WITHDRAWN BY THE
UNIVERSITY OF MICHIGAN

A Guide to Probability Theory and Application

A Guide to Probability Theory and Application

Cyrus Derman
Columbia University

Leon J. Gleser
Purdue University

Ingram Olkin
Stanford University

HOLT, RINEHART AND WINSTON, INC.
*New York · Chicago · San Francisco · Atlanta
Dallas · Montreal · Toronto · London · Sydney*

Copyright © 1973 by Holt, Rinehart and Winston, Inc.
All Rights Reserved
Library of Congress Catalog Card Number: 72-88276
AMS 1970 Subject Classifications: 60-01, 98A25
ISBN: 0-03-0-78885-4
Printed in the United States of America
3 4 5 6   038   1 2 3 4 5 6 7 8 9

To Martha, Marilyn, and Anita

PREFACE

Probabilistic concepts and probability models have been an integral part of the theoretical structure of the physical sciences (physics, chemistry, engineering) since the turn of the century. As a result, an exposure to the theory of probability has become an essential part of the professional training of almost every physical scientist. In recent years, fruitful models have been proposed and used within the social and environmental sciences (psychology, education, sociology, economics, political science, biology, geography, and ecology), and the professions (medicine, law, business). Such models have related to learning, conformity, social mobility, language development, cognition, environmental change, organic growth, geological composition, contagion of disease, organizational decision-making, and monetary growth, to name only a few. Since these models also rest on a framework of probability theory, a knowledge of the essentials of probability theory is a necessary prerequisite for the understanding and development of models in these areas.

The purpose of the present book is to provide an extensive treatment of the concepts and models of probability theory at an elementary mathematical level, with specific attention to the needs and interests of researchers in the social and environmental sciences, and of professionals in medicine, law, and business. Our intent has been to present various topics of probability theory with a minimum of mathematics and a maximum of illustrative examples taken from scientific and professional practice. In addition to a thorough coverage of the basic probabilistic concepts and methods, a great variety of useful probability models are presented, including some (continuous models, multivariate models, Markov chains) not usually covered in a book at this level. We believe that the present book can serve those individuals who are interested in surveying the general principles of probability theory, and also those who are more interested in detailed discussions of aspects of this theory that can be applied directly to their specific interests.

RELATION TO STATISTICS

Probability and statistics are integrally related. Probability theory provides a conceptual framework for many phenomena, while statistical inference provides the methods by which such models may be verified. It is virtually impossible to present probability theory in a setting of applications without giving supporting data. When using such data, we have had to gloss over the statistical principles of analysis in order not to obscure the underlying probability models. For example, when we mention fitting a particular probability model to data, we use the most elementary method of fitting (the method of moments) and we do not discuss alternative methods. Our concern in such discussions is with the underlying structure of the model—and the type of data for which this model might be applicable—rather than with a comparison of methods of fitting, which is a statistical problem. There is a large body of literature dealing with principles of fitting probability models to data and evaluating the fit, but such topics are outside the scope of this book.

PREREQUISITES

Most of the material in this book will be comprehensible to a reader who has had only a course in mathematics at the level of college algebra. Differential or integral calculus is not needed. Readers with a higher level of mathematical sophistication can still appreciate this book, however. The essential prerequisite is the ability (and willingness) to use and understand mathematical notation and formulas. Specific topics in mathematics which will be helpful to the reader of this book are logarithms, general functional notation, and exponential and power functions. Some understanding of the use of mathematical logic (for example, the logic used in the proofs of plane geometry or in algebra) will also be helpful, particularly in Chapters 2–5. All of this background can be obtained either from a college algebra course or from reading one of the recent high-school texts in mathematics (for example, texts published by the School Mathematics Study Group).

HOW TO USE THIS BOOK

This book is divided into two parts as follows. Part A—Elementary Concepts—develops the basic probability calculus. By design, the pace is slow in order to permit those with the minimum of mathematical background to learn the material thoroughly. We hope that the negative aspects of a very slow pace will be offset by the increased accessibility of the theory. Part B—Probability Models—deals with material not generally found in other texts. Here the style is more brisk, and some of the topics (in particular, Chapter 12) are relatively advanced. These topics are included because they are important to the social and biological sciences, and have been previously unavailable at an elementary mathematical level. However, they also provide a wide spectrum of material so that the book can serve students at all levels of mathematical sophistication. The more difficult sections (or exercises) are starred to indicate advanced material that may be ignored on a first reading.

Because the book has multiple goals, it may not be desirable for every reader to follow the order of presentation here. The following are some suggested combinations for different courses.

For a One-Semester Course Taken by Students with a Minimal Mathematical Background:
    All of Part A, except, perhaps, for isolated sections;
    Some parts of Chapters 6 or 11 might be included as an alternative to some of the material in Part A.

For a One-Semester Course Taken by Students with More Than a Minimal Mathematical Background:
    Part A plus Chapters 6, 7, and 8. If there is more material than can be covered in the allotted time, some sections in Chapter 6 and 8 can be omitted.

For a Two-Semester Course Taken by Students with a Minimal Mathematical Background:
    (a) Part A plus Chapters 6, 7, 8, 9, and 10,
    or
    (b) Part A plus Chapters 6, 7, 9, 10, and 11. The section in Chapter 9 on the bivariate normal distribution can be deleted if time is short.

For a Two-Semester Course Taken by Students with More Than the Minimal Mathematical Background:
    Parts A and B together can be covered. More time should be devoted to Part B.

EXERCISES

The exercises in this book were chosen not only to illustrate the ideas in the text, but also to provide additional learning material and motivation for study. The more standard problems of dice throwing, coin tossing, and games of chance are very limited here. Instructors who wish may supplement the materials here with such exercises. Answers are provided to those exercises marked with an [A]. The most difficult problems are indicated by an asterisk.

NUMBERING SYSTEM

In order to achieve simplicity in numbering, the sections and equations are numbered sequentially throughout each chapter only. This has an advantage in simplicity, but requires a chapter reference whenever one chapter refers to an equation or other material from another chapter.

TABLES

The necessary tables are provided in the text. Any entries in a table which are left blank are either equal to 0 or correspond to undefined quantities.

As with all tables, the roundoff problem has been unavoidable. For example, if we have $6.64 + 3.34 = 9.98$, the answer rounded to one decimal after summing is 10.0, whereas rounding before summing yields $6.6 + 3.3 = 9.9$. As a consequence, totals which should be 1.00 may be 0.98, 0.99, or 1.01, and similarly for other totals. Where these discrepancies are minor and recognizable, we have made no attempt to correct them.

ACKNOWLEDGMENTS

No book of this size can be completed without the help of many persons. To name only a few, we owe thanks to Del Clear, Peter Husen, and Mary Lou Koran, for pointing out ambiguities in an early draft; to William Ericson and Burton Singer for their many helpful suggestions; to Judith Tanur for her editorial help; to John Petkau and Bert Wiser for discovering errors we missed; to Henrietta Gallagher for help with some of the computations. Vera Bertrand, Betty Clausen, Donna Lembeck, Beverly Ralphs Ross, Mary Evelyn Runyon, Claire Smith, and Betty Springsteen, deserve the credit for the typing of the many drafts, while withstanding our constant harassment. A very special thanks is due to the editor, Lois Wernick, for her help, advice, and patience through the many stages of the preparation of this book. Finally, we wish to express our sincere appreciation to Educational Testing Service for making their facilities available to us.

New York City Cyrus Derman
Lafayette, Indiana Leon J. Gleser
Stanford, California Ingram Olkin
September 1972

CONTENTS

Preface ix

PART I FUNDAMENTAL CONCEPTS

1 Introduction 3

   1 History, 3
   2 Models, 5
   3 The Empirical Basis of Probability Theory, 8
   4 Notes and References, 15

2 The Elements of Probability Theory 25

   1 Preliminaries, 25
   2 Formal Structure: Outcomes and Event Sets, 25
      Equality of Event Sets, 28
      Inclusion of Event Sets, 29
      The Union of Event Sets, 30
      The Intersection of Event Sets, 31
      The Complement of an Event Set, 31
      Mutually Exclusive Event Sets, 34
      Some Useful Rules, 35
      The Correspondence between the Probability Model and Random Experiments, 38
   3 Formal Structure: Probabilities, 40
   4 The Probability Calculus, 42
      Law of Inclusion, 42
      Law of Complementation, 43
      Law of Addition, 45

xii CONTENTS

    5 Fitting a Probability Model, 51
    6 Notes and References, 54

3 Further Concepts of Probability Theory   65

    1 Finite Probability Models, 66
    2 Simple Random Sampling, 69
    3 Composite Random Experiments, 72
        A Comment on Notation, 74
    4 Conditional Probability, 79
        The Probability Calculus for Conditional Probabilities, 83
        Law of Multiplication, 86
        Law of Multiplication for Nested Events, 89
        Partitions, 93
        Bayes' Rule, 95
    5 Independence, 97
    6 Appendix: Combinatorial Analysis, 100
        Orderings and Permutations, 100
        Orderings with Similar Elements, 101
        Ordering of r Items from n, 104
        Counting of Sets of Items, 105
    7 Notes and References, 108

4 Random Variables   125

    1 Random Variables, 125
    2 Discrete and Continuous Random Variables, 130
    3 Cumulative Distribution Functions, 133
        Estimating the Cumulative Distribution Function from Data, 143
        Mixed Discrete and Continuous Variables and their Cumulative Distribution Functions, 144
    4 Probability Mass Functions of Discrete Random Variables, 146
        Estimating a Probability Mass Function: Bar Graphs, 149
    5 Density Functions, 150
        *Rectangular Probability Approximations, 153
        Empirical Distributions: Histograms, 160

5 Descriptive Properties of Distributions   175

    1 Introduction, 175
    2 Measures of Location for a Distribution: The Expected Value, 177
        The Expected Value of a Discrete Random Variable, 177
        *The Expected Value of a Discrete Random Variable Having an Infinite Number of Possible Values, 185
        The Expected Value of a Continuous Random Variable, 186
        *Use of the Rectangular Method of Approximation, 188
        The Expected Value as a Measure of Location, 191
        Notation and Terminology, 193
    3 Other Measures of Location for a Distribution: Median, Quantile, Mode, 193

Medians, 193
Quantiles, 197
Modes, 203
Choice of a Measure of Location, 205
4   Measures of Dispersion for a Distribution: The Variance, 210
5   Higher Moments, 217
6   Inequalities for Probabilities, 223
Proof of the Markov Inequality, 224
Use of the Markov Inequality, 225
Supplementary Reading, 231
7   Appendix, 233

PART II   PROBABILITY MODELS

6   Special Distributions: Discrete Case   247

1   Bernoulli and Binomial Distributions, 247
Bernoulli Distribution, 248
Binomial Distribution, 249
Tables, 255
2   Hypergeometric Distribution, 256
Derivation of the Hypergeometric Distribution, 258
Approximation of the Hypergeometric Distribution by the Binomial Distribution, 262
Uses of the Hypergeometric Distribution, 262
Description of Tables, 264
3   Poisson Distribution, 265
A Method for Fitting the Poisson Distribution, 266
Applications of the Poisson Distribution, 269
The Poisson Distribution as an Approximation of the Binomial Distribution, 272
Tables of the Poisson Distribution, 273
4   Geometric Distribution, 274
A Method for Fitting the Geometric Distribution, 277
Description of Tables of the Geometric Distribution, 279
Notes and References, 279
5   Negative Binomial Distribution, 280
*Generalization of the Negative Binomial Distribution, 284
*A Method for Fitting the Negative Binomial Distribution, 287
Applications of the Negative Binomial Distribution, 290
Description of Tables of the Negative Binomial Distribution, 292
6   Discrete Uniform Distribution, 293
Lotteries, 294
Generalized Discrete Uniform Distribution, 295
7   *Other Discrete Distributions, 296
The Zeta Distribution, 296
The Truncated Poisson Distribution, 299
References to Other Discrete Distributions, 304
8   Summary of Discrete Distribution, 304

xiv CONTENTS

7 The Normal Distribution — 324

1. Introduction, 324
   - History, 326
2. Probability Calculations for the Normal Distributions, 332
   - Tail Probabilities, 336
   - Probabilities of Intervals of Numbers, 339
3. Fitting a Normal Distribution, 341
4. The Central Limit Theorem, 345
   - *A Justification of the Normal Distribution by the Central Limit Theorem, 349

8 Special Distributions: Continuous Case — 358

1. The Exponential Distribution, 359
   - Fitting the Exponential Distribution to Data, 361
   - Probability Calculations, 364
   - The Survival Function, 365
   - An Interesting Property of the Exponential Distribution, 367
   - Notes and References, 368
2. The Gamma Distribution, 369
   - The Standard Gamma Density Function, 370
   - Fitting the Gamma Distribution to Data, 372
   - Probability Calculations, 374
3. *The Weibull Distribution, 378
   - A Theoretical Explanation, 385
   - Fitting the Weibull Distribution to Data, 385
   - The Survival Function, 389
4. The Uniform Distribution, 391
   - Probability Calculations, 393
   - Applications, 396
5. *The Beta Distribution, 398
   - Fitting the Beta Distribution to Data, 404
   - Probability Calculations, 405
   - Notes and References, 409
6. *Other Continuous Distributions, 410
   - The Lognormal Distribution, 410
   - The Cauchy Distribution, 414
   - Notes and References, 418

9 Multivariate Distributions — 432

1. Introduction, 432
2. Discrete Bivariate Distributions, 433
   - Marginal Distributions, 439
   - Example of the Construction of a Bivariate Probability Model, 443
   - Conditional Distributions, 446
   - The Regression Function, 450
3. Descriptive Properties of Bivariate Distribution, 453
   - Rescaling Techniques, 460
   - The Correlation Coefficient, 464

Correlation and Statistical Independence, 469
Correlation and Statistical Dependence, 472
4   Continuous Bivariate Distributions, 475
    Marginal Probability Density Functions, 479
    Conditional Probability Density Functions, 482
    Statistical Independence, 483
    Descriptive Properties of Bivariate Continuous Distributions, 484
5   The Bivariate Normal Distribution, 486
    The Probability Contours of a Bivariate Normal Distribution, 490
    The Standard Bivariate Normal Distribution, 493
6   *Some Multivariate Generalizations: Trivariate Discrete Distributions, 501
    Conditional Distributions and Partial Correlation Coefficients, 506
    Statistical Independence, 508
    The Multinomial Distribution, 509

10 Weighted Sum of Random Variables — 519

1   Standardization of Random Variables, 519
2   Weighted Sums of Random Variables, 523
    Some Special Cases, 526
    Expected Values and Variances, 527
3   The Law of Large Numbers, 530
4   Further Comments on the Central Limit Theorem, 533
    The Continuity Correction, 539
5   *Appendix: Moment-Generating Functions, 541

11 Markov Chains — 550

1   Introduction, 550
2   Markov Chains, 552
3   Applications, 559
    An Example from Psychology, 559
    Social Processes, 563
    Genetics, 566
4   Calculation of k-Step Transition Probabilities, 571
    2-Step Transition Probabilities, 572
    3-Step Transition Probabilities, 578
    General k-Step Transition Probabilities, 582
    An Alternative Method for Calculating k-Step Transition Probabilities, 587
5   Absolute or Unconditional Probabilities, 588

12 *Markov Chains II: First Passage Times, Recurrent States, and Long-Run Probabilities — 597

1   First-Passage Times, 597
    Calculation of the Probability Distribution of the First-Passage Time K_{ij}, 600
    Infinite First-Passage Times, 611
2   Recurrent and Transient States, 615

3   Expected First-Passage Times, 620
        Alternate Computation for Expected First-Passage Times, 623
4   Probabilistic Equilibrium of a Markov Chain, 630
5   Classes of States, 634
        Classes of Communicating States, 636
6   Aperiodic, Irreducible Markov Chains, 640
        Periodic States and Periodic Chains, 642
        Aperiodic States, 644
        Steady-State Equations, 648
7   Absorbing Markov Chains, 651
8   Notes and References, 657

Bibliography 663

Tables 671

Answers 717

Name Index 743

Subject Index 745

INTERNATIONAL SERIES IN DECISION PROCESSES

INGRAM OLKIN, Consulting Editor

A Basic Course in Statistics with Sociological Application, 2d ed., T. R. Anderson and M. Zelditch, Jr.
Probability Theory and Elements of Measure Theory, H. Bauer
Probability and Statistics for Decision Making, Ya-lun Chou
A Guide to Probability Theory and Application, C. Derman, L. J. Gleser, and I. Olkin
Statistics: Probability, Inference, and Decision, Volumes I and II and Combined ed., W. L. Hays and R. L. Winkler
Introduction to Statistics, R. A. Hultquist
Introductory Statistics with Fortran, A. Kirch
Reliability Handbook, B. A. Kozlov and I. A. Ushakov (edited by J. T. Rosenblatt and L. H. Koopmans)
An Introduction to Probability, Decision, and Inference, I. H. LaValle
Elements of Probability and Statistics, S. A. Lippman
Modern Mathematical Methods for Economics and Business, R. E. Miller
Applied Multivariate Analysis, S. J. Press
Fundamental Research Statistics for the Behavioral Sciences, J. T. Roscoe
Applied Probability, W. A. Thompson, Jr.
Elementary Statistical Methods, 3d ed., H. M. Walker and J. Lev
An Introduction to Bayesian Inference and Decision, R. L. Winkler

FORTHCOMING TITLES

Time Series, R. Brillinger
Decision Analysis for Business, R. Brown
Introduction to Statistics and Probability, E. Dudewicz
Statistics: Probability, Inference, and Decision, 2d ed., W. L. Hays and R. L. Winkler
Fundamentals of Decision Analysis, I. H. LaValle
Quantitative Methods and Operations Research for Business, R. E. Trueman

SERIES IN QUANTITATIVE METHODS FOR DECISION-MAKING

ROBERT L. WINKLER, Consulting Editor

Probability and Statistics for Decision Making, Ya-lun Chou
Statistics: Probability, Inference, and Decision, Volumes I and II and Combined ed., W. L. Hays and R. L. Winkler
An Introduction to Probability, Decision, and Inference, I. H. LaValle
Elements of Probability and Statistics, S. A. Lippman
Modern Mathematical Methods for Economics and Business, R. E. Miller
Applied Multivariate Analysis, S. J. Press
An Introduction to Bayesian Inference and Decision, R. L. Winkler

FORTHCOMING TITLES

Decision Analysis for Business, R. Brown
Statistics: Probability, Inference, and Decision, 2d ed., W. L. Hays and R. L. Winkler
Fundamentals of Decision Analysis, I. H. LaValle
Quantitative Methods and Operations Research for Business, R. E. Trueman

part   I

Fundamental Concepts

1
INTRODUCTION

It is remarkable that a science which began with the consideration of games of chance should have become the most important object of human knowledge. ... The most important questions of life are, for the most part, really only problems of probability.
    Pierre Simon de Laplace (Théorie analytique des probabilités)

1. HISTORY

In French society during the middle of the seventeenth century, games of chance were a popular, even fashionable, pastime. The proliferation and variety of such games, together with the large sums of money involved, led many members of the nobility to spend their free hours seeking a rational method for assessing the fortunes of a gambler engaged in games of chance. A particularly interested individual, the Chevalier de Méré, consulted the mathematician and philosopher Blaise Pascal (1623–1662) concerning this problem. By so doing, he stimulated the birth of a unified theory of chance phenomena. This theory is today called the *theory of probability*.

Pascal and his fellow mathematicians, most notably Pierre de Fermat (1601–1665), restricted their interest in chance phenomena almost entirely to the small class of such phenomena involved in games of chance. They considered each game separately, and appear to have worked out no connected theory. However, starting from about the year 1700, the Swiss mathematician Jakob (James) Bernoulli (1654–1705) and the French mathematician Abraham de Moivre (1667–1754) led a vigorous development in the direction of a more general theory.

In addition, various scientific and practical investigators of the eighteenth and nineteenth centuries began to note analogies between the laws of uncertainty involved with games of chance and the laws of variation observed in other apparently uncontrollable phenomena. This led some of these investigators to conclude that the variability of their observations could be described by the laws governing the results of games of chance. For example, students of human populations (and geneticists) noted that the sequential record of births at city hospitals exhibited patterns (alternations) of male and female births similar to corresponding patterns of "heads" and "tails" resulting from successive tosses of a coin. This observation suggested the hypothesis that Nature chooses the sex of a child in much the same way as she chooses the side of a tossed coin—that is, by chance. Laws governing the sex of a child should then be the same as laws governing the result of a coin-toss. From such arguments came models of reality which treated observed phenomena as if they resulted from a game of chance played by Nature. These *probabilistic (or stochastic) models* have played increasingly important roles in almost every scientific and practical activity in our society.

The word *stochastic* derives from Greek στόχοσ—stochos, a target; in ancient Greece a *stochastiches* was a person who forecast a future event (in the sense of aiming at the truth). In this sense the word occurs in sixteenth-century writing. Bernoulli in the *Ars conjectandi* (1713) refers to the "ars conjectandi sive stochastice." The translation of the relevant paragraph is:

"We are said to *know* or to *understand* those things which are certain and beyond doubt; all other things we are said merely to *conjecture* or *guess about*.

To *conjecture about* something is to measure its probability; and therefore, the *art of conjecturing* or the *stochastic art* is defined by us as the art of measuring as exactly as possible the probabilities of things with this end in mind: that in our decisions or actions we may be able always to choose or to follow what has been perceived as being superior, more advantageous, safer, or better considered; in this alone lies all the wisdom of the philosopher and all the discretion of the statesman." (From M. G. Kendall and W. R. Buckland, *A Dictionary of Statistical Terms*, Oliver and Boyd, 1960, and B. Sung, *Translations from James Bernoulli*, with a preface by A. P. Dempster, Technical Report No 2, Harvard University, February, 1966.)

As more and more opportunities for the use of probabilistic models were discovered, an increasing number of outstanding scientists became interested in probability theory. As a result, starting with the eighteenth century and continuing to the present time, many of the greatest names in science have devoted their energies to solving problems in this field. As a partial list of these contributors, we should mention Laplace (1749–1827), Gauss (1777–1855), Maxwell (1831–1879), Bertrand (1822–1900), Poincaré (1854–1912), Einstein (1879–1955), and Wiener (1894–1964). One consequence of the concentration of so much genius and effort on problems concerning chance phenomena has been the production of several unified approaches to a theory of probability. Among these, the axiomatic approach, based on "the stability properties of frequency ratios" (Section 3), has proved to be the most popular and pervasive. It is this approach that is emphasized in the following chapters.

To close this very brief historical introduction to the theory of probability, it may be of interest to cite some examples to show how widely stochastic models are used today. Insurance policies are written taking into account the chance that a person with a certain medical history will die within the year (actuarial tables); public opinion experts gauge the reactions of the total populace by questioning a few people selected through a lottery (random sampling); theorists in physics conceive of elementary particles moving, colliding, and splitting according to chance; psychologists analyze learning behavior as if it were a chance phenomenon; some sociologists view mobility of populations as being governed by a probabilistic mechanism; hereditary characteristics of biological organisms are hypothesized to be assigned by chance; inventory systems are devised to meet demands fluctuating in a random fashion; and so forth. From this short list, it should be apparent that today the use of stochastic models (and thus probability theory) pervades most sophisticated attempts to explore, explain, and control our physical and social environment.

2. MODELS

Does Nature gamble with the universe? This question bothered many of the scientists of the twentieth century and was partly responsible for their reluctance to adopt stochastic models as a fundamental part of a theory of reality for our universe. To clarify the problem that these men faced, it is first necessary to explain what a scientific model is, how it is used, and how it is chosen.

A scientific model is an abstract and simplified description of a given phenomenon. Certain basic aspects of this phenomenon are isolated as being of primary interest, and an analogy is drawn between these aspects and some logical (or phenomenological) structure concerning which we already have (or can obtain) detailed information. Scientific models are most often based on mathematical structures; but organic models have been used in sociology [for example, see Spencer (1877)], physical models in psychology and economics, and economic models in engineering. If the analogy between model and phenomenon is strong enough, knowledge about the model (in the form of theorems or laws) can suggest new research or explain already obtained results concerning the phenomenon. Figure 2.1 illustrates how a scientific model may deepen our understanding of natural phenomena.

When an investigator builds a mathematical model for a particular natural phenomenon (say, the motion of an asteroid), he identifies the

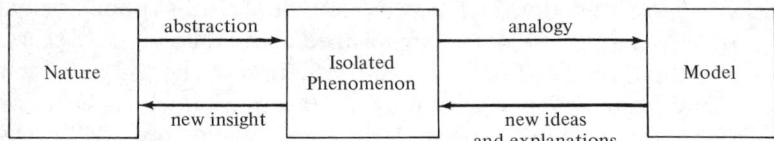

Figure 2.1 Use and development of a scientific model.

important elements of this phenomenon (the position, mass, shape, and speed of the asteroid) with the basic elements of some mathematical structure (numbers). Certain fundamental facts relating the important elements of the phenomenon are restated as axioms relating the analogous mathematical entities. Finally, the more complex relationships between the basic elements of the natural phenomenon are made to correspond to laws or theorems in the mathematical structure. If this correspondence is reasonably valid, the investigator does not have to haphazardly experiment with the phenomenon in order to find new facts; he can instead argue logically from his mathematical axioms to a theorem that presumably has an analogy to a law of nature. Experimentation can now be directed toward verifying this law. Thus, in the example of the asteroid, an application of that part of mathematical logic known as the integral calculus to the numbers representing the position, mass, and velocity of the asteroid at a certain time t_0 enables us to predict its position, mass, and velocity at some later time t_1. Still another application of mathematical logic may lead us to expect the asteroid to orbit our sun. Astronomical observations can now be guided by these predictions.

The fact that an investigator need only concentrate on the few axioms that define the mathematical structure of his model leads to a simplification and unification of his knowledge concerning the natural phenomenon of interest. Every fact known to him can be reproduced by starting from the axioms and using mathematical logic. Thus, his discipline becomes a cohesive whole in which all facts are logically interrelated, rather than merely a list of isolated facts. For example, the problem of the motion of an asteroid is one of the problems treated by classical Newtonian mechanics, a body of knowledge dealing with the motion of all physical bodies in the universe. The known facts of this discipline fill volumes; yet all of these facts can be derived, more or less accurately, from just three axioms through the use of mathematical logic.

[*Remark:* When we try to fit a mathematical model to Nature, we sometimes deliberately force the analogies. In particular, relationships which in Nature only hold approximately are assumed to hold exactly in the mathematical model. When using the model to gain insight into the natural phenomenon of interest, we must keep this point in mind. One particular pitfall that must be avoided is the belief that every fact in Nature corresponds to a statement which is true about the model. This belief is not necessarily true, since the model refers to an abstraction of the natural phenomenon in which many features peculiar to the phenomenon have been ignored. On the other hand, it is also true that facts in the model need not always correspond to facts about Nature. We do not say, "Nature *is* such-and-such a model," but only that "Nature has some *features* like such-and-such a model." The test of a model is its usefulness in giving us insights and explanations of Nature.]

Having discussed what a scientific model is, how such a model is interpreted, and why we choose a given scientific model, we are ready to return to the question that opened this section: Does Nature play dice with the universe? On one level, this question is metaphysical, implying that the very existence of chance phenomena is open to doubt. On another level, however, our question concerns the methodology and logic of science. At this level, we are asking whether stochastic methods provide a useful framework for explaining and predicting certain classes of phenomena.

For more than a century after Newton's time, the prevalent scientific and philosophical viewpoint was that Nature is *mechanistic*. That is, once all relevant causal variables were identified and measured, the eighteenth century scientist believed that any phenomenon could be *exactly* predicted. Even a tossed coin, he might have argued, could be predicted, provided the magnitude and direction of all forces acting on the coin were known. Such a view of Nature provided a *mechanistic model*. To eighteenth century scientists, all variability in an observed phenomenon was due to our inability to control certain causal variables, and an investigation should be directed toward identifying and controlling (or at least measuring) these variables in order to reduce (or explain) their variability, and thus improve their predictability. In physics, chemistry, and engineering, such models proved very fruitful and aided the rapid expansion of knowledge that marked the eighteenth and nineteenth centuries.

However, even in the eighteenth century it had become apparent that no matter how mechanistic one believes Nature is, models incorporating uncertainty (stochastic models) are needed. Some of the reasons for this need are:

(i) Even if Nature is mechanistic, to list and measure all of the variables involved in the complex phenomena found in, say, micro-physics, micro-biology, psychology, or social behavior would be an impossible task.
(ii) Some scientists and philosophers [for example, the German physicist Heisenberg (1901–)] argue that there is a limit to man's ability to measure natural phenomena. Certain of these thinkers assert further

that Nature may not even *be* mechanistic, in the sense that our measurements of natural phenomena themselves produce unpredictable variations in these same phenomena.

(iii) Strangely enough, it has been discovered in some of the social and biological sciences that a deliberate *introduction* of uncertainty provides a cheaper (and often more accurate) way of investigating natural phenomena. For example, in taking a census, it has been found that interviewing everyone in the population is not only expensive and time consuming, it is also inaccurate (because of mobility of the population, fatigue effect in the interviewer, and so forth). On the other hand, almost the same information can be obtained more cheaply, in less time, and more accurately, by interviewing a *random sample* (a sample chosen by lottery) of the population, and then extrapolating to the population by means of probability theory.

In consequence, starting with the work of Gauss and Laplace on errors in measurement, and continuing with the nineteenth century actuaries and demographers, stochastic models of natural phenomena have been used in preference to mechanistic models *even by those who believe that Nature is inherently mechanistic.* (This viewpoint seems to have been adopted by Einstein. Although he is supposed to have said, "I shall never believe that God plays dice with the world," he nonetheless made significant contributions to the development of stochastic models in physics.) A model need not be completely true to be useful; the decisive criteria for choosing a model are practical, not metaphysical: Does the model provide a simple, yet comprehensive, explanation of known phenomena, and at the same time have a strong potentiality for providing further insight into the natural world? Stochastic models meet these criteria. As a consequence, the assumption that chance phenomena exist and can be described, whether true or not, has proved valuable in almost every discipline. When stochastic models are no longer useful, or when better models are found, then we will cease to make use of stochastic models. Judging by the current proliferation of stochastic models in almost every branch of science and technology, such a prospect is unlikely.

3. THE EMPIRICAL BASIS OF PROBABILITY THEORY

The applicability of probability theory is a consequence of the observed fact that chance phenomena exhibit statistical regularities. A clear example of this fact is the ancient game of dice-throwing. Although we cannot say what the result of any given throw of a die will be, we can predict in a *large number* of throws of the die what proportion of the throws will result in (say) a six. We use this information to place our bets for each throw. A more practical application of the statistical regularity of chance phenomena occurs in the field of insurance. Although it may not be possible to predict the life span of a given individual, precise statements concerning longevity can be made about large populations of individuals. Such a statement may be, for example, that 50 percent of the present population will die before age 65. For the purposes of life insurance or annuities, these statements are entirely adequate to enable the insurance company to "place its bet"—that is, decide what premium to charge for a given amount of insurance. In offering a policy to a large group of individuals, it is not important to the company to know what will happen to any one individual; only what happens to the group as a whole is of concern.

Let us discuss this subject in a slightly more formal way. A basic concept for all discussions of chance phenomena is the notion of a *random experiment*. A random experiment is a procedure that involves certain actions under specified conditions and that has as its outcome one (and only one) of a collection of possible *simple results*. When the experiment is run, the particular simple result that occurs is unpredictable a priori (that is, before the experiment is performed). The following are some common examples of random experiments.

Example 3.1 (Sample Survey). A phone number from a phone book is chosen by means of a lottery. The number is called and the person who answers is asked whether he or she is listening to a certain radio station. The possible simple results are "Yes" or "No" responses (or perhaps "No Answer"). Before the person is called, there is no way of accurately predicting what his or her answer will be.

Example 3.2 (Psychological Measurement). A test consisting of 50 true-false questions is given to a specified individual. Each possible simple result of this experiment can be described as the pattern of "Correct" and "Incorrect" responses to the questions. A typical single result might be represented by (C, C, I, C, I, I, C, ..., I, C), which says that the first and second questions were correctly answered, the third question was incorrectly answered, and so forth. Alternatively, we might agree to restrict our interest to the *number* of questions answered correctly. In this case, the simple results are the numbers 0, 1, 2, ..., or 50, of correctly answered questions. The above experiment is potentially random both because individuals differ as to their performance on the test, and because a given individual may, himself, vary with time in performance on the test.

Example 3.3 (Physical Measurement). A physicist attempts to find the weight of a given object. Here the possible simple results are all possible weights. The experiment is random for two reasons: because minute changes in moisture content, air pressure, and temperature in the room may affect the scale and the actual weight of the object in unknown and unpredictable (a priori) ways, and because there may be slight visual or adjustment errors on the part of the physicist.

Example 3.4 (Quality Control). An electric light bulb is allowed to burn until it burns out. The possible simple results are all the possible times to extinction.

Example 3.5 (Games of Chance). A die is placed in a cup, shaken vigorously, and hurled against a wall so that it bounces on the floor. The number of dots appearing on the upturned face of the die is counted. Possible simple results are 1, 2, 3, 4, 5, and 6 dots.

We can think (at least ideally) of repeating each of the above experiments under identical conditions, a so-called *repeated trial*. Since these experiments are random, the result of each trial is still unpredictable a priori. However, let us assume that we make many, say N, repeated trials of a particular experiment, and for each trial record the (simple) result. For example, in the phone survey (Example 3.1) we might have a list such as

Trial	1	2	3	4	...	$N-1$	N
Result	Yes	No	No	Yes	...	No	Yes

.

In the physical measurement example (Example 3.3), we might have

Trial	1	2	3	4	...	$N-1$	N
Result (in pounds)	4.5	4.4	4.6	4.4	...	4.2	4.4

,

while for the die throws in Example 3.5 we might have

Trial	1	2	3	4	...	$N-1$	N
Result (number of dots)	5	3	2	1	...	3	4

.

THE EMPIRICAL BASIS OF PROBABILITY THEORY 11

Consider some statement that describes or summarizes the result of an individual trial. In Example 3.1, this statement might be, "The person called says that he is listening to the radio station"; in Example 3.3, the statement might be, "The weight of the object is less than 4.4 pounds"; and in Example 3.5, it might be, "The number of dots shown by the die is an odd number." Each individual trial has as its outcome a single simple result, but out of all possible simple results, more than one may be describable by the statement in which we are interested. Thus, in Example 3.3 any weight (such as 2.2 pounds, 3.1 pounds, 4.0 pounds, and so on) which is less than 4.4 pounds will satisfy the description of the statement "The weight of the object is less than 4.4 pounds," while in Example 3.5 we can say "The number of dots shown by the die is an odd number" if we observe 1, 3, or 5 dots. Let us call the collection of all possible simple results that satisfy (or are described by) the statement of interest to us the *composite result* defined by this statement, and denote this composite result by the letter E. If the result of the trial is describable by the statement that defines the composite result E, then we say that E *has occurred*. Among the N trials that we observe, we count the number of times (denoted by $\#E$) that the composite result E occurs, and divide this number by N. The result of this operation is a fraction $\#E/N$ which gives the proportion of times in N trials that the composite result E occurs. We call this fraction the *relative frequency* (in N trials) of E, and denote this fraction by r.f.(E).

Let us, as we continue to make more and more trials, keep a record to see how r.f.(E) varies with the number of trials. Thus after the first trial, we compute r.f.(E) (it is 0 if E did not occur, or 1 otherwise); after the second trial, we again compute r.f.(E) over the *two* trials (this number is 0 if E has not occurred on either trial, $\frac{1}{2}$ if E has occurred on one trial and not on the other, or 1 if E occurred on both trials); after the third trial, we compute r.f.(E) for the three trials; and so on. We graph the relative frequencies so obtained against the number of trials for which they were obtained. Thus if we have

Trial	1	2	3	4	5
Result	E	not E	not E	E	not E
r.f.(E)	1	$\frac{1}{2}$	$\frac{1}{3}$	$\frac{2}{4}$	$\frac{2}{5}$

(where not E indicates that the composite result E did not occur), our graph would look like Figure 3.1 and over many trials might look like

12 INTRODUCTION 1

Figure 3.2. Note that a logarithmic scale is used in order to more easily exhibit a large number of trials. (Table 3.1 gives the data that have been used to construct Figure 3.2.)

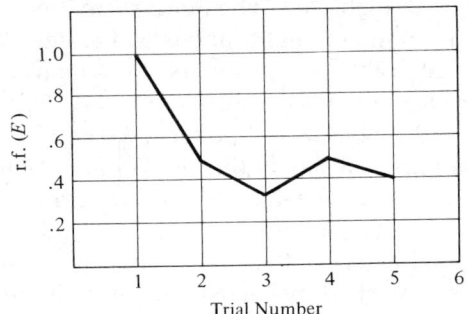

Figure 3.1 Relative frequency of the composite result E.

Table 3.1. A record of how the relative frequency r.f.(E) of the composite result E varies with the number N of trials.

Trial	Relative Frequency	Trial	Relative Frequency	Trial	Relative Frequency	Trial	Relative Frequency
1	1.0000	30	0.3333	300	0.3567	1700	0.3912
2	0.5000	35	0.3143	350	0.3514	1800	0.3961
3	0.3333	40	0.3250	400	0.3625	1900	0.3974
4	0.5000	45	0.2889	450	0.3689	2000	0.3985
5	0.4000	50	0.2800	500	0.3760	2500	0.4032
6	0.5000	55	0.2727	550	0.3927	3000	0.4003
7	0.4286	60	0.3167	600	0.4000	3500	0.3986
8	0.5000	65	0.3231	650	0.3985	4000	0.3973
9	0.4444	70	0.3143	700	0.3929	4500	0.3953
10	0.5000	75	0.3067	750	0.3947	5000	0.3956
11	0.4545	80	0.3125	800	0.4025	5500	0.3958
12	0.5000	85	0.3529	850	0.4059	6000	0.3978
13	0.4615	90	0.3778	900	0.4078	6500	0.3983
14	0.4286	95	0.3789	950	0.4084	7000	0.3994
15	0.4000	100	0.3900	1000	0.4060	7500	0.4019
16	0.3750	120	0.4000	1100	0.3973	8000	0.4000
17	0.3529	140	0.3786	1200	0.3917	8500	0.4016
18	0.3889	160	0.3625	1300	0.3938	9000	0.4016
19	0.3684	180	0.3667	1400	0.3893	9500	0.3994
20	0.3500	200	0.3700	1500	0.3880	10000	0.4001
25	0.3600	250	0.3680	1600	0.3919		

3. THE EMPIRICAL BASIS OF PROBABILITY THEORY 13

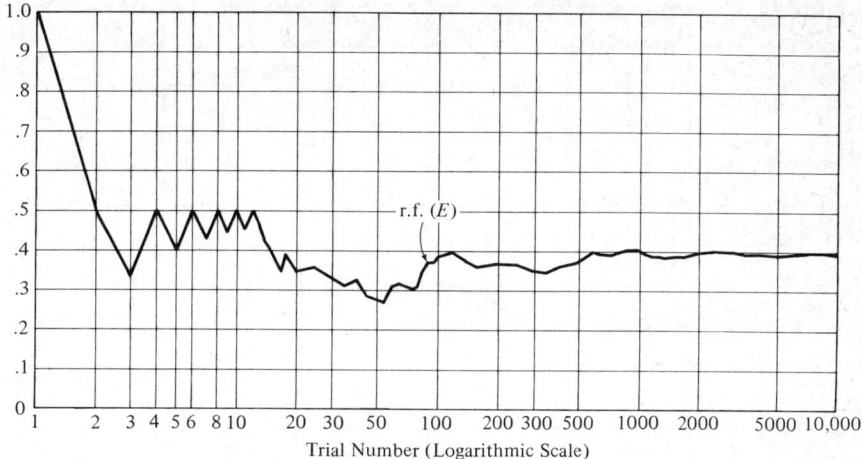

Figure 3.2 Relative frequency of the composite result E.

As N grows larger and larger, the number r.f.(E), that is, the ratio $\#E/N$, should grow closer and closer to a certain value which we shall call p. In Figure 3.2, $p = \frac{2}{5}$. Let us illustrate (Figure 3.3) this same phenomenon for Examples 3.1, 3.3, and 3.5 (where the composite result for Example 3.1 is, "the person called is listening to the radio station," for Example 3.3 is, "the weight of the object is less than 4.4 pounds," and for Example 3.5 is, "the number of dots on the die is an odd number").

Notice that the value of p in each of these examples is different; what is common to these three examples is the convergence of r.f.(E) as N grows large. That is, for small N, r.f.(E) may vary widely over the range of numbers from 0 to 1, but as N gets large, r.f.(E) becomes less and less different from p. This phenomenon is what we call the *statistical regularity of chance phenomena*; it is also known as the *stability of relative frequencies*. Such stability may be all that we need in order to describe and utilize chance phenomena (a point noted earlier in connection with life insurance policies). When the property of stability of relative frequencies is assumed to hold for *all* composite results of interest in a given experiment, we can use probability theory as a mathematical model for this chance phenomenon.

Before we turn to an exposition of probability theory, we should make two remarks. First, we do not always test the stability of relative frequencies in order to justify the use of probability theory. In many random experiments, it may be impossible or impractical to perform many repeated trials (although we can perhaps conceive of such trials in our imagination). Example 3.4 is an example of a random experiment where more than one

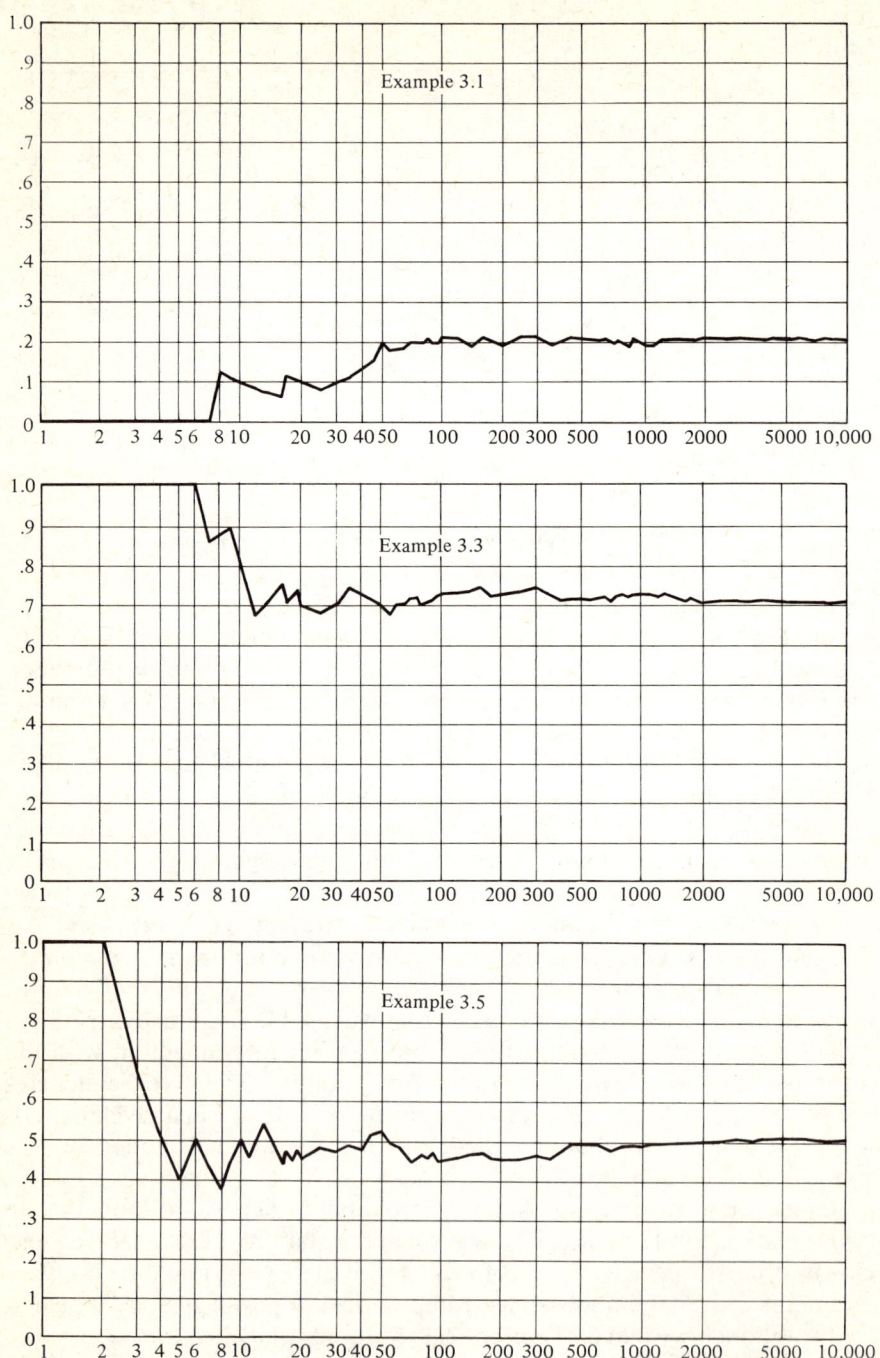

Figure 3.3 Relative frequency of the composite result E. (Trial number is on a logarithmic scale.)

trial is impossible (once a bulb burns out, it is gone forever), while in all of Examples 3.1 to 3.5, we might find that taking repeated trials is an expensive or time-consuming task. Even when repeated trials are not performed, however, we may have theoretical reasons for assuming that the relative frequencies are stable. The only test of this assumption then becomes the success of the model in describing and explaining the phenomenon. This brings us to our second point: *probability theory has proved itself successful* whether or not relative frequencies ever really stabilize.

NOTES AND REFERENCES

The following books and journal articles provide an opportunity for the reader interested in the history of probability to read further in this area.
1. Pearson, E. S., and Kendall, M. G. (1970). *Studies in the History of Statistics and Probability*. Hafner Publishing Company, New York.

This book contains a collection of historical articles. The articles dealing with probability are:

(i) David, F. N. Dicing and gaming (a note on the history of probability), pp. 1–17.

This article traces the history of dicing and gaming from prehistory, through classic Greece, and up to the late Renaissance. It includes comments on the early literature on gambling. Written for nonmathematicians with illustrated photographs.

(ii) Kendall, M. G. The beginnings of a probability calculus, pp. 19–34.

This article contains further history of dicing and gaming, emphasizing the development of the mathematical theory of probability as applied to games of chance.

(iii) Kendall, M. G. A note on playing cards, pp. 35–36.

The history and combinatorial aspects of card playing are discussed.

(iv) Kendall, M. G. The book of fate, pp. 37–38.

This article discusses the ancient practice of divination through chance mechanisms. Special attention is paid to a sixteenth century Italian book which told fortunes by means of playing cards, and an eighteenth century system for composing music according to the results of dice throws.

(v) Hasofer, A. M. Random mechanisms in Talmudic literature, pp. 39–43.

Biblical examples of lotteries used to determine boundaries, in civil law, in division of sacrificial offerings, and in the allocation of jobs are provided in this article.
2. David, F. N. (1962). *Games, Gods, and Gambling*. Hafner Publishing Company, New York.

This is a more extensive and inclusive treatment of the history of dicing and gaming than the above article, Reference 1(i).
3. Klein, M. (1962). *Mathematics: A Cultural Approach*. Addison-Wesley Publishing Company, Inc., Reading, Massachusetts.

The author devotes two chapters (Chapters 29 and 30) to the history and use of stochastic models and probability theory in science. The question of the application of other mathematical models in science is also given extensive discussion. The book is written for people who have had some exposure to mathematics at the level of the calculus, but most of the book can be easily read by students with a less advanced background.

4. Ore, Oyestein (1953). *Cardano, the Gambling Scholar*. Princeton University Press, Princeton, New Jersey.

Of primary interest is the material contained on pp. 143–241, which discusses the contributions of this great sixteenth century mathematician to the probabilistic theory of gambling.

5. Ore, Oyestein (1960). Pascal and the invention of probability theory. *American Mathematical Monthly*, vol. 67, pp. 409–419.

The story of Chevalier de Méré's problem, and how it stimulated Pascal to begin work on the mathematical study of games of chance.

6. Todhunter, Isaac (1865). *A History of the Mathematical Theory of Probability from the Time of Pascal to That of Laplace*. Reprinted by Chelsea Publishing Company, London, 1949.

The classical history of early developments in probability theory (1650–1825). For a brief biography of Todhunter himself, see M. G. Kendall, "Isaac Todhunter's history of the mathematical theory of probability," in *Biometrika*, vol. 50 (1963), pp. 204–205.

[*Note:* The reader who enjoys reading original source material may find interesting the series of essays on probability by Pierre Laplace, Charles S. Peirce, John Maynard Keynes, Henri Poincaré, and Ernest Nagel reproduced in *The World of Mathematics*, vol. 2, edited by James R. Newman (Simon and Schuster, New York, 1956). Another interesting original source is Laplace's popular exposition of probability theory available in English translation as *A Philosophical Essay on Probabilities* (with an introduction by E. T. Bell, Dover Publications, Inc., New York, 1951).]

EXERCISES

1. Give an example of a model that is used to describe phenomena in one field of science, but uses concepts and intuitions derived from an entirely different field of science. State some advantages and some limitations of such a model.

A2.[1] Consider the following experiment: A coin is tossed in the air and allowed to fall onto a flat, hard surface. The visible face of the coin is then recorded.
   (a) Is this experiment a *random* experiment? Give arguments to support your assertion.
   (b) What are the simple results of this experiment? Does the concept of a "simple result" of an experiment have meaning if the experiment is not a random experiment? Why or why not?

[1] Exercises preceded by A denote problems for which answers are provided; exercises preceded by * denote the more difficult problems.

(c) How can the coin-toss experiment described above be used to serve as a model for determining sex in a newborn child? Explain.

3. Give an example of a random experiment that has been of interest in some area of scientific research.
   (a) Why do you assert that the experiment you have chosen is a random experiment?
   (b) Describe the simple results of your experiment.
   (c) Describe one composite result of your experiment and, if possible, give the value p at which the relative frequency of this composite result stabilizes (or would stabilize) in repeated trials of the experiment.

A4. A study of purchasing intentions and actions of members of a consumers organization was made by Columbia University in 1962. As part of this study, the annual family income (in hundreds of dollars) of each member interviewed was recorded. Table E.1 gives the annual family income reported by each person in a sample of 49 members of the organization. This sample of people was taken only from the "young marrieds" in the organization (married adults, age 35–39, who had only one working adult in their family). Regard the results reported in this table as if they arose from 49 repeated trials of the random experiment in which a "young married" member of the consumers organization is chosen by lottery from among all "young married" members of the organization, and in which the annual family income of the person so chosen is recorded.

Table E.1. Annual family income (in hundreds of dollars) for a sample of 49 members of a consumers union.

Member Number	Annual Family Income	Member Number	Annual Family Income	Member Number	Annual Family Income
1	$085	18	$065	34	$075
2	135	19	147	35	083
3	200	20	084	36	092
4	050	21	120	37	060
5	070	22	092	38	105
6	076	23	090	39	088
7	105	24	100	40	055
8	065	25	074	41	085
9	121	26	070	42	110
10	105	27	110	43	096
11	130	28	100	44	100
12	065	29	075	45	066
13	120	30	069	46	072
14	100	31	067	47	060
15	127	32	060	48	095
16	150	33	120	49	090
17	105				

(a) What are the simple results of the random experiment?

(b) What is the relative frequency (in these 49 trials) of the composite result E that the annual family income of the "young married" individual chosen in the experiment is not greater than \$10,000 and not less than \$8500? What simple results are contained in this composite result?

5. Fifty phone numbers are drawn, one at a time, from the phone book of a very large city. Each such selection is made by means of a lottery in which *all* phone numbers are eligible to be chosen. Each phone number is called, and the person who answers is asked whether he or she is listening to a certain radio station.

(a) Do you see any relationships between this random experiment and the experiment described in Example 3.1? If so, what are the relationships?

(b) Do you see any relationships between this random experiment and the experiment described in Example 3.2? If so, what are the relationships?

A6. Consider the following experiment which is part of the conceptual basis of a theory of statistical linguistics [see Herdan (1964)]. In writing a letter, essay, book, or other document, an individual chooses one linguistic form (for example, a letter) at a time from among a collection of the permissible linguistic forms in his language (for example, the letters of the English language). Seen in this fashion, a sample of literary output from an individual is composed of repeated trials of the experiment which consists of choosing one such linguistic form from among all permissible linguistic forms. Without knowledge of the context in which a given choice of linguistic form is made, the particular form which is chosen will be unpredictable. For example, without knowing what is being said in a given sentence, we could not say

Table E.2. Frequency of occurrence of the letters of the English alphabet as taken from the initial letters of the nouns in a sample of the writing of John Bunyan.

Initial Letter of Noun	Frequency of Occurrence	Initial Letter of Noun	Frequency of Occurrence
A	111	N	40
B	147	O	41
C	210	P	188
D	153	Q	7
E	69	R	133
F	112	S	256
G	72	T	112
H	110	U	16
I	72	V	43
J	22	W	100
K	18	X	0
L	84	Y	5
M	124	Z	1

with certainty what letter begins the first noun in that sentence. In a long literary selection, a sufficient variety of contexts is present so that an individual's style, rather than the subject matter, determines the relative frequency with which such linguistic forms appear. Table E.2 [taken from Herdan (1964)] gives the frequency with which the letters A–Z of the English alphabet occur as the first letter of the nouns used in a sample from the works of the English author John Bunyan (1628–1688).
(a) What is the relative frequency of the composite result "The letter chosen is a vowel"?
(b) What letter has the largest relative frequency?
(c) Construct a table (similar to Table E.2) based on a sample (two or three typewritten pages in length) of your own writing, and answer questions (a) and (b) for your own writing style.

7. Consider the following experiment: With a tape measure, a person measures the length of the longest wall in a given room to the nearest $(\frac{1}{8})$th of an inch.
(a) Do you believe that this experiment is a random experiment? Why or why not?
(b) What are the possible simple results of this experiment?
(c) Twenty-five people were asked to perform this experiment. Each person who performed the measurement was not told of the result obtained by any of the other measurers. The results obtained are shown in Table E.3. Regard each person's measurement as a repeated trial of the measuring experiment described above. Find the relative frequency of each simple result that you actually observe in these repeated trials. Based on these observations, what conclusions do you draw about the variability of the measurements which people obtain for the length of a given object?

Table E.3. Lengths of a given wall in a given room obtained by 25 different people.

Person	Length Obtained	Person	Length Obtained
1	20 ft. $8\frac{1}{8}$ in.	14	20 ft. $7\frac{3}{8}$ in.
2	20 ft. $8\frac{1}{2}$ in.	15	20 ft. $8\frac{1}{8}$ in.
3	20 ft. $8\frac{3}{8}$ in.	16	20 ft. 8 in.
4	20 ft. $7\frac{7}{8}$ in.	17	20 ft. $8\frac{1}{4}$ in.
5	20 ft. 8 in.	18	20 ft. $7\frac{7}{8}$ in.
6	20 ft. $8\frac{1}{8}$ in.	19	20 ft. $7\frac{3}{4}$ in.
7	20 ft. $7\frac{3}{4}$ in.	20	20 ft. 8 in.
8	20 ft. $7\frac{7}{8}$ in.	21	20 ft. $8\frac{1}{8}$ in.
9	20 ft. $8\frac{1}{2}$ in.	22	20 ft. $7\frac{1}{2}$ in.
10	20 ft. $8\frac{1}{4}$ in.	23	20 ft. $8\frac{1}{4}$ in.
11	20 ft. $7\frac{5}{8}$ in.	24	20 ft. 8 in.
12	20 ft. 8 in.	25	20 ft. $8\frac{3}{8}$ in.
13	20 ft. $7\frac{3}{4}$ in.		

(d) Verify your conclusions in part (c) by performing a similar experiment using the members of your class (or at least 10 of your acquaintances).

^{A}8. A standard deck of cards is shuffled several times and cut by a blindfolded individual. After the deck has been cut, the top card in the deck is turned up and the suit (clubs, spades, hearts, and diamonds) and the value (ace, deuce, ..., jack, queen, king) of the card are noted. If the shuffling is well done, the result of this process is one trial of a certain random experiment.
(a) What are the simple results of this experiment?
(b) Consider the composite result E_1 consisting of simple results in which the card drawn is a heart. List the simple results belonging to E_1.
(c) If you conducted many trials of the above experiment, at what number p do you think the relative frequency r.f.(E_1) would stabilize? (That is, to what number p does r.f.(E_1) tend as the number of trials becomes large?) Why?
(d) Consider the composite result E_2 consisting of simple results in which the card drawn is a "face card" (king, queen, jack). List the simple results belonging to E_2.
(e) If you conducted many trials of the above experiment, to what number p do you think r.f.(E_2) would tend as the number of trials becomes large? Why?

9. Consider the following random experiment: You stand on the corner of an intersection of two streets, and face along one of the intersecting streets. You observe the last digit (number) on the license plate of the first car to pass you headed in the direction toward which you are facing.
(a) What are the simple results of this experiment?
(b) Repeat this experiment 100 times. What is the relative frequency of the composite result "The last digit on the license plate is k"? Answer this question for each of the numbers $k = 0, 1, 2, 3, 4, 5, 6, 7, 8$, and 9.

10. Consider the same random experiment as in Question 9 above, but assume that for every car observed, we record the last digit (number) on the license plate, and *also* record whether the car is of American make or foreign make.
(a) What are the simple results of this random experiment?
(b) If you observed a large number of cars, at what number p do you think the relative frequency of the composite result "The car observed is of foreign make" would stabilize? Explain your answer. Do you think the relative frequency of this composite result would ever stabilize at a given number? If not, explain.

^{A}11. Ten cards numbered 0, 1, 2, ..., 9, respectively, are put into a hat and shuffled. One card is drawn and its number is recorded. It is then replaced, the hat shaken again, and another card is drawn and recorded. This is done 100 times, resulting in the following array of numbers:

```
2 3 1 5 7 5 4 8 5 9 0 1 8 3 7 2 5 9 9 3
7 6 2 4 9 7 0 8 8 6 9 5 2 3 0 3 6 7 4 4
0 5 5 4 5 5 5 0 4 3 1 0 5 3 7 4 3 5 0 8
9 0 6 1 1 8 3 7 4 4 1 0 9 6 2 2 1 3 4 3
1 4 8 7 1 6 0 3 5 0 3 2 4 0 4 3 6 2 2 3
```

(a) Summarize the data by making a table of frequencies and relative frequencies for the results "The card chosen has the number k written on it," where k can equal 0, 1, 2, 3, 4, 5, 6, 7, 8, or 9. That is, fill in the blank spaces in the following table:

k	Frequency of: THE RESULT "THE CARD CHOSEN HAS THE NUMBER k WRITTEN ON IT"	Relative Frequency of:
0		
1		
2		
3		
4		
5		
6		
7		
8		
9		

(b) Which of the experiments described in Questions 1–10 does the experiment described in this question most closely resemble in form? Explain your answer.

12. A meteorologist is conducting a study of the effectiveness of cloud-seeding techniques for producing rain. He has selected a certain type of cloud to seed. Before beginning his experiments, however, he wants to determine the natural tendency of this type of cloud to produce rain. Thus, he chooses one cloud of this type at a time, using a method which, he believes, produces the same result as if he had sampled the chosen cloud by lottery from among all clouds of the given type. Each cloud that he observes is assigned to one of the following three nonoverlapping categories:

Category A: The cloud dissipates before producing rain.
Category B: The cloud produces rain somewhere else, but not on the area over which it was first observed by the meteorologist.

Category C: The cloud produces rain on the area over which it was first observed.

The category of each cloud selected by the meteorologist can be regarded as the simple result of a trial of the experiment in which a cloud of the given type is selected, and the cloud is assigned to one of the above three categories. Table E.4 gives the results of the meteorologist's observations on 100 clouds of the given type, listed in the order in which these clouds were observed.

Table E.4. Results of observation of 100 clouds.

Trial Number	1	2	3	4	5	6	7	8	9	10
Category of Observed Cloud	A	A	A	A	B	B	C	B	A	B

Trial Number	11	12	13	14	15	16	17	18	19	20
Category of Observed Cloud	C	A	C	C	B	B	A	A	B	C

Trial Number	21	22	23	24	25	26	27	28	29	30
Category of Observed Cloud	B	A	A	A	A	B	A	A	A	C

Trial Number	31	32	33	34	35	36	37	38	39	40
Category of Observed Cloud	A	A	B	C	A	B	C	A	A	B

Trial Number	41	42	43	44	45	46	47	48	49	50
Category of Observed Cloud	A	B	B	A	B	A	A	C	C	A

Trial Number	51	52	53	54	55	56	57	58	59	60
Category of Observed Cloud	C	A	A	A	B	B	B	A	A	A

Trial Number	61	62	63	64	65	66	67	68	69	70
Category of Observed Cloud	C	C	A	B	A	A	A	B	C	C

Trial Number	71	72	73	74	75	76	77	78	79	80
Category of Observed Cloud	C	A	A	A	A	B	A	A	A	B

Trial Number	81	82	83	84	85	86	87	88	89	90
Category of Observed Cloud	C	B	A	A	B	A	B	A	B	B

Trial Number	91	92	93	94	95	96	97	98	99	100
Category of Observed Cloud	C	A	B	B	A	B	A	C	A	B

(a) Consider the composite result "The observed cloud produces rain somewhere." Construct a table showing how the relative frequency of this composite result varies with the number of trials. Graph the relative frequencies so obtained against the number of trials for which they were obtained (see Figures 3.1 and 3.2). Do the relative frequencies seem to stabilize? If so, around what number p do the relative frequencies stabilize?

(b) Consider the result "The cloud does not produce rain on the area over which it was first observed." Construct a table and a graph for this result similar to the table and graph constructed in part (a) of this question. Do the relative frequencies seem to stabilize? If so, around what number p do the relative frequencies stabilize?

(c) Based on the given data, do you believe that we can use probability theory as a mathematical model for the experiment described in this question? Explain your answer. When can we use probability theory to serve as a model for a chance phenomenon? Are these conditions met for the present experiment?

13. (a) Using the data on annual family incomes of 49 "young married" members of a consumers organization given in Table E.1, construct a table showing how the relative frequency of the composite result "The annual family income of the member is $10,000 or higher" varies with the number of trials. Graph the relative frequencies so obtained against the number of trials for which they were obtained. [*Note:* Let member number 1 correspond to the first trial, member number 2 correspond to the second trial, and so on.] Do the relative frequencies stabilize?

(b) Construct a table and graph [similar to the table and graph described in part (a)] showing how the relative frequency of the composite result "The annual family income of the member is no less than $8500 and no greater than $10,000" varies with the number of trials. Do the relative frequencies stabilize?

*^14. A medical researcher has studied the performance of a diagnostic computer. Such a computer receives a medical history and a list of symptoms from a patient, compares this information with summaries of the histories and symptoms of previous patients who had diseases of known type, and on this basis makes a diagnosis of the new patient's problem. For each patient diagnosed by the computer, a team of physicians also makes a diagnosis (the physicians are not informed of the computer's diagnosis). If the computer's diagnosis for the patient agrees with the diagnosis of the doctors, an "A" (for agreement) is recorded. Otherwise, a "D" is recorded (for disagreement). Eighty patients are diagnosed both by the computer and by the doctors. The results obtained are shown (in order of their occurrence) in Table E.5.

(a) The table can be regarded as giving the results of 80 trials of a certain random experiment. What is that experiment, and what are the simple results of that experiment?

(b) List *all* of the different composite results for this experiment. How many such composite results are there?
(c) On the basis of the data given in Table E.5, verify by an explicit argument that the random experiment which you described in step (a) can be modeled using probability theory. What do you need to show?

Table E.5. Results of the diagnosis of 80 patients.

Patient Number	1	2	3	4	5	6	7	8	9	10	11	12
Result	A	A	A	D	A	A	A	A	A	D	A	A

Patient Number	13	14	15	16	17	18	19	20	21	22	23	24
Result	A	A	A	A	A	D	D	A	A	A	A	A

Patient Number	25	26	27	28	29	30	31	32	33	34	35	36
Result	A	A	A	A	A	D	A	A	A	A	A	A

Patient Number	37	38	39	40	41	42	43	44	45	46	47	48
Result	A	A	A	D	A	A	A	A	D	A	A	D

Patient Number	49	50	51	52	53	54	55	56	57	58	59	60
Result	A	A	A	A	A	A	D	A	A	A	A	A

Patient Number	61	62	63	64	65	66	67	68	69	70	71	72
Result	A	A	A	A	A	A	D	A	A	A	A	A

Patient Number	73	74	75	76	77	78	79	80
Result	D	A	A	A	D	A	A	A

2

THE ELEMENTS OF PROBABILITY THEORY

> "That's not the regular rule: you invented it just now."
> "It's the oldest rule in the book," said the King.
> "Then it ought to be Number One," said Alice.
> Lewis Carroll (Alice in Wonderland)

> ... it is clear that nobody who does not understand insurance and comprehend in some degree its enormous possibilities is qualified to meddle in national business. And nobody can get that far without at least an acquaintance with the mathematics of probability....
> George Bernard Shaw (The Vice of Gambling and the Virtue of Insurance).

1. PRELIMINARIES

We have indicated how the understanding of chance phenomena is an important part of almost all scientific and technological disciplines. We have also noted that it is the long-run (that is, over many repeated trials) relative frequencies of certain composite results of random experiments that are useful in describing and understanding chance phenomena. Thus, our discussion has isolated the four basic elements which are of major common interest in all chance phenomena: random experiments, simple results, composite results, and relative frequencies. In line with our previous remarks on the construction of mathematical models (Chapter 1, Section 2) we now seek a mathematical structure with entities that can be made to correspond to these basic elements of chance phenomena, and with axioms that describe the properties of these basic elements.

2. FORMAL STRUCTURE: OUTCOMES AND EVENT SETS

Corresponding to the simple results of our random experiment, we conceive of basic mathematical entities in our model called *outcomes*. Symbolically, we denote an outcome by the lower case Greek letter omega (ω). When we want to distinguish among several outcomes related to the same random experiment, we enumerate them by means of subscripts (that is, ω_1, ω_2, ω_3, ω_4, and so on). The collection of all outcomes related to a given random experiment is called the *sample space*, denoted by the capital Greek letter omega (Ω).

A more intuitive and apt term to denote the collection of all outcomes would be *sample universe*, since it is indeed the universe of discourse which is being defined. However, the terms "point" and "space" have a long history in the development of mathematics, particularly in geometry where these words have physical meaning. These terms have thus become somewhat standard in mathematical systems with "points" referring to the undefined notions of a theory, and "space" to the totality of "points." In keeping with this tradition, we call the collection of all outcomes the sample space.

In mathematics, an important concept is that of a *set*. A set is a well-defined collection of entities (such as, for example, names or numbers). When we speak of a collection of outcomes, we call such a collection an *event set*. Since simple results in the random experiment correspond to outcomes in the probability model, and since composite results are collections of simple results, event sets in the model must correspond to composite results of a random experiment. Consequently, we will use the same symbol E to represent both event sets and composite results.

Event sets E are frequently defined by giving a proposition or property (say, property $\mathscr{P}$) which is satisfied by every outcome in the event set E, and which is not satisfied by every outcome not in E. We write such event sets as follows:

(2.1) $$E = \{\omega: \omega \text{ satisfies property } \mathscr{P}\}.$$

Here the braces are a symbolic shorthand meaning "set of," and the colon means "such that." Equation (2.1) may be rephrased in words by saying, "E is the event set of outcomes ω such that ω has property $\mathscr{P}$." Sometimes we adopt a succinct equivalent to (2.1), namely,

(2.2) $$E = \{\omega \text{ satisfies property } \mathscr{P}\}.$$

Still another way of describing an event set E is by enumeration of all the outcomes that belong to E. Thus, if E is the event set consisting of the outcomes ω_1, ω_2, and ω_6, we write this fact as

$$E = \{\omega_1, \omega_2, \omega_6\}.$$

[*Note:* We have used the capital letter E to denote an event set. If we use this letter to denote every event set that comes up in discussion, our ability to distinguish among event sets decreases rapidly as the number of problems that we consider increases. Consequently, it is necessary to use different symbols (such as the letters A, B, C) or numerical subscripts (for example, E_1, E_2, and so on) to denote event sets. If the choice of symbols is clearly indicated, no confusion should result, and the added flexibility of notation provides dividends in brevity and convenience. Thus, in the following examples, we utilize different letter symbols in order to keep the event sets in each example notationally distinct from the event sets of the other examples. Further, within a given example, we subscript event sets to distinguish them from one another.]

Example 2.1. Consider the sample space (universe) Ω whose outcomes consist of the first names of children in a given school class. For specificity, let the outcomes be $\omega_1 =$ John, $\omega_2 =$ Sue, $\omega_3 =$ Helen, $\omega_4 =$ Jane, $\omega_5 =$ Harry, $\omega_6 =$ Elizabeth. Although each of the outcomes ω_1, ω_2, ..., ω_6 is eventually to be associated with a simple result of a random experiment (say, the experiment in which one name is drawn from a hat), we think of them now as mathematical entities entirely divorced from any experimental context.[1] A particular event set that we can consider is the event set A_1 consisting of those outcomes that are boys' names; that is,

$$A_1 = \{\omega: \omega \text{ is a boy's name}\} = \{\omega_1, \omega_5\}.$$

Another event set A_2 consists of names beginning with the letter "J"; that is,

$$A_2 = \{\omega: \omega \text{ begins with a "J"}\} = \{\omega_1, \omega_4\}.$$

The sample space $\Omega = \{\omega_1, \omega_2, \omega_3, \omega_4, \omega_5, \omega_6\}$ of this example is itself an event set. Still another event set is the set of names beginning with the letter "K." This is an event set that has *no* outcomes as members, and is called the *null event set*. For the null event set we reserve the special symbol $\emptyset$. Notice that in the context of this example $\emptyset$ can be constructed in many ways. We have already observed that $\emptyset$ is the event set consisting of outcomes (names) beginning with "K"; it is also the event set of names having more than ten letters, the event set of names starting and ending with the same letter, the event set of names ending with the letter "q," and so on.

[1] The terminology "outcome" and "event set" may confuse the reader somewhat at this point since these terms seem to refer to a happening outside the mathematical universe under discussion. If it helps the reader, he can substitute the less connotative terms "element" and "set" for "outcome" and "event set," respectively. This substitution is appropriate for Sections 2 through 4, since in these sections we are really talking about the more general mathematical theory of sets. We have chosen not to introduce set theory terminology in this book, because we believe that such generality is unnecessary and potentially distracting. After working a few examples, the reader should begin to feel more comfortable with the present terminology. Although at present we are distinguishing in our terminology between the mathematical model (outcomes, event sets) and the random experiment (simple results, composite results), we intend to drop the distinction once the analogy between these structures has been completely exposited.

Example 2.2. Another sample space Ω might consist of the letters of the English alphabet. Here the outcomes would be $\omega_1 = $ a, $\omega_2 = $ b, $\omega_3 = $ c, ..., $\omega_{25} = $ y, $\omega_{26} = $ z. One event set B_1 is the event set consisting of outcomes that are vowels. Thus,

$$B_1 = \{a,e,i,o,u\} = \{\omega_1, \omega_5, \omega_9, \omega_{15}, \omega_{21}\}.$$

Another event set B_2 is the event set consisting of the letters (outcomes) that make up the word "probability"; that is,

$$B_2 = \{a,b,i,l,o,p,r,t,y\} = \{\omega_1, \omega_2, \omega_9, \omega_{12}, \omega_{15}, \omega_{16}, \omega_{18}, \omega_{20}, \omega_{25}\}.$$

Finally, we can consider the event set B_3 made up of the letters (outcomes) which, when capitalized, look the same right-side-up or upside-down. Then

$$B_3 = \{h,i,n,o,s,x,z\} = \{\omega_8, \omega_9, \omega_{14}, \omega_{15}, \omega_{19}, \omega_{24}, \omega_{26}\}.$$

In this example, the null event set $\emptyset$ can be described as the event set of letters *not* appearing in the sentence, "The quick brown fox jumps over the lazy dog."

Example 2.3. As a third example of a sample space, consider the sample space consisting of possible rankings on a scale of ten units (a rank of 1 is high and a rank of 10 is low). The outcomes of this sample space are $\omega_1 = 1$, $\omega_2 = 2$, $\omega_3 = 3$, $\omega_4 = 4$, ..., $\omega_9 = 9$, $\omega_{10} = 10$. One event set C_1 consists of all ranks (outcomes) larger than 5; that is,

$$C_1 = \{\omega : \omega \text{ is larger than } 5\} = \{\omega_6, \omega_7, \omega_8, \omega_9, \omega_{10}\}.$$

Another event set C_2 consists of the outcomes (ranks) that are even integers. Thus,

$$C_2 = \{\omega : \omega \text{ is an even integer}\} = \{\omega_2, \omega_4, \omega_6, \omega_8, \omega_{10}\}.$$

Among other event sets are C_3, the event set consisting of ranks between 3 and 6, and C_4, the event set of ranks divisible by 4.

Equality of Event Sets

It is often possible to describe the same event set in many ways. For instance, in Example 2.3, the event set C_2 is described as $\{\omega : \omega \text{ is an even integer}\}$. We could have also described C_2 as $\{\omega : \omega \text{ is divisible without remainder by 2}\}$ or as $\{\omega : \omega \text{ is not an odd integer}\}$. Suppose that we mistakenly think that the event set $\{\omega : \omega \text{ is divisible without remainder by 2}\}$ is an event set C_2^* different from the event set C_2. One way of correcting our error and showing that C_2 and C_2^* are the same event set (that is, C_2 equals C_2^*) is to show that C_2 and C_2^* have the same outcomes as members: *Two event sets are equal if and only if they contain the same outcomes.*

To demonstrate that C_2 and C_2^* have the same outcomes as members, we may directly verify that every outcome which belongs to C_2 also belongs to C_2^*, and that every outcome which belongs to C_2^* also belongs to C_2. However, it is often more convenient to verify that every outcome which belongs to one of the event sets (say, C_2) also belongs to the other, and that every outcome which does *not* belong to C_2 also does not belong to C_2^*. Using this latter approach, we note that since all even integers are divisible (without remainder) by 2, it follows that every outcome ω in C_2 also is an outcome of C_2^*. On the other hand, since each odd integer is not divisible (without remainder) by 2, it follows that every outcome ω which does not belong to C_2 also does not belong to C_2^*. Hence, we have shown that the event sets C_2 and C_2^* contain the same outcomes, and, in consequence, are the same event set.

Inclusion of Event Sets

The argument used above to show that C_2 and C_2^* are equal (are the same event set) illustrates another important concept, that of inclusion. *An event set E_1 is said to include another event set E_2 if every outcome belonging to E_2 also belongs to E_1.* This relation is illustrated in Figure 2.1 [called a *Venn diagram* in honor of the English logician John Venn (1834–1923)]. In the figure, the dots represent outcomes ω, the event set E_1 is represented by the larger oval, and the event set E_2 is represented by the smaller oval (contained in E_1). Notice that every dot (outcome) contained in E_2 is also contained in E_1.

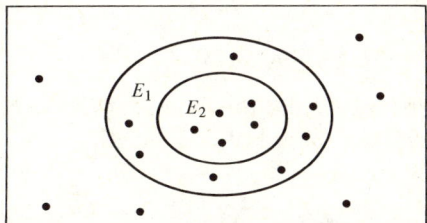

Figure 2.1 Venn diagram illustrating inclusion.

If two event sets each include the other, then these event sets are equal. This fact has already been used above in showing that the event sets C_2 and C_2^* (from Example 2.3) are equal. If the event set E_1 includes the event set E_2, but E_2 does not include E_1, then we say that E_2 is a *proper subset* (sub-event set) of E_1. (This situation is illustrated in Figure 2.1.) Thus, in Example 2.3, the event set C_4 is a proper subset of the event set C_2.

30 THE ELEMENTS OF PROBABILITY THEORY 2

It is possible for two event sets to be defined for a given random experiment, and for neither event set to include the other. Example 2.3 again provides us with an example: event sets C_1 and C_2. Other examples are A_1 and A_2 in Example 2.1, and B_1 and B_2 in Example 2.2.

A special notation is used to denote inclusion of event sets. To represent the relation "E_2 is included in E_1," we write $E_2 \subset E_1$. It is sometimes convenient to say "E_1 includes E_2"; the notation for this statement is then $E_1 \supset E_2$.

The Union of Event Sets

In elementary algebra, we learn that there is an algebra of numbers that allows us to construct new numbers from given numbers. Starting from the number 1, for example, we can obtain the remaining positive integers (2, 3, 4, and so on) by the operation of addition. Similarly, there is an algebra of event sets. If we start with any given list of event sets, we can perform certain operations on these event sets in order to make new event sets. Three operations on event sets that are of importance in probability theory are *union*, *intersection*, and *complementation*.

Consider two event sets, say E_1 and E_2. The *union* $E_1 \cup E_2$ (in words, E_1 *union* E_2) of these two event sets is defined to be the event set consisting of all outcomes that are members of E_1, of E_2, or of both E_1 and E_2. A visual picture (Venn diagram) of E_1, E_2, and their union $E_1 \cup E_2$ is given in Figure 2.2. Here the vertically lined area represents E_1, the horizontally lined area represents E_2, and all the lined areas together represent $E_1 \cup E_2$. In terms of our symbolic conventions, we can write the definition of the union of E_1 and E_2 as

(2.3) $\quad E_1 \cup E_2 = \{\omega : \omega \text{ is a member of } E_1, \text{ of } E_2, \text{ or of both } E_1 \text{ and } E_2\}$.

Notice that the event set E_1 and the event set E_2 are both subsets of $E_1 \cup E_2$; that is, $E_1 \subset (E_1 \cup E_2)$ and also $E_2 \subset (E_1 \cup E_2)$. This fact may be perceived visually in Figure 2.2.

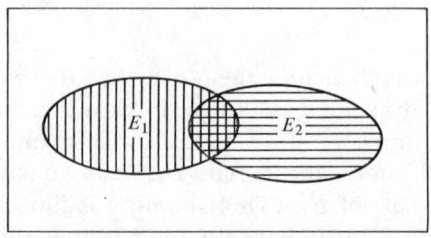

Figure 2.2 Venn diagram illustrating union of two sets.

2. FORMAL STRUCTURE: OUTCOMES AND EVENT SETS 31

[*Note*: Other notations for the operation $E_1 \cup E_2$ are used in certain textbooks of probability theory. Since the union of two sets has many (*but not all*) of the properties that the addition of two numbers has, some authors use the notation "$E_1 + E_2$." On the other hand, since $E_1 \cup E_2$ consists of all outcomes in E_1 *or* in E_2 (or both), other authors use the notation "E_1 *or* E_2." The first of these notations is convenient, but can be misleading; the second can become cumbersome. Nevertheless, because the notation "E_1 *or* E_2" may be of intuitive aid, we use both "$E_1 \cup E_2$" and "E_1 *or* E_2" in the early sections.]

The Intersection of Event Sets

Given two event sets E_1 and E_2, we can consider the event set which is the *intersection* of E_1 and E_2, namely, the event set $E_1 \cap E_2$ (E_1 *intersection* E_2) whose elements are members of *both* E_1 and E_2. A Venn diagram of E_1, E_2, and $E_1 \cap E_2$ is shown in Figure 2.3, where again E_1 is vertically lined, E_2 is horizontally lined, and $E_1 \cap E_2$ is the cross-hatched region.

In our set notation conventions, we can write $E_1 \cap E_2$ as

(2.4) $\qquad E_1 \cap E_2 = \{\omega: \omega \text{ is a member of } E_1 \text{ } and \text{ of } E_2\}$.

Note that $E_1 \cap E_2$ is a subset of E_1 and of E_2; that is, $(E_1 \cap E_2) \subset E_1$ and $(E_1 \cap E_2) \subset E_2$. This fact can be seen pictorially in Figure 2.3.

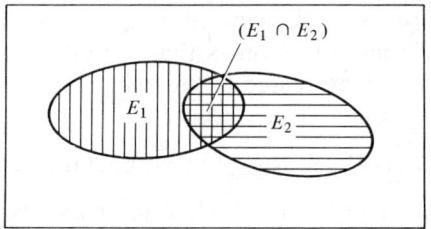

Figure 2.3 Venn diagram illustrating intersection of two sets.

[*Note:* Two alternative notations that are frequently used for the intersection of two event sets are "$E_1 E_2$" and "E_1 *and* E_2." The former notation is used to suggest the similarity of set intersection and numerical multiplication; the justification for the latter notation comes directly from the definition of set intersection. We shall use the notations "$E_1 \cap E_2$" and "E_1 *and* E_2" interchangeably during the early sections.]

The Complement of an Event Set

The *complement* of an event set E (denoted E^c; in words, "E-complement") is the event set consisting of all outcomes in the sample space Ω which are not members of E. A Venn diagram of E and E^c is given in Figure 2.4, where the entire box represents the sample space Ω and the

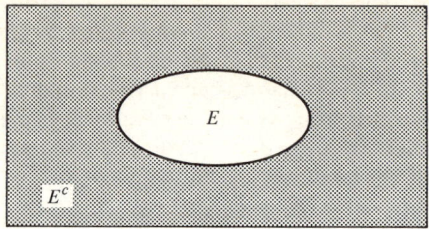

Figure 2.4 Venn diagram illustrating the complement of a set.

shaded region represents E^c. In our set notation, we write E^c by saying

(2.5) $$E^c = \{\omega: \omega \text{ is not a member of } E\}.$$

[*Note:* Alternative notations found in probability literature are "$\sim E$," "$\bar{E}$," and "*not E*." Of these, the last will be used together with "E^c" during the early sections.]

As an example of the application of these three operations on event sets, let us refer back to Example 2.1 where the sample space consisted of $\omega_1 =$ John, $\omega_2 =$ Sue, $\omega_3 =$ Helen, $\omega_4 =$ Jane, $\omega_5 =$ Harry, $\omega_6 =$ Elizabeth. There we discussed the event sets $A_1 = \{\omega: \omega \text{ is a boy's name}\} = \{\omega_1, \omega_5\}$, and $A_2 = \{\omega: \omega \text{ starts with the letter "J"}\} = \{\omega_1, \omega_4\}$. If we want the event set consisting of outcomes that are either boys' names or start with a "J," then we are interested in the event set $A_1 \cup A_2$ ("A_1 or A_2"). This event set is

$$A_1 \cup A_2 = \{\omega_1, \omega_4, \omega_5\} = \{\text{John, Jane, Harry}\}.$$

On the other hand, if we want those outcomes (names) that are boys' names beginning with a "J," then we are interested in the event set $A_1 \cap A_2$ ("A_1 *and* A_2"), where

$$A_1 \cap A_2 = \{\omega_1\} = \{\text{John}\}.$$

Finally, since all names that are not boys' names in our example are girls' names, to find the event set consisting of outcomes that are girls' names, we need only consider the complement of A_1 ("*not* A_1"), namely,

$$A_1^c = \{\omega_2, \omega_3, \omega_4, \omega_6\} = \{\text{Sue, Helen, Jane, Elizabeth}\}.$$

The operations of union and intersection on the event sets C_1 and C_2 described in Example 2.3 yield

$$C_1 \cup C_2 = \{\omega: \omega \text{ is greater than 5 } or \text{ } \omega \text{ is even}\}$$
$$= \{\omega_2, \omega_4, \omega_6, \omega_7, \omega_8, \omega_9, \omega_{10}\},$$

and

$$C_1 \cap C_2 = \{\omega: \omega \text{ is greater than 5 } and \text{ } \omega \text{ is even}\}$$
$$= \{\omega_6, \omega_8, \omega_{10}\}.$$

The complement of C_2 ("*not C_2*") is

$$C_2{}^c = \{\omega: \omega \text{ is not even}\} = \{\omega: \omega \text{ is odd}\} = \{\omega_1, \omega_3, \omega_5, \omega_7, \omega_9\}.$$

We can also consider the union of three event sets. The union $E_1 \cup E_2 \cup E_3$ ("*E_1 or E_2 or E_3*") of the event sets E_1, E_2, E_3 is the event set whose outcomes each belong to one (or more) of these event sets (see the shaded area of Figure 2.5). Notice that we can regard $E_1 \cup E_2 \cup E_3$ as being the union of E_1 with $E_2 \cup E_3$, the union of E_2 with $E_1 \cup E_3$, or the union of E_3 with $E_1 \cup E_2$. Thus, if we have four event sets E_1, E_2, E_3, and E_4, we could define the union $E_1 \cup E_2 \cup E_3 \cup E_4$ ("*E_1 or E_2 or E_3 or E_4*") as the union of E_4 with $E_1 \cup E_2 \cup E_3$, or as the union of E_1 with $E_2 \cup E_3 \cup E_4$, and so on. However, a more convenient, but equivalent, definition of $E_1 \cup E_2 \cup E_3 \cup E_4$ is to say that the outcomes of this event set are members of one (or more) of the event sets E_1, E_2, E_3, E_4. In general, the union of k event sets $E_1 \cup E_2 \cup \cdots \cup E_k$ ("*E_1 or E_2 or $\cdots$ or E_k*") is defined to be the event set whose outcomes belong to one or more of the event sets $E_1, E_2, \ldots, E_k$.

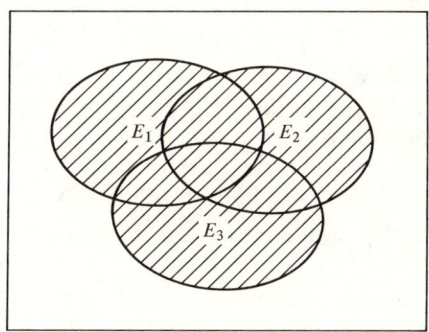

Figure 2.5 Venn diagram illustrating union of three sets.

Analogously, we can define the intersection of three event sets. Given E_1, E_2, E_3, their intersection $E_1 \cap E_2 \cap E_3$ ("*E_1 and E_2 and E_3*") is the event set whose outcomes are members of *all* of the event sets E_1, E_2, E_3 (the shaded area in Figure 2.6). The event set $E_1 \cap E_2 \cap E_3$ can be regarded as being the intersection of E_1 with $E_2 \cap E_3$, of E_2 with $E_1 \cap E_3$, or of E_3 with $E_1 \cap E_2$. Thus, if we have another event set E_4 in addition to E_1, E_2, and E_3, we could think of defining $E_1 \cap E_2 \cap E_3 \cap E_4$ to be the intersection of E_4 and $E_1 \cap E_2 \cap E_3$. However, an equivalent and more

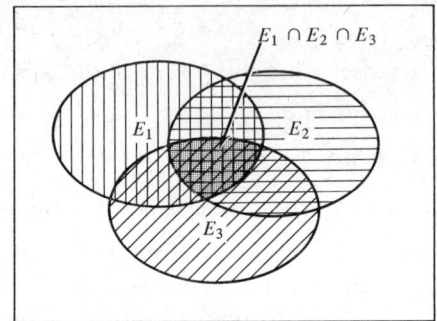

Figure 2.6 Venn diagram illustrating intersection of three sets.

flexible definition is to define $E_1 \cap E_2 \cap E_3 \cap E_4$ as the event set whose outcomes belong to all of the event sets E_1, E_2, E_3, E_4. The intersection $E_1 \cap E_2 \cap \cdots \cap E_k$ ("E_1 and E_2 and $\cdots E_k$") of $E_1, E_2, \ldots, E_k$ is correspondingly defined to be the event set whose outcomes are members of all of the event sets $E_1, E_2, \ldots, E_k$.

In Example 2.2, the union of B_1, B_2, and B_3 is the event set consisting of all letters (outcomes) that are either vowels, form part of the word "probability," or look the same (when capitalized) right-side-up or upside-down. Recalling that $\omega_1 = a$, $\omega_2 = b$, ..., $\omega_{26} = z$, we conclude that

$$B_1 \cup B_2 \cup B_3 = \{\omega_1, \omega_2, \omega_5, \omega_8, \omega_9, \omega_{12}, \omega_{14}, \omega_{15}, \omega_{16}, \omega_{18}, \omega_{19}, \omega_{20}, \omega_{21}, \omega_{24}, \omega_{25}, \omega_{26}\}.$$

On the other hand, the event set consisting of all letters that are vowels appearing in the word "probability," and that look the same right-side-up as upside-down, is the event set

$$B_1 \cap B_2 \cap B_3 = \{\omega_9, \omega_{15}\}.$$

That is, $B_1 \cap B_2 \cap B_3$ contains only the outcomes (letters) "i" and "o."

Mutually Exclusive Event Sets

If two event sets have no outcomes in common (their intersection is the null event set ∅), then we say that they are *mutually exclusive* or *disjoint* (see Figure 2.7).

Example 2.4. Suppose our sample space Ω consists of all of the chemical elements. Since this sample space is quite large, we content ourselves with listing only three such outcomes (chemical elements): ω_1 = hydrogen, ω_2 = helium, ω_3 = lithium. Let D_1 be the event set consisting of outcomes that are chemical elements belonging to the family of halogens; that is,

$$D_1 = \{\text{bromine, chlorine, flourine, iodine, astatine}\}.$$

2. FORMAL STRUCTURE: OUTCOMES AND EVENT SETS 35

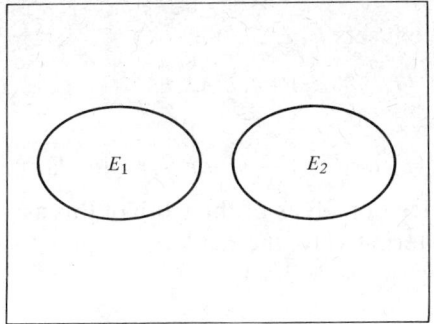

Figure 2.7 Venn diagram illustrating disjoint sets.

Let D_2 be the event set consisting of all metals. Then D_1 and D_2 have no outcomes (chemical elements) in common. The event sets D_1 and D_2 are thus mutually exclusive—a chemical element cannot be both a halogen and a metal. On the other hand, the event set D_3 consisting of chemical elements that are solids has (at least) one outcome (chemical element) in common with D_1, namely the outcome iodine. Thus, D_1 and D_3 are not mutually exclusive.

Some Useful Rules

From the definitions of union, intersection, and complementation, we can derive the following frequently useful rules:

Rule 2.1. The union of an event set and its complement is the sample space. That is, for any event set E, we have

(2.6) $$E \cup E^c = \Omega.$$

It is perhaps easier to see this fact if we say: "E or 'not E'" $= \Omega$.

Rule 2.2. The complement of Ω is the null event set $\emptyset$. The complement of $\emptyset$ is Ω.

Rule 2.3. For any event set E we have

(a) $E \cap \Omega = E$ (E and $\Omega = E$),
(b) $E \cap \emptyset = \emptyset$ (E and $\emptyset = \emptyset$),
(c) $E \cup \Omega = \Omega$ (E or $\Omega = \Omega$),
(d) $E \cup \emptyset = E$ (E or $\emptyset = E$),
(e) $(E^c)^c = E$ (not (not E) $= E$).

The reader should try to convince himself of the truth of Rules 2.1, 2.2, and 2.3 either by drawing appropriate Venn diagrams or by the use of examples.

Rule 2.4. For any event sets E, E_1, E_2,

(2.7) $\qquad E \cup (E_1 \cap E_2) = (E \cup E_1) \cap (E \cup E_2).$

In words,
$$E \text{ or } (E_1 \text{ and } E_2) = (E \text{ or } E_1) \text{ and } (E \text{ or } E_2).$$

Let us try to convince ourselves of the truth of this assertion by considering an example. (Alternatively, the reader may find it simpler to construct a Venn diagram to verify Rule 2.4.)

Example 2.5. Consider the sample space of all diseases of human beings. As in Example 2.4, there are too many outcomes for us to list all possible outcomes, but we can give a few: ω_1 = Pneumonia, ω_2 = German Measles, ω_3 = Scarlet Fever, ω_4 = Typhoid, and so on. Let the event set E consist of all dangerous diseases (those that kill a significant proportion of humans who catch them). Let the event set E_1 consist of all tropical diseases, and let the event set E_2 consist of all diseases carried by flies. Then the event set $E \cup (E_1 \cap E_2)$ contains diseases which are either dangerous or which are tropical diseases carried by flies. On the other hand, $(E \cup E_1) \cap (E \cup E_2)$ consists of diseases that are both "dangerous or tropical" *and* "dangerous or carried by flies." In particular, if a disease ω is an outcome contained in this latter event set, ω must be dangerous or ω must be tropical (since ω is both "dangerous or tropical" and "dangerous or carried by flies"). If ω is dangerous, then ω belongs to E and thus belongs to "E or $(E_1$ and $E_2)$." If ω is not dangerous, then since ω is either dangerous or tropical, ω must be tropical. Similarly, if ω is not dangerous, it must be carried by flies (since ω is both "dangerous or carried by flies" and "dangerous or tropical"). Thus, either ω is dangerous or it is a tropical disease carried by flies. We therefore see that if ω belongs to the event set $(E \cup E_1) \cap (E \cup E_2)$, it must also belong to $E \cup (E_1 \cap E_2)$. A corresponding argument shows that if ω belongs to $E \cup (E_1 \cap E_2)$, it must also belong to $(E \cup E_1) \cap (E \cup E_2)$. The two event sets $E \cup (E_1 \cap E_2)$ and $(E \cup E_1) \cap (E \cup E_2)$ have been shown to contain the same outcomes. These event sets must consequently be the same (see the previous discussion on the equality of event sets).

A generalization of Equation (2.7) is the following: For any event sets $E, E_1, E_2, \ldots, E_k$,

(2.8) $\quad E \cup (E_1 \cap E_2 \cap \cdots \cap E_k)$
$$= (E \cup E_1) \cap (E \cup E_2) \cap \cdots \cap (E \cup E_k);$$

that is,
$$E \text{ or } (E_1 \text{ and } E_2 \text{ and } \cdots \text{ and } E_k)$$
$$= (E \text{ or } E_1) \text{ and } (E \text{ or } E_2) \text{ and } \cdots \text{ and } (E \text{ or } E_k).$$

2. FORMAL STRUCTURE: OUTCOMES AND EVENT SETS

Rule 2.5. For any event sets E, E_1, E_2,

(2.9) $\qquad E \cap (E_1 \cup E_2) = (E \cap E_1) \cup (E \cap E_2).$

In words,

$$E \text{ and } (E_1 \text{ or } E_2) = (E \text{ and } E_1) \text{ or } (E \text{ and } E_2).$$

Let us try to verify this result in our example of human diseases.

Example 2.5 (Continued). We have already considered the event sets, E consisting of all dangerous diseases, E_1 consisting of all tropical diseases, and E_2 consisting of all diseases carried by flies. The event set $E \cap (E_1 \cup E_2)$ consists of all diseases that are dangerous *and* are either tropical or carried by flies. An outcome ω in $E \cap (E_1 \cup E_2)$ is always dangerous, but it may be tropical (in which case it is a dangerous tropical disease and is contained in $E \cap E_1$), or it may be carried by flies (in which case it belongs to $E \cap E_2$). Thus, ω belongs to $E \cap E_1$ or to $E \cap E_2$, and therefore ω belongs to $(E \cap E_1) \cup (E \cap E_2)$. Similarly, if ω is either a dangerous tropical disease or a dangerous disease carried by flies, then it is always dangerous, and may be either tropical or carried by flies; that is, ω belongs to $E \cap (E_1 \cup E_2)$. We conclude that $(E \cap E_1) \cup (E \cap E_2)$ and $E \cap (E_1 \cup E_2)$ have the same outcomes (in this special case) and are thus the same event set. Rule 2.5 has consequently been verified in this case.

Rule 2.5 has the following generalization: For any event sets $E, E_1, E_2, \ldots, E_k$, it is true that

(2.10) $\qquad E \cap (E_1 \cup E_2 \cup \cdots \cup E_k)$
$$= (E \cap E_1) \cup (E \cap E_2) \cup \cdots \cup (E \cap E_k),$$

or in words,

$$E \text{ and } (E_1 \text{ or } E_2 \text{ or } \cdots \text{ or } E_k)$$
$$= (E \text{ and } E_1) \text{ or } (E \text{ and } E_2) \text{ or } \cdots \text{ or } (E \text{ and } E_k).$$

Rule 2.6. The complement of the union of two event sets E_1 and E_2 is the intersection of their complements. In symbols,

(2.11) $\qquad (E_1 \cup E_2)^c = E_1^c \cap E_2^c.$

Example 2.6. Consider the sample space of all Peace Corps volunteers. The outcomes in this sample space are the individual volunteers. Let E_1 be the event set consisting of all volunteers (outcomes) who speak Urdu. Let E_2 be the event set consisting of all volunteers who speak Tamil. Then $(E_1 \cup E_2)^c$ is the event set of all volunteers who speak neither Urdu nor Tamil [*not* $(E_1 \text{ or } E_2)$]. However, another way of describing volunteers who speak neither Urdu nor Tamil is to say that they do not speak Urdu

($E_1{}^c$) and they do not speak Tamil ($E_2{}^c$). Thus, every volunteer who belongs to $(E_1 \cup E_2)^c$ also belongs to $E_1{}^c \cap E_2{}^c$. But volunteers who do not speak Urdu and do not speak Tamil, speak neither Urdu nor Tamil. Therefore, every volunteer who belongs to $E_1{}^c \cap E_2{}^c$ also belongs to $(E_1 \cup E_2)^c$. Consequently, $E_1{}^c \cap E_2{}^c$ and $(E_1 \cup E_2)^c$ are the same event set. This verifies Equation (2.11) in this special case.

Equation (2.11) generalizes to the following result: The complement of the union of any number of event sets $E_1, E_2, \ldots, E_k$ is the intersection of the complements of these event sets. In symbols,

(2.12) $\qquad (E_1 \cup E_2 \cup \cdots \cup E_k)^c = E_1{}^c \cap E_2{}^c \cap \cdots \cap E_k{}^c,$

or

$\quad$ *not* $(E_1 \text{ or } E_2 \text{ or } \cdots \text{ or } E_k) = (\text{not } E_1) \text{ and } (\text{not } E_2) \text{ and } \cdots \text{ and } (\text{not } E_k).$

Rule 2.7. The complement of the intersection of two event sets E_1 and E_2 is the union of their complements. That is,

(2.13) $\qquad\qquad\qquad (E_1 \cap E_2)^c = E_1{}^c \cup E_2{}^c.$

The reader who is fond of logical arguments might try to prove this result, given that Rules 2.3(c) and 2.6 above are true. Or, he might prove Rule 2.7 in the special case of Example 2.6 of the Peace Corps volunteers, using arguments similar to those used to prove Rule 2.6 in that case. A generalization of Rule 2.7 is

(2.14) $\qquad (E_1 \cap E_2 \cap \cdots \cap E_k)^c = E_1{}^c \cup E_2{}^c \cup \cdots \cup E_k{}^c,$

or in words,

$\quad$ *not* $(E_1 \text{ and } E_2 \text{ and } \cdots \text{ and } E_k) = (\text{not } E_1) \text{ or } (\text{not } E_2) \text{ or } \cdots \text{ or } (\text{not } E_k).$

This equation is true for all event sets $E_1, E_2, \ldots, E_k$.

[*Note:* Rules 2.6 and 2.7 are known as *de Morgan's Laws*, named after the English mathematician Augustus de Morgan (1806–1871).]

The Correspondence between the Probability Model and Random Experiments

So far, we have discussed the probabilistic concepts "outcome" and "event set." These concepts in the probability model are analogous, respectively, to the concepts "simple result" and "composite result" discussed in connection with random experiments. Through such a correspondence, we can apply the algebra of event sets to composite results.

Example 2.7. A sentence is chosen by lottery from a book (written in English) and the first word of this sentence is observed. The simple results

2. FORMAL STRUCTURE: OUTCOMES AND EVENT SETS 39

of this random experiment are all of the possible words in the English language. One possible composite result A is that the word observed is a verb. Another possible composite result B is that the word is three letters in length. A third possible composite result C is that the word begins with a vowel. From these three composite results, more complicated composite results can be formed. For example, the composite result $A \cap B$ ("A and B") occurs if the word observed is a three-letter verb; the composite result $A \cup C$ ("A or C") occurs if the word is either a verb or begins with a vowel; and the composite result $A \cap B \cap C$ occurs if the word is a three-letter verb beginning with a vowel. The important point to note here is that we are able to express quite complicated composite results by combining simple results into composite results and then combining these composite results in turn through the operations of the algebra of sets.

Example 2.8. A doctor measures the blood pressure of a human patient. Here the simple results are not denumerable[2] (as they were in Examples 2.1 through 2.7); they are all possible positive numbers X. One possible composite result A is that the systolic blood pressure (highest blood pressure during a cardiac cycle) is over 180 millimeters mercury. Another possible composite result B is that the diastolic blood pressure (lowest blood pressure during a cardiac cycle) is under 60 millimeters mercury. A patient's condition is serious if his systolic pressure is over 180 or if his diastolic pressure is under 60, and we can write this composite result as $A \cup B$ ("A or B").

Let us summarize the correspondences we have found so far between our mathematical model and random experiments:

Random Experiment	*Probability Model*
Simple Result: The basic occurrences of the experiment	*Outcome:* The basic elements of the model
Totality of Simple Results: All simple results that can occur in our experiment	*Sample Space:* All possible outcomes
Composite Result: A collection of simple results having some property	*Event Set (Event):* A collection of outcomes obeying some proposition

[2] A collection is *denumerable* if we can assign to each one of its elements a different positive integer. For example, the collection of all fractions is denumerable, as is the collection of all even integers. Note that both of these collections have an infinite number of elements. On the other hand, any finite collection (such as the sample space of letters in Example 2.2) is also denumerable. A synonym that is often used in mathematics for the word "denumerable" is the word "countable."

(*continued*)

Random Experiment	Probability Model
Algebra of Composite Results: The algebra of sets applied to composite results	*Algebra of Event Sets:* The algebra of sets applied to event sets
Relative Frequency: The proportion of times a given composite result is observed in repeated trials of a random experiment	?

We need a mathematical entity to fill the blank opposite "relative frequency." This entity is the probability of an event set.

3. FORMAL STRUCTURE: PROBABILITIES

We have already assumed that in a large number of repeated trials of our random experiment the relative frequency of every composite result is close to a certain number, p. Correspondingly, we assume that for every event set, there exists a number (called the *probability* of the event set) that is the idealization of the long-run relative frequency of the composite result that corresponds to the event set. In order to make the analogy between the concepts of probability and relative frequency useful, we would like the probability of an event set to have all of the important properties of the relative frequency of a composite result.

We first recall that the relative frequency r.f.(E) of a composite (or simple) result E in repeated trials is the number of times $\#E$ that the composite result occurs divided by the number, N, of repeated trials. Because the number, $\#E$, of times that the composite result E occurs in N trials cannot be less than 0 or greater than N, it follows that the relative frequency r.f.(E), of E is a number greater than or equal to 0 and less than or equal to 1 (that is, $0 \leq$ r.f.$(E) \leq 1$). Further, since one and only one simple result must occur at every trial, the relative frequency of the composite result consisting of all possible simple results is equal to 1, while the relative frequency of the composite result consisting of no simple results (the impossible result) is equal to 0. Finally, if two composite results E_1 and E_2 have no simple results in common, then the number of trials in which a simple result from either E_1 or E_2 is observed equals the number of trials in which a simple result from E_1 is observed plus the number of trials in which a simple result from E_2 is observed. (Because E_1 and E_2 have no simple results in common, they cannot occur together at the same trial.) Thus, when E_1 and E_2 have no simple results in common,

(3.1) $$\#(E_1 \cup E_2) = \#(E_1) + \#(E_2).$$

Dividing both sides of (3.1) by the number N of trials, we find that

$$\frac{\#(E_1 \cup E_2)}{N} = \frac{\#E_1}{N} + \frac{\#E_2}{N},$$

or

(3.2) $$\text{r.f.}(E_1 \cup E_2) = \text{r.f.}(E_1) + \text{r.f.}(E_2).$$

[*Note:* Equations (3.1) and (3.2) are not, in general, true if the two composite results E_1 and E_2 have simple results in common.]

The three properties just listed are some basic properties obeyed by relative frequencies. Corresponding to these basic properties, we demand that the probabilities of event sets obey the following axioms:

Axiom 1. For every event set E, the probability of E is a number $P(E)$ between 0 and 1. That is, $0 \leq P(E) \leq 1$.

Axiom 2. The sample space Ω has probability 1, whereas the null event $\emptyset$ has probability 0. Thus,

$$P(\Omega) = 1 \quad \text{and} \quad P(\emptyset) = 0.$$

Axiom 3. If E_1 and E_2 are two mutually exclusive event sets, then

$$P(E_1 \cup E_2) = P(E_1) + P(E_2).$$

It is important to realize that $\emptyset$ need not be the only event set that has probability 0. Certain events may be such that they correspond, loosely speaking, to composite results that "happen only once in a lifetime." The relative frequencies of such composite results in a large number of trials will be almost indistinguishable from 0. Thus a probability of 0 does not necessarily mean that the event set corresponds to an impossible result, but only that it corresponds to a very rarely observed result. Similarly, a probability of 1 does not necessarily mean that the event set corresponds to a sure result, but only that it corresponds to a result whose nonoccurrence is very rarely observed.

*Remark:[3] For probability models such as that needed to describe the blood pressure measurements in Example 2.8, the axiomatic formulation given above is not entirely sufficient. However, the extra assumptions needed are of a mathematically more difficult nature since they arise from

---

[3] Paragraphs and sections preceded by an asterisk are of greater difficulty; they may be omitted upon a first reading.

the necessity of handling event sets containing an infinite number of outcomes. The understanding of these points of mathematical rigor demands a more advanced mathematical background than we have assumed in this book. Fortunately, the part of probability theory that the reader can expect to use in his own research can be utilized without excessive concern about such topics. If the reader only deals with sample spaces consisting of a finite number (no matter how large) of outcomes, such problems of theory need never concern him.

4. THE PROBABILITY CALCULUS

Suppose that a sample space consists of 15 outcomes. How many event sets are there in the probability model? The answer is somewhat staggering: 32,768 (or 2^{15}) event sets. The task of determining the probabilities for each of these event sets thus appears to be quite extensive. Fortunately, Axioms 1, 2, and 3 assure that certain mathematical relationships hold between the probabilities. Indeed, from the axioms we can construct a probability calculus that allows us to evaluate the probability of *every* event set given only the probabilities of certain basic event sets. For the sample space of 15 outcomes mentioned above, there are only 15 such basic event sets (the event sets consisting of only one outcome); thus, the number of probabilities to be determined is reduced from 32,768 to 15.

Although there are many rules of the probability calculus, the following are among the most useful for practical calculations.

Law of Inclusion

Recall that an event set E_2 is said to be included in an event set E_1 (denoted by $E_2 \subset E_1$) if every outcome which belongs to E_2 also belongs to E_1. If $E_2 \subset E_1$, then $P(E_2)$ is less than or equal to $P(E_1)$. We write this fact in the following manner.

Rule 4.1. If $E_2 \subset E_1$, then $P(E_2) \leq P(E_1)$.

We may derive Rule 4.1 from the axioms through use of the Venn diagram shown in Figure 4.1. Event set E_1 contains both the event set E_2 and the shaded area outside of E_2 (but in E_1). The shaded area represents the event set consisting of outcomes in E_1 and not in E_2; that is, the event set "E_1 *and not* E_2" or equivalently $E_1 \cap E_2^c$. The Venn diagram shows that E_2 and $E_1 \cap E_2^c$ have no outcomes in common (they are mutually exclusive) and that the union $E_2 \cup (E_1 \cap E_2^c)$ of these two event sets is equal

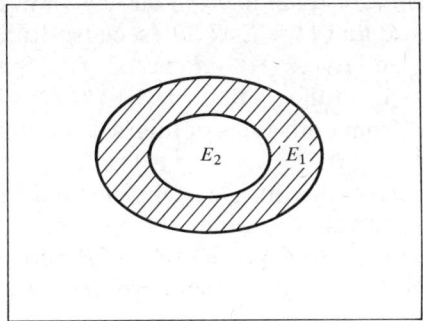

Figure 4.1 Comparison of probabilities for included event sets.

to E_1. Thus, Axiom 3 applies, and

$$P(E_1) = P(E_2 \cup (E_1 \cap E_2^c)) = P(E_2) + P(E_1 \cap E_2^c).$$

But by Axiom 1, $P(E_1 \cap E_2^c)$ is a nonnegative number. Therefore, $P(E_1)$ is the sum of $P(E_2)$ and a nonnegative number; this implies that $P(E_2) \leq P(E_1)$, which is the assertion of Rule 4.1.

Law of Complementation

The number of repeated trials resulting in the composite result E in a total of N trials is equal to N minus the number of trials resulting in the occurrence of "not E"; that is, $\#E = N - \#$ (not E). From this fact, we have r.f.$(E) = 1 - $ r.f.(E^c). The corresponding fact for probabilities is:

Rule 4.2. For any event set E, $P(E) = 1 - P(E^c)$.

The verification of this rule again follows easily if we refer to a Venn diagram (Figure 4.2). The whole box represents the sample space Ω; the

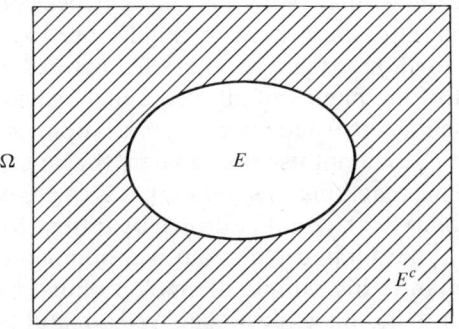

Figure 4.2 The probability of the complement of an event set.

shaded area represents E^c. The diagram clearly shows that E and E^c are mutually exclusive and that $\Omega = E \cup E^c$ (see also Rule 2.1 of Section 2). Applying Axiom 3, we have $P(\Omega) = P(E) + P(E^c)$. Since we know from Axiom 2 that $P(\Omega) = 1$, it follows that $P(E) + P(E^c) = 1$. Rule 4.2 follows by subtracting $P(E^c)$ from both sides of this last equality.

Example 4.1. Police statistics in a certain city indicate that of all cars stolen, 9 percent are 10 years or older, 11 percent are from 5 to 9 years old, 15 percent are from 3 to 4 years of age, 20 percent are 2 years old, 20 percent are 1 year old, and 25 percent are less than 1 year old. A court reporter, hearing that a car has been reported stolen, bets that the car is less than a year old. What is the probability that he loses his bet?

In this example, our sample space can either consist of all possible ages for the stolen car, or of the outcomes ω_1 = older than 10 years, ω_2 = 5 to 9 years, ω_3 = 3 to 4 years, ω_4 = 2 years, ω_5 = 1 year, and ω_6 = less than 1 year. We choose to use the latter sample space because the police statistics give us probabilities for the event sets

$$E_1 = \{\omega_1\}, \quad E_2 = \{\omega_2\}, \quad E_3 = \{\omega_3\}, \quad E_4 = \{\omega_4\}, \quad E_5 = \{\omega_5\},$$
$$\text{and} \quad E_6 = \{\omega_6\}.$$

To use these police statistics as probabilities, we must assume, of course, that the set of circumstances that created previous car thefts (giving rise to the police statistics) has not changed (that is, we are still observing the same random experiment). Under this assumption, we can say that the probability that the reporter is right is $P(E_6) = 0.25$, since the reporter is right only if the event E_6 occurs. The reporter is wrong if the event E_6 does not occur, that is, if E_6^c occurs. By Rule 4.2, $P(E_6^c) = 1 - P(E_6) = 1 - 0.25 = 0.75$, so that the probability that the reporter is wrong and loses his bet is 0.75.

Example 4.2. A scientist studying the influence of smoking on the health of adult males in the United Kingdom finds it necessary to distinguish between those cigarette smokers who inhale and those who do not inhale. He is going to choose male smokers for his experiments by lottery from among a list of the names of all of the male smokers in the United Kingdom. Since he must pay a fee to everyone whom he selects, and since for his experiments he cannot use noninhalers (because noninhalers differ from inhalers in their reactions to smoking), it is of great interest to him to find out the probability that a smoker drawn by lottery from the population of male smokers in the United Kingdom is a noninhaler. The larger this probability is, the larger his sample of smokers must be if he is to select enough inhalers to make his experiments meaningful to other scientists, and thus the larger the amount of money he must expect to spend to pay his subjects (he must pay inhaler and noninhaler alike). From studies carried out by the Tobacco Research Council of London in 1966 [G. F. Todd, *Statistics of Smoking in the United Kingdom* (fourth edition)], the scientist finds that the probability that a smoker drawn by lottery from the population of male smokers in the United Kingdom inhales when he smokes is equal to 0.87. Thus, by Rule 4.2, the probability that a male smoker does not inhale when he smokes is equal to $1 - 0.87 = 0.13$.

Law of Addition

Suppose that we know the probabilities of two event sets E_1 and E_2. Is this enough information to let us compute the probability of the event set $E_1 \cup E_2$ ("E_1 or E_2")? The answer, in general, is "No." This answer may be surprising to the reader since Axiom 3 seems to assert that the answer is always "Yes" and that, indeed, $P(E_1 \cup E_2) = P(E_1) + P(E_2)$. However, there is a condition to be met before Axiom 3 may be applied to calculate $P(E_1 \cup E_2)$; namely, that the event sets E_1 and E_2 be mutually exclusive (have no outcomes in common). Not all event sets E_1 and E_2 are mutually exclusive. A Venn diagram of one case where E_1 and E_2 have outcomes in common appears in Figure 4.3. The cross-hatched region $E_1 \cap E_2$ consists (by definition; see Section 2) of outcomes that are common to E_1 and

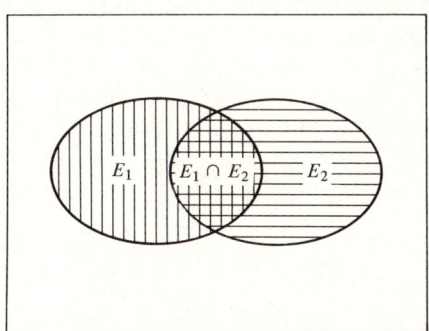

Figure 4.3 The probability of the intersection of two event sets.

E_2. The region that is vertically striped but not cross-hatched represents the event set of outcomes that belong to E_1 and not to E_2 (that is, the event set $E_1 \cap E_2^c$). Similarly, the region with only horizontal stripes is $E_2 \cap E_1^c$. From the diagram we see that any two of the event sets $A = E_1 \cap E_2^c$, $B = E_1 \cap E_2$, and $C = E_2 \cap E_1^c$ have no outcomes in common, and that

$$E_1 = A \cup B,$$
$$E_2 = B \cup C,$$
$$E_1 \cup E_2 = E_1 \cup C = A \cup B \cup C.$$

Because A and B are mutually exclusive and $E_1 = A \cup B$, we may apply Axiom 3 to obtain $P(E_1) = P(A \cup B) = P(A) + P(B)$.

Similarly, because B and C are mutually exclusive and $E_2 = B \cup C$, we may apply Axiom 3 to obtain $P(E_2) = P(B \cup C) = P(B) + P(C)$. If we add $P(E_1)$ and $P(E_2)$, we obtain

(4.1) $$\begin{aligned} P(E_1) + P(E_2) &= [P(A) + P(B)] + [P(B) + P(C)] \\ &= P(A) + 2P(B) + P(C). \end{aligned}$$

However, since $E_1 \cup E_2 = E_1 \cup C$, and since C and E_1 are mutually exclusive (C has no outcomes in common with A or B, and thus has no outcomes in common with $E_1 = A$ or B), we may apply Axiom 3 a third time to obtain

(4.2) $$\begin{aligned} P(E_1 \cup E_2) &= P(E_1 \cup C) = P(E_1) + P(C) \\ &= P(A) + P(B) + P(C). \end{aligned}$$

Comparing (4.1) and (4.2) we see that $P(E_1 \cup E_2)$ is not equal to $P(E_1) + P(E_2)$ unless $P(B)$ is zero. The error in attempting to assert that $P(E_1 \cup E_2) = P(E_1) + P(E_2)$ lies in the fact that $P(E_1) + P(E_2)$ counts $P(B) = P(E_1 \cap E_2)$ one too many times. A correct formula, of course, can always be obtained from (4.2), but the quantities $P(A) = P(E_1 \cap E_2^c)$ and $P(C) = P(E_2 \cap E_1^c)$ are not always available to us. If, however, we subtract $P(B) = P(E_1 \cap E_2)$ from the expression for $P(E_1) + P(E_2)$ given in (4.1), we see from (4.2) that we obtain $P(E_1 \cup E_2)$. Indeed, we find that $P(E_1 \cup E_2) = P(E_1) + P(E_2) - P(E_1 \cap E_2)$. Since the probabilities of $P(E_1)$, $P(E_2)$, and $P(E_1 \cap E_2)$ are often known, the result we have obtained gives a useful formula for computing $P(E_1 \cup E_2)$. We summarize this result in the following rule:

Rule 4.3(a). For any two event sets E_1 and E_2, the probability of their union $E_1 \cup E_2$ can be found from the equation:

(4.3) $$P(E_1 \cup E_2) = P(E_1) + P(E_2) - P(E_1 \cap E_2).$$

The most general case where we can correctly assert that $P(E_1 \cup E_2) = P(E_1) + P(E_2)$ is when $P(E_1 \cap E_2)$ is zero. We state this as a separate rule.

Rule 4.3(b). We can assert that $P(E_1 \cup E_2) = P(E_1) + P(E_2)$ if and only if the event set $E_1 \cap E_2$ has zero probability, that is, $P(E_1 \cap E_2) = 0$.

Note that when E_1 and E_2 are mutually exclusive (disjoint), then $(E_1 \cap E_2) = \emptyset$. In this case, we know from Axiom 2 that $P(E_1 \cap E_2) = P(\emptyset) = 0$. Thus Rule 4.3(b) and Axiom 3 are consistent with one another [this is hardly surprising since we used Axiom 3 to verify Rules 4.3(a) and 4.3(b)].

However, recalling (see Section 3) that an event set can have zero probability and yet not be the null event set $\emptyset$, we see that Rule 4.3(b) is slightly more general than Axiom 3.

Extensions of Rules 4.3(a) and 4.3(b) to the case of three (or more) event sets are possible. We state and verify here only the rule for three event sets.

Rule 4.4. For any three event sets E_1, E_2, and E_3,

$$\begin{aligned}P(E_1 \cup E_2 \cup E_3) = & P(E_1) + P(E_2) + P(E_3) - P(E_1 \cap E_2) \\ & - P(E_1 \cap E_3) - P(E_2 \cap E_3) \\ & + P(E_1 \cap E_2 \cap E_3).\end{aligned} \quad (4.4)$$

We may verify Rule 4.4 in two ways. One way is by using a Venn diagram (Figure 4.4). Here E_1 is represented by the area with horizontal stripes, E_2 by the area with vertical stripes, and E_3 by the dotted area. By breaking up the total area into seven disjoint (mutually exclusive) areas and using an argument similar to that used to derive Rule 4.3(a), we may verify Rule 4.4. This argument is left to the reader.

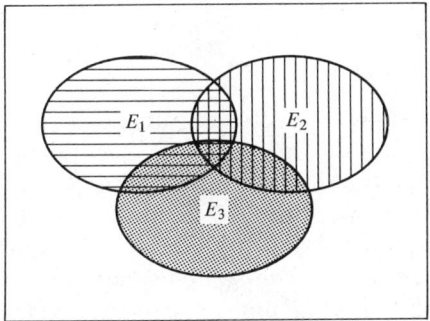

Figure 4.4 The probability of the union of three event sets.

*An alternative method, which is of a type frequently adopted in mathematical arguments, is to proceed by induction. Since we know that Rule 4.3(a) is true, let us apply this rule wherever appropriate. First, consider the two event sets $A = (E_1 \cup E_2)$ and E_3. Applying Rule 4.3(a), we obtain

$$(4.5) \quad P(E_1 \cup E_2 \cup E_3) = P(A \cup E_3) = P(A) + P(E_3) - P(A \cap E_3).$$

Applying Rule 4.3(a) again to $P(A) = P(E_1 \cup E_2)$, we obtain

$$(4.6) \quad P(A) = P(E_1 \cup E_2) = P(E_1) + P(E_2) - P(E_1 \cap E_2).$$

Now we note that $A \cap E_3 = E_3 \cap A = E_3 \cap (E_1 \cup E_2)$, which by Rule 2.5 is equal to $(E_3 \cap E_1) \cup (E_3 \cap E_2)$ or equivalently to $(E_1 \cap$

$E_3) \cup (E_2 \cap E_3)$. If we let $B = E_1 \cap E_3$ and $C = E_2 \cap E_3$, then $B \cap C = E_1 \cap E_2 \cap E_3$. By applying Rule 4.3(a) once again, we obtain

(4.7) $\quad P(A \cap E_3) = P(B \cup C) = P(B) + P(C) - P(B \cap C)$
$\quad\quad\quad = P(E_1 \cap E_3) + P(E_2 \cap E_3) - P(E_1 \cap E_2 \cap E_3).$

Substituting (4.7) and (4.6) into (4.5) gives us (4.4) and verifies Rule 4.4.

A very special case of Rule 4.4 is when the event sets E_1, E_2, and E_3 have no outcomes in common. Then each of $E_1 \cap E_2$, $E_1 \cap E_3$, $E_2 \cap E_3$, and $E_1 \cap E_2 \cap E_3$ is the null set $\emptyset$. Since $P(\emptyset) = 0$ by Axiom 2, we can substitute 0 for $P(E_1 \cap E_2)$, $P(E_1 \cap E_3)$, $P(E_2 \cap E_3)$, and $P(E_1 \cap E_2 \cap E_3)$ in Equation (4.4). From this substitution we find that if E_1, E_2, and E_3 have, pairwise, no outcomes in common, then

$$P(E_1 \cup E_2 \cup E_3) = P(E_1) + P(E_2) + P(E_3).$$

We can generalize this result to the following rule, using inductive arguments similar to those used to obtain Rule 4.4 from Rule 4.3(a).

Rule 4.5. If $E_1, E_2, \ldots, E_k$ are event sets, no two of which have outcomes in common, then

(4.8) $\quad P(E_1 \cup E_2 \cup \cdots \cup E_k) = P(E_1) + P(E_2) + \cdots + P(E_k).$

Example 4.3. A geologist inspects ore samples from ground that he suspects contains uranium. From previous experience he knows that if uranium is present at the drill site, the probability that the ore samples contain continental sediment is 0.90. He also knows that when uranium is present at the site, there is a probability of 0.80 that non-red clastic sediments are present in the ore samples, and a probability of 0.78 that both continental sediments and non-red clastic sediments are present. The geologist decides that if he observes either continental sediments or non-red clastic sediments in his ore samples, he will order drilling to continue. [Note that in order to find a good rule for deciding when to continue drilling, the geologist should also consider probabilities of finding these sediments if uranium is not present.] Assuming that there is uranium in the ground and that probabilities given above are correctly stated, what is the probability that the geologist will correctly order drilling to continue? Let E_1 be the event set corresponding to the (composite) result that the geologist's ore samples contain continental sediments. Let E_2 be the event set corresponding to the result that non-red clastic sediments are present in the ore samples. Then $P(E_1)$ equals 0.90, $P(E_2)$ equals 0.80, and $P(E_1 \cap E_2)$ equals 0.78. The geologist will order drilling to continue if either the result corresponding to event set E_1 or the result corresponding to event set E_2 occur. This decision will be the correct one provided that uranium is present at the drill site. Hence, if uranium is present, the probability that the geologist correctly orders drilling to continue is equal to $P(E_1 \cup E_2)$. From Rule 4.3(a), we find that

$P\{\text{the geologist correctly orders drilling to continue}\}$
$= P(E_1) + P(E_2) - P(E_1 \cap E_2)$
$= 0.90 + 0.80 - 0.78 = 0.92 \ .$

In an attempt to obtain an even larger probability of correctly drilling for uranium, the geologist also decides to look for the presence of carbonaceous matter in his ore samples. Let the event set E_3 correspond to the (composite) result that carbonaceous matter is discovered in the geologist's samples. The geologist reads various technical papers on the subject, recalls his own experiences, and arrives at the conclusion that if uranium is present, the probability, $P(E_3)$, that carbonaceous matter will be found in the ore samples is 0.50. He further finds that if uranium is present, $P(E_1 \cap E_3) = 0.42$, $P(E_2 \cap E_3) = 0.39$, and the probability that continental sediments, non-red clastic sediments, and carbonaceous matter are all found in the ore samples is 0.38 [that is, $P(E_1 \cap E_2 \cap E_3) = 0.38$]. The geologist decides to continue drilling if continental sediments, non-red clastic sediments, *or* carbonaceous matter is found in the ore samples. Therefore, if uranium is present, the probability that he correctly decides to continue drilling is equal to $P(E_1 \cup E_2 \cup E_3)$ and can be found by applying Rule 4.4. That is,

$P\{\text{the geologist correctly decides to continue drilling for uranium}\}$
$= P(E_1) + P(E_2) + P(E_3) - P(E_1 \cap E_2) - P(E_1 \cap E_3) - P(E_2 \cap E_3)$
$+ P(E_1 \cap E_2 \cap E_3)$
$= 0.90 + 0.80 + 0.50 - 0.78 - 0.42 - 0.39 + 0.38 = 0.99 .$

Thus, if uranium is present, the geologist's new rule for deciding when to continue drilling for uranium will almost always lead him to make the correct decision.

Example 4.4. An engineer who works for the natural gas industry in a certain European country is concerned with the possibility that pipeline breaks might occur and shut off the flow of gas to certain sections of the country. Figure 4.5 illustrates a network of pipelines servicing various areas of the country. From a consideration of the previous experience of both his own country and other countries, and from consideration of the strengths of the pipes and the pressures of gas flow in each segment of the system, the engineer has calculated for each segment the probability that a break will occur in that segment, given that a break occurs somewhere in the network. In constructing his probability model, the engineer assumes

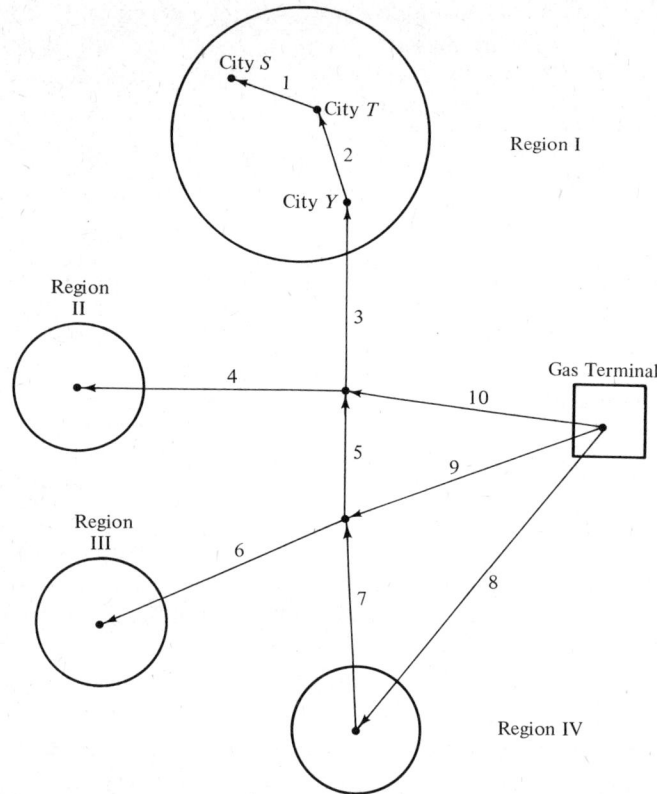

Figure 4.5 Network of pipelines serving the various regions of the country. The branch servicing Region I consists of segments 1, 2, and 3.

that a break in the network, when it occurs, can occur in one and only one segment of the network (these segments are shown and numbered in Figure 4.5). Thus, the segments of the network can be identified with the outcomes of the probability model; outcome ω_i corresponds to a break in segment i, $i = 1, 2, 3, \ldots, 10$. The engineer's calculations give the probabilities of the event sets $E_1 = \{\text{pipeline segment number 1 breaks}\} = \{\omega_1\}$, $E_2 = \{\text{pipeline segment number 2 breaks}\} = \{\omega_2\}$, $E_3 = \{\omega_3\}$, and so on. These probabilities appear in the following table:

Event Set	E_1	E_2	E_3	E_4	E_5	E_6	E_7	E_8	E_9	E_{10}
	0.096	0.066	0.109	0.049	0.027	0.149	0.147	0.163	0.092	0.102

Since the event sets $E_1, E_2, \ldots, E_{10}$ are mutually exclusive (E_i and E_j have no outcomes in common, $i \neq j$), Rule 4.5 can be applied when finding the probability of the union of any k of the event sets $E_1, E_2, ..., E_{10}$. Thus, for example, if the engineer is interested in the probability that a break in the network, when one occurs, occurs in the branch of the network feeding Region I (this branch consists of segments 1, 2, and 3; see Figure 4.5), then we can find this probability by adding up the probabilities of event set E_1, event set E_2, and event set E_3. That is, from Rule 4.5, with $k = 3$,

$P\{$the break occurs in the branch of the network feeding Region I$\}$
$= P\{$the break occurs in segment numbers 1, 2, or 3$\}$
$= P(E_1 \cup E_2 \cup E_3)$
$= P(E_1) + P(E_2) + P(E_3) = 0.096 + 0.066 + 0.109 = 0.271.$

Calculations of probabilities, such as that given above, may help repairmen search for breaks when they occur by indicating portions of the network where the break is most likely to have occurred.

5. FITTING A PROBABILITY MODEL

With the introduction of the notion of the probability of an event set, we have completed the general construction of probability models. To correspond to the simple results, composite results, totality of simple results, and long-run relative frequencies of a random experiment, we have assigned the outcomes, event sets (or, for short, *events*), sample space, and probabilities, respectively, to our probability model. Therefore, once the totality of simple results has been specified, we can cease to distinguish between the random experiment and the probability model and can refer to the random experiment completely in terms of the probability model. We then have at our disposal various mathematical tools (for example, the algebra of sets and the probability calculus) that enable us to develop more and more complicated descriptions (propositions) about the chance phenomenon, and to compute the probabilities with which these propositions hold.

It should always be remembered when using a probability model that the simple results that are abstracted by the outcomes of the model must have the two properties listed below. If these requirements are not met, the probability model may be useless as an aid to understanding the chance phenomena of interest.

(i) The simple results must be sufficiently detailed so that their totality contains all the information of interest to us in the random phenomenon. If, for example, in an investigation of peaks in the Dow-Jones stock market average we have decided to record just the month in which the peak occurred, then the simple results (outcomes) would

be the 12 months of the year. If we later wanted more detailed information about when the peaks occurred (such as what day of the month) or about what concomitant phenomena also occurred at that time, we would be unable to extract this information from our model.
(ii) The collection of simple results must have the property that on a given trial of the experiment *one and only one* of the simple results can occur.

Having agreed to study random experiments only in terms of the corresponding probability models, we now need to determine the probability associated with each *event* (event set) in the model.[4] The rule or list that assigns a probability to every event is called the *probability measure* for the model. Except for some mathematical details (which are somewhat complicated), we can summarize the goals of an experimenter investigating a chance phenomenon by saying that he needs to (a) identify all important outcomes of the phenomenon, and (b) specify the probability measure associated with this phenomenon.

In Section 4 we noted that it is not necessary to determine the probabilities for *every* event in order to determine the probability measure. Only the values of the probability measure for certain basic events need to be found. In practice, such values are determined in one of two different ways. We may actually take a large number of trials of the random experiment, calculate the resulting relative frequencies, and set the probabilities equal to these relative frequencies. Alternatively, we may employ theoretical arguments to specify that certain mathematical relationships are to hold between the probabilities of the basic events. That is, we look at either the experiment or the mathematical model. To conclude this section, we give an example of each of these two methods for determining the probability measure of a random experiment.

Example 5.1 (Experimental Method). Consider the experiment that consists of observing the length of human life. As we remarked in Chapter 1, insurance companies have traditionally been concerned with the data resulting from such experiments, and over the years have accumulated records of human life spans. These are usually summarized in *life tables*, the earliest example of which appears to have been given by Captain John Graunt (1620–1674) in 1662. (The foundation of a theory of life annuities is due to the English astronomer Edmund Halley in 1693.) An example of a life table is given in Table 5.1. The simple results are the ages at death; the probabilities are actually relative frequencies calculated over thousands of observed life spans.

[4] Since "event" is a more intuitively meaningful term than "event set," we use the former term from now on.

5. FITTING A PROBABILITY MODEL

Table 5.1. Life table for individuals living in the city of Breslau (Wroclaw), Poland, 1693.

Age	Probability	Age	Probability	Age	Probability	Age	Probability
1	0.145	22	0.007	43	0.010	64	0.010
2	0.057	23	0.006	44	0.010	65	0.010
3	0.038	24	0.006	45	0.010	66	0.010
4	0.028	25	0.007	46	0.010	67	0.010
5	0.022	26	0.007	47	0.010	68	0.010
6	0.018	27	0.007	48	0.010	69	0.010
7	0.012	28	0.007	49	0.011	70	0.011
8	0.010	29	0.008	50	0.011	71	0.011
9	0.009	30	0.008	51	0.011	72	0.011
10	0.008	31	0.008	52	0.011	73	0.011
11	0.007	32	0.008	53	0.011	74	0.010
12	0.006	33	0.008	54	0.010	75	0.010
13	0.006	34	0.009	55	0.010	76	0.010
14	0.006	35	0.009	56	0.010	77	0.010
15	0.006	36	0.009	57	0.010	78	0.009
16	0.006	37	0.009	58	0.010	79	0.008
17	0.006	38	0.009	59	0.010	80	0.007
18	0.006	39	0.009	60	0.010	81	0.006
19	0.006	40	0.009	61	0.010	82	0.005
20	0.006	41	0.009	62	0.010	83-	0.023
21	0.006	42	0.010	63	0.010		

Example 5.2 (Theoretical Approach). Suppose we are concerned with the number of automobile accidents that occur over a typical weekend. This is a random experiment whose outcomes (simple results) are positive integers $0, 1, 2, \ldots$. It is possible to assign probabilities to the simple results of this random experiment, by making use of certain theoretical assumptions about the nature of automobile accidents. That is, we may assume that for every motorist on the road, Nature tosses a biased coin. If it comes up heads, the motorist has a fatal accident over the weekend. However, we assume that the chance that the toss results in a head is infinitesimally small. In addition, we may hypothesize that one motorist's chances of dying are not affected by the chances of any other motorist's death. Since there are many motorists on the road, even though the chance is almost zero that any one motorist will die, among all motorists a few will die. The number of those few who die becomes the simple result of our random experiment; namely, the number ω of people who die over the weekend. The logic of mathematics is now utilized to show that if our assumptions are true, then the probability of no deaths on the weekend is

approximately a certain number c, and the probability of one death is approximately $c\lambda$ where λ is another fixed number related to the number c. Indeed, it can be shown that the probability of two deaths is $c\lambda^2/2$, the probability of three deaths is $c\lambda^3/6$, and the probability of d deaths on the weekend is

$$\frac{c\lambda^d}{d(d-1)(d-2)\cdots(3)(2)(1)}.$$

The numbers c and λ are related to one another in such a way that the sum of the above probabilities is equal to one.

Thus, rather than trying to find a probability for each event set (an impossible task since the number of outcomes ω is infinite), we need determine only the numbers c and λ and in order to completely determine the probability model. The probability measure that we have obtained is called the *Poisson distribution* (see Chapter 6, Section 2).

Of course, someone might challenge the theoretical arguments of Example 5.2 and defy us to prove that they are indeed a description of Nature. Our only recourse then is to take experimental data and show that the probabilities of our model agree to a close enough approximation with the observed relative frequencies.

NOTES AND REFERENCES

The following are general articles giving some idea of the meaning and use of probability theory:
1. Ayer, A. J. (1965). "Chance." *Scientific American*, vol. 213, October, p.44.
2. Kac, M. (1964). "Probability." *Scientific American*, vol. 211, September, p. 92.
3. Weaver, W. (1950). "Probability." *Scientific American*, vol. 183, October, p. 44.
4. Weaver, W. (1952). "Statistics" *Scientific American*, vol. 186, January, p. 60.

Elementary expositions of probability theory, which can serve as supplementary reading to the present book, can be found in the following textbooks:
5. Dwass, M. (1967). *First Steps in Probability*. McGraw-Hill, Inc., New York.
6. Goldberg, S. (1960). *Probability, An Introduction*. Prentice-Hall, Inc., Englewood Cliffs, New Jersey.
7. Guenther, W. C. (1968). *Concepts of Probability*, McGraw-Hill, Inc., New York.
8. Hodges, J. L., Jr., and Lehmann, E. L. (1970). *Elements of Finite Probability* (2nd ed.). Holden-Day, Inc., San Francisco.
9. Mosteller, F., Rourke, R. E. K., and Thomas, G. B. (1961). *Probability with Statistical Applications*. Addison-Wesley Publishing Company, Inc., Reading, Massachusetts.

More advanced treatises dealing with the foundations of probability are the following:
10. Feller, W. (1968). *An Introduction to Probability Theory and Its Applications* (3rd ed.). John Wiley & Sons, Inc., New York.
11. Kolmogorov, A. N. (1933). *Foundations of the Theory of Probability*. Translation of original published by Chelsea Publishing Company, New York, 1950, 1956.
12. Von Mises, Richard (1957). *Probability, Statistics, and Truth* (second revised English edition prepared by Hilda Geiringer). The Macmillan Company, New York.

One of the advantages of the mathematical conceptualization of a real phenomenon is that the same conceptualization may serve, with suitable translations in language, as a model for a different reality. Probabilities were conceptualized to correspond to relative frequencies in repeated trials of a random experiment. Such probabilities are sometimes called *objective probabilities*. However, a number of present-day statisticians believe that the probability calculus can also serve as a model for the way in which a "rational" man assesses his own belief in the truth of a given statement. In most cases the statement is, objectively, either true or false. However, the rational man may not know which is the case (because he has not been informed, because the event to which the statement applies will happen in the future, and so on). Nevertheless, he may be able to make an assessment as to whether the statement is true or false. His assessment is a personal number (a probability), based on the evidence available to him, signifying the degree to which he *believes* the given statement to be true. When probabilities are given this kind of interpretation, they are referred to as *subjective probabilities*. The probability calculus to be presented in the remainder of this chapter, and in the succeeding chapters, is valid *no matter which interpretation of the probabilities* (objective or subjective) *is used*.

Some idea of the history, philosophy, and theoretical foundations of the subjectivist approach to probability theory can be obtained by reading:
13. Kyburg, H. E., and Smokler, H. E. (1964). *Studies in Subjective Probability*. John Wiley & Sons, Inc., New York.
14. Savage, L. J. (1954). *The Foundations of Statistics*. John Wiley & Sons, Inc., New York.

EXERCISES

1. Suppose the sample space Ω of a probability model consists of integers from 1 to 10; that is, $\Omega = \{1,2,3,4,5,6,7,8,9,10\}$, and $\omega_1 = 1$, $\omega_2 = 2$, $\omega_3 = 3$, ..., $\omega_{10} = 10$. Let E_1 be the event set consisting of the numbers (outcomes) 1, 2, 3, 4, and 5. Let E_2 be the event set consisting of the even numbers between 1 and 10 (that is, 2, 4, 6, 8, and 10). Let E_3 be the event set consisting of the odd numbers between 1 and 10.
   (a) Draw a Venn diagram showing Ω, E_1, E_2, and E_3. Use points to repre-

sent the outcomes and clearly indicate on the diagram which outcomes belong to which event sets.
(b) Describe the event sets $E_1 \cap E_2$, $E_1 \cap E_3$, $E_2 \cap E_3$, and $E_1 \cap E_2 \cap E_3$ in terms of the outcomes which are members of these event sets.
(c) Describe the event sets $E_1 \cup E_2$, $E_1 \cup E_3$, $E_2 \cup E_3$, and $E_1 \cup E_2 \cup E_3$ in terms of the outcomes which are members of these event sets. Prove that for this example $E_2 \cap E_3 = \emptyset$ and $E_2 \cup E_3 = \Omega$.
(d) Describe the event sets $(E_1 \cap E_2) \cup E_3$ and $(E_1 \cup E_3) \cap (E_2 \cup E_3)$ in terms of the outcomes which are members of these event sets. Show directly (using the definition of the equality of event sets) that these two event sets are equal. This example illustrates the truth of what rule?
(e) Describe the event sets $(E_1 \cup E_2)^c$ and $E_1^c \cap E_2^c$ in terms of the outcomes which are members of these event sets. Show directly that these two event sets are equal. What rule does this example illustrate?
(f) Describe the event set $E_1^c \cup E_2^c \cup E_3^c$ in terms of the outcomes which are members of this event set. Is there another way to describe this event set?
(g) Draw Venn diagrams for each of the event sets described in parts (b) through (f).

A2. A sample space Ω consists of the 52 outcomes $\omega_1, \omega_2, \omega_3, \ldots, \omega_{52}$. The event set E_1 consists of outcomes $\omega_1, \omega_2, \ldots, \omega_{13}$. The event set E_2 consists of outcomes $\omega_{11}, \omega_{12}, \omega_{13}, \omega_{24}, \omega_{25}, \omega_{26}, \omega_{37}, \omega_{38}, \omega_{39}, \omega_{50}, \omega_{51}, \omega_{52}$.
(a) What outcomes belong to $E_1 \cup E_2$?
(b) What outcomes belong to $E_1 \cap E_2$?
(c) What outcomes belong to $E_1^c \cap E_2$?
(d) What outcomes belong to $E_1 \cap E_2^c$?
(e) For which of the examples of random experiments described in Chapter 1 and in the exercises of Chapter 1 could the abstract sample space described above serve as sample space? Explain your answer.

3. Consider an experiment in which a single family is selected by lottery from among all U.S. families which have exactly 3 children. The sexes of the 3 children from this family are recorded in sequence. For example, MFM denotes that the first-born child is a male, the second-born child is a female, and the third-born child is a male. Altogether, assuming only 1 child can be born at a time, there are 8 simple results of this experiment, namely, MMM, MFF, MMF, MFM, FMM, FFM, FMF, and FFF. Let the outcomes $\omega_1, \omega_2, \omega_3, \omega_4, \omega_5, \omega_6, \omega_7, \omega_8$ correspond respectively to these simple results (for example, ω_5 corresponds to the simple result FMM).
(a) What event set E_1 corresponds to the composite result that exactly 2 of the 3 children are male? Answer this question by listing the outcomes that belong to that event set.
(b) What event set E_2 corresponds to the composite result that the first-born child is a female?
(c) Suppose that when you observe the family chosen in your random experiment you only count the *number* of male children. Is this the same random experiment as the one described above? What are the

simple results of this experiment? What information do you lose by only counting the number of male children?

A4. Let E_1, E_2, and E_3 be 3 event sets associated with a random experiment. Express the following verbal statements both in set notation *and* by means of Venn diagrams.
   (a) At least 1 of the events occurs.
   (b) Exactly 1 of the events occurs.
   (c) Exactly 2 of the events occur.
   (d) Not more than 2 of the events occur simultaneously.
   (e) All of the events occur.
   (f) None of the events occur.

A5. Use Venn diagrams or logical arguments to establish the following relationships. You may argue abstractly or through a particular example, whichever is more convenient for you. If possible, however, try to argue abstractly.
   (a) Let E_1, E_2, and E_3 be event sets. If $E_1 \subset E_2$ and $E_2 \subset E_3$, then $E_1 \subset E_3$.
   (b) If $E_1 \subset E_2$, then $E_1 \cap E_2 = E_1$.
   (c) If $E_1 \subset E_2$, then $E_2^c \subset E_1^c$.
   (d) If $E_1 \subset E_2$, then $(E_1 \cap E_3) \subset (E_2 \cap E_3)$.
   (e) If $E_1 \cap E_2 = \emptyset$ and if $E_3 \subset E_1$, then $E_3 \cap E_2 = \emptyset$.
   (f) If E_1 and E_2 are event sets, then $(E_1 \cap E_2) \subset E_1$, $(E_1 \cap E_2) \subset E_2$, $E_1 \subset (E_1 \cup E_2)$, and $E_2 \subset (E_1 \cup E_2)$.

6. Use Venn diagrams or an explicit example of a sample space (such as the sample spaces described in Exercises 1, 2, or 3) to verify the truth of Rules 2.1, 2.2, and 2.3.

A7. Which of the following pairs of event sets A and B are mutually exclusive?
   (a) A: being the son of a lawyer,
       B: being born in Chicago;
   (b) A: being under 18 years of age,
       B: voting in a Presidential election in 1968;
   (c) A: owning a Chevrolet,
       B: owning a Ford.

8. Let the sample space of an experiment have as outcomes all of the adult individuals (over 18 years of age) in the United States. Construct a Venn diagram indicating your perception of the relationship among the following 5 event sets: the event set A consisting of all registered Democrats, the event set B consisting of all registered Republicans, the event set C consisting of all farmers, the event set D consisting of all members of labor unions, and the event set E consisting of all Naval personnel.

9. Suppose that in a survey concerning the smoking habits of adult males, it is found that: (i) 50 percent of the men smoke cigarettes; (ii) 35 percent of the men smoke pipes; and (iii) 30 percent of the men smoke both cigarettes and pipes. Which of the following assertions is false?
   (a) 45 percent of the adult males smoke neither cigarettes nor pipes.

(b) 25 percent of the adult males smoke exactly one of the choices: cigarettes or pipes.
(c) 85 percent of the adult males either smoke cigarettes or smoke pipes.
(d) 70 percent of the adult males do not smoke *both* cigarettes *and* pipes.

A10. Let the sample space of a probability model have as outcomes all schools in the United States. According to the 1965 *Statistical Abstract* (p. 106), there are 81,910 public elementary schools, 14,762 nonpublic elementary schools, 25,350 public secondary schools, 4129 nonpublic secondary schools, 721 public schools of higher education, and 1316 nonpublic schools of higher education. Use Venn diagrams to answer the following questions:
(a) How many schools are public schools?
(b) How many schools are elementary schools?
(c) How many schools are either private schools or schools of higher education?
(d) How many schools are either private schools or elementary schools, but not both?
(e) How many schools are *not* secondary schools?
(f) How many schools have no more than 1 of the following characteristics: private, school of higher education, elementary school?
(g) How many schools have *none* of the following characteristics: private, elementary, secondary?

11. A certain city of size 100,000 has 3 newspapers: A, B, and C. The proportions of people who read these papers are:

A: 0.10 A and B: 0.08 A and B and C: 0.01
B: 0.30 A and C: 0.02
C: 0.05 B and C: 0.04

(a) Find the proportion and the number of people who read only one newspaper.
(b) How many people read at least two newspapers?
(c) If A and C are morning papers, and B is an evening paper, how many people read *at least* one morning paper and also an evening paper?
(d) How many people read only *one* morning paper and *one* evening paper?

A12. Let C be the event set that a college student belongs to a fraternity or sorority and D the event set that a college student has a scholastic grade-point average of 2.5 or higher. Express each of the following symbolically:
(a) the probability that a college student belongs to a fraternity or sorority;
(b) the probability that a college student has a grade-point average of 2.5 or higher;
(c) the probability that a college student has a grade-point average less than 2.5;
(d) the probability that a college student does not belong to a fraternity or sorority *and* has a grade-point average of 2.5 or higher;
(e) the probability that a college student does not belong to a fraternity or sorority *and* has a grade-point average less than 2.5.

A13. Prove the following assertions. In so doing, state what axioms and rules of the probability calculus you are using to support your arguments.
  (a) For any two event sets A and B,
  $$P(A \cap B) = P(A) + P(B) - P(A \cup B).$$
  (b) For any two event sets A and B, $P(A \cap B^c) = P(A) - P(A \cap B)$.
  [*Hint:* Note that the event sets $A \cap B$ and $A \cap B^c$ have no outcomes in common, and that $(A \cap B) \cup (A \cap B^c) = A$. Now use the axioms of probability theory.]
  (c) Suppose that the event set A is included in the event set B, and that $P(B) = 0$. Then $P(A) = 0$.

A14. Suppose that we are given a probability model in which we are only told that $P(A) = 0.3$, $P(B) = 0.4$, and $P(A \cap B) = 0.2$. Establish the truth or the falsity of each of the following assertions. State explicitly what rules or axioms of the probability calculus you are using in verifying your assertions.
  (a) $P(A \cup B) = 0.6$;   (d) $P(A^c) = 0.7$;
  (b) $P(A^c \cap B) = 0.2$;   (e) $P(B^c) = 0.4$;
  (c) $P(A \cap B^c) = 0.2$;   (f) $P(A^c \cup B^c) = 0.8$.

15. Suppose that we are given a probability model in which two mutually exclusive event sets A and B have the respective probabilities $P(A) = 0.30$ and $P(B) = 0.40$. Find
  (a) the probability that either A or B occur;
  (b) the probability that both A and B occur;
  (c) the probability that exactly one of the event sets A and B occurs;
  (d) the probability that neither A nor B occurs;
  (e) the probability that B occurs but A does not occur;
  (f) the probability that either B does not occur or A does not occur.
  For each case state the axioms or rules of the probability calculus that you use to obtain your answer.

16. Let A be the event that a person is a college graduate and B the event that a person is wealthy. State in words what probabilities are expressed by
  (a) $P(A \text{ or } B)$;   (d) $1 - P(A)$;
  (b) $P(A \text{ and } B)$;   (e) $1 - P(B)$;
  (c) $P(A \text{ or } B) - P(A \text{ and } B)$; (f) $1 - P(A \text{ or } B)$.

17. At a doctor's office, we can read *Reader's Digest*, *Life Magazine*, *New Yorker*, or *Time*, whereas at the dentist's, we can read *Life Magazine*, *Reader's Digest*, *Look Magazine*, or *Sports Illustrated*. Suppose that only one magazine is available to us at each place at each visit, and that each pairing of magazines (one from each office) has an equal chance of being available. If we visit both offices, what is the probability that *Life Magazine* will be available for us to read at least in one of the two offices?

A18. In a certain psychological experiment, 2 individuals, I and II, are allowed to wait in a room unaware that they are being observed through a one-way mirror. The individuals are strangers to one another, and both think they are waiting in this room until the psychologist is ready to run them through an entirely different experiment. The interest in this experiment is the vocal interaction between the individuals.

(a) Let A be the event set corresponding to the result that individual I vocalizes at a given instant of time. Let B be the event set corresponding to the result that individual II vocalizes at that given instant of time. Draw a Venn diagram to indicate the event sets A, B, $A \cap B$, $A \cup B$. What is the event set corresponding to the result that both individuals are silent?

(b) Experimentally, the following probabilities are determined:
  (i) the probability that both individuals are silent is 0.3;
  (ii) the probability that both individuals speak simultaneously is 0.1;
  (iii) the probability that individual II speaks and individual I is silent is 0.4.

Determine $P(A)$, $P(B)$, and $P(A \cup B)$.

A19. A biologist is experimenting with cross-fertilization of two varieties of rose. He observes two characteristics of the resulting progeny: color of petal and shape of leaf. Suppose that genetic theory tells him that the probability of observing red petals is 0.37, the probability of observing a certain shape of leaf ("type Z") is 0.48, and that the probability of observing either red petals *or* a "type Z" shape of leaf is 0.68.

(a) What is the probability that the progeny has red petals *and* a "type Z" shape of leaf?

(b) What is the probability that the progeny has neither red petals nor a "type Z" shape of leaf?

(c) What is the probability that the progeny has red petals but does not have a "type Z" shape of leaf?

(d) What is the probability that the progeny has a "type Z" shape of leaf but does not have red petals?

(e) What is the probability that the progeny has *exactly one* of the two characteristics: red petals or "type Z" shape of leaf?

[*Hint:* Let the event set E_1 correspond to the composite result "progeny has red petals," and let the event set E_2 correspond to the composite result "progeny has 'type Z' shape of leaf." The information we are given is $P(E_1) = 0.37$, $P(E_2) = 0.48$, and $P(E_1 \cup E_2) = 0.68$. Part (a) asks for $P(E_1 \cap E_2)$. Part (b) asks for $P(E_1^c \cap E_2^c)$. Part (c) asks for $P(E_1 \cap E_2^c)$, and so on.]

20. Consider an experiment in which a single family is selected by lottery from among all U.S. families which have exactly 3 children. The simple results of this experiment are MMM, MFF, MMF, MFM, FMM, FFM, FMF, and FFF (see Exercise 3). A certain demographic theory states that the probability of the event set E_1 corresponding to the composite result "the first-born child is a male" is 0.50. This theory also states that the probability of the event set E_2 corresponding to the composite result "the second-born child is a male" is 0.51, and that the probability of the event set E_3 corresponding to the composite result "the third-born child is a male" is 0.52. The

demographic theory also gives you the following information: $P(E_1 \cap E_2) = 0.25$, $P(E_1 \cap E_3) = 0.25$, $P(E_2 \cap E_3) = 0.26$, and $P(E_1 \cap E_2 \cap E_3) = 0.13$.
- (a) What is the probability of the event set corresponding to the composite result "at least 1 of the children is male"?
- (b) What is the probability of the event set corresponding to the composite result "all 3 children are females"?
  [*Hint:* Use your answer to part (a).]
- (c) What is the probability of the event set corresponding to the composite result "exactly 2 of the children are male"?

21. A study is being made of the tendency of husbands and wives to come from similar social groups. A hierarchy of 4 social groups is defined; these groups are numbered in order of increasing prestige within the community. Thus, social group I is the most prestigious and social group IV is the least prestigious. For each married couple interviewed, the social groups of origin of the husband and of the wife are determined.
- (a) For this experiment identify all of the possible simple results of the experiment. For each possible simple result, assign this result to an outcome. How many outcomes are there in the sample space?
- (b) As the result of one study, it was found that the probability that both husband and wife come from the same social group is 0.45. What is the probability that a husband and wife will come from different social groups? What rule of the probability calculus did you use to answer this question?
- (c) In the same study it was found that the probabilities that the husband comes from social groups I, II, III, and IV are 0.07, 0.25, 0.45, and 0.23, respectively. What is the probability that the husband will come from one of the extreme social groups (social group I or social group IV)? What rule of the probability calculus did you use to answer this question?
- (d) It is stated that in this same study, it was found that the probability that husband and wife both come from social group I is 0.08. Explain why this statement must be false if the assertions made in parts (b) and (c) are true.

*^22. A study is being made of the relationship between the sex and school class of students in the School of Education and their choice of a field of specialization. For each student studied, the following information is recorded: sex (male or female), class year (freshman, sophomore, junior, or senior), and field of specialization (social studies, psychological studies, secondary education, or science and mathematics). Each student studied corresponds to one trial of a certain random experiment.
- (a) List all of the possible simple results of this random experiment and identify these results with corresponding outcomes in a probability model for the experiment. Verify that there are a total of 32 outcomes, $\omega_1, \omega_2, \omega_3, \ldots, \omega_{32}$, in the sample space of this probability model.
- (b) Assign probabilities to the event sets $E_1 = \{\omega_1\}$, $E_2 = \{\omega_2\}$, ..., $E_{32} = \{\omega_{32}\}$ which seem reasonable to you in the context of the given experi-

ment. What checks do you need to make in order to assure yourself that you have a legitimate probability model for this experiment? Is it enough to make sure that $P(E_1)+P(E_2)+\cdots+P(E_{32})=1$ and that each $P(E_i)$ is nonnegative? Explain.

(c) Let M be the event set corresponding to the composite result that the student you study is a male. Use your probability model to find $P(M)$. What rule do you use? From your answer for $P(M)$, directly obtain the probability that the student studied is a female. What rule do you use?

(d) Let A_1, A_2, A_3, and A_4 be event sets corresponding respectively to the composite results that the student is a freshman, sophomore, junior, and senior. Use your probability model to find $P(A_1)$, $P(A_2)$, $P(A_3)$, and $P(A_4)$. Do the probabilities $P(A_1)$, $P(A_2)$, $P(A_3)$, and $P(A_4)$ sum to 1? Should they? Explain why, or why not.

(e) Let B_1, B_2, B_3, and B_4 correspond to the composite results that the student's field of specialization is social studies, psychological studies, secondary education, and science and mathematics, respectively. Use your probability model to find $P(B_1)$, $P(B_2)$, $P(B_3)$, and $P(B_4)$. Do the probabilities $P(B_1)$, $P(B_2)$, $P(B_3)$, and $P(B_4)$ sum to 1? Should they?

(f) Use your probability model to find $P(M \cap A_1)$, $P(M \cap B_1)$, $P(A_1 \cap B_1)$, and $P(M \cap A_1 \cap B_1)$. Now use the values of $P(M)$, $P(A_1)$, and $P(B_1)$ which you obtained earlier, plus the values of $P(M \cap A_1)$, $P(M \cap B_1)$, $P(A_1 \cap B_1)$, and $P(M \cap A_1 \cap B_1)$ just obtained, to find $P(M \cup A_1 \cup B_1)$. What rule do you use? Check your answer by finding $P(M \cup A_1 \cup B_1)$ directly from your probability model.

(g) Does $P(M \cup A_1) = P(M) + P(A_1)$ for your probability model? If so, should this result also hold for any probability model for this experiment? Why? If not, why does this result not hold?

(h) Does $P(M \cup (M^c \cap A_1)) = P(M) + P(M^c \cap A_1)$ for your probability model? If so, should this result also hold for any probability model for this experiment? Why? If not, why does this result not hold?

(i) Suppose you know only the values of $P(M \cap A_1)$, $P(M \cap A_2)$, $P(M \cap A_3)$, and $P(M \cap A_4)$. From this information (and *only* this information), verify that you can obtain the value of $P(M)$.

*A23. A biologist is studying a certain species of insect. Each insect in the species can have either marked wings or unmarked wings, each insect can have either red eyes or pale eyes, and each insect can have either long or short antennae.

(a) There are 8 possible simple results for the experiment which consists of observing 1 insect of the given species and recording the characteristics of its wings, eyes, and antennae. List these simple results and identify each of their results with corresponding outcomes $\omega_1, \omega_2, \omega_3, \ldots, \omega_8$ for a probability model for this experiment.

(b) A researcher who previously studied this species of insect reported probabilities for the following events: E_1, E_2, E_3, $E_1 \cup E_2$, $E_1 \cup E_3$, $E_2 \cup E_3$, and $E_1^c \cap E_2^c \cap E_3^c$, where

$E_1 =$ the event set corresponding to the composite result "the insect has marked wings,"

E_2 = the event set corresponding to the composite result "the insect has red eyes,"

E_3 = the event set corresponding to the composite result "the insect has long antennae."

Show that the reported probabilities give sufficient information for you to calculate the probabilities of the event sets $\{\omega_1\}, \{\omega_2\}, \{\omega_3\}, ..., \{\omega_8\}$. Do this by giving formulas for $P\{\omega_1\}, P\{\omega_2\}, P\{\omega_3\}, P\{\omega_4\}, P\{\omega_5\}, P\{\omega_6\}, P\{\omega_7\},$ and $P\{\omega_8\}$ in terms of the given possibilities $P(E_1), P(E_2), P(E_3), P(E_1 \cup E_2), P(E_1 \cup E_3), P(E_2 \cup E_3),$ and $P(E_1^c \cap E_2^c \cap E_3^c)$. [*Hint:* A Venn diagram showing the events E_1, E_2, and E_3 and the outcomes $\omega_1, \omega_2, ..., \omega_8$ will be of help to you in answering this question.]

(c) Show that knowledge of $P\{\omega_1\}, P\{\omega_2\}, ..., P\{\omega_8\}$ allows you to calculate the probability of any event set for this probability model.
[*Hint:* Every event set E contains outcomes. Given the outcomes in E and the probabilities of each of the event sets $\{\omega_j\}$ which contain one and only one outcome, how would you calculate $P(E)$?]

(d) The probabilities reported by the previous researcher are $P(E_1) = \frac{1}{2}$, $P(E_2) = \frac{1}{2}$, $P(E_3) = \frac{1}{2}$, $P(E_1 \cup E_2) = \frac{3}{4}$, $P(E_1 \cup E_3) = \frac{3}{4}$, $P(E_2 \cup E_3) = \frac{3}{4}$, and $P(E_1^c \cap E_2^c \cap E_3^c) = \frac{1}{8}$. Find $P\{\omega_1\}, P\{\omega_2\}, P\{\omega_3\}, ..., P\{\omega_8\}$ using the formulas you gave in part (b).

^24. Suppose in the context of Exercise 23 that the probabilities reported by the previous researcher are $P(E_1) = \frac{1}{3}$, $P(E_2) = \frac{1}{3}$, $P(E_3) = \frac{1}{2}$, $P(E_1 \cup E_2) = \frac{1}{2}$, $P(E_1 \cup E_3) = \frac{2}{3}$, $P(E_2 \cup E_3) = \frac{2}{3}$, and $P(E_1^c \cap E_2^c \cap E_3^c) = \frac{1}{2}$. Show that the researcher must have made a mistake in obtaining these probabilities. That is, show that these probabilities do not produce a valid probability model. What axiom or rule of the probability calculus is violated?

25. Use the life table (Table 5.1) of Example 5.1 to answer the following questions:
   (a) What is the probability that a randomly chosen individual living in the city of Breslau, Poland in 1693 lived to be at least 60 years of age?
   (b) What is the probability that a randomly chosen individual living in Breslau, Poland in 1693 did not live past 18 years of age?
   (c) What is the probability that a randomly chosen individual living in Breslau, Poland in 1693 lived at least to 15 years of age, but no longer than 70 years of age?
   (d) Can you use the probabilities you found in parts (a), (b), and (c) to calculate the probability that a randomly chosen individual living in Breslau, Poland in 1693 lived past 15 years of age? If so, calculate this probability and check your answer by comparing it to the probability of this event as calculated from Table 5.1. If not, explain why not.

*26. You and a friend of yours are arguing over whether or not a certain result E can ever happen. You believe that a certain probability model gives a valid description of the random phenomenon being discussed. Under this probability model the event set corresponding to the result E has a probability of 0. Your friend asserts that E can occur and is willing to bet that in 1000 observations of the given random phenomenon, the result E will occur at least once. Assuming that your friend can and will match any bet, how much money would you be willing to bet him? Explain your answer. If your answer depends on the result E, describe a result E that is not logically impossible (that is, the result E cannot be described by a self-contradictory proposition) but which you believe has a probability of 0, and give your answer in that context. What does your answer tell you about *your* use of probability models as a description of naturally occurring random phenomena?

27. You want to construct a probability model for a given random experiment. The experiment has been run to answer a specific theoretical question in some area of science. What information do you need to know about the experiment in order to construct your probability model? If your probability model has been constructed, how would you check its "fit" (descriptive value) for your experiment?

3
FURTHER CONCEPTS OF PROBABILITY THEORY

If a man will begin with certainties, he shall end in doubts; but if he will be content to begin with doubts, he shall end in certainties.
                                  Francis Bacon (Advancement of Learning)

In this chapter, we go deeper into some aspects of probability theory that have already been introduced in Chapter 2; in addition, we discuss new concepts. In Chapter 2 we mentioned that the probability calculus enables us to calculate the probability of each possible event (event set) for a given probability model if we are given the probabilities of only certain basic events. This fact is illustrated in Section 1 of the present chapter, where we discuss a class of probability models whose sample spaces contain only a finite number of outcomes (finite models), and introduce a particularly useful probability model: the uniform probability model. One of the most important uses of the uniform probability model occurs in survey sampling. Section 2 provides a brief introduction to the probability theory of sample surveys, while some helpful methods of computation for this theory (and for probability theory in general) appear in Section 6.

Three new concepts introduced in this chapter are (1) composite random experiments, (2) conditional probabilities, and (3) stochastic independence. Since, as noted in Chapter 1, the word "stochastic" refers to prediction, we might expect a stochastic (that is, probability) model to aid in predicting the occurrence of events. The way in which the probability model does aid in prediction is made clear by these three new concepts. Discussion of composite experiments appears in Section 3. Conditional probabilities are introduced in Section 4, while Section 5 takes up the subject of stochastic independence.

The reader should note that starting in this chapter, we begin to use the language of probability theory (sample space, outcome, event, probability) for both the random experiment and its associated probability model.

1. FINITE PROBABILITY MODELS

There are a variety of random experiments whose sample spaces are finite. Among such experiments are most problems of population sampling (for example, sample surveys, lot inspections, and so on) most problems in genetics, almost all of the classical games of chance, and interesting problems in many other disciplines.

Specific finite models are discussed in Chapter 6. For the moment, we restrict ourselves to one or two brief general remarks about finite models, and then turn to a discussion of one special finite model: the uniform probability model.

Assume that the sample space Ω of our model has a finite number M of outcomes, that is, $\Omega = \{\omega_1, \omega_2, \ldots, \omega_M\}$. We define the *simple events* $S_1 = \{\omega_1\}$, $S_2 = \{\omega_2\}$, $\ldots$, $S_M = \{\omega_M\}$ of the probability model to be the events consisting of one and only one outcome. Figure 1.1 illustrates the sample space of one finite model, the corresponding outcomes and simple events, and a typical event set. There are seven outcomes $\omega_1, \omega_2, \ldots, \omega_7$ in the sample space pictured by Figure 1.1; these outcomes are represented by points. Corresponding to these seven points are seven simple events $S_1, S_2, \ldots, S_7$ (which are the small ovals in the figure). The entire box represents the sample space $\Omega = \{\omega_1, \omega_2, \ldots, \omega_7\}$. The event set E (the large oval) contains the outcomes $\omega_1, \omega_2, \omega_4$. Note that

$$E = \{\omega_1, \omega_2, \omega_4\} = S_1 \cup S_2 \cup S_4.$$

Figure 1.1 illustrates the following facts about finite models (facts true for the model whose sample space appears in Figure 1.1, and for *all* finite

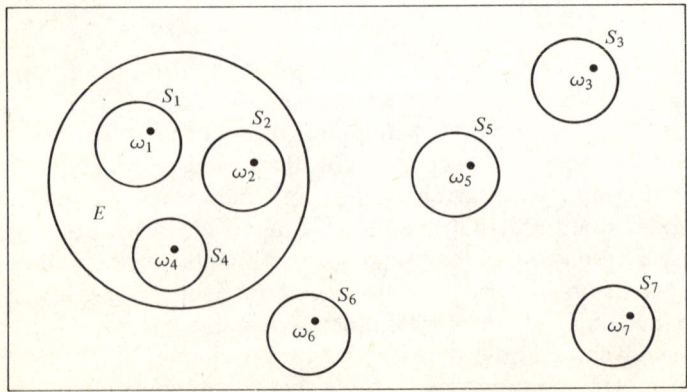

Figure 1.1 Sample space of a finite model.

models):

(i) Any two (different) simple events have no outcomes in common. That is, any two simple events are mutually exclusive (disjoint).
(ii) Any event E is the union of simple events (namely, the union of those simple events corresponding to outcomes in E).

These two facts, together with Rule 4.5 of Chapter 2, lead to a fundamental rule. This rule enables us to calculate the probability of any event E in the sample space of a finite model once we are given the probabilities for the simple events $S_1, S_2, \ldots, S_M$ of that model.

Rule 1.1. The probability of any event E of a finite probability model is the sum of the probabilities of the simple events whose union is E.

Thus, if, in the example illustrated in Figure 1.1, the probabilities of the simple events $S_1, S_2, \ldots, S_7$ are

$$P(S_1) = \frac{1}{12}, \quad P(S_2) = \frac{1}{4}, \quad P(S_3) = \frac{1}{4}, \quad P(S_4) = \frac{1}{12},$$

$$P(S_5) = \frac{1}{6}, \quad P(S_6) = \frac{1}{12}, \quad P(S_7) = \frac{1}{12},$$

then the probability of the event $E = S_1 \cup S_2 \cup S_4$ pictured in Figure 1.1 is

$$P(E) = P(S_1) + P(S_2) + P(S_4) = \frac{1}{12} + \frac{1}{4} + \frac{1}{12} = \frac{5}{12}.$$

From Rule 1.1, it follows that any finite probability model can be completely described by listing its simple events and their respective probabilities; thus

Simple Event	S_1	S_2	S_3	$\cdots$	S_{M-1}	S_M
Probability	p_1	p_2	p_3	$\cdots$	p_{M-1}	p_M

where $p_1 = P(S_1), p_2 = P(S_2), \ldots, p_M = P(S_M)$.

Example 1.1. In the random experiment which consists of spinning a standard American roulette wheel, the outcomes are the numbers 1, 2, 3, 4, ..., 36, 0, 00 (38 outcomes in all). The simple events are $S_1 = \{1$ is observed$\}$, $S_2 = \{2$ is observed$\}$, $S_3 = \{3$ is observed$\}$, ..., $S_{37} = \{0$ is observed$\}$, $S_{38} = \{00$ is observed$\}$. Assuming that the wheel is balanced any number on the wheel should have an equal probability of being observed. Therefore, we have the following list of probabilities:

Simple Event	S_1	S_2	S_3	S_4	$\cdots$	S_{37}	S_{38}
Probability	$\dfrac{1}{38}$	$\dfrac{1}{38}$	$\dfrac{1}{38}$	$\dfrac{1}{38}$	$\cdots$	$\dfrac{1}{38}$	$\dfrac{1}{38}$

From these probabilities we can find the probabilities of any event set. For example, if E is the event consisting of outcomes that are even numbers (excluding the zeros), then

$$P\{E\} = P(S_2) + P(S_4) + P(S_6) + \cdots + P(S_{36}) = \frac{18}{38} = \frac{9}{19}.$$

Example 1.1. illustrates a probability model that has validity whenever there is a basic homogeneity in the outcomes with respect to the natural forces which produce variation in the random experiment. Such homogeneity comes about, for example, when a balanced roulette wheel is spun (Example 1.1), when cards are chosen from a well-mixed deck, when two dice are thrown, and so on. Such homogeneity is often assumed for the choice of sex in a newborn child (although actually approximately 51 percent of all newborn children are boys), for guessing in multiple choice situations, for the choice of magnetic polarity in a given electron, and in many other experiments of importance in science and technology. It is a *logical* consequence of this assumption of homogeneity that the simple events of the model have *equal* probabilities. Any finite probability model for which the simple events have equal probability is called a *uniform probability model*.

Let us continue to assume that there are M outcomes in Ω. Then in a uniform probability model, each of the simple events has the same probability p [that is, $P(S_1) = P(S_2) = \cdots = P(S_M) = p$]. Since $P(\Omega) = 1$ and $\Omega = S_1 \cup S_2 \cup \cdots \cup S_M$, we have

$$1 = P(S_1) + P(S_2) + \cdots + P(S_M) = p + p + \cdots + p = Mp,$$

so that $p = 1/M$.

Rule 1.2. In a uniform probability model, the probability of any simple event equals $1/M$, where M is the number of outcomes in Ω.

If the event E consists of K outcomes, then E is the union of K simple events, each having probability $1/M$. Thus, the probability of E is K times $1/M$ or K/M.

Rule 1.3. In a uniform probability model, the probability of any event E is

$$P(E) = \frac{\text{number outcomes in } E}{\text{number outcomes in } \Omega}.$$

Example 1.2. In a learning experiment, a rat is placed in a cell. There are ten identical doors leading from the cell. One of the doors leads to food and water, one to food alone, eight lead to no rewards. The rat has never been placed in this cell before; we want to see if he can learn to choose any of the doors leading to rewards. At the initial stage it is reasonable to assume that the rat can only guess which door to choose (that is, he picks each door with equal probability). Thus the probability that the rat is rewarded at the first stage is

$$P\{\text{rat is rewarded}\} = \frac{(\text{number of doors to rewards})}{10} = \frac{2}{10} = \frac{1}{5}.$$

2. SIMPLE RANDOM SAMPLING

Perhaps the single most important example of the use of the uniform probability model is in survey sampling. Here the randomness in the experiment is deliberately introduced by the experimenter. Rather than examine every unit in the population of interest to him, he finds it more convenient to sample a population in a controlled random way (that is, choose some members of the population by means of a lottery), examine these chosen members, and then obtain inferences about the entire population through the use of probability theory.

The basic probability model for survey sampling is that of *random sampling*. Suppose that we have a population of N units $u_1, u_2, ..., u_N$. This population may consist of people, domiciles, radio tubes, chemical distillates, items on an examination, numbers, and so on. We decide to take a sample of n (n less than or equal to N) units from the population by means of some probabilistic mechanism. That is, we decide to perform a random experiment whose outcomes are *samples* of n units, each particular sample being chosen with a certain probability. Sampling schemes are distinguished from one another according to the types of samples that can be outcomes of the experiment and according to the probabilities with which those outcomes (samples) are drawn. Regardless of the sampling scheme, any sample which is the outcome of random sampling is called a *random sample*.

Formally, a *sample* is an array of units listed in a definite order. For this reason, it is perhaps more precise to use the phrase *ordered sample*. For example, one possible sample of five units from the population consisting of the units $u_1, u_2, \ldots, u_N$ is the sample $(u_1, u_4, u_5, u_7, u_3)$. This sample would be distinguished from the sample $(u_5, u_1, u_4, u_7, u_3)$ because even though the same units appear in both samples, the *order* in which they appear is different. The definition of a sample should be interpreted to allow a given unit (say u_2) to appear $0, 1, 2, \ldots, n$ times in an (ordered) sample of size n. For example, the ordered array of units $(u_3, u_2, u_2, u_7, u_1)$ and the ordered array of units $(u_1, u_1, u_1, u_2, u_2)$ are permissible samples of five units.

Example 2.1. In the population consisting of the four units u_1, u_2, u_3, u_4, the following are all possible ordered samples of two units:

$\omega_1 = (u_1, u_1)$, $\omega_5 = (u_2, u_1)$, $\omega_9 = (u_3, u_1)$, $\omega_{13} = (u_4, u_1)$,
$\omega_2 = (u_1, u_2)$, $\omega_6 = (u_2, u_2)$, $\omega_{10} = (u_3, u_2)$, $\omega_{14} = (u_4, u_2)$,
$\omega_3 = (u_1, u_3)$, $\omega_7 = (u_2, u_3)$, $\omega_{11} = (u_3, u_3)$, $\omega_{15} = (u_4, u_3)$,
$\omega_4 = (u_1, u_4)$, $\omega_8 = (u_2, u_4)$, $\omega_{12} = (u_3, u_4)$, $\omega_{16} = (u_4, u_4)$.

One possible type of random sampling is *simple random sampling with replacement*. This is a random experiment in which the sample spaces Ω consist of all possible ordered samples, and in which each sample has equal probability of being selected. Another possible type of random sampling is *simple random sampling without replacement*. This random experiment is described by a probability model which gives zero probability to all ordered samples in which any unit appears more than once, and equal probability to all other ordered samples.

Example 2.2. For the population of four units discussed in Example 2.1 above, the probability model for *simple random sampling with replacement* can be described in the following manner (see Section 1):

Simple Event	$\{\omega_1\}$	$\{\omega_2\}$	$\{\omega_3\}$	$\{\omega_4\}$	$\{\omega_5\}$	$\{\omega_6\}$	$\{\omega_7\}$	$\{\omega_8\}$
Probability	$\frac{1}{16}$	$\frac{1}{16}$	$\frac{1}{16}$	$\frac{1}{16}$	$\frac{1}{16}$	$\frac{1}{16}$	$\frac{1}{16}$	$\frac{1}{16}$
Simple Event	$\{\omega_9\}$	$\{\omega_{10}\}$	$\{\omega_{11}\}$	$\{\omega_{12}\}$	$\{\omega_{13}\}$	$\{\omega_{14}\}$	$\{\omega_{15}\}$	$\{\omega_{16}\}$
Probability	$\frac{1}{16}$	$\frac{1}{16}$	$\frac{1}{16}$	$\frac{1}{16}$	$\frac{1}{16}$	$\frac{1}{16}$	$\frac{1}{16}$	$\frac{1}{16}$

In Example 2.1, the probability model for *simple random sampling without replacement* can be described in the following manner:

Simple Event	$\{\omega_1\}$	$\{\omega_2\}$	$\{\omega_3\}$	$\{\omega_4\}$	$\{\omega_5\}$	$\{\omega_6\}$	$\{\omega_7\}$	$\{\omega_8\}$
Probability	0	$\frac{1}{12}$	$\frac{1}{12}$	$\frac{1}{12}$	$\frac{1}{12}$	0	$\frac{1}{12}$	$\frac{1}{12}$
Simple Event	$\{\omega_9\}$	$\{\omega_{10}\}$	$\{\omega_{11}\}$	$\{\omega_{12}\}$	$\{\omega_{13}\}$	$\{\omega_{14}\}$	$\{\omega_{15}\}$	$\{\omega_{16}\}$
Probability	$\frac{1}{12}$	$\frac{1}{12}$	0	$\frac{1}{12}$	$\frac{1}{12}$	$\frac{1}{12}$	$\frac{1}{12}$	0

where we recall that $\omega_1 = (u_1,u_1)$, $\omega_6 = (u_2,u_2)$, $\omega_{11} = (u_3,u_3)$, and $\omega_{16} = (u_4,u_4)$ are samples in which a unit appears more than once.

Simple random sampling with replacement gives rise to a uniform probability model for the sample space whose outcomes are all ordered samples (*ordered samples with replacement*). Since, under simple random sampling *without* replacement, all samples in which a given unit appears more than once have zero probability, we might agree to eliminate these outcomes from the sample space. Then, over the remaining samples (*ordered samples without replacement*), simple random sampling without replacement defines a uniform probability model.

Under either sampling model, in order to apply Rule 1.3 to find the probability that an observed sample has a given property, we must find the number K of possible samples having this property, find the total number M of possible samples under the sampling scheme, and then take the ratio K/M.

Rule 2.1. The total number M_R of ordered samples (with replacement) of size n from a population of N units is $M_R = N^n$.

Rule 2.2. The total number M_W of ordered samples of size n from a population of N units in which no unit appears more than once (that is, ordered samples without replacement) is

$$M_W = N(N-1)(N-2) \cdots (N-n+1).$$

The subscripts R and W on M_R and M_W are mnemonic devices to remind us that M_R refers to sampling with replacement and M_W to sampling without replacement. For the moment, these two rules must be taken on faith; their validity is demonstrated in Section 6.

Under either of the sampling models discussed above, to find the probability that an observed sample will have a given property, we need only find K, the number of possible samples (which have nonzero probabilities under the model) having that property. The number K gives the numerator of the fraction which gives the desired probability (see Rule 1.3); the denominator of that fraction comes from either Rule 2.1 or Rule 2.2 of this section, depending on which sampling model is appropriate. There are several rules that simplify the process of calculating K. These rules are called the rules of *combinatorial analysis* (see Section 6).

Example 2.3. Of 50 people who are members of a given population (say, a given city block), 25 (say, units $u_1, u_2, u_3, \ldots, u_{25}$) are Democrats and 25 (units $u_{26}, u_{27}, \ldots, u_{50}$) are Republicans. Random sampling without replacement is used to draw a sample of two individuals. (How this might be done in practice is indicated in Section 3 when we discuss lotteries.) What is the probability that both people drawn are Republicans?

The total number M_W of ordered samples without replacement of size $n = 2$ from a population of $N = 50$ units is, by Rule 2.2, equal to $50(49) = 2450$. The number of possible samples in which both people drawn are Republicans can also be calculated from Rule 2.2 if we note that such samples may be conceived of as actually having been drawn only from the population consisting of the 25 people (in the city block) who are Republicans. Thus, the sample being drawn can now be viewed as being an ordered sample without replacement of size $n = 2$ from a population of $N = 25$ Republicans; Rule 2.2 tells us that $(25)(24) = 600$ such samples are possible.

Applying Rule 1.3, we conclude that

$P\{$sample drawn without replacement consists of two Republicans$\}$

$$= \frac{600}{2450} = \frac{12}{49}.$$

If we had drawn samples from the city block using sampling with replacement, we could apply Rule 2.1 twice (once for the denominator with $n = 2, N = 50$; once for the numerator with $n = 2, N = 25$) and then apply Rule 1.3 to obtain the result

$P\{$sample drawn with replacement consists of two Republicans$\}$

$$= \frac{(25)^2}{(50)^2} = \frac{1}{4}.$$

Comparison of the above probabilities demonstrates how the probability of an event depends on the method of sampling.

3. COMPOSITE RANDOM EXPERIMENTS

In practice, random sampling is not usually performed by drawing the entire sample by a single probabilistic mechanism. Instead, the sample is constructed in stages (or *draws*), and each stage can be considered to be a random experiment. Thus, the total sample results from a collection (or composite) of random experiments.

A typical example of a sampling experiment that can be described by random sampling models is a lottery (for example, the lottery that is used to select winners at a raffle). One way of setting up a lottery is to write the names of the participating individuals on homogeneous slips of paper. These names (or, equivalently, the individual slips of paper) now become the units of our population; if there are N slips of paper (names), our

3. COMPOSITE RANDOM EXPERIMENTS

population has size N. Suppose n prizes are to be given out. The names of the winners of these prizes then comprise a sample of size n from the population of names (slips) participating in the lottery. A frequently used device for choosing the winners in such a lottery is to place the slips of paper containing names of the participants in a big urn. The urn is thoroughly shaken, a slip is drawn, and the bearer of the name on that slip receives a prize. The winning slip is placed aside, once again the urn is shaken, another slip is drawn, and the bearer of the name on that slip is awarded a prize. The drawing proceeds in such stages until n slips have been chosen. The collection of such slips arranged in the order in which they were drawn from the urn is an ordered sample without replacement (the nature of the drawing assures that a given slip can only be drawn once). If the shaking of the urn at each stage of the draw thoroughly mixes the slips, and if the slips are chosen in as uniform and impartial a manner as possible, we have good reason to believe that any possible arrangement of n slips (in which no slip can appear more than once in the array) has an equal chance of being observed, and thus the lottery is a way of generating a random sample (of size n) without replacement.

In a lottery, however, each draw itself results in a sample of size 1 from the population of slips remaining in the urn after previous draws. For example, the first slip drawn is a random sample of size 1, with or without replacement (there is no distinction between with or without replacement when we only draw one unit), from the complete population of N slips. That is, the shaking of the urn and the impartiality of the method of drawing the slip means that every slip (every sample of size 1) has an equal chance of being chosen. Once this first slip is drawn, the remaining population of slips in the urn has $N-1$ members. The second slip drawn is thus a random sample of size 1 from a population of $N-1$ units. Drawing this second slip reduces the population of slips in the urn to $N-2$ units, so that the third slip drawn is a random sample of size 1 from a population of $N-2$ units, and so on. From this discussion, we not only see that random sampling without replacement can be regarded as being a collection of smaller component random sampling experiments, but also that the outcomes of certain of these component random sampling experiments (for example, the outcome of the first draw in the lottery) can affect the conditions under which others of the component random sampling experiments are performed (for example, the population of units from which the second draw in the lottery is made depends on what unit is chosen on the first draw). Such observations motivate a study of the concept of a composite random experiment, both because such a study can allow us to break down random experiments (such as lotteries) into components that may be easier to handle individually, and because causal relationships among such component random experiments may allow us to better predict the results of one component random experiment if we know the results of some other component random experiment(s).

A lottery is only one example of a random experiment that conceptually can be broken down into component random experiments. Another example occurs when we take concurrent measurements on several different variables in the same random experiment. This *multivariable random experiment* can be thought of as the simultaneous running of several component random experiments—one for each variable measured. Usually random experiments that are carried out in practical contexts have several identifiable components, each of which can be considered to be a random experiment.

Mathematically, a *composite random experiment* is any random experiment whose outcomes can be represented as *n-tuples* for some fixed integer n. An n-tuple is an ordered array $(a^{(1)}, a^{(2)}, \ldots, a^{(n)})$ of entities which has n positions, the entity $a^{(1)}$ filling the first position, the entity $a^{(2)}$ filling the second position, ..., and the entity $a^{(n)}$ filling the nth position. For example, a 2-tuple or *pair* is an ordered array having two positions, distinguished by calling one the first position and the other the second position. It should be noted that the positions of two pairs may be filled by the same entities, but yet the pairs may be different because the entities are placed in different positions. Thus, the pair (Eve,Apple) is not the same as the pair (Apple,Eve), even though both pairs contain the same entities. Example 2.1 lists all pairs that can be formed from the entities (units) u_1, u_2, u_3, u_4. We have already seen that such pairs comprise all possible ordered samples of size 2 from the population consisting of the units u_1, u_2, u_3, u_4. Similarly, any ordered sample of size n from a population of N units can be represented as an n-tuple of these units.

A Comment on Notation: The reader who has read other mathematics books may already be familiar with the concept of an n-tuple, but may find our use of superscripts to denote the positions of entities in an n-tuple confusing. In many mathematics books, n-tuples are written in the form $(a_1, a_2, \ldots, a_n)$. We would have preferred this notation, and indeed would have used it here, except that we have already reserved subscripts for enumeration of outcomes, event sets, and so on, *within* a given random experiment. Now, however, we are talking about composites of random experiments, and so we must have a notation that distinguishes *among* these random experiments. Thus, we have chosen to use superscripts to identify particular components of a composite random experiment, reserving subscripts for their role of distinguishing among the outcomes, event sets, and so on, of each particular component random experiment. Because, in the next paragraph, we start writing n-tuples in which the ith position is filled by an outcome from a component random experiment, we have, at least for this chapter, chosen to represent n-tuples using the superscript notation of the previous paragraph [that is, $(a^{(1)},a^{(2)},...,a^{(n)})$] in order to keep a consistent notation.

If a random experiment $\mathscr{E}$ has outcomes that can be represented as an n-tuple $\omega = (\omega^{(1)},\omega^{(2)},...,\omega^{(n)})$ for some fixed integer n, we can regard the entities $\omega^{(1)}$, $\omega^{(2)}$, ..., $\omega^{(n)}$ which fill, respectively, positions $1, 2, ..., n$ of the n-tuple ω as being *aspects* or *component outcomes* of the random experiment $\mathscr{E}$. Thus, in the lottery mentioned earlier, $\omega^{(1)}$ may represent the name chosen on the first draw, $\omega^{(2)}$ may represent the name chosen on the second draw, ..., and $\omega^{(n)}$ may represent the name chosen on the nth draw of the lottery. The outcome $\omega = (\omega^{(1)},\omega^{(2)},...,\omega^{(n)})$ is then an ordered sample (drawn without replacement) of size n of names drawn from the population consisting of the names of the people participating in the lottery. In the context of Example 1.1 we might run n rats through the experiment instead of just 1; then $\omega^{(1)}$ might be the observed choice of door for the first rat run through the experiment, $\omega^{(2)}$ might be the observed choice for the second rat run through that experiment, ..., and $\omega^{(n)}$ might be the observed choice of door for the nth rat. The outcome ω of the entire random experiment now consists in observing the choice of doors for all n rats; this outcome ω can be represented as the n-tuple $\omega = (\omega^{(1)},\omega^{(2)},...,\omega^{(n)})$. Still a third example is a random experiment in which we observe n different aspects of an individual; say, $\omega^{(1)} = $ his height, $\omega^{(2)} = $ his weight, ..., $\omega^{(n)} = $ his I.Q. This experiment is a multivariate random experiment with outcomes $\omega = $ (height, weight, ..., I.Q.).

A component outcome $\omega^{(i)}$ [that is, the entry in the ith position of the outcome $\omega = (\omega^{(1)},\omega^{(2)},...,\omega^{(i-1)},\omega^{(i)},\omega^{(i+1)},...,\omega^{(n)})$] may itself be considered to be the outcome of a random experiment $\mathscr{E}^{(i)}$. The collection of all possible entries $\omega^{(i)}$ that can fill the ith position of the outcome ω is then the sample space $\Omega^{(i)}$ for a random experiment which consists of running the entire composite random experiment $\mathscr{E}$, observing the outcome ω of $\mathscr{E}$, and recording only the ith position in the n-tuple representation of that outcome. This random experiment $\mathscr{E}^{(i)}$ is called a *component random experiment* of the composite random experiment $\mathscr{E}$. A very important fact to remember is the following: *Given a composite random experiment $\mathscr{E}$ which has component random experiments $\mathscr{E}^{(1)}, \mathscr{E}^{(2)}, ..., \mathscr{E}^{(n)}$, the probability measures of these component random experiments are completely determined by the probability measure of the composite random experiment $\mathscr{E}$.* We illustrate the validity of this assertion in the following example.

Example 3.1. As our composite random experiment $\mathscr{E}$, consider observing the sex of the children in a family that has exactly 2 children. The sample space of this experiment has outcomes that can be represented as $\omega =$ (sex of first-born, sex of second-born). The components of the pair $\omega = (\omega^{(1)},\omega^{(2)})$ are $\omega^{(1)} =$ sex of first-born and $\omega^{(2)} =$ sex of second-born. Thus the sample space Ω of the composite random experiment $\mathscr{E}$ consists of the outcomes

(3.1)
$$\begin{aligned}\omega_1 &= (\omega_1^{(1)},\omega_1^{(2)}) = \text{(male,male)},\\ \omega_2 &= (\omega_1^{(1)},\omega_2^{(2)}) = \text{(male,female)},\\ \omega_3 &= (\omega_2^{(1)},\omega_1^{(2)}) = \text{(female,male)},\\ \omega_4 &= (\omega_2^{(1)},\omega_2^{(2)}) = \text{(female,female)}.\end{aligned}$$

We remind the reader that in our notation, the superscript serves to identify the component while the subscript identifies the particular outcome. Thus $\omega_1^{(i)}$ represents a male, and $\omega_2^{(i)}$ represents a female for the ith component random experiment. The lack of a superscript means that we are referring to an outcome of the entire composite experiment; thus, $\omega_3 = (\omega_2^{(1)},\omega_1^{(2)})$ represents the outcome that the first-born child is a female and the second-born child is a male, that is, (female,male). Using this notation, we see that the sample space $\Omega^{(1)}$ for the first component random experiment is $\Omega^{(1)} = \{\omega_1^{(1)},\omega_2^{(1)}\} = \{$first-born is a male, first-born is a female$\}$, and the sample space for the second component random experiment is $\Omega^{(2)} = \{\omega_1^{(2)},\omega_2^{(2)}\} = \{$second-born is a male, second-born is a female$\}$.

Suppose that the probability measure for the composite random experiment $\mathscr{E}$ is given by

Simple Event	$\{\omega_1\}$	$\{\omega_2\}$	$\{\omega_3\}$	$\{\omega_4\}$
Probability	$\frac{1}{3}$	$\frac{1}{6}$	$\frac{1}{6}$	$\frac{1}{3}$

We recall (Section 1) that for a finite probability model (such as we have in the present example), the above list of probabilities enables us to calculate the probabilities of all of the more complicated events that may be considered in such a model. To give a probability model for the first component random experiment $\mathscr{E}^{(1)}$, we need only fill in the blanks in the following table (since $\Omega^{(1)}$ has a finite number of outcomes):

Simple Event	$\{\omega_1^{(1)}\}$	$\{\omega_2^{(1)}\}$
Probability		

But, looking at the definitions of the outcomes of the composite random experiment as given in Equation (3.1), we see that

$P(\{\omega_1^{(1)}\}) = P(\{\text{first-born child is a male}\}) = P(\{\omega_1\} \cup \{\omega_2\}) = \frac{1}{3}+\frac{1}{6}=\frac{1}{2},$
$P(\{\omega_2^{(1)}\}) = P(\{\text{first-born child is a female}\}) = P(\{\omega_3\} \cup \{\omega_4\}) = \frac{1}{6}+\frac{1}{3}=\frac{1}{2}.$

3. COMPOSITE RANDOM EXPERIMENTS

Similar considerations produce the equations

$$P(\{\omega_1^{(2)}\}) = P(\{\omega_1\} \cup \{\omega_3\}) = \tfrac{1}{3} + \tfrac{1}{6} = \tfrac{1}{2},$$
$$P(\{\omega_2^{(2)}\}) = P(\{\omega_2\} \cup \{\omega_4\}) = \tfrac{1}{6} + \tfrac{1}{3} = \tfrac{1}{2},$$

so that the probability model for $\mathscr{E}^{(2)}$ is given by the list

Simple Event	$\{\omega_1^{(2)}\}$	$\{\omega_2^{(2)}\}$
Probability	$\tfrac{1}{2}$	$\tfrac{1}{2}$

A Further Notational Remark: In reading through Example 3.1, the reader may be confused between the notations $\{a,b\}$ and (a,b). The brackets $\{\cdot,\cdot\}$ mean "collection of" and are used to indicate what entities are in the collection of which we are speaking. With the notation $\{\cdot,\cdot\}$, the order in which we list the entities does not matter; thus, $\{a,b\}$ is the same as $\{b,a\}$ since both expressions mean "the collection containing the entities 'a' and 'b'." On the other hand, with the parentheses $(\cdot,\cdot)$ we have to be very careful about the order in which we list the entities because (a,b) is not, in general, the same as (b,a). The former expression means "the pair in which a is in the first position and b is in the second position," while (b,a) is "the pair in which b is in the first position and a is in the second position." The reader who has been confused between these notations should now reread Example 3.1.

Example 3.1 illustrates our previous assertion that knowledge of the probability measure for the composite random experiment $\mathscr{E}$ determines the probability measure of each of its component random experiments $\mathscr{E}^{(i)}$. We have shown (at least for finite models) how to obtain the probability measures for the component random experiments when given the probability measure P for the entire composite random experiment $\mathscr{E}$; namely, to find the probability measure of the jth simple event $\{\omega_j^{(i)}\}$ in the ith component random experiment $\mathscr{E}^{(i)}$, simply add the probabilities of all simple events $\{\omega\}$ in the composite random experiment $\mathscr{E}$ for which the ith component outcome of ω is $\omega_j^{(i)}$. Let us now use the same example to show that many probability measures for a composite random experiment $\mathscr{E}$ can yield the *same* probability measure for each given component random experiment $\mathscr{E}^{(i)}$.

Example 3.1 (Continued). Consider again the random experiment $\mathscr{E}$ in which we observe the sexes of the children in a family that has exactly 2 children. We have already considered one probability measure for this experiment. Consider now the following probability measure:

Simple Event	$\{\omega_1\}$	$\{\omega_2\}$	$\{\omega_3\}$	$\{\omega_4\}$
Probability	$\frac{1}{4}$	$\frac{1}{4}$	$\frac{1}{4}$	$\frac{1}{4}$

Call this probability measure "probability measure number 2" and the probability measure previously given for this example "probability measure number 1." Let us find the probability measure for the component random experiment $\mathscr{E}^{(1)}$ (in which the sex of the first-born child is observed) under probability measure number 2 for the composite random experiment $\mathscr{E}$. To do this, we need only find the probabilities of the simple events $\{\omega_1^{(1)}\}$ and $\{\omega_2^{(1)}\}$. (Note that these events are *not* simple events for the composite random experiment $\mathscr{E}$.) These probabilities are

$$P(\{\omega_1^{(1)}\}) = P(\{\omega_1\} \cup \{\omega_2\}) = \tfrac{1}{4} + \tfrac{1}{4} = \tfrac{1}{2},$$
$$P(\{\omega_2^{(1)}\}) = P(\{\omega_3\} \cup \{\omega_4\}) = \tfrac{1}{4} + \tfrac{1}{4} = \tfrac{1}{2}.$$

We recognize these probabilities as being the same as those previously found for $\{\omega_1^{(1)}\}$ and $\{\omega_2^{(1)}\}$ under probability measure number 1. Similarly,

$$P(\{\omega_1^{(2)}\}) = P(\{\omega_1\} \cup \{\omega_3\}) = \tfrac{1}{4} + \tfrac{1}{4} = \tfrac{1}{2},$$
$$P(\{\omega_2^{(2)}\}) = P(\{\omega_2\} \cup \{\omega_4\}) = \tfrac{1}{4} + \tfrac{1}{4} = \tfrac{1}{2},$$

are the same probabilities for $\{\omega_1^{(2)}\}$ and $\{\omega_2^{(2)}\}$ as previously found under probability measure number 1. Thus probability measures numbers 1 and 2 for the composite random experiment $\mathscr{E}$ each result in the same probability measure for component random experiment $\mathscr{E}^{(1)}$, and the same probability measure for component random experiment $\mathscr{E}^{(2)}$.

Since two different probability measures for a composite random experiment $\mathscr{E}$ can yield the same probability measure for each of the component random experiments $\mathscr{E}^{(i)}$, *knowledge of the probability measures for each of the component random experiments of a composite random experiment is not enough to determine the probability measure for the entire composite experiment.* The extra information about the composite random experiment which is carried by the probability measure of the entire composite and which is not carried by the probability measures of the component experiments is information on how the components tend to influence one another. For example, in Example 3.1 above, belief in probability measure number 1 allows us to assert that families in which both children have the same sex are more probable than families in which the 2 children are of different sex. In contrast, belief in probability measure number 2 forces us to assert that all combinations of sexes are equally probable.

Suppose that the correct model in Example 3.1 turns out to be probability model number 1. An acquaintance tells you that he has exactly two children and that the older is a boy. What sex would you predict for his second child? The new concept that tells you how to use the composite probability model and the information you have about the sex of your acquaintance's first-born is the concept of *conditional probability*. This concept is introduced in the next section.

4. CONDITIONAL PROBABILITY

One type of composite random experiment is the sequential experiment. Here, the experiment is completed in stages with the experimenter observing and recording the results of each stage of the experiment as it is performed. To know the outcome of the entire experiment, the experimenter must wait for its completion; however, the information obtained at each stage of the experiment often permits the experimenter to narrow down the list of possible outcomes that can finally occur and reassess the probabilities of those events that interest him. For example, if an engineer takes observations on the output of a given factory at various times t_1, t_2, and so on, he can use this infomation to reassess the probabilities for the reliability of future outputs of that factory; if the reassessed probabilities indicate declining quality of production, he can then modify the production process.

Alternatively, there are many situations where we have partial knowledge concerning the outcome of a random experiment before the complete result becomes known, and where this partial knowledge enables us to re-evaluate the probabilities with which events of interest will occur. For example, consider the following condensed (and fictitious) life table:

	Probability of Dying Before Age 65	*Probability of Living to Age 65*
Men	0.26	0.74
Women	0.20	0.80
Both Men and Women	0.23	0.77

If actuaries in an insurance company do not know the sex of a particular applicant, they will undoubtedly base their insurance premium on a probability of death for both sexes before age 65 of 0.23. However, if the actuaries know that the applicant is a woman, they should lower their premium since the chance that a person will die before age 65, given that the person is a female, is only 0.20.

This process of re-evaluating probabilities in the light of partial information about the outcomes of an experiment is made precise by the concept of *conditional probability*. Formally, let A be an event such that $P(A) > 0$, and let B be another event whose occurrence or nonoccurrence is of concern. The conditional probability of B occurring *given* that A has occurred [denoted by $P(B|A)$] is defined by the equation

$$(4.1) \qquad P(B|A) = \frac{P(A \cap B)}{P(A)}.$$

If $P(A) = 0$, the conditional probability of B given A is undefined.

To distinguish the conditional probability $P(B|A)$ of the event B *given* the event A from the probability $P(B)$ of B evaluated before the experiment begins, we sometimes speak of $P(B)$ as being the *unconditional*, *marginal*, or *absolute* probability of B. Of course, *all* probabilities are conditional in the sense that we assign (or calculate) these probabilities in the light of our present knowledge of the properties of whatever random experiment we are observing. If a change occurs in the circumstances which guided our choice of the probability model for the random experiment (the conditions under which the experiment is performed, our theoretical knowledge about underlying processes, and so on), a similar change in the probabilities $P(B)$ which we assign to events B may be in order. The difference between the above sense of the word "conditional" and the sense in which this word is used when we talk about the conditional probability of an event B given that an event A is known to have occurred [that is, $P(B|A)$] is that the use of the word "conditional" in the latter sense refers to a change in our knowledge provided by the random experiment itself.

The frequency interpretation for $P(E)$ that has been presented previously states that the proportion of trials, r.f.(E), in which the event E occurs in a large number of repetitions of the experiment tends to be close to $P(E)$. In a similar manner, a frequency interpretation can be given for $P(B|A)$. Let $\#A$ denote the number of times A occurs in a large number N of repeated trials. Let $\#(A \cap B)$ be the number of times that A and B both occur in the N trials. Then the ratio $\#(A \cap B)/\#A$ is the proportion of times that B occurs among those trials for which A also occurs. However,

$$\frac{\#(A \cap B)}{\#A} = \frac{\#(A \cap B)/N}{\#A/N} = \frac{\text{r.f.}(A \cap B)}{\text{r.f.}(A)},$$

which by the frequency interpretation of the measure $P(E)$ tends to be approximately $P(A \cap B)/P(A)$. Thus, the conditional probability of B given A is the abstraction of the long-run proportion of times that B occurs among those trials for which A also occurs.

Example 4.1. Consider the following table of relative frequencies (Table X of A. B. Hollingshead, *Elmtown's Youth: The Impact of Social Classes on Adolescents*. John Wiley, New York, 1949) which resulted from a study of the impact of social class structures on 390 high school students in a middle western community.

Table 4.1. Frequency of enrollment of five social classes of Elmtown youth in three alternative high school curricula.

	Social Class			
Curriculum	I and II	III	IV	V
College Preparatory	23/390	40/390	16/390	2/390
General	11/390	75/390	107/390	14/390
Commercial	1/390	31/390	60/390	10/390

Table 4.1 can be regarded as resulting from 390 replications of the random experiment in which a high school age youth (from Elmtown) is observed and his social class[1] and choice of high school curriculum are recorded. The outcomes (simple results) of this experiment are the pairs in which, say, the first position gives the social class of the individual and the second position gives the choice of curriculum for the individual. One such outcome, for example, is the pair (Social Class V, General Curriculum). Let A be the event (composite result) that an observed individual is in Social Class V, and let B be the event that an observed individual chooses a general high school curriculum. Note that the relative frequency of event A is given by the sum of the entries in the last column of Table 4.1; that is, r.f.$(A) = 26/390$. The proportion of times that an individual chooses a general high school curriculum among the trials in which the individual's social class is observed to be Class V is then

$$\frac{\#(A \cap B)}{\#(A)} = \frac{\text{r.f.}(A \cap B)}{\text{r.f.}(A)} = \frac{14/390}{26/390} = \frac{7}{13}.$$

If we interpret the above table of relative frequencies as being a table of probabilities for the random experiment described above, then we would say that the probability that an individual Elmtown youth (of high school age) chooses a general high school curriculum, *given* that he belongs to Social Class V, is 7/13.

[1]Based on a rating scale developed in the study — Social Class I is the highest status class; Social Class V is the lowest.

Another useful way of thinking of conditional probabilities is the following: Knowing that event A has occurred changes the sample space of our experiment. Rather than expecting the outcome ω of the random experiment to belong to the sample space Ω, we now know that ω must belong to the subset (subevent set) A. Thus, we can consider a new probability model in the light of our knowledge that the event A has occurred. In this new probability model, the sample space is A, outcomes that can occur must belong to A, and event sets B occur if and only if an outcome in $B \cap A$ occurs. The sample space of the new model is shown pictorially (through a Venn diagram) in Figure 4.1. There the old sample space Ω is replaced by a new sample space A, and each event set B is replaced by the event set $B \cap A$. Because the relative frequency of the event B given that the event A has occurred is proportional to r.f.$(A \cap B)$, the new model must have a probability measure which assigns a probability to the event B proportional to the probability that its replacement event $B \cap A$ had under the old (unconditional) model. That is, if P^* is the new probability measure, then

$$P^*(B) = P^*(B \cap A) = cP(B \cap A),$$

where c is a constant. Now, $P^*(A)$ must equal 1 because A is the new sample space and Axiom 2 of Chapter 2 demands that the probability of the sample space must be equal to 1 no matter what probability measure we are talking about. Thus,

(4.2) $\qquad 1 = P^*(A) = cP(A \cap A) = cP(A),$

because $A \cap A = A$. Solving for c in (4.2), we obtain $c = 1/P(A)$. Thus for any event B, $P^*(B) = P(B \cap A)/P(A) = P(A \cap B)/P(A) = P(B|A)$. The probability measure for the new probability model is thus the conditional probability defined in Equation (4.1).

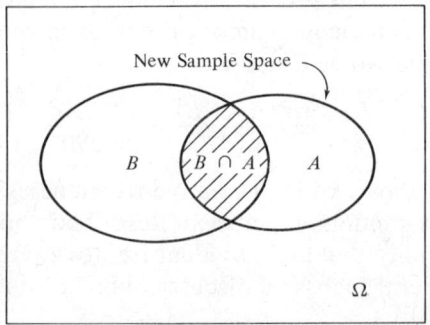

Figure 4.1 The event A replaces the sample space Ω, and $B \cap A$ replaces B when we know that A occurs.

Example 4.2. Let us return to the question asked at the end of Section 3. If you have an acquaintance who has exactly 2 children, if that acquaintance tells you that his older child is a boy, and if you believe that your acquaintance's family can be regarded as the result of a random experiment in which the probability model is given by the list

Simple Event	$\{\omega_1\}$	$\{\omega_2\}$	$\{\omega_3\}$	$\{\omega_4\}$
Probability	$\frac{1}{3}$	$\frac{1}{6}$	$\frac{1}{6}$	$\frac{1}{3}$

where
$\omega_1 = \{\text{first-born is a male, second-born is a male}\}$,
$\omega_2 = \{\text{first-born is a male, second-born is a female}\}$,
$\omega_3 = \{\text{first-born is a female, second-born is a male}\}$,
$\omega_4 = \{\text{first-born is a female, second-born is a female}\}$,

what probability should you assign to the event (assertion) that your acquaintance's second child is a male? Let A be the event {the first-born is a male} and let B be the event {the second-born is a male}. The probability that we require is, by Equation (4.1), equal to

$$P(B|A) = \frac{P(A \cap B)}{P(A)} = \frac{P(\{\omega_1\})}{P(\{\omega_1\} \cup \{\omega_2\})} = \frac{1/3}{(1/3)+(1/6)} = \frac{2}{3},$$

since $B = \{\omega_1\} \cup \{\omega_3\}$, $A = \{\omega_1\} \cup \{\omega_2\}$, and $A \cap B = \{\omega_1\}$. Notice that $P(B|A)$ is greater than the probability $P(B) = \frac{1}{2}$ which we assigned to event B before we were told about the sex of our acquaintance's oldest child (see Example 3.1).

The Probability Calculus for Conditional Probabilities

We have already mentioned that the conditional probability $P(B|A)$ of event B given that event A has occurred can be interpreted as being a new probability measure defined on a sample space which consists only of the outcomes belonging to event A. To be able to make this assertion, however, we must show that $P(B|A)$ satisfies Axioms 1, 2, and 3 of Chapter 2.

To verify that $P(B|A)$ satisfies Axiom 1, we must show that for every event E, that $0 \leq P(E|A) \leq 1$. Since $P(E|A)$ is equal to the ratio of the nonnegative numbers $P(A \cap E)$ and $P(A)$, it follows that $P(E|A) \geq 0$. On the other hand, since the event $A \cap E$ is included in event A (since every outcome in A and E is an outcome of A), the Law of Inclusion (Rule 4.1 of Chapter 2) implies that $P(A \cap E) \leq P(A)$. Dividing by $P(A)$ on both sides of this last inequality, we obtain $P(E|A) = P(A \cap E)/P(A) \leq P(A)/P(A) = 1$. Thus $P(B|A)$ satisfies Axiom 1. We have already mentioned that $P(A|A) = 1$. Because $A \cap \emptyset = \emptyset$ [see Rule 2.3(b) of Chapter

2], it follows that $P(A \cap \emptyset) = P(\emptyset) = 0$, and thus $P(\emptyset|A) = P(A \cap \emptyset)/P(A) = 0$. Hence, $P(B|A)$ satisfies Axiom 2. Finally, to show that $P(B|A)$ satisfies Axiom 3, we need to show that if the events E_1 and E_2 have no outcomes in A in common [that is, $(E_1 \cap E_2) \cap A = \emptyset$], then $P(E_1 \cup E_2|A) = P(E_1|A) + P(E_2|A)$. By Rule 2.5 of Chapter 2, we know that $(E_1 \cup E_2) \cap A = (E_1 \cap A) \cup (E_2 \cap A)$. Hence, applying Rule 4.3(a) of Chapter 2, we find that

$$P[(E_1 \cup E_2) \cap A] = P(E_1 \cap A) + P(E_2 \cap A) - P[(E_1 \cap A) \cap (E_2 \cap A)]$$
$$= P(E_1 \cap A) + P(E_2 \cap A),$$

since $P[(E_1 \cap A) \cap (E_2 \cap A)] = P[(E_1 \cap E_2) \cap A] = P(\emptyset) = 0$. Dividing every term of the equation $P[(E_1 \cup E_2) \cap A] = P(E_1 \cap A) + P(E_2 \cap A)$ by $P(A)$ allows us to conclude that

$$P(E_1 \cup E_2|A) = P(E_1|A) + P(E_2|A).$$

Thus, $P(B|A)$ satisfies Axiom 3. Since $P(B|A)$ satisfies Axioms 1, 2, and 3, it is a legitimate probability measure.

From the fact that the conditional probability measure $P(B|A)$ satisfies Axioms 1, 2, and 3, we can show (by a repetition of the arguments in Section 4 of Chapter 2) that $P(B|A)$ obeys the following rules:

Rule 4.1 (Law of Inclusion for Conditional Probabilities). If $E_2 \subset E_1$, then $P(E_2|A) \leq P(E_1|A)$.

Rule 4.2 (Law of Complementation for Conditional Probabilities). For any event set E, $P(E|A) = 1 - P(E^c|A)$.

Rule 4.3(a) (Law of Addition for Conditional Probabilities). For any two event sets E_1 and E_2, the conditional probability of their union given that the event A has occurred can be found from the equation

$$P(E_1 \cup E_2|A) = P(E_1|A) + P(E_2|A) - P(E_1 \cap E_2|A).$$

Rules for calculating conditional probabilities of more than two events can be obtained by paralleling the form of Rules 4.4 and 4.5 of Chapter 2. Rules 4.1, 4.2, and 4.3(a) (plus the rules applicable for calculating conditional probabilities of more than two events) form the basis for a probability calculus for conditional probabilities.

Example 4.3. A stock market research firm has done extensive research on the power of a certain market indicator to predict changes in the prices of stocks. Their studies show that the conditional probability of a rise in the Dow-Jones industrial average of at least 3 percent during the coming week, *given* that their market indicator rose at least 5 percent during the previous week, is 0.53. One of their staff members S prepares a memo in which he states that the conditional probability of a rise in the Dow-Jones industrial average of at least 5 percent during the coming week, given that the market indicator rose at least 5 percent during the previous week, is 0.55. The memo is reviewed by a colleague R in the research section, who notes that S has made a mistake in his probability calculations. How did R know of S's mistake? Let A be the event that the research company's market indicator rises at least 5 percent during the previous week, let E_1 be the event that the Dow-Jones industrial average rises at least 3 percent during the coming week, and let E_2 be the event that the Dow-Jones average rises at least 5 percent during the coming week. Then, the research company knows that $P(E_1|A) = 0.53$. On the other hand, S has claimed that $P(E_2|A) = 0.55$, which is a number strictly larger than 0.53. Since a rise of at least 5 percent certainly implies a rise of at least 3 percent, all outcomes of event E_2 must be outcomes of event E_1, and by Rule 4.1, $P(E_2|A)$ must be less than or equal to $P(E_1|A)$. Since S had assigned to $P(E_2|A)$ a number strictly larger than $P(E_1|A)$, R knows that S made an error.

The staff member S prepares another note correcting his error. He states that instead of the Dow-Jones industrial average, he meant rather to refer to an average kept by the firm of the prices of certain highly speculative stocks (the speculative stock average). Thus, the event to which he was actually referring is the event E_3 that the speculative stock average rises at least 5 percent during the coming week, and it is this event which has a conditional probability of 0.55 given that event A occurs. Further, he states that the conditional probability that both stock averages will rise at least the stated percentages (that is, both event E_1 *and* event E_3 occur) given that event A has occurred is 0.40; that is, $P(E_1 \cap E_3|A) = 0.40$. In the review, R notes that if the company can correctly forecast the rise of *either* the Dow-Jones industrial average *or* the speculative stock average with a probability (actually, relative frequency) appreciably better than that which could be obtained by guessing (that is, a probability of 0.50), the resulting publicity will bring in more business. In order to find out whether the market indicator developed by the firm can provide such a high probability of correct forecasting, R computes the conditional probability of either a rise of at least 3 percent in the Dow-Jones industrial average or a rise of at least 5 percent in the speculative stock average during the coming week *given* that the market indicator has risen at least 5 percent during the previous week; that is, R computes $P(E_1 \cup E_3|A)$. From the figures available, and from Rule 4.3(a), R finds that

$$P(E_1 \cup E_3|A) = P(E_1|A) + P(E_3|A) - P(E_1 \cap E_3|A)$$
$$= 0.53 + 0.55 - 0.40 = 0.68.$$

Since 0.68 is appreciably greater than 0.50, R decides to make use of the firm's market indicator in forecasting stock prices.

Law of Multiplication

Equation (4.1) tells us how to compute $P(B|A)$ when we know $P(B \cap A)$ and $P(A)$. In many instances, we may know $P(A)$ and $P(B|A)$, but we may not know $P(A \cap B)$. However, by multiplying both sides of Equation (4.1) by $P(A)$ and noting that $B \cap A = A \cap B$, we obtain the following:

Rule 4.4 (Law of Multiplication). Given $P(B|A)$ and $P(A)$, we can compute $P(A \cap B)$ through the formula

(4.3) $$P(A \cap B) = P(A) P(B|A).$$

Example 4.4. Our long-term records of the performance of a particular type of electrical generator leads us to assume that the probability $P(A)$ of failure for that generator is 0.80. Simulated repair runs have indicated that if that particular type of generator fails (that is, if event A occurs), we are able to correctly find the breakdown and successfully repair it with probability $P(B|A)$ equal to 0.90. Our interest is in the probability that the electrical generator fails (event A) and we find and repair the failure (event B). Assuming that only one generator of the given type is being used, we find from Rule 4.4 that the probability that the generator fails and we find and repair the failure (event $A \cap B$) is

$$P(A \cap B) = P(A) P(B|A) = (0.80)(0.90) = 0.72.$$

We may generalize Rule 4.4 to any number of events. For example, if we have 3 events E_1, E_2, E_3, and are given the probabilities $P(E_3|E_2 \cap E_1)$, $P(E_2|E_1)$, and $P(E_1)$, we can compute $P(E_1 \cap E_2 \cap E_3)$ as follows. First, we apply Rule 4.4 to the events E_3 and $C = E_1 \cap E_2$, obtaining

(4.4) $$P(E_1 \cap E_2 \cap E_3) = P(C \cap E_3) = P(E_3|C)P(C).$$

Next, we apply Rule 4.4 to E_1 and E_2 and find that

(4.5) $$P(C) = P(E_1 \cap E_2) = P(E_2|E_1)P(E_1).$$

Remembering that $C = E_1 \cap E_2 = E_2 \cap E_1$, we combine Equations (4.4) and (4.5) and conclude that

(4.6) $$P(E_1 \cap E_2 \cap E_3) = P(E_3|C)P(C) = P(E_3|E_1 \cap E_2)P(E_2|E_1)P(E_1)$$
$$= P(E_1)P(E_2|E_1)P(E_3|E_1 \cap E_2).$$

This same type of argument produces a similar result for k events $E_1, E_2, \ldots, E_k$; namely,

(4.7) $P(E_1 \cap E_2 \cap \cdots \cap E_k)$
$= P(E_1)P(E_2|E_1)P(E_3|E_1 \cap E_2) \cdots P(E_k|E_1 \cap E_2 \cap E_3 \cap \cdots \cap E_{k-1}).$

Example 4.5. In Example 2.3 we evaluated the probability of obtaining two Republicans in a random sample of size 2 taken without replacement from a population of 50 individuals, 25 of whom are Republicans. Previously we obtained this probability using Rule 2.2. Here, we obtain it through the use of conditional probabilities. Let A be the event that the first individual drawn is a Republican, and B be the event that the second individual drawn is a Republican. Since every individual in the population has an equal chance of being drawn on the first draw and since 25 of these 50 individuals are Republicans, application of Rule 1.3 yields $P(A) = 25/50$. Suppose now that a Republican is drawn on the first draw. Then for the second draw there are 49 individuals left in the population of whom 24 are Republicans. Because every one of the 49 remaining individuals has an equal probability of being selected on the second draw, the probability $P(B|A)$ of a Republican on the second draw *given* a Republican on the first draw is (by Rule 1.3) equal to 24/49. Thus by Rule 4.4,

$P\{\text{both individuals chosen are Republicans}\} = P(A \cap B) = P(A)P(B|A)$
$$= \left(\frac{25}{50}\right)\left(\frac{24}{49}\right) = \frac{12}{49},$$

which is the result obtained earlier in Section 2.

Suppose, instead of selecting 2 individuals, we take a random sample of 5 individuals without replacement. By repeating the arguments given above, and using Equation (4.7), where E_i is the event that a Republican appears on the ith draw, $i = 1, 2, \ldots, 5$ we obtain the result

$P\{\text{all 5 individuals chosen are Republican}\}$
$= P(E_1 \cap E_2 \cap E_3 \cap E_4 \cap E_5)$
$= P(E_1)P(E_2|E_1)P(E_3|E_1 \cap E_2)P(E_4|E_1 \cap E_2 \cap E_3)$
$\cdot P(E_5|E_1 \cap E_2 \cap E_3 \cap E_4)$
$= \left(\frac{25}{50}\right)\left(\frac{24}{49}\right)\left(\frac{23}{48}\right)\left(\frac{22}{47}\right)\left(\frac{21}{46}\right) = \frac{33}{1316}.$

Another event that might be of interest in connection with this example is the event that we choose Republicans on the first two draws and Democrats on the last three draws; that is, the event $E_1 \cap E_2 \cap E_3^c \cap E_4^c \cap E_5^c$. We have previously argued that $P(E_1) = 25/50$ and that $P(E_2|E_1) = 24/49$, so that the probability, $P(E_1 \cap E_2)$ of drawing Republicans on the first two draws is equal to $P(E_1)P(E_2|E_1) = (25)(24)/(50)(49) = 12/49$.

Now, given that Republicans have been drawn on the first two draws, we know that before we begin the third draw there are $(25-2) = 23$ Republicans and 25 Democrats remaining in the population, each having equal probability of being chosen on the third draw. From this fact, we can argue that the probability that we draw a Republican on the third draw *given* that Republicans were drawn on the first two draws is $P(E_3|E_1 \cap E_2) = (23)/(23+25) = 23/48$. Hence, from Rule 4.2, $P(E_3^c|E_1 \cap E_2) = 1 - P(E_3|E_1 \cap E_2) = 25/48$. A similar argument, based on the fact that after two Republicans and one Democrat have been drawn, 23 Republicans and 24 Democrats remain in the population, allows us to state that $P(E_4^c|E_1 \cap E_2 \cap E_3^c) = 1 - P(E_4|E_1 \cap E_2 \cap E_3^c) = 24/47$. One final repetition of this type of argument yields the fact that $P(E_5^c|E_1 \cap E_2 \cap E_3^c \cap E_4^c) = 23/46 = 1/2$. Thus, applying Equation (4.7), we find that

$P\{$Republicans are drawn on the first two draws and Democrats are drawn on the last three draws$\}$

$$= P(E_1 \cap E_2 \cap E_3^c \cap E_4^c \cap E_5^c)$$
$$= P(E_1 \cap E_2) P(E_3^c|E_1 \cap E_2) P(E_4^c|E_1 \cap E_2 \cap E_3^c)$$
$$\cdot P(E_5^c|E_1 \cap E_2 \cap E_3^c \cap E_4^c)$$
$$= \left(\frac{12}{49}\right)\left(\frac{25}{48}\right)\left(\frac{24}{47}\right)\left(\frac{1}{2}\right) = \frac{75}{2303}.$$

Probabilities of other similar events can be obtained in an analogous fashion. For example, the probability that we draw first a Democrat, second a Republican, third a Democrat, fourth a Republican, and fifth a Democrat [that is, $P(E_1^c \cap E_2 \cap E_3^c \cap E_4 \cap E_5^c)$] is equal to

$P(E_1^c \cap E_2 \cap E_3^c \cap E_4 \cap E_5^c)$
$$= P(E_1^c) P(E_2|E_1^c) P(E_3^c|E_1^c \cap E_2) P(E_4|E_1^c \cap E_2 \cap E_3^c)$$
$$\cdot P(E_5^c|E_1^c \cap E_2 \cap E_3^c \cap E_4)$$
$$= \left(\frac{25}{50}\right)\left(\frac{25}{49}\right)\left(\frac{24}{48}\right)\left(\frac{24}{47}\right)\left(\frac{23}{46}\right)$$
$$= \frac{75}{2303}.$$

Notice that both the event $E_1 \cap E_2 \cap E_3^c \cap E_4^c \cap E_5^c$ and the event $E_1^c \cap E_2 \cap E_3^c \cap E_4 \cap E_5^c$ have the same probability, 75/2303, of occurring. If either of these events occur, our sample will contain three Democrats and two Republicans. Does this suggest any rule for calculating the probabilities of other events for which three Democrats and two Republicans are drawn? What is the probability of $E_1 \cap E_2^c \cap E_3 \cap E_4^c \cap E_5^c$ or of $E_1^c \cap E_2 \cap E_3 \cap E_4^c \cap E_5^c$? We take up this question again in Chapter 6 when we discuss the hypergeometric probability distribution.

Law of Multiplication for Nested Events

Equation (4.7) takes on a somewhat simpler form in the special case where the events $E_1, E_2, \ldots, E_k$ are decreasing or nested events. The events $E_1, E_2, \ldots, E_k$ are nested if all outcomes of the event E_k are outcomes of the event E_{k-1}, all outcomes of E_{k-1} are outcomes of E_{k-2}, ..., all outcomes of E_3 are outcomes of E_2, and all outcomes of E_2 are outcomes of E_1. In the notation developed in Chapter 2 for set inclusion this definition becomes the following: The events $E_1, E_2, \ldots, E_k$ are said to be nested if $E_k \subset E_{k-1} \subset \cdots \subset E_3 \subset E_2 \subset E_1$. A Venn diagram of the nested events $E_1, E_2,$ and E_3 is shown in Figure 4.2.

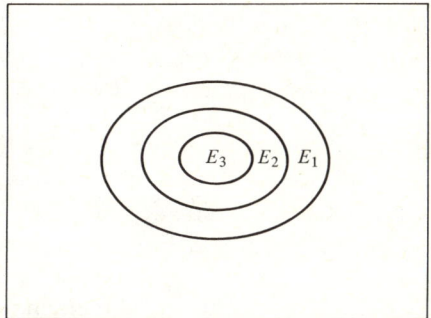

Figure 4.2 Nested events. $E_3 \subset E_2 \subset E_1$.

Looking at Figure 4.2, it is easy to see that when the event sets $E_1, E_2,$ and E_3 are nested, then $E_1 \cap E_2 = E_2$ and $E_1 \cap E_2 \cap E_3 = E_3$. Substituting these facts into Equation (4.6), we find that if $E_3 \subset E_2 \subset E_1$, then

$$P(E_3) = P(E_1 \cap E_2 \cap E_3) = P(E_1)P(E_2|E_1)P(E_3|E_1 \cap E_2)$$
$$= P(E_1)P(E_2|E_1)P(E_3|E_2).$$

In general, if $E_1, E_2, \ldots, E_k$ are nested events, then $E_1 \cap E_2 = E_2$, $E_1 \cap E_2 \cap E_3 = E_3, \ldots,$ and $E_1 \cap E_2 \cap \cdots \cap E_k = E_k$. Substituting these facts into Equation (4.7) yields the following rule:

Rule 4.5. If $E_1, E_2, \ldots, E_k$ are nested events, that is, $E_k \subset E_{k-1} \subset \cdots \subset E_2 \subset E_1$, then

(4.8) $P(E_k) = P(E_1)P(E_2|E_1)P(E_3|E_2) \cdots P(E_{k-1}|E_{k-2})P(E_k|E_{k-1}).$

Example 4.6. A systems engineer is interested in assessing the reliability of the latest moon rocket. This rocket is composed of three stages. At takeoff, the engine of the first stage of the rocket must lift the rocket off the ground. If that engine accomplishes its task, it and its accompanying stage are jettisoned and the engine of the second stage takes over the task of lifting the rocket into orbit. Once the engine of the second stage is finished it is discarded, and the engine of the third stage is used to complete the rocket's mission. The systems engineer has tested engines similar to the engines of all three stages, and has concluded that the engine of the first stage has a probability of 0.99 of performing successfully, the engine of the second stage has a probability of 0.97 of performing successfully, and the engine of the third stage has a probability of 0.98 of performing successfully. He now wants to calculate the probability that the rocket will complete its mission. Let E_1 be the event that the rocket successfully completes the first stage of its ascension. Since the event that the first stage of ascension is successful is the same as the event that the engine of the first stage performs successfully, $P(E_1) = 0.99$. Let E_2 be the event that the rocket successfully completes the first *two* stages of its ascension. Event E_2 occurs when event E_1 occurs *and* the engine of the second stage of the rocket performs successfully. Thus, $E_2 \subset E_1$, and $E_1 \cap E_2 = E_2$. Given that event E_1 occurs, event E_2 occurs if and only if the engine of the second stage performs successfully, so that the probability of E_2 *given* E_1 is $P(E_2|E_1) = P\{\text{engine of second stage performs successfully}\} = 0.97$. Finally, let event E_3 be the event that the rocket successfully completes all *three* stages of its ascension (and completes its mission). Event E_3 occurs when events E_1 *and* E_2 occur *and* the engine of the third stage of the rocket performs successfully, so that $E_3 \subset (E_1 \cap E_2) = E_2$ and $E_1 \cap E_2 \cap E_3 = E_3$. Because given the occurrence of event E_2, event E_3 occurs when the engine of the third stage of the rocket performs successfully, we can conclude that $P(E_3|E_2) = 0.98$. Since we have demonstrated that $E_3 \subset E_2 \subset E_1$, the events E_1, E_2, E_3 are nested events and Rule 4.5 applies. Thus,

$$P\{\text{the rocket successfully completes its mission}\}$$
$$= P(E_3)$$
$$= P(E_1)P(E_2|E_1)P(E_3|E_2)$$
$$= (0.99)(0.97)(0.98) = 0.941.$$

Example 4.7. An anthropologist is trying to account for the unusually large number of families in which there are more female than male children in a certain primitive tribe. He postulates that when a child is born, the probability that its sex will be female is $\frac{1}{2}$ (and thus the probability that its sex will be male is also $\frac{1}{2}$). As a first attempt to account for the high number of families observed which have more female children than male children, he hypothesizes that families continue to try to have children until a male child is born. If this model is a correct one, then the events

$$E_1 = \{\text{first child is a girl}\},$$
$$E_2 = \{\text{first and second children are girls}\},$$
$$E_3 = \{\text{first 3 children are girls}\},$$
$$\vdots \qquad \vdots$$
$$E_k = \{\text{first } k \text{ children are girls}\},$$

are nested.

To demonstrate this fact, note that the outcomes of the random experiment under consideration here are lists of the sexes of the children born to a family, where the lists are ordered by birth. For an outcome to have the property that the first two sexes listed are female, that outcome must also have the property that the first sex listed is female. Thus, $E_2 \subset E_1$. For an outcome to have the property that the first three sexes listed are female, the outcome must have the property that the first two sexes listed are female. Thus, $E_3 \subset E_2$. Continuing the argument in this fashion verifies that

$$E_k \subset E_{k-1} \subset \cdots \subset E_2 \subset E_1.$$

Since the anthropologist's model tells him the probability $P(E_{i+1}|E_i)$ that a family that has produced i successive girl children will (if they give birth to another child) have an $(i+1)$st girl child, Equation (4.8) enables him to calculate the probability that the first k children of a given family will be girls:

$$P\{\text{first } k \text{ children are girls}\}$$
$$= P(E_k) = P(E_1)P(E_2|E_1)P(E_3|E_2) \cdots P(E_k|E_{k-1})$$
$$= (\tfrac{1}{2})(\tfrac{1}{2})(\tfrac{1}{2}) \cdots (\tfrac{1}{2}) = (\tfrac{1}{2})^k.$$

Thus, for example,

$$P\{\text{first 4 children are girls}\} = P(E_4) = (\tfrac{1}{2})^4 = \tfrac{1}{16}.$$

Note that under the anthropologist's model, the following list of the sexes of children born to a given family would have zero probability: (girl,boy,girl,...). This is because under the model, the family does not have more children after a boy is born. The model is an inadequate one not only because of this rather unreasonable assumption, but also because it postulates no mechanism other than the birth of a boy for stopping the growth of a family. The anthropologist can calculate the probability that a family has *exactly* m children by the following argument: To have exactly m children, a family must have had $m-1$ girls and then a boy. The probability that the first $m-1$ children of a family are girls is $P(E_{m-1})$ which by our previous calculation (with $k=m-1$) is equal to

$(\frac{1}{2})^{m-1}$. The probability that the mth child is a boy given that the first $m-1$ children have been girls is, by the model, equal to $\frac{1}{2}$. Thus,

$$P\{\text{exactly } m \text{ children}\} = P(E_{m-1})P(\{m\text{th child a boy}\}|E_{m-1})$$
$$= (\tfrac{1}{2})^{m-1}\tfrac{1}{2} = (\tfrac{1}{2})^m.$$

Consequently, the probability that a family has exactly 5 children is

$$P\{\text{exactly 5 children}\} = (\tfrac{1}{2})^5 = \tfrac{1}{32}.$$

The probability that a family has more than 4 children is 1 minus the probability that the family has 1, 2, 3, or 4 children (by the Law of Complementation; see Rule 4.2 of Chapter 2). Hence, since the events {exactly 1 child}, {exactly 2 children}, and so on, are mutually exclusive,

$$P\{\text{more than 4 children}\} = 1 - P\{\text{exactly 1, 2, 3, or 4 children}\}$$
$$= 1 - P\{\text{exactly 1 child}\} - P\{\text{exactly 2 children}\}$$
$$\quad - P\{\text{exactly 3 children}\} - P\{\text{exactly 4 children}\}$$
$$= 1 - \tfrac{1}{2} - \tfrac{1}{4} - \tfrac{1}{8} - \tfrac{1}{16} = \tfrac{1}{16}.$$

At the beginning of this section, we motivated interest in conditional probabilities by considering a certain (fictitious) life table. This table provided the conditional probability that a person will live to age 65 given that the person is a male, as well as the conditional probability that a person will live to age 65 given that the person is a female. We pointed out that if an insurance company knew the sex of an applicant, it would undoubtedly base its premiums for insurance on the information provided by such conditional probabilities, and would thus not be very interested in the unconditional probability that a person of either sex lives to age 65. Suppose, however, that we are interested in the unconditional probability that a person of either sex lives to age 65. Suppose further that an actuarial table provides us with the conditional probabilities mentioned above. Finally, assume that the probability that a randomly selected individual is a male is equal to 0.50. How can we find the unconditional probability that a (randomly selected) person of either sex lives to age 65?

To answer this question, let B be the event whose unconditional probability is of interest to us; that is, B is the event that a (randomly selected) person of either sex lives to age 65. Let E_1 be the event that a (randomly selected) person is a male, and let E_2 be the event that a (randomly selected) person is female. Note that E_1 and E_2 are mutually exclusive, and that the union, $E_1 \cup E_2$, of E_1 and E_2 is the sample space Ω. We are given the value of $P(B|E_1)$, the value of $P(B|E_2)$, and we know that the value of $P(E_1)$ is equal to 0.50. By the Law of Complementation (Rule 4.2 of Chapter 2) and the fact that $E_1{}^c = E_2$, we find that $P(E_2) = P(E_1{}^c) = 1 -$

$P(E_1) = 0.50$. Next, note that because $E_1 \cup E_2 = \Omega$, it follows [using Rules 2.3(a) and 2.5 of Chapter 2] that

$$B = B \cap \Omega = B \cap (E_1 \cup E_2) = (B \cap E_1) \cup (B \cap E_2).$$

Since $E_1 \cap E_2 = \emptyset$ (E_1 and E_2 are mutually exclusive), it follows from Rule 2.3(b) of Chapter 2 that $(B \cap E_1) \cap (B \cap E_2) = B \cap E_1 \cap E_2 = B \cap \emptyset = \emptyset$, so that $(B \cap E_1)$ and $(B \cap E_2)$ are mutually exclusive. Thus, if we want $P(B) = P[(B \cap E_1) \cup (B \cap E_2)]$, we know from Axiom 3 of Chapter 2 that

$$P(B) = P(B \cap E_1) + P(B \cap E_2).$$

However, the Law of Multiplication (Rule 4.4) tells us that $P(B \cap E_1) = P(E_1)P(B|E_1)$ and that $P(B \cap E_2) = P(E_2)P(B|E_2)$. Therefore,

$$\begin{aligned}P(B) &= P(B \cap E_1) + P(B \cap E_2) = P(E_1)P(B|E_1) + P(E_2)P(B|E_2) \\ &= (0.50)P(B|E_1) + (0.50)P(B|E_2).\end{aligned}$$

If the values of $P(B|E_1)$ and $P(B|E_2)$ are obtained from the fictitious life table given at the beginning of this section, then $P(B|E_1) = 0.74$, $P(B|E_2) = 0.80$, and thus

$$P(B) = (0.50)(0.74) + (0.50)(0.80) = 0.77.$$

The basic idea behind the exercise in calculation illustrated above can be formalized by use of the notion of a partition of a sample space.

Partitions

The events $E_1, E_2, ..., E_k$ are said to *partition* the sample space Ω if every outcome ω of Ω belongs to one and only one of the events $E_1, E_2, ..., E_k$. In other words, $E_1, E_2, ..., E_k$ partition the sample space if $\Omega = E_1 \cup E_2 \cup \cdots \cup E_k$ and if no two of the events $E_1, E_2, ..., E_k$ have an outcome in common. One such circumstance was illustrated by the events E_1 and E_2 in the actuarial example discussed above. Another such circumstance is pictured in Figure 4.3. Here, the sample space Ω is the entire box, and E_1, E_2, E_3, and E_4 partition Ω. Notice that the event B can be broken up into a part common with E_1 (that is, $E_1 \cap B$), a part common with E_2 (that is, $E_2 \cap B$), a part common with E_3 (that is, $E_3 \cap B$), and a part common with E_4 (that is, $E_4 \cap B$), and that these parts have no outcomes in common with one another. That is,

$$B = (E_1 \cap B) \cup (E_2 \cap B) \cup (E_3 \cap B) \cup (E_4 \cap B)$$

and $(E_i \cap B)$ and $(E_j \cap B)$ are mutually exclusive for $i \neq j$. Hence, *a partition of the sample space partitions any event B*. Using arguments similar to those presented in the actuarial example given above, we find

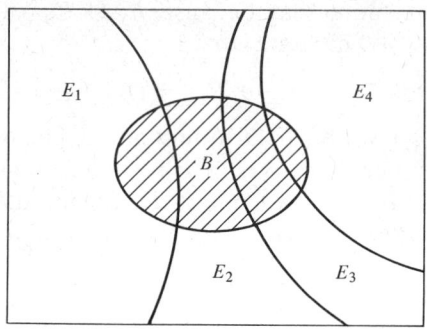

Figure 4.3 The event sets E_1, E_2, E_3, E_4 partition Ω.

that

(4.9) $P(B) = P[(E_1 \cap B) \cup (E_2 \cap B) \cup (E_3 \cap B) \cup (E_4 \cap B)]$
$= P(E_1 \cap B) + P(E_2 \cap B) + P(E_3 \cap B) + P(E_4 \cap B),$

and applying Rule 4.4, we obtain

(4.10) $\quad P(E_1 \cap B) = P(E_1)P(B|E_1),$
$P(E_2 \cap B) = P(E_2)P(B|E_2),$
$P(E_3 \cap B) = P(E_3)P(B|E_3),$
$P(E_4 \cap B) = P(E_4)P(B|E_4).$

Substituting the results of Equation (4.10) into Equation (4.9), we conclude that

(4.11) $P(B) = P(E_1)P(B|E_1) + P(E_2)P(B|E_2) + P(E_3)P(B|E_3)$
$+ P(E_4)P(B|E_4).$

Thus, if we know the probabilities of the events E_1, E_2, E_3, E_4 (the events that partition the sample space) and if we know the conditional probabilities of the event B given E_1, given E_2, given E_3, and given E_4, we can use such information to calculate $P(B)$. This assertion can be stated generally:

Rule 4.6. If the events $E_1, E_2, ..., E_k$ partition the sample space, then the probability of any event B can be calculated as follows:

(4.12) $P(B) = P(E_1 \cap B) + P(E_2 \cap B) + \cdots + P(E_k \cap B)$
$= P(E_1)P(B|E_1) + P(E_2)P(B|E_2) + \cdots + P(E_k)P(B|E_k).$

Remark A: Note that Equation (4.12) expresses $P(B)$ as a weighted average of the conditional probabilities of B given E_1, B given E_2, ..., and B given E_k. The weights used are the probabilities of the events $E_1, E_2, ..., E_k$.

Remark B: Of course, there can be many partitions of the sample space. The partition that should be chosen in order to use Rule 4.6 to calculate $P(B)$ is the one for which $P(B|E_1)$, $P(B|E_2)$, and so on, can be most easily determined.

Remark C: Rule 4.6 generalizes to nonfinite partitions as follows: Let $E_1, E_2, \ldots$, be a denumerable collection of events, having (pairwise) no outcomes in common, such that $\Omega = \text{``}E_1 \cup E_2 \cup E_3 \cdots\text{.''}$ For any event B,

$$P(B) = P(E_1 \cap B) + P(E_2 \cap B) + \cdots$$
$$= P(E_1)P(B|E_1) + P(E_2)P(B|E_2) + \cdots.$$

Example 4.8. A social scientist decides to study in depth one individual to be randomly sampled from a given population. The social scientist is worried about the possibility that the chosen individual will refuse to cooperate. A previous study provides data on the probability of noncooperation for individuals in different income classes:

High income	0.08
Medium income	0.10
Low income	0.04

The social scientist knows that in the population he is studying the proportion of high income people is 0.25, of medium income people is 0.50, and of low income people is 0.25. Let B be the event that the person drawn is noncooperative, E_1 be the event that the person has high income, E_2 be the event that the person has medium income, and E_3 be the event that the person has low income. Applying Rule 4.6 we have

$$P\{\text{draw noncooperative person}\} = P(B)$$
$$= P(E_1)P(B|E_1) + P(E_2)P(B|E_2) + P(E_3)P(B|E_3)$$
$$= (0.25)(0.08) + (0.50)(0.10) + (0.25)(0.04)$$
$$= 0.08.$$

Bayes' Rule

Suppose that in Example 4.8 the social scientist actually draws an individual at random. He knows immediately that the probability that this individual comes from the low income group is 0.25. The social scientist begins to examine the individual and finds him uncooperative. Should this finding cause the social scientist to reassess the probability that the individual comes from the low income group? The answer is "yes," and the method of reassessment was known to probabilists as long ago as the seventeenth century, when the Reverend Thomas Bayes (1702–1761) first employed the now famous rule.

Rule 4.7 (Bayes' Rule). If the events $E_1, E_2, \ldots, E_k$ partition the sample space Ω and if B is any other event set for which $P(B) > 0$, then the probability of the particular partitioning event E_i, given that event B has occurred, can be calculated as follows:

$$(4.13) \quad P(E_i|B) = \frac{P(E_i \cap B)}{P(B)}$$

$$= \frac{P(E_i)P(B|E_i)}{P(E_1)P(B|E_1) + P(E_2)P(B|E_2) + \cdots + P(E_k)P(B|E_k)},$$

for $i = 1, i = 2, \ldots, i = k$.

As Equation (4.13) indicates, Bayes' Rule is derived by first writing down the definition of $P(E_i|B)$, then substituting $P(E_i)P(B|E_i)$ for $P(E_i \cap B)$ in the numerator using Rule 4.4, and finally substituting $P(E_1) \cdot P(B|E_1) + P(E_2)P(B|E_2) + \cdots + P(E_k)P(B|E_k)$ for $P(B)$ in the denominator using Rule 4.6. Note that if $P(B) = 0$, then since $E_i \cap B$ is included in B, the Rule of Inclusion and Axiom 1 for probabilities implies that $0 \leq P(E_i \cap B) \leq P(B) = 0$, or $P(E_i \cap B) = 0$. Thus, if $P(B) = 0$, we would be dividing 0 by 0 in Equation (4.13) — an undefined operation.

Example 4.8 (Continued). Using Bayes' Rule, we can, for example, calculate

$P\{\text{low income person drawn} \,|\, \text{the person drawn is uncooperative}\}$

$$= \frac{P(E_3)P(B|E_3)}{P(E_1)P(B|E_1) + P(E_2)P(B|E_2) + P(E_3)P(B|E_3)}$$

$$= \frac{(0.25)(0.04)}{(0.25)(0.08) + (0.50)(0.10) + (0.25)(0.04)} = \frac{1}{8} = 0.125.$$

Before the individual drawn by random sample was observed, the probability that he would come from the low income group was 0.25. Because of the information the social scientist has gained about this individual (that he is uncooperative), he has reassessed that probability and now calculates it to be 0.125 (or half the former number). However, the original probability referred to a sample space in which the outcomes (individual characteristics such as income group and willingness to cooperate) referred to individuals who could be either cooperative or uncooperative. The new probability refers to the new conditional sample space (see the discussion following Example 4.1 of this section) of outcomes having the property that they refer to uncooperative individuals. Thus, the probability that the individual comes from the low-income group has been reassessed in the light of the knowledge that one characteristic of the individual actually chosen is that he is uncooperative.

5. INDEPENDENCE

Conditional probabilities are used to reassess probabilities of events in the light of partial information concerning the outcome of a random experiment. The degree to which the probabilities change as a result of the use of this partial information depends on the events in question. For example, if the event A is included in the event B (every outcome of A is an outcome of B), then the conditional probability of B given A is equal to 1. On the other hand, if A and B have no outcomes in common (that is, they are mutually exclusive), the conditional probability of B given A is equal to 0. In these two examples, our knowledge that A has occurred allows us to know with certainty whether or not event B has occurred.

On the other hand, there may be no new probabilistic information about B that can be obtained from knowing that A has occurred. In this case,

(5.1) $$P(B|A) = P(B);$$

that is, the conditional probability of B given A is the same as the unconditional probability of B. If (5.1) holds, then

(5.2) $$P(A|B) = \frac{P(A \cap B)}{P(B)} = \frac{P(A \cap B)}{P(B|A)} = \frac{P(A \cap B)}{P(A \cap B)/P(A)} = P(A)$$

also holds. In other words, if knowledge that A has occurred does not alter the probability that we assign to the occurrence of B, neither does knowledge that B has occurred alter the probability that we assign to the occurrence of A. Any two events A and B for which either Equation (5.1) or Equation (5.2) hold are said to be *statistically independent*.

An equivalent definition of the statistical independence of two events A and B is the following: A and B are said to be statistically independent if

(5.3) $$P(A \cap B) = P(A)P(B).$$

Although definition (5.3) is less intuitive than definition (5.1) or definition (5.2), it is often more useful for computation.

We note that definition (5.3) holds if and only if A and B are statistically independent. If A and B are not statistically independent (in which case we say they are *statistically dependent*), then to find $P(A \cap B)$ we may either use the formula

$$P(A \cap B) = P(B|A)P(A) = P(A|B)P(B),$$

or we may calculate $P(A \cap B)$ in terms of the outcomes in $A \cap B$.

A third characterization of the statistical independence of the events A

and B is given by either of the formulas

$$P(A|B) = P(A|\text{ not } B), \quad P(B|A) = P(B|\text{ not } A).$$

Thus, the events A and B are independent if the probability of A given B is the same as the probability of A given the complement of B.

At first glance there is a tendency to equate the concepts "statistically independent" and "mutually exclusive." This is a fallacy. Indeed, if A and B are mutually exclusive, they are necessarily statistically dependent; on the other hand, if A and B are statistically independent, they are never mutually exclusive. [We have excluded the trivial cases where either $P(A)$ or $P(B)$ is 0.] Suppose A and B are mutually exclusive and A occurs, then B cannot occur and thus the conditional probability that B occurs is 0 [that is, $P(B|A) = 0$]. Therefore, since we have assumed that $P(B) \neq 0$, it cannot be the case that $P(B|A) = P(B)$, and thus A and B cannot be independent. On the other hand, suppose A and B are independent and neither $P(A)$ nor $P(B)$ are zero; then $P(A \cap B) = P(A)P(B) \neq 0$, so that $A \cap B$ cannot be the null set $\emptyset$ [since $P(\emptyset)$ is always 0].

Example 5.1. Consider the experiment of randomly selecting two units with replacement from a population consisting of the two units u_1 and u_2. The probability model for this experiment is

Simple Event	$\{(u_1,u_1)\}$	$\{(u_1,u_2)\}$	$\{(u_2,u_1)\}$	$\{(u_2,u_2)\}$
Probability	$\frac{1}{4}$	$\frac{1}{4}$	$\frac{1}{4}$	$\frac{1}{4}$

Recall that (u_1,u_2) is the sample in which u_1 is the result on the first draw, u_2 is the result on the second draw. Let A be the event that u_1 is the result on the first draw, so that $A = \{(u_1,u_1),(u_1,u_2)\}$, and let B be the event that u_2 is the result on the second draw, so that $B = \{(u_1,u_2),(u_2,u_2)\}$. Then $A \cap B = \{(u_1,u_2)\}$, and thus

$$P(B|A) = \frac{P(A \cap B)}{P(A)} = \frac{\frac{1}{4}}{\frac{1}{4}+\frac{1}{4}} = \frac{1}{2} = P(B).$$

Hence the events A and B are independent.

In the above example, an heuristic argument can be given to explain why the events A and B are independent. Note that event A refers to the first draw, and event B refers to the second draw. Because the sampling is *with replacement*, knowledge of the result of the first draw yields no information as to the probabilities governing what will take place on the second draw. This lack of information is what is reflected in the mathematical notion of statistical independence. In most practical applications, the assumption that events A and B are statistically independent arises from the nature of the experiment (as it does in the above example). Frequently, the random experiment is performed in a number of physically independent stages, where the event A and the event B relate to what takes place at different stages. That is, it is possible to know whether A occurs by observing only what happens when a certain stage of the experiment is performed, and, similarly, whether B occurs by observing what happens at a different stage of the experiment. In such cases, it is reasonable to assume that the two events A and B are statistically independent.

Although statistically independent events usually arise from experiments performed in physically independent stages, from time to time examples do arise in other contexts. The point of importance here is that although each of the defining conditions (5.1), (5.2), or (5.3) of statistical independence are mathematical conditions which reflect the statistical properties of physically independent events, this does not imply that statistically independent events *must* also be physically independent.

The definition of the statistical independence of two events generalizes to the case of more than two events. Events $E_1, E_2, \ldots, E_n$ are said to be *mutually (statistically) independent* if the conditional probability of any one of them (say, event E_k) given *any* subcollection of the events $E_1, E_2, \ldots, E_{k-1}, E_{k+1}, \ldots, E_n$ is equal to the unconditional probability of that one event, E_k. For example, E_1, E_2, and E_3 are mutually (statistically) independent if and only if *all* of the following equations are satisfied:

(5.4) $\quad P(E_1) = P(E_1|E_2) = P(E_1|E_3) = P(E_1|E_2 \cap E_3),$
$\quad\quad\quad P(E_2) = P(E_2|E_1) = P(E_2|E_3) = P(E_2|E_1 \cap E_3),$
$\quad\quad\quad P(E_3) = P(E_3|E_1) = P(E_3|E_2) = P(E_3|E_1 \cap E_2).$

In words, knowledge about the simultaneous occurrence of one or more of the events does not alter the probabilities that we assign to the occurrence of any of the other events.

Recall that a composite random experiment is made up from two or more random experiments $\mathscr{E}^{(1)}, \mathscr{E}^{(2)}, \ldots, \mathscr{E}^{(k)}$ (called *components, factors,* or *trials*). In such an experiment, we say that an event $E^{(j)}$ *belongs to component* $\mathscr{E}^{(j)}$ if the occurrence or nonoccurrence of the event $E^{(j)}$ can be investigated simply by observing only the component $\mathscr{E}^{(j)}$ of the composite experiment. Two components $\mathscr{E}^{(1)}$ and $\mathscr{E}^{(2)}$ of a composite experiment are said to be statistically independent if for any event A belonging to component $\mathscr{E}^{(1)}$ and any event B belonging to component $\mathscr{E}^{(2)}$, A and B are statistically independent events. For example, in Example 5.1 the two draws (that is, the two trials) are statistically independent.

The k components $\mathscr{E}^{(1)}, \mathscr{E}^{(2)}, \ldots, \mathscr{E}^{(k)}$ of a composite random experiment are said to be *mutually statistically independent components* if for every collection $E_1, E_2, \ldots, E_k$ of events, where E_i belongs to component $\mathscr{E}^{(i)}$,

$i = 1, 2, \ldots, k$, we have that $E_1, E_2, \ldots, E_k$ are mutually statistically independent events.

If a composite random experiment consists of k mutually statistically independent components, each component being a repetition of the same experiment $\mathscr{E}$, then we say that the composite random experiment consists of *independent and identically distributed* trials. Indeed, it is often the case that the components of a composite random experiment are assumed to be independent and identically distributed trials.

6. APPENDIX: COMBINATORIAL ANALYSIS

For some simple experiments we can enumerate all the possible outcomes by counting. For example, if we wish to choose a committee of 2 people from a group of 5 people [whom, for convenience, we label as A, B, C, D, E], the 10 possible committees are $\{A,B\}$, $\{A,C\}$, $\{A,D\}$, $\{A,E\}$, $\{B,C\}$, $\{B,D\}$, $\{B,E\}$, $\{C,D\}$, $\{C,E\}$, and $\{D,E\}$. Now suppose that we need to choose a committee of 10 from a group of 100; the enumeration of all possible committees in this case, while possible, becomes a long and tedious task.

The above problem is one of a variety of counting problems for which *counting rules* (or *algorithms*) would be helpful. In particular, counting rules are extremely useful in dealing with uniform probability models (Section 1), because the probability of any event can be obtained by counting the number of outcomes contained in that event and dividing by the total number of outcomes in the sample space. In the present section we present some rules which aid in the task of enumeration in such counting problems. These rules are shortcuts to computations, and are not intrinsic to the understanding of the theory of probability.

Orderings and Permutations

Recall that in the definition of a set we do not take note of the order in which the elements of the set are listed. Thus the set $\{A,B,C\}$ is the same set as the set $\{A,C,B\}$ because both sets contain the same elements. We now wish to take into account the order in which the elements of the set are arranged.

An arrangement of n distinct objects in a given order is called a *permutation* (this concept may be defined equivalently in terms of an n-tuple, as in Section 3). Suppose that we have n distinct objects; in how many distinguishable ways can we order them (that is, how many permutations can we form)?

To answer this question, we start by considering the case when we have

6. APPENDIX: COMBINATORIAL ANALYSIS

2 distinct objects A and B (that is, the case $n = 2$). The possible permutations are "A followed by B," denoted by AB, and "B followed by A," denoted by BA. Thus, in the case $n = 2$, there are 2 distinguishable permutations. If we have 3 distinct objects A, B, and C ($n = 3$), we may obtain the number of distinguishable permutations by appending the third item C in all possible ways to all possible permutations of A and B. That is, C is placed at any of the positions indicated by the arrows:

$$\downarrow AB \quad A \downarrow B \quad AB \downarrow \quad\quad \downarrow BA \quad B \downarrow A \quad BA \downarrow,$$

resulting in the permutations

$$CAB \quad ACB \quad ABC \quad\quad CBA \quad BCA \quad BAC.$$

For each of the 2 permutations of A and B, we have three choices of where to put C, thus giving a total of 6 permutations (that is, 3 choices times 2 permutations of A and B). For 4 distinct objects, A, B, C, and D, we may repeat the argument. That is, for each of the $2 \times 3 = 6$ possible permutations of A, B, and C, we have 4 positions in which to place D. Thus, we have $2(3)(4) = 24$ permutations of 4 distinct objects. If we continue arguing in the above way, we find that there is 1 permutation of 1 distinct object, there are 2 permutations of 2 distinct objects, there are $2(3) = 6$ permutations of 3 distinct objects, there are $(2)(3)(4) = 24$ permutations of 4 distinct objects, there are $(2)(3)(4)(5) = 120$ permutations of 5 distinct objects, and so on.

Counting Rule 6.1. The number of distinguishable permutations (arrangements) of n distinct objects is $1 \times 2 \times 3 \times \cdots \times (n-1) \times n$.

Because the product $1 \times 2 \times 3 \times \cdots \times (n-1) \times n$ recurs frequently in counting problems, it is assigned the special notation $n!$. (This symbol is read "n factorial.") For mathematical convenience in writing formulas, we define $0!$ to be 1.

Example 6.1. Six diplomats are asked to line up for a group picture. The number of ways in which they could order themselves is

$$6! = 1 \times 2 \times 3 \times 4 \times 5 \times 6 = 720.$$

Orderings with Similar Elements

Consider a shelf upon which we are to place 5 books each having a dust jacket with a different design. If we wish to pick the most aesthetic arrangement of dust jackets, we have to view $5! = 1 \times 2 \times 3 \times 4 \times 5 = 120$ distinguishable arrangements of the books. Now, however, suppose that 3 of the books have identical dust jackets (so that they are, for aesthetic purposes, indistinguishable). How does this indistinguishability alter the above result? That is, how many arrangements of the books can we distinguish now?

Assume that books A, B, and C have identical jackets, whereas books D and E have jackets different from those of A, B, and C, and from each other. Now, consider any single permutation of the books, say $ABDCE$. Keeping books D and E fixed in place, rearrange A, B, and C in this ordering in all possible ways. We obtain

$$\begin{array}{ccc} ABDCE & BADCE & CADBE \\ ACDBE & BCDAE & CBDAE \end{array}$$

corresponding to the 3! possible orderings of A, B, C. But A, B, and C have the same dust jackets and cannot be distinguished from each other. If J denotes their common jacket design, then all 6 of the above arrays look to us like $JJDJE$. This same argument can be carried out for every permutation of A, B, C, D, and E. Thus, rather than 120 distinguishable orderings (the number we obtain when the books have different jackets), we have here only $\frac{1}{6}$th (that is, $1/3!$) as many distinguishable orderings of the jacket designs. That is, when we counted the orderings of *books* and obtained the number $5! = 120$, we counted every distinguishable ordering of the jackets 6 times—a *redundancy factor* of 6. Thus, the true number of distinguishable orderings of book jackets equals the number of orderings of the books divided by the redundancy factor arising from our counting book orderings as being different when actually they looked the same; namely, $5!/3! = 120/6 = 20$.

Now suppose that D and E have dust jackets identical to one another, but different from the common dust jacket of books A, B, and C. By the same argument as above, every arrangement in which D and E are permuted (for example, $JJDJE$ and $JJEJD$) appears alike to us. Consequently, there is an additional redundancy factor of $2!$ (the number of permutations of D and E) which must be accounted for if we want to know the number of distinguishable orderings of the book jackets. This number of distinguishable orderings is thus $(5!/3!)$ divided by $2!$ or $5!/(3!2!) = 10$.

Extending this argument to cover more general situations we obtain the following rule.

Counting Rule 6.2. The number of distinguishable arrangements of n items of which n_1 are of type 1 and n_2 are of type 2 is $n!/(n_1!n_2!)$. Here $n_1 + n_2 = n$, and types 1 and 2 are assumed to be different.

The expression $n!/(n_1!n_2!)$ occurs in so many contexts that we would like to be able to denote it by a shorthand symbol. One possible symbol is $\binom{n}{n_1,n_2}$. However, we know that $n = n_1 + n_2$; hence, once we know n and n_2, we know n_1. Thus, either of the symbols $\binom{n}{n_1}$ or $\binom{n}{n_2}$ serves equally well to denote the number $n!/(n_1!n_2!)$.

The expression $\binom{n}{k} = n!/[k!(n-k)!]$ is called the kth *binomial coefficient*. The terminology "binomial coefficient" arose historically because $\binom{n}{k}$ is the coefficient of a^k in the expansion

$$(a+1)^n = \binom{n}{0}a^0 + \binom{n}{1}a^1 + \binom{n}{2}a^2 + \cdots + \binom{n}{n-1}a^{n-1} + \binom{n}{n}a^n$$

of $(a+1)^n$ in powers of a (the so-called *binomial expansion*). Values of $\binom{n}{k}$ for $n = 1, 2, \ldots, 10$ and various values of k are presented in Table 6.1.

Table 6.1 Values of the kth binomial coefficient $\binom{n}{k}$ for $k \leq n, n = 1, 2, \ldots, 10$.

n \ k	0	1	2	3	4	5	6	7	8	9	10
1	1	1									
2	1	2	1								
3	1	3	3	1							
4	1	4	6	4	1						
5	1	5	10	10	5	1					
6	1	6	15	20	15	6	1				
7	1	7	21	35	35	21	7	1			
8	1	8	28	56	70	56	28	8	1		
9	1	9	36	84	126	126	84	36	9	1	
10	1	10	45	120	210	252	210	120	45	10	1

The binomial coefficients can be arranged to form a triangle-like figure called Pascal's triangle, where each coefficient is obtained as the sum of the two coefficients directly above it (see Figure 6.1).

One of the most important examples of the use of Counting Rule 6.2 is the case in which each of the n items belongs to one and only one of two distinct and exhaustive categories (the case of dichotomous classification), and we are interested in the number of distinguishable ways in which these items can be arranged. For example, we may have n machine parts which are either "defective" or "not defective," n students who receive either a "pass" or a "fail" grade, n patients who either have or do not have

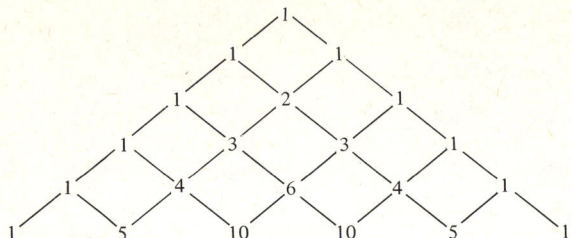

Figure 6.1 Pascal's triangle.

a certain disease, and so on. In general, we have n items, n_1 of which are classified as "success" (S) and $n_2 = n - n_1$ of which are classified as "failure" (F). If we want to know the number of distinguishable ways in which these items can be arrayed, Counting Rule 6.2 tells us that the number we require is $n!/(n_1!n_2!) = \binom{n}{n_1} = \binom{n}{n_2}$.

Counting Rule 6.2 can be extended to situations where the n items to be arranged belong to k distinct types, and consist of n_1 items of type 1, n_2 items of type 2, n_3 items of type 3, ..., and n_k items of type k.

Counting Rule 6.3. The number of distinguishable arrangements of n items, n_1 of which are of type 1, n_2 of which are of type 2, ..., n_k of which are of type k is

$$\binom{n}{n_1, n_2, \ldots, n_k} = \frac{n!}{n_1! n_2! \cdots n_k!}.$$

Here $n_1 + n_2 + \cdots + n_k = n$ and the types 1, 2, ..., k are distinct.

Example 6.2. In attempting to find a pleasing abstract design, we have available 20 positions in which to place patterned tiles. We have available 4 distinct tile patterns, A, B, C, and D. Seven of our tiles are of pattern A, 5 are of pattern B, 4 are of pattern C, and 4 are of pattern D. From Counting Rule 6.3, we thus have a total of $20!/(7!5!4!4!)$ designs from which to choose. Although the number $20!/(7!5!4!4!)$ is the required answer, it is no small feat to actually carry out this tedious computation. In fact, the number $20!/(7!5!4!4!)$ equals 6,983,776,800.

Ordering of r Items from n

Consider again the problem of ordering 5 distinct books on a shelf. However, now suppose we may choose these 5 books from a bin holding 12 different books. This problem may be viewed as one in which we attempt to fill the 5 vacant places [1] [2] [3] [4] [5] with items chosen from among 12 distinct items. There are 12 ways in which we can fill the first vacant place. Once this place is filled, there are 11 ways of filling the second vacant place. Then, after the second place is filled, there are 10 items to choose from to fill the third place, then 9 items for the fourth place, and finally 8 items for the fifth place. For each of the 12 ways we can fill the first vacant place, there are 11 ways to fill the second. For each of the $(12)(11) = 132$ ways in which we can fill the first 2 places, there are 10 ways to fill the third. Continuing in this way, we see that there are $(12)(11)(10)(9)(8)$ or 95,040 ways of filling the 5 vacant places. Thus, there are 95,040 distinct ways of placing 5 books on the shelf when we have 12 books to choose from.

This calculation in general is given by the rule:

Counting Rule 6.4. The number of ways of ordering r items from among n items is $(n)(n-1)(n-2) \cdots (n-r+2)(n-r+1) = n!/(n-r)!$.

The quantity $n!/(n-r)!$ is often denoted by the symbol $n_{(r)}$. The asserted equality between $(n)(n-1)(n-2) \cdots (n-r+2)(n-r+1)$ and $n_{(r)} = n!/(n-r)!$ in Counting Rule 6.4 is verified by noting that

$$\frac{n!}{(n-r)!} = \frac{(n)(n-1)(n-2) \cdots (n-r)(n-r-1) \cdots (3)(2)(1)}{(n-r)(n-r-1)(n-r-2) \cdots (3)(2)(1)}$$
$$= (n)(n-1)(n-2) \cdots (n-r+1),$$

since some of the terms in the numerator of the fraction (for example, $n-r$, $n-r-1$, and so on) are cancelled by the terms in the denominator.

The drawing of an ordered sample of size n without replacement from a population consisting of the N units $u_1, u_2, \ldots, u_N$ (note the change of indices from n to N) is directly analogous to the ordering of n of the units from among all N of the units. In choosing this sample, we have n places to fill in the ordered array (n-tuple) of units which is to be our ordered sample. Place number 1 can be filled by any one of the N units; given the choice of that unit, there are $N-1$ units remaining to fill the second place. Thus if $n = 2$, there are $N(N-1)$ possible samples, since for each of the N ways of filling the first place we can assign any of the $N-1$ ways of filling the second place. This manner of arguing is the same as we used to verify Counting Rule 6.4. Thus Rule 2.2 of Section 2 has been verified: *The total number M_W of ordered samples without replacement of size n from a population of N units is $M_W = N_{(n)} = N(N-1)(N-2) \cdots (N-n+1)$.*

Counting of Sets of Items

In a variety of counting problems we do not distinguish between orderings containing the same items; that is, we are only interested in *sets* of items. For example, when we ask for the number of ways in which a jury of 12 can be chosen from a panel of 300 eligible jurors, we are not concerned with the order in which the jurors are chosen, but only with *which* jurors are chosen. To answer this question, we first ask how many ways the jurors can be chosen if we *do* distinguish the order in which they are chosen. This is the problem of ordering $r = 12$ items from among $n = 300$ items considered in Counting Rule 6.4 and the answer is that there are $(300)(299)(298)(297) \cdots (290)(289) = 300!/288!$ ways. However, if we do *not* wish to distinguish order, then the above counting process has counted each set of jurors 12! times (one count for each way this set of jurors can be ordered; see Counting Rule 6.1). Thus, if we divide the number of ordered ways the jurors can be selected (that is, $300!/288!$) by the number of times (12!) each set has been counted, we obtain the number of ways a *set* (unordered sample) of 12 jurors can be chosen from among a panel of 300; the number that we so obtain (the number of possible juries) is

$$\frac{(300)_{(12)}}{12!} = \frac{300!}{288!\,12!} = \binom{300}{288} = \binom{300}{12}.$$

Counting Rule 6.5. The number of ways of choosing a set (combination, unordered sample) of r items from n distinct items is $\binom{n}{r}$.

Again consider the problem of choosing a jury, but suppose there is the additional complication that the law requires that a jury consist of 8 men and 4 women. There are 180 men and 120 women on the panel of 300 eligible jurors. How many possible juries can we now choose? The answer to this problem is a product of the answers for the two separate problems: (i) In how many ways can we choose 8 men from 180, and (ii) In how many ways can we choose 4 women from 120? The answers to these two problems are given by Counting Rule 6.5; namely, $\binom{180}{8}$ and $\binom{120}{4}$, respectively. We multiply the results, because for every group of 8 men we choose, we can choose one of $\binom{120}{4}$ different groups of women.

We can extend these arguments to the following more general result.

Counting Rule 6.6. If we must choose a set of r items from k distinct categories in such a way that r_1 items belong to category 1, r_2 items to category 2, ..., r_k items to category k $(r_1 + r_2 + \cdots + r_k = r)$, and if we have n items to choose from, of which n_1 are in category 1, n_2 in category 2, ..., n_k in category k $(n_1 + n_2 + \cdots + n_k = n)$, then the number of distinct ways to choose the set of r items is

$$\binom{n_1}{r_1}\binom{n_2}{r_2}\cdots\binom{n_k}{r_k}.$$

Example 6.3. From a university student council, in which there is a representation of 4 freshmen, 6 sophomores, 6 juniors, and 14 seniors, it is desired to form a subcommittee consisting of 1 freshman, 1 sophomore, 2 juniors, and 3 seniors. Using Counting Rule 6.6 we find that there are

$$\binom{4}{1}\binom{6}{1}\binom{6}{2}\binom{14}{3}$$

distinct ways in which this subcommittee can be formed. As in Example 6.2 although we can easily write down this answer in symbols, it takes some effort to compute the actual number, which in the present case is 131,040.

There are several special cases of counting Rule 6.6 which are of interest. Consider, for example, the number of possible ways in which a voting slate for a political party can be compiled. In a certain election district for the state primary, there may be 3 candidates for Governor, 2 candidates for Senator, 2 for Representative, and 3 for Mayor. We have to choose 1 candidate for Governor, 1 for Senator, 1 for Representative, and 1 for Mayor in order to compile the voting slate. By Counting Rule 6.6 (with $n_1 = 3$, $n_2 = 2$, $n_3 = 2$, $n_4 = 3$, and $r_1 = r_2 = r_3 = r_4 = 1$), we find that the total number of possible voting slates for the political party is

$$\binom{3}{1}\binom{2}{1}\binom{2}{1}\binom{3}{1} = 3\times 2\times 2\times 3 = 36.$$

This problem of choosing a voting slate is typical of problems where we wish to choose k items, *one* from each of k distinct collections of items. If there are n_1 items in the first collection, n_2 items in the second collection, ..., and n_k items in the kth collection, then by Counting Rule 6.6 (with $r_1 = r_2 = \cdots = r_k = 1$), the total number of distinct ways in which we can choose these items is

$$\binom{n_1}{1}\binom{n_2}{1}\cdots\binom{n_k}{1} = n_1\times n_2\times\cdots\times n_k.$$

A further simplification of this problem occurs when each of the k categories has the same number N of items; here $n_1 = n_2 = \cdots = n_k = N$ and $n_1 \times n_2 \times \cdots \times n_k = N^k$. Thus, in such a case there are N^k different ways of choosing k items, one from each category. For example, if we toss a coin k times, there are 2^k possible outcomes (each outcome consists of choosing one out of $N = 2$ "items" — either a "Head" or a "Tail" —

for each of the k "categories" (or trials). For $k=3$, the 8 possible outcomes are (H,H,H), (H,H,T), (H,T,H), (T,H,H), (T,T,H), (T,H,T), (H,T,T), and (T,T,T). A second example is when we toss a die k times; here the number of possible outcomes is 6^k (k trials or "categories" for each of which we "choose" one of the $N=6$ possible faces 1, 2, 3, 4, 5, or 6).

NOTES AND REFERENCES

There are various symbols used to denote factorials and binomial coefficients. In older books, the notation $\underline{|n}$ is frequently used in place of the notation $n!$. The most common alternative notations for the binomial coefficient $\binom{n}{k}$ are C_k^n and $C(n,k)$.

A number of books contain extensive tables of binomial coefficients. For example, in the *Handbook of Mathematical Functions*, edited by M. Abramowitz and I. A. Stegun, binomial coefficients for $n = 1(1)50$ are provided. The notation $n = 1(1)50$ means that the first value of n is 1, the last value of n is 50, and the steps are in units of 1.

A table of the factorials is given in *Tables of $n!$ and $\Gamma(n+\frac{1}{2})$ for the First Thousand Values of n*, National Bureau of Standards, Applied Mathematics, Series 16.

EXERCISES

A1. In the context of Example 1.1, find the following probabilities:
  (a) The probability of the event E_1 that the roulette wheel stops at a number which exceeds 18.
  (b) The probability of the event E_2 that the roulette wheel stops at a number (other than 0 and 00) which is divisible without remainder by 3.
  (c) The probability of the event $E_1 \cap E_2$.
  (d) Find the probability of the event $E_1 \cup E_2$ in two different ways. What outcomes belong to the event $E_1 \cup E_2$?
  (e) Find the probability that exactly 1 of the events E_1, E_2 occurs. Do this (i) by using the results you obtained in parts (a), (b), and (c); and (ii) by counting the number of outcomes that belong to this event set.

2. A botanist is interested in studying the process of fertilization in a certain type of flowering plant. He will be using time-lapse photography for his studies, so he decides to study only 1 plant in order to cut his costs for the experiment. There are 30 plants of the desired type in the botanist's greenhouse. However, he must choose 1 of these plants before it produces flowers. The night before the botanist chooses the plant, a power failure lowers the temperature in the greenhouse sufficiently so that 3 of the plants suffer damaged buds. Assume that this damage can only be ascertained after the

plant to be studied has been chosen, and that the botanist chooses the plant to be studied at random (that is, with equal probability) from among the 30 plants available to him. What is the probability that the botanist chooses a plant that does not flower (has damaged buds)?

A3. Suppose that in the context of Exercise 2, the botanist chooses 2 plants to study by means of a simple random sample without replacement from among the 30 plants available to him.
(a) What is the probability that neither of the plants chosen will flower?
(b) What is the probability that exactly 1 of the plants chosen will flower?
(c) What is the probability that both plants will flower?

4. There are 10 children in a certain class. Two members of the class are chosen by the teacher to clean the blackboard after class. They protest that they have been chosen for this duty because they are the only 2 members of the class who write for an underground newspaper. The teacher denies this and states that the students were chosen by a random sample of size 2 without replacement from among all the children in the class. If the teacher is telling the truth, what is the probability that the 2 persons chosen to clean the blackboard are the 2 members of the class who write for the underground newspaper?

A5. There is an urn on the table containing 5 red balls and 5 black balls. Jimmy takes a simple random sample of 2 balls *with* replacement from the urn. He records the colors of the balls he chooses and then replaces them in the urn. Harry takes a simple random sample of 2 balls from the urn *without* replacement. Let P_J be the probability that Jimmy's sample contains 1 red ball and 1 black ball. Let P_H be the probability that Harry's sample contains 1 red ball and 1 black ball. What is the value of $P_J - P_H$?

6. A population of 50 registered voters contains 30 Democrats and 20 Republicans. An opinion survey selects a simple random sample of 4 voters without replacement from this population.
(a) What is the probability that no Republicans will be represented in the sample?
(b) What is the probability that at least 1 Republican will be represented in the sample?
(c) What is the probability that exactly 1 Republican will appear in the sample?
(d) What is the probability that a majority of the sample are Republicans? (A majority is *more* than half of the voters represented.)
(e) What is the probability that one of the parties (either Democratic or Republican) will have a majority representation in the sample?

A7. In a uniform probability model with 33 possible outcomes, what is the common probability of each simple event? What is the probability of an event that consists of 6 outcomes? Suppose that an event E has probability $P(E) = 1/11$. How many outcomes are in the event E? Another event F has probability $P(F) = 2/11$, and there are 2 outcomes common to the events E and F. What is the probability of the event $E \cup F$?

8. Suppose that we run 2 rats through the experiment described in Example 1.2.
   (a) How many outcomes are there for this composite random experiment?
   (b) How many outcomes correspond to results in which every rat receives a reward?
   (c) How many outcomes correspond to results in which none of the rats receive rewards?
   (d) How many outcomes correspond to results in which at least 1 rat receives a reward?
   Suppose that this composite random experiment is a uniform probability model.
   (e) Find the probability that every rat receives a reward.
   (f) Find the probability that none of the rats receive rewards.
   (g) Find the probability that at least 1 rat receives a reward.

A9. Answer parts (a) through (g) of Exercise 8, but now assume that 3 rats are run through the experiment described in Example 1.2.

10. Families with 3 sons are studied, and it is asked whether each son is colorblind or not.
    (a) List the possible outcomes of this composite random experiment.
    (b) Does your list of outcomes take account of the order in which the sons were born? If not, list a new set of outcomes that take account of the order of birth.
    (c) Suppose that we have a uniform probability model for this experiment (using a list of outcomes that take account of order of birth). What is the probability that the oldest son is colorblind? What is the probability that the middle son is colorblind? What is the probability that the youngest son is colorblind? What is the probability that all 3 sons are colorblind?
    (d) Suppose that we have a probability model for this composite experiment in which the probability that the oldest son is colorblind is 0.5, the probability that the middle son is colorblind is 0.5, and the probability that the youngest son is colorblind is also 0.5. Do you have enough information to enable you to calculate the probability that all 3 sons are colorblind? Why, or why not? If you do have enough information, calculate the desired probability.

A11. Consider the experiment in which a pebble is selected at random from among all of the pebbles on the shore of Lake Michigan. Suppose that the color of this pebble is observed, and that the pebble can be either brown, black, or green in color. Based on a group of 1000 pebbles studied by Miller and Kahn (1962), an appropriate probability model for this experiment is the following:
   (i) The outcomes are the colors $\omega_1 =$ brown, $\omega_2 =$ black, and $\omega_3 =$ green. The simple events are $S_1 = \{\omega_1\} = \{\text{brown}\}$, $S_2 = \{\omega_2\} = \{\text{black}\}$, and $S_3 = \{\omega_3\} = \{\text{green}\}$.

(ii) The probability model is determined by

Simple Event	S_1 (brown)	S_2 (black)	S_3 (green)
Probability	0.852	0.093	0.055

Under this model,
(a) What is the conditional probability that a pebble is brown *given* that the pebble is either black or brown?
(b) What is the unconditional probability that the pebble is brown? How does the conditional probability found in part (a) compare with the unconditional probability found here? Which is the larger probability? Are the probabilities close to one another in magnitude? If there is a big difference in these probability values, explain why this should be the case. If these probabilities are nearly equal, explain why they should be nearly equal.

12. Based on studies described by Gregory (1963, p. 78), the following probability model describes the number of floods in a given wet season:
   (i) The outcomes are the numbers, 0, 1, 2, 3, 4, 5, ... of floods. The simple events are $S_1 = \{0 \text{ floods}\}$, $S_2 = \{1 \text{ flood}\}$, $S_3 = \{2 \text{ floods}\}$, and so on.
   (ii) The probability model is determined by

Simple Event	{0 floods}	{1 flood}	{2 floods}	{3 floods}
Probability	0.2466	0.3452	0.2417	0.1127
Simple Event	{4 floods}	{5 floods}	{6 floods}	
Probability	0.0395	0.0110	0.0033	

Simple events corresponding to more than 6 floods have probability 0.
(a) What is the probability of 2 or more floods in a year?
(b) Given that a particularly wet year is being experienced (in which 2 floods have already been observed), what is the conditional probability that 4 or more floods will be observed? That is, what is the conditional probability of 4 or more floods *given* that 2 or more floods have occurred?
(c) If at the beginning of a given wet season we observe a flood, what probability (conditional probability) do we assign to the event that we will observe at least 1 *more* flood before the wet season is over? Explain your answer. What is the new sample space for the (conditional) probability model if we know that at least 1 flood will occur this season? What is the event whose conditional probability we want to assess?

A13. (a) Evaluate $P(A|A \cap B)$ and $P(A|A^c)$ without carrying out any computations. Support your answers by logical arguments.

(b) If the event A is included in the event B, what is the value of $P(B|A)$? Why?

(c) If the events A and B are mutually exclusive, what is $P(B|A)$? What is $P(A|B)$?

14. Table E.1 describes data obtained by Lazarsfeld and Thielens (1958) on the relationships between age, productivity, and party vote in 1952 for a sample of 2117 social scientists studied in 1954–1955.

Table E.1. Social scientists, classified by age, productivity score, and party vote in 1952.

Age	Productivity Score					
	Low		Middle		High	
	DEMO-CRATS	OTHERS	DEMO-CRATS	OTHERS	DEMO-CRATS	OTHERS
40 or younger	260	118	226	60	224	60
41 to 50	60	60	78	46	231	91
51 or older	43	60	59	60	206	175

Calculate the following (and show the steps in your calculations):

(a) The percentage of social scientists who voted Democratic in 1952.

(b) The percentage of social scientists in the age group of 41 to 50 who are classified in the middle of the productivity score.

(c) The percentage of social scientists 51 years or older and with a low productivity score among all who voted Democratic in 1952.

(d) The percentage of social scientists 40 years or younger among all who did not vote Democratic in 1952.

(e) The percentage of social scientists 51 years or older and with a low productivity score among all who did not vote Democratic in 1952.

(f) The percentage of social scientists low on the productivity scale and voting Democratic in 1952 among all who are in the age group 41–50 years.

15. Using the data of Table 4.1 of Example 4.1 as probabilities, compute the conditional probability that a student will choose a commercial curriculum *given* that the student is a member of social class III. Compute the conditional probability that a student will choose a commercial curriculum *given* that the student belongs to one of the social classes I, II, or III.

16. Studies of the number of children in families provide experimental evidence to support the following table of probabilities [*Statistical Abstract* (1965), p. 36]:

Event	{0 children}	{1 child}	{2 children}	{3 children}	{4 or more children}
Probability	0.431	0.173	0.174	0.113	0.109

(a) What is the probability that a randomly chosen family has 1 or more children?

(b) Given that a randomly chosen family has children (1 or more), what is the conditional probability that they have exactly 1 child?

(c) You have chosen the name of a family at random from the phone book. You are now going to interview that family about the sibling relationships among their children. For the questions to be meaningful, the family must have at least 2 children. What is the probability that the family you are going to interview has 2 or more children? As you approach the house of the family to be interviewed, you see that the house is too small to house more than 3 children, so that you hypothesize that the family you are going to interview has no more than 3 children. Assuming that you are right, what is the conditional probability that the family has at least 2 children (*given* that they have no more than 3 children)? How does your answer compare with the unconditional probability that the family has 2 or more children?

A17. A person takes a 3-part examination. The parts of the examination are set up in such a way that the answers to parts of the exam taken previously to a given part of the exam provide clues to the correct answer to that given part of the exam. The person has probability $\frac{1}{2}$ of correctly answering the first part of the exam. His probability of correctly answering the second part of the exam given that he correctly answered the first part of the exam is $\frac{3}{4}$; while his probability of correctly answering the second part of the exam given that he missed the first part of the exam is $\frac{1}{4}$. Finally, his probability of correctly answering part 3 of the exam is

$\frac{1}{5}$, given that he missed both previous parts of the exam;
$\frac{2}{5}$, given that he missed part 2 but correctly answered part 1;
$\frac{3}{5}$, given that he missed part 1 but correctly answered part 2;
$\frac{4}{5}$, given that he correctly answered parts 1 and 2.

(a) What is the probability that the person misses all 3 parts of the exam?

(b) What is the probability that the person misses the first 2 parts, but answers the third part correctly?

(c) What is the probability that the person misses the first and last parts, but answers the second part correctly?

(d) What is the probability that the person misses the last 2 parts, but answers the first part correctly?

(e) What is the probability that the person answers exactly 1 part correctly?

(f) What is the probability that the person answers exactly 2 parts correctly?

(g) Given that the person answers the last part correctly, what is the conditional probability that he answers 2 or more parts correctly?

18. Consider all persons born in the United States and now living in the United States. Suppose that 25 percent of these were born west of the Mississippi River and that: (i) the probability that a person is now living west of the Mississippi given that he was born west of the Mississippi is $\frac{2}{3}$; (ii) the probability that a person is now living west of the Mississippi given that he was

born east of the Mississippi is $\frac{1}{5}$. A person is chosen at random from among all persons born in and now living in the United States.
  (a) What is the probability that the person is now living west of the Mississippi River? [*Hint:* Let W be the event that the person is now living west of the Mississippi, let A be the event that the person was born west of the Mississippi. Then A^c is the event that the person was born east of the Mississippi and $W = (W \cap A) \cup (W \cap A^c)$.]
  (b) What is the conditional probability that the person chosen was born west of the Mississippi *given* that he now lives west of the Mississippi? [*Hint:* Use Bayes' Rule.]
  (c) What is the conditional probability that the person chosen was born east of the Mississippi *given* that he now lives west of the Mississippi?
  (d) What is the conditional probability that the person chosen was born east of the Mississippi *given* that he is now living east of the Mississippi?
  (e) What is the unconditional probability that the person chosen is now living in the region (west or east of the Mississippi) where he was born?

^A19. Suppose that we have carried out a study relating education level to income level, and find that the population can be adequately represented by two income levels, "high" and "low," having relative sizes $\frac{1}{3}$ and $\frac{2}{3}$, respectively. The proportion of people in the high income level who have a college education is 0.75, while the proportion of people in the low income level who are college educated is 0.36. (a) Making use of Bayes' Theorem, find the conditional probability that a person chosen at random will be from the high income level, given that he is college educated. (b) If the person is college educated, what is the probability that he is from the low income level?

20. Smith and Suchman (1955) are concerned with the question "Do people know why they buy?" In a telephone survey made in March 1938 in Syracuse, New York, 764 people were called up during the time Boake Carter (who was then advertising Philco radios) was on the air, and asked for the program to which they were listening at the moment of the call, and also the make of the radio they owned. The following relative frequencies were obtained:

		Owned a Philco Radio	
		YES	NO
Listened to	YES	0.136	0.170
Boake Carter	NO	0.272	0.422

Assume that these relative frequencies are actually probabilities for the events described.
  (a) Given that someone owns a Philco radio, what is the conditional probability that this person listened to Boake Carter?
  (b) Given that someone listened to Boake Carter, what is the conditional probability that this person owns a Philco radio?

^A21. There are 2 diagnostic tests T_1 and T_2 for a certain disease D. Test T_1 has a (conditional) probability of 0.8 of being negative if the person tested does

not have the disease D, and a (conditional) probability of 0.1 of being negative if the person tested does have the disease. Test T_2 has a (conditional) probability of 0.7 of being negative if the person tested does not have the disease D, and a (conditional) probability of 0.05 of being negative if the person tested does have the disease. The 2 diagnostic tests have outcomes that are statistically independent of one another whether or not the person tested has the disease D.
- (a) What is the (conditional) probability that both diagnostic tests are negative given that the person tested does not have the disease?
- (b) What is the (conditional) probability that at least 1 of the diagnostic tests will be positive if the person tested does have the disease?
- (c) It is known that a person who presents himself for testing for the given disease D has (unconditional) probability 0.2 of actually having the disease. What is the (unconditional) probability that at least 1 of the tests is positive? Given that at least 1 of the tests is positive, what is the conditional probability that the person tested has disease D?
- (d) What is the unconditional probability that both diagnostic tests are negative? Given that both tests are negative, what is the conditional probability that the person tested has the disease?

*22. Three prisoners, A, B, and C, with apparently equally good records have applied for parole. The parole board has decided to release 2 of the 3, and the prisoners know this but not *which* 2. A warden friend of prisoner A knows who are to be released. Prisoner A realizes that it would be unethical to ask the warden if he, A, is to be released, but thinks of asking for the name of *one* prisoner *other than* himself who is to be released. He thinks that before he asks, his chances of release are $\frac{2}{3}$. He thinks that if the warden says "B will be released" his own chances have now gone down to $\frac{1}{2}$ because either A and B or B and C are to be released. And so A decides not to reduce his chances by asking. Clearly, A is mistaken in his reasoning. (If you accept his reasoning, merely asking the question "Who besides me is to be paroled?" reduces his chances of parole even if he never hears the answer to the question.) What was his mistake? [*Hint:* What is the proper sample space for this problem? What kind of probabilities is A calculating? Is not the warden's answer part of the sample space?] As part of your answer, show the correct calculations needed. [See Mosteller (1965).]

A23. The probability that a married man will vote in a given election is 0.40. The probability that a married woman will vote in that election is 0.35. The probability that a married woman will vote given that her husband votes is 0.60.
- (a) What is the probability that both husband and wife will vote in the election?
- (b) Are the actions of husband and wife with respect to voting statistically independent? Why or why not?
- (c) What is the probability that a married man will vote given that his wife votes?

24. Students of political behavior are often interested in the point in time at which a voter decides on the candidate for whom he is going to cast his ballot [Campbell and Kahn (1952)]. Following the 1948 presidential election a national sample of 390 individuals who had voted in the election was taken. Each individual was asked, among other things, whether he had voted for Dewey or Truman and at what point he made up his mind. The results were:

When Decision Was Made	Truman	Dewey	Total
Knew all along	76	75	151
Decided at time of convention	47	60	107
Decided during campaign	30	23	53
Decided within 2 weeks of election	30	5	35
Decided on election day	6	4	10
Do not know	2	4	6
Time not ascertained	21	7	28
Total	212	178	390

For present purposes regard these 390 individuals as constituting a finite population from which we draw a random sample of size 1.
   (a) If a single individual is drawn at random from this population, what is the probability that he:
      (i) knew all along for whom he was going to vote?
      (ii) knew all along for whom he was going to vote and voted for Truman?
      (iii) knew all along for whom he was going to vote and voted for Dewey?
      (iv) voted for Truman?
      (v) voted for Dewey?
   (b) If it is known that the individual voted for Truman, what is the (conditional) probability that he knew all along for whom he was going to vote?
   (c) If it is known that the individual voted for Dewey, what is the (conditional) probability that he knew all along for whom he was going to vote? What conclusions do you draw about the comparative characteristics of Truman and Dewey voters by comparing your answers in parts (b) and (c)? Explain.
   (d) Compare the probability that the voter knew all along for whom he was going to vote and voted for Truman, with the product of the probabilities obtained in (i) and (iv) of part (a). Are the events "voted for Truman" and "knew all along for whom he was going to vote" statistically independent? Why or why not?

25. The following is a slightly edited newspaper account of a trial in which the issue is suicide versus murder. Comment on this excerpt in terms of concepts of probability theory. How many errors can you find in the testimony of the expert witness?
   "The expert witness, Mr. W's testimony was, briefly, that the odds against

J.B. having committed suicide were 300,000,000 to 1; that is, about the same as the chance an honest poker player has of drawing to an inside straight flush 6 times in a row.

D has maintained that J.B. killed his wife, C.B., and then stabbed himself in murder and suicide. The state holds that D killed both in an effort to collect C.B.'s $100,000 estate.

The knife, W said, is an unpopular weapon in suicides; it is used only once in 60 times.

As a murder weapon, however, the knife is used in 1 out of 4 killings.

The clothing of suicides by knife is stabbed through only once in every 10 cases; it was stabbed through in the case of J.B.

In only 1 out of 10 cases does a knife suicide occur without "hesitation wounds" resulting from tentative jabs; there were no such wounds on J.B.

In 9 out of 10 knife suicides, the would would be in the left chest, instead of in the right, as in J.B.'s case.

In only 1 out of 100 knife suicides is the knife thrust vertical instead of horizontal; the thrust was vertical in J.B.'s body.

In W's experience, only 1 person in 50 would be able to coordinate his bodily action with the amount of alcohol found in the blood of J.B.

Multiplying these factors together according to the laws of probability, Mr. W reached the stated odds of 300,000,000 to one."

26. An important classification of individuals is in terms of blood groups O, A, B, and AB. In the United States, the proportion of people falling into each category are

Blood Group	O	A	B	AB
Proportion	0.45	0.42	0.10	0.03

Assume that both sexes have the same proportions of people falling into these categories, and assume that choice of a marriage partner is (statistically) independent of blood group. Finally, assume that the proportions stated above actually hold for the subclasses of males (females) who are newly married today. We perform the following experiment: Among all couples newly married today, we choose one couple by a simple random sample (that is, every newly married couple has an equal chance to be observed). We observe the blood groups of the man and of his wife. The simple results (outcomes) of this experiment are OO, OA, OB, $O\overline{AB}$, AO, AA, AB, and so on, where the first letter (or letter pair) refers to the blood group of the husband, the second letter (or letter pair) refers to the blood group of the wife.

(a) What is the probability that in the observed couple, the husband has blood group A and the wife has blood group B?

(b) What is the probability that in the observed couple one spouse has blood group A and the other spouse has blood group B?

(c) What is the probability that both spouses have the same blood group?

(d) What is the probability that each spouse has a different blood group?
(e) Given that the husband has blood group O, what is the probability that the wife has blood group B?

A27. An adult who has had neither Mumps nor Measles suddenly discovers (too late for him to protect himself) that 1 of his children has Mumps and a second has Measles. Suppose that the probability of his contracting Mumps after exposure is 0.15 and of contracting Measles after exposure is 0.20 and that it is possible to have both diseases at once. What is the probability that:
(a) He has both diseases?
(b) He has neither?
(c) He has Mumps and not Measles?
(d) He has Measles and not Mumps?
In making your calculations assume that the event that the adult contracts Measles is statistically independent of the event that he contracts Mumps.

28. In a study of marriages, both husbands and wives were classified as being either good tempered (G) or bad tempered (B). The following probability model (based on observations made on 300 couples) was found to hold.

		Temper of Wife	
		G	B
Temper of Husband	G	0.22	0.24
	B	0.31	0.23

Thus, for example, the probability that in a given couple the husband is bad tempered and the wife is good tempered is 0.31.
(a) What is the conditional probability that a husband is good tempered *given* that his wife is good tempered?
(b) What is the conditional probability that a husband is good tempered *given* that his wife is bad tempered?
(c) Are the tempers of husband and wife statistically independent of one another? Why or why not? Based on the above table of probabilities, if you want to find a good tempered husband should you look for a good tempered wife? Or does it matter?

29. We have seen that if A and B are independent, then $P(A|B) = P(A)$. Describe an example of dependent events A and B for which $P(A|B) > P(A)$. Describe an example for which $P(A|B) < P(A)$.

A30. If A and B are statistically independent events, evaluate $P(A \text{ and } B|B)$. To help you make the evaluation, use a Venn diagram.

31. Which of the following are pairs of statistically independent events?
(a) Getting a 6 on a roll of a 6-sided balanced die, and getting a 6 on the very next roll of that same die.
(b) Being intoxicated, and having an auto accident.
(c) Studying tonight, and not studying tonight.
(d) Being on time for class, and the day being sunny and warm.

(e) Winning the first door prize on a raffle in which there are 2 door prizes, and winning the second door prize. (Assume that winning raffle tickets are not replaced for future draws.)
Explain your answer.

32. In a certain area, 10 percent of the homes have color television sets and 25 percent have automatic dishwashers. A single home is drawn at random. Your laboratory assistant says that the probability that you will draw a home having both a color television set and an automatic dishwasher is $(0.10)(0.25) = 0.025$. Do you agree? Why or why not?

*A33. In electrical systems, components may be inserted into a circuit in "series" or in "parallel." If 2 or more components appear in parallel in a circuit, electric current will flow through the circuit if any 1 (or if more than 1) component is capable of conducting the current. If 2 or more components appear in series in a circuit, all components must conduct current if electricity is to flow through the circuit. The figure below shows an electric circuit in which the components B, D, and E, and the subcircuits A and C, are inserted in series. Subcircuit A consists of 2 components, A_1 and A_2, in parallel; while subcircuit C consists of 3 components, C_1, C_2, and C_3, in parallel. The abilities to conduct electricity of the components A_1, A_2, B, C_1, C_2, C_3, D, and E are assumed to be mutually statistically independent.

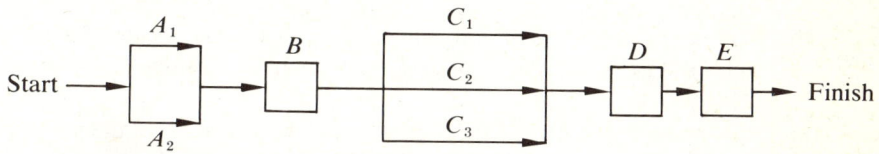

(a) Write out a list of the ways in which the circuit can conduct electric current. Two such ways are:
(A_1 conducts, B conducts, C_2 conducts, D conducts, E conducts),
(both A_1 and A_2 conduct, B conducts, C_1 conducts, D conducts, E conducts).

(b) Assume that the following probabilities hold:
$P(A_1 \text{ conducts}) = P(A_2 \text{ conducts}) = 0.7$,
$P(C_1 \text{ conducts}) = P(C_2 \text{ conducts}) = P(C_3 \text{ conducts}) = 0.6$,
$P(B \text{ conducts}) = P(D \text{ conducts}) = P(E \text{ conducts}) = 0.9$.
Find the probability that the entire circuit conducts.

(c) Find the probability that current passes at least through subcircuit A and component B.

(d) Find the probability that current can pass through the part of the circuit consisting of subcircuit C and components D and E.

(e) If we already had the answers to parts (c) and (d) of this question, how could we get the answer to part (b)? Explain your answer.

(f) Suppose that the entire circuit does not conduct electricity. What is the conditional probability that component B is not conducting electricity?

34. Miller and Kahn (1962, p. 6) report an experiment in which samples of rocks were collected and classified both according to type and according to modal

size. Based on the data they report (2000 rock samples), the following probability model can be constructed:

(i) The outcomes are (Arkose,Coarse), (Arkose,Medium), and so forth. That is, the outcomes can be represented as pairs, where the first component of the pair gives the type of rock sample chosen, and the second component of the pair gives the modal size of the rock sample.

(ii) The probabilities of the simple events of the experiment can be summarized in the following table:

Rock Type \ Model Size	Coarse	Medium	Fine	Very Fine
Arkose	0.160	0.075	0.065	0.025
L. R. Gray Wocke	0.025	0.250	0.225	0.050
Quartzite	0.040	0.050	0.010	0.025

(a) Note that this experiment is a composite random experiment in which there are two component random experiments: the observation of the rock type of the chosen rock sample, and the observation of the modal size of the rock sample. Find the probability models for each of the component experiments (considered separately) of this composite random experiment.

(b) Are the two component experiments statistically independent? That is, is rock type statistically independent of rock size? Support your assertion.

A35. Sir Francis Galton in his book *Finger Prints* (1892) discusses the prints of the right forefingers of 101 pairs of school children chosen at random from a large collection. In addition, 105 pairs of children having a fraternal relation are also chosen. The contingency tables for the two sets of data are as follows:

Nonfraternal Children
A Children

B Children	ARCHES	LOOPS	WHORLS	TOTALS
ARCHES	5	12	8	25
LOOPS	8	18	8	34
WHORLS	9	20	13	42
TOTALS	22	50	29	101

Fraternal Children
A Children

B Children	ARCHES	LOOPS	WHORLS	TOTALS
ARCHES	5	12	2	19
LOOPS	4	42	15	61
WHORLS	1	14	10	25
TOTALS	10	68	27	105

If the measurements on the A and B children were actually statistically independent experiments, what distributions would we expect?

^{A}36. A fair coin is tossed 100 times, the tosses being stochastically independent. (Remember that a "fair" coin has an equal chance of landing "Heads" or of landing "Tails" at any given toss.) Given that the first 99 tosses result in "Heads," what is the probability of obtaining a "Head" on the 100th toss?

In each of the following problems state explicitly the counting rules which you use to solve the problem.

37. How many 5-place numbers can be made using each of the digits
    (a) 3, 4, 5, 8, 9;
    (b) 0, 4, 5, 8, 9;
    (c) 3, 4, 5, 5, 9?
    How many *even* 5-place numbers can be made using each of the digits 2, 3, 5, 7, 9?

^{A}38. If 13 diplomats are asked to line up for a group picture with the senior diplomat always to be in the center, in how many distinguishable ways can they be arranged?

39. A housewife has 7 types of meat available to her for dinners, 1 meat for every day of the week. She would like to vary her choices of meats so that the meats appear in a different order each week of the year. Can she achieve her goal?

^{A}40. How many different sets of initials can be formed if every person has exactly 1 surname and
    (a) exactly 2 given names?
    (b) at most 2 given names?
    (c) at most 3 given names?
    [Note that a person may have no given names.]

41. A wine taster claims he can discriminate among 5 different varieties of wine by taste. An experiment is set up to test his ability. The wine taster is blindfolded and given the wine varieties one at a time.
    (a) In how many different possible orders could the 5 wine varieties be presented to the wine taster?
    (b) The wine taster was really boasting. When the actual experiment is run, he guesses his answers at random. That is, the order in which he names the 5 varieties is a random sample of one order from among all possible orders in which he could name the 5 varieties. What is the probability that the wine taster guesses correctly?
    (c) What is the probability that the wine taster guesses 3 right out of 5?

42. A radiologist is to look at 7 chest X-rays which his assistant has arranged in order. Two of the X-rays show a tumor at the identical position on the lung, while the remaining X-rays are normal. The radiologist only pays attention to the distinction between diseased and normal X-rays. In how many

possible distinguishable orders could the X-rays be presented to the radiologist?

A43. A child is playing with black and white blocks. He has 3 black blocks and 3 white blocks, and he arranges the 6 blocks in one line. (a) How many possible distinguishable patterns can he make? The child arranges the blocks in 2 rows of 3 blocks each. (b) Now, how many possible distinguishable patterns can the child make? (c) Compare your answers in parts (a) and (b). What do you conclude?

44. In a certain random experiment, there are only 2 possible outcomes: "success" and "failure." We repeat this experiment 5 times, and distinguish the repetitions from one another by labeling the initial run of the experiment as "trial 1," the second run of the experiment as "trial 2," and so forth.
    (a) Describe the outcomes of this composite random experiment.
    (b) In how many outcomes of this experiment do we observe 2 "successes" and 3 "failures"?
    (c) In how many outcomes of this experiment do we observe 3 "successes" and 2 "failures"?
    (d) Compare your answers in parts (b) and (c). What do you conclude? Do you have any explanation for this result?
    (e) In how many outcomes of this experiment is there exactly 1 failure?
    (f) In how many outcomes of this experiment is there exactly 1 success?
    (g) Compare your answers in parts (e) and (f). What do you conclude? Do you have an explanation for this result?

A45. Nine skaters are competing in a team figure-skating contest. Of the skaters, 3 are from the United States, 3 are from the U.S.S.R., and 3 are from mainland China. At the end of the contest, the skaters will be ranked from best to worst (no ties are permitted), but the scoring will only take account of what countries these skaters represent, and not their individual identities.
(a) For the purpose of scoring, how many possible distinguishable outcomes are there to this contest? (b) How many outcomes correspond to results in which the United States skaters are ranked 1, 2, 3?

A46. A certain stained-glass window is made of 25 panes (5 rows of 5 panes each). The artist who constructs the window has 7 blue panes, 5 red panes, 6 green panes, and 7 yellow panes. In how many different distinguishable ways can he arrange the panes in order to make an abstract design for the window? [*Hint:* Does it make any difference to your answer whether the panes are arranged in a row of 25 panes, or arranged in 5 rows of 5 panes each?]

47. A family has a choice of 5 vacation spots. They decide to visit 2 of these spots on their vacation, spending part of their time at each. How many different choices of vacations do they have if

(a) we distinguish the order in which they visit the vacation spots that they choose?
(b) we do not distinguish the order in which they visit the chosen vacation spots, but only name which vacation spots were chosen?

48. The refrigerator holds 2 cartons of ice cream. The dairy has 5 flavors, and the family always likes to buy 2 different flavors. How many times can the family go to the dairy and bring home a different pair of flavors?

^{A}49. A telephone dial has a finger hole for each of the 10 digits.
(a) How many telephone numbers, each with 7 digits but with no digit repeated, are possible?
(b) How many telephone numbers, each with 7 digits, are possible?

50. A signalman has 6 flags. The emblems on the various flags are: a stripe, a dot, a triangle, a rectangle, a bar, and a circle. By showing 2 different flags, one after the other, the signalman can send a signal. How many different signals are possible?

51. Suppose that we have a population of 12 units $u_1, u_2, ..., u_{12}$. We draw a simple random sample of size 5 *with replacement*. What is the probability that we draw a sample in which no unit appears more than once?

*^{A}52. Assume that 5 people gather for a dinner party. It is reasonable to hypothesize that the month of birth for each person is the simple result of a random sample (lottery) in which the month of birth is chosen from among all of the months of the year, and that the choice of birth-month for each of the 5 persons is statistically independent of the choice of birth-months for the other people at the party. What is the probability that everyone at the party has a different birth-month?

53. A certain restaurant's menu offers a choice of soup or orange juice for an appetizer; a choice of steak, chicken, or fish for the entree; and a choice of pie, cake, or ice cream for the dessert course. A complete 3-course dinner consists of one choice in each course (appetizer, entree, and dessert). How many different complete 3-course dinners are offered by the restaurant?

^{A}54. From 8 men and 5 women, in how many ways can one select a committee of
(a) 3 men and 2 women?
(b) 5 people, subject to the condition that a committee not contain more than 2 men?

*^{A}55. A certain jury venire consists of 300 individuals of whom 135 are opposed to the death penalty and 165 favor the death penalty. (a) If an initial list of 12 individuals are picked from this venire by a simple random sample without replacement, what is the probability that 6 of them will favor the death penalty, and 6 will oppose the death penalty? (b) What is the probability that a majority (7 or more) of these 12 individuals will oppose the death penalty?

56. Bicycle locks frequently are arranged with 4 disks, each disk having 5 numbers. The correct combination is an ordered list of 4 numbers. (a) If we wish to open the lock, how many combinations do we have to try? (b) Suppose the digit on the third disk represents a geographical area where the lock is made. Thus 1 may denote the North Eastern states, 2 the South Eastern states, 3 the Midwest, 4 the Northwest, and 5 the Southwest. If we know the bicycle comes from a western state, how many combinations do we have to try?

4
RANDOM VARIABLES

Defendit numeris: There is safety in numbers.
Anonymous

1. RANDOM VARIABLES

By and large, most experiments of scientific interest are complex, with outcomes that are likely to be difficult to describe completely. A common device for reducing the complexity of an experiment and its outcomes is that of confining attention to one or more quantitative aspects of the experiment. As a matter of fact, an essential characteristic of modern science and technology is an interest in the quantitative aspects of natural phenomena.

Example 1.1. As a random experiment, a ticket window at a bus terminal is observed for a given period of time by an employee of the bus company. The outcome of this random experiment might be defined as the various happenings that can occur during the period. However, for the purposes of the employee, not every incident that takes place is of interest. Rather, his concern is with those quantities that are relevant to the affairs of the bus company. Specifically, the company is considering the possibility of adding other ticket windows to be used during the peak hours. During the time he observes the ticket window, the employee notes the number X of people who arrive at the window, the lengths of time $T_1, T_2, \ldots$ required for each person to be served (starting from the time of arrival), the length of the waiting line at different times, and so on. The employee thus has converted the happenings resulting from the random experiment into a list of quantitative aspects (that is, the list $X, T_1, T_2, \ldots,$ and so on).

The quantification of the outcomes of a random experiment in such a way that each outcome is represented by exactly one numerical value involves the notion of a *random variable* (*chance variable, stochastic variable*). A random variable Y, then, is a prescription by which every outcome ω in the sample space Ω of the random experiment is assigned a number $Y(\omega)$. The number $Y(\omega)$ so assigned is called the *value* of the random variable Y for the outcome ω. We may view a random variable as being a coding whereby every outcome ω is given a numerical code $Y(\omega)$. Alternatively, stated in mathematical terms, a random variable is a *function* which assigns a real number $Y(\omega)$ to each outcome ω in the sample space Ω. In the present book, capital letters X, Y, Z are used to denote different random variables, while possible numerical values that may be assigned by these random variables to a particular outcome are denoted by corresponding lower case letters x, y, z.

Example 1.1 (Continued). The outcomes ω of the random experiment that consists of observing the ticket window at the bus terminal can be described by lists giving in minute detail every happening that occurred (could have occurred) during the period of observation. One random variable X, described earlier, abstracts from the list of minutia corresponding to a given outcome ω the number $X(\omega) = x$ of people served during the period. Every time the bus company employee reruns the experiment, a different outcome ω may be observed, and a different number $X(\omega)$ of people may be served during the period of observation. If 5 people are served during one observation period, then the value of X for that run of the experiment is $X(\omega) = 5$; if 17 people are served, then the value of X is $X(\omega) = 17$; if x people are served, then the value of X is $X(\omega) = x$.

Example 1.2. Three infants are given a simple puzzle, and are kept under observation for a prescribed length of time. The possible outcomes of this random experiment could be quite detailed descriptions of the behavior of each infant during the period of observation. However, the psychologist running the experiment decides to record only whether or not each of the 3 infants completes the solution of the puzzle during the period of observation. Any outcome of the experiment can be described in terms of a triple (see Chapter 3, Section 3). For example, one outcome ω_1 is the triple $\omega_1 =$ (first infant succeeds, second infant succeeds, third infant succeeds), which we abbreviate for convenience to $\omega_1 =$ (S,S,S), where S stands for "success" and F stands for "failure." Similarly, the other possible outcomes are $\omega_2 =$ (F,S,S), $\omega_3 =$ (F,F,S), $\omega_4 =$ (F,F,F), $\omega_5 =$ (S,F,S), $\omega_6 =$ (S,S,F), $\omega_7 =$ (S,F,F), and $\omega_8 =$ (F,S,F). Now, in the process of summarizing his results, the psychologist decides that he is not interested in which infants succeeded in solving the puzzle, but only in *how many* infants succeeded. Let X be the random variable defined as the total number of infants who succeeded in solving the puzzle. Then by inspection we can see that $X(\omega_1) = 3, X(\omega_2) = 2, X(\omega_3) = 1, X(\omega_4) = 0$, and so on.

The psychologist later changes his mind and decides to summarize the result of his experiment by the proportion Y of infants who succeeded in solving the puzzle. Since the proportion of infants who succeeded in solving the puzzle is equal to the number who succeeded divided by the total number, 3, of infants involved in the experiment, we see by direct inspection of each outcome ω that $Y(\omega_1) = 1$, $Y(\omega_2) = \frac{2}{3}$, $Y(\omega_3) = \frac{1}{3}$, and so on. Alternatively, since X is the number of infants who succeeded in solving the puzzle, $Y = X/3$. Thus, we can either calculate $Y(\omega)$ by inspection of the outcome ω (as above), or we can calculate $Y(\omega)$ directly from knowledge of $X(\omega)$. For example, since $X(\omega_7) = 1$, it follows that $Y(\omega_7) = X(\omega_7)/3 = \frac{1}{3}$.

Construction of the random variable Y in Example 1.2 above illustrates two important points: (i) *given one random variable X we can form any number of new random variables Y by means of mathematical operations upon* (the values of) X, and (ii) *if the random variable Y is known to be a function of the random variable X, $Y(\omega)$ can be calculated directly from knowledge of $X(\omega)$* without any need to refer to a more complete description of the outcome ω. For example, another random variable that the psychologist might have used to summarize his results is $Z = X - (3 - X) = 2X - 3$, the difference between the number, X, of infants who solved the puzzle and the number, $3 - X$, who did not. If we want to know the value $Z(\omega_6)$, we could either note that in $\omega_6 = $ (S,S,F) there are two S's and one F, so that $Z = 2 - 1 = 1$; or we could remember that $X(\omega_6) = 2$, and obtain $Z(\omega_6) = 2X(\omega_6) - 3 = 1$.

Notice that in Example 1.2 above there are 8 possible outcomes of the experiment (namely, $\omega_1, \omega_2, ..., \omega_8$), but that the random variable X takes on only 4 different values (namely, 0, 1, 2, or 3). This example illustrates a consequence of the quantization of the outcome of a random experiment by means of a random variable: *The total number of possibilities is decreased or kept the same* (but never increased) *when we confine attention only to that aspect of the outcome ω expressed by the value $X(\omega)$ of the random variable X.*

Suppose that we have already identified a probability measure P appropriate for the random experiment as originally defined in terms of the outcomes ω and the events E, which are collections of those outcomes. Once we have decided to summarize the random experiment only by the observed value of the random variable X, however, it seems overly cumbersome to keep the very detailed model based on the outcomes ω and the probability measure P. We look instead for a simpler model whose sample space (denoted, say, by $\mathscr{X}$) consists of all possible values x of the random variable X. The event sets of such a model are subcollections of the possible numerical values x (usually described by such mathematical statements as "the collection of all values x that are less than or equal to 5"). The probability measure P_X of such a model gives the probability of these event sets as calculated from the original probability measure P. This new probability model does not give us all of the detailed information about the random experiment obtainable from our original model; however, if we are only interested in the value of the random variable X and in no other aspect of the observed outcome of our random experiment, the information provided by the new model is sufficient for our purposes. Making use of Example 1.2 above, and recalling that the random variable X is the number of infants who successfully solved the puzzle, we illustrate how the new probability model for a random variable X can be derived from knowledge of the original probability model for the experiment.

Example 1.2 (Continued). Recall that the outcomes of this random experiment were $\omega_1 = (S,S,S)$, $\omega_2 = (F,S,S)$, $\omega_3 = (F,F,S)$, $\omega_4 = (F,F,F)$, $\omega_5 = (S,F,S)$, $\omega_6 = (S,S,F)$, $\omega_7 = (S,F,F)$, and $\omega_8 = (F,S,F)$, where, for example, (S,F,S) is a shorthand expression that means that the first and third infants solved the puzzle that was presented to them, while the second infant did not solve the puzzle. The different possible values of the random variable X (which tells us the number of infants who solved the puzzle) are 0, 1, 2, and 3. We denote these possible values by x_1, x_2, x_3, and x_4, respectively. Each of the possible values of X arises as a result of the occurrence of one or more of the outcomes ω. For example, $X(\omega) = 2$ for $\omega = \omega_2$, $\omega = \omega_5$, and $\omega = \omega_6$. The event "$X = 2$" is thus the same as the event $\{\omega_2, \omega_5, \omega_6\}$. Similarly, the event "$X = 0$" is the same as the event $\{\omega_4\}$, the event "$X = 1$" is the same as the event $\{\omega_3, \omega_7, \omega_8\}$, and the event "$X = 3$" is the same as the event $\{\omega_1\}$. In our new probability model based on the random variable X, the outcomes are the different possible values $x_1 = 0, x_2 = 1, x_3 = 2, x_4 = 3$, the sample space is $\mathscr{X} = \{0,1,2,3\}$, and we can determine the new probability measure P_X by finding the probabilities of the simple events $\{x_1\}, \{x_2\}, \{x_3\}, \{x_4\}$. (See Chapter 3, Section 1.) Suppose that the following probabilities were assigned to the simple events $\{\omega_1\}, \{\omega_2\}, ..., \{\omega_8\}$ in our original probability model:

Simple Event	$\{\omega_1\}$	$\{\omega_2\}$	$\{\omega_3\}$	$\{\omega_4\}$	$\{\omega_5\}$	$\{\omega_6\}$	$\{\omega_7\}$	$\{\omega_8\}$
Probability	p_1	p_2	p_3	p_4	p_5	p_6	p_7	p_8

For example, we might have $p_1 = p_2 = p_3 = p_4 = p_5 = p_6 = p_7 = p_8 = \frac{1}{8}$, the *uniform probability model*. To find the probabilities of the simple events $\{x_1\}$, $\{x_2\}$, $\{x_3\}$, $\{x_4\}$ in the new model, we simply add the probabilities of the simple events $\{\omega_1\}$, $\{\omega_2\}$, and so on, for which $X(\omega)$ takes on the appropriate value. Thus,

$$P_N\{x_2\} = P\{\omega: X(\omega) = x_2\} = P\{\omega_3\} + P\{\omega_7\} + P\{\omega_8\} = p_3 + p_7 + p_8,$$

which under the uniform probability model mentioned above would be $P_X\{x_2\} = \frac{3}{8}$. Our new probability model can be summarized in the following table:

Simple Event	$\{x_1\}$	$\{x_2\}$	$\{x_3\}$	$\{x_4\}$
Probability	p_4	$p_3 + p_7 + p_8$	$p_2 + p_5 + p_6$	p_1

or equivalently in the form

x	0	1	2	3
$P_X\{X = x\}$	p_4	$p_3 + p_7 + p_8$	$p_2 + p_5 + p_6$	p_1

Since we have agreed that we are only interested in that aspect of the original outcome ω which is given by the value $X(\omega)$ of the random variable X, we can now drop the old model from consideration and keep only the new model.

Example 1.3. A criminologist is investigating hand guns seized by the police in a large North American city as evidence in homicide cases. Since the overwhelming majority of these weapons are either Wessons, Mossbergs, or Colts, and either of 22, 38, or 45 caliber, the criminologist decides to limit the outcomes of his probability model (for the random experiment of observing a randomly chosen gun which has been used in a homicide) to the following list: $\omega_1 =$ Wesson 22 caliber, $\omega_2 =$ Wesson 38 caliber, $\omega_3 =$ Wesson 45 caliber, $\omega_4 =$ Mossberg 22 caliber, $\omega_5 =$ Mossberg 38 caliber, $\omega_6 =$ Mossberg 45 caliber, $\omega_7 =$ Colt 22 caliber, $\omega_8 =$ Colt 38 caliber, and $\omega_9 =$ Colt 45 caliber. Using the experimental method for forming the probability model of a random experiment (see Chapter 2, Section 5), the criminologist constructs the following table of the probabilities of simple events:

Simple Event	$\{\omega_1\}$	$\{\omega_2\}$	$\{\omega_3\}$	$\{\omega_4\}$	$\{\omega_5\}$	$\{\omega_6\}$	$\{\omega_7\}$	$\{\omega_8\}$	$\{\omega_9\}$
Probability	0.11	0.13	0.10	0.05	0.07	0.08	0.12	0.13	0.21

Thus, the probability that a hand gun used in a homicide in that large North American city is a Colt 45 caliber weapon is $P\{\omega_9\} = 0.21$.

The criminologist is asked to testify in a certain homicide case where the murder weapon was not found, but a bullet believed to have been fired from the murder weapon is of 22 caliber. Since the murder weapon was not found, he assumes (and we assume) that the make of the hand gun cannot be ascertained, and thus decides to reduce consideration of the outcomes of his random experiment to that quantitative aspect $X(\omega)$ of the outcome ω which is the caliber of the gun. All other aspects of the outcomes of his original probability model are irrelevant to the testimony he must give in this particular case. The possible values of the random variable X are then $x_1 = 22$, $x_2 = 38$, and $x_3 = 45$. Using his original table of probabilities, the criminologist calculates that

$$P_X\{X = x_1\} = P\{\text{Wesson 22, Mossberg 22, or Colt 22}\}$$
$$= P\{\omega_1\} + P\{\omega_4\} + P\{\omega_7\} = 0.11 + 0.05 + 0.12 = 0.28,$$
$$P_X\{X = x_2\} = P\{\omega_2\} + P\{\omega_5\} + P\{\omega_8\} = 0.13 + 0.07 + 0.13 = 0.33,$$
$$P_X\{X = x_3\} = P\{\omega_3\} + P\{\omega_6\} + P\{\omega_9\} = 0.10 + 0.08 + 0.21 = 0.39.$$

Hence, the criminologist records in his notebook only the new probability model summarized by the following table:

x	22	38	45
Probability	0.28	0.33	0.39

From this table, he can calculate the probabilities of any events about which he is likely to be asked during his testimony at the trial.

Remark: The method just described above for changing from a probability model based on possibly nonquantitative outcomes to a new model based on the values of a given random variable X is appropriate only for random variables that take on a denumerable (countable) number of values. If we are interested in a random variable that takes on a nondenumerable number of different values (for example, all numbers expressible in decimal form), the construction of a probability model for that random variable requires more advanced mathematical concepts. In such cases, it is usually easier to form the model starting from the beginning with numerical values as outcomes; indeed, this is usually what is done in practice.

2. DISCRETE AND CONTINUOUS RANDOM VARIABLES

For the most part, random variables fall into two categories: discrete and continuous. *Discrete random variables* are random variables that take on only "isolated" values. That is, if we mark the *possible values*

2. DISCRETE AND CONTINUOUS RANDOM VARIABLES

(values taken on with nonzero probability) on the real line, there are gaps between the marks (see Figure 2.1). Examples of discrete random variables include: (i) the number of traffic accidents over a given weekend, (ii) the number of children born to a given low-income family, (iii) the number of dots on the upper face of a tossed die, and (iv) the income of a randomly chosen farm family in 1965 (to the nearest dollar).

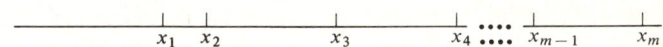

Figure 2.1 Values of a discrete random variable.

A *continuous random variable* is a random variable that (at least conceptually) can be measured to any desired degree of accuracy. That is, between every two possible values x_1 and x_2 of the random variable there is another possible value x_3 not equal to either x_1 or x_2. The collection of all possible values of a continuous random variable usually consists of one or more intervals of real numbers. Such collections are not denumerable and, as noted previously, require a more complicated mathematical theory for their analysis.

The concept of a continuous random variable is a mathematical abstraction. To illustrate the concept, imagine a man being weighed on a scale that can give his weight to whatever number of decimal places we care to specify. Obviously, such accuracy can never be achieved, but our conception (or model) of such accuracy turns out to provide us with great mathematical simplification in many real problems. Examples of such problems are measurements of the amount of oil obtained from an oil well, the amount of alcohol distilled in a given distillation process, a child's intelligence, the distance a car can travel on 1 gallon of gas, the blood pressure of a patient, and so on.

One important distinction between discrete and continuous variables may prove troublesome. For discrete random variables it is reasonable to consider the probability of the event that X equals some number x, and for some numbers x this probability is not zero; however, for continuous random variables, it is necessary to assume that for all possible values x, the probability of the event $\{X = x\}$ is zero.

How do we justify such a statement? Doesn't this mean that every event must have probability zero? First, recall that a probability of zero for an event does *not* mean that the event is impossible, only that it is very very rare. (See Chapter 2, Section 3.)

Next consider weighing a man on the perfectly accurate scale described above; we obtain the value x for him once, but the next time our measurement might differ by a very small amount from x (too small perhaps to pick up on most scales, but measurable on our perfectly accurate scale); the time after that we might differ from both previous values, and in a large number of weighings it is unlikely that we would ever obtain the same weight twice. This is because such an accurate scale would be sensitive to infinitesimal changes in the environment (including changes in the person weighed) and could potentially take on a nondenumerable (see Chapter 2, Section 2) infinity of possible values due to such changes. Out of such an infinity of values, no matter how many weighings we perform, we can only observe a finite number of values, so that if every one of those infinity of values has some chance of occurring at least once, we would not expect to see the same value twice, and, of course, there would be an infinity of possible values that we could not see at all. If we calculate, therefore, the relative frequency of any one value, that relative frequency would appear to be zero, or a very small number that would get smaller as we performed more and more weighings. On the other hand, if we ask for the man's weight accurate only to the first decimal place (such information describes an *interval* of possible values), we are likely to see certain readings fairly frequently (those readings that are very close to the person's real weight) and would assign these intervals of values positive probability. Thus, although individual values of a continuous random variable are assumed to have zero probability of occurring, *intervals* of possible values can have positive probability.

An analogy to geometry may help to further clarify this point. Consider a two-dimensional region as in Figure 2.2. The portion of the object between lines *a* and *b* has a certain area which is not zero. However, if line *a* is moved toward line *b*, the area between line *a* and line *b* decreases, becoming equal to zero when the two lines (lines *a* and *b*) meet and become one line. Since any given line can be considered as resulting from moving two distinct lines together (as above), all lines have area equal to zero. On the other hand, any figure (such as the portion of the object between lines *a* and *b* in Figure 2.2) with nonzero area is composed of lines "squeezed" together. Thus, any figure (interval) of nonzero area (probability) results from the composition of lines (values), each having zero area (zero probability).

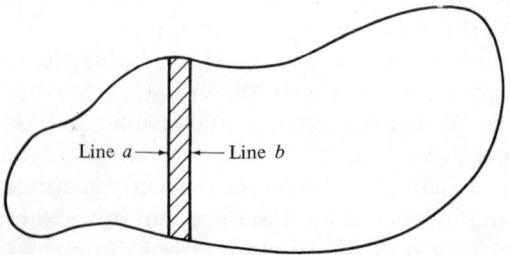

Figure 2.2 A two-dimensional object.

We emphasize once again: Although the assumptions connected with a continuous random variable may seem paradoxical and unreal, such models often provide a close approximation to observed reality. Indeed, because we tend to think of certain variables (such as length, weight, and time) as possessing a potential for infinitely refined measurement, continuous models frequently correspond more closely to our conception of measurement than our measuring ability enables us to produce practically. Even when we believe that a certain variable is discrete, if that variable has a large number of possible values, we may decide to think of the variable as being described by a continuous model, since such models are handled more easily mathematically than are alternative discrete models.

Because of the distinction that we have shown exists between discrete and continuous random variables, each of these types of model requires a different mathematical apparatus for its description and analysis. One descriptive tool common to both models is the notion of the cumulative distribution function. We discuss this concept in Section 3 and then turn in the following sections to the methods of analysis appropriate respectively to discrete and to continuous random variables.

3. CUMULATIVE DISTRIBUTION FUNCTIONS

Suppose that we have restricted ourselves to some single quantitative aspect of a given random experiment. Thus, we are interested in the probability model defined by a random variable X, the totality of whose possible values x is the sample space $\mathscr{X}$. Through empirical or theoretical constructions we have determined the probability measure P_X for X.

Example 3.1. A linguist is studying syllabication in the German language. He focuses his interest on that random variable X which is equal to the number of syllables in a German word. To construct a probability model for the random variable X, he first determines all possible values for X. Although theoretically a German word could exist with an arbitrarily large number of syllables, in practice a word of more than nine syllables is almost never seen. For reasons of convenience (so as not to make his model too cumbersome), the linguist decides to assume that X takes only the values 1, 2, 3, ..., 9. He now needs to find probabilities for the simple events $\{X = 1\}$, $\{X = 2\}$, ..., $\{X = 9\}$, and these he finds by the experimental method (Chapter 2, Section 5) using data taken from Viëtor-Meyer's *Deutches Aussprachewörterbuch* (P. Menzerath and W. Meyer-Eppler, *Die Architektonik des Deutschen Wortschatzes*, Dümmler,

Bonn, 1954). The probability model that he obtains for X is given in Table 3.1.

Table 3.1. X denotes the number of syllables of a randomly chosen German word. The probabilities given are based on 20,453 words taken from Viëtor-Meyer's *Deutches Aussprachewörterbuch*.

x	1	2	3	4	5	6	7	8	9
$P_X\{X=x\}$	0.110	0.313	0.341	0.178	0.045	0.010	0.002	0.001	0.000[a]

[a] If calculation had been carried out to five decimal places, $P_X\{X = 9\}$ would have been seen to equal a nonzero number. Since the values in Table 3.1 were rounded off to three decimal places, $P_X\{X = 9\}$ appears to be equal to zero.

From Table 3.1 we can determine (by means of the probability calculus) the probability of any event concerning the random variable X. Two events that may be of interest are the event "the number of syllables in a (randomly chosen) German word is greater than or equal to 2 and less than or equal to 5" and the event "a (randomly chosen) German word is polysyllabic (has 2 or more syllables)." Using Table 3.1, we can quickly find that the probability of the former event equals the probability that $X = 2, 3, 4$, or 5; this probability is thus equal to $0.313 + 0.341 + 0.178 + 0.045 = 0.877$. Similarly, the probability that a German word is polysyllabic can be found either by summing the probabilities that X equals 2, 3, 4, 5, 6, 7, 8, and 9, or, more simply, by subtracting from 1 the probability that $X = 1$ (see the Law of Complementation, Chapter 2, Section 4). Using the latter method of calculation, we find that the probability that a (randomly chosen) German word is polysyllabic is $1 - 0.110 = 0.890$.

Example 3.1 is an instance of a random variable that assumes a finite number of possible values. For every random variable of this type, a table similar to Table 3.1 can be constructed. Such a table allows us to calculate probabilities of events of the form "X lies between the numbers a and b $(a \leq X \leq b)$," of the form "X is greater than or equal to a $(a \leq X)$," or, indeed, of any form whatsoever. Hence, probability calculations for random variables that assume only a finite number of values are, at least in principle, directly and easily accomplished, once we have a table of the probabilities of simple events similar to Table 3.1.

If X is a discrete random variable with an infinite number of possible values, or if X is a continuous random variable, the problem of describing the probability model for X is not as obviously solvable as it is for discrete random variables with only a finite number of possible values. Although the construction of a table of probabilities for simple events is *conceptually* possible when X is a discrete random variable with an infinite number of possible values, in practice no such table could be given, since

3. CUMULATIVE DISTRIBUTION FUNCTIONS

such a table would have to have an infinite number of entries. The problem is even more complex when X is a continuous random variable, since for such a random variable the probabilities of events of the form "$X = x$" are all zero. Thus a table similar to Table 3.1 would be useless for calculating probabilities of more complicated events, even if such a table could be constructed. Hence, if X is not a discrete random variable with a finite number of possible values, some way other than a table of probabilities of simple events (such as Table 3.1) must be found to describe the probability model for X.

One way to describe the probability model of *any* random variable is the *cumulative distribution function*. The usefulness of the cumulative distribution function arises from the fact that the *probability measure of a random variable X can be completely specified once we know the probabilities assigned to all events of the form "X is less than or equal to x."* A rigorous proof of this fact requires advanced mathematical techniques. We demonstrate the validity of the above assertion in the case of discrete random variables having only a finite number of possible values. Before doing this, we need to introduce some notation.

Let us designate the event "the random variable X is less than or equal to the number τ" by the symbol $\{X \leq \tau\}$, and let

(3.1) $$F_X(\tau) = P_X\{X \leq \tau\}$$

be the probability of the event $\{X \leq \tau\}$. Here the symbol τ is the Greek letter tau.

For every number τ, $F_X(\tau)$ assigns to τ a number $F_X(\tau)$ which is the probability $P_X\{X \leq \tau\}$ of the event $\{X \leq \tau\}$. Hence, $F_X(\tau)$ is a *function* of τ; we call $F_X(\tau)$ the *cumulative distribution function* (abbreviated *c.d.f.*) of the random variable X. When we discuss more than one random variable at a time, say the random variables X, Y, and Z, we write the c.d.f. (cumulative distribution function) of X as $F_X(\tau)$ so as to distinguish it from the c.d.f.'s $F_Y(\tau)$ and $F_Z(\tau)$ of Y and Z, respectively. *Every random variable has its associated cumulative distribution function.* When it is clear from the context which random variable and cumulative distribution function we are considering, the subscript X on the c.d.f. $F_X(\tau)$ will often be omitted.

We now show that when the random variable X is a discrete random variable having only a finite number of possible values, then the c.d.f. $F_X(\tau)$ of X determines its probability measure. Suppose that the possible values of X are $x_1, x_2, \ldots, x_M$, and that the x's are ordered $x_1 < x_2 < \cdots < x_M$ (as in Figure 2.1). To determine the probability measure of X, we need only calculate the probabilities of the simple events $\{X = x_1\}$, $\{X = x_2\}$, ..., $\{X = x_{M-1}\}$, $\{X = x_M\}$; these probabilities can then be used, together with the probability calculus, to calculate the probabilities of all other events concerning X (Chapter 3, Section 1). Since

$$x_1 < x_2 < \cdots < x_{i-1} < x_i < x_{i+1} < \cdots < x_{M-1} < x_M,$$

it follows that the random variable X is less than or equal to the value x_i if and only if X takes on one of the values $x_1, x_2, \ldots, x_{i-1}, x_i$ (since each of these values is less than or equal to x_i, while all other possible values $x_{i+1}, x_{i+2}, \ldots, x_M$ of X are strictly greater than x_i). Consequently,

$$F_X(x_i) = P_X\{x_1, x_2, \ldots, x_{i-1}, x_i\} = P_X\{x_1\} + P_X\{x_2\} + \cdots + P_X\{x_{i-1}\} + P_X\{x_i\}.$$
(3.2)

By a similar argument,

(3.3) $\quad F_X(x_{i-1}) = P_X\{x_1, x_2, \ldots, x_{i-2}, x_{i-1}\} = P_X\{x_1\} + P_X\{x_2\} + \cdots + P_X\{x_{i-1}\}.$

Hence we find by subtraction that

(3.4) $$F_X(x_i) - F_X(x_{i-1}) = P_X\{x_i\}.$$

What we have shown here is that if X is a discrete random variable having only a finite number of possible values, the probability of the simple event $\{X = x_i\}$ for any possible value x_i of the random variable X can be found from the c.d.f. $F_X(x)$ of X by taking the difference between the value of the c.d.f. $F_X(x_i)$ at x_i and the value of the c.d.f. $F_X(x_{i-1})$ at that value x_{i-1} next smallest to x_i. Thus, from the c.d.f. of X, we can determine the probabilities $P_X\{X = x_i\}$ of all simple events. Since the probabilities of simple events determine the probability model of any random experiment (or random variable) having only a finite number of outcomes, knowledge of the c.d.f. $F_X(\tau)$ of the random variable X for all values of τ allows us to determine the probability model of X.

Of course, if X is a discrete random variable having only a finite number of possible values, and if we know the probabilities of the simple events $\{X = x_1\}, \{X = x_2\}, \ldots, \{X = x_M\}$ as given by the probability model for X, then the c.d.f. of X is easily obtained. For, if τ is any number, then either τ is strictly less than x_1, or $x_1 \leq \tau < x_2$, or $x_2 \leq \tau < x_3$, ..., or $x_{M-1} \leq \tau < x_M$, or τ is greater than x_M. If τ is strictly less than x_1, then the event $\{X \leq \tau\}$ is the null event $\emptyset$ (since X cannot be less than or equal to τ) and $F_X(\tau) = P_X\{X \leq \tau\} = P_X(\emptyset) = 0$. If $x_{i-1} \leq \tau < x_i$ for some i, $i = 1, 2, \ldots, M$, then

(3.5) $\quad F_X(\tau) = P_X\{X \leq \tau\} = P_X\{x_1, x_2, \ldots, x_{i-1}\}$
$\qquad\qquad = P\{x_1\} + P_X\{x_2\} + \cdots + P\{x_{i-1}\},$

since if X equals any of the possible values $x_1, x_2, \ldots, x_{i-1}$, then $X \leq x_{i-1} \leq \tau$, while if X equals any of the values $x_i, x_{i+1}, \ldots, x_M$, then $X \geq x_i > \tau$ (X is *not* less than or equal to τ). Finally, if τ is greater than x_M, then

$\{X \le \tau\} = \{x_1, x_2, \ldots, x_M\}$ which is the whole sample space $\mathcal{X}$, and thus $F_X(\tau) = 1$.

Example 3.1 (Continued). Consider the random experiment discussed earlier in which we were interested in the random variable X which is the number of syllables in a (randomly chosen) German word. The probability model for X has already been given in Table 3.1. There, for example, we see that $P_X\{X = 1\} = 0.110$, $P_X\{X = 2\} = 0.313$, and so on. Since the possible values for X are 1, 2, 3, 4, 5, 6, 7, 8, and 9, no outcome of X can be strictly less than 1. Thus, $F(-1) = 0$, $F(-\frac{1}{2}) = 0$, $F(0) = 0$, $F(\frac{1}{2}) = 0$, and so on. Indeed, for all numbers τ that are strictly less than 1, $F(\tau) = 0$. On the other hand, the event that X is less than or equal to 1 is the same as the event that $X = 1$ (since X cannot be strictly less than 1), and hence $F(1) = P_X\{X \le 1\} = P_X\{1\} = 0.110$. If τ is a number that is strictly less than 2 but greater than or equal to 1 (that is, $1 \le \tau < 2$), then the random variable X can be less than or equal to τ only when $X = 1$. Hence when $1 \le \tau < 2$, $\{X \le \tau\}$ is the same event as $\{X = 1\}$ and $F(\tau) = P_X\{1\} = 0.110$. Now, if $2 \le \tau < 3$, then X is less than or equal to τ only when $X = 1$ or when $X = 2$. It follows that when $2 \le \tau < 3$, the event $\{X \le \tau\}$ is the same as the event $\{X = 1 \text{ or } X = 2\}$, so that $F(\tau) = P_X\{1,2\} = P_X\{1\} + P_X\{2\} = 0.110 + 0.313 = 0.423$. Proceeding in a similar fashion, we can determine the mathematical form of the c.d.f. $F(\tau)$ of X. Perhaps the easiest way of displaying $F(\tau)$ is to plot the graph of $y = F(\tau)$ as shown in Figure 3.1.

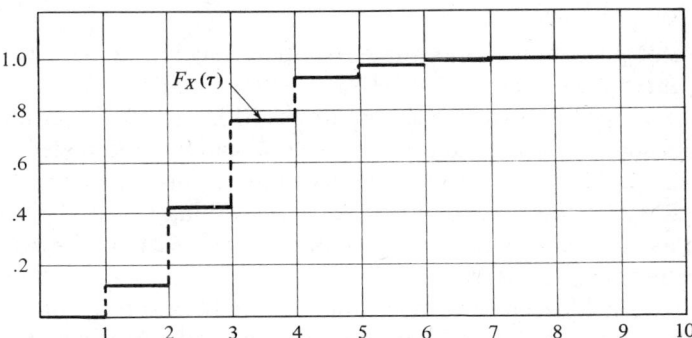

Figure 3.1 Graph of the c.d.f. $F_X(\tau)$ of the random variable X, which is the number of syllables in a randomly chosen German word.

Notice that the graph of $F(\tau)$ jumps at the numbers $\tau = 1, \tau = 2, \ldots, \tau = 9$. The total graph looks like a series of steps rising from left to right, and for this reason it is called a *step-function*. Notice also that the height of the jump of the graph at $\tau = 1$ is equal to $F(1) - F(0) = F(1)$, which is the same as $P_X\{X = 1\}$ [see Equation (3.4)]. Similarly, the height of the jump of the graph at $\tau = 2$ is $F(2) - F(1) = 0.313 = P_X\{X = 2\}$, the height of the jump of the graph at $\tau = 3$ is $F(3) - F(2) = P_X\{X = 3\}$, and so on. In general, *for any discrete random variable X, the cumulative distribution function of X is a step-function, and for any number τ at which the graph jumps, the height of the jump of the graph at τ is equal to $P_X\{X = \tau\}$.*

For a discrete random variable X which takes on only a finite number of possible values, we have, as noted before, a choice as to which representation for the probability model of X we wish to use: (i) a table of probabilities of simple events such as Table 3.1 or (ii) a graph of the c.d.f. $F(x)$ of X as in Figure 3.1. Either of these representations gives us all the information we need to know if we wish to calculate probabilities of events associated with the random variable X. Now suppose that X is a discrete random variable that can take on any one of an infinite number of possible values.

Example 3.2. An astronomer is interested in counting the number of separate radio emissions received during a given observation period from a newly discovered galaxy. Let X be the (random) number of radio emissions recorded. Note that X could take on any of the values 0, 1, 2, ... (that is, any nonnegative integer). The possible values 0, 1, 2, ..., have spaces between them if they are marked off as points on a line, and yet there are an infinite number of possible values for X. Hence, X is an example of a discrete random variable having an infinite number of possible values.

A table of probabilities of simple events for such a random variable X could in theory be constructed. However, as noted before, such a table would have an infinite number of entries. More information about the probability model for the random variable X can be obtained visually by means of a graph of the cumulative distribution function $F(\tau)$ of X (see Figure 3.2). It is still not possible to see everything about $F(\tau)$, since to plot it in its entirety would require a horizontal axis of infinite length, but we can infer much more from the shape of the graph of $F(\tau)$ than we can assimilate from looking at a long table of numbers. Thus, for a discrete random variable X having an infinite number of possible values, calculation of the cumulative distribution function of X is frequently a useful way of gaining information about the probability model for X.

Note that the graph of the c.d.f. of the discrete random variable X in Example 3.2 (see Figure 3.2) rises in steps just like the graph of the c.d.f. of the discrete random variable presented in Figure 3.1. The cumulative distribution function of a discrete random variable X is always a step-function regardless of the number of possible values that X may take on.

3. CUMULATIVE DISTRIBUTION FUNCTIONS

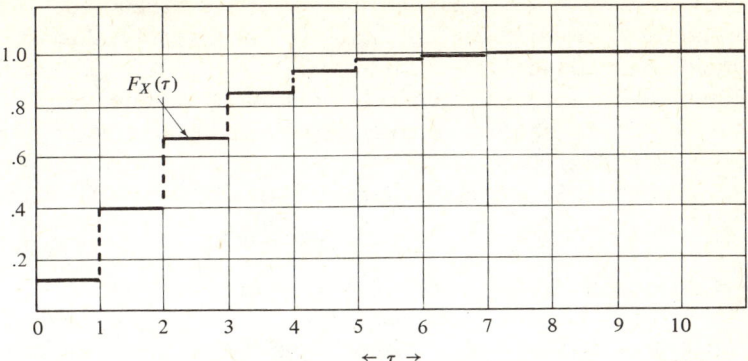

Figure 3.2 The cumulative distribution function for the number X of radio emissions from a newly discovered galaxy.

This assertion is a direct consequence of the fact, discussed above, that the probabilities for discrete random variables are "lumped" at individual numbers.

In contrast to the c.d.f.'s of discrete random variables, *the c.d.f.'s of continuous random variables are smooth* (see Figure 3.3), reflecting the fact that all events of the form "$X = x$" (for a given number x) are assigned zero probability by the probability model for a continuous random

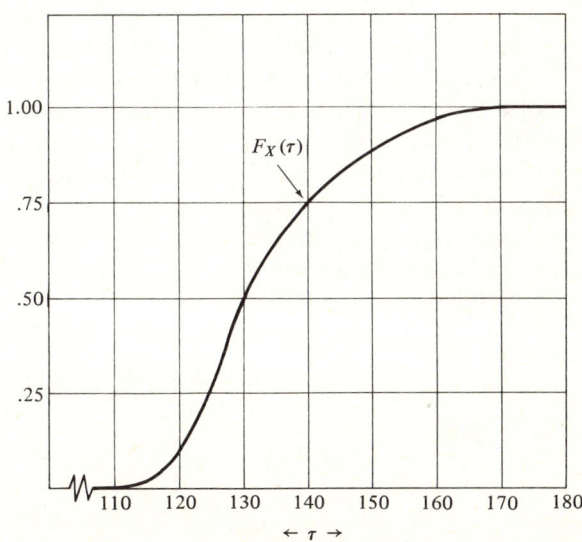

Figure 3.3 A cumulative distribution function of a continuous random variable X.

variable (see Section 2). Speaking intuitively, the probabilities of continuous random variables are not "lumped" at distinct values, but are "smoothed" over all numbers.

Whether they arise from discrete or continuous random variables, all c.d.f.'s have some properties in common. Note that if τ_1 and τ_2 are two numbers and $\tau_1 \leq \tau_2$, then the event $\{X \leq \tau_1\}$ is included in the event $\{X \leq \tau_2\}$. Thus, the Law of Inclusion (Rule 4.1 of Chapter 2) tells us that

$$P_X\{X \leq \tau_1\} \leq P_X\{X \leq \tau_2\}.$$

Because $F(\tau_1) = P_X\{X \leq \tau_1\}$ and $F(\tau_2) = P_X\{X \leq \tau_2\}$, it follows that $F(\tau_1) \leq F(\tau_2)$ when $\tau_1 \leq \tau_2$. Hence, *the c.d.f. $F(\tau)$ of a random variable is a nondecreasing function* [that is, as τ increases, $F(\tau)$ does not decrease but either increases or stays constant]. Further, if the random variable X has a largest value x_L, then for all $\tau \geq x_L$, the event "$X \leq \tau$" is the whole sample space $\mathscr{X}$, and $F(\tau) = P_X\{X \leq \tau\} = P_X(\mathscr{X}) = 1$. If X has a smallest value x_S, then the event "$X \leq \tau$" for any number $\tau < x_S$ is impossible and $F(\tau) = 0$. Even if X has neither a smallest nor a largest value, it is still the case that the graph of the cumulative distribution function $F(\tau)$ of a random variable X increases from 0 to 1 as we move from left to right with increasing τ (see Figures 3.1, 3.2, and 3.3).

We often need to find the probabilities of such events as: (i) the event that "X is strictly greater than the number a (that is, $\{a < X\}$)," (ii) the event that "X is strictly greater than the number a and less than or equal to the number b (that is, $\{a < X \leq b\}$)," (iii) the event "X is greater than or equal to the number a and less than or equal to the number b (that is, $\{a \leq X \leq b\}$)," and so forth. Although we know from the previous discussion that all such probabilities are determined by the c.d.f. $F(\tau)$ of X, we still need formulas that tell us how to calculate such probabilities from knowledge of the values of the c.d.f. $F(\tau)$.

Let us start with the first case [case (i)] mentioned above. How do we find $P_X\{a < X\}$, the probability that the random variable X is strictly greater than the number a, from a knowledge only of the values of the c.d.f. $F(\tau)$? Note first that the event $\{a < X\}$ is the complement of the event $\{X \leq a\}$. Hence, by the Law of Complementation (Rule 4.2 of Chapter 2), it follows that

$$P_X\{a < X\} = 1 - P_X\{X \leq a\}$$
$$= 1 - F(a).$$

To find the probability of the event $\{a < X \leq b\}$ using the values of $F(\tau)$, observe that the event $\{a < X \leq b\}$ is the intersection of the event $A = \{a < X\}$ and the event $B = \{X \leq b\}$. Since the Law of Addition [Rule 4.3(a) of Chapter 2] tells us that

$$P_X(A \cup B) = P_X(A) + P_X(B) - P_X(A \cap B),$$

it follows that

$$P_X\{a < X \leq b\} = P_X(A \cap B) = P_X(A) + P_X(B) - P_X(A \cup B).$$

But $P_X(A) = P_X\{a < X\} = 1 - F(a)$, $P_X(B) = P_X\{X \leq b\} = F(b)$, and $P_X(A \cup B) = P_X\{a < X \text{ or } X \leq b\} = P_X(\mathcal{X}) = 1$ (see Figure 3.4). Hence it follows that

$$P_X\{a < X \leq b\} = P_X(A) + P_X(B) - P_X(A \cup B)$$
$$= F(b) - F(a).$$

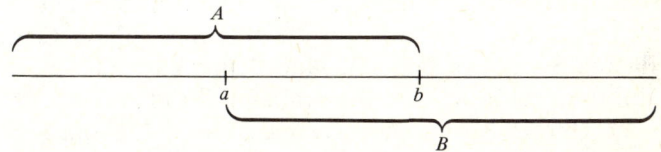

Figure 3.4 The event $\{a < X \leq b\}$ is represented by all points on the line between a and b except the point a. Note that $A \cup B = \mathcal{X}$.

Given the result that we have just obtained in case (ii), it is easy to obtain an answer for case (iii). Here, we want to calculate $P_X\{a \leq X \leq b\}$. Since the event $\{a \leq X \leq b\}$ is the union of the mutually exclusive events $\{a < X \leq b\}$ and $\{X = a\}$, and since $P_X\{a < x \leq b\} = F(b) - F(a)$, it follows from the special Law of Addition [Rule 4.3(b) of Chapter 2] that

$$P_X\{a \leq X \leq b\} = P_X\{a < X \leq b\} + P_X\{X = a\}$$
$$= F(b) - F(a) + P_X\{X = a\}.$$

In the case when X is a continuous random variable, $P_X\{X = a\} = 0$, and $P_X\{a \leq X \leq b\} = P_X\{a < X \leq b\} = F(b) - F(a)$. If X is a discrete

Table 3.2. Formulas for calculating probabilities of certain events given the values of the c.d.f. $F(\tau)$. Note that the above calculations assume that $a \leq b$.

Event	Formula for Probability of Event
$\{X = a\}$	Height of jump of graph of $F(\tau)$ at $\tau = a$
$\{a < X\}$	$1 - F(a)$
$\{a \leq X\}$	$1 - F(a) + P_X\{X = a\}$
$\{X \leq b\}$	$F(b)$
$\{X < b\}$	$F(b) - P_X\{X = b\}$
$\{a < X \leq b\}$	$F(b) - F(a)$
$\{a < X < b\}$	$F(b) - F(a) - P_X\{X = b\}$
$\{a \leq X \leq b\}$	$F(b) - F(a) + P_X\{X = a\}$
$\{a \leq X < b\}$	$F(b) - F(a) + P_X\{X = a\} - P_X\{X = b\}$

random variable, to find $P_X\{a \leq X \leq b\}$ we simply add to $F(b) - F(a)$ the height of the jump of the graph of $F(\tau)$ at the point $\tau = a$.

Table 3.2 gives formulas for calculating the probabilities of other events (intervals of numbers) of interest when we know only the values of the c.d.f. $F(\tau)$ of the random variable X.

Example 3.1 (Continued). Recall that in this example, the random variable X is the number of syllables in a randomly chosen German word. The c.d.f. $F(\tau)$ of X is graphed in Figure 3.1. Suppose that we want to know the probability that a randomly chosen German word has strictly more than one syllable (that is, we want to know the probability that a German word is polysyllabic). When we first introduced this example, we calculated this probability using a table of the probabilities of simple events of the form $\{X = x\}$. Here, we use Figure 3.1 and Table 3.2. From Table 3.2, we see that we need to find $1 - F(1)$. From the graph in Figure 3.1 $F(1)$ is equal to 0.110. Hence the probability that a German word is polysyllabic is 0.890. Similarly, using Table 3.2 and Figure 3.1, we find that

$$P_X\{X = 2\} = \text{height of jump of graph of } F(\tau) \text{ at } \tau = 2$$
$$= 0.423 - 0.110 = 0.313,$$

which is the value for $P_X\{X = 2\}$ appearing in Table 3.1. Also

$$P_X\{5 < X \leq 7\} = F(7) - F(5) = 0.999 - 0.987 = 0.012,$$
$$P_X\{5 \leq X \leq 7\} = 0.999 - 0.987 + 0.045 = 0.057.$$

The problem of calculating probabilities of other events of the form listed in Table 3.2 is handled in a similar fashion.

Example 3.2 (Continued). The astronomer would like to delegate to an assistant the work of counting radio emissions from the newly discovered galaxy. He does not want to be left ignorant of any unusual activity (or lack of activity) in the new galaxy. He tells the assistant to notify him if either 0 or more than 5 counts are recorded in a given observation period. Assuming that the cumulative distribution function shown in Figure 3.2 describes the distribution of the number X of radio emissions from the galaxy, the astronomer calculates the probability that during any given observation period his assistant will observe either 0 or more than 5 counts. Since the event $\{X = 0 \text{ or } X > 5\}$ is the complement of the event $\{0 < X \leq 5\}$, the astronomer uses Figure 3.2 to calculate

$$P_X\{0 < X \leq 5\} = F_X(5) - F_X(0) = 0.98 - 0.12 = 0.86,$$

and then subtracts this probability from 1. The probability that the astronomer will be notified is then

$$P_X\{X = 0 \text{ or } X > 5\} = 1 - P_X\{0 < X \leq 5\} = 1 - 0.86 = 0.14.$$

Estimating the Cumulative Distribution Function from Data

Suppose we make N independent trials of the random experiment that gives rise to a random variable X which is of interest to us. We do not know the probability measure that describes the random variation of X, and so we decide to use the experimental method for determining this probability measure (see Chapter 2, Section 5). Since the probability measure of a random variable X is completely specified once we know the values $F(\tau)$ of the cumulative distribution function of X for all numbers τ, we can confine our efforts to using our data to estimate the values of $F(\tau)$. The observations of X which result from the N trials of our random experiment are numbers that we can order according to size; thus, our observations are given as the numbers $t_1 \leq t_2 \leq t_3 \leq \cdots \leq t_N$. (Note that the subscripts $1, 2, \ldots, N$ on the t's refer to their order in terms of size, not to the order in which they were observed.) Since there are N trials, the relative frequency of events of the form $\{X \leq \tau\}$ equals

$$\text{r.f.}\{X \leq \tau\} = \frac{\text{number of } t\text{'s that are } \leq \tau}{N}.$$

For example, if we took $N = 10$ trials and obtained the numbers (arranged in order of size) $-3.1, -1.9, -0.8, 0.1, 0.3, 0.3, 0.9, 1.3, 1.3,$ and 1.7, then

$$\text{r.f.}\{X \leq 0.0\} = 3/10, \qquad \text{r.f.}\{X \leq 1.0\} = 7/10,$$

and so forth. We can graph r.f.$\{X \leq \tau\}$ against τ, obtaining a graph such as in Figure 3.5.

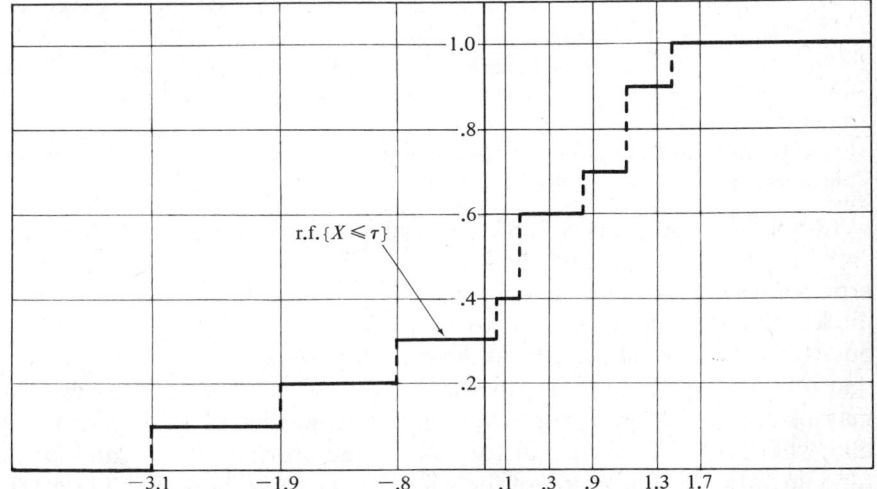

Figure 3.5 Sample or empirical cumulative distribution function obtained from $N = 10$ trials.

The function r.f.$\{X \leq \tau\}$ which assigns to each number τ the relative frequency of the event $\{X \leq \tau\}$ in our sample, and which is graphed in a fashion similar to that displayed in Figure 3.5, is called the *sample* (or *empirical*) *cumulative distribution function*. As the number of trials, N, of the random experiment gets larger and larger, the frequency interpretation of probability (see Chapter 1) leads us to expect the graph of r.f.$\{X \leq \tau\}$ to more and more closely resemble the theoretical c.d.f. $F_X\{\tau\}$. (See Figure 3.6.) Hence, if we have observed a large number of trials of the random experiment and have no compelling theoretical reason to expect the graph of the c.d.f. $F_X(\tau)$ of X to take a particular shape, a reasonable approximation to the c.d.f. $F_X(\tau)$ of X is the sample cumulative distribution function r.f.$\{X \leq \tau\}$.

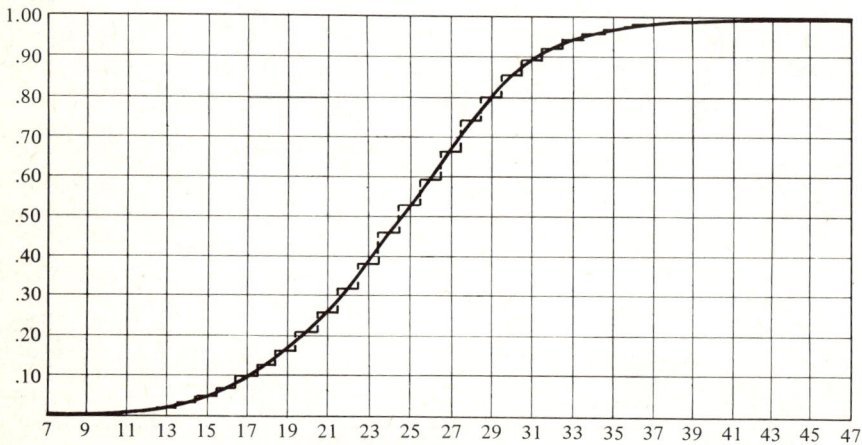

Figure 3.6 The smooth curve is the theoretical cumulative distribution function. The step curve is the sample or empirical cumulative distribution function for large N.

Mixed Discrete and Continuous Variables and Their Cumulative Distribution Functions

Although in most cases random variables are either continuous or discrete, occasionally we must deal with a random variable having the properties of both. For example, consider a random variable X which denotes the length of time that a person arriving at a bus terminal must wait for the bus. It may be reasonable to assume that there is a positive probability that the bus is already there; that is, $P_X\{X = 0\}$ is not zero. However, it may be appropriate to assume that any other waiting time is possible, so that whenever X is not equal to 0, it has the properties of a continuous random variable. The graph of the c.d.f. $F(\tau)$ of X is shown in Figure 3.7. Note that $F(\tau)$ jumps at $\tau = 0$, but rises smoothly thereafter.

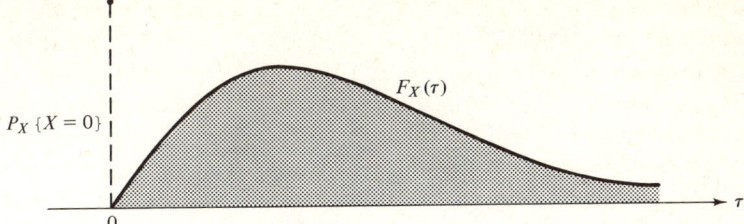

Figure 3.7(a) The mixed probability and density function of the length of waiting time for a bus. The area under the shaded part equals $1 - P_X\{X = 0\}$.

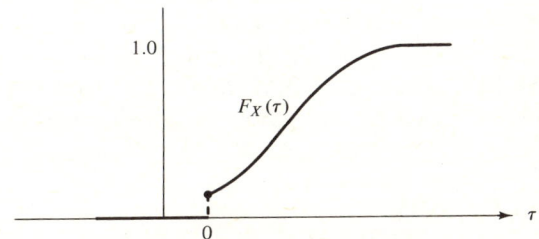

Figure 3.7(b) The mixed cumulative distribution function of the length of waiting time for a bus.

As another example, suppose that the random variable X denotes the amount of sales of a certain product by a manufacturer. Sales are limited by the amount a manufacturer can produce. It is possible that with non-zero probability the demand for the product exceeds the maximum amount b that can be produced so that all of b is sold, and $P_X\{X = b\}$ is positive. Similarly, there may be nonzero probability of zero demand, so that $P_X\{X = 0\}$ is also positive. On the other hand, all other numbers τ between 0 and b (but not equal to 0 or to b) may be possible values for X, but over these values the probability model for X is that of a continuous variable. The graph of the c.d.f. $F(\tau)$ of X is shown in Figure 3.8; $F(\tau)$ jumps twice, at $\tau = 0$ and at $\tau = b$, and otherwise rises smoothly.

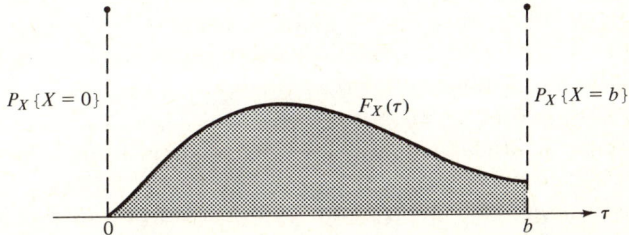

Figure 3.8(a) The mixed probability and density function of the amount of sales. The area under the shaded part equals $1 - P_X\{X = 0\} - P_X\{X = b\}$.

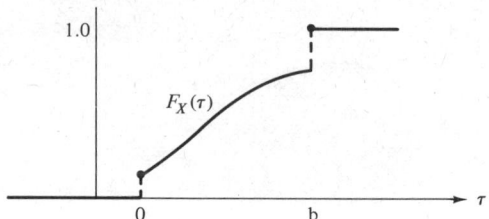

Figure 3.8(b) The mixed cumulative distribution function of the amount of sales.

In general, it is characteristic of the graphs of c.d.f.'s of mixed discrete and continuous random variables that they rise from left to right in an alternate series of jumps and smooth rises. The techniques of handling purely discrete and purely continuous variables must first be mastered before any attempt is made to deal with mixed discrete and continuous variables.

4. PROBABILITY MASS FUNCTIONS OF DISCRETE RANDOM VARIABLES

As we have said earlier, every random variable X has an associated cumulative distribution function $F(\tau)$ which contains the essential probabilistic information concerning X. However, the c.d.f. $F(\tau)$ is not the only function that contains such probabilistic information about X. Indeed, in certain cases other functional (or graphical) representations may be more directly interpretable. A useful function for describing the probability model of discrete random variables is the *probability mass function* (or *probability function*); for continuous random variables, corresponding information about the probability model for X is provided by the *probability density function* (or *density function*).

The *probability mass function* of a discrete random variable X conveys the same information about the probability model for X as does a table of probabilities of the simple events $\{X = x_1\}$, $\{X = x_2\}$, and so on, where $x_1, x_2, \ldots$ are the possible values (outcomes) of X. We denote the probability mass function of X by $p_X(\tau)$; when no confusion with probability mass functions of other random variables is likely to arise, we drop the subscript X and write $p(\tau)$. The probability mass function $p_X(\tau)$ of X assigns to every number the probability of the event $\{X = \tau\}$; that is,

$$p_X(\tau) = P_X\{X = \tau\}.$$

For every possible value $x_1, x_2, x_3, \ldots$ of X, the probability mass function

$p_X(\tau)$ evaluated at $\tau = x_1, x_2, \ldots$ is equal to a positive number; for all numbers τ that are not possible values of X, $p_X(\tau) = 0$. Thus, the probability mass function conveys all of the information provided by a table of probabilities of simple events (such as Table 3.1), but also conveys (somewhat redundantly) information concerning the probabilities of events $\{X = \tau\}$ corresponding to numbers τ that are not possible values of X. This extra information is really only used to graph the probability mass function $p_X(\tau)$, as in Figure 4.1.

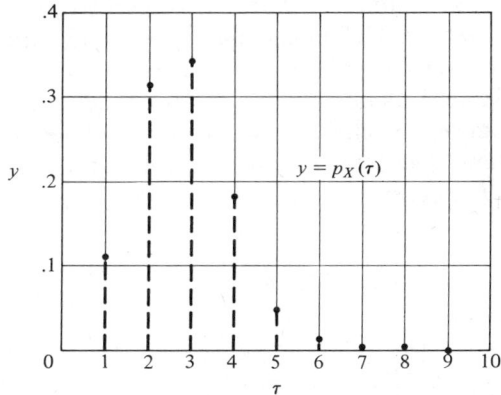

Figure 4.1 Graph of the probability mass function corresponding to the probability model for the random variable X in Example 3.1. (X is the number of syllables in a randomly chosen German word.)

Compare Table 3.1 and Figure 4.1. (Note again that although $P_X\{X = 9\}$ appears to be zero in Table 3.1, this is a consequence of the fact that we rounded off to three decimal places.) From Figure 4.1 we gain a visual impression of the information conveyed by Table 3.1; this impression allows us to obtain a more comprehensive picture of the probability model for the random variable X. Notice that the points which are raised over the horizontal axis in Figure 4.1 correspond to probabilities for the simple events $\{X = 1\}$, $\{X = 2\}$, ..., $\{X = 9\}$ of the random variable X. To display these points more clearly, and to indicate which value of τ is associated with each point, we have drawn dotted lines from each point down to (and perpendicular to) the horizontal axis. The height of each one of these dotted lines equals the probability of the simple event $\{X = x\}$, where $\tau = x$ is the point on the horizontal axis at which the dotted line meets the axis. Thus, at $\tau = 3$, we know that $p_X(3) = P_X\{X = 3\} = 0.341$, and this is the height of the dotted line which rises from the point $\tau = 3$ on the horizontal axis. Since the sum of the probabilities of all simple events must add to 1 (Chapter 3, Section 1), it follows that the sum of the heights of all dotted lines in Figure 4.1 is equal to 1.

The probability mass function and its graph may be thought of as a group of clumps, one clump placed at each of several points $x_1, x_2, x_3, \ldots$ on a line. The weight of the clump placed at x_j corresponds to the probability that X equals x_j; we represent the magnitude of this weight by the height of the line at x_j.

The graph of the probability mass function is useful as a way of quickly gaining a meaningful intuition about the probability characteristics of a random variable. For example, Figure 4.2 has only one line, which must, therefore, have height 1. This means that there is no variation in the random variable X having this probability mass function, and that every time this random variable X is observed, we can expect X to be equal to the number x_1. In Figure 4.3, each of the five dotted lines has equal height; this graph represents a Uniform Probability Model (Chapter 3, Section 1; also, see Chapter 6, Section 6).

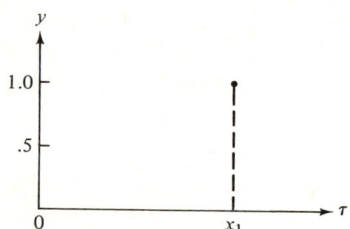

Figure 4.2 Graph of the probability function for a distribution concentrated at one point.

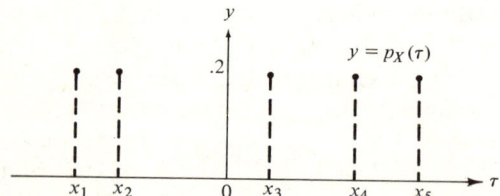

Figure 4.3 Graph of the probability function for a uniform probability model.

In Figure 4.4, the probabilities are shown to be clustered near the center. Here we expect some variation in repeated experiments; but values of X near zero (where the dotted lines are the tallest) will tend to occur most frequently.

4. PROBABILITY MASS FUNCTIONS   149

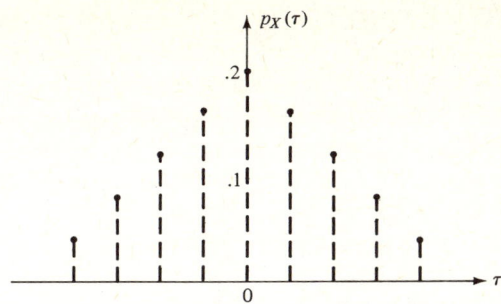

Figure 4.4 Graph of the probability function for a symmetric and clustered model.

Estimating a Probability Mass Function: Bar Graphs

If we make N independent trials of the random experiment which gives rise to the random variable X, and if we compute for every τ the observed relative frequency of X values that equal τ (that is, the number of times τ is observed divided by N), then this *sample* (or *empirical*) *probability mass function* can also be represented by a bar graph (as in Figures 4.2, 4.3, and 4.4). When N is large, the graph of the empirical probability mass function closely resembles the graph of $p_X(\tau)$. (This fact is a consequence of the relative frequency interpretation of probability theory.)

Example 4.1. To illustrate construction of a sample probability mass function, consider a study of literary style in which we tabulate the number of letters in each of 2000 words in Thackeray's *Vanity Fair*. The result of this tabulation is discussed by Williams (1956) in his statistical study of literary style. The tabulation itself follows:

Numbers of Letters in the Word	Frequency of Words Having That Number of Letters (Out of 2000)	Relative Frequency	Numbers of Letters in the Word	Frequency of Words Having That Number of Letters (Out of 2000)	Relative Frequency
1	58	0.029	8	100	0.050
2	315	0.1575	9	63	0.0315
3	480	0.240	10	43	0.0215
4	351	0.1755	11	16	0.008
5	244	0.122	12	15	0.0075
6	154	0.077	13	4	0.002
7	152	0.076	14	5	0.0025

If we let X equal the number of letters in a randomly chosen word, and for each of the above values of X we plot a dotted line whose height equals

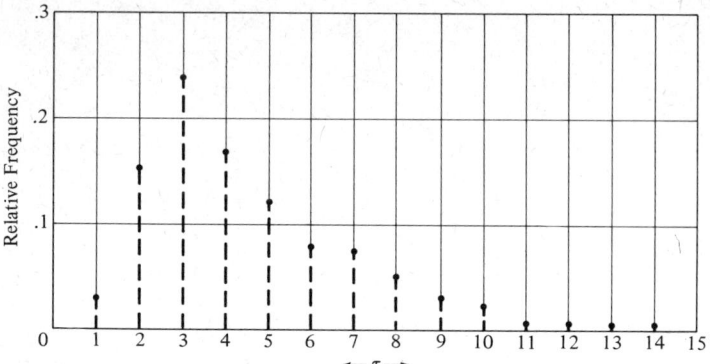

Figure 4.5 Frequency distribution of the number of letters in words from Thackeray's *Vanity Fair*.

the relative frequency of that value (of X), we obtain Figure 4.5. The height of the dotted line at each possible value is the observed relative frequency of that value, and is also approximately equal to the value of the probability mass function $p_X(\tau)$ evaluated at that value.

5. DENSITY FUNCTIONS

For continuous random variables the function analogous to the probability mass function (Section 4) is called the *density function*. To denote the density function of the random variable X, we use the symbol $f_X(\tau)$, or, when no confusion is possible, simply $f(\tau)$. The density function of the random variable X may be given a graphical representation as in Figure 5.1. Here the *area* under the graph between the numbers a and b represents the probability of the event $\{a \leq X \leq b\}$. Since the probability is 1

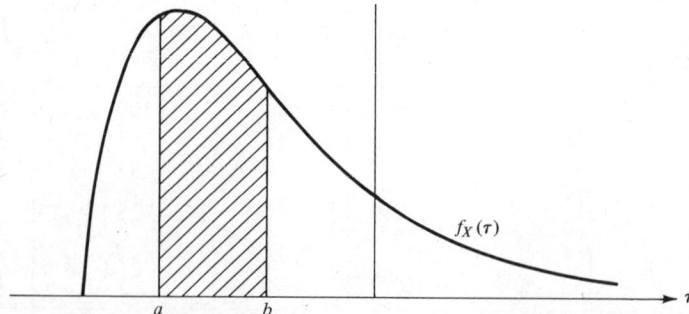

Figure 5.1 A density function. The area of the shaded part is equal to the probability that X lies between a and b.

that X is equal to some real number, it follows that *the area under the graph of $f(\tau)$ over the entire horizonal axis is always equal to 1*.

We can think of the shaded part of the graph in Figure 5.1 as being a section cut out of a uniformly thick sheet of metal of total weight equal to one unit. The value of $f_X(\tau)$ at the number τ is then the *density* of the metal sheet at τ. Note that the *mass* of the sheet at any point τ must be zero (the metal over that point is infinitesimally thin, and thus has no weight), but the mass over any line segment may not be zero. The probability of the event $\{a \leq X \leq b\}$ is then the weight of the sheet between a and b.

From the density function, certain qualitative conclusions concerning the variation of the random variable X over repeated independent and identical trials are evident. If the density function is nonzero over a certain line segment from x_1 to x_2 (see Figure 5.2) and zero elsewhere, then no values outside the range x_1 to x_2 may occur. If, in addition, the density function is constant over the line segment from x_1 to x_2, then all subintervals of equal length are equally likely (that is, the probability that X takes on a value in an interval I_1 is equal to the probability that X takes on a value in some other interval I_2 of the same length as I_1, provided both intervals I_1 and I_2 fall with the range from x_1 to x_2; see Figure 5.2).

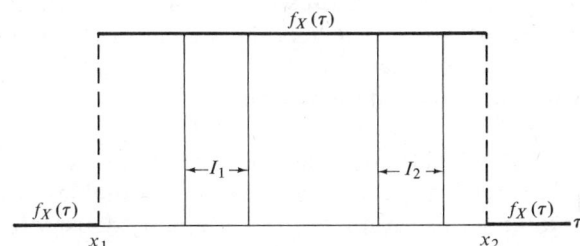

Figure 5.2 Constant density between the values x_1 and x_2 implies equal probability for the intervals I_1 and I_2 both of length L.

On the other hand, if the density function is such that most of the area beneath it is concentrated within a very narrow range (see Figure 5.3), then repeated experiments on X tend to yield values of the random variable X mostly within the range of numbers where the area is concentrated.

The area between a and b under the graph of a density $f(x)$ equals $P\{a < X < b\}$. Note that if a approaches b, the area between a and b approaches 0; this is still another way of visualizing the fact that the probability that X equals b is zero for any number b. It follows from this that when X is a continuous random variable, all the probabilities $P\{a \leq X \leq b\}$, $P\{a < X \leq b\}$, $P\{a \leq X < b\}$, and $P\{a < X < b\}$ are equal. This last result is, in general, not true for discrete random variables.

We emphasize that $f(\tau)$ is *not* to be interpreted as being equal to the

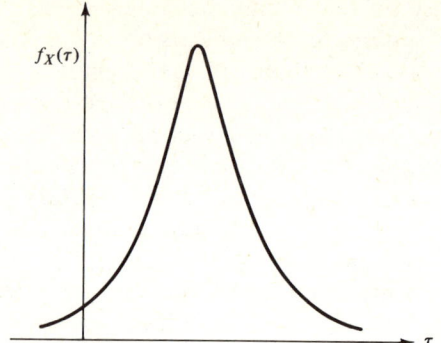

Figure 5.3 A highly centralized density function.

probability that X equals τ (as would be the case for discrete random variables and their probability mass functions), since for continuous random variables, the probability that $X = \tau$ (for any τ) is zero. Instead, probabilities are computed by finding areas under the density function.

Example 5.1. A delicatessen has 400 pounds of sandwich meat delivered every day. The owner of the delicatessen has hired a marketing specialist whose studies have shown that if X is the weight of the lunch meat sold in a morning (assumed measured *exactly*), then the probability model of the random variable X can be described by the density function $f(\tau)$ graphed in Figure 5.4.

The only possible weights for X are between 0 and 400 pounds (since 400 pounds is all the meat that the store has available). Note that the shape of the area under the graph of the probability density function $f_X(\tau)$ is that of a right triangle with base of length 400 and altitude of height $f_X(400)$. Since the total area under the graph of $f_X(\tau)$ must be equal to 1,

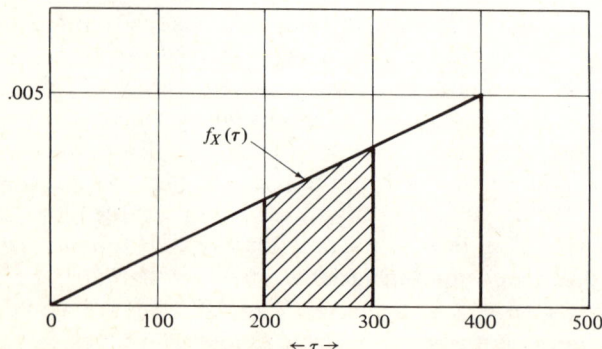

Figure 5.4 A density function that has constant growth.

and since the area of a triangle is half the product of the length of its base and the height of its altitude, the height $f_X(400)$ of the altitude of this triangle must equal $1/200 = 0.005$. Thus, $f_X(400) = 0.005$. If we want the probability that between 200 and 300 pounds of lunch meat are sold on a given morning, we must calculate the area of the shaded part of Figure 5.4. This may be determined by subtracting the area of the triangle whose base is the line from 0 to 200 from the area of the triangle whose base is the line from 0 to 300. That is,

$$P_X\{200 \leq X \leq 300\} = \tfrac{1}{2}(300-0)f_X(300) - \tfrac{1}{2}(200-0)f_X(200)$$
$$= 150f_X(300) - 100f_X(200),$$

since the height of the altitude of the triangle whose base is from 0 to 200 is $f_X(200)$, while the height of the altitude of the triangle whose base is from 0 to 300 is $f_X(300)$. But, using a result from analytic geometry concerning the sides of similar triangles, we know that

$$\frac{f_X(200)}{200} = \frac{f_X(300)}{300} = \frac{f_X(400)}{400}.$$

Since we have earlier calculated that $f_X(400) = 0.005$, it follows that $f_X(200) = (200/400)(0.005) = 0.0025$, $f_X(300) = (300/400)(0.005) = 0.00375$, and hence, $P\{200 \leq X \leq 300\} = 150(0.00375) - 100(0.0025) = 0.3125 = 5/16$.

*Rectangular Probability Approximations

The geometrical type of argument illustrated above in Example 5.1 for calculating the *exact* probability that a continuous random variable X lies in the interval of numbers between (and including) the numbers a and b (that is, $P_X\{a \leq X \leq b\}$) is rarely applicable since density functions do not usually come in nice triangular shapes. To obtain exact answers, it is generally necessary to make use of the results of the theory of integral calculus (which gives techniques and formulas for finding areas under curves). Such calculations can become impossibly complex or exhaustingly tedious. Thus when forming tables of probabilities, statisticians often employ various approximate numerical methods for finding areas under curves. These approximate methods, in the context of probability calculations, make use of the device of approximating the continuous variable X by a discrete random variable Y.

Suppose that our random experiment produces a continuous random variable X having the density function $f_X(\tau)$ shown in Figure 5.5. In this example, values that are strictly less than the number x_S (the smallest possible value of X) or strictly greater than the number x_L (the largest possible value of X) have zero density, so that $P_X\{x_S \leq X \leq x_L\} = 1$. Let

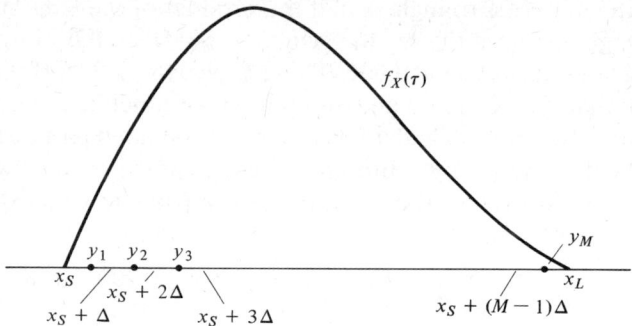

Figure 5.5 Density function $f_X(\tau)$ of a continuous random variable X.

us divide the interval of numbers between x_S and x_L into M subintervals of equal length Δ (see Figure 5.5). Then since the total length $x_L - x_S$ of the interval between x_S and x_L is equal to the sum of the lengths of the M subintervals, and since each subinterval has length Δ, it follows that $\Delta = (x_L - x_S)/M$. Now suppose that before we are given the results of the random experiment in which X is observed, a laboratory assistant rounds off the observed value of X as follows: If the value of X is between x_S and $x_S + \Delta$ (strictly speaking $x_S \leq X < x_S + \Delta$), the assistant rounds off X to the number $y_1 = x_S + (\frac{1}{2})\Delta$, midway between x_S and $x_S + \Delta$. Similarly, if X is between $x_S + \Delta$ and $x_S + 2\Delta$, the assistant rounds off X to the number $y_2 = x_S + (\frac{3}{2})\Delta$, midway between $x_S + \Delta$ and $x_S + 2\Delta$. In general, if the observed value of X is between $x_S + (i-1)\Delta$ and $x_S + i\Delta$, $i = 1, 2, \ldots, M$, the assistant rounds off X to the value $y_i = x_S + (i - \frac{1}{2})\Delta$. Thus, the information that the assistant relays to us is not the value of the continuous random variable X, but instead a value of the discrete random value Y which takes on one of the possible values $y_1, y_2, \ldots, y_m$. Note that Y is a function of X, since $Y = y_i$ if and only if $x_S + (i-1)\Delta \leq X < x_S + i\Delta$. Since the value of Y is always equal to the midpoint of the particular subinterval of numbers into which X falls, and since the length of this subinterval is Δ, it follows that Y and X never differ by more than $(\frac{1}{2})\Delta$. By making M, the number of subintervals, large, or equivalently, by making Δ small, we can make the continuous random variable X and the discrete random variable Y as close to one another as we wish, so that little information is lost by "rounding" in this fashion. (One may ask why we even consider continuous random variables if we can always round off to a discrete random variable Y arbitrarily close to a given continuous random variable X. To get close agreement between X and Y, we may have to take M, the number of possible values of Y, to be a very large integer. If M is very large, the probability model for Y becomes very cumbersome to work with. Further, we must still calculate $P_Y\{Y = y_i\}$ for each y_i.)

From the way we constructed the random variable Y, the probability that $Y = y_i$, $i = 1, 2, \ldots, M$, is the same as the probability of the event $\{x_S + (i-1)\Delta \leq X < x_S + i\Delta\}$. Thus, even though we have replaced the continuous variable X by the discrete random variable Y, it seems that we must still calculate areas under the graph of the probability density function of X in order to find the probability model for Y. Thus, it appears that we have replaced our original problem of calculating the area under the graph of the probability density function $f_X(\tau)$ of X by M similar problems!

However, if the graph of $f_X(\tau)$ is smooth (there are no jumps or jagged edges) and if Δ is small enough, then over each of the M subintervals that we constructed (Figure 5.5), $f_X(\tau)$ is very nearly flat, and we can approximate the area under the graph of $f_X(\tau)$ over each subinterval by the area of the rectangle whose base is the given subinterval and whose height is equal to the value of $f_X(\tau)$ at the midpoint of the subinterval (see Figure 5.6). Since the area of a rectangle is equal to the product of the lengths of

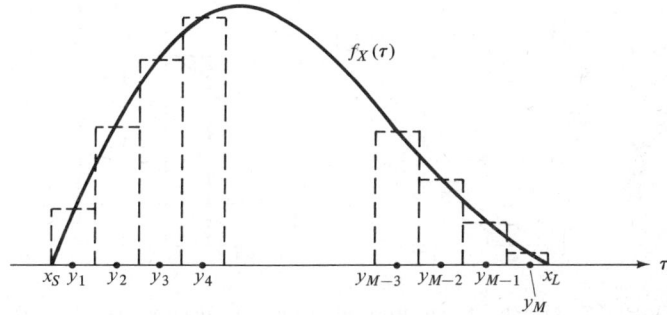

Figure 5.6 Graph of the probability density function, $f_X(\tau)$, of a continuous random variable, X, showing the method for approximating the area under the curve.

the base and altitude, the area of the rectangle constructed over the subinterval from $x_S + (i-1)\Delta$ to $x_S + i\Delta$ is equal to Δ times the value $f_X(x_S + (i - \frac{1}{2})\Delta)$ of the density function at the midpoint of the subinterval. The approximation we use to $P_Y\{Y = y_i\} = P_X\{x_S + (i-1)\Delta \leq X < x_S + i\Delta\}$ is therefore

(5.1) $$P_Y\{Y = y_i\} \cong \Delta f_X(x_S + \tfrac{1}{2}(i-1)\Delta),$$

where $\cong$ denotes approximate equality. Now, to approximate the probability $P_X\{a \leq X \leq b\}$ that X lies in the interval between the number a and the number b, we approximate X by Y, and then we find $P_Y\{a \leq Y \leq b\}$ by use of the probability calculus, using the approximation (5.1) instead of the exact values of $P_Y\{Y = y_i\}$ to make our calculations. This way of approximating $P_X\{a \leq X \leq b\}$ will give answers close to the exact value only when Δ is small enough. If Δ is not chosen to be small enough, unreasonable answers will often be obtained. For example, if Δ is not small enough, our approximation of $P_X\{x_S \leq X \leq x_L\}$ by

$$P_Y\{x_S \leq Y \leq x_L\} = P_Y\{Y = y_1\} + P_Y\{Y = y_2\} + \cdots + P\{Y = y_M\}$$
$$\cong \Delta f_X(x_S + (\tfrac{1}{2})\Delta) + \Delta f_X(x_S + (\tfrac{3}{2})\Delta) + \cdots + \Delta f_X(x_S + [M - \tfrac{1}{2}]\Delta)$$

may yield a value not equal to or even very close to the correct value of $P_X\{x_S \leq X \leq x_L\}$, which is 1. The more accurate we want our approximation of $P_X\{a \leq X \leq b\}$ to be, the smaller we must choose Δ, the larger we must choose M, and the more terms we must sum to find $P_Y\{a \leq Y \leq b\}$.

Example 5.1 (Continued). Recall that in Example 5.1 the random variable X of interest is the weight of lunch meat sold by a delicatessen in a morning. The probability density function $f_X(\tau)$ of X is shown in Figure 5.7. Note

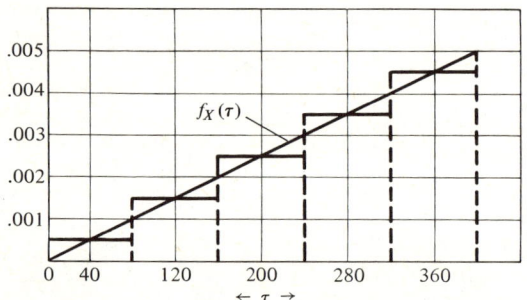

Figure 5.7 Graph of the probability density function, $f_X(\tau)$, of the weight, X, of luncheon meat sold in a given morning, showing rectangular approximation with $M = 5$ subintervals.

that $P_X\{0 \leq X \leq 400\} = 1$. We wish to find $P_X\{200 \leq X \leq 300\}$. Let us divide the interval between 0 and 400 into $M = 5$ subintervals. Then $\Delta = (x_L - x_S)/5 = 400/5 = 80$ and we obtain the division shown in Figure 5.7. Note that $y_1 = 40$, $y_2 = 120$, $y_3 = 200$, $y_4 = 280$, $y_5 = 360$, and that

$$P_Y\{200 \leq Y \leq 300\} = P_Y\{Y = y_3 = 200\} + P_Y\{Y = y_4 = 280\}$$
$$\cong \Delta f_X(200) + \Delta f_X(280).$$

We earlier showed that $f_X(200) = 0.0025$. Using the method of similar triangles (or simply reading the graph in Figure 5.7) we find that $f_X(280) = (280/400)(0.005) = 0.0035$. Thus,

$$P_X\{200 \leq X \leq 300\} \cong P_Y\{200 \leq Y \leq 300\} \cong 80(0.0025) + 80(0.0035)$$
$$= 0.480.$$

5. DENSITY FUNCTIONS 157

When we calculated $P_X\{200 \leq X \leq 300\}$ exactly, we found that $P_X\{200 \leq X \leq 300\} = 0.315$. In this case, our approximation greatly overestimates the exact value of the probability that we want. If we double M and make $M = 10$, then $\Delta = 40$, $y_1 = 20$, $y_2 = 60$, $y_3 = 100$, $y_4 = 140$, $y_5 = 180$, $y_6 = 220$, $y_7 = 260$, $y_8 = 300$, $y_9 = 340$, $y_{10} = 380$ (see Figure 5.8), and our approximation now is

$$\begin{aligned}
P_X\{200 \leq X \leq 300\} &= P_Y\{200 \leq Y \leq 300\} \\
&= P_Y\{Y = y_6 = 220\} + P_Y\{Y = y_7 = 260\} + P_Y\{Y = y_8 = 300\} \\
&\cong 40 f_X(220) + 40 f_X(260) + 40 f_X(300) \\
&= 40(0.00275 + 0.00325 + 0.00375) \\
&= 0.3900,
\end{aligned}$$

which is still an over-approximation, but closer to the exact answer. If we double M, the number of subintervals, again and make $M = 20$ (so that $\Delta = 20$), a repetition of the above method of calculation gives

$$\begin{aligned}
P_X\{200 \leq X \leq 300\} &\cong P_Y\{Y = 210, 230, 250, 270, 290\} \\
&\cong 20[f_X(210) + f_X(230) + \cdots + f_X(290)] \\
&= 0.3125,
\end{aligned}$$

which is the same as the exact probability of the event $\{200 \leq X \leq 300\}$ which we calculated previously; therefore, for this value of M (that is, $M = 20$), the approximation to $P_X\{200 \leq X \leq 300\}$ is precisely equal to the exact probability.

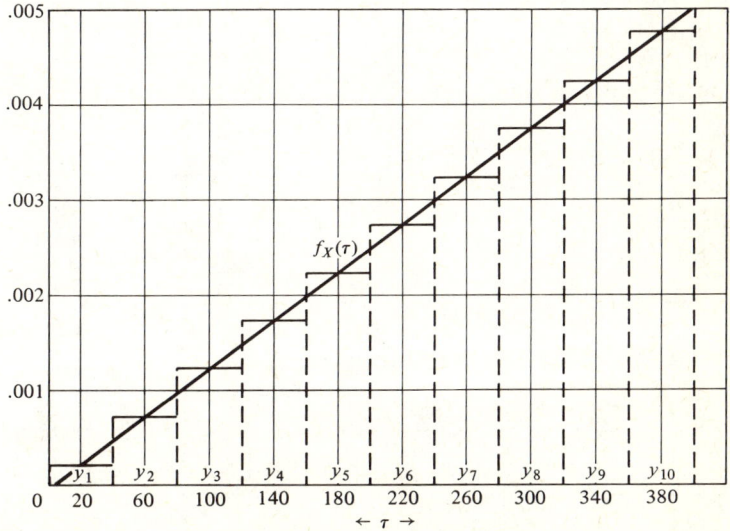

Figure 5.8 Graph of the probability density function, $f_X(\tau)$, of the weight, X, of luncheon meat sold in a given morning, showing rectangular approximation with $M = 10$ subintervals.

158 RANDOM VARIABLES 4

Note: When X is a continuous random variable having no smallest possible value x_S and/or no largest possible value x_L, the method of approximation to probabilities of the form $P_x\{a \leq X \leq b\}$ discussed above can still be made to work. Before constructing the discrete random variable Y to approximate X, we look for two numbers x_1 and x_2 such that $P_X\{x_1 \leq X \leq x_2\}$ is so close to 1 that the error made in saying that $P_X\{x_1 \leq x \leq x_2\}$ is approximately equal to 1 is very small. If we are interested in finding $P_X\{a \leq X \leq b\}$ for some two numbers a and b, then x_1 and x_2 should also be chosen so that $x_1 \leq a$ and $x_2 \geq b$. Once we have found numbers $x_1 \leq x_2$ that satisfy us, set $x_S = x_1$ and $x_L = x_2$, and repeat the approximation process for $P_X\{a \leq X \leq b\}$ that we have described.

*Example 5.2. The lifetime X (in playing hours) of a diamond phonograph needle produced by a certain company, Company C, is known to have the probability density function $f_X(\tau)$ which is described by the following equation:

$$f_X(\tau) = \begin{cases} (1/100)\, e^{-\tau/100}, & \text{if } \tau \geq 0, \\ 0, & \text{if } \tau \leq 0, \end{cases}$$

where e is a certain constant. The constant e is the base of the so-called natural logarithms, and is approximately equal to 2.718. However, for the present example it is not necessary to know the exact value of the constant e. The probability density function $f_X(\tau)$ is graphed in Figure 5.9. Some values of $f_X(\tau)$ for certain values of τ appear in a table directly to the right of Figure 5.9.

We might wish to determine the probability that if we buy a diamond

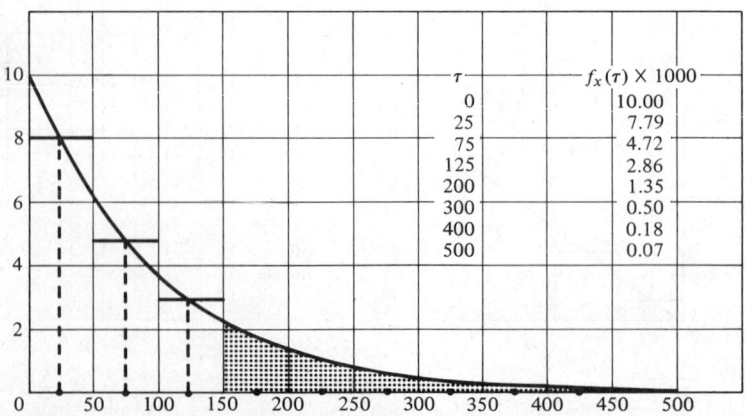

Figure 5.9 The probability density function for the lifetime of a diamond phonograph needle.

needle from Company C, it will last (strictly) more than 150 hours. In terms of the graph $f_X(\tau)$ in Figure 5.9, this probability is equal to the unshaded area to the right of $\tau = 150$ under the graph of $f_X(\tau)$. Alternatively, since we know that the total area under the graph of $f_X(\tau)$ is equal to 1, we can find the desired probability by calculating the area of the shaded region in Figure 5.9 and subtracting this area from 1. To approximate this area by our method (or equivalently to approximate $P_X\{0 \leq X \leq 150\}$), we must first choose a number x_L so large that $P_X\{0 \leq X \leq x_L\}$ is very close to 1. Since $P_X\{X \leq 0\} = 0$, $x_S = 0$. Looking at Figure 5.9, it appears that we might choose $x_L = 500$, since $f_X(500) = 0.00007$ and $f_X(\tau)$ continues to decrease for all $\tau \geq 500$, and since the area under the graph of $f_X(\tau)$ to the right of $\tau = 500$ appears to the naked eye to be nearly zero. (Neither of the above assertions guarantees that we have made a good choice, but as long as x_L is considerably larger than $\tau = 150$, a bad choice of x_L will not greatly affect our approximation of $P_X\{0 \leq X \leq 150\}$ if we choose the number of subintervals M in our method of approximation to be large.) As it happens, it can be shown that $P_X\{0 \leq X \leq 500\} = 0.993$, so that we have not made a bad choice at all. Now, let $M = 10$ and divide up the interval of numbers between 0 and 500 into 10 equal subintervals. Then $\Delta = (500-0)/10 = 50$, and $y_1 = 25$, $y_2 = 75$, $y_3 = 125$, $y_4 = 175$, $y_5 = 225$, and so forth. Also,

$$P_X\{0 \leq X \leq 150\} \cong P_Y\{0 \leq Y \leq 150\}$$
$$= P_Y\{Y = y_1 = 25\} + P_Y\{Y = y_2 = 75\} + P_Y\{Y = y_3 = 125\}$$
$$\cong 50 f_X(25) + 50 f_X(75) + 50 f_X(125) = 0.7685.$$

It can be shown, using the theory of integral calculus, that the exact value of $P_X\{0 \leq X \leq 150\}$ is about 0.777. Hence, even with $M = 10$ intervals, in this case we have obtained a pretty close approximation. Our approximate value for the probability that a diamond needle bought from Company C would last more than 150 playing hours is thus $1 - 0.769 = 0.231$, while the exact probability of this event is equal to $1 - 0.777 = 0.223$. Of course, we could have carried out a finer approximation by increasing M.

The method which we have discussed above for approximating the area, $P_X\{a \leq X \leq b\}$, under the graph of $f_X(\tau)$ between $\tau = a$ and $\tau = b$ is known (for obvious reasons, see Figures 5.6 to 5.9) as the *rectangular method* of approximation. This method is a very coarse method usually requiring a large amount of computation for even a moderate amount of accuracy in approximation. Better methods of approximation exist, but all of these methods agree in two ways: (i) Replacing areas under curves of irregular shape with areas that are easier to calculate, and (ii) replacing calculation of one entire area by the summation of a great number of smaller areas. In situations where probabilities are desired for a continuous random variable X which has a probability density function for which no tables have already been computed, consulting a numerical analyst before starting computation frequently can lead to large savings in effort and computation time.

We have not introduced the rectangular method of numerical approximation in order to recommend its use in calculating probabilities, but rather to illustrate that there are a great variety of discrete random variables Y which closely approximate a given continuous random variable X and whose probability models (particularly the probability mass functions) can give some insight into properties of the probability model of X. More use of this fact will be made in the next chapter (Chapter 5).

Empirical Distributions: Histograms

Just as we found empirical c.d.f.'s and empirical probability mass functions for discrete random variables, so we can also find empirical density functions (called histograms) for continuous random variables. There are several methods for constructing histograms and no general criteria of goodness seem to exist. However, some rules of thumb seem to have survived the test of time.

Perhaps the biggest problem involved in constructing histogram representations of probability densities is that we are trying to show probabilistic properties of an infinite number of possible values with only a finite amount of data. Our data appear to be discrete, but the random variable observed is continuous. To resolve this problem, we try to construct representations of probabilities for certain intervals; these intervals are chosen beforehand in such a way that, hopefully, we get some idea of the way probability is spread over the line. The choice of the intervals is a problem over which there is occasional disagreement—in part because there is no criterion that tells us which choice is best.

To discuss the construction of histograms let us consider a concrete example. We observe 417 40-watt, internally frosted incandescent lamps that are kept lit continuously in a life test [Davis (1952)]. We find that the lifetimes of the lamps vary between a minimum of 225 hours and a maximum of 1690 hours. If X is the random variable which equals the lifetime of such a lamp, we seek a sample representation of the probability density of X. Obviously, since X does not have a discrete distribution, we gain little information by plotting a bar graph for the observed values (because we do not learn anything about the infinite number of unobserved values). Thus, we decide to find the relative frequency of certain intervals of numbers; this tells us more about the density of X.

Before choosing such intervals, we must decide (1) how many intervals are needed, (2) how big these intervals are to be, and (3) where these intervals are to be placed. Statisticians usually agree that the intervals to

5. DENSITY FUNCTIONS

be chosen should be nonoverlapping, and, when this is feasible, should be of equal length. The intervals should also be chosen so that all of the intervals together include every one of the observed values.

Looking first at the problem of how many intervals to choose in our example, we note that we could choose one single interval from 225 to 1690 hours. All that we would learn from this choice, however, is that the probability of this interval is approximately 1. If we choose two equal intervals (one interval from 225 to 957; the other from 958 to 1690), we would find that a little over $\frac{1}{3}$ of the observed values fall in one interval, and a little more than $\frac{2}{3}$ of the observed values fall in the other interval. This is slightly more information than we obtained from using only one interval, but still very vague. The greater the number of intervals we choose, the more information we obtain about the probability of these intervals, as long as the number of intervals remains small relative to the number of observations. If we choose too many intervals, many of these intervals will contain few or no observations, and we will not know whether this is due to the fact that these intervals have low probability or to the fact that we do not have enough observations (with more observations, we might have seen observations in such intervals). Thus, to have a representative picture, we should have neither too few nor too many intervals. The rule of thumb advocated by many statisticians is that the number of intervals should be between 5 and 15. Another suggestion is that the number of intervals should be approximately twice the ratio of the range of values observed (in our case, the range of values is $1690 - 225 = 1465$) to the standard deviation (this term is defined in the next chapter; for our example, the standard deviation is 190). Thus, this second suggestion would advocate our using $2(1465)/190 = 15.4$ intervals. Since the number of intervals used must be an integer, this means that we should use either 15 or 16 intervals.

If we can decide where to start the first interval, and if our intervals are to be of equal length and include all of the observed data, then the answer to the problem of how big the intervals should be [problem (2) above] will be determined. We might decide to make the left endpoint of the first interval (that is, the interval farthest to the left) equal to the smallest observed value, and the right endpoint of the last interval (the one farthest to the right) equal to the largest observed value. Suppose we decide to use 15 equal intervals, then this choice would dictate that our interval lengths would be $(1690-225)/15$; that is, somewhere between 90 and 100 hours. However, this choice generally gives a misleading picture since our choice of intervals depends on the observed data in much too direct a fashion. The minimum and maximum values in a sample are usually highly variable from sample to sample; this variability tends to distort the representation. A more reasonable practice is to choose the left endpoint of the first interval small enough so that it is almost certain to be smaller than the smallest value that might be observed in any sample, and to choose the right endpoint of the last interval so that it is almost certain to be larger than the largest value that might be observed in any sample. For example, in our life testing problem the left endpoint of the first interval is chosen to be 200 hours and the right endpoint of the last interval is chosen to be 1700 hours (in part, this choice is also made so as to make the length of the intervals 100 hours). The data of the lifetime of the bulbs are now collected and summarized as follows:

Table 5.1. Lifetime of 417 40-watt bulbs.

Interval	*Frequency*
200 hours to 300 hours	1
300 hours to 400 hours	0
400 hours to 500 hours	0
500 hours to 600 hours	3
600 hours to 700 hours	10
700 hours to 800 hours	21
800 hours to 900 hours	45
900 hours to 1000 hours	91
1000 hours to 1100 hours	85
1100 hours to 1200 hours	80
1200 hours to 1300 hours	44
1300 hours to 1400 hours	23
1400 hours to 1500 hours	9
1500 hours to 1600 hours	3
1600 hours to 1700 hours	2
	417

From a frequency distribution such as this, it is a direct process to form the *histogram*. At the midpoint of each interval, a bar is constructed with height equal to the ratio of the relative frequency (frequency of the interval divided by the total number of observations) of the interval to the length of the interval. The relative frequency of the interval is now "spread" over the entire interval by constructing a rectangle over the interval with base equal to the length of the interval and height equal to the height of the bar (see Figure 5.10). The resulting construction is called a (relative frequency) histogram. The tops of the constructed rectangles form a curve; the area under this curve over any interval of values is equal approximately to the probability of the event corresponding to that interval of values.

5. DENSITY FUNCTIONS 163

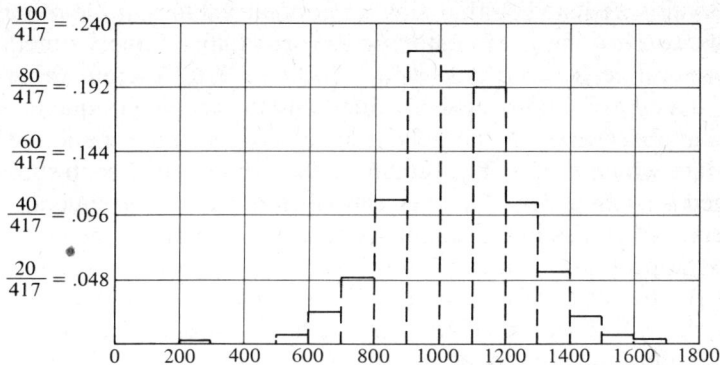

Figure 5.10 Relative frequency histogram of data on life testing.

For convenience, most statisticians construct *frequency histograms* where the frequency (and not the relative frequency) of an interval determines the height of the rectangle constructed over that interval. The areas under the resulting curve (see Figure 5.11) formed by the tops of the rectangles are all inflated by a constant factor (the number of observations) as compared to the corresponding areas for the relative frequency histogram, so that the same information is conveyed by either representation.

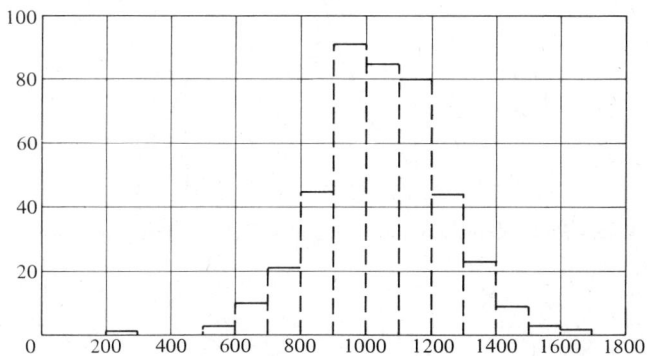

Figure 5.11 Frequency histogram of data on life testing.

It should be remarked that the construction of a histogram is, in a certain sense, the reverse of the rectangular method for estimating probabilities. In a rectangular method, a discrete random variable (and its probability mass function) is constructed to approximate a continuous random variable X. In this construction each subinterval of values for X is identified with a single number—the midpoint of the interval—and the total probability (or an approximation thereof by a rectangular area) of that subinterval is assigned to that single midpoint value. On the other hand, when we use histograms to estimate the probability density function of a random variable X, each observed value t_i of X (a discrete value) is assigned to a certain subinterval of values, and the relative frequency of that particular observation t_i (namely, $1/n$) is assigned to the subinterval of values into which t_i falls. The rectangle constructed over each subinterval is formed in such a way that the relative frequency (or frequency) of the subinterval of values in the sample is equal to the area of the rectangle. If it seems appropriate, we can then draw some appropriate smooth curve though the top of each rectangle of the histogram at the midpoint of the subinterval (see Figure 5.12). If this curve is carefully chosen, the area under the curve can be made to equal 1, and this curve can become our estimate of the probability density function of the observed continuous random variable X. Since many such curves can be drawn for a given histogram, this procedure is not always a good practice, but it does illustrate how the construction of a histogram in a sense reverses the method of construction used in the rectangular approximation method.

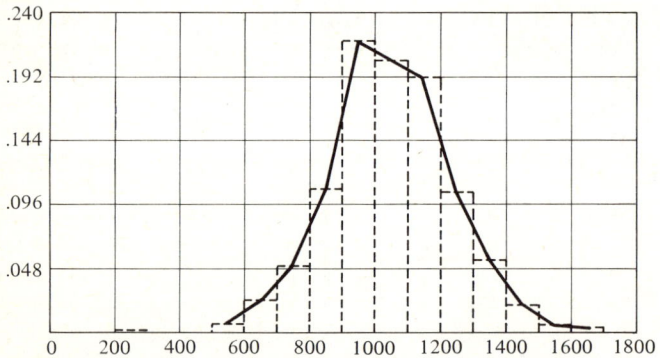

Figure 5.12 Representation of data as a relative frequency curve.

EXERCISES

1. At a certain hospital, the progress of 4 patients suffering from a certain disease is observed. After each patient has been in the hospital for 1 month, he is observed and a prognosis made as to whether he will recover from the disease or not. If the prognosis is favorable, an "R" (for recovery) is recorded for the patient; if the prognosis is unfavorable, an "N" (for nonrecovery) is recorded for the patient.
   (a) What is the sample space for this experiment? List all possible outcomes. How many outcomes are there?
   (b) The hospital assumes that the prognoses for the 4 patients are mutually statistically independent. Previous experience leads them to also as-

sume that for each patient the probability of a favorable prognosis is 0.75. For each simple event of the random experiment, find the corresponding probability.

(c) From the outcomes listed in part (a) and the probabilities obtained in part (b), find the probability model of the discrete random variable X, where X is the number of patients who receive a favorable prognosis. What are the possible values of X? What information would we lose if we change the sample space of the experiment from the sample space described in part (a) to a sample space in which the outcomes are the possible values of X?

A2. In the game "two-finger Morra," 2 players each show 1 or 2 fingers and simultaneously guess the number of fingers their opponent will show. If only 1 player guesses correctly, he wins an amount (in dollars) equal to the sum of the number of fingers shown by him and his opponent. If both players guess correctly, or if neither player guesses correctly, the game is a draw.

(a) Write all of the possible outcomes of one play of this game. Your outcomes should allow us to determine how many fingers each player holds up, and how many fingers each player guesses that his opponent will hold up.

(b) Let the random variable X equal the amount of money (in dollars) won by a specified player. What are the possible values of X? For each possible value x of X indicate what outcomes belong to the event $\{X = x\}$.

(c) Suppose that each player acts independently of the other players, and that each player makes his choice of the number of fingers he will hold up and the number of fingers he guesses his opponent will hold up by a random mechanism that assigns equal probability to each of his possible alternatives. Find the resulting probability model for the random variable X.

(d) Suppose that each player acts independently of the other player, but that both players hold up the *same* number of fingers that they guess their opponent will hold up. If each player guesses at random the number of fingers his opponent will hold up, find the resulting probability model for X.

3. In the context of Example 1.2, find the probability model for the number X of infants who solved the puzzle if $p_1 = 0.10$, $p_2 = 0.31$, $p_3 = 0.22$, $p_4 = 0.05$, $p_5 = 0.07$, $p_6 = 0.15$, $p_7 = 0.03$, and $p_8 = 0.07$. (Recall that $p_i = P\{\omega_i\}$.) Find the probability model for the difference Z between the number of infants who solved the puzzle and the number of infants who did not solve the puzzle.

A4. Which of the following random variables are discrete?
(a) The time required to answer this question.
(b) The height of mercury in a barometer at this moment.
(c) The number of "heads" in 6 flips of a coin.
(d) The highest temperature of the air during the daylight hours today.

A5. Which of the following random variables are continuous?
   (a) The volume of air breathed in by an individual when that individual is asked to "take a deep breath."
   (b) The annual income to the nearest dollar of a randomly chosen wage-earner.
   (c) The population of a randomly chosen town in the United States.
   (d) The population density (number of individuals per square mile) of a randomly chosen census tract in New York City.

6. In Exercise 12 of Chapter 3, we described a probability model for the number X of floods in a wet season. This probability model is

Simple Event	$\{X=0\}$	$\{X=1\}$	$\{X=2\}$	$\{X=3\}$	$\{X=4\}$	$\{X=5\}$	$\{X=6\}$
Probability	0.2466	0.3452	0.2417	0.1127	0.0395	0.0110	0.0033

   (a) Is X a discrete random variable or a continuous random variable?
   (b) Find the cumulative distribution function $F_X(\tau)$ of X and graph this function. What is the shape of the graph of $F_X(\tau)$? [Note: In graphing $F_X(\tau)$, you may round off the values of $F_X(\tau)$ to two decimal places.]

7. (a) Determine and graph the cumulative distribution function $F_X(\tau)$ of the random variable X described in Exercise 1.
   (b) Determine and graph the cumulative distribution function $F_Z(\tau)$ for the random variable Z described in Example 1.2 when the probabilities $p_1, p_2, \ldots, p_8$ are as described in Exercise 3.

A8. The following is the cumulative distribution function $F_X(\tau)$ of a random variable X:

$$F_X(\tau) = \begin{cases} 0.0, & \text{if } \tau < 8.0, \\ 0.1, & \text{if } 8.0 \leq \tau < 9.0, \\ 0.4, & \text{if } 9.0 \leq \tau < 10.0, \\ 0.7, & \text{if } 10.0 \leq \tau < 11.0, \\ 0.9, & \text{if } 11.0 \leq \tau < 12.5, \\ 1.0, & \text{if } 12.5 \leq \tau. \end{cases}$$

   (a) Is X a continuous random variable or a discrete random variable?
   (b) What are the possible values of X?
   (c) What is the probability model for X?
   (d) Using the values of the cumulative distribution function find the probabilities of the following events:
      (i) $\{X \leq 9.5\}$,
      (ii) $\{X > 10.8\}$,
      (iii) $\{9.0 < X \leq 11.5\}$,
      (iv) $\{9.0 \leq X \leq 11.5\}$.
   (e) Graph $F_X(\tau)$.

9. The cumulative distribution function $F_X(\tau)$ of a random variable X has the form

$$F_X(\tau) = \begin{cases} 0, & \text{if } \tau < 0, \\ \tau, & \text{if } 0 \leq \tau \leq 1, \\ 1, & \text{if } \tau > 1. \end{cases}$$

(a) Is X a discrete random variable or a continuous random variable?
(b) What are the possible values of X?
(c) Using the values of $F_X(\tau)$, find $P_X\{\frac{1}{4} \leq X \leq \frac{3}{4}\}$.
(d) Using the values of $F_X(\tau)$, find $P_X\{X > \frac{1}{2}\}$.
(e) Graph $F_X(\tau)$.

^10. The cumulative distribution function $F_Y(\tau)$ of a random variable Y has the form

$$F_Y(\tau) = \begin{cases} 0, & \text{if } \tau < 0, \\ \tau^2, & \text{if } 0 \leq \tau \leq 1, \\ 1, & \text{if } \tau > 1. \end{cases}$$

(a) Is Y a discrete random variable or a continuous random variable?
(b) What are the possible values of Y?
(c) Using the values of $F_Y(\tau)$, find $P_Y\{\frac{1}{4} \leq Y \leq \frac{3}{4}\}$.
(d) Using the values of $F_Y(\tau)$, find $P_Y\{Y > \frac{1}{2}\}$.
(e) Graph $F_Y(\tau)$ and compare this graph to the graph of the cumulative distribution function $F_X(\tau)$ described in Exercise 9.
(f) On the basis of the comparison that you made in part (e), which probability do you think is the larger, $P_X\{X \leq \tau\}$ or $P_Y\{Y \leq \tau\}$? Answer for $\tau = -1.0$, 0.0, 0.2, 0.4, 0.6, 0.8, 1.0, and 2.0. Does your answer depend on the value of τ? Explain.

^11. The cumulative distribution function $F_W(\tau)$ of a random variable W has the form

$$F_W(\tau) = \begin{cases} 0, & \text{if } \tau < 0, \\ \tau, & \text{if } 0 \leq \tau < \frac{1}{2}, \\ \frac{1}{2}\tau + \frac{1}{2}, & \text{if } \frac{1}{2} \leq \tau \leq 1, \\ 1, & \text{if } \tau > 1. \end{cases}$$

(a) What kind of random variable is W? Graph the cumulative distribution function $F_W(\tau)$. What shape does it have?
(b) What are the possible values of W?
(c) Find $P_W\{W = \frac{1}{2}\}$, $P_W\{W > \frac{1}{2}\}$, $P_W\{W < \frac{1}{4}\}$, and $P_W\{\frac{1}{3} \leq W < \frac{2}{3}\}$.

12. Use the data on the annual family income X of "young married" members of a consumers union described in Exercise 4 of Chapter 1 to determine and graph the sample (empirical) cumulative distribution function of X.

^13. Use the data on the length X of a wall in a given room described in Exercise 7 of Chapter 1 to determine and graph the sample cumulative distribution function of X.

14. Suppose that you obtain data on a continuous random variable X from another researcher (or from an organization). Quite frequently you will obtain these data in *grouped* form (for example, in the form of an empirical density function, or in the form of a histogram). The following is a frequency tabulation of 370 scores X on a portion of the College Entrance Board Advanced Placement Examination in English Composition. Each interval is assumed to include its right end point, but not its left end point. Thus, 9 observations were observed which had values strictly greater than 5.5, but less than or equal to 7.5.

Interval of Values of X	Observed Frequency
1.5 to 3.5	1
3.5 to 5.5	1
5.5 to 7.5	9
7.5 to 9.5	24
9.5 to 11.5	49
11.5 to 13.5	67
13.5 to 15.5	51
15.5 to 17.5	49
17.5 to 19.5	40
19.5 to 21.5	44
21.5 to 23.5	19
23.5 to 25.5	11
25.5 to 27.5	5
	370

Note that because only the frequencies of *intervals* of values are given, we cannot determine what the original observations were, and hence cannot say what the value of r.f.$\{X \leq \tau\}$ is for *every* number τ. However, we can evaluate r.f.$\{X \leq \tau\}$ when τ is any one of the right end points of the intervals of values reported for X. Thus, for example,

$$\text{r.f.}\{X \leq 1.5\} = 0,$$
$$\text{r.f.}\{X \leq 3.5\} = \frac{1}{370} = 0.0027,$$
$$\text{r.f.}\{X \leq 5.5\} = \frac{1+1}{370} = \frac{2}{370} = 0.0054;$$

and, in general, when τ is the right end point of an interval,

$$\text{r.f.}\{X \leq \tau\} = \frac{\text{sum of observed frequencies of all intervals to the left of } \tau}{\text{number of observations}}.$$

(a) Compute r.f.$\{X \leq \tau\}$ from the above data for every value τ which is a right end point of one of the reported intervals. That is, find r.f.$\{X \leq \tau\}$ for $\tau = 3.5, 5.5, 7.5, 9.5, 11.5, \ldots, 23.5, 25.5,$ and 27.5.

(b) To construct a sample cumulative distribution from the grouped data shown above, at least 3 strategies are possible:

(i) Assume that all of the observations in each interval are concentrated at the right end point of that interval. In this case, a graph of the sample cumulative distribution would resemble a series of rising steps with jumps only at the right end points of the intervals.

(ii) Assume that all of the observations in each interval are concentrated at the midpoint of that interval. In this case, a graph of the sample cumulative distribution function would resemble a series of rising steps with jumps only at the midpoints of the intervals.

(iii) Assume that the data in each interval are equally spaced within the interval. In this case, a graph of the sample cumulative distribution function would again be a step function, but each of the jumps would be of the same height. An interval with many observations would have many jumps of the graph within the interval.

Using the given data on the English Composition scores X, graph a sample cumulative distribution for X using each of the strategies (i), (ii), and (iii) described above. Are the graphs similar? For each such graph, verify whether or not the graph correctly indicates the value of r.f.$\{X \leq \tau\}$ when τ is a right end point of an interval (that is, $\tau = 3.5$, 5.5, and so on). Which one of the strategies do you prefer? Why?

[*Note:* In most of the situations in this book in which we will need to construct a sample cumulative distribution function, we will be comparing this sample cumulative distribution function to a theoretical cumulative distribution function derived from some hypothesized probability model. In such cases, greatest interest will center on whether the values of the theoretical cumulative distribution function at the right end points of the intervals agree closely with the values of r.f.$\{X \leq \tau\}$ calculated for these end points, since the agreement of the sample and theoretical cumulative distribution functions at other values of τ will depend on which strategy is used to calculate the sample cumulative distribution function. For this reason, when making comparisons between sample and theoretical cumulative distribution functions in the remainder of this book, if the data are given in grouped form, *we will graph only the values of r.f.$\{X \leq \tau\}$ at the right end points of the grouping intervals*, and will not graph the entire cumulative distribution function.]

15. Consider the data given in Table 5.1 of Chapter 4 on the lifetimes X of incandescent lamps which are kept lit continuously in a life test.
    (a) Find r.f.$\{X \leq \tau\}$ for the right end points $\tau = 300, 400, 500, ..., 1700$ of the grouping intervals in the table. Note that r.f.$\{X \leq 200\} = 0.0$.
    (b) Use each of the 3 strategies (i), (ii), and (iii) described in Exercise 14 to construct and graph a sample cumulative distribution function for X. Do the 3 graphs closely resemble one another? For what values of τ are values of all 3 sample cumulative distribution functions the same? Explain.

A16. In a large urban shopping mall, the stores are arranged as shown in the figure below. Each store contains an underground garage. If a shopper wants to remain indoors, he can pass from store to store (and garage to garage) by the routes shown in the figure. The distances between adjacent stores are the same; let the distance between any two adjacent stores be the unit of distance. Thus, store A and store B are 1 unit apart.

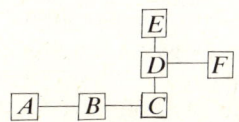

(a) On a given shopping day, a shopper parks at the garage under a randomly chosen store (that is, the garage under each store has an equal probability of being chosen). The shopper finishes shopping at another randomly chosen store. The choice of the store under which to park and the choice of the store at which the shopper finishes shopping are statistically independent. When the shopper finishes shopping, it is raining; thus, the shopper elects to reach his or her car via the routes shown in the figure. Let X denote the number of units of distance that the shopper has to walk to reach his or her car. Find the probability mass function of X.

(b) What is the probability that the shopper must walk more than 1 unit of distance to reach his or her car?

(c) Graph the probability mass function and cumulative distribution function of X.

(d) Suppose that on the given shopping day, the parking garages C and D were closed for repair and resurfacing, so that the shopper could park only in the garages under stores A, B, E, and F. Under the assumptions of random choice of parking garage and of random choice of the store at which shopping ceases made in part (a), find the probability mass function of the number Y of units of distance that the shopper has to walk to reach his or her car. Graph the probability mass function and cumulative distribution function of Y. Would you say that the probability distributions of X and of Y are quite different? Explain your answer.

*A 17. Consider the discrete random variable X which has possible values 0, 1, 2, 3, 4, 5, 6, 7, 8, 9, and for which each of the events $\{X=1\}$, $\{X=2\}$, $\{X=3\}$, ..., $\{X=8\}$, $\{X=9\}$ has the same probability, 1/10.

(a) Let Y be the remainder obtained after dividing X^2 by 10. For example, if $X=9$, then $X^2=81$, and when we divide 81 by 10, we obtain the remainder $Y=1$. Similarly, if $X=7$, then $X^2=49$ and $Y=9$. Using this definition of the random variable Y, find the probability mass function of Y.

(b) Let Z be the remainder obtained after dividing Y^2 by 10. Find the probability mass function of Z.

(c) Let W be the remainder obtained after dividing Z^2 by 10. Find the probability mass function of W. Compare the probability mass functions of Z and of W. Are they the same?

(d) If, starting with the random variable W, we continue to define new random variables in terms of old random variables by the process described in parts (a), (b), and (c), do we ever obtain a probability mass function different from the probability mass function of W? Explain.

*A 18. The following problem is a prototype of a set of problems known as "matches," "coincidences," or "rencontre." It has many variants, some of which date from the early eighteenth century. Suppose we have N pairs of distinguishable tickets, divided into 2 decks of N tickets each. Each deck contains 1 ticket from each of N pairs. Both decks of N tickets are shuffled, and 1 ticket at a time is taken from the top of each deck and matched against a corresponding ticket from the top of the other deck. If the tickets agree, we say a *match* has occurred. Among the N tickets compared, we count the number X_N of matches. Assuming that both decks of tickets are well shuffled, find the probability mass function of X_N when (a) $N=2$, (b) $N=3$, (c) $N=4$, and (d) $N=5$.

A19. The probability mass function $p_X(\tau)$ of a discrete random variable X has the form

$$p_X(\tau) = \begin{cases} \left(\frac{1}{2}\right)^\tau \left(\frac{32}{31}\right), & \text{for } \tau = 1, 2, 3, 4, 5, \\ 0, & \text{otherwise.} \end{cases}$$

(a) Show that $p_X(1) + p_X(2) + p_X(3) + p_X(4) + p_X(5) = 1$.
(b) Graph $p_X(\tau)$.
(c) Determine and graph the cumulative distribution function $F_X(\tau)$ of X.
(d) From the values of $p_X(\tau)$, find
    (i) $P_X\{X \geq 3\}$, (ii) $P_X\{2 \leq X \leq 4\}$, (iii) $P_X\{X < 4\}$.

20. In a study of the incidence of dental caries in adults, a random sample of 100 adults was selected from the population of a rural Southern town. These adults were offered free dental examinations at the beginning and end of a given year. A total of 80 adults accepted the offer. For each adult studied, the number X of new caries that developed over the year were counted. The raw (untabled) data obtained from this experiment appear below:

Patient Number	1	2	3	4	5	6	7	8	9	10
Number X of Caries	2	1	3	0	1	2	2	1	1	1

Patient Number	11	12	13	14	15	16	17	18	19	20
Number X of Caries	0	0	3	1	3	0	1	0	2	1

Patient Number	21	22	23	24	25	26	27	28	29	30
Number X of Caries	0	0	1	2	2	1	0	2	1	5

Patient Number	31	32	33	34	35	36	37	38	39	40
Number X of Caries	0	0	1	2	1	3	1	4	1	1

Patient Number	41	42	43	44	45	46	47	48	49	50
Number X of Caries	0	1	0	1	0	0	0	0	1	1

Patient Number	51	52	53	54	55	56	57	58	59	60
Number X of Caries	4	0	0	1	1	0	1	3	2	0

Patient Number	61	62	63	64	65	66	67	68	69	70
Number X of Caries	0	0	1	0	1	1	2	0	0	1

Patient Number	71	72	73	74	75	76	77	78	79	80
Number X of Caries	0	1	2	0	0	1	0	1	0	1

Determine and graph the sample probability mass function of X. Determine and graph the sample cumulative distribution function of X.

21. (a) From any (white) page of a telephone directory, note the *last digit* for each of the first 100 telephones on the page. Treat these 100 observations as repeated trials of the random experiment which consists of observing the last digit X of a randomly chosen personal telephone number. Based on these data, construct and graph the sample probability mass function for X.
    (b) Repeat the above random experiment using the classified (yellow) pages of the phone directory. Regard your data as 100 observations on the last digit Y of a randomly chosen business number. Based on these data, construct and graph the sample probability mass function for Y.
    (c) Compare the sample probability mass functions of X and of Y. What differences in these mass functions do you note? Which event is apparently more probable, the event $\{X=0\}$ or the event $\{Y=0\}$? Or are these events apparently equally probable? If there is a difference between r.f.$\{X=0\}$ and r.f.$\{Y=0\}$, can you explain why this difference exists? What is the nature of most business numbers?

*A22. The probability density function $f_X(\tau)$ of a continuous random variable X has the form
$$f_X(\tau) = \begin{cases} 0.4, & \text{if } 0 \leq \tau < 1, \\ 0.2, & \text{if } 1 \leq \tau < 2, \\ 0.3, & \text{if } 2 \leq \tau < 3, \\ 0.1, & \text{if } 3 \leq \tau < 4, \\ 0.0, & \text{if } \tau < 0 \text{ or } \tau \geq 4. \end{cases}$$

(a) Graph $f_X(\tau)$.
(b) Find $P_X\{0 < X \leq 2\}$.
(c) Find $P_X\{2 \leq X < 3\}$.
(d) Find $P_X\{\frac{1}{2} \leq X \leq \frac{3}{2}\}$.
(e) Find $P_X\{X \geq 1\}$.
(f) Find $P_X\{X > 2\}$.
(g) (Optional) Find and graph the cumulative distribution function $F_X(\tau)$ of X.

*23. Suppose that the continuous random variable Y has a probability density function $f_Y(\tau)$ of the form
$$f_Y(\tau) = \begin{cases} 0, & \text{if } \tau < 0, \\ 3\tau^2, & \text{if } 0 \leq \tau \leq 1, \\ 0, & \text{if } \tau > 1. \end{cases}$$

(a) Graph $f_Y(\tau)$ on graph paper that has a fine grid.
(b) Using the method of rectangular approximation (with $M = 10$ intervals and $M = 20$ intervals) estimate $P_Y\{0 \leq Y \leq \frac{1}{2}\}$ and $P_Y\{\frac{1}{4} \leq Y \leq \frac{3}{4}\}$.
(c) The exact values of $P_Y\{0 \leq Y \leq \frac{1}{2}\}$ and $P_Y\{\frac{1}{4} \leq Y \leq \frac{3}{4}\}$ are $\frac{1}{8}$ and $\frac{13}{32}$, respectively. How close are the approximate values which you obtained in part (b) for these probabilities?

*^24. The probability density function $f_Z(\tau)$ of a continuous random variable Z has the form

$$f_Z(\tau) = \begin{cases} 0, & \text{if } \tau < 1, \\ \tau - 1, & \text{if } 1 \leq \tau < 2, \\ 3 - \tau, & \text{if } 2 \leq \tau \leq 3, \\ 0, & \text{if } \tau > 3. \end{cases}$$

(a) Graph $f_Z(\tau)$.
(b) Find the exact values of $P_Z\{Z \leq \frac{3}{2}\}$, $P_Z\{\frac{3}{2} \leq Z \leq \frac{3}{2}\}$, and $P_Z\{Z > \frac{3}{2}\}$. [*Hint:* The area of a triangle equals $\frac{1}{2}$ times the product of the lengths of its base and altitude.]
(c) Use the rectangular method of approximation with $M = 10$ intervals to approximate $P_Z\{\frac{3}{2} \leq Z \leq \frac{3}{2}\}$. How good an approximation do you obtain?

25. Example 5.2 deals with the lifetime X (in hours) of a diamond phonograph needle. Using the rectangular method of approximation, approximate the probability that a needle will last at least 50 hours. What is the probability that the needle lasts *exactly* 50 hours? Does your answer upset you? Explain.

^26. The following data were obtained [Gregory (1963)] by observing the total amount X of rainfall (in inches) at Bidsten Observatory, Birkenhead, England from 1901–1930:

Year	1901	1902	1903	1904	1905	1906	1907	1908	1909	1910
Amount X	25.19	25.57	34.42	25.18	24.01	28.08	26.57	28.90	28.45	28.59

Year	1911	1912	1913	1914	1915	1916	1917	1918	1919	1920
Amount X	25.27	30.17	25.78	26.02	26.83	24.87	30.59	31.93	29.12	33.34

Year	1921	1922	1923	1924	1925	1926	1927	1928	1929	1930
Amount X	22.47	25.97	30.92	32.87	28.00	28.95	34.81	29.11	25.15	36.50

(a) Find and graph the sample cumulative distribution function of X.
(b) Starting with a smallest value of 21.00 and using 8 intervals of length 2.00, construct a relative frequency histogram (sample probability density function) for X.
(c) (Optional) Use each of the 3 strategies [(i), (ii), and (iii)] described in Exercise 14 to construct and graph a sample cumulative distribution function for X based on the *grouping* of the data obtained in part (b). Which of the 3 strategies seems to do the best job of approximating the sample cumulative distribution function for X which you obtained in part (a) from the raw (ungrouped) data? Explain your conclusion. Do you think you would obtain the same conclusion if you used other groupings of these data? Explain. Do you think that the same strategy would do the best job for all cases in which continuous data are observed? Explain.

27. One measure of air pollution is the amount Y of beta radioactivity concentration in the air (measured in micro microcuries per cubic meter). Data for those states which have air sampling stations are collected by the Federal government [*Statistical Abstracts* (1965)]. The results, listed alphabetically by states, are: 2.2, 6.5, 5.8, 3.9, 5.9, 4.0, 4.6, 4.9, 4.5, 2.6, 5.7, 4.7, 4.8, 3.5, 4.2, 5.0, 4.0, 4.1, 4.5, 5.0, 5.2, 3.9, 4.6, 4.1, 4.8, 3.9, 8.8, 5.9, 4.2, 4.8, 4.6, 5.5, 4.3, 5.0, 3.3, 5.2, 4.0, 4.6, 4.0, 3.7, 5.7, 4.6, 5.6, 5.2, 4.4, 3.6, 4.9, 5.4, 6.2, 2.2. Prepare a frequency histogram and a relative frequency histogram from these data.

5
DESCRIPTIVE PROPERTIES OF DISTRIBUTIONS

I collected statistics, I worked out the golden mean, and never understood that extremes join hands, that the man who goes to bed very late meets the man who gets up very early, and he who chooses to take his seat on the golden mean, risks falling between two stools.

André Gide (Prometheus Misbound)

1. INTRODUCTION

Cumulative distribution functions, and probability mass functions in the case of discrete variables, or density functions in the case of continuous variables, are profiles that contain all of the relevant information about the statistical properties of a random variable. We may refer to these profiles, whichever is used, as the *distribution* of the random variable.

For many purposes it is not necessary (or possible) to obtain all of the information contained in the distribution; rather, several descriptive properties may (or must) suffice. These descriptive properties serve to summarize various important aspects of a distribution. Two of the most frequently useful pieces of information about a distribution are its *location* and its *dispersion*.

To explain what we mean by the location of a distribution, consider Figure 1.1. Here we have two densities that have exactly the same graphical shape except that one $[f_X(\tau)]$ concentrates about $\tau = x_0$, and the other $[f_X^*(\tau)]$ concentrates about $\tau = x_1$. Thus, we can see that $f_X^*(\tau)$ is, in some sense, $f_X(\tau)$ *relocated* from x_0 to x_1 (we may think of the horizontal axis as a track upon which we can slide the distribution), and this relocation is the only difference between these two distributions. If, therefore, we had a descriptive measure for a distribution that changed value by (approximately) d whenever the distribution changed location by d, we would have a *measure of location* for the distribution.

On the other hand, a *measure of dispersion* (measure of variation) measures how strongly a distribution concentrates about a central value.

176 DESCRIPTIVE PROPERTIES OF DISTRIBUTIONS **5**

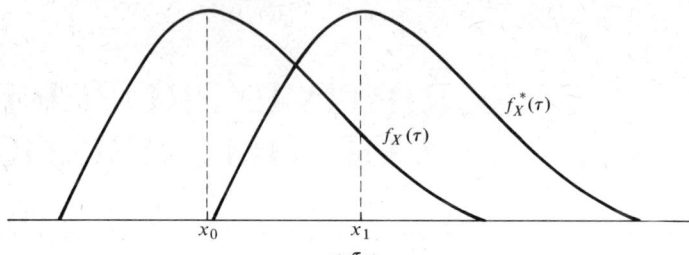

Figure 1.1 Two densities differing in location.

A measure of dispersion is large when the spread of the distribution about a central value is large; and is small when this spread is negligible, becoming zero when all of the probability is concentrated at a single point. For example, the distribution corresponding to the discrete random variable X whose probability mass function is graphed in Figure 1.2 has small spread around the central point c, and thus its measure of dispersion should be small. On the other hand, the distribution corresponding to the probability mass function graphed in Figure 1.3 is more widely spread about the central point c, so that its measure of dispersion should be larger. Similarly, the distribution of the random variable X whose probability density function is graphed in Figure 1.4 should have a lower measure of dispersion than that of the distribution of the continuous random variable Y whose probability density function is shown in Figure 1.5.

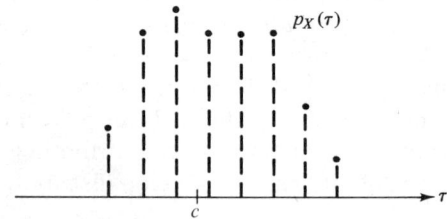

Figure 1.2 Probability mass function with low dispersion.

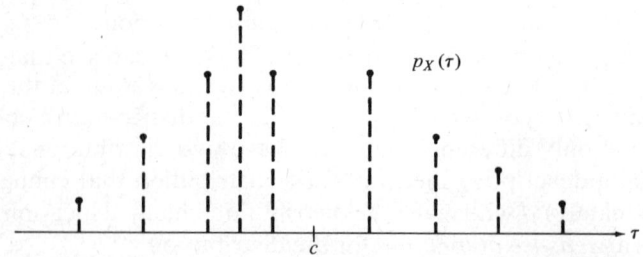

Figure 1.3 Probability mass function with a larger dispersion than that of the mass function in Figure 1.2.

2. MEASURES OF LOCATION FOR A DISTRIBUTION

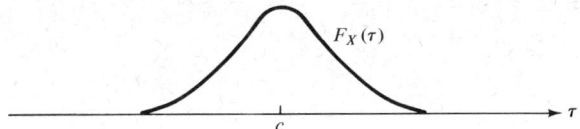

Figure 1.4 Probability density function with low dispersion.

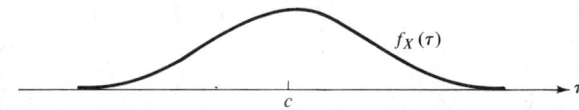

Figure 1.5 Probability density function with larger dispersion than that of the probability density function in Figure 1.4.

In this chapter, we introduce several measures of location for a distribution (expected value, median, quantile, mode), each of which can be used to answer interesting questions about probability distributions. We also discuss one or two measures of dispersion, one of which (the variance) plays a prominent role in statistical analysis. Other descriptive measures (the moments) and descriptive terminology (skewness, peakedness or kurtosis, and so on) are briefly mentioned. Finally, we show how a knowledge of a certain measure of location (the expected value) and/or a certain measure of dispersion (the variance) for a random variable X in many cases allows us to state upper or lower limits for the values of the probabilities of certain events of interest concerning X, even when nothing else is known about the distribution of X.

2. MEASURES OF LOCATION FOR A DISTRIBUTION: THE EXPECTED VALUE

One of the most useful characteristics of a distribution is its *expected value* (sometimes called the *mean* or the *expectation*). The expected value of a distribution may be viewed as being the fulcrum, the center of gravity, or the balance point of the distribution of probability on the real line.

The Expected Value of a Discrete Random Variable

In mechanics, the center of gravity of a distribution of discrete masses is determined as the sum of the products of the magnitude of each weight times its distance from a given origin (see Figure 2.1). The expected value of a discrete random variable X, denoted by $E(X)$, is defined analogously as the sum of the products of the possible values for X by their respective

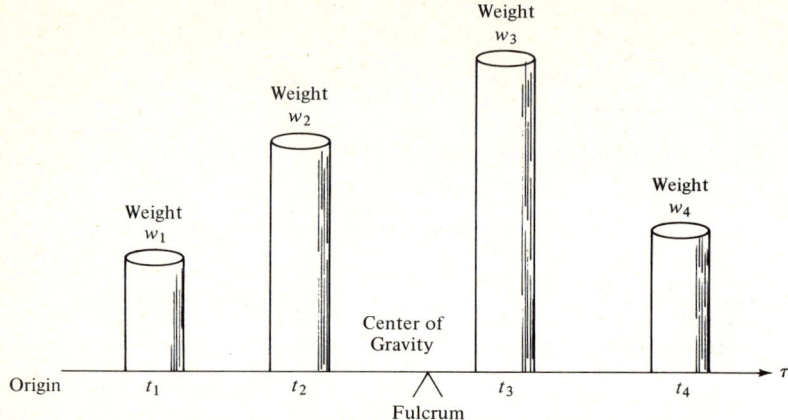

Figure 2.1 Center of gravity of four weights w_1, w_2, w_3, w_4, placed respectively at distances t_1, t_2, t_3, t_4 from the origin. The center of gravity is defined to be $w_1 t_1 + w_2 t_2 + w_3 t_3 + w_4 t_4$. If the horizontal axis is a see-saw of no weight supported by a fulcrum at the center of gravity, the weights on one side of the fulcrum balance the weights on the other side.

probabilities of occurrence. For the probability model

x	x_1	x_2	x_3	...
Probability	p_1	p_2	p_3	...

,

the expected value $E(X)$ is defined by the equation

(2.1) $$E(X) = x_1 p_1 + x_2 p_2 + x_3 p_3 + \cdots.$$

In the case where the discrete random variable X has possible values $x_1, x_2, \ldots, x_k$, and $P_X\{X = x_j\} = p_j$, $j = 1, 2, 3, \ldots, k$, then formula (2.1) says that

$$E(X) = x_1 p_1 + x_2 p_2 + x_3 p_3 + \cdots + x_k p_k.$$

The probability mass function $p_X(\tau)$ and the expected value $E(X)$ of a discrete random variable X having five possible values are shown in Figure 2.2.

Example 2.1. There have been many studies comparing family sizes in lower-class urban and upper-class suburban environments. After one such study, a sociologist uses the experimental method (see Chapters 1 and 4) to construct probability models for the number X of children born to a ran-

2. MEASURES OF LOCATION FOR A DISTRIBUTION

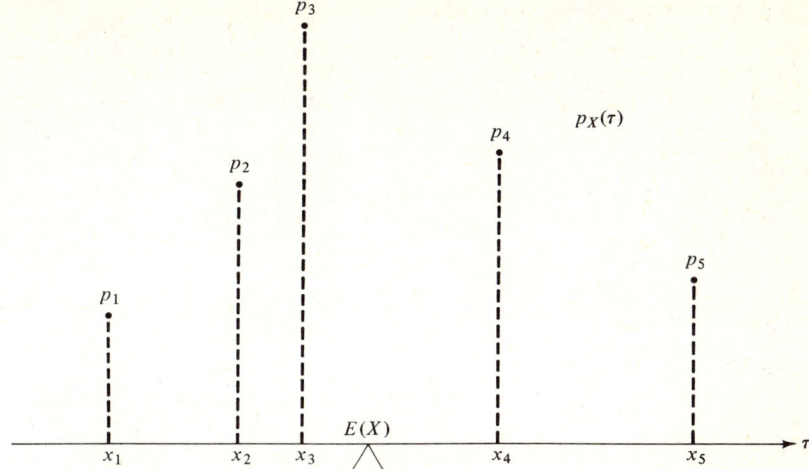

Figure 2.2 The probability mass function $p_X(\tau)$ of a discrete random variable X having possible values x_1, x_2, x_3, x_4, x_5 with $p_X(x_j) = p_j$, $j = 1, 2, \ldots, 5$. $E(X) = x_1 p_1 + x_2 p_2 + x_3 p_3 + x_4 p_4 + x_5 p_5$ is also shown.

domly chosen lower-class urban family and for the number Y of children born to a randomly chosen family from an upper-class suburb. Suppose that his probability model for X is given by the following table:

Number of Children x	0	1	2	3	4	5	6	7	8	9	10	11
Probability $P_N\{X = x\}$	0.05	0.05	0.10	0.15	0.20	0.20	0.15	0.03	0.02	0.02	0.02	0.01

From this table, we can calculate that the expected value of X is

$$\begin{aligned} E(X) &= (0)(0.05) + (1)(0.05) + (2)(0.10) + (3)(0.15) + (4)(0.20) \\ &\quad + (5)(0.20) + (6)(0.15) + (7)(0.03) + (8)(0.02) + (9)(0.02) \\ &\quad + (10)(0.02) + (11)(0.01) \\ &= 4.26. \end{aligned}$$

The expected value $E(X) = 4.26$ gives the sociologist some idea as to how many children "on the average" are born in a lower-class urban family.

In contrast, suppose that the sociologist's probability model for the number Y of children born to a randomly chosen upper-class suburban family is given by the following table:

Number of Children y	0	1	2	3	4	5	6	7	8	9	10	11
Probability $P_Y\{Y = y\}$	0.12	0.15	0.19	0.16	0.12	0.10	0.07	0.02	0.02	0.02	0.02	0.01

From this table, the expected value $E(Y)$ of Y is

$$\begin{aligned} E(Y) &= (0)(0.12) + (1)(0.15) + (2)(0.19) + (3)(0.16) + (4)(0.12) \\ &\quad + (5)(0.10) + (6)(0.07) + (7)(0.02) + (8)(0.02) + (9)(0.02) \\ &\quad + (10)(0.02) + (11)(0.01) \\ &= 3.20. \end{aligned}$$

Although by themselves the two expected values $E(X)$ and $E(Y)$ do not allow us to compare the probability distributions of X and Y in detail, a comparison of $E(X) = 4.26$ and $E(Y) = 3.20$ does suggest that lower-class urban families tend to be larger than upper-class suburban families. This last assertion can be directly verified if we have graphs of the probability mass functions of X and Y (see Figure 2.3), but these are more complicated and harder to obtain than the two numbers $E(X)$ and $E(Y)$. Certainly if we can arrive at a conclusion about the relative family sizes of lower-class urban and upper-class suburban families by merely comparing two numbers $E(X)$ and $E(Y)$, this is a preferable alternative to the construction and comparison of the graphs of the two probability mass functions. The information provided by the numbers $E(X)$ and $E(Y)$ is also easier to transmit to others than is the more complex and detailed

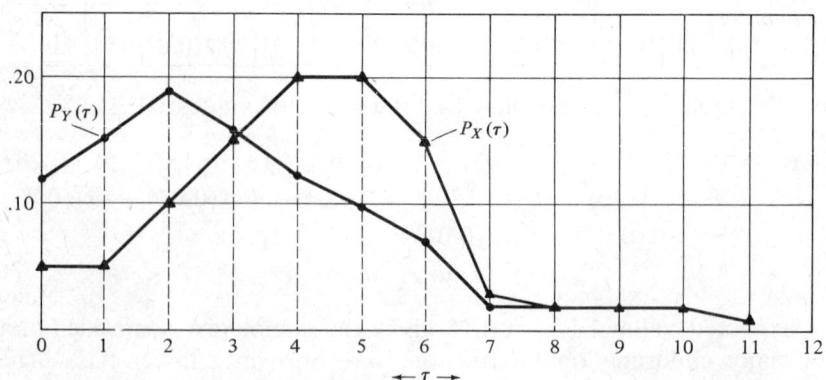

Figure 2.3 Probability mass functions of X and Y. The triangles represent values of $p_X(\tau)$, while the circles represent values of $p_Y(\tau)$.

2. MEASURES OF LOCATION FOR A DISTRIBUTION

information provided by Figure 2.3. On the other hand, the disadvantage of keeping only the information about X and Y provided by their expected values $E(X)$ and $E(Y)$ is that we do not obtain a complete picture of the differences between their distributions, and thus miss important and interesting facts. (In this case, for example, we miss the fact that both lower-class urban and upper-class suburban families have equal probability of having 8 or more children.) It should be noted that similar advantages and disadvantages accrue to any other single index which is reported in place of the entire distribution of a random variable.

Example 2.2. In an investigation of the theory of Brownian motion in physics, the Swedish physicist T. Svedberg (*Existenz der Moleküle*, Leipzig, 1912, p. 160) counted the number of particles X in an optically isolated volume of a colloidal solution of gold at a certain instance of time. If the particular theory which Svedberg was investigating holds true (we are simplifying greatly here), the following table of probabilities gives the probability mass function of X:

Number of Particles x	0	1	2	3	4	5	6	7
$P_X\{X=x\}$	0.182	0.310	0.264	0.150	0.064	0.022	0.006	0.002

The expected value $E(X)$ of X under this probability model is

$$E(X) = (0)(0.182) + (1)(0.310) + (2)(0.264) + (3)(0.150) + (4)(0.064)$$
$$+ (5)(0.022) + (6)(0.006) + (7)(0.002)$$
$$= 1.704.$$

Thus, the expected value of the number of particles is 1.704.

The measure $E(X)$ is called the expected value of the distribution of X, but this terminology should not be interpreted as literally meaning that $E(X)$ is the value expected as a result of performing the random experiment and observing X. Indeed, the expected value of X need not even be a possible value of X! As examples, notice that in Example 2.1 neither $E(X) = 4.26$ nor $E(Y) = 3.20$ are possible values of the corresponding random variables X or Y, respectively. Similar remarks hold for Example 2.2. We would like $E(X)$ to be close to those possible values of X which occur with high probability, so that $E(X)$ reflects "likely" values of X, but there is no reason to expect that $X = E(X)$, even when $E(X)$ *is* a possible value of the random variable X.

The expected value $E(X)$ of X is, however, a number which, in a large number of independent trials of the experiment that produces X, would be very close to the "average" of the observed X values. This fact is an interpretation of the expected value $E(X)$ that follows directly from the frequency interpretation of probability (Chapter 2). To be more precise, suppose that we perform N independent trials of the random experiment from which the discrete random variable X arises, and suppose the resulting (observed) values of X in these N trials are $t_1, t_2, t_3, \ldots, t_N$. For each distinct value t of X that we have observed, multiply t by the relative frequency r.f.$\{X = t\}$ of this value in the N trials, then sum all of these products. The result equals the *arithmetic average* $(t_1 + t_2 + \cdots + t_N)/N = \bar{t}$ of the observed values $t_1, t_2, \ldots, t_N$; that is

(2.2) $$\bar{t} = \frac{t_1 + t_2 + t_3 + \cdots + t_N}{N} = \text{sum } [t \times \text{r.f.}\{X = t\}],$$

where the sum is over all distinct values t that are observed. To illustrate the summing process in the above equation, assume that $N = 10$ and that $t_1 = 2, t_2 = 4, t_3 = 1, t_4 = 5, t_5 = 2, t_6 = 1, t_7 = 3, t_8 = 9, t_9 = 1, t_{10} = 1$. Then the distinct values observed are $t = 1, 2, 3, 4, 5,$ and 9 and the following table gives the relative frequencies of these values:

t	1	2	3	4	5	9
r.f.$\{X = t\}$	$\frac{4}{10}$	$\frac{2}{10}$	$\frac{1}{10}$	$\frac{1}{10}$	$\frac{1}{10}$	$\frac{1}{10}$

Now

$$\bar{t} = \frac{t_1 + t_2 + t_3 + \cdots + t_{10}}{10} = \frac{2+4+1+5+2+1+3+9+1+1}{10} = \frac{29}{10}$$

and

$$\text{sum } [t \times \text{r.f.}\{X = t\}] = \left(1 \times \frac{4}{10}\right) + \left(2 \times \frac{2}{10}\right) + \left(3 \times \frac{1}{10}\right) + \left(4 \times \frac{1}{10}\right)$$
$$+ \left(5 \times \frac{1}{10}\right) + \left(9 \times \frac{1}{10}\right) = \frac{29}{10},$$

so that Equation (2.2) is indeed correct in this case.

From the frequency interpretation of probability, we know r.f.$\{X = t\}$ tends to be close to $P_X\{X = t\}$ as the number of trials N becomes very large. Thus, looking at Equation (2.2) we see that if N, the number of trials of our random experiment, is very large, the arithmetic average (also called the *sample average* or *sample mean*) of the observed collection of values for X tends to be close to the expected value $E(X)$ of X. That is, when the number N of trials is large,

(2.3) $$\bar{t} = \frac{t_1 + t_2 + \cdots + t_N}{N} \cong E(X),$$

where the symbol "$\cong$" means "approximately equal to."

2. MEASURES OF LOCATION FOR A DISTRIBUTION

Example 2.3. The approximation given by Equation (2.3) is frequently used in practice to predict the magnitude of the sum $t_1 + t_2 + \cdots + t_N = N\bar{t}$ of the observed values of X before they are actually observed. This fact often aids scientists and administrators in planning. For example, suppose that the Board of Education in a lower-class urban district is planning for future school construction. As part of their considerations, they must estimate how many children to plan for. As yet, few families with children are living in the district, but a new housing project for 1000 families is being built, and the Board needs to predict how many children from this housing project will be using the schools. Each of these 1000 families can be considered a trial of a random experiment from which the number X of children is the obtained random variable. Thus, the observations $t_1, t_2, \ldots, t_{1000}$ of such a random experiment will be counts of the number of children in each of the 1000 families. From Example 2.1, we know that $E(X) = 4.26$, and from the relative frequency interpretation of $E(X)$ we know that

$$\frac{t_1 + t_2 + \cdots + t_{1000}}{1000} \cong E(X) = 4.26.$$

Thus (perhaps to a lesser degree of accuracy),

$$t_1 + t_2 + \cdots + t_{1000} = 1000 \left(\frac{t_1 + t_2 + \cdots + t_{1000}}{1000} \right) \cong 1000(4.26) = 4260.$$

The Board of Education decides to plan for adequate schooling for approximately 4260 children.

Example 2.4. A man who has just become 44 years of age and who is in good health wants to take out insurance on his life over a period of 1 year (a so-called "term insurance"). He has a friend A who is well-to-do and likes to take risks. Taking along the mortality table shown in Table 2.1, he approaches his friend A and offers him a financial proposition. He will pay A a premium of $100. In turn, A will sign a contract guaranteeing to pay the man's heirs $25,000 if the man dies before he is 45 years of age. From the mortality table, we find that the probability that a man of 44 will *live* to be 45 is $p = 1 - 0.00319 = 0.99681$. Let X be the random variable which is equal to the amount of profit (or loss) in dollars that A makes from the transaction. That is, if his friend lives to the age of 45, A obtains $X = \$100$, while if his friend dies before he becomes 45, $X = \$100 - \$25,000$. The expected value $E(X)$ is then

$$\begin{aligned}
E(X) &= (\$100)p + (\$100 - \$25,000)(1-p) \\
&= (\$100)(0.99681) + (-\$24,900)(0.00319) \\
&= \$99.681 - \$79.431 \\
&= \$20.25.
\end{aligned}$$

Table 2.1. Mortality table for males.

Age	Probability	Age	Probability	Age	Probability	Age	Probability
0	0.00404	28	0.00090	55	0.01056	82	0.10244
1	0.00158	29	0.00095	56	0.01149	83	0.11211
2	0.00089	30	0.00100	57	0.01246	84	0.12267
3	0.00072	31	0.00107	58	0.01348	85	0.13418
4	0.00063	32	0.00114	59	0.01454	86	0.14671
5	0.00057	33	0.00121	60	0.01566	87	0.16033
6	0.00053	34	0.00130	61	0.01687	88	0.17512
7	0.00050	35	0.00139	62	0.01820	89	0.19115
8	0.00049	36	0.00149	63	0.01967	90	0.20849
9	0.00048	37	0.00161	64	0.02128	91	0.22719
10	0.00048	38	0.00173	65	0.02307	92	0.24733
11	0.00049	39	0.00187	66	0.02503	93	0.26896
12	0.00050	40	0.00203	67	0.02719	94	0.29212
13	0.00051	41	0.00222	68	0.02958	95	0.31683
14	0.00052	42	0.00248	69	0.03220	96	0.34312
15	0.00054	43	0.00280	70	0.03509	97	0.37097
16	0.00055	44	0.00319	71	0.03827	98	0.40035
17	0.00057	45	0.00363	72	0.04177	99	0.43120
18	0.00058	46	0.00412	73	0.04562	100	0.46341
19	0.00060	47	0.00466	74	0.04985	101	0.49687
20	0.00062	48	0.00525	75	0.05450	102	0.53139
21	0.00065	49	0.00588	76	0.05961	103	0.56676
22	0.00067	50	0.00656	77	0.06522	104	0.60271
23	0.00070	51	0.00728	78	0.07137	105	0.63896
24	0.00073	52	0.00804	79	0.07811	106	0.67514
25	0.00077	53	0.00884	80	0.08550	107	0.71090
26	0.00081	54	0.00968	81	0.09359	108	0.74582
27	0.00085					109	1.00000

(Entries give the probability that a male of a given age does not survive an additional year.)
SOURCE: 1967 Life Insurance Fact Book, Institute of Life Insurance, New York.

Thus, the expected value of X indicates a profit (positive amount of money) for A. The frequency interpretation of the expected value of X says that if A entered into a very large number N of statistically independent transactions of the kind described above (all with men of 44 years of age in good health), then "on the average" he would make $20.25 profit on each person. The larger the number N of people A deals with, the more accurately $E(X)$ indicates what his average profit per person will be. Thus, if A goes into the insurance business on a larger scale and takes $N = 100$ customers, and if he can assume that all other costs are negli-

2. MEASURES OF LOCATION FOR A DISTRIBUTION

gible (this is not actually the case in practice), he can anticipate making a very nice total profit on this kind of transaction. However, suppose instead that A has exactly $25,000, and suppose he makes only the one transaction with his friend mentioned above. Does $E(X) = \$20.25$ reflect the actual amount of money that A can "expect" to get out of the transaction? Does either person have an advantage in this transaction? If you were A, would you agree to insure your friend? Some thought on these questions may help us to obtain a clearer perspective of the information which the expected value, $E(X)$, conveys about the variation from trial to trial of the random variable X.

*The Expected Value of a Discrete Random Variable Having an Infinite Number of Possible Values

Most of our discussion thus far has been about discrete random variables with a finite number of possible values. Our discussion, however, holds equally well for discrete random variables with an infinite number of possible values. Nevertheless, there are two technical points about the definition of the expected value of a discrete random variable X with an infinite number of possible values which need mentioning before turning our attention to the definition of the expected value of a continuous random variable.

When a discrete random variable X has an infinite number of possible values $x_1, x_2, \ldots$, then the summation (2.1) defining $E(X)$ must extend over an infinite number of terms. In such cases, it is possible for $E(X)$ to be infinitely large [that is, $E(X) = \infty$, where ∞ is the mathematical symbol for infinity] or for $-E(X)$ to be infinitely large. As an example of the former situation, suppose that the probability mass function $p_X(\tau)$ of X is nonzero only for values of τ which are squares of positive integers; that is, the possible values of X are $x_1 = (1)^2$, $x_2 = (2)^2$, $x_3 = (3)^2$, ..., and so on. Suppose that

$$P_X\{X = k^2\} = p_X(k^2) = c(1/k^2), \quad k = 1, 2, \ldots,$$

where c is a number chosen so that the probability of the whole sample space

$$\mathscr{X} = \{(1)^2, (2)^2, (3)^2, (4)^2, \ldots\} = \{1, 4, 9, 16, \ldots\}$$

is 1; that is,

$$1 = P_X\{\mathscr{X}\} = P_X\{1, 4, 9, 16, \ldots\} = c\left[\left(\frac{1}{1}\right) + \left(\frac{1}{4}\right) + \left(\frac{1}{9}\right) + \cdots\right],$$

or $c^{-1} = [1 + (1/4) + (1/9) + (1/16) + \cdots] = 1.645$. Then using Equation

(2.1) to calculate $E(X)$, we find that
$$E(X) = x_1 p_1 + x_2 p_2 + x_3 p_3 + \cdots$$
$$= (1)^2 c \frac{1}{(1)^2} + (2)^2 c \frac{1}{(2)^2} + (3)^2 c \frac{1}{(3)^2} + \cdots$$
$$= c + c + c + \cdots = \infty.$$

When X is discrete and has an infinite number of possible values, it is also possible for the sum in Equation (2.1) not to be well-defined. This is the case, for example, when the possible values x_i and the probabilities $p_i = p_X(x_i)$ are such that $x_i p_i = (-1)^i$, so that $E(X) = 1 - 1 + 1 - 1 + 1 - 1 + \cdots$. In such a case, it may appear that $E(X) = 1$ or that $E(X) = 0$.

In both cases mentioned above [where $E(X)$ or $-E(X)$ equals ∞, or when $E(X)$ is not well-defined], the expected value $E(X)$ of X is said *not to exist*, and the frequency interpretations of $E(X)$ given earlier need not be valid. Although such mathematical technicalities are of concern to theoretical probabilists, and warn us not to carry analogies between observed random phenomena and our probability models too far, only rarely are such technicalities of practical importance. Further, most discrete random variables X which have an infinite number of possible values also have quite well-defined and finite expected values.

The Expected Value of a Continuous Random Variable

We have already mentioned the analogy existing between the center of gravity of the distribution of a discrete set of masses on a line and the expected value $E(X)$ of a discrete random variable. In physics (mechanics), we can also speak of the center of gravity for a continuous distribution of mass on a line, where the distribution of mass is expressed as a function assigning to every point on the line the density of mass at that point. Analogously, if a random variable X is continuous, we define the expected value of X to be the center of gravity of the distribution of probability on the line [where the distribution of probability is given by the density function $f_X(\tau)$ of X]. Explicitly, the expected value $E(X)$ of the continuous random variable X is defined as follows:

(2.4) $E(X) =$ "total area between the graph of the function $\tau f_X(\tau)$ and the horizontal axis."

Notice the similarity of Equation (2.4) to the defining equation [Equation (2.1)] for the expected value of a discrete random variable. The difference between the definition of the expected value of a continuous random variable and the definition of a discrete random variable is that the former definition involves finding the total area of a geometrical figure, while the latter definition involves finding the sum of a collection of discrete numbers.

2. MEASURES OF LOCATION FOR A DISTRIBUTION

Because $f_X(\tau)$ is a density function, $f_X(\tau)$ is always nonnegative; hence the graph of the function $\tau f_X(\tau)$ lies below the horizontal axis when τ is negative and above the horizontal axis when τ is positive. Thus, the area under the graph of $\tau f_X(\tau)$ in Equation (2.4) should be interpreted algebraically; that is, when the graph of $\tau f_X(\tau)$ is below the horizontal axis, the area between the graph and the horizontal axis should be taken to be a negative number, while when the graph of $\tau f_X(\tau)$ is above the horizontal axis, the area between the graph and the horizontal axis is taken to be positive.

Example 2.5. Suppose the graph of the density function $f_X(\tau)$ of the continuous random variable X is given by Figure 2.4. That is, $f_X(\tau)$ is 0 for values of τ less than $\tau = -1$, $f_X(\tau) = \frac{1}{2}$ for values of τ between (and including) $\tau = -1$ and $\tau = 1$, and $f_X(\tau)$ is 0 for values of τ greater than $\tau = 1$. The graph of $\tau f_X(\tau)$ is displayed in Figure 2.5.

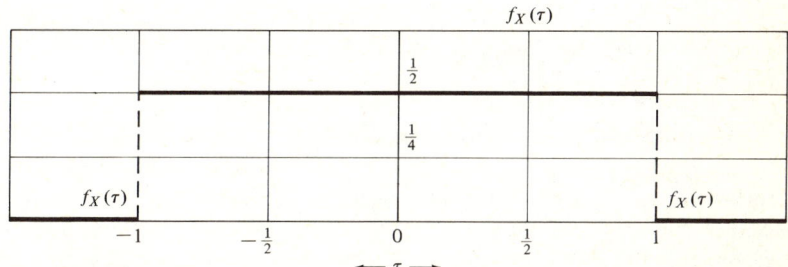

Figure 2.4 Graph of the density function $f_X(\tau)$, which is flat from $\tau = -1$ to $\tau = 1$ and is zero elsewhere.

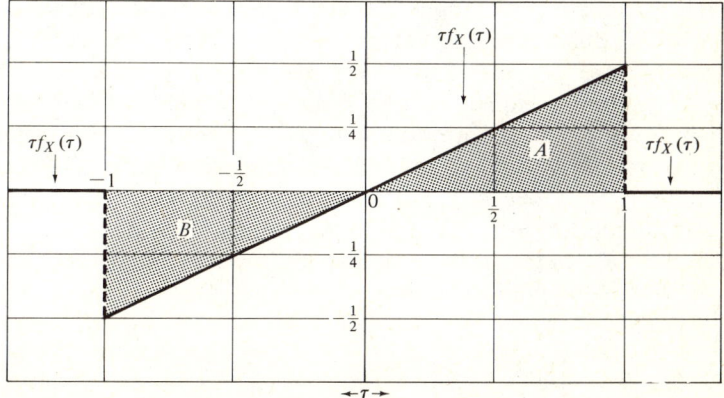

Figure 2.5 Graph of $\tau f_X(\tau)$ for the density function $f_X(\tau)$ graphed in Figure 2.4.

Notice that the total area between the graph of $\tau f_X(\tau)$ and the horizontal axis is the sum of the areas of the two shaded triangles, A and B, shown in Figure 2.5. The area of triangle A is equal to $\frac{1}{2}$ times the product of the length of the base (which is equal to 1) and the height of the altitude (which is equal to $[(1)f_X(1) - 0] = \frac{1}{2}$). Thus, the area of triangle A equals $\frac{1}{4}$. Similarly, the area of triangle B is equal to $\frac{1}{2}$ the product of the length of the base (which is equal to 1) and the height of the altitude (which is equal to $[0 - (-1)f_X(-1)] = \frac{1}{2}$). Hence, the area of triangle B is $-\frac{1}{4}$ [we attach the minus sign since the graph of $\tau f_X(\tau)$ is below the horizontal axis for τ between $\tau = -1$ and $\tau = 0$]. Consequently,

$E(X) =$ "total area between graph of $\tau f_X(\tau)$ and horizontal axis"
$=$ area of triangle $B +$ area of triangle A
$= (-\frac{1}{4}) + (\frac{1}{4}) = 0.$

The obtained result [that $E(X) = 0$] is intuitively reasonable because of the symmetry of the graph of $\tau f_X(\tau)$ about the point 0.

*Use of the Rectangular Method of Approximation

In Chapter 4 we pointed out that geometrical methods (such as we have used in Example 2.5) for calculating the area between a graph and the horizontal axis are seldom applicable for obtaining exact values. Unless the graph has a nice form (such as in Figure 2.5), the geometrical figure whose area must be calculated generally has a shape for which formulas for calculating the area are not available. However, various numerical methods can be employed to approximate an exact answer. We here illustrate the so-called "rectangular method" of approximation, earlier discussed in Chapter 4; in the process, we also show how Equations (2.1) and (2.4), for the expected values of discrete and continuous random variables, are analogous to one another.

Suppose we are interested in finding the expected value $E(X)$ of the continuous random variable X whose probability density function $f_X(\tau)$ is graphed in Figure 2.6. To approximate $E(X)$, we first approximate the continuous random variable X by a discrete random variable Y using the method earlier described in Chapter 4, Section 5. We choose an integer $M > 0$ and divide the interval of possible values for X into M equal subintervals each of length $\Delta = (x_L - x_S)/M$ as shown in Figure 2.6. The random variable Y is then defined in terms of the random variable X by letting $Y = y_i$, where $y_i = x_S + (i - \frac{1}{2})\Delta$ is the midpoint of the ith subinterval whenever X is between $x_S + (i-1)\Delta$ and $x_S + i\Delta$ [that is, $Y = y_i$ if and only if $x_S + (i-1)\Delta \leq X < x_S + i\Delta$], $i = 1, 2, ..., M$. Next we approximate $P_Y\{Y = y_i\} = P_X\{x_S + (i-1)\Delta \leq X < x_S + i\Delta\}$, a number that is equal to the area under the graph of the probability density function $f_X(\tau)$ between

2. MEASURES OF LOCATION FOR A DISTRIBUTION

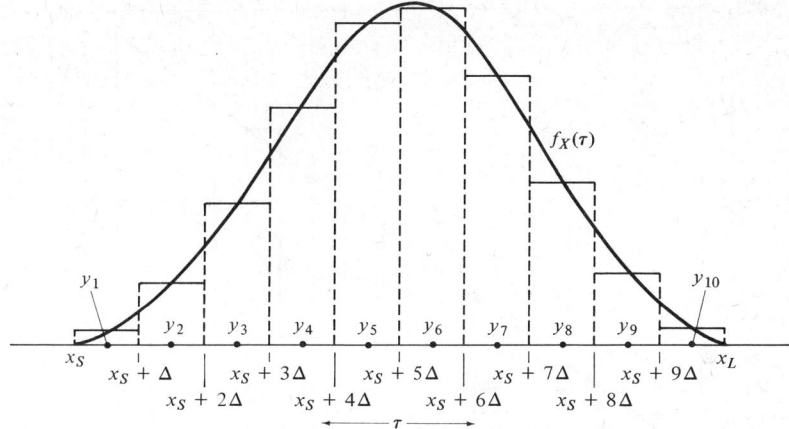

Figure 2.6 Graph of the probability density function $f_X(\tau)$ of the continuous random variable X showing construction of the discrete random variable y approximating X ($M = 10$ subintervals).

$x_S + (i-1)\Delta$ and $x_S + i\Delta$, by the area of a rectangle whose base is of length Δ, and whose height is the value of $f_X(\tau)$ at the midpoint $y_i = x_S + (i-\frac{1}{2})\Delta$ of the base. That is,

$$P_Y\{Y = y_i\} \cong \Delta f_X(x_S + (i-\tfrac{1}{2})\Delta).$$

Using this two-step method of approximation [first approximating X by Y, and then approximating $P_Y\{Y = y_i\}$ by $\Delta f_X(x_S + (i-\frac{1}{2})\Delta)$, $i = 1, 2, \ldots, M$], we find that

$$(2.5) \quad E(X) \cong E(Y) = y_1 P_Y\{Y = y_1\} + y_2 P_Y\{Y = y_2\} + \cdots + y_M P_Y\{Y = y_M\}$$
$$\cong y_1(\Delta f_X(y_1)) + y_2(\Delta f_X(y_2)) + \cdots + y_M(\Delta f_X(y_M)),$$

where the symbol $\cong$ stands for "approximately equal." Equation (2.5) shows how the expected value $E(X)$ of a continuous random variable X can be approximated by the expected value (actually, an approximation to the expected value) of a discrete random variable Y. The larger we choose the number M of subintervals to be, the better is the approximation given in Equation (2.5).

Example 2.6. In Chapter 4, Section 5, we have given an example of a delicatessen that receives an order of 400 pounds of meat every day. The random variable X considered in this example is the total weight of meat (assumed measured exactly) sold in a day. This random variable is assumed to have the probability density function $f_X(\tau)$ shown in Figure 2.7. In order to approximate the total amount of meat that they can expect to sell

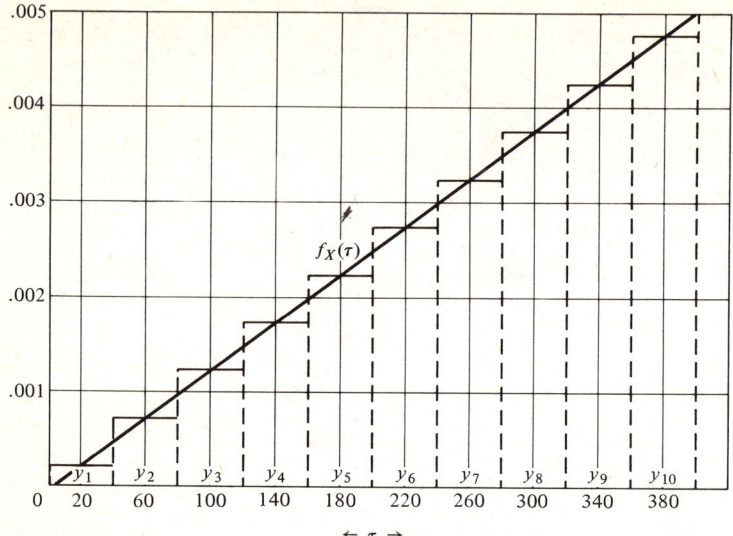

Figure 2.7 The probability density function $f_X(\tau)$ of the weight X of meat sold by the delicatessen in one day.

over the period of a year (365 days), the delicatessen owners use the frequency interpretation of probability (and the arguments used already in Examples 2.3 and 2.4) to estimate that approximately $(365)[E(X)]$ pounds of meat will be sold. To complete their calculations, the owners now need to calculate $E(X)$. They decide to use the rectangular method to provide them with an approximation, and choose to approximate X with the discrete random variable Y based on a division of the interval of possible values of X into $M = 10$ subintervals. Here, $x_S = 0$, $x_L = 400$, and thus $\Delta = (x_L - x_S)/10 = 40$. The possible values of Y are $y_1 = 20$, $y_2 = 60$, $y_3 = 100$, $y_4 = 140$, $y_5 = 180$, $y_6 = 220$, $y_7 = 260$, $y_8 = 300$, $y_9 = 340$, and $y_{10} = 380$. From Figure 2.7 (and from use of the theory of similar triangles),

$$f_X(y_i) = \frac{y_i}{400}(0.005),$$

so that using Equation (2.5),

$$E(X) \cong (20)[\Delta f_Y(y_1)] + 60[\Delta f_Y(y_2)] + \cdots + 380[\Delta f_Y(y_{10})]$$
$$= (20)(40)\left[\frac{20}{400}(0.005)\right] + (60)(40)\left[\frac{60}{400}(0.005)\right]$$
$$+ \cdots + 380(40)\left[\frac{380}{400}(0.005)\right] = 266.00 \text{ pounds},$$

2. MEASURES OF LOCATION FOR A DISTRIBUTION

and the total amount of meat the delicatessen will sell in a year is approximately $(365)(266.00) = 96{,}506$ pounds.

The exact value of $E(X)$ can be calculated by means of the calculus and is 266.66 pounds. Thus, by using the rectangular method of approximation in this case, the delicatessen made an error in calculating $E(X)$ of only 0.66 pounds, and an error in calculating $(365)[E(X)]$ of about 241 pounds (a small error compared to the total quantity being estimated).

Despite the apparent success of the rectangular method of approximation in Example 2.6, it is worth repeating the warning given in Chapter 4 that the rectangular method is not always this accurate, nor is it usually the best available method of numerical approximation in any particular case.

Note: Another warning about the use of the approximate method for calculating $E(X)$ described above should be given here. If the collection of possible values for the random variable X has no smallest value x_S and/or no largest value x_L, the method described in Chapter 4 [which involved finding numbers x_1 and $x_2 > x_1$ so that the area between x_1 and x_2 under the graph of $f_X(\tau)$ is nearly equal to 1 and setting $x_S = x_1$, $x_L = x_2$] does not always give good results when used to approximate $E(X)$. The reason for this inaccuracy is that even though the probability that X is less than x_1 or greater than x_2 is small (so that possible values of X less than x_1 or greater than x_2 can be ignored when calculating the probabilities of events of the form $\{a \leq X \leq b\}$), these small probabilities when weighted by numbers much smaller than x_1 or much larger than x_2 can add an appreciable amount to the overall area between the graph of $\tau f_X(\tau)$ and the horizontal axis, [that is to the value of $E(X)$], and thus cannot be ignored when using the "rectangular method" to approximate $E(X)$. There are several possible ways of handling this problem, but since these are not directly relevant to our purposes here (to introduce and motivate the expected value of a continuous random variable), we do not pursue discussion of this matter further.

The Expected Value as a Measure of Location

As the title of this section asserts, the expected value is a measure of location for a probability distribution. We have not proven this assertion, although hopefully the interpretation of the expected value as the center of gravity of a probability distribution gives some intuitive support to our claim. Another way of seeing that $E(X)$ is a measure of location for the distribution of the random variable X is to investigate what would result if we formed a new random variable $X + c$ based on X and took its expected value. Let $Z = X + c$. Suppose that X is a discrete random variable with possible values $x_1, x_2, \ldots$, and that $P_X\{X = x_i\} = p_i$, $i = 1, 2, \ldots$. Now note that $X = x_i$ if and only if $Z = x_i + c$. Thus, the events $\{X = x_i\}$ and $\{Z = x_i + c\}$ are equivalent, and

$$p_i = P_X\{X = x_i\} = P_Z\{Z = x_i + c\}.$$

Let $z_i = x_i + c$, $i = 1, 2, \ldots$. From the definition [Equation (2.1)] of $E(Z)$,

$$\begin{aligned} E(Z) &= z_1 P_Z\{Z = z_1\} + z_2 P_Z\{Z = z_2\} + \cdots \\ &= (x_1 + c) P_Z\{Z = x_1 + c\} + (x_2 + c) P_Z\{Z = x_2 + c\} + \cdots \\ &= (x_1 + c) p_1 + (x_2 + c) p_2 + \cdots \\ &= (x_1 p_1 + x_2 p_2 + \cdots) + c(p_1 + p_2 + \cdots). \end{aligned}$$

Because $E(X) = x_1 p_1 + x_2 p_2 + \cdots$ and $p_1 + p_2 + \cdots = P_X\{\mathscr{X}\} = 1$, it follows that

(2.6) $$E(X + c) = E(Z) = E(X) + c.$$

Hence if X is relocated to a new random variable $X + c$ by addition of a constant, c, $E(X)$ is also relocated in the same manner into $E(X + c) = E(X) + c$. Equation (2.6) is a property of the expected value of a random variable that is often of use. Another relationship that is frequently of use in dealing with expected values is

(2.7) $$E(aX + c) = aE(X) + c,$$

where a and c are any constants. We can verify Equation (2.7) for discrete random variables X by calculations similar to those by which we verified Equation (2.6). To verify that Equations (2.6) and (2.7) also hold for continuous random variables X, the following *heuristic* approach can be followed:

(i) First use the rectangular method described above to form a discrete random variable Y which approximates X. Remember that we can make Y approximate X as closely as we like by dividing the range of possible values of X into a large enough number M of subintervals.

(ii) Apply Equations (2.6) and (2.7) to Y and obtain the results that

$$E(Y + c) = E(Y) + c, \qquad E(aY + c) = aE(Y) + c.$$

(iii) Note that $Y + c$ and $aY + c$ approximate $X + c$ and $aX + c$, respectively.

(iv) Finally, put the above remarks together to obtain the results

$$E(X + c) \cong E(Y + c) = E(Y) + c \cong E(X) + c,$$
$$E(aX + c) \cong E(aY + c) = aE(Y) + c \cong aE(X) + c.$$

By taking M large enough, we can change the above approximate equali-

3. OTHER MEASURES OF LOCATION FOR A DISTRIBUTION

ties into equalities, thus verifying Equations (2.6) and (2.7) in the case of continuous random variables.

Notation and Terminology

The expected value of a random variable X is also commonly referred to as the *mean* of the distribution of the random variable X. This latter terminology has arisen because the value of the expected value $E(X)$ is usually a number in the middle range of possible values of the random variable X. Because in physics the center of gravity of a distribution is called the first moment of inertia of the distribution, the expected value $E(X)$ of a probability distribution is also called the *first moment* of that distribution.

Partly as a mnemonic for the above two synonyms (mean, moment) for the expected value of a random variable, and partly as a notational convenience, the expected value of a random variable X is often denoted by the symbol μ_X (μ is the Greek letter mu corresponding to the Latin letter M). Depending on notational and pedagogical convenience, we will use the symbols $E(X)$ and μ_X interchangeably throughout the remainder of this book. When the context is clear as to what random variable is being considered, we use the symbol μ instead of the subscripted symbol μ_X.

3. OTHER MEASURES OF LOCATION FOR A DISTRIBUTION: MEDIAN, QUANTILE, MODE

The expected value $E(X)$ of a random variable X is not the only measure of location that we could use to describe the distribution of X. Other measures of location that are often used are medians, quantiles, and modes.

Medians

A *median* of the distribution of a random variable X is any number $\text{Med}(X)$ having the property that the probability that X is strictly less than $\text{Med}(X)$ is no greater than $\frac{1}{2}$, and the probability that X is strictly greater than $\text{Med}(X)$ is no greater than $\frac{1}{2}$. That is,

(3.1) $\quad P_X\{X < \text{Med}(X)\} \leq \frac{1}{2} \quad$ and $\quad P_X\{X > \text{Med}(X)\} \leq \frac{1}{2}$.

Because $P_X\{X > \text{Med}(X)\}$ is less than or equal to $\frac{1}{2}$ if and only if $1 - P_X\{X > \text{Med}(X)\}$ is greater than or equal to $(1 - \frac{1}{2}) = \frac{1}{2}$, and because $1 - P_X\{X > \text{Med}(X)\} = P_X\{X \leq \text{Med}(X)\}$, it follows from Equation (3.1) that an equivalent way of defining a median of the random variable

X is to say that Med(X) must be a number such that

(3.2) $$P_X\{X < \text{Med}(X)\} \le \tfrac{1}{2} \le P_X\{X \le \text{Med}(X)\}.$$

Since we know that the cumulative distribution function (c.d.f.) $F_X(\tau)$ evaluated at any number τ equals the probability of the event $\{X \le \tau\}$, we see from Equation (3.2) that a median of X is any number Med(X) for which $F_X(\tau) \le \tfrac{1}{2}$ for all numbers τ which are strictly less than Med(X), and for which $F_X(\text{Med}(X)) \ge \tfrac{1}{2}$. That is,

(3.3) $$F_X(\tau) \le \tfrac{1}{2} \le F_X(\text{Med}(X))$$

for all numbers $\tau < \text{Med}(X)$. This last way of defining a median of the distribution of X suggests a convenient way of finding Med(X) in situations where the c.d.f. $F_X(\tau)$ of X is available in either graphical or tabular form. Note that we say "a median" rather than the "median" because more than one number may satisfy Equations (3.1), (3.2), or (3.3).

Suppose first that we have available the graph of $F_X(\tau)$. Draw a line from the point 0.5 on the vertical axis parallel to the horizontal axis (see Figures 3.1 and 3.2), and note where this line intersects the graph of $F_X(\tau)$. From any point of intersection between the line we have constructed and the graph of $F_X(\tau)$, drop a perpendicular to the horizontal axis. The value $\tau = \text{Med}(X)$ where this perpendicular meets the horizontal axis is a median of the distribution of X.

Figures 3.1 and 3.2 illustrate cases where only one number, $\tau = \text{Med}(X)$, satisfies Equations (3.1), (3.2), and (3.3), and hence there is a

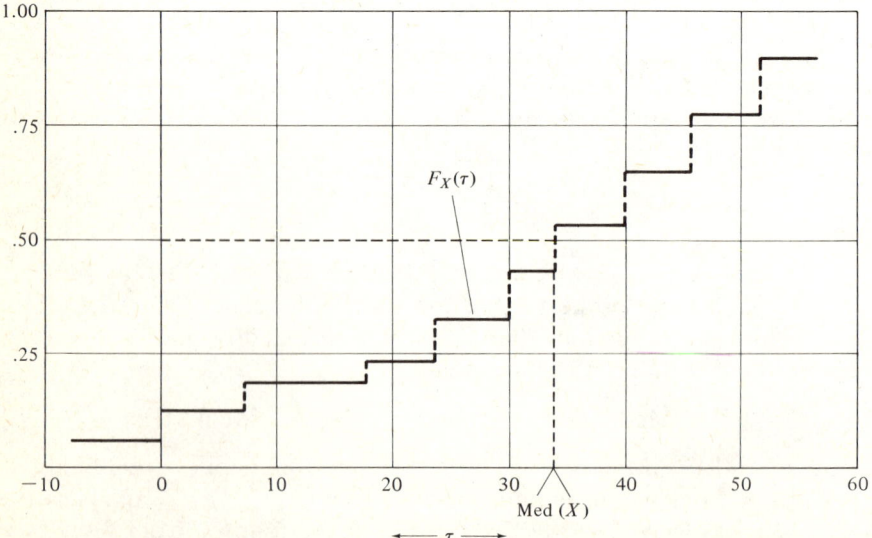

Figure 3.1 Finding a median, Med(X), of a discrete random variable X.

3. OTHER MEASURES OF LOCATION FOR A DISTRIBUTION 195

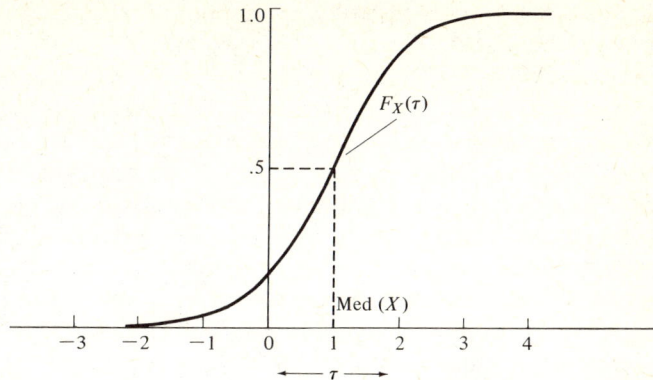

Figure 3.2 Finding a median, Med(X), of a continuous random variable X.

unique value for the median. Notice that in both Figure 3.1 and Figure 3.2, the value of the median is a possible value of X. [In Figure 3.1, Med(X) = 34 and is a point at which the graph of the c.d.f. jumps. Hence, Med(X) is a possible value of X. A similar comment holds for Figure 3.2, where Med(X) = 1.0 is also a possible value of X.] In general, whenever there is a unique value for the median, that value is a possible value for the random variable X. In contrast, recall that the expected value of X need not be a possible value of X.

As we have mentioned, there are cases where more than one number satisfies Equations (3.1), (3.2), or (3.3). In such cases (see Figure 3.3), $F_X(\tau)$ equals $\frac{1}{2}$ for values of τ in an interval of numbers ranging from a lower value τ_L to an upper value τ_U [that is, $F_X(\tau) = \frac{1}{2}$ for $\tau_L \leq \tau \leq \tau_U$]. All

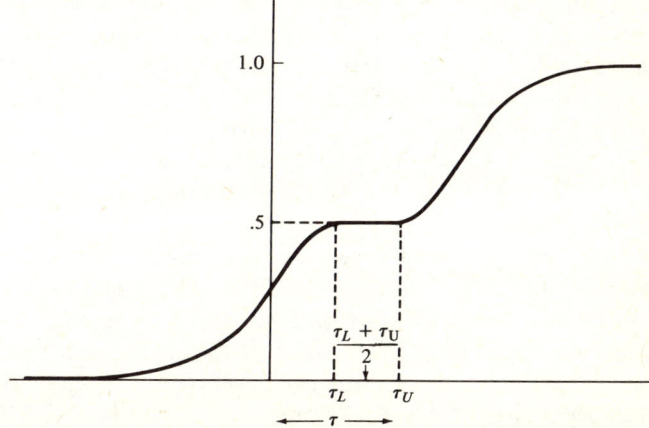

Figure 3.3 A case where there is not a unique value for the median.

numbers τ such that $F_X(\tau) = \frac{1}{2}$ (that is, all numbers τ such that $\tau_L \leq \tau \leq \tau_U$) are medians of the distribution of X. The graphical method of finding the median from the c.d.f. $F_X(\tau)$ allows us to find τ_L, the smallest number that is a median of X. We could say in this case that $\text{Med}(X) = \tau_L$; however, common usage in statistics is to report instead that $\text{Med}(X) = (\tau_L + \tau_U)/2$, the midpoint of the interval of numbers that are medians of X. Either method of assigning a unique number to be reported as "the median of X" is acceptable [since both τ_L and $(\tau_L + \tau_U)/2$ are medians of X], but the method of calculation should always be indicated. The reason why we do not usually report that all numbers between τ_L and τ_U are medians is that it is awkward to combine, compare, or algebraically manipulate intervals of numbers. For this reason, in cases where there is no unique value for the median, a single value for the median is nevertheless reported.

If no graph of the c.d.f. $F_X(\tau)$ is available, but we do have a table of values for $F_X(\tau)$ corresponding to selected values of τ (as in Table 3.1), we can proceed in the following manner to find an *approximate* value of a median of the random variable X:

(i) We look in the table and see if there exists any number τ^* for which $F_X(\tau^*) = \frac{1}{2}$. If one and only one such value τ^* exists for which $F_X(\tau^*) = \frac{1}{2}$, we say that $\text{Med}(X) = \tau^*$. If in the table there is more than one value τ for which $F_X(\tau) = \frac{1}{2}$, we set $\text{Med}(X)$ to be equal to the average of the smallest and the largest values of τ in the table for which $F_X(\tau) = \frac{1}{2}$.

(ii) If there is no number τ^* in the table for which $F_X(\tau^*) = \frac{1}{2}$, we look in the table and find the largest value of τ, say $\tau = \tau_1$, for which $F_X(\tau) < \frac{1}{2}$. We also find the smallest value of τ, say $\tau = \tau_2$, for which $F_X(\tau) > \frac{1}{2}$. Of necessity (by the way we defined τ_1 and τ_2), $\tau_2 > \tau_1$ and the values τ_1 and τ_2 are adjacent to one another in the table (that is, there is no value of τ represented in the table which is between τ_1 and τ_2). Once we have τ_1 and τ_2, we use the method of *linear interpolation* to obtain an (approximate) value for $\text{Med}(X)$. That is,

$$\text{Med}(X) \cong \left[\frac{\frac{1}{2} - F_X(\tau_1)}{F_X(\tau_2) - F_X(\tau_1)}\right]\tau_2 + \left[\frac{F_X(\tau_2) - \frac{1}{2}}{F_X(\tau_2) - F_X(\tau_1)}\right]\tau_1.$$

As an example, consider Table 3.1 which gives values of the c.d.f. $F_X(\tau)$ of a certain random variable X corresponding to selected values of τ (in this case values of τ increasing by steps of 0.5 from -3.0 to 3.0). Glancing at this table, we see that there does not exist a value of τ in the table for which $F_X(\tau) = \frac{1}{2}$. Hence, we turn to step (ii) and notice that $\tau_1 = 0.0$ is the largest value for τ in the table for which $F_X(\tau) < \frac{1}{2}$. Similarly, $\tau_2 = 0.5$ is the smallest value for τ in the table for which $F_X(\tau) > \frac{1}{2}$. Using the method

3. OTHER MEASURES OF LOCATION FOR A DISTRIBUTION

Table 3.1. Table of values for the c.d.f. $F_X(\tau)$ corresponding to selected values of τ.

τ	$F_X(\tau)$
−3.0	0.001
−2.5	0.019
−2.0	0.057
−1.5	0.111
−1.0	0.189
−0.5	0.298
0.0	0.476
0.5	0.583
1.0	0.692
1.5	0.783
2.0	0.857
2.5	0.954
3.0	0.999

of linear interpolation, we find that

$$\text{Med}(X) \cong \left[\frac{\frac{1}{2} - F_X(0.0)}{F_X(0.5) - F_X(0.0)}\right](0.5) + \left[\frac{F_X(0.5) - \frac{1}{2}}{F_X(0.5) - F_X(0.0)}\right](0.0)$$

$$= \left[\frac{0.024}{0.107}\right](0.5) + \left[\frac{0.083}{0.107}\right](0.0) = 0.112.$$

It should be emphasized that using the method of linear interpolation in a table of values for the c.d.f. $F_X(\tau)$ of X need not result in a true median of X, but only an approximation to a median. On the other hand, if there is a value of τ^* in the table for which $F_X(\tau^*) = \frac{1}{2}$, then τ^* is a median of the distribution of X. Even here, however, it is possible that the value we report for Med(X) from looking at the table will differ from the number we would have reported if we had a graph of the c.d.f. $F_X(\tau)$. The reason for this is that the table gives only selected values of $F_X(\tau)$, while the graph displays all values of $F_X(\tau)$.

Quantiles

Quantiles arise from a straightforward generalization of the concept of a median. Let p be a number between (and including) 0 and 1. The *pth quantile* of the distribution of the random variable X is any number $Q_p(X)$ such that

(3.4) $\quad P_X\{X < Q_p(X)\} \leq p \quad$ and $\quad P_X\{X > Q_p(X)\} \leq 1-p,$

or equivalently such that

(3.5) $\quad\quad\quad\quad\quad\quad F_X(\tau) \leq p \leq F_X(Q_p(X))$

for all numbers $\tau < Q_p(X)$. When $p = \frac{1}{2}$, the definitions of a pth quantile of X and a median of X coincide [compare Equations (3.1) and (3.4)], so that *a median is a $\frac{1}{2}$th quantile* of the distribution of the random variable X.

For a given distribution, and for fixed p, $0 \leq p \leq 1$, there may be more than one number which is a pth quantile. In such cases, there exists an interval of values for τ, say $\tau_{pL} \leq \tau \leq \tau_{pU}$, for which $F_X(\tau) = p$. We arbitrarily set

$$Q_p(X) = \frac{\tau_{pL} + \tau_{pU}}{2}$$

in such cases, in order to always have a single number as our reported value of the pth quantile. (Our reasons for requiring a single number here are the same as the reasons we have given for arbitrarily choosing one number to be the median of X in cases where there are many medians.)

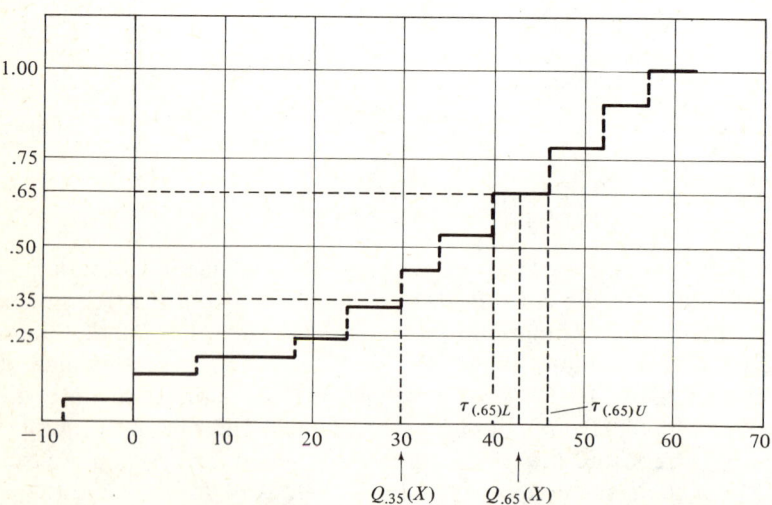

Figure 3.4 Finding the pth quantile $Q_p(X)$ of the distribution of a discrete random variable X. Two cases, $p_1 = 0.35$ and $p_2 = 0.65$, are shown. In the second case, there is more than one (0.65)th quantile, and we choose $Q_{0.65}(X)$ to be the midpoint $(\tau_{(0.65)L} + \tau_{(0.65)U})/2$ of the interval of numbers which are (0.65)th quantiles.

3. OTHER MEASURES OF LOCATION FOR A DISTRIBUTION

The pth quantile $Q_p(X)$ can be obtained from a graph of the c.d.f. $F_X(\tau)$ of X by a method similar to that by which we obtain $\text{Med}(X)$. The only difference is that the line that we draw parallel to the horizontal axis is drawn from the value p on the vertical axis. The method is illustrated in Figures 3.4 and 3.5.

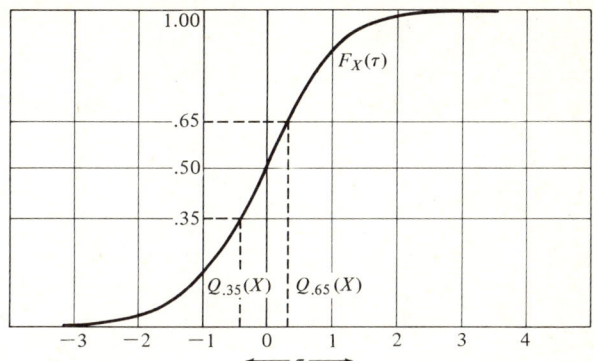

Figure 3.5 Finding a pth quantile $Q_p(X)$ of the distribution of a continuous random variable X. Two cases, $p_1 = 0.35$ and $p_2 = 0.65$, are shown.

Example 2.1 (Continued). Earlier we discussed probability models for the number X of children born to a randomly chosen lower-class urban family and the number Y of children born to a randomly chosen upper-class suburban family. From the tables of probabilities of simple events given in Example 2.1 for these two discrete random variables, we can calculate and then graph the cumulative distribution functions $F_X(\tau)$ and $F_Y(\tau)$. From the graphs (see Figure 3.6) of $F_X(\tau)$ and $F_Y(\tau)$, we can calculate the quantiles of these distributions that are of interest to us. Thus,

$Q_{0.10}(X) = 1.5,$ $Q_{0.10}(Y) = 0.0,$
$Q_{0.25}(X) = 3.0,$ $Q_{0.25}(Y) = 1.0,$
$Q_{0.50}(X) = \text{Med}(X) = 4.0,$ $Q_{0.50}(Y) = \text{Med}(Y) = 3.0,$
$Q_{0.75}(X) = 5.5,$ $Q_{0.75}(Y) = 5.0,$
$Q_{0.90}(X) = 6.5,$ $Q_{0.90}(Y) = 6.0,$

and so on. Notice that all of the quantiles calculated above satisfy the inequality $Q_p(X) \geq Q_p(Y)$. This inequality agrees in direction with the inequality $E(X) \geq E(Y)$, which we earlier observed in connection with Example 2.1.

The pth quantile $Q_p(X)$ of a random variable X may be approximated from a table of values of the c.d.f. $F_X(\tau)$ of X by using the method

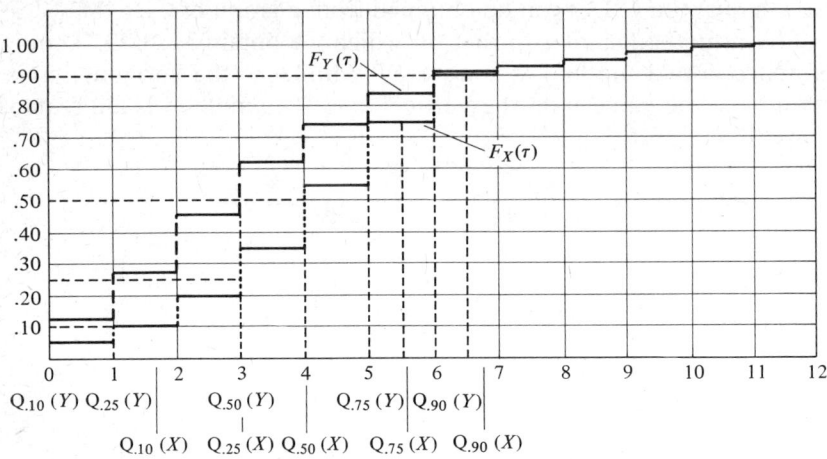

Figure 3.6 The c.d.f.'s $F_X(\tau)$ and $F_Y(\tau)$, respectively, of the number X of children born to a randomly chosen lower-class urban family, and the number Y of children born to a randomly chosen upper-class suburban family.

of linear interpolation. We proceed as follows:

(i) We look in the table and see if there is a value τ^* of τ for which $F_X(\tau^*) = p$. If there is one and only one such value τ^* in the table, we set $Q_p(X) = \tau^*$. If in the table there is more than one value τ for which $F_X(\tau) = p$, we set $Q_p(X)$ to be equal to the average of the smallest and the largest values of τ in the table for which $F_X(\tau) = p$.

(ii) If there is no number τ^* in the table for which $F_X(\tau^*) = p$, we find the largest value τ_1 of τ (in the table) for which $F_X(\tau) < p$ and the smallest value τ_2 of τ for which $F_X(\tau) > p$. Of necessity, $\tau_1 < \tau_2$ and the values τ_1 and τ_2 are adjacent to one another in the table. Once we have τ_1 and τ_2, we set

$$Q_p(X) \cong \left[\frac{p - F_X(\tau_1)}{F_X(\tau_2) - F_X(\tau_1)}\right]\tau_2 + \left[\frac{F_X(\tau_2) - p}{F_X(\tau_2) - F_X(\tau_1)}\right]\tau_1.$$

Thus, for example, if we are interested in the (0.25)th quantile of the random variable X whose c.d.f. $F_X(\tau)$ is tabulated in Table 3.1 (for selected values of τ), then using the method of linear interpolation we find that $\tau_1 = -1.0$, $\tau_2 = -0.5$, and

$$Q_{0.25}(X) \cong \left[\frac{0.25 - F_X(-1.0)}{F_X(-0.5) - F_X(-1.0)}\right](-0.5)$$
$$+ \left[\frac{F_X(-0.5) - 0.25}{F_X(-0.5) - F_X(-1.0)}\right](-1.0)$$
$$= \frac{0.061}{0.109}(-0.5) + \frac{0.048}{0.109}(-1.0) = -0.720.$$

3. OTHER MEASURES OF LOCATION FOR A DISTRIBUTION

Note: We reiterate our earlier warning that the method of linear interpolation is an approximate method and does not always yield the same value of $Q_p(X)$ as does the (accurate) graphical method based on the graph of the c.d.f. $F_X(\tau)$ of X. For example, if we tabulated the c.d.f. $F_X(\tau)$ of the number X of children born to a randomly chosen lower-class urban family (see Example 2.1), we might obtain the following table (where the only values of τ included are possible values of X):

τ	$F_X(\tau)$	τ	$F_X(\tau)$
0	0.05	6	0.90
1	0.10	7	0.93
2	0.20	8	0.95
3	0.35	9	0.97
4	0.55	10	0.99
5	0.75	11	1.00

If, for example, we use the method of linear interpolation in this table to obtain $Q_{0.25}(X)$, we find that $Q_{0.25}(X) \cong 2.33$, whereas the precise value of $Q_{0.25}(X) = 3.0$. The number $\tau = 2.33$ is not a (0.25)th quantile for the distribution of X, but rather a (0.20)th quantile.

The approximation obtained from a table of values for the c.d.f. $F_X(\tau)$ improves as we tabulate $F_X(\tau)$ for more and more values of τ. For example, if we tabulate $F_X(\tau)$, the c.d.f. of the random variable X discussed in Example 2.1, for $\tau = 0.0, 0.5, 1.0, 1.5, 2.0, 2.5$, and so on, we find from the method of linear interpolation that $Q_{0.25}(X) \cong 2.67$. This approximation is closer to the correct value $Q_{0.25}(X) = 3.0$ than is our previous approximation [which was based on a table of $F_X(\tau)$ for $\tau = 0.0, 1.0, 2.0$, and so on].

A pth quantile is a measure of location for a probability distribution. $Q_p(X)$ is a number that divides the distribution of mass into two parts. The total probability mass corresponding to values of X that are less than $Q_p(X)$ is no greater than p, and the total probability mass corresponding to values of X that are larger than $Q_p(X)$ is no greater than $(1-p)$. If we graph the mass function of a discrete random variable X (Figure 3.7), then the point $Q_p(X)$ on the horizontal axis has the property that there is no more than $(100p)$ percent of the total probability mass to the left of $Q_p(X)$ and no more than $[100(1-p)]$ percent of the total probability mass to the right of $Q_p(X)$. Similarly, if we graph the density function $f_X(\tau)$ of a continuous random variable X (Figure 3.8), then the point $Q_p(X)$ on the horizontal axis divides the total area under the graph of $f_X(\tau)$ into a part to the left of $Q_p(X)$ of area p and a part to the right of $Q_p(X)$ of area

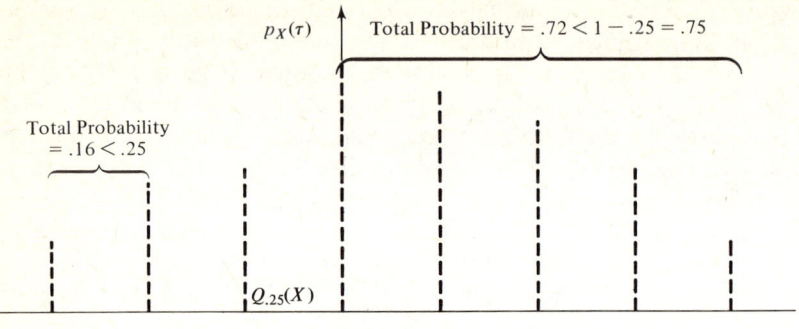

Figure 3.7 Graph of the probability mass function $p_X(\tau)$ of a discrete random variable X showing how the (0.25)th quantile $Q_{0.25}(X)$ divides the probability mass into two parts.

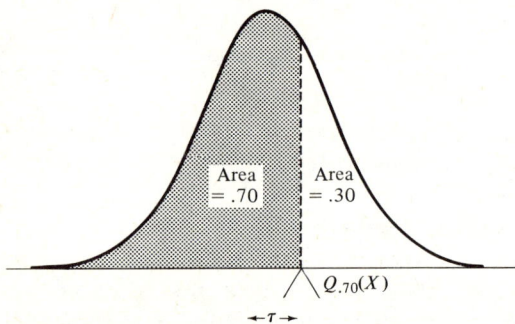

Figure 3.8 Graph of the probability density function $f_X(\tau)$ of a continuous random variable X showing how the (0.70)th quantile $Q_{0.70}(X)$ divides the total area under the graph of $f_X(\tau)$ into two parts.

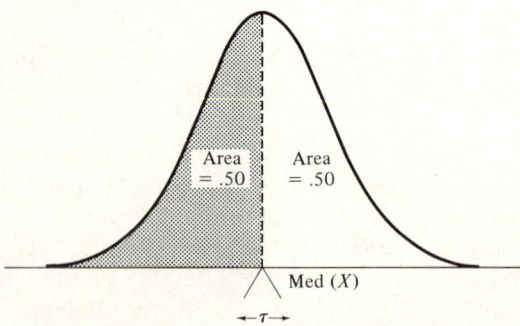

Figure 3.9 Graph of the probability density function $f_X(\tau)$ of a continuous random variable X showing how the median. Med(X), divides the probability mass function into two equal (or nearly equal) parts.

3. OTHER MEASURES OF LOCATION FOR A DISTRIBUTION

$(1-p)$. It follows from the above interpretation of the pth quantile $Q_p(X)$ that for any constants a and b,

(3.6) $$Q_p(aX+b) = aQ_p(X) + b.$$

Since the median $\text{Med}(X)$ is a (0.50)th quantile, the median is a measure of location, and

(3.7) $$\text{Med}(aX+b) = a\text{Med}(X) + b,$$

for any constants a and b. The median is a number that divides the distribution of mass into two equal or nearly equal parts (see Figure 3.9).

Before turning to a discussion of modes, we should remark that quantiles are known by other names. We have already noted that $Q_{0.50}(X)$ is also a median of X. The quantiles $Q_{0.25}(X)$ and $Q_{0.75}(X)$ are sometimes called the *lower quartile* and the *upper quartile* of the distribution of X, respectively. The quantiles $Q_{0.10}(X)$, $Q_{0.20}(X)$, $Q_{0.30}(X)$, ..., $Q_{0.80}(X)$, $Q_{0.90}(X)$ are also known as the *deciles* (first decile, second decile, third decile, ..., eighth decile, ninth decile) of X. Finally, for any p, $Q_p(X)$ is often called the $(100p)$th *percentile* or the (p)th *fractile* of the distribution of X.

Modes

A *mode* of a random variable is (one of) its "most probable" value(s). A mode of the distribution of a discrete random variable X is that possible value, $\text{Mode}(X)$, of X at which the probability mass function $p_X(\tau)$ of X has its largest value. There can be one, or more than one, mode of the distribution of a discrete random variable (see Figures 3.10 and 3.11).

Since every possible value x of a continuous random variable X has zero probability of occurrence, the definition of a mode of a continuous

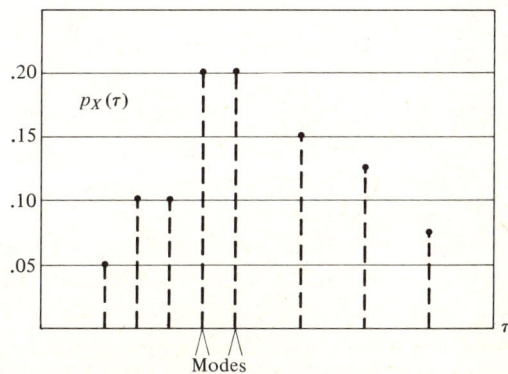

Figure 3.10 A discrete unimodal distribution.

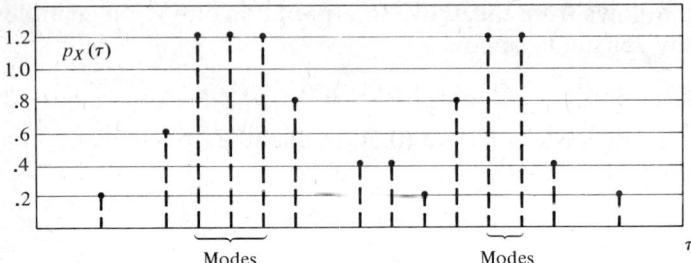

Figure 3.11 A discrete bimodal distribution.

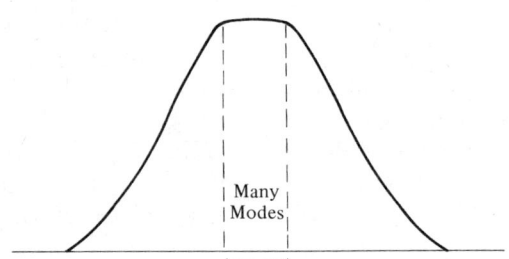

Figure 3.12 A continuous unimodal distribution.

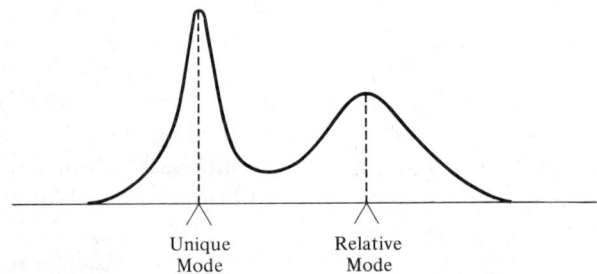

Figure 3.13 A continuous bimodal distribution.

random variable cannot refer to the value of X which has the largest probability. Otherwise every number would be a mode of a continuous random variable; clearly this would not be a helpful definition! Instead, a mode of the distribution of a continuous random variable X with probability density function $f_X(\tau)$ is that possible value of X at which $f_X(\tau)$ is largest. Again, there can be one, or more than one, mode of the distribution of a continuous random variable (see Figures 3.12 and 3.13).

Many books define a *relative mode* of a discrete (continuous) random variable X to be any possible value of X for which the probability mass function (probability density function) is *locally* the highest. Examples of

3. OTHER MEASURES OF LOCATION FOR A DISTRIBUTION

this concept are shown in Figures 3.10 to 3.13. A mode, of course, is always a relative mode.

Example 2.1 (Continued). In Figure 2.2 we graphed the probability mass functions of X, the number of children born to a randomly chosen lower-class urban family, and Y, the number of children born to a randomly chosen upper-class suburban family. From this figure, we see that the largest value of $p_X(\tau)$ is 0.20, and that $p_X(\tau) = 0.20$ for $\tau = 4$ and $\tau = 5$. Hence $\tau = 4$ and $\tau = 5$ are modes of the distribution of X. On the other hand, the largest value of $p_Y(\tau)$ is 0.19. This largest value of $p_Y(\tau)$ is achieved only for $\tau = 2$. Thus $\tau = 2$ is the unique mode of the distribution of Y. From Figure 2.2, we see that the modes of X and Y are the only relative modes of the respective distributions of these random variables.

Note: The descriptive terms "unimodal" and "bimodal" are often used in connection with distributions. A *unimodal distribution* is a distribution having the characteristic that the shape of its probability mass function or probability density function is that of a single "hump" (see Figures 3.10 and 3.12). In contrast, a *bimodal distribution* has a probability mass function or probability density function with two "humps" (see Figures 3.11 and 3.13). A bimodal distribution has two different separated relative modes, but may have a single unique mode. A unimodal distribution may have several modes or only one mode. The terms "unimodal" and "bimodal" thus do *not* refer to the *number* of modes (or of relative modes), but rather are terms describing the "humpiness" of the probability mass function or probability density function (whichever is appropriate).

Choice of a Measure of Location

There are many potential measures of location for a distribution. In Sections 2 and 3, we have discussed a few of these: the expected value $E(X)$, the median $\text{Med}(X)$, the quantiles $Q_p(X)$, and the mode (or modes). Note that there are as many quantiles as there are numbers p between 0 and 1. Thus, even if we restrict ourselves to choosing a particular quantile as a measure of location, we must choose among an infinite number of possible alternatives! In addition to the measures of location that we have described, there are infinitely many other possible measures of location, many of which have been used at one time or another to describe a probability distribution. Hence we see that there is a staggeringly rich choice of measures of location available to us in any given situation. Which measure of location do we choose?

There is no conclusive answer to the question just raised. Each measure of location that we have discussed in Sections 2 and 3 (and each of the alternative measures that we have not discussed) has its advantages and disadvantages. It is not possible to say that one measure is the "best" in every case. Ultimately, our choice of a measure of location for a distribution depends both on what facet of that distribution we want to describe, and on what measure of location most unambiguously conveys the information we wish to transmit. In making our choice, we should also give some consideration to such properties of a measure of location as: (i) the ease with which the measure can be computed, (ii) the ease with which such a measure can be compared to (or combined with) other measures, and (iii) the sensitivity of the value of the measure to slight variations in the shape of the distribution. Property (iii) of a measure of location must be considered since we often try to determine a probability distribution of a random variable X from a finite number of observations of X. Since only a finite number of trials are observed, the stability property of relative frequencies discussed in Chapter 1 may not be totally in effect, and several different probability models may describe (or fit) the data equally well. If possible, in such situations we would like our summary measure for a distribution (such as a measure of location) to be relatively insensitive to our selection of a probability distribution.

Because the choice of a measure of location depends so heavily both on what we want to say about a distribution and on the distribution we actually have in hand, guidelines for such a choice are difficult to state, and subject to frequent exceptions. In defining each of the measures of location that we have introduced, we have tried to indicate what information these measures convey about a distribution. The following example illustrates how the information that a measure of location gives about a distribution is influenced by the shape of the distribution, and why the choice of a measure of location is largely determined by the goals of the investigator.

Example 3.1. Two economists are studying the distribution of income in the United States. Economist A is preparing his study for a congressional committee which is considering poverty-oriented legislation. Economist B is reporting the results of his work to a market research firm. Both economists decide to investigate the probability distribution of the total pre-tax income, X, of a randomly chosen U.S. family. Here, X is given in dollars, and an unmarried person counts as a single family. Although X is admittedly discrete (since income is at best measured to the nearest 1/100 of a dollar), because X has thousands of possible values, it is both mathematically and conceptually more convenient to treat X as if it were a continuous random variable with a probability density function $f_X(\tau)$. Suppose X has the density function $f_X(\tau)$ graphed in Figure 3.14. (This graph is somewhat crudely based on information taken from the 1959 *Statistical Abstract of the United States*.) In Figure 3.14, we have

3. OTHER MEASURES OF LOCATION FOR A DISTRIBUTION

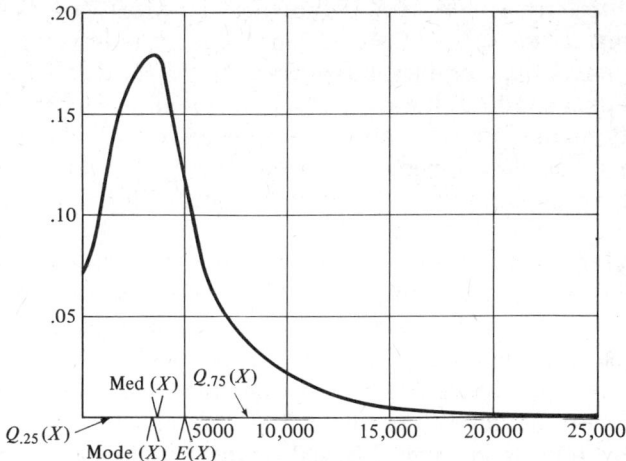

Figure 3.14 Probability density function of the income (in dollars) of a randomly chosen U.S. family.

indicated on the horizontal axis the (approximate) values of $Q_{0.25}(X)$, $Q_{0.50}(X) = \text{Med}(X)$, $Q_{0.75}(X)$, the (unique) mode of X, and the expected value $E(X)$.

Since Economist A is studying the distribution of income in the United States in order to provide guidance to a congressional committee as to the extent of poverty (or the prevalence of low incomes) in the United States, he decides not to use the expected value $E(X)$ to describe the location of the distribution of X, since the value of this measure can be greatly influenced by the incomes of the very few wealthy families in the country (say, those with incomes over \$50,000). Even though the probability that a randomly chosen family is wealthy is quite small, in calculating $E(X)$ that small number is multiplied by a very large value for X [see Equation (2.1)] and the resulting product may add an appreciable quantity to the value of $E(X)$. In other words, a few large possible values of X, even though they have low probability, can inflate $E(X)$ to such an extent that its value does not reflect the incomes of the large majority of families [in the present case $P_X\{X \leq E(X)\}$ is about 0.60]. Consequently, the majority of families appear to be better off financially than is actually the case.

On the other hand, the value of $\text{Med}(X)$ is not influenced by large incomes occurring with small probability (see Figure 3.14). Indeed, we can calculate $\text{Med}(X)$ without ever looking at the right-hand tail of the graph in Figure 3.14. [We simply find a number τ^* such that the area under the graph of $f_X(\tau)$ between $\tau = 0$ and $\tau = \tau^*$ is equal to $\frac{1}{2}$; this method does not require that we know how much of the probability to the right of $\text{Med}(X)$

is assigned to large values of X.] Knowing that $\text{Med}(X) = \$3500$ [which is approximately the value of $\text{Med}(X)$ in this case], tells us that 50 percent of the probability mass of the random variable X is associated with income values less than $3500. If the congressional committee decides that $3500 is a poverty income for a family and is considering legislation to supplement such incomes, knowing that $\text{Med}(X) = \$3500$ tells them that 50 percent of the population of the United States will be affected by their legislation.

Similarly, as long as p is not too close to 1, $Q_p(X)$ is not affected by extremely large incomes that have small probability. Knowing that $Q_{0.25}(X) = \$2100$ [which is approximately the value of $Q_{0.25}(X)$ in this case], the congressmen on the committee would know that 25 percent of the families in the United States have incomes less than $2100. [*Note:* Our figures are intended to be realistic, but remember that they refer to 1959 pre-inflationary incomes!] Based on these arguments, Economist *A* decides to report one or more quantiles (including the median) of the distribution of X to the congressional committee.

On the other hand, Economist *B* is as much (or more) interested in the very wealthy families as he is in the mass of low and middle income families. The market research firm for which he works has come up with a theory that says that family consumer purchases for any given year can be predicted from knowledge of family income in the immediately preceding year. They want to know how much consumer spending to expect for the coming year. If Economist *B* reports $\text{Med}(X)$ instead of $E(X)$ as the "average" family income, he will be ignoring the incomes of the very rich families, and this ultimately will lead his firm to underpredict the amount of consumer spending for the coming year. The measure $E(X)$ allows us to estimate the total income for *all* families in the United States [using the frequency interpretation of $E(X)$; see Section 2]. Since $\text{Med}(X)$, or indeed any quantile $Q_p(X)$, takes little or no account of the incomes of the very wealthy, use of the quantiles does not allow us to recapture total U.S. income, and thus is of questionable help in predicting total consumer purchases for the coming year. Thus, Economist *B* decides to report the value of $E(X)$ to his firm.

Neither Economist *A* nor Economist *B* would give much consideration to the mode of X as a measure of location. Economist *A* would not be interested because the mode, reflecting as it does the income (or interval of incomes) with highest probability, tells him little about how low the incomes of the lower income families actually are. The mode of X could be at a very low income (one exceeded by the incomes of almost all families), a middle income, or a high income. Thus, the mode tells little about the need for poverty legislation in the United States. On the other hand, Economist *B* is not interested in the mode because from the mode of the

3. OTHER MEASURES OF LOCATION FOR A DISTRIBUTION 209

distribution of incomes, he has no way of calculating even approximately the total income of all U.S. families. Even if the mode were relevant to their goals, however, both investigators would hesitate to use the mode as a measure of location for the following reasons:

(i) the mode is sensitive to slight and relatively unimportant changes in the shape of the probability distribution;
(ii) in contrast to the expected value (which is unique) or any (p)th quantile (which even when not unique, has values which are always adjacent to one another), there may be two widely separated values for the mode (see Figure 3.13). If this is the case, it is not clear which value of the mode is being reported. Thus, whenever a mode is reported, we cannot be sure (unless we are told) that there is not another quite different possible value of X that is also a mode of the distribution of X;
(iii) whereas the expected value $E(X)$, the median $\text{Med}(X)$, and each of the pth quantiles $Q_p(X)$ are usually intermediate between the largest and smallest possible values of X (when these exist), a mode of a random variable X can be either the largest or the smallest possible value of X. Further, a mode of a random variable X can be a number which is quite far away from the interval of values over which most of the probability mass of the distribution lies.

Although the above failings of the mode as a measure of location eliminate the mode from consideration in most problems of scientific or practical interest, there is one situation where the mode is the preferred measure of location. This occurs when we want to know a possible value (or interval of possible values) for the random variable X which will allow us to have the highest probability of making a correct guess. We then guess that an observation of X will equal this value (or fall in that interval of values).

In most situations where a measure of location for a distribution is desired, choice of such a measure is usually made between the expected value $E(X)$ and the median $\text{Med}(X)$. As we have seen, each one of these measures of location is preferred to the other in certain circumstances. However, even when the median is believed to be a more meaningful or useful measure than the expected value, the expected value is often computed along with the median. There are many reasons why investigators remain interested in the expected value even in situations where other measures of location are preferable. The influence of statistical tradition is one reason for their interest in the expected value. Another reason is the role that the expected value plays in the definition of two very important measures of dispersion: the variance and standard deviation. We discuss the variance and standard deviation in the next section.

4. MEASURES OF DISPERSION FOR A DISTRIBUTION: THE VARIANCE

The *variance* of a random variable X is a measure of the dispersion of the distribution of X around the expected value (center of gravity) of the distribution. To define the variance of X, it is convenient to consider the function $(X-\mu_X)^2$ of X which measures how close X is to its expected value $E(X) = \mu_X$. (Notice that we are now using the alternate notation for the expected value which we mentioned in Section 2.) If the observed value of X is far from the center of gravity μ_X, then the observed value of $(X-\mu_X)^2$ is a large number; if the observed value of X is near μ_X, then the observed value of $(X-\mu_X)^2$ is small. Thus, if X tends to vary widely about its expected value μ_X, there are quite a few outcomes for which X deviates from μ_X by a large amount, and, consequently, large values of $(X-\mu_X)^2$ have high probability. If X does not vary much about μ_X, then most outcomes of X yield values close to the expected value, and small values of $(X-\mu_X)^2$ have high probability. A way to quantify just how large the deviation $(X-\mu_X)^2$ is "on the average" is to consider the expected value of $(X-\mu_X)^2$. Because $(X-\mu_X)^2$ is a function of the random variable X, it is itself a random variable and hence may have an expected value, just as the random variable X may have an expected value. We can write this expected value as $E(X-\mu_X)^2$. This expectation is called the *variance* of X (or the variance of the distribution of X) since it quantifies how extensively X varies "on the average" about its expected value μ_X. Large values of the variance are associated with large dispersions (or variation) of X about μ_X, and small values of the variance are associated with small dispersions. If the variance of X is equal to zero, then X does not vary at all about μ_X, and $P_X\{X = \mu_X\} = 1$.

To denote the variance of X, we follow tradition in probability and statistics and use the symbol σ_X^2. (σ is the Greek letter sigma.) Thus,

$$(4.1) \qquad \sigma_X^2 = E(X-\mu_X)^2.$$

When it is clear from the context that we are talking about the variance σ_X^2 of a particular random variable X, we frequently drop the subscript X on σ_X^2, and write σ^2.

The term *standard deviation of X* refers to the square root, $\sigma_X = \sqrt{\sigma_X^2}$ of the variance of X. The standard deviation is expressed in terms of the same units of measurement as X, while the variance is not. For this reason, the standard deviation is frequently used as a measure of dispersion in preference to the variance. Once we have obtained the variance σ_X^2, we can easily obtain the standard deviation σ_X; therefore, we concentrate on properties of the variance in the remainder of this section.

How do we calculate the variance σ_X^2 of a particular random variable X? From the definition we have given of σ_X^2, and from the way we defined

4. MEASURES OF DISPERSION FOR A DISTRIBUTION

the expected value of a random variable in Section 2, it might appear necessary to first find the probability distribution of $Y = (X - \mu_X)^2$ from the probability distribution of X by the means described in Chapter 4, Section 1, and then compute from that distribution the expected value of Y. Fortunately, this roundabout way of calculating $\sigma_X^2 = E(X - \mu_X)^2$ is not really necessary (although it sometimes proves useful) *since the expected value of any function $g(X)$ of a random variable X can be calculated directly from knowledge of the distribution of X*. In the case where X is a discrete random variable with possible values $x_1, x_2, \ldots,$

(4.2) $\quad E[g(X)] = g(x_1)P_X\{X = x_1\} + g(x_2)P_X\{X = x_2\} + \cdots,$

while when X is a continuous random variable with probability density function $f_X(\tau)$,

(4.3) $\quad E[g(X)] =$ "the area between the graph of $g(\tau)f_X(\tau)$ and the horizontal axis."

We illustrate the truth of this assertion in the special case where the random variable X is discrete, $g(X) = (X - \mu_X)^2$, and the probability distribution of X is given by the following table of probabilities of simple events:

x	0	1	2	3	4	5	6
$P_X\{X = x\}$	$\frac{1}{16}$	$\frac{2}{16}$	$\frac{3}{16}$	$\frac{3}{16}$	$\frac{5}{16}$	$\frac{1}{16}$	$\frac{1}{16}$

From the definition [Equation (2.1)] of $E(X) = \mu_X$, we find that

$$\mu_X = (0)\frac{1}{16} + (1)\frac{2}{16} + (2)\frac{3}{16} + (3)\frac{3}{16} + (4)\frac{5}{16} + (5)\frac{1}{16} + (6)\frac{1}{16} = 3.$$

Let us first calculate $\sigma_X^2 = E(X - \mu_X)^2 = E(X - 3)^2$ by the roundabout way described above. That is, let us find the distribution of $Y = (X - 3)^2$ and then calculate $E(Y)$. Note that Y is a discrete random variable having only four possible values: $y_1 = 0, y_2 = 1, y_3 = 4, y_4 = 9$. Note also that $Y = 0$ if $X = 3$, that $Y = 1$ if $X = 2$ or $X = 4$, that $Y = 4$ if $X = 1$ or $X = 5$, and that $Y = 9$ if $X = 0$ or $X = 6$. Thus,

(4.4) $\quad P_Y\{Y = 0\} = P_X\{X = 3\} = \frac{3}{16},$

$P_Y\{Y = 1\} = P_X\{X = 2 \text{ or } X = 4\} = P_X\{X = 2\} + P_X\{X = 4\} = \frac{8}{16},$

$P_Y\{Y = 4\} = P_X\{X = 1 \text{ or } X = 5\} = P_X\{X = 1\} + P_X\{X = 5\} = \frac{3}{16},$

$P_Y\{Y = 9\} = P_X\{X = 0 \text{ or } X = 6\} = P_X\{X = 0\} + P_X\{X = 6\} = \frac{2}{16}.$

and hence

$$E(Y) = (0)P_Y\{Y=0\} + (1)P_Y\{Y=1\} + (4)P_Y\{Y=4\} + (9)P_Y\{Y=9\}$$
$$= (0)\left(\frac{3}{16}\right) + (1)\left(\frac{8}{16}\right) + (4)\left(\frac{3}{16}\right) + (9)\left(\frac{2}{16}\right) = \frac{19}{8}.$$
(4.5)

On the other hand, if we use Equation (4.2) to calculate $E[g(X)] = E(X-3)^2$, we find that

$$E(X-3)^2 = (0-3)^2 P_X\{X=0\} + (1-3)^2 P_X\{X=1\} + (2-3)^2 P_X\{X=2\}$$
$$+ (3-3)^2 P_X\{X=3\} + (4-3)^2 P_X\{X=4\}$$
(4.6)
$$+ (5-3)^2 P_X\{X=5\} + (6-3)^2 P_X\{X=6\}$$
$$= (0)P_Y\{Y=0\} + (1)P_Y\{Y=1\} + (4)P_Y\{Y=4\}$$
$$+ (9)P_Y\{Y=9\} = \frac{19}{8}.$$

The results shown in Equation (4.6) were obtained by grouping terms for which $(X-3)^2$ is the same, using the relations (4.4) between the probabilities for X and the probabilities for Y, and then using Equation (4.5). Comparison of Equation (4.5) and (4.6) demonstrates, at least when $g(X) = (X-3)^2$ and X has the special discrete distribution given above, that we can calculate $E[g(X)]$ directly from the distribution of X by use of Equation (4.2). A look at how we obtained Equation (4.6) also indicates how we might prove that Equation (4.2) holds for general discrete random variables X and general functions $g(X)$.

From Equation (4.2), we conclude that if X is a discrete random variable having expected value $E(X) = \mu_X$, then $\sigma_X^2 = E(X - \mu_X)^2$ can be calculated from the formula

(4.7) $\quad \sigma_X^2 = (x_1 - \mu_X)^2 P_X\{X = x_1\} + (x_2 - \mu_X)^2 P_X\{X = x_2\} + \cdots,$

where $x_1, x_2, \ldots$ are the possible values of X. Similarly, if X is a continuous random variable having expected value μ_X and density function $f_X(\tau)$, then σ_X^2 can be calculated from the formula

$$\sigma_X^2 = \text{"area between the graph of } (\tau - \mu_X)^2 f_X(\tau) \text{ and the horizontal axis."}$$

We now give some examples of distributions of discrete random variables X to show how the variance σ_X^2 of X is related to the spread, dispersion, or variation of X around its expected value. In Figure 4.1, the probability mass of the distribution of the discrete random variable X is equally distributed at the numbers $x_1 = \mu - \Delta$, $x_2 = \mu + \Delta$, where Δ is some constant. That is, $P_X\{X = x_1\} = P_X\{X = x_2\} = \frac{1}{2}$. The expected

4. MEASURES OF DISPERSION FOR A DISTRIBUTION

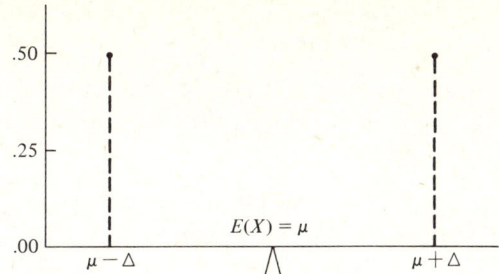

Figure 4.1 Probability distributed equally between two values of the discrete random variable X equally spaced about the expected value $E(X) = \mu$ of X. The variance σ_X^2 of X is Δ^2.

value of this discrete distribution is $E(X) = \mu$. The variance σ_X^2 of X is

$$\sigma_X^2 = (x_1 - \mu)^2 P_X\{X = x_1\} + (x_2 - \mu)^2 P_X\{X = x_2\}$$
$$= (-\Delta)^2 \tfrac{1}{2} + (\Delta)^2 \tfrac{1}{2} = \Delta^2.$$

The larger the value of Δ, the larger the spread of the distribution about its expected value becomes, and the larger the variance becomes.

If one random variable Y has a larger expected value than another random variable X, there are many situations in which Y would be considered preferable to X. However, small variances may also be desirable. The next example illustrates that large expected values and small variances do not necessarily go together.

Example 4.1. Two bowlers, Don and Mike, are having a dispute. Don admits that Mike is the better bowler in that Mike has a higher score "on the average," but insists that he himself is better in that he is a more consistent bowler (that is, has less variation in his bowling scores). Each bowler has a record of his performance, and from these records (using the experimental method described in Chapter 2, Section 5) they determine the probability mass functions for their bowling scores. Let X be Don's bowling score for a randomly chosen game, and let Y be Mike's bowling score. Suppose that the following tables describe the distributions of X and of Y:

Don's Score x	139	140	141	142	143	144	145	146	147	148	149
$P_X\{X = x\}$	0.01	0.03	0.05	0.10	0.18	0.26	0.18	0.10	0.05	0.03	0.01

Mike's Score y	145	146	147	148	149	150	151	152	153	154	155
$P_Y\{Y = y\}$	0.02	0.05	0.08	0.12	0.14	0.18	0.14	0.12	0.08	0.05	0.02

For each bowler, the probability model only reflects scores that were assigned nonzero probabilities. From the above tables, we find that

$$\begin{aligned}\mu_{\text{Don}} = E(X) &= (139)(0.01) + (140)(0.03) + (141)(0.05) + (142)(0.10) \\ &+ (143)(0.18) + (144)(0.26) + (145)(0.18) + (146)(0.10) \\ &+ (147)(0.05) + (148)(0.03) + (149)(0.01) \\ &= 144,\end{aligned}$$

$$\begin{aligned}\mu_{\text{Mike}} = E(Y) &= (145)(0.02) + (146)(0.05) + (147)(0.08) + (148)(0.12) \\ &+ (149)(0.14) + (150)(0.18) + (151)(0.14) + (152)(0.12) \\ &+ (153)(0.08) + (154)(0.05) + (155)(0.02) \\ &= 150,\end{aligned}$$

so that the expected value $E(Y) = \mu_{\text{Mike}}$ of Mike's bowling score is larger than the expected value $E(X) = \mu_{\text{Don}}$ of Don's bowling score by 6 points. On the other hand, the variances of Don's bowling score and of Mike's bowling score are respectively:

$$\begin{aligned}\sigma^2_{\text{Don}} &= (139-144)^2(0.01) + (140-144)^2(0.03) + \cdots + (148-144)^2(0.03) \\ &+ (149-144)^2(0.01) \\ &= (25)(0.01) + (16)(0.03) + (9)(0.05) + (4)(0.10) + (1)(0.18) \\ &+ (0)(0.26) + (1)(0.18) + (4)(0.10) + (9)(0.05) + (16)(0.03) \\ &+ (25)(0.01) \\ &= 3.52,\end{aligned}$$

and

$$\begin{aligned}\sigma^2_{\text{Mike}} &= (145-150)^2(0.02) + (146-150)^2(0.05) + \cdots + (154-150)^2(0.05) \\ &+ (155-150)^2(0.02) \\ &= (25)(0.02) + (16)(0.05) + (9)(0.08) + (4)(0.12) + (1)(0.14) \\ &+ (0)(0.18) + (1)(0.14) + (4)(0.12) + (9)(0.08) + (16)(0.05) \\ &+ (25)(0.02) \\ &= 5.28.\end{aligned}$$

A comparison of the variances of Don's score and of Mike's score indicates that Don is right; he is the more consistent bowler.

Some caution should be noted with respect to the use of σ_X^2 as a measure of variation. The relationship of the degree of clustering of the distribution to the size of σ_X^2 holds in most practical cases; however, some mathematical examples can be given where the variance is not a good indicator of the shape of the distribution. For example, for the random variable X having the probability model

x	0	d
$P_X\{X = x\}$	$1 - \dfrac{1}{d}$	$\dfrac{1}{d}$

,

4. MEASURES OF DISPERSION FOR A DISTRIBUTION

where d is a constant, $d > 1$, we have

$$\mu_x = (0)\left[1 - \left(\frac{1}{d}\right)\right] + (d)\left(\frac{1}{d}\right) = 1,$$

and

$$\sigma_x^2 = (0-1)^2\left[1 - \left(\frac{1}{d}\right)\right] + (d-1)^2\left(\frac{1}{d}\right)$$
$$= d - 1.$$

When $d = 10$, the probability model for X is given by

x	0	10
$P_X\{X = x\}$	0.9	0.1

and $\mu_X = 1$, $\sigma_X^2 = 10 - 1 = 9$. When $d = 100$, the probability model for X becomes

x	0	100
$P_X\{X = x\}$	0.99	0.01

and $\mu_X = 1$, $\sigma_X^2 = 100 - 1 = 99$. Finally, when $d = 1000$, the probability model for X becomes

x	0	1000
$P_X\{X = x\}$	0.999	0.001

and $\mu_X = 1$, $\sigma_X^2 = 1000 - 1 = 999$. Thus, as d becomes larger, almost all of the probability mass of X becomes located at zero, with less and less probability being placed at d, and thus X is less and less variable (but more and more dispersed). On the other hand, the variance σ_X^2 increases as d grows larger, implying that X is more and more variable. The reason for this contradiction is that the variance is unduly sensitive to a small amount of probability placed far out from the main body of the probability mass of the distribution. In Section 3, we noted that the expected value $E(X)$ had a similar sensitivity to extreme values assumed with small probability, and that because of this sensitivity it is sometimes preferable to use one of the quantiles $Q_p(X)$ in place of $E(X)$ as a measure of location. Similarly, many probabilists and statisticians would use a measure of dispersion based on the quantiles in place of the variance in cases where extreme values may be assumed with small probability. One such measure of dispersion based on the quantiles is the *semi-interquartile range* $R = [Q_{0.75}(X) - Q_{0.25}(X)]/2$. A justification for using this as an index of dispersion follows from the fact that the interval of numbers between and including $Q_{0.25}(X)$ and $Q_{0.75}(X)$ has a length of $2R$ and is associated with at least 50 percent of the total probability mass of the distribution of X. Likewise, 50% of the total probability mass is centered within $\pm R$ units of a certain central value, namely the midpoint $[Q_{0.25}(X)+Q_{0.75}(X)]/2$ or *mid-range* of the interval of numbers between $Q_{0.25}(X)$ and $Q_{0.75}(X)$. Because it is not as mathematically convenient a measure of dispersion as is the variance, the semi-interquartile range is infrequently used in the analysis of probability distributions. This measure and other similar measures [such as the *semi-interdecile range* $\frac{1}{2}[Q_{0.90}(X) - Q_{0.10}(X)]$ and so on], however, are often useful in statistical inference.

We will have more to say about calculation of the variance in the next section (Section 5), where we show how the variance is related to a certain collection of descriptive measures for a distribution called the *moments*. In particular, we provide one example of the calculation of the variance of a continuous random variable. In Section 6, we show how knowledge of the expected value and/or variance of a random variable X enables us to make certain statements concerning the magnitude of the probabilities of extreme (extremely large or extremely small) values of X even when we know nothing else about the distribution of X.

Before we close Section 4, however, it is worth mentioning (and proving) one very important property of the variance σ_X^2 of a random variable X. This property is the following: *if X is a random variable with variance σ_X^2, and if for any two numbers a and c we let $Z = aX+c$, then Z is a random variable with variance σ_Z^2 given by*

$$\sigma_Z^2 = a^2 \sigma_X^2.$$

That is, for any two numbers a and c, it is the case that

(4.9) $$\sigma_{aX+c}^2 = a^2 \sigma_X^2.$$

The above property (4.9) of the variance of a random variable X is useful in many ways. In particular, suppose that we wish to change the units of measurement in terms of which we are measuring a given random variable X. For example, X may be a measurement of temperature in degrees Fahrenheit which we wish to convert to a measurement Z in degrees Centigrade. Similarly, X may be a measurement of height in feet which we wish to convert to a measurement Z in inches. In the former case, $Z = (5/9)X + (-160/9)$, so that $a = 5/9$ and $c = -160/9$. In the latter case, $Z = 12X$, so that $a = 12$, $c = 0$. Suppose that we know the variance σ_X^2 of X. We could, of course, obtain the distribution of Z from knowledge of the distribution of X (if this knowledge is available!), and then from the distribution of Z compute the variance σ_Z^2 of Z by the methods described in this section. However, Equation (4.9) provides an alternative, and far easier, way of obtaining σ_Z^2.

To verify Equation (4.9), we make use of Equation (2.7). Since $Z = aX + c$, it follows from Equation (2.7) that

$$\mu_Z = E(aX + c) = aE(X) + c = a\mu_X + c,$$

so that $\mu_{aX+c} = \mu_Z = a\mu_X + c$. Thus,

(4.10) $$\sigma_Z^2 = E(Z - \mu_Z)^2 = E[(aX + c) - (a\mu_X + c)]^2$$
$$= E[a(X - \mu_X)]^2$$
$$= E[a^2(X - \mu_X)^2].$$

Let $d = a^2$ and $W = (X - \mu_X)^2$. Since $E(W)$ is a measure of location for the distribution of the random variable W, we know from Equation (2.7) that

$$E(dW + 0) = dE(W) + 0 = dE(W).$$

However, from Equation (4.10), remembering that $d = a^2$ and $W = (X - \mu_X)^2$,

$$\sigma_Z^2 = E(dW + 0) = dE(W) = dE(X - \mu_X)^2 = a^2\sigma_X^2,$$

which proves Equation (4.9). Further use will be made of this property of the variance of a random variable in Chapter 9.

5. HIGHER MOMENTS

The expected value or mean of the random variable X is sometimes referred to as the *first moment* of the distribution of X. A generalization of this concept is the expected value of the random variable $(X - c)^r$ ($r = 1, 2, \ldots$) which is referred to as the *rth moment about the point c* of the distribution of X. When c is the expected value μ_X, these are called *central moments*. [*Note:* Given the first r moments about any point c, it is straightforward, although tedious, to find the first r moments about any other point b.]

The moments of a random variable play a role in a method of approximating the distribution of a random variable. That is, occasionally a number of moments of a distribution are known, but the distribution itself is unknown. Mathematical techniques can be employed to find a convenient distribution which possesses the given moments. Using this convenient distribution, we can calculate the probabilities of events that are of interest to us. These probabilities are often reasonably close to the probabilities that we could have calculated if we knew the true distribution.

How do we calculate the rth moment of X about the constant c? Note that $(X - c)^r$ is a function of X so that we can apply Equations (4.2) and

(4.3). Thus, when X is a discrete random variable with possible values $x_1, x_2, \ldots,$

(5.1) $\quad E(X-c)^r = (x_1-c)^r P_X\{X=x_1\} + (x_2-c)^r P_X\{X=x_2\} + \cdots,$

while when X is a continuous random variable with probability density function $f_X(\tau)$,

(5.2) $\quad E(X-c)^r =$ "area (taken algebraically) between the graph of $(\tau-c)^r f_X(\tau)$ and the horizontal axis."

As we noted in Section 2, the first moment $\mu_X = E(X)$ of X about $c = 0$ is a measure of location for the distribution of X. The first moment of X around $c = \mu_X$ is [from Equation (2.6)],

(5.3) $\qquad E(X-\mu_X) = E(X) + (-\mu_X) = \mu_X - \mu_X = 0.$

The second moment of X around $c = \mu_X$ is the variance σ_X^2 of X, and (as we pointed out in Section 4) is a measure of dispersion for X. Indeed, the arguments that we used to justify $\sigma_X^2 = E(X-\mu_X)^2$ as a measure of dispersion for X can also serve to justify the second moment of X around any constant c as a measure of dispersion for X. Why, then, do we choose $c = \mu_X$ and $\sigma_X^2 = E(X-\mu_X)^2$ to provide us with a measure of dispersion for X? Why not choose $c = 0$, or $c = 5$, or any other value of c? Part of the justification for choosing σ_X^2 comes from the physical interpretation of σ_X^2 as the second moment of inertia of the probability mass about its center of gravity; and part of the justification for σ_X^2 comes from the usefulness of μ_X as a center or measure of location for the distribution of X. However, a mathematical justification for choosing σ_X^2 is the following: *Among all quantities of the form $E(X-c)^2$, choosing $c = \mu_X$ results in the smallest value.* That is, σ_X^2 is the smallest second moment about a central number c, and

(5.4) $\qquad\qquad\qquad \sigma_X^2 \leq E(X-c)^2,$

for all constants c. Inequality (5.4) holds as a result of the very useful formula

(5.5) $\qquad\qquad E(X-c)^2 = \sigma_X^2 + (\mu_X-c)^2,$

which relates any second moment of X to the variance of X. Since $(\mu_X-c)^2$ is nonnegative, $\sigma_X^2 + (\mu_X-c)^2 \geq \sigma_X^2$. Thus if we can show that (5.5) is true, (5.4) will follow as an immediate consequence.

In general, the proof of the validity of formula (5.5) requires much cumbersome mathematics. The idea behind the proof, however, can be revealed by proving (5.5) for the case when X is a discrete random variable having two possible values x_1 and x_2. Let $p_1 = P_X\{X=x_1\}$ and $p_2 =$

$P_X\{X = x_2\}$. Note that
$$(x_i-c)^2 = [(x_i-\mu_x)+(\mu_X-c)]^2 = (x_i-\mu_X)^2 + 2(x_i-\mu_x)(\mu_x-c)+(\mu_X-c)^2,$$
so that
$$\begin{aligned}(5.6)\quad E(X-c)^2 &= (x_1-c)^2 p_1 + (x_2-c)^2 p_2 \\ &= [(x_1-\mu_X)^2 + 2(x_1-\mu_X)(\mu_X-c)+(\mu_X-c)^2]p_1 \\ &\quad + [(x_2-\mu_X)^2 + 2(x_2-\mu_X)(\mu_X-c)+(\mu_X-c)^2]p_2.\end{aligned}$$

Now from Equation (4.7),
$$\sigma_X^2 = (x_1-\mu_X)^2 p_1 + (x_2-\mu_X)^2 p_2,$$
and from Equation (5.3),
$$\begin{aligned}&2(x_1-\mu_X)(\mu_X-c)p_1 + 2(x_2-\mu_X)(\mu_X-c)p_2 \\ &= 2(\mu_X-c)[(x_1-\mu_X)p_1 + (x_2-\mu_X)p_2] \\ &= 2(\mu_X-c)E(X-\mu_X) = 0.\end{aligned}$$

Thus it follows from Equation (5.6) that
$$\begin{aligned}E(X-c)^2 &= \sigma_X^2 + 0 + (\mu_X-c)^2 p_1 + (\mu_X-c)^2 p_2 \\ &= \sigma_X^2 + (\mu_X-c)^2(p_1+p_2),\end{aligned}$$
and since $p_1 + p_2 = 1$,
$$E(X-c)^2 = \sigma_X^2 + (\mu_X-c)^2,$$
proving Equation (5.5) in this special case. The method of proof in the general case is similar.

Let $c = 0$ in Equation (5.5). We obtain the equation
$$E(X)^2 = \sigma_X^2 + (\mu_X)^2,$$
and solving for σ_X^2, we obtain
$$(5.7)\qquad \sigma_X^2 = E(X)^2 - (\mu_X)^2.$$

This formula often provides a convenient shortcut for computing the variance.

Example 5.1. Suppose the continuous random variable X has density function $f_X(\tau)$ defined as follows:
$$f_X(\tau) = \begin{cases}(1.4430)/\tau, & \text{if } 1 \leq \tau \leq 2, \\ 0, & \text{if } \tau < 1 \text{ or } \tau > 2.\end{cases}$$

A graph of $f_X(\tau)$ appears in Figure 5.1. Recall that $\mu_X = E(X)$ is found by computing the area (taken algebraically) between the horizontal axis and the graph of $\tau f_X(\tau)$. The graph of $\tau f_X(\tau)$ rises above the horizontal axis only for values of τ between 1 and 2 (see Figure 5.2) and is constant in height over that interval of τ values. Thus, the area we need to compute to

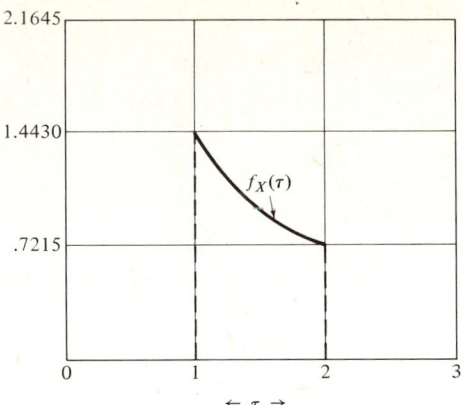

Figure 5.1 Graph of the probability density function $f_X(\tau)$.

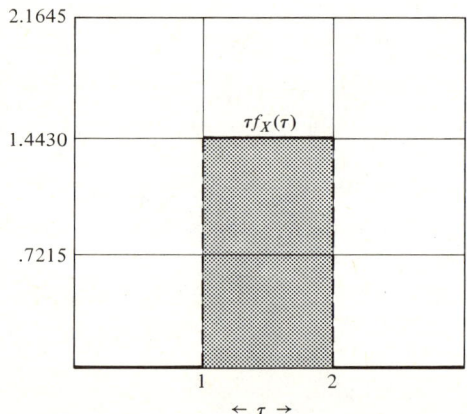

Figure 5.2 Graph of $\tau f_X(\tau)$. The area of the shaded part is equal to $\mu_X = E(X)$.

find μ_X is that of a rectangle with base of length equal to 1, and height equal to 1.4430. Thus $\mu_X = (1)(1.4430) = 1.4430$.

To find σ_X^2, we could use Equation (4.8) and find the area between the graph of

$$f_x(\tau) = \begin{cases} (\tau - 1.4430)^2(1.4430/\tau), & \text{if } 1 \leq \tau \leq 2, \\ 0, & \text{if } \tau < 1 \text{ or } \tau > 2, \end{cases}$$

and the horizontal axis, but this is a difficult task without the aid of integral calculus. However, let us first compute $E(X)^2$, which is the area between $\tau^2 f(\tau)$ and the horizontal axis. The graph of $\tau^2 f_X(\tau)$ is given in Figure 5.3. From this graph, we see that $E(X)^2$ is equal to the area of a trapezoid whose altitude is the interval of numbers between $\tau = 1$ and

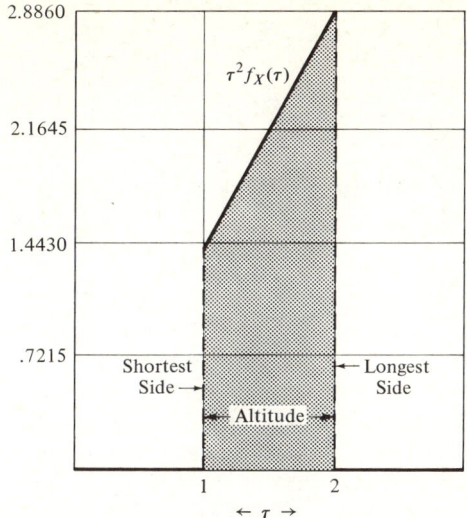

Figure 5.3 Graph of $\tau^2 f_X(\tau)$. The area of the shaded part is equal to $E(X)^2$.

$\tau = 2$, of length equal to 1, whose smallest side has length equal to $(1)^2 f_X(1) = 1.4430$, and whose largest side has length equal to $(2)^2 f_X(2) = 2.8860$. Since the area of a trapezoid equals the length of its altitude multiplied by the average of the lengths of the longest and smallest sides,

$$E(X)^2 = (1)\left(\frac{1.4430 + 2.8860}{2}\right) = 2.1645.$$

Using Equation (5.7), we conclude that

$$\sigma_X^2 = E(X)^2 - (\mu_X)^2 = 2.1645 - (1.4430)^2 = 0.0823.$$

Despite the fact that this example of calculating the variance of a continuous random variable works out nicely using results from geometry and Equation (5.7), in general we need the theory of integral calculus to compute the variance of a continuous random variable.

We have already discussed the usefulness of the first moment about 0, namely $E(X)$, and the second central moment $E(X - \mu_X)^2 = \sigma_X^2$. Two more central moments, the third central moment and the fourth central moment, are also often used to describe properties of a distribution. The *third central moment*, $E(X - \mu_X)^3$, is a measure of the symmetry of the distribution of a random variable about its expected value. The third central moment of X is zero when the distribution of X is symmetric about μ_X (see Figure 5.4); it tends to be negative when the distribution is *skewed*

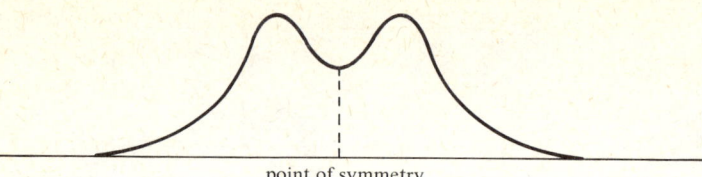

Figure 5.4 Graph of a probability density function which is symmetric.

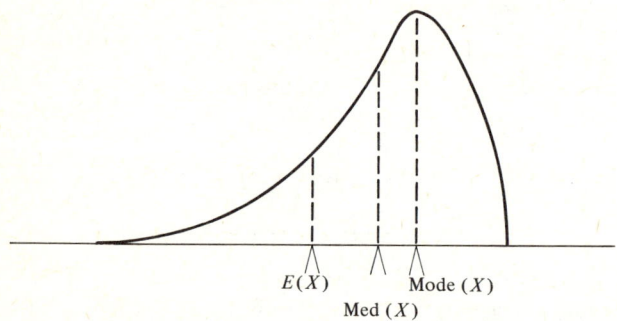

Figure 5.5 Graph of a probability density function which is skewed to the left. Note the relative positions of the expected value, median, and mode.

to the left (or negatively skewed; see Figure 5.5); and it tends to be positive when the distribution is *skewed to the right* (or positively skewed; see Figure 5.6). The fourth central moment $E(X-\mu_X)^4$ gives an indication of the "peakedness" or "kurtosis" of a distribution. The very famous *Normal Distribution* (see Chapter 7) serves as a standard. For this distribution, the ratio of $E(X-\mu_X)^4$ to the square of the variance [that is, $E(X-\mu_X)^4/(\sigma_X^2)^2$] is equal to 3. If the ratio $E(X-\mu_X)^4/(\sigma_X^2)^2$ exceeds 3, we say that the distribution of X is "less peaked" (or has greater kurtosis)

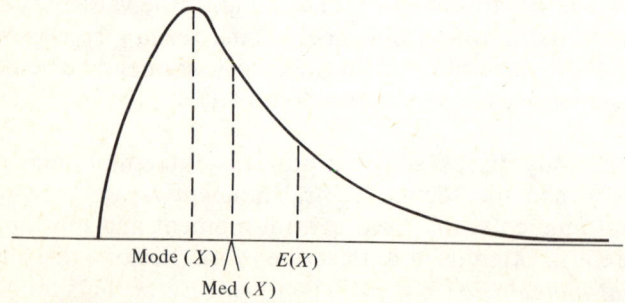

Figure 5.6 Graph of a probability density function which is skewed to the right. Note the relative positions of the expected value, median, and mode.

than the normal distribution; if this ratio is less than 3, we say that the distribution of X is "more peaked" (or has less kurtosis) than the normal distribution (see Figures 5.7 and 5.8). Knowledge of the location, dispersion, skewness, and peakedness of a distribution is usually enough to give a coarse, but useful, picture of the distribution. For this reason, investigators seldom go beyond computing the expected value, variance, third central moment, and fourth central moment of a distribution.

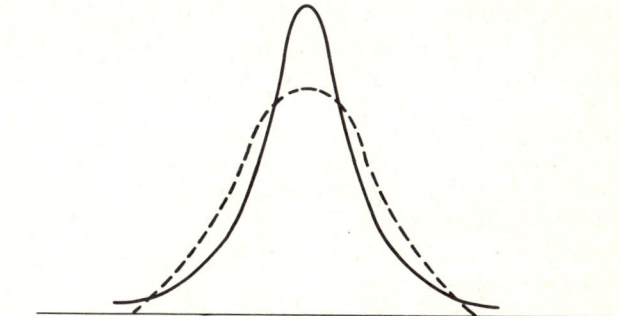

Figure 5.7 Graph of the probability density function of the normal distribution (dotted line) and a more peaked distribution (solid line).

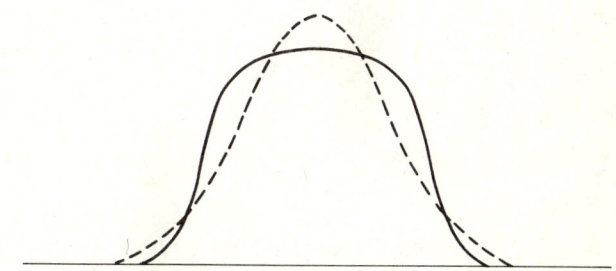

Figure 5.8 Graph of the probability density functions of the normal distribution (dotted line) and a less peaked distribution (solid line).

6. INEQUALITIES FOR PROBABILITIES

In the previous sections, various descriptive characteristics of distributions were discussed. Now suppose we are given the values of some of these descriptive measures, but are not informed as to the exact distribution which these values describe. How much of the probability distribution can we reproduce with only this information?

If only the expected value μ of the distribution is given to us (and the value of no other descriptive measure), we cannot, in general, make any statements about the probability distribution. If, however, we also know that the random variable X, which has the unknown distribution in which we are interested, can only assume positive or zero values (and is not always equal to zero), then from this fact and from the value of $\mu_X = \mu$, we can obtain information about probabilities of certain events. Regardless of the distribution of X, it can be asserted that *for any positive number c ($c > 0$), the probability, $P_X\{X \geq c\mu\}$, that the random variable X is greater than c times its expected value $\mu_X = \mu$ does not exceed $1/c$*. In mathematical notation,

(6.1) $$P_X\{X \geq c\mu\} \leq \frac{1}{c}.$$

This inequality is known as the Markov Inequality [after the Russian mathematician A. A. Markov (1903–) who first obtained the result].

*Proof of the Markov Inequality

To prove this assertion when X is a discrete random variable with possible values $0 \leq x_1 < x_2 < x_3 < \ldots$, note that

(6.2) $$\mu = E(X) = x_1 P_X\{X = x_1\} + x_2 P_X\{X = x_2\} + \cdots.$$

Let us separate the possible values $x_1, x_2, \ldots$ into two groups. One group consists of those x's which are less than $c\mu$, the other group consists of those x's which are greater than or equal to $c\mu$. Consequently, there is an integer $m \geq 1$ such that

(6.3) $$x_1 < x_2 < \cdots < x_m < c\mu \leq x_{m+1} < x_{m+2} < \cdots,$$

and from Equation (6.2),

(6.4) $$\mu = [x_1 P_X\{X = x_1\} + x_2 P_X\{X = x_2\} + \cdots + x_m P_X\{X = x_m\}]$$
$$+ [x_{m+1} P_X\{X = x_{m+1}\} + x_{m+2} P_X\{X = x_{m+2}\} + \cdots].$$

Since $x_{m+1} \geq c\mu$, $x_{m+2} \geq c\mu$, and so on, it must be the case that

$$(c\mu) P_X\{X = x_{m+1}\} \leq x_{m+1} P_X\{X = x_{m+1}\},$$
$$(c\mu) P_X\{X = x_{m+2}\} \leq x_{m+2} P_X\{X = x_{m+2}\}, \quad \cdots.$$

Hence,

$$x_{m+1} P_X\{X = x_{m+1}\} + x_{m+2} P_X\{X = x_{m+2}\} + \cdots$$
$$\geq (c\mu)[P_X\{X = x_{m+1}\} + P_X\{X = x_{m+2}\} + \cdots],$$

and substituting this last inequality into Equation (6.4), we find that

(6.5) $$\mu \geq [x_1 P_X\{X = x_1\} + x_2 P_X\{X = x_2\} + \cdots + x_m P_X\{X = x_m\}]$$
$$+ (c\mu)[P_X\{X = x_{m+1}\} + P_X\{X = x_{m+2}\} + \cdots].$$

Now since $0 \leq x_1 < x_2 < \cdots < x_m$,

$$[x_1 P_X\{X = x_1\} + x_2 P_X\{X = x_2\} + \cdots + x_m P_X\{X = x_m\}] \geq 0,$$

and therefore from Equations (6.3) and (6.5), it follows that

(6.6) $$\mu \geq c\mu [P_X\{X = x_{m+1}\} + P_X\{X = x_{m+2}\} + \cdots]$$
$$= c\mu P_X\{X \geq c\mu\}.$$

Finally, since X is never negative (and not always zero), it must be the case that $\mu = E(X) > 0$. Hence, if we divide both sides of Equation (6.6) by $c\mu$, we obtain Equation (6.1).

Use of the Markov Inequality

Inequality (6.1) is of no value if $1/c$ exceeds 1, because then it merely tells us that the probability of the event $\{X \geq c\mu\}$ never equals or exceeds a number larger than 1, a fact we know is true for any event under any distribution. However, for c greater than 1 (so that $1/c$ is less than 1), the information supplied by Inequality (6.1) is meaningful, since the inequality tells us that the probability of the event $\{X \geq c\mu\}$ can never exceed $1/c$, a number less than 1. For very large c, the inequality tells us that X is greater than or equal to $c\mu$ (that is, X is large) with very small probability. It must be noted that as long as we know that $P_X\{X \geq 0\} = 1$, Inequality (6.1) is true *no matter what distribution* governs the variability of the random variable X. Table 6.1 shows the largest possible prob-

Table 6.1. Largest possible values of the probability, $P_X\{X \geq c\mu\}$, in the tail of the distribution of a nonnegative random variable X.

c	Largest Possible Value $1/c$ of $P_X\{X \geq c\mu\}$
1.0	1.000
1.5	0.667
2.0	0.50
2.5	0.40
3.0	0.333
3.5	0.286
4.0	0.25
4.5	0.222
5.0	0.20
10.0	0.10
30.0	0.0333
50.0	0.02
100.0	0.01

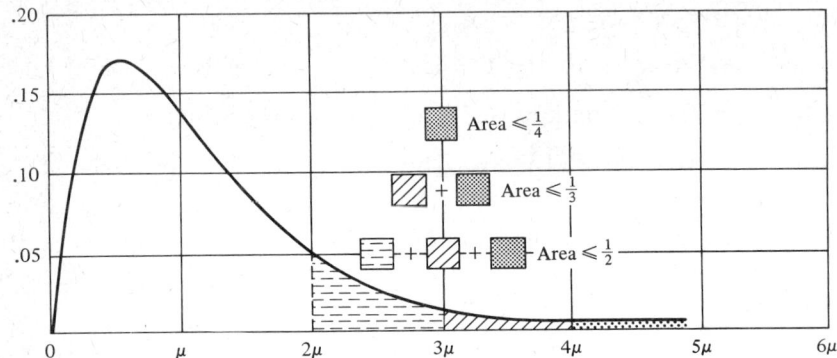

Figure 6.1 Inequality on area based on the value of $\mu = E(X)$.

abilities of the event $\{X \geq c\mu\}$ for various values of c (see Figure 6.1).

Example 6.1. A car rental company announces in its advertisements that the expected time that a customer must wait from the time he reaches their counter until the time he drives off with one of their cars is 5 minutes. Let X be the random variable which is the waiting time (in minutes) for a customer; then the company's announcement says that $\mu = E(X) = 5$ minutes. A salesman who is in a hurry to keep an appointment with an important client figures that he has a maximum of 1 hour (60 minutes) that he can spend waiting to get a rental car; after that time, he will have been delayed so much that he will miss his appointment. He can call now and make his appointment for a later time, but he knows this may make his client less receptive. On the other hand, if he is late his client may be pressed by other engagements and may not be able to see him. The salesman tries to estimate the probability that he will be delayed so badly as to be late for his appointment. He decides that if this probability is larger than 0.10, he will call and move his appointment to a later time, preferring to risk facing a less receptive customer than a busy one. Otherwise, he will keep to his schedule. Since $60 = 12\mu$, the value of c is 12. It follows from Inequality (6.1) that the probability that the salesman will be delayed more than 1 hour could possibly be as large as $1/c = 1/12 = 0.083$, but no larger. Although the salesman knows that $P_X\{X \geq 60\} = P_X\{X \geq 12E(X)\}$ may be smaller than $1/c = 0.083$ ($1/c$ is only the *largest* possible value), he knows that it cannot be larger than this number. Since 0.083 is less than 0.10, the salesman decides that he can keep to his schedule.

Inequality (6.1) is useful when we know that the random variable X of interest to us is never negative. Unfortunately, not all random variables are nonnegative. Furthermore, we may not always be interested in events of the form $\{X \geq c\mu\}$. Sometimes, events of the form $\{X \leq a \text{ or } X \geq b\}$

6. INEQUALITIES FOR PROBABILITIES

are of greater interest to us. If we know both the expected value μ and the variance σ^2 of a random variable X, then we can derive an inequality for the probability of certain events of this last form, regardless of whether X takes on only nonnegative values or not. Indeed, we can assert that for any number $d > 0$, *the probability that X is less than or equal to $\mu - d\sigma$ or is greater than or equal to $\mu + d\sigma$ cannot exceed $1/d^2$*. In mathematical notation,

(6.7) $$P_X\{X \leq \mu - d\sigma \text{ or } X \geq \mu + d\sigma\} \leq \frac{1}{d^2}.$$

Inequality (6.7) is known as the Bienaymé-Chebyshev Inequality [after the French mathematician I. J. Bienaymé (1796–1878) and the Russian mathematician P. L. Chebyshev (Čebyšev) (1821–1894), both of whom independently discovered the result]. This inequality may be proved in a manner similar to the proof of Inequality (6.1). However, it may also be obtained as a consequence of Inequality (6.1) as follows: Note that $Y = (X - \mu)^2$ is a nonnegative random variable with expected value $E(Y) = \sigma^2$. From Inequality (6.1), letting $c = d^2$, we know that $P_Y\{Y \geq d^2 E(Y)\} \leq 1/d^2$. However, $Y \geq d^2 E(Y)$ if and only if $\sqrt{Y^2} = |Y| \geq d\sigma$, where $|Y|$ is the magnitude (or absolute value, or unsigned value) of Y. In turn, $|Y| \geq d\sigma$ if and only if $Y \geq d\sigma$ or $Y \leq -d\sigma$. Since $Y = (X - \mu)^2$, putting all of the above facts together tells us that the events $\{Y \geq d^2 E(Y)\}$ and $\{X \leq \mu - d\sigma \text{ or } X \geq \mu + d\sigma\}$ are the same event. Thus,

$$P_X\{X \leq \mu - d\sigma \text{ or } X \geq \mu + d\sigma\} = P_Y\{Y \geq d^2 E(Y)\} \leq \frac{1}{d^2},$$

which verifies Inequality (6.7).

If $d \leq 1$, Inequality (6.7) furnishes no information about the distribution of X that we do not already know *without* knowing the values of μ and σ^2. If $d \leq 1$, Inequality (6.7) says that the probability of a certain event concerning X is less than a number greater than 1; since the probabilities of all events have this property, we have learned nothing new. On the other hand, if $d > 1$, then we gain some useful information about the probability of the event $\{X \leq \mu - d\sigma \text{ or } X \geq \mu + d\sigma\}$. Without knowing μ and σ^2, we could only say that this probability is a number between 0 and 1. However, by knowing the values of μ and σ^2, we can say that the probability is less than a certain positive number which is itself less than 1. In Table 6.2, this positive number (which is the largest value that the probability $P_X\{X \leq \mu - d\sigma \text{ or } X \geq \mu + d\sigma\}$ can possibly be) is tabulated for various values of d.

Suppose that from an inequality of the form of Inequality (6.1) or Inequality (6.7) we know that the probability of a certain event E involving a random variable X cannot exceed a certain number p; that is,

$$P_X(E) \leq p.$$

Table 6.2. Largest possible probability of the event $\{X \leq \mu - d\sigma \text{ or } X \geq \mu + d\sigma\}$.

d	Largest Possible Probability That Either $X \leq \mu - d\sigma$ or $X \geq \mu + d\sigma$
1	1·000
1.5	0.444
2	0.250
2.5	0.160
3	0.111
3.5	0.082
4	0.063
4.5	0.049
5	0.040
10	0.010
30	0.0011
50	0.0004
100	0.0001

Then it must follow (from facts about inequalities) that

$$1 - P_X(E) \geq 1 - p.$$

Since $1 - P_X(E) = P_X(E^c)$ by the Law of Complementation, we conclude that if $P_X(E) \leq p$, then it must be the case that $P_X(E^c) \geq 1 - p$. Hence, if we know that the probability of an event E cannot exceed p, we also know that the probability of the complement of E cannot be less than $1 - p$. Applying this result to the event $E = \{X \leq \mu - d\sigma \text{ or } X \geq \mu + d\sigma\}$ and using Inequality (6.7), we can conclude that

(6.8) $$P_X\{\mu - d\sigma < X < \mu + d\sigma\} = P_X(E^c) \geq 1 - \frac{1}{d^2}.$$

If $d \leq 1$, this statement furnishes no information about the distribution that we do not already know *without* knowing μ and σ^2. (If $d \leq 1$, the inequality says that the probability of a certain event is bigger than a negative number; but we already know that the probability of any event is always greater than or equal to 0, and thus greater than any negative number.) However, if $d > 1$, then we gain some nontrivial information about the probability of the event $\{\mu - d\sigma < X < \mu + d\sigma\}$. Without knowing μ and σ^2, we could only say that this probability is between 0 and 1. However, by knowing the values of μ and σ^2, we can say that the prob-

6. INEQUALITIES FOR PROBABILITIES

ability is greater than a certain positive number; in Table 6.3, this smallest possible probability is tabulated for various values of d.

Table 6.3. Smallest possible probability of the event $\{\mu - d\sigma < X < \mu + d\sigma\}$.

d	Smallest Possible Probability That X Is Between $\mu - d\sigma$ and $\mu + d\sigma$
1	0.000
1.5	0.556
2	0.750
2.5	0.840
3	0.889
3.5	0.918
4	0.937
4.5	0.951
5	0.960
10	0.990
30	0.9989
50	0.9996
100	0.9999

Thus, we know almost certainly (almost with probability 1) that X will be within 50 standard deviations of its mean. The huge bulk of *any* distribution (over 95 percent) lies within $4\frac{1}{2}$ standard deviations of the mean, and so on. Figure 6.2 illustrates this bound for $d = 1$, 2, and 3 and X a continuous random variable.

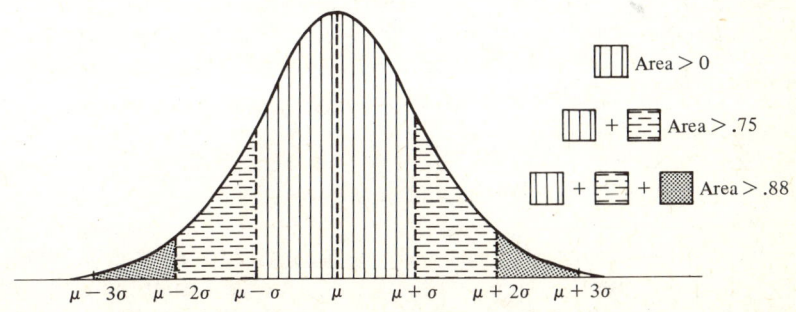

Figure 6.2 Inequality on area based on the values μ and σ.

Remark: Since the event $\{\mu - d\sigma \leq X \leq \mu + d\sigma\}$ includes the event $\{\mu - d\sigma < x < \mu + d\sigma\}$, the inequality

$$P\{\mu - d\sigma \leq X \leq \mu + d\sigma\} \geq 1 - \frac{1}{d^2}$$

holds for all random variables X that have expected value μ and variance σ^2.

Example 6.2. A multiple choice reading comprehension test (in which incorrect answers receive negative scores) has been constructed by a psychologist in such a way that the score X of a randomly chosen high school senior has expected value $E(X) = \mu_X = 0$ and variance $\sigma_X^2 = 100$. The psychologist who constructed the reading test has reported μ_X and σ_X^2 to high school advisors, but has not provided more detailed information about the probability distribution of scores on this test. A student asks his high school advisor about the chance that he can get into a certain highly selective college. Since he knows that this college is only interested in students who show above average reading ability, the advisor gives the student the reading comprehension test. Suppose that the student makes a score of 30. The advisor now wants to determine where the student stands in reading ability relative to all high school seniors. One way in which he can measure his advisee's standing is to ask for the probability that a randomly chosen high school senior obtains a score which equals or exceeds his advisee's score of 30; that is, the advisor is interested in the probability of the event $\{X \geq 30\}$. Since this event is in the form of the event shown in Inequality (6.1), he starts to use that inequality to obtain some idea of the magnitude of $P_X\{X \geq 30\}$. Then he recalls that the random variable X can be negative, so that Inequality (6.1) is inapplicable. However, (6.7) concerns an event almost of the form of the event $\{X \geq 30\}$, and indeed, the event $\{X \geq 30\}$ is included in the event $\{X \leq -30 \text{ or } X \geq 30\}$. Thus, by the Law of Inclusion,

$$P_X\{X \geq 30\} \leq P_X\{X \leq -30 \text{ or } X \geq 30\}.$$

Since $\mu_X = 0$, $\sigma_X = \sqrt{\sigma_X^2} = 10$, it follows that if we want $\mu_X + d\sigma_X = 30$, then d must be equal to $30/\sigma_X = 30/10 = 3$. Using Inequality (6.7) or Table 6.2 the advisor calculates that

$$\begin{aligned} P_X\{X \geq 30\} &\leq P_X\{X \leq -30 \text{ or } X \geq 30\} \\ &= P_X\{X \leq \mu_X - 3\sigma_X \text{ or } X \geq \mu_X + 3\sigma_X\} \\ &\leq \frac{1}{(3)^2} = 0.111. \end{aligned}$$

The advisor concludes that his advisee has a reading ability which is exceeded by at most 11.1 percent of the American high school senior population (and possibly by an even smaller percentage of such seniors).

Since this result certainly implies that his advisee has above average reading ability, the advisor can recommend with confidence that the student go ahead and apply to the college of his choice.

Supplementary Reading

The problem of finding largest or smallest possible values of probabilities of various events when we have available only a few descriptive indices (such as the mean and variance, or the median, and so on) of a distribution has been of renewed interest during the last 20 years.

The inequality for probabilities most often quoted is the Bienaymé-Chebyshev Inequality (6.7). There are other inequalities similar to Inequality (6.7) or (6.8) in that they make use of the values of μ and σ^2. As an example, when X is a *nonnegative* random variable and we know $E(X) = \mu$ and $\sigma_X^2 = \sigma^2$, then

(6.9) $$P_X\{X \geq \mu + b\sigma\} \leq \frac{1}{1+b^2}.$$

Inequality (6.9) provides information about the probability that X equals or exceeds $\mu + b\sigma$ *regardless of the value of b* (provided only that b is positive). Table 6.4 gives values of the largest possible probability that X equals or exceeds $\mu + b\sigma$ for various values of b.

Since Inequalities (6.1) and (6.9) are similar in form, it is of interest to compare Tables 6.1 and 6.4. To make this comparison, we need to equate $c\mu$ to $\mu + b\sigma$ (otherwise Tables 6.1 and 6.4 refer to probabilities of different events). Suppose that $\sigma = k\mu$ (remember that since X is always nonnegative, so is μ). To obtain a comparison between Inequalities (6.1) and (6.9), if we look up a fixed b in Table 6.4, we must look up $c = 1 + bk$ in Table 6.1. For example, if $k = 1$, $b = 3$, we must look up $c = 4$. From Table 6.4, the entry opposite $b = 3$ is 0.100, while from Table 6.1, the entry opposite $c = 4$ is 0.250. Thus, if $\sigma = \mu$, Inequality (6.9) tells us that

$$P_X\{X \geq \mu + 3\sigma\} = P_X\{X \geq 4\mu\} \leq 0.100,$$

while Inequality (6.1) gives us the less precise result that

$$P_X\{X \geq 4\mu\} \leq 0.250.$$

Suppose instead that $\sigma = 3\mu$. Then the entry in Table 6.1 corresponding to $b = 3$ in Table 6.4 is for $c = 1 + (3)(3) = 10$. In this case, both tables tell us that

$$P_X\{X \geq 10\mu\} \leq 0.10.$$

Finally, suppose that $k = \frac{1}{2}$; then the entry in Table 6.1 corresponding to $b = 3$ is for $c = 1 + 3(\frac{1}{2}) = 2.5$. From Table 6.1, we obtain the bound

$$P_X\{X \geq 2.5\mu\} \leq 0.40,$$

Table 6.4. Largest possible value of the probability, $P_X\{X \geq \mu + b\sigma\}$, in the tail of the distribution of a nonnegative random variable X,

b	Largest Possible Value, $1/(1+b^2)$, of $P_X\{X \geq \mu + b\sigma\}$
0.5	0.800
1	0.500
1.5	0.308
2	0.200
2.5	0.138
3	0.100
3.5	0.075
4	0.059
4.5	0.047
5	0.039
10	0.010
30	0.001
50	0.0004
100	0.0001

while from Table 6.4,

$$P_X\{X \geq \mu + 3(\tfrac{1}{2})\mu\} = P_X\{X \geq 2.5\mu\} \leq 0.1.$$

Again, we obtain more information from Table 6.4 than from Table 6.1. In general, Table 6.4 [referring to Inequality (6.9)] tells us more about probabilities of events of the form $\{X \geq a\}$ than does Table 6.1 [referring to Inequality (6.1)]. This fact is not surprising, since to use Table 6.4 we need to have more information (namely, the value of the variance of X) than we do to use Table 6.1. *The more information we have concerning a distribution, the better we are able to bound the values of the probabilities of events concerning random variables which have that distribution.* Thus, if we have more information (in terms of the expected value, variance, median, higher moments, and so on) about a distribution than we have used in Inequalities (6.1), (6.7), (6.8), or (6.9), we can get a better idea about probabilities of events of the form $\{X \geq a\}$, $\{a < X < b\}$, or $\{X \leq a \text{ or } X \geq b\}$ by looking up a probability inequality that uses all of the information which we have available. There are two useful expository papers designed to aid the user in finding the appropriate probability inequality to fit his needs. They are the following:

Godwin, H. J. (1955). On generalizations of Tchebychef's inequality. *Journal of the American Statistical Association*, vol. 50, pp. 923–945.

Savage, I. R. (1961). Probability inequalities of the Tchebycheff type. *Journal of Research of National Bureau of Standards*, vol. 65B, pp. 211–222.

7. APPENDIX

In Chapters 6, 7, and 8, where we discuss various special probability distributions for random variables, we will need a way of approximating (estimating) the expected value μ_X and variance σ_X^2 of a random variable X in situations where we do not know the exact distribution of X, but have available observations on X taken from N statistically independent trials (see Chapter 3) of the random experiment in which X is observed. In the present Appendix, we indicate how observations upon X can be used to estimate (approximate) the values of μ_X and σ_X^2.

Let us begin with the case in which X is a discrete random variable. In this case the N observations $t_1, t_2, \ldots, t_N$ upon X are either reported to us in the form of a list of the observations $t_1, t_2, \ldots, t_N$, or in the form of an empirical probability mass function (Chapter 4, Section 4). For example, in Section 2 we considered a situation in which values $t_1 = 2$, $t_2 = 4$, $t_3 = 1$, $t_4 = 5$, $t_5 = 2$, $t_6 = 1$, $t_7 = 3$, $t_8 = 9$, $t_9 = 1$, and $t_{10} = 1$ for X are observed. The empirical probability mass function based on these observations can be represented in the form of a table such as Table 7.1.

Table 7.1 An empirical probability mass function based on the observations $t_1 = 2$, $t_2 = 4$, and so on.

t	1	2	3	4	5	9
r.f.$\{X = t\}$	$\frac{4}{10}$	$\frac{2}{10}$	$\frac{1}{10}$	$\frac{1}{10}$	$\frac{1}{10}$	$\frac{1}{10}$

Here the top row of Table 7.1 lists the distinct values t of X that have been observed (that is, $t = 1, 2, 3, 4, 5,$ and 9), and the bottom row lists the observed relative frequency of the events $\{X = t\}$, for $t = 1, 2, 3, 4, 5,$ and 9. An alternative, and equivalent, way of displaying the observations in a table is to give a frequency distribution for the observations $t_1, t_2,$ and so on. The frequency distribution is usually given in a table similar to Table 7.2. Here the left-hand column lists the distinct values of X that are ob-

Table 7.2 Frequency distribution based on the observations $t_1 = 2$, $t_2 = 4$, and so on.

t	Observed Frequency of the Event $\{X = t\}$
1	4
2	2
3	1
4	1
5	1
9	1
	10

served (in this case, $t = 1, 2, 3, 4, 5$, and 9), and the right-hand column gives the frequency of the event $\{X = t\}$, $t = 1, 2, 3, 4, 5$, and 9, among the observations (for example, the entry in Table 7.2 in the right-hand column corresponding to the second row, $t = 2$, of the left-hand column gives the number of times that an observation of X has the value 2).

If we are given N observations on a discrete random variable X (either in the form of a list $t_1, t_2, \ldots, t_N$, of the observed values, or in the form of an empirical probability mass function or frequency distribution), then from the frequency interpretation of probability, we have argued in Section 2 that

(7.1) $\bar{t} = \dfrac{t_1 + t_2 + \cdots + t_N}{N} = \dfrac{1}{N}$ sum $[t \times \text{frequency of } \{X = t\}]$

$= \text{sum } [t \times \text{r.f.} \{X = t\}]$

should be approximately equal to μ_X if the number N of trials is sufficiently large. In the example given above in which values $t_1 = 2$, $t_2 = 4, \ldots, t_{10} = 1$ are observed for X,

$$\bar{t} = \frac{t_1 + t_2 + \cdots + t_{10}}{10} = \frac{2 + 4 + 1 + 5 + 2 + 1 + 3 + 9 + 1 + 1}{10} = \frac{29}{10},$$

or alternatively,

$\bar{t} = \dfrac{1}{10}$ sum$[t \times \text{frequency of } \{X = t\}]$

$= \text{sum}[t \times \text{r.f.} \{X = t\}]$

$= (1)(\text{r.f.}\{X = 1\}) + (2)(\text{r.f.}\{X = 2\}) + (3)(\text{r.f.}\{X = 3\})$
$+ (4)(\text{r.f.}\{X = 4\}) + (5)(\text{r.f.}\{X = 5\}) + (9)(\text{r.f.}\{X = 9\})$

$= (1)\left(\dfrac{4}{10}\right) + (2)\left(\dfrac{2}{10}\right) + (3)\left(\dfrac{1}{10}\right) + (4)\left(\dfrac{1}{10}\right) + (5)\left(\dfrac{1}{10}\right) + (9)\left(\dfrac{1}{10}\right) = \dfrac{29}{10}.$

To remind us that the value of $\bar{t}$ is an approximation to the value of μ_X, we adopt a special notation for the quantity $\bar{t}$; namely, $\hat{\mu}_X = \bar{t}$. Thus

(7.2) $\quad \hat{\mu}_X = \dfrac{t_1 + t_2 + \cdots + t_N}{N} = \dfrac{1}{N} \text{sum}[t \times \text{frequency of } \{X = t\}]$

$\qquad\qquad = \text{sum}[t \times \text{r.f.}\{X = t\}],$

and for our example, $\hat{\mu}_X = 29/10 = 2.9$.

Let us now turn to the problem of estimating (approximating) the variance σ_X^2 of X from observations taken on X. If the value μ_X of the expected value of X is known, then since σ_X^2 is equal to the expected value of $W = (X - \mu_X)^2$, the frequency interpretation of probabilities suggests that we estimate $\sigma_X^2 = E(W)$ by

(7.3) $\quad$ estimate of $\sigma_X^2 = \dfrac{1}{N} \text{sum} \, [(t - \mu_X)^2 \times \text{frequency of } \{X = t\}]$

$\qquad\qquad\qquad\quad = \text{sum} \, [(t - \mu_X)^2 \times \text{r.f.}\{X = t\}],$

where the "sum" is taken over all distinct values t that are observed. Since we usually do not know the value of μ_X, to estimate σ_X^2 we may use formulas (7.3) or (7.4), but with $\hat{\mu}_X$ substituted in place of μ_X. This substitution yields the formula

(7.4) $\quad \hat{\sigma}_X^2 = \dfrac{(t_1 - \hat{\mu}_X)^2 + (t_2 - \hat{\mu}_X)^2 + \cdots + (t_N - \hat{\mu}_X)^2}{N}$

$\qquad\qquad = \dfrac{1}{N} \text{sum} \, [(t - \hat{\mu}_X)^2 \times \text{frequency of } \{X = t\}]$

$\qquad\qquad = \text{sum} \, [(t - \hat{\mu}_X)^2 \times \text{r.f.}\{X = t\}],$

where the notation $\hat{\sigma}_X^2$ reminds us that formula (7.4) gives us an estimate (approximate value) of σ_X^2. In our above example,

$\hat{\sigma}_X^2 = \dfrac{1}{10} [(2 - \hat{\mu}_X)^2 + (4 - \hat{\mu}_X)^2 + (1 - \hat{\mu}_X)^2 + (5 - \hat{\mu}_X)^2 + (2 - \hat{\mu}_X)^2$

$\qquad\quad + (1 - \hat{\mu}_X)^2 + (3 - \hat{\mu}_X)^2 + (9 - \hat{\mu}_X)^2 + (1 - \hat{\mu}_X)^2 + (1 - \hat{\mu}_X)^2]$

$\quad = \dfrac{1}{10} \left[\left(2 - \dfrac{29}{10}\right)^2 + \left(4 - \dfrac{29}{10}\right)^2 + \cdots + \left(9 - \dfrac{29}{10}\right)^2 + \left(1 - \dfrac{29}{10}\right)^2 + \left(1 - \dfrac{29}{10}\right)^2 \right]$

$\quad = \dfrac{589}{100} = 5.89.$

Alternatively, using either Table 7.1 or Table 7.2, we can calculate $\hat{\sigma}_X^2$ by the formula

$$\hat{\sigma}_X^2 = \frac{1}{10} \text{sum} \left[(t - \hat{\mu}_X)^2 \times \text{frequency of } \{X = t\} \right]$$

$$= \text{sum} \left[(t - \hat{\mu}_X)^2 \times \text{r.f. of } \{X = t\} \right]$$

$$= \left(1 - \frac{29}{10}\right)^2 \left(\frac{4}{10}\right) + \left(2 - \frac{29}{10}\right)^2 \left(\frac{2}{10}\right) + \left(3 - \frac{29}{10}\right)^2 \left(\frac{1}{10}\right)$$

$$+ \left(4 - \frac{29}{10}\right)^2 \left(\frac{1}{10}\right) + \left(5 - \frac{29}{10}\right)^2 \left(\frac{1}{10}\right) + \left(9 - \frac{29}{10}\right)^2 \left(\frac{1}{10}\right)$$

$$= \frac{5890}{1000} = 5.89.$$

Hence, if the observed values of X are $t_1 = 2$, $t_2 = 4$, $t_3 = 1$, $t_4 = 5$, $t_5 = 2$, $t_6 = 1$, $t_7 = 3$, $t_8 = 9$, $t_9 = 1$, $t_{10} = 1$, then we can estimate the expected value μ_X of X by $\hat{\mu}_X = 2.9$, and we can estimate the variance, σ_X^2, of X by $\hat{\sigma}_X^2 = 5.89$.

If X is a continuous random variable, and if we are given the actual observed values $t_1, t_2, \ldots, t_N$ of X which are obtained from N independent trials of the random experiment in which X is observed, then we may estimate the expected value μ_X of X by

(7.5) $$\hat{\mu}_X = \frac{t_1 + t_2 + \cdots + t_N}{N},$$

and the variance σ_X^2 of X by

(7.6) $$\hat{\sigma}_X^2 = \frac{(t_1 - \hat{\mu}_X)^2 + (t_2 - \hat{\mu}_X)^2 + \cdots + (t_N - \hat{\mu}_X)^2}{N},$$

just as we would do if X were a discrete random variable. However, it is often the case that the result of taking N observations upon X are reported to us in the form of a table of frequencies. For example, we may obtain the results of observing X in the form of a table such as Table 7.3. In this case, we cannot recreate from the table what the actual observed values $t_1, t_2, \ldots, t_{40}$ of X are. All observed values of X which lie between 1.5 and 2.5, for example, are counted in the interval of values whose midpoint is 2.0. The actual observed values of X counted in this interval might have been 2.25 and 1.97; all that we know from Table 7.3 is that two observations on X fall into the interval of values between 1.5 and 2.5 (corresponding to the event $\{1.5 \leq X < 2.5\}$). Because the information that we obtain from Table 7.3 is less than the information that we would have if we knew the actual values of the observations on X, we cannot expect the estimates of μ_X and σ_X^2 obtained from Table 7.3 to be the same as the estimates of μ_X and σ_X^2 obtained from the actual observed values on X. (Note that this fact is in contrast to the case when X is discrete, where the estimates of

Table 7.3 Frequency distribution based on $N = 40$ observations of the random variable X.

Midpoint x of Interval of Values	Observed Frequency of the Event $\{x-0.5 \leq X < x+0.5\}$
1.0	1
2.0	2
3.0	4
4.0	7
5.0	12
6.0	8
7.0	5
8.0	$\dfrac{1}{40}$

μ_X and σ_X^2 are the same regardless of whether the estimates are computed from the actual observed values of X or from tables similar to Tables 7.1 and 7.2.)

To estimate μ_X and σ_X^2 from a frequency histogram such as Table 7.3, we may act as if the observations whose values are included in a given interval of possible values are all equal to the midpoint of that interval of possible values. For example, in Table 7.3 we act as if the two observations in the interval of possible values for X between 1.5 and 2.5 are both equal to the value $x = 2.0$ of the midpoint of that interval. Having done this, we can then estimate μ_X by multiplying the value of each midpoint x by the relative frequency of the corresponding interval, and summing these products for all intervals. This operation is most easily illustrated by showing how it can be applied to Table 7.3:

$$\hat{\mu}_X = \text{sum}\left[\text{midpoint} \times \left(\frac{\text{frequency of interval}}{N}\right)\right]$$

$$= (1.0)\left(\frac{1}{40}\right) + (2.0)\left(\frac{2}{40}\right) + (3.0)\left(\frac{4}{40}\right) + (4.0)\left(\frac{7}{40}\right) + (5.0)\left(\frac{12}{40}\right)$$

$$+ (6.0)\left(\frac{8}{40}\right) + (7.0)\left(\frac{5}{40}\right) + (8.0)\left(\frac{1}{40}\right)$$

$$= \frac{196}{40} = 4.90.$$

To estimate σ_X^2 the same idea is employed. We square the difference $(x - \hat{\mu}_X)$ between the value of each midpoint x and $\hat{\mu}_X$, multiply this square by the relative frequency of that interval of possible values whose mid-

238 DESCRIPTIVE PROPERTIES OF DISTRIBUTIONS 5

point is x, and then sum these products over all intervals. For Table 7.3,

$$\hat{\sigma}_X^2 = \text{sum}\left[(\text{midpoint } x - \hat{\mu}_X)^2 \times \left(\frac{\text{frequency of interval}}{N}\right)\right]$$

$$= (1.0 - \hat{\mu}_X)^2\left(\frac{1}{40}\right) + (2.0 - \hat{\mu}_X)^2\left(\frac{2}{40}\right) + (3.0 - \hat{\mu}_X)^2\left(\frac{4}{40}\right)$$
$$+ (4.0 - \hat{\mu}_X)^2\left(\frac{7}{40}\right) + (5.0 - \hat{\mu}_X)^2\left(\frac{12}{40}\right) + (6.0 - \hat{\mu}_X)^2\left(\frac{8}{40}\right)$$
$$+ (7.0 - \hat{\mu}_X)^2\left(\frac{5}{40}\right) + (8.0 - \hat{\mu}_X)^2\left(\frac{1}{40}\right)$$

$$= (-3.9)^2\left(\frac{1}{40}\right) + (-2.9)^2\left(\frac{2}{40}\right) + (-1.9)^2\left(\frac{4}{40}\right) + (-0.9)^2\left(\frac{7}{40}\right)$$
$$+ (0.1)^2\left(\frac{12}{40}\right) + (1.1)^2\left(\frac{8}{40}\right) + (2.1)^2\left(\frac{5}{40}\right) + (3.1)^2\left(\frac{1}{40}\right)$$

$$= 2.34.$$

We emphasize again that the estimates of μ_X and of σ_X^2 obtained from a frequency histogram are usually not equal to estimates of μ_X and σ_X^2 obtained from the actual observed values on X [using Equations (7.5) and (7.6)]. For example, the actual observations on X that gave rise to Table 7.3 are given below:

0.78	4.53	5.61
1.97	4.61	5.83
2.25	4.67	5.89
2.66	4.83	5.91
2.92	4.91	6.03
3.12	4.99	6.14
3.25	5.05	6.25
3.51	5.08	6.51
3.69	5.20	6.62
3.81	5.27	6.67
3.99	5.31	6.82
4.13	5.44	7.15
4.21	5.58	7.56
4.43		

From these observations and Equation (7.5), we find that

$$\hat{\mu}_X = \frac{0.78 + 1.97 + \cdots + 6.82 + 7.15 + 7.56}{40}$$
$$= \frac{193.18}{40} = 4.83,$$

whereas from Table 7.3 we obtained an estimate of μ_X of 4.90. Similarly, from the actual observed values and Equation (7.7), we find that

$$\hat{\sigma}_x^2 = \frac{(0.78-4.83)^2+(1.97-4.83)^2+\cdots+(7.15-4.83)^2+(7.56-4.83)^2}{40}$$

$$= \frac{87.9192}{40} = 2.20,$$

whereas from Table 7.3 we obtained an estimate of 2.34 for σ_x^2. The discrepancy between the value of the estimate of μ_x obtained from the actual observations and the value of the estimate of μ_x obtained from Table 7.3 is due to the fact that in calculating the latter estimate we act as if every observation in a given interval is equal to the value of the midpoint of that interval. Since for each observation the error made by adopting this assumption is no more than half the common length ℓ of each such interval, the discrepancy between the two estimates of μ_x *can never be greater than $\ell/2$* (and frequently is less than this amount). Hence, the smaller is the common length ℓ of each interval used to construct the frequency histogram, the smaller is the discrepancy between the two estimates of μ_x. Similar comments hold for the discrepancy that may exist between the estimate of σ_x^2 based on the actual observations and the estimate of σ_x^2 based on a frequency distribution formed from these observations.

EXERCISES

1. Let the random variable X have the following probability model:

x	0	1	2
Probability	$\frac{1}{3}$	$\frac{1}{2}$	$\frac{1}{6}$

   (a) Find the expected value of X.
   (b) Find the variance of X.
   (c) Find the median of X.
   (d) Is the expected value of X a possible value of X? Is the median of X a possible value of X? Find a mode of the distribution of X. Is the mode a possible value of X?

A2. An integer is chosen at random (that is, with equal probability) from among the integers 1, 2, 3, 4, 5, 6, 7, 8, 9, and 10. Let X be the number of different positive integers which divide the chosen integer without remainder. (For example, the integer 6 has $X = 4$ divisors since 6 can be divided without remainder by the positive integers 1, 2, 3, and 6.)
   (a) Find the probability mass function of the random variable X.
   (b) Find the expected value of X.
   (c) Find the median of X.

(d) Find the mode of X.
(e) If you were to play the "Guess the Number of Divisors Game," in which your opponent chooses an integer at random from among the integers 1, 2, 3, ..., 9, 10, and you have to guess the number X of divisors it has, which value would you guess for X? Why? What is the probability that you would correctly guess the number of divisors of the integer chosen by your opponent?
(f) Find the variance of X.

3. An insurance company wishes to establish the premium C for selling a 1-year term insurance policy of $1000 to a man of age 45. Suppose that the probability that a 45-year-old man will die during the next year is 0.00363 (see Table 2.1). Let the random variable X denote the gain (or loss) to the insurance company as a result of selling a 1-year term insurance policy of $1000 to a 45-year-old man. Thus, the probability model for X is

x	C	$C - 1000$
Probability	0.99637	0.00363

(a) Determine C so that the company's expected gain is equal to 0.
(b) What premiums allow the company to make a profit "on the average"?
(c) If the company sets a premium of $5.00 for every $1000 of term insurance and sells $500,000 worth of this term insurance to 45-year-old men (that is, 500 thousand-dollar policies), approximately what total profit (or loss) will they receive?
(d) Does it make sense for the insurance company to choose the premium C so that their *median* gain (or loss) is greater than 0? Is knowledge of the median of the distribution of X useful to the company? Explain.

A4. In the problem on "matches" or "rencontres" (Exercise 18 of Chapter 4), we obtained the probability mass functions for the number X_N of matches among N pairs of identical tickets. For $N = 2, 3,$ and 4, use these probability mass functions to find the expected value and variance of X_N. For convenience, the needed probability mass functions are as follows:

X_2:
x	0	1	2
Probability	$\frac{1}{2}$	0	$\frac{1}{2}$

X_3:
x	0	1	2	3
Probability	$\frac{2}{6}$	$\frac{3}{6}$	0	$\frac{1}{6}$

X_4:
x	0	1	2	3	4
Probability	$\frac{9}{24}$	$\frac{8}{24}$	$\frac{6}{24}$	0	$\frac{1}{24}$

5. A gambler is playing roulette at Monte Carlo (see Figure 4.1 of Chapter 6 for a picture of the roulette wheel). He bets 1 chip on his lucky number, number 7. The wheel is spun and if his number comes up, he receives 35 chips plus his 1 chip; otherwise he loses his 1 chip. Let X denote the gambler's gain (or loss) on a single bet. Find the expected value of X. Does the gambler win or lose "on the average"? The gambler bets on 400 spins of the wheel (each time betting on a "7"). Assuming that the results of the spins of the wheel are mutually statistically independent, approximately how many chips will the gambler win (or lose)?

A6. An experiment consists in selecting 2 balls (without replacement) from an urn which contains 2 white and 3 red balls, and then selecting 3 more balls (without replacement) from a second urn which contains 3 white and 2 red balls. If X denotes the number of white balls among the 5 balls selected from the two urns, calculate the expected value of X.

7. Plates of cells are observed and the number of cells in the process of reproduction are counted. Suppose that each plate has 5 cells on it. Let X be the number of cells observed in reproduction on a given plate. The following is a possible probability model for this experiment:

x	0	1	2	3	4	5
Probability of x Cells	$\frac{1}{243}$	$\frac{10}{243}$	$\frac{40}{243}$	$\frac{80}{243}$	$\frac{80}{243}$	$\frac{32}{243}$

Find the expected value and variance of the random variable X.

8. In a discussion concerning the wealth of two underdeveloped countries, A argues that the total wealth of his country is twice the total wealth of B's country. B retorts that the average wealth per capita in his country is twice the per capita wealth of A's country. With which protagonist would you agree? Why? (In your answer tell how you would define wealth.) Which country is larger in population? How do you know?

A9. Studies have been made by quantitative linguists of the syllabification of words by various authors. For example, let X be the number of syllables in a word chosen at random from the works of the English author and lexicographer Samuel Johnson. In 2000 such words, the following sample probability mass function was obtained for X:

Number x of Syllables	1	2	3	4	5	6
Observed Relative Frequency of the Event $\{X = x\}$	$\frac{1268}{2000}$	$\frac{423}{2000}$	$\frac{195}{2000}$	$\frac{77}{2000}$	$\frac{29}{2000}$	$\frac{8}{2000}$

(a) Find the sample average (sample mean) $\hat{\mu}_X$ of X.
(b) Find the sample variance $\hat{\sigma}_X^2$ of X.

(c) Estimate the median of X and the quartiles $Q_{0.25}(X)$ and $Q_{0.75}(X)$ of X by assuming that the above sample probability mass function for X is the actual probability mass function for X.

(d) Graph the sample cumulative distribution function of X and locate on this graph (i) the sample average $\hat{\mu}_X$, (ii) the sample median [as calculated in part (c)], (iii) $\hat{\mu}_X + \hat{\sigma}_X$, and (iv) $\hat{\mu}_X + 2\hat{\sigma}_X$.

10. A study of the syllabification in the works of the German poet and dramatist Johann Wolfgang von Goethe has also been made. Let Y be the number of syllables in a word chosen at random from Goethe's works. In 1200 such words, the following sample probability mass function for Y was obtained.

Number y of Syllables	1	2	3	4	5
Observed Relative Frequency of the Event $\{Y = y\}$	$\frac{587}{1200}$	$\frac{410}{1200}$	$\frac{146}{1200}$	$\frac{49}{1200}$	$\frac{8}{1200}$

(a) Calculate the sample average $\hat{\mu}_Y$ of Y.
(b) Calculate the sample variance $\hat{\sigma}_Y^2$ of Y.
(c) Estimate the median of Y and the quartiles $Q_{0.25}(Y)$ and $Q_{0.75}(Y)$ by the method described in Exercise 9, part (c).
(d) Compare graphs of the sample cumulative distribution functions of the number X of syllables in a randomly chosen word in the works of Johnson (see Exercise 9) and the number Y of syllables in a randomly chosen word in the works of Goethe.
(e) Compare graphs of the sample probability mass functions of X and Y.
(f) Use the comparisons in (d) and (e) to compare the syllabification styles of Johnson and Goethe. How much of this information can you obtain by just comparing the sample averages and sample variances of X and Y? Explain your answer.

*11. Let X be the yearly income (in thousands of dollars) of a person who earns more than K thousand dollars and who pays some federal income tax. [K is the minimum income (in thousands of dollars) for which an individual must report his income to the Internal Revenue Service.] The density function $f_X(\tau)$ of the random variable X is assumed to be of the form

$$f_X(\tau) = \begin{cases} aK^a/\tau^{a+1}, & \text{for } \tau \geq K, \\ 0, & \text{for } \tau < K. \end{cases}$$

Suppose $K = 2$ and $a = 3$.
(a) Graph the probability density function $f_X(\tau)$.
(b) Approximate the expected value of μ_X of X using the rectangular method of approximation. Compare with the exact value $\mu_X = 3$.
(c) Approximate the variance σ_X^2 of X using the method of rectangular approximation and compare with the exact value $\sigma_X^2 = 3$. [*Hint:* Use the rectangular method to approximate $E(X^2)$, and then use the fact that $\sigma_X^2 = E(X^2) - (\mu_X)^2$.]

[*Note:* The distribution of X is a special case of a *Pareto distribution*. This application of Pareto distributions is discussed by Hagstroem (1960).]

A12. Let X be the distance (in miles) traveled by a passenger on a certain bus route. The bus route one way covers 2 miles. Studies by the bus company have shown that the probability density function $f_X(\tau)$ of X is given by

$$f_X(\tau) = \begin{cases} \frac{3}{8}(2-\tau)^2, & \text{for } 0 \leq \tau \leq 2, \\ 0, & \text{for } \tau < 0 \text{ or } \tau > 2, \end{cases}$$

and $f_X(\tau) = 0$ for all other values of τ.
(a) Graph the probability density function $f_X(\tau)$.
(b) Is this density function unimodal? If so, what is the mode? Where do you believe the expected value and median of X occur with respect to the mode? Why?
(c) The cumulative distribution function of X is given by

$$F_X(\tau) = \begin{cases} 0, & \text{if } \tau < 0, \\ 1 - (2-\tau)^3/8, & \text{if } 0 \leq \tau \leq 2, \\ 1, & \text{if } 2 < \tau. \end{cases}$$

Graph the function $F_X(\tau)$.
(d) Find the median and 90th percentile of the distribution of X.
(e) The expected value of X is actually equal to $\frac{1}{2}$. Is this what you anticipated?

13. Let X be a random variable. It is known that the expected value of $Y = 3X - 5$ is equal to 0. What is the expected value of X? The variance of Y is 9. What is the variance of X?

A14. Let X be a random variable with $E(X) = -1$, $\sigma_X^2 = 4$. For what positive values of a and b does $Z = aX + b$ have expected value equal to 0 and variance equal to 1?

A15. Suppose that the (unique) median of the distribution of the *positive* random variable X is equal to 4 and that the expected value of X is equal to 5.
(a) What is the value of the median of the distribution of $Y = 3X - 5$?
(b) What is the value of the expected value of Y?
(c) Can you calculate the median of the random variable $Z = X^2$? If so, what is this value? If not, what information do you need to calculate the median of Z?
(d) Can you calculate the expected value of X^2? If so, what is this value? If not, what information do you need?

A16. In Exercise 1 find $E(X^2)$, $E(X^3)$, and $E(X^4)$.

17. A random variable X has a continuous distribution. The largest possible value of X is 1, and the smallest possible value of X is 0. The upper quartile $Q_{0.75}(X)$ of X is 0.29. A new random variable $Y = 1 - X$ is constructed from the random variable X. Find the lower quartile $Q_{0.25}(Y)$ of Y.

A18. Scores X on the verbal scale of the Scholastic Aptitude Test (that is, the "SAT Verbal" scores) are nonnegative and have an expected value μ_X of 650. Even if we do not make any other assumptions about the distribution of X, the various probability inequalities discussed in Section 6 allow us to make estimates of the probabilities of various events that may be of interest to us. For example, use Inequality (6.1) to find the largest possible values of the probabilities
   (a) $P\{X \geq 715.0\}$,
   (b) $P\{X \geq 780.0\}$,
   when the only fact that we know about the distribution of X is that $\mu_X = 650$.
   (c) If someone claims that 30 percent of all SAT Verbal scores are larger than 780.0, would you believe their claim? Answer first on the basis of the result you obtained in part (b), and then on the basis of your own personal knowledge. Is Inequality (6.1) very helpful to you in this case? Explain.

19. In the context of Exercise 18, suppose that we also know that the standard deviation σ_X of X is 64. Note that $780 = 650 + (65/32)64$. Now use Inequality (6.9) to find the largest possible value of $P\{X \geq 780\}$. Do you now believe the claim that 30 percent of all SAT Verbal scores X exceed 780? Explain. Would you believe the claim that 80 percent of all SAT Verbal scores X exceed 715? Explain.

A20. Still in the context of Exercises 18 and 19, suppose you know that the SAT Verbal scores X have an expected value μ_X of 650 and a standard deviation σ_X of 64. Use the Bienaymé-Chebyshev Inequalities to answer the following questions:
   (a) Find smallest values for $P\{650 - 64d < X < 650 + 64d\}$ for $d = 1.1$, 1.5, 1.8, 2, and 2.2.
   (b) Find largest values for $P\{X \geq 650 + 64b\}$ for $b = 0.5, 0.7, 1, 1.2,$ and 1.5.
   (c) What do the answers in parts (a) and (b) tell you about the probability distribution of SAT Verbal scores?

21. Suppose that X is a discrete random variable with the following probability mass function:

x	$\mu - k\sigma$	μ	$\mu + k\sigma$
Probability	$\dfrac{1}{2k^2}$	$1 - \dfrac{1}{k^2}$	$\dfrac{1}{2k^2}$

where μ is any number, σ is any positive number, and where $k \geq 1$. Show that:
   (a) the expected value of X is μ,
   (b) the variance of X is σ^2,
   (c) $P\{X \geq \mu + k\sigma \text{ or } X \leq \mu - k\sigma\} = \dfrac{1}{k^2}$.

Relate the result obtained in part (c) to Inequality (6.7). What have you shown by the example in this exercise?

part II

Probability Models

6
SPECIAL DISTRIBUTIONS: DISCRETE CASE

Things occur by chance which we dare not hope for.
Terence (Phormio)

Order is Heav'n's first law.
Alexander Pope (An Essay on Man)

In any analysis of quantitative data, it is a major step when we think we know the form of the underlying probability distributions of those random variables that are of interest. Fortunately, there are certain basic probability distributions that are applicable in many diverse contexts, and thus repeatedly arise in practice. In this and in the next two chapters, we describe some frequently encountered discrete and continuous distributions.

1. BERNOULLI AND BINOMIAL DISTRIBUTIONS

Whenever a random experiment has two possible outcomes, we have what is called a *Bernoulli trial* [after Jacob (James) Bernoulli, 1654–1705]. Thus, for example, we have a Bernoulli trial when we are interested in whether a student's grade is pass or fail, whether a person has or has not contracted a given disease, whether a stock price rises or does not rise, and so on. Because, in a Bernoulli trial, there are two possible outcomes, we might, somewhat arbitrarily, call one outcome a "success" and the other outcome a "failure." On the other hand, we might perhaps call one outcome a "head" and the other outcome a "tail." The choice, as we have said, is arbitrary. Since the terminology "success" and "failure" has meaning in many contexts where Bernoulli trials occur, we adopt it in the present section.

Bernoulli Distribution

We can define a random variable associated with every Bernoulli trial as follows: If the outcome ω is a "success," set $X(\omega) = 1$; whereas if the outcome ω is a "failure," set $X(\omega) = 0$. Consequently, we have the probability model

(1.1)

x	0	1
$p_X(x)$	$1-p$	p

where p is the probability of "success." Any random variable having this probability mass function for some probability p is said to be a *Bernoulli random variable*, and the distribution (1.1) is called a *Bernoulli distribution*. The probability mass function (1.1) varies with changing values of p. We call p the *parameter* of the Bernoulli distribution; if we know that a random variable has a Bernoulli distribution, then if we can determine p, we have completely determined the distribution.

Just as our choice of names for the outcomes of a Bernoulli trial is arbitrary, so is our choice of possible values for the random variable X. Any other choice of numbers a and b to assign to the outcomes "success" and "failure" of a Bernoulli trial would do as well, provided that a and b are unequal. For example, suppose that instead of the Bernoulli random variable X we had decided to use the random variable Y defined as $Y(\omega) = b$ if ω is a "success," and $Y(\omega) = a$ if ω is a "failure." Then Y would have the probability model

y	a	b
$p_Y(y)$	$1-p$	p

If we now let $X = (Y-a)/(b-a)$, we see that the event $\{X=1\}$ occurs if and only if the event $\{Y=b\}$ occurs (and both of these events occur if the outcome ω which occurs is a "success"). Similarly, the event $\{X=0\}$ occurs if and only if the event $\{Y=a\}$ occurs. Thus, $P_X\{X=1\} = P_Y\{Y=b\} = p$, $P_X\{X=0\} = P_Y\{Y=a\} = 1-p$, and so X has the distribution (1.1). If we are given the random variable Y, we can transform Y to the Bernoulli random variable $X = (Y-a)/(b-a)$ without losing the basic information which the probability models of both random variables convey—namely, that the probability of the event {"success"} is equal to p. Consequently, we may restrict our study of the quantitative aspects of Bernoulli trials to analysis of the Bernoulli random variable X. Results about other random variables having the form of the random variable Y discussed above will then follow without difficulty.

If X has a Bernoulli distribution with parameter p, then the expected

1. BERNOULLI AND BINOMIAL DISTRIBUTIONS

value μ_X and variance σ_X^2 of X are given by

$$\mu_X = 0(1-p) + 1(p) = p,$$
$$\sigma_X^2 = (0-p)^2(1-p) + (1-p)^2 p = p(1-p).$$

Binomial Distribution

If we make n independent trials of the random experiment which gives rise to a Bernoulli random variable (and a Bernoulli distribution), it is natural to consider a random variable Z associated with this composite random experiment (see Chapter 3) that records the *number of successes* in the n trials. This random variable Z can then assume any of the possible values $0, 1, 2, 3, \ldots, n$. For example, if $n = 3$, we may observe any one of the following 8 outcomes: (S,S,S), (S,S,F), (S,F,S), (F,S,S), (F,F,S), (F,S,F), (S,F,F), (F,F,F), where S denotes "success" and F denotes "failure." Corresponding respectively to these outcomes are the following values of Z: 3, 2, 2, 2, 1, 1, 1, 0. To find the appropriate probability model for Z, we start by finding probabilities for the simple events of the composite random experiment composed of the 3 Bernoulli trials. Let E_1 be the event {"success" on the first trial}, E_2 be the event {"success" on the second trial}, and E_3 be the event {"success" on the third trial}. Since the trials are statistically independent (see Chapter 3), and since the simple event $\{(S,S,S)\} = E_1 \cap E_2 \cap E_3$, the simple event $\{(S,S,F)\} = E_1 \cap E_2 \cap E_3^c$, and so on, it follows that

$$P\{(S,S,S)\} = P(E_1)P(E_2)P(E_3) = p^3,$$
$$P\{(S,S,F)\} = P(E_1)P(E_2)P(E_3^c) = p^2(1-p),$$
$$\ldots \quad \ldots \quad \ldots$$
$$P\{(F,F,F)\} = P(E_1^c)P(E_2^c)P(E_3^c) = (1-p)^3.$$

We summarize these findings in Table 1.1. In the third column of Table 1.1, we indicate the number of successes, Z, associated with each simple event of the probability model which describes $n = 3$ independent Bernoulli trials.

Using the methods described in Chapter 4, Section 1, we find that the probability model for Z is given by the following table:

z	0	1	2	3
$P_Z\{Z=z\}$	$(1-p)^3$	$3p(1-p)^2$	$3p^2(1-p)$	p^3

Notice from Table 1.1 that all simple events that have the same associated number Z of successes, have the same probability. For example, the simple events $\{(S,S,F)\}$, $\{(S,F,S)\}$, and $\{(F,S,S)\}$ all correspond to the

Table 1.1 Probability model for $n = 3$ independent Bernoulli trials.

Simple Event	Probability	Number of Successes Z
{(S,S,S)}	p^3	3
{(S,S,F)}	$p^2(1-p)$	2
{(S,F,S)}	$p^2(1-p)$	2
{(F,S,S)}	$p^2(1-p)$	2
{(F,F,S)}	$p(1-p)^2$	1
{(F,S,F)}	$p(1-p)^2$	1
{(S,F,F)}	$p(1-p)^2$	1
{(F,F,F)}	$(1-p)^3$	0

value $Z = 2$. Because

$$P\{(S,S,F)\} = P(E_1 \cap E_2 \cap E_3^c) = P(E_1)P(E_2)P(E_3^c) = p^2(1-p),$$
$$P\{(S,F,S)\} = P(E_1 \cap E_2^c \cap E_3) = P(E_1)P(E_2^c)P(E_3)$$
$$= p(1-p)p = p^2(1-p),$$
$$P\{(F,S,S)\} = P(E_1^c \cap E_2 \cap E_3) = P(E_1^c)P(E_2)P(E_3)$$
$$= (1-p)p^2 = p^2(1-p),$$

each of these simple events have the same probability $p^2(1-p)$. To find $P_Z\{Z = 2\}$, we have added the probabilities of these three events, or, what is the same, we have multiplied the common probability $p^2(1-p)$ of these three events by the number 3 of such events. Similarly, to find $P_Z\{Z = z\}$, for $z = 0$, 1, or 3, we can count the number $C(3,z)$ of simple events which correspond to $Z = z$, and then multiply this number by the common probability $p^z(1-p)^{3-z}$ of these simple events; that is, $P_Z\{Z = z\} = C(3,z)p^z(1-p)^{3-z}$.

In general, if we take n independent Bernoulli trials and count the number Z of successes, then we can find the appropriate probability model for Z as follows. First, we note that the event $\{Z = k\}$ is the same as the event that there are k "successes" and $(n-k)$ "failures." To find all simple events for which $Z = k$, we look for those simple events corresponding to outcomes in which k "successes" [that is, k S's] and $(n-k)$ "failures" [that is, $(n-k)$ F's] appear. Then, to find the probability of each such simple event, we multiply together n terms, k of which equal the probability p of "success" on a single trial, and $n-k$ of which equal the probability $(1-p)$ of a "failure" on a single trial. Thus, the probability of any simple event corresponding to an outcome in which k "successes" and $(n-k)$ "failures" appear is equal to $p^k(1-p)^{n-k}$. The probability of the event $\{Z = k\}$ is now found by multiplying the common probability,

$p^k(1-p)^{n-k}$, of all such simple events by the number, $C(n,k)$, of such simple events; that is, $P_Z\{Z=k\} = C(n,k)p^k(1-p)^{n-k}$. To evaluate $C(n,k)$ we note that to have an outcome of n Bernoulli trials which corresponds to k "successes" and $(n-k)$ "failures," we must assign each of the k "successes" and the $(n-k)$ "failures" to one of the n trials. From Rule 6.2 of Chapter 3, we know that there are $n!/k!(n-k)!$ ways of making this assignment, and that each way corresponds to one and only one outcome. Thus,

$$C(n,k) = \frac{n!}{k!(n-k)!} = \binom{n}{k}.$$

[Recall that for any integer m, $m! = m(m-1)(m-2) \ldots (2)(1)$, and $0! = 1$.] Consequently, the probability mass function, $p_Z(k)$, for Z is given by the equation

(1.2) $$p_Z(k) = P_Z\{Z=k\} = \binom{n}{k} p^k (1-p)^{n-k},$$

for values $k = 0, 1, 2, \ldots, n-1, n$.

For example, if $n = 4$, we obtain the probability model

k	0	1	2	3	4
$p_Z(k)$	$(1-p)^4$	$4p(1-p)^3$	$6p^2(1-p)^2$	$4p^3(1-p)$	p^4

and if $n = 6$, we obtain

k	0	1	2	3	4	5	6
$p_Z(k)$	$(1-p)^6$	$6(1-p)^5 p$	$15(1-p)^4 p^2$	$20(1-p)^3 p^3$	$15(1-p)^2 p^4$	$6(1-p)p^5$	p^6

The distribution given by (1.2) is called the *binomial distribution*. It has two parameters n and p, where n is the number of repeated Bernoulli trials, and p is the probability of success on each one of these trials. Knowledge of both n and p determines the probabilities $p_Z(k)$ given in Equation (1.2). If Z has a binomial distribution with parameters n and p, then the expected value and variance of Z are, respectively,

(1.3) $$\mu_Z = np, \quad \sigma_Z^2 = np(1-p).$$

We note that the Bernoulli distribution is a special case of the binomial distribution that occurs when $n = 1$. The expressions given earlier for the expected value and variance of a Bernoulli random variable are in accord with Equation (1.3).

In Table 1.2, we illustrate how values of the parameter p affect probabilities for the binomial distribution in the particular case when the parameter n equals 10.

Table 1.2. Values of $p_Z(k)$ for the binomial distribution when $n = 10$ and p varies.

k \ p	0	1	2	3	4	5
0.1	0.3487	0.3874	0.1937	0.0574	0.0112	0.0015
0.2	0.1074	0.2684	0.3020	0.2013	0.0881	0.0264
0.3	0.0282	0.1211	0.2335	0.2668	0.2001	0.1029
0.4	0.0060	0.0403	0.1209	0.2150	0.2508	0.2007
0.5	0.0010	0.0098	0.0439	0.1172	0.2051	0.2461
0.6	0.0001	0.0016	0.0106	0.0425	0.1115	0.2007
0.7	0.0000	0.0001	0.0014	0.0090	0.0368	0.1029
0.8	0.0000	0.0000	0.0001	0.0008	0.0055	0.0264
0.9	0.0000	0.0000	0.0000	0.0000	0.0001	0.0015

k \ p	6	7	8	9	10
0.1	0.0001	0.0000	0.0000	0.0000	0.0000
0.2	0.0055	0.0008	0.0001	0.0000	0.0000
0.3	0.0368	0.0090	0.0014	0.0001	0.0000
0.4	0.1115	0.0425	0.0106	0.0016	0.0001
0.5	0.2051	0.1172	0.0439	0.0098	0.0010
0.6	0.2508	0.2150	0.1209	0.0403	0.0060
0.7	0.2001	0.2668	0.2335	0.1211	0.0282
0.8	0.0881	0.2013	0.3020	0.2684	0.1074
0.9	0.0112	0.0574	0.1937	0.3874	0.3487

Note that for any fixed value of the parameter p, the probabilities $P_Z\{Z = k\} = p_Z(k)$ are smaller at the extremes (that is, when $k = 0$ or $k = 10$) and are large in the center. In fact, as k increases from $k = 0$ to $k = 10$, the probabilities increase, reach a maximum, and then decrease.

Observe also that in Table 1.2, the probability values for $p = 0.2$ and $p = 0.8 = 1.0 - 0.2$ have a certain symmetry. For a fixed value of n, the probability of the event $\{Z = k\}$, when p is the probability of "success" on a single (Bernoulli) trial, is equal to the probability of the event $\{Z = n - k\}$, when $(1 - p)$ is the probability of "success" on a single trial. This relationship holds in general, regardless of the values of n, p, and k. To see that this must be the case, note that the event $\{Z = k\}$ is the same as the event $\{n - Z = n - k\}$, and that $n - Z$ is the number of "failures" in the n trials. Recall that the labels "success" and "failure" were arbitrarily assigned to the outcomes of a Bernoulli trial. Thus, we can in this case think of temporarily switching the labels "success" and "failure" and regarding $W = n - Z$ as the random variable counting the number of "successes" in the n trials. Thus, W has a binomial distribution with parameters n and $(1-p)$, where we recall that $(1-p)$ is the probability of a "failure" (which we are temporarily calling a "success") on a single trial. Since $W = n - Z$, it follows that $P_Z\{Z = k\} = P_W\{W = n - k\}$, thus proving the general assertion made above.

As a consequence, we need to prepare a table of binomial probabilities only for values of the parameter p which are less than or equal to 0.5. If we want the probability $P_Z\{Z = k\}$ for a random variable Z which has a binomial distribution with parameters n and p, and if p is greater than 0.5, we can find $P_Z\{Z = k\}$ by looking up the probability $P_W\{W = n - k\}$ that a random variable W having a binomial distribution with parameters n and $(1-p)$ is equal to $n - k$. Thus, if Z has the binomial distribution with parameters $n = 9$ and $p = 0.8$, and we want $P_Z\{Z = 7\}$, we can find this probability by looking up the probability in Table T.1[1] corresponding to $k = 9 - 7 = 2$ and $p = 1.0 - 0.8 = 0.2$; namely, $P_Z\{Z = 7\} = P_W\{W = 2\} = 0.3020$.

The binomial distribution can arise whenever we select a random sample of n units *with replacement*. Each unit in the population is classified into one of two categories according to whether it possesses a certain characteristic or property. For example, the unit may be a person and the property may be whether he intends to vote "Yes" on a certain bond issue, or whether he has ever been arrested. If the unit is a machine part, this property may be whether the part is defective; if the unit is a leaf, the property may be whether the leaf has worm damage; and so on. Suppose the proportion of units in the population possessing the property of interest is p. Then, if Z denotes the number of units in the sample (of size n) which possess the given property, Z has a binomial distribution with parameters n and p.

Example 1.1. An agent of a consumer protection agency is investigating the incidence of consumer fraud in advertising. He is particularly interested in bait-and-switch tactics used by appliance stores. (This is a practice where a desirable brand is advertised at an attractively low price, but when the customer comes to take advantage of the bargain, he is "switched" over to a less desirable brand at a higher price by an aggressive salesman.) The agent decides to sample 10 stores with replacement from the population of all appliance stores advertising bargains that week, and to have one of his representatives visit each of these stores posing as a customer. (If a store is selected more than once, it is visited more than once, each time by a different representative.) The agent decides that if he finds more than 2 cases of "bait-and-switch" tactics, he will call for a full-scale investigation by his agency. Since such investigations are very expensive, the agent wants to be sure before starting his survey that he has a low probability of

[1]Tables T.1 through T.11 are contained in the tables chapter, pp. 671–716.

a false alarm. He feels that if only 10 percent ($p = 0.10$) of all stores practice "bait-and-switch" techniques, it is not worthwhile to the public to run an investigation, so he does not want the probability of a false alarm to be large if $p = 0.10$. He notes that the event $\{Z > 2\}$, where Z is the number of cases of "bait-and-switch" tactics found in the survey of 10 stores, is the complement of the event $\{Z = 0, 1, \text{or } 2\}$. Hence the probability of a false alarm is

$$P_Z\{Z > 2\} = 1 - P_Z\{Z = 0, \text{or } Z = 1, \text{or } Z = 2\}$$
$$= 1 - p_Z(0) - p_Z(1) - p_Z(2).$$

Since Z has a binomial distribution with parameters $n = 10$ and $p = 0.10$, Table 1.2 yields $p_Z(0) = 0.3487$, $p_Z(1) = 0.3874$, $p_Z(2) = 0.1937$, so that

$$P_Z\{Z > 2\} = 1 - 0.3487 - 0.3874 - 0.1937 = 0.0702.$$

The agent decides that this is a tolerable probability for a false alarm, and orders the survey to begin.

The binomial distribution also arises in contexts other than in random sampling experiments. Whenever a random variable Z can be hypothesized to have arisen as the number of occurrences of a certain characteristic or property in n independent trials of some random experiment, Z has a binomial distribution. The hypothesis as to how Z has arisen may spring from a scientific theory, and data may then be collected to see how well the hypothesis explains the data. That is, we may collect data to see how well the observed relative frequencies of the simple events $\{Z = 0\}$, $\{Z = 1\}$, $\{Z = 2\}$, and so on, agree with the probabilities assigned to these events by the probability model for Z which comes from the scientific hypothesis.

Example 1.2 (Genetics). The inheritance of biological characteristics depends on special carriers, called genes. These genes appear in pairs. In the simplest genetic model, each gene of a pair may assume one of two forms: a or A. Consequently, three combinations, called genotypes, may be formed: aa, Aa, AA. (Note that aA and Aa are indistinguishable.)

Fisher and Mather (1936) describe an experiment (called a "backcross") in which the gene controlling straight (W) or wavy (w) hair in mice was segregated. The female parents were all wavy-haired of genotype (ww), and the male parents were all straight-haired of genotype (wW). If the Mendelian laws of inheritance are true, each mouse in the offspring of these parents has a probability 0.5 of being straight-haired (wW), and a probability of 0.5 of being wavy-haired (ww). We assume now that the hair character of one offspring is statistically independent of the hair character of any other offspring. Thus, in litters of 8 mice, the theoretical

1. BERNOULLI AND BINOMIAL DISTRIBUTIONS

distribution of Z, the number of straight-haired offspring, is the binomial distribution with $n = 8$ and $p = 0.5$. This distribution has the following probability mass function:

k	0	1	2	3	4	5	6	7	8
$p_Z(k)$	0.0039	0.0312	0.1094	0.2188	0.2734	0.2188	0.1094	0.0312	0.0039

In the genetic experiment described by Fisher and Mather, 32 such litters of 8 mice were observed. A comparison of the observed and theoretical frequencies are given in Table 1.3. [The theoretical frequencies are obtained by multiplying each theoretical probability $p_Z(k)$ by 32.] Note that there are some differences between the observed and theoretical frequencies in this table. Thus, it might be asked whether or not the binomial model truly "fits" the observed data. Answers to such a question are discussed in statistics textbooks under the topic of "goodness-of-fit tests."

Table 1.3 Comparison of observed and theoretical frequency distributions for the number of straight-haired mice in a litter of 8 mice.

Number of Straight-Haired Mice in a Litter of 8 Mice	Observed Frequency	Theoretical Frequency
0	0	0.1
1	1	1.0
2	2	3.5
3	4	7.0
4	12	8.7
5	6	7.0
6	5	3.5
7	2	1.0
8	0	0.1
	32	31.9

SOURCE: Fisher, R. A., and Mather, K. (1936). A linkage test with mice. *Annals of Eugenics*, vol. 7, pp. 265–280. Reprinted with permission.

Description of Tables

Table T.1 gives the probability mass functions $p_Z(k)$ of binomially distributed random variables Z for the following values of the parameter

n and p:

$$p = 0.01, 0.05, 0.10, 0.15, 0.20, 0.25, 0.30, 0.35, 0.40, 0.45, 0.50$$
$$n = 1, 2, 3, 4, 5, 6, 7, 8, 9, 10.$$

More detailed tabulations may be found in the following books.

1. National Bureau of Standards (1950). *Applied Mathematics Series 6*, United States Government Printing Office, Washington, D.C.

This book gives the probability mass functions and cumulative distribution functions (evaluated only at possible values k) of binomially distributed random variables for $n = 2(1)49$ and $p = 0.01(0.01)0.50$. (By this notation we mean that n ranges from 2 to 49 by increments of 1, and p ranges from 0.01 to 0.50 by increments of 0.01.)

2. Romig, H. (1953). *50–100 Binomial Tables*. John Wiley & Sons, Inc., New York.

This book gives the probability mass functions and cumulative distribution functions (evaluated only at possible values k) of binomially distributed random variables for $n = 50(5)100$ and $p = 0.01(0.01)0.50$.

3. Pearson, E. J. and Hartley, H. O. (1958). *Biometrika Tables for Statisticians*, vol. I, Cambridge University Press.

This book gives the probability mass functions and cumulative distribution functions (evaluated only at possible values k) of binomially distributed random variables or $n = 5(5)30$, and $p = 0.01, 0.02(0.02)0.10(0.01)0.50$.

2. HYPERGEOMETRIC DISTRIBUTION

In Section 1, we noted that the binomial distribution quite frequently arises from a random experiment in which there is sampling with replacement. In contrast, the *hypergeometric distribution* typically results from a random experiment in which we select a random sample of units without replacement from a population of N units. In such experiments, each unit in the population can be classified into one of two categories (which we call, arbitrarily, "success" and "failure") according to whether the unit does or does not possess a certain property of interest. We count the number X of "successes" in our sample; it is this random variable that has the hypergeometric distribution.

As an example, consider the problem in which we choose 2 individuals from a population of $N = 5$ individuals by means of random sampling without replacement. Of the 5 individuals in the population, 3 of them, H_1, H_2, H_3, rate high on a certain achievement scale, and 2 of them, L_4, L_5,

2. HYPERGEOMETRIC DISTRIBUTION

rate low on that scale. In Table 2.1, we list the possible outcomes of the experiment; there are a total of 20 such outcomes (see Chapter 3, Section 2).

Table 2.1. Outcomes of a simple random sample of size 2 without replacement from a population consisting of the units H_1, H_2, H_3, L_4, L_5.

$\omega_1 = (H_1, H_2)$	$\omega_6 = (H_2, H_3)$	$\omega_{11} = (H_3, L_4)$	$\omega_{16} = (L_4, L_5)$
$\omega_2 = (H_1, H_3)$	$\omega_7 = (H_2, L_4)$	$\omega_{12} = (H_3, L_5)$	$\omega_{17} = (L_5, H_1)$
$\omega_3 = (H_1, L_4)$	$\omega_8 = (H_2, L_5)$	$\omega_{13} = (L_4, H_1)$	$\omega_{18} = (L_5, H_2)$
$\omega_4 = (H_1, L_5)$	$\omega_9 = (H_3, H_1)$	$\omega_{14} = (L_4, H_2)$	$\omega_{19} = (L_5, H_3)$
$\omega_5 = (H_2, H_1)$	$\omega_{10} = (H_3, H_2)$	$\omega_{15} = (L_4, H_3)$	$\omega_{20} = (L_5, L_4)$

In this example, we are interested in the number X of persons in our (random) sample who have a high achievement rating. From Table 2.1, we have the following assignment of the possible values of X to the outcomes ω_i:

$X(\omega_1) = 2$ $X(\omega_6) = 2$ $X(\omega_{11}) = 1$ $X(\omega_{16}) = 0$
$X(\omega_2) = 2$ $X(\omega_7) = 1$ $X(\omega_{12}) = 1$ $X(\omega_{17}) = 1$
$X(\omega_3) = 1$ $X(\omega_8) = 1$ $X(\omega_{13}) = 1$ $X(\omega_{18}) = 1$
$X(\omega_4) = 1$ $X(\omega_9) = 2$ $X(\omega_{14}) = 1$ $X(\omega_{19}) = 1$
$X(\omega_5) = 2$ $X(\omega_{10}) = 2$ $X(\omega_{15}) = 1$ $X(\omega_{20}) = 0$.

To find the probability distribution of X, note that by the definition of random sampling without replacement, the probability $P\{\omega_i\}$ of each simple event $\{\omega_i\}$, $i = 1, 2, \ldots, 20$, is equal to $1/20$. Thus,

$$P_X\{X = 0\} = P\{\omega_{16}, \omega_{20}\} = P\{\omega_{16}\} + P\{\omega_{20}\} = 2(\tfrac{1}{20}) = \tfrac{1}{10},$$

$$P_X\{X = 1\} = P\{\omega_3, \omega_4, \omega_7, \omega_8, \omega_{11}, \omega_{12}, \omega_{13}, \omega_{14}, \omega_{15}, \omega_{17}, \omega_{18}, \omega_{19}\}$$
$$= 12(\tfrac{1}{20}) = \tfrac{3}{5},$$

$$P_X\{X = 2\} = P\{\omega_1, \omega_2, \omega_5, \omega_6, \omega_9, \omega_{10}\} = 6(\tfrac{1}{20}) = \tfrac{3}{10},$$

so that the probability distribution of X is given by

x	0	1	2
$P_X\{X = x\}$	$\tfrac{1}{10}$	$\tfrac{6}{10}$	$\tfrac{3}{10}$

In the above example, we found the distribution of X through an explicit enumeration of the outcomes ω resulting from random sampling without replacement. This is not a difficult task in the example above, because the size N of the population and the sample size n are both small. However, when N and n are large, enumeration of the outcomes ω_i becomes quite cumbersome. (For example, when $N = 100$, $n = 3$, there are 970,200 outcomes to list.) Fortunately, the distribution of X can be obtained without the need to enumerate the outcomes.

Derivation of the Hypergeometric Distribution

Assume that we draw a random sample of size n without replacement from a population of N units. Let p denote the proportion of units in the population of N units which possess the property of interest (that is, p is the proportion of "successes" in the population). Since the number of "successes" in the population is an integer, p is one of the fractions $0/N$, $1/N$, $2/N$, ..., $(N-1)/N$, N/N. Let $q = 1 - p$; then q is the proportion of "failures" in the population. Thus, in the population of N units there are Np "successes" and Nq "failures." (Remember that a unit which is a "success" possesses the property of interest to us, while a unit which is a "failure" does not possess the property of interest.) In the sample of size n which we have drawn, we are interested in the number X of "successes" that appear. To obtain the distribution of the random variable X, we calculate the probabilities of the simple events $\{X = k\}$, $k = 0, 1, 2, ..., n$. Since the definition of random sampling without replacement implies that every ordered sample of size n in which no unit appears more than once has the same probability of being drawn, we know from Rules 1.3 and 2.2 of Chapter 3 that

$$P_X\{X = k\} = \frac{\text{number of ordered samples of size } n \text{ containing } k \text{ "successes"}}{\text{number of ordered samples of size } n}$$

$$= \frac{\text{number of ordered samples of size } n \text{ containing } k \text{ "successes"}}{[N!/(N-n)!]}.$$

(2.1)

Thus, to find $P_X\{X = k\}$, we need to find the number A_k of ordered samples in which no unit appears more than once and in which k units are "successes."

Note that any (ordered) sample of size n which contains k "successes" must also contain $(n - k)$ "failures." Keeping in mind that we are talking about a sample in which no unit appears more than once, we now determine the number A_k of ordered samples of size n in which k units are "successes" and $(n - k)$ units are "failures." We may think of forming each such sample in the following three steps: (i) First, we select a collection of k units from among the Np units in the population that are "successes"; (ii) second, we select a collection of $(n - k)$ units from among the Nq units in the population that are "failures"; (iii) finally, we combine the two collections chosen in steps (i) and (ii) into a single collection of n units, and then choose the order (permutation) in which we will list these

2. HYPERGEOMETRIC DISTRIBUTION

n units in our ordered sample. The result of these three steps is an ordered sample of n units in which k "successes" and $(n-k)$ "failures" appear.

If $k > Np$, there is no way in which we can select k units from among the Np units in the population which are "successes," since we exhaust the Np "successes" in the population before we complete our selection of k "successes." Thus, if $k > Np$, we cannot complete step (i) above and thus cannot form a sample containing k "successes"; in this case, the number A_k of ordered samples of size n in which k "successes" appear (and no unit appears more than once) is equal to 0. Similarly, if $n-k > Nq$, step (ii) cannot be completed, and again $A_k = 0$. When $k \leq Np$ and $n-k \leq Nq$, we can perform step (i) in one of $\binom{Np}{k}$ distinguishable ways (see Counting Rule 6.2 of Chapter 3). For each way in which we complete step (i), there are $\binom{Nq}{n-k}$ ways in which we can complete step (ii). Hence, we can select the collection of n units (k "successes" and $n-k$ "failures") which will appear in our sample in one of $\binom{Np}{k}\binom{Nq}{n-k}$ distinguishable ways (see Counting Rule 6.6 of Chapter 3). For every collection of units which we draw in steps (i) and (ii), there are $n!$ distinguishable ways (see Counting Rule 6.1 of Chapter 3) of ordering these units, so as to produce the ordered sample of size n and complete step (iii). Consequently, by another application of Counting Rule 6.6 of Chapter 3, we conclude that when $k \leq Np$, $(n-k) \leq Nq$, there are

$$(2.2) \qquad A_k = \binom{Np}{k}\binom{Nq}{n-k} n!$$

possible ordered samples of size n in which k units are "successes" and $(n-k)$ units are "failures" (and in which no unit appears more than once). From Equations (2.1) and (2.2), it follows that

$$(2.3) \qquad P_X\{X = k\} = \frac{A_k}{[N!/(N-n)!]} = \begin{cases} 0, & \text{when } k > Np \text{ or } n-k > Nq, \\ \dfrac{\binom{Np}{k}\binom{Nq}{n-k}}{\binom{N}{n}}, & \text{when } k \leq Np \text{ and } n-k \leq Nq. \end{cases}$$

In order to give an explicit illustration of the arguments that we have used to obtain Equations (2.1), (2.2), and (2.3), suppose that a radio supply house has an inventory (population) of 200 transistor radios (which we label $R_1, R_2, \ldots, R_{200}$). In their inventory, they know that a proportion

$p = 0.02$ of the radios have been improperly soldered. Since $N = 200$ and $p = 0.02$, it follows that $Np = 4$ and $Nq = N(1-p) = 196$. Let R_1, R_2, R_3, R_4 be the improperly soldered radios in the inventory. Suppose the supply house draws a random sample of 5 radios without replacement and sends them to a customer. To construct a particular ordered sample for which the event $\{X = 2\}$ would occur, we need to find an ordered sample in which 2 improperly soldered radios ("successes") and 3 properly soldered radios ("failures") appear. We can select such a sample by (i) choosing 2 radios, say R_1 and R_3, from among the 4 improperly soldered radios, (ii) choosing 3 radios, say R_7, R_{54}, and R_{127}, from among the properly soldered radios, and (iii) arranging the collection $\{R_1, R_3, R_7, R_{54}, R_{127}\}$ in some particular order, say $(R_7, R_3, R_{127}, R_1, R_{54})$. The total number A_2 of ordered samples of size 5 in which 2 improperly soldered radios appear (and no radio appears more than once) is the product of: (a) the number $\binom{4}{2}$ of ways of choosing 2 radios from among the 4 improperly soldered radios [step (i)]; (b) the number $\binom{196}{3}$ of ways of choosing 3 radios from among the properly soldered radios $R_5, R_6, \ldots, R_{199}, R_{200}$ [step (ii)]; and (c) the number 5! of ways of ordering the collection of 5 radios chosen in steps (i) and (ii). That is, $A_2 = \binom{4}{2}\binom{196}{3}5!$. The probability that the supply house sends $X = 2$ improperly soldered radios to their customer is

$$P_X\{X = 2\} = \frac{A_2}{[(200)!/(200-5)!]} = \frac{\binom{4}{2}\binom{196}{3}}{\binom{200}{5}} = \frac{1261}{431{,}233} = 0.002924.$$

On the other hand, the probability that the supply house sends 5 improperly soldered radios to its customer is equal to 0, since the radio supply house only has $Np = 4$ improperly soldered radios to send.

Any random variable X possessing a distribution of the form (2.3) is said to have a hypergeometric distribution with parameters N, n, and p. The parameters n and N are frequently known to us, but the parameter p may have to be determined either by theory or by observing repeated trials of X. Table 2.2 illustrates, for $N = 10$, $n = 1(1)5$, and various values of p, how the probabilities $P_X\{X = k\}$ depend on the value of the parameter p.

If X has the hypergeometric distribution (2.3), then the expected value μ_X and the variance σ_X^2 of X are

(2.4) $\qquad \mu_X = np, \qquad \sigma_X^2 = \left(\frac{N-n}{N-1}\right)np(1-p).$

2. HYPERGEOMETRIC DISTRIBUTION

Table 2.2. Selected values of the hypergeometric distribution.

				$N = 10$				
n	Np	k	$p_X(k)$		n	Np	k	$p_X(k)$
1	1	0	0.9000		4	4	0	0.0714
		1	0.1000				1	0.3810
							2	0.4286
2	1	0	0.8000				3	0.1143
		1	0.2000				4	0.0048
2	2	0	0.6222		5	1	0	0.5000
		1	0.3556				1	0.5000
		2	0.0222					
					5	2	0	0.2222
3	1	0	0.7000				1	0.5556
		1	0.3000				2	0.2222
3	2	0	0.4667		5	3	0	0.0833
		1	0.4667				1	0.4167
		2	0.0667				2	0.4167
							3	0.0833
3	3	0	0.2917					
		1	0.5250		5	4	0	0.0238
		2	0.1750				1	0.2381
		3	0.0083				2	0.4762
							3	0.2381
4	1	0	0.6000				4	0.0238
		1	0.4000					
					5	5	0	0.0040
4	2	0	0.3333				1	0.0992
		1	0.5333				2	0.3968
		2	0.1333				3	0.3968
							4	0.0992
4	3	0	0.1667				5	0.0040
		1	0.5000					
		2	0.3000					
		3	0.0333					

Note that for given values of N, n, and p, the expected value of the hypergeometric distribution agrees with the expected value (see Section 1) of the binomial distribution with parameters n and p. On the other hand, the variances of the hypergeometric and binomial distributions differ. The ratio of the variance of the hypergeometric distribution to the variance $np(1-p)$ of the binomial distribution is equal to $(N-n)/(N-1)$. This number is sometimes called the *finite population factor* (or *finite population correction*). As the population size N becomes infinitely large, the factor $(N-n)/(N-1)$ tends to 1, so that we might say that in infinite populations there is no difference (at least in terms of location and dispersion) between the binomial and hypergeometric distributions. When N is finite, the factor $(N-n)/(N-1)$ loosely indicates how much difference (in dispersion) there is between the binomial and hypergeometric distributions.

Approximation of the Hypergeometric Distribution by the Binomial Distribution

If N is large with respect to n (say, 10 times as large as n), the finite population factor $(N-n)/(N-1)$ is nearly equal to 1. It seems reasonable in this case to expect the binomial distribution to serve as an approximation to the hypergeometric distribution.

For example, suppose $N = 100$, $p = 0.1$, $n = 10$, so that $(N-n)/(N-1) = 90/99 = 0.909$. Then the exact probability mass function for the hypergeometrically distributed random variable X is

x	0	1	2	3	4	5	6	7	8	9	10
$p_X(x)$	0.330	0.408	0.202	0.052	0.008	0.001	0.000	—	—	—	—

while the binomial approximation yields the values

x	0	1	2	3	4	5	6	7	8	9	10
Approximate Probability	0.349	0.387	0.194	0.057	0.011	0.002	0.000	—	—	—	—

As can be seen, the comparison of exact with approximate probabilities is quite favorable; the maximum difference being 0.021. Since the binomial probabilities are simpler to compute and more readily available in tabular form, they are frequently used to approximate the hypergeometric probabilities.

Uses of the Hypergeometric Distribution

The hypergeometric distribution is very important in the analysis of opinion surveys. Most opinion surveys involve drawing a random sample of individuals from a population of individuals (voters, taxpayers, school children, and so on) and asking them one or more questions. The questions asked are usually of a kind admitting only one of two possible answers. The interviewer is interested in the number X of people in his sample who answer his question in a particular way (for example, by saying "yes" to a yes-no type of question), since X allows him to make inferences about the proportion p of people in the population who would answer his question in that particular manner. How these inferences are made is a subject for a book on statistical inference; we illustrate here a probability calculation that might be made in the course of such an investigation.

Example 2.1. A psychologist is interested in finding out how many students in the senior class at a certain high school have tried a particular drug. There are 300 members in the senior class at the high school. The psychologist draws a random sample of 5 students without replacement from among the 300 senior class members and asks each of these 5 students if he has tried the particular drug. Two of the students interviewed say that they have tried the drug. Assuming that the students have all told him the truth, the psychologist asks himself how probable this result is if, as he believes, 50 percent of the seniors have tried the drug. Thus, he wants to know the probability, $P_X\{X=2\}$, when X has a hypergeometric distribution with parameters $p=0.50$, $n=5$, $N=300$. From Equation (2.3), noting that $Np = Nq = 150$,

$$P_X\{X=2\} = \frac{\binom{150}{2}\binom{150}{3}}{\binom{300}{5}} = \frac{18{,}625}{59{,}202} = 0.3146.$$

Note that the probability $P_X\{X=2\} = 0.3125$ when X is binomially distributed with parameters $n=5$, $p=0.5$. The finite population factor $(N-n)/(N-1)$ in this example is equal to $(300-5)/(300-1) = 0.987$.

Another application of the hypergeometric distribution arises in the area of quality control. From an inventory of N items, a random sample of n items without replacement is sampled and the number X of defective items is noted. From knowledge of the value $X=x$, and from knowledge of the form of the distribution (2.3) for X for various values of the parameter p, statistical inferences may be made concerning the proportion p of defective items in the inventory.

There are many other examples of applications of the hypergeometric distribution, not all of which are concerned with inferences based on a single random sample.

Example 2.2 (Card Playing). Pearson (1924) reported on the records of 25,000 actual deals of the card game whist. His concern was in the determination of whether the shuffling was "perfect." This was done by comparing actual and theoretical results. Table 2.3 gives observed and theoretical observed frequencies for the event that X trump cards appear in the first hand of a deal of whist. The data are taken from a sample of 3400 deals of the cards. The theoretical observed frequencies are obtained by multiplying the theoretical probability $P_X\{X = x\}$ by 3400.

Table 2.3. Comparison of observed and theoretical frequencies for the event that X trump cards appear in the first hand of a deal of the card game whist.

Number x of Trump Cards in the Hand	Observed Frequency of X Trump Cards	Theoretical Observed Frequency of X Trump Cards Computed from the Hypergeometric Distribution with $N = 52, n = 13, p = \frac{1}{4}$
0	35	43.4
1	290	272.3
2	697	700.2
3	937	973.4
4	851	811.2
5	444	424.0
6	115	141.4
7	21	29.9
8	10	3.7
9 and over	0	0.5
	3400	3400.0

SOURCE: Pearson, K. (1924). On a certain double hypergeometrical series and its representation by continuous frequency surfaces. *Biometrika*, vol. 16, pp. 172–188. Reprinted with permission.

Description of Tables

Table T.2 provides values for the probability mass function of the hypergeometric distribution for $N = 2(1)9$, and selected values of n and p.

A more detailed tabulation may be found in the following book:

Lieberman, G. J., and Owen, D. B. (1961). *Tables of the Hypergeometric Distribution*, Stanford University Press, Stanford, California.

This book gives the probabiliy mass functions and cumulative distribution functions of hypergeometrically distributed random variables for $N = 2(1) 50(10) 100$ (that is, N ranges from 2 to 50 by increments of 1 and then to 100 by increments of 10).

3. POISSON DISTRIBUTION

If we study data collected on accidents by insurance companies, an interesting fact emerges. Suppose, for example, that we are interested in the number X of individuals injured while getting out of an automobile in a given year. For any particular individual, the probability of such an accident in a given year is quite small; in fact, it is almost zero. However, data reveal that a number of such accidents do occur every year, and that the number X of such accidents varies from year to year. Over many years, the relative frequencies of the events $\{X = k\}$, for k an integer, exhibit the stability property mentioned in Chapter 1. When we search for a discrete probability distribution which appears to fit this type of statistically regular behavior, we find that the relative frequency of the event that the number of accidents X equals a given number k is closely approximated by the probability assigned to this event by the *Poisson distribution* (named after the French mathematician Simeon D. Poisson, 1781–1840). This discrete distribution has a probability mass function of the form

$$(3.1) \qquad p_X(k) = \frac{e^{-\lambda}\lambda^k}{k!},$$

for values $k = 0, 1, 2, \ldots$. In Equation (3.1), the symbol e stands for the absolute constant 2.7183... (this number is not a fraction and can be expanded indefinitely in decimal form), λ denotes a certain positive number, and $e^{-\lambda} = (1/e)^\lambda$. Some representative values of the mass function are

k	0	1	2	3	...
$p_X(k)$	$e^{-\lambda}$	$\lambda e^{-\lambda}$	$\frac{1}{2}\lambda^2 e^{-\lambda}$	$\frac{1}{6}\lambda^3 e^{-\lambda}$	...

The constant λ is the only parameter of the Poisson distribution; λ can be any positive number, depending on the random phenomenon being observed. The effect of the parameter λ on the behavior of the probability mass function $p_X(k)$ of the Poisson distribution can be observed from Table 3.1, where the values of $p_X(k)$, for $k = 0, 1, 2, \ldots$, are given corresponding to the following choice of values for λ: $\lambda = 0.5, 1.0, 2.0, 3.0, 5.0,$ and 10.0.

In Table 3.1, note that for any value of λ greater than 1 (that is, $\lambda = 2, 3, 5, 10$), the probabilities $P_X(k) = p_X\{X = k\}$ increase until k is approximately equal to λ, and then decrease. Indeed, for integer values of λ, both λ and $\lambda - 1$ are modes of the Poisson distribution. (When λ is not an integer, the mode of the Poisson distribution is the integer k which lies between $\lambda - 1$ and λ.) In general, regardless of whether or not λ is an integer, when $\lambda > 1$, the values of $p_X(k)$ increase as k increases, reach a maximum for k close to λ, and then decrease as k further increases.

Table 3.1. Values of $p_X(k)$ for the Poisson distribution when $\lambda = 0.5, 1, 2, 3, 5, 10$.

k \ λ	0.5	1	2	3	5	10
0	0.6065	0.3679	0.1353	0.0498	0.0067	0.0000
1	0.3033	0.3679	0.2707	0.1494	0.0337	0.0005
2	0.0758	0.1839	0.2707	0.2240	0.0842	0.0023
3	0.0126	0.0613	0.1804	0.2240	0.1404	0.0076
4	0.0016	0.0153	0.0902	0.1680	0.1755	0.0189
5	0.0002	0.0031	0.0361	0.1008	0.1755	0.0378
6		0.0005	0.0120	0.0504	0.1462	0.0631
7		0.0001	0.0034	0.0216	0.1044	0.0901
8			0.0009	0.0081	0.0653	0.1126
9			0.0002	0.0027	0.0363	0.1251
10				0.0008	0.0181	0.1251
11				0.0002	0.0082	0.1137
12				0.0001	0.0034	0.0948
13					0.0013	0.0729
14					0.0005	0.0521
15					0.0002	0.0347
16						0.0217
17						0.0128
18						0.0071
19						0.0037
20						0.0019
21						0.0009
22						0.0004
23						0.0002
24						0.0001

It can be shown that when X has a Poisson distribution, the expected value μ_X and the variance σ_X^2 of X are both equal to λ. That is,

(3.2) $$\mu_X = \lambda, \qquad \sigma_X^2 = \lambda.$$

Hence, a property of the Poisson distribution is that the ratio of its expected value μ_X to its variance σ_X^2 is always unity (that is, $\mu_X/\sigma_X^2 = 1$).

A Method for Fitting the Poisson Distribution

The wide variety of random phenomena which give rise to random variables X having a Poisson distribution is truly astonishing. Some classical

3. POISSON DISTRIBUTION

examples for which the Poisson distribution provides a reasonable model are given in this section. These examples serve to illustrate the various magnitudes of the parameter λ that may be appropriate in applications of the Poisson distribution. Before giving these examples, we need to explain how the Poisson distribution might be "fit" to data resulting from repeated trials of a random experiment.

In many biological examples a surface is subdivided into equal sections, from which counts of specimens are made. This occurs, for example, in studies of blood samples, yeast cultures, and so on.

Example 3.1 (Zoology). A horizontal quarry surface was divided into 30 squares about 1 meter on a side. In each square the number X of specimens of the extinct mammal *Litolestes notissimus* was counted. The results are shown in Table 3.2.

Table 3.2. Observed relative frequencies of the events $\{X = k\}$ in 30 trials (squares) of the random experiment in which we count the number X of specimens of *Litolestes notissimus* in a square of horizontal quarry surface.

$k =$ Number of Specimens Per Square	Observed Frequency of Squares with k Specimens	Observed Relative Frequency of Squares with k Specimens
0	16	8/15
1	9	3/10
2	3	1/10
3	1	1/30
4	1	1/30
5 and over	0	0
	30	1

SOURCE: From *Quantitative Zoology* by G. G. Simpson and A. Roe. Copyright 1939 by McGraw-Hill Book Company, New York. Used with permission of McGraw-Hill Book Company.

To fit a Poisson distribution to these data, we need to determine an appropriate value for the parameter λ. Recall that λ is the expected value of the Poisson distribution; this fact suggests estimating λ using the arithmetic average of the data. This estimate is given by

$$\text{estimate of } \lambda = (0)(\text{r.f.}\{X = 0\}) + (1)(\text{r.f.}\{X = 1\}) + (2)(\text{r.f.}\{X = 2\})$$
$$+ (3)(\text{r.f.}\{X = 3\}) + (4)(\text{r.f.}\{X = 4\}) + \cdots$$
$$= (0)\left(\frac{8}{15}\right) + (1)\left(\frac{3}{10}\right) + (2)\left(\frac{1}{10}\right) + (3)\left(\frac{1}{30}\right) + (4)\left(\frac{1}{30}\right) + 0$$
$$= \frac{11}{15} = 0.73.$$

With this estimated value of λ (that is, $\lambda = 0.73$), we can compute values of $p_X(k)$, for $k = 0, 1, 2, \ldots$, using the formula for $p_X(k)$ given in Equation (3.1). Thus, for example, $p_X(2)$ is given by

$$p_X(2) = \frac{e^{-0.73}(0.73)^2}{2!},$$

and $p_X(4)$ is given by

$$p_X(4) = \frac{e^{-0.73}(0.73)^4}{4!}.$$

The calculations required to find $p_X(2)$ and $p_X(4)$ are not easy. Fortunately, tables exist (see Table T.3) which provide answers for such computations. From one such table, the following probability mass function has been obtained for X:

k	0	1	2	3	4	5 and over
$p_X(k)$	0.48	0.35	0.13	0.03	0.01	0.00

This probability mass function gives the theoretical probabilities of the simple events $\{X = 0\}$, $\{X = 1\}$, $\{X = 2\}$, $\{X = 3\}$, and so on, assuming that λ equals its estimated value of 0.73. Using this table, we may also determine the theoretical *number* of squares having k specimens that we could expect in our sample of size $n = 30$. This we can do by noticing that for a given value of k, the random variable Y_k, which gives the number of times in our sample of 30 trials that the simple event $\{X = k\}$ occurs, has the binomial distribution with parameter $p = P_X\{X = k\}$ and $n = 30$ (see Section 1). Since the expected value of a random variable having the binomial distribution with parameters p and n is $\mu = np$ [see Equation (1.3)], the expected number of times in our sample of 30 trials that the simple event $\{X = k\}$ occurs is equal to the expected value of Y_k or $\mu_k = 30 P_X\{X = k\}$. Thus, from the theoretical probability mass function of X shown above, the expected number of times that $X = 0$ in our sample is equal to $\mu_0 = (30) P_X\{X = 0\} = (30)(0.48) = 14.4$, the expected number of times that $X = 1$ in our sample is equal to $\mu_1 = (30) P_X\{X = 1\} = (30)(0.35) = 10.5$, and so on. From such calculations, we obtain a table (Table 3.3) which compares the observed number of squares in which we find k specimens to the expected number of such squares, for $k = 0, 1, 2$, and so on.

Judging by the above table, a Poisson distribution with parameter $\lambda = 0.73$ agrees closely (but not perfectly) with the observed frequencies (or relative frequencies). If the size n of our sample of squares (that is, the number of trials of our random experiment) had been larger than 30, the fit might have been better.

Table 3.3. Comparison of observed and theoretical frequencies, and of observed and theoretical relative frequencies, of the events $\{X = k\}$ in 30 trials (squares) of the random experiment in which we count the number X of specimens of *Litolestes notissimus*.

$k =$ Number of Specimens Per Square	Observed Number of Squares in Which k Specimens Are Observed	Theoretical Probability of Observing k Specimens: $P_X\{X = k\}$	Theoretical Expected Number of Squares in Which k Specimens Are Observed
0	16	0.48	14.4
1	9	0.35	10.5
2	3	0.13	3.9
3	1	0.03	0.9
4	1	0.01	0.3
5 and larger	0	0.00	0.0

SOURCE: From *Quantitative Zoology* by G. G. Simpson and A. Roe. Copyright 1939 by McGraw-Hill Book Company, New York. Used with permission of McGraw-Hill Book Company.

[*Remark:* In "fitting" a Poisson distribution in the example above, we used the sample average from the data to determine the parameter λ of the distribution. Actually the theory of fitting and of determining goodness-of-fit for probability distributions is well developed, and many alternative techniques are available. Because these techniques are peripheral to our main interest, we arbitrarily have selected a method here which appears to be both useful and easily understood.]

Applications of the Poisson Distribution

We have already remarked upon the versatility of application of the Poisson distribution. In industry, engineers have applied the Poisson distribution to the distribution of the number of flaws in capacitors, to the number of defects per linear unit of wire and of rope, and to the number of strands in a cross section of thread. In agriculture and biology, applications have been made to the number of beetle larvae, to the number of fish caught in a day, to the number of photons reaching the retina, and to bacteria counts. In medicine, we find the Poisson distribution applied to the number of defective teeth per individual, and to the number of victims suffering from various specific diseases. In sociology, the Poisson distribution has been applied to the numbers of university graduates, to the number of vacancies in the Supreme Court, and to the number of labor strikes. Applications of the Poisson distribution have also been made to the number of words misread in a text, to the number of misprints, and to the frequency of earthquakes.

The following three examples are classical examples of the use of the Poisson distribution. In each case, the appropriate value of the parameter λ is estimated by taking the sample average of the observations. Because in these examples the sample average is computed from the data *before* the data is condensed for the tables, there may be discrepancies between the stated value of the average and the average that can be obtained from the results shown in the tables.

Example 3.2 (Bombing Hits). The distribution of the daily number X of hits by flying bombs in London during World War II has been approximated by a Poisson distribution with $\lambda = 0.943$. The data used to fit the Poisson distribution came from dividing London into 580 different regions, each of nearly equal area, and recording the number of flying bomb hits in each given area. The number of bomb hits in a given area thus becomes a trial of the random experiment which consists of observing the daily number of bomb hits in a given region of London. See Table 3.4.

Table 3.4. Comparison of observed and theoretical frequencies of k flying bomb hits in 580 trials of the random experiment in which one region (out of 580 regions) in London is observed over a given day.

Number of Hits k	Observed Number of Areas (Trials) N_k	Theoretical Number of Areas (Trials) $Np_X(k)$ $\lambda = 0.943$
0	229	225.62
1	211	212.86
2	93	100.34
3	39	31.32
4	7	7.54
5 or more	1	2.32
	580	580.00

SOURCE: Clarke, R. D. (1946) An application of the Poisson distribution. *Journal of the Institute of Actuaries*, vol. 72, p. 48. Reprinted with permission.

Example 3.3 (Telephone Connections). Groups of coin-box telephones in a large transportation terminal were observed. The calls made on several statement of ages by persons under 21 years is quite common in marriage statistics; however, no adjustment of the data was made. The results of the fit are shown in Table E.7 [Pretorius (1930)]. Comment on the use of the lognormal model for these data. Does the lognormal distribution provide a good fit? Do you think any of the other families of continuous distributions which have been discussed in Chapter 8 (or in Chapter 7) would provide a better fit? Explain your answer.

9
MULTIVARIATE DISTRIBUTIONS

Order and simplification are the first steps toward the mastery of a subject—the actual enemy is the unknown.

Thomas Mann (The Magic Mountain)

Happy the man, who, studying Nature's laws, through known effects can trace the secret cause.

Virgil (Georgics)

1. INTRODUCTION

Many of the most important problems in science and technology are concerned with the question of how the values of one measurable quantity (or several measurable quantities) can be used to explain, predict, or control another quantity (or quantities). Thus, physicists may be interested in how the motion of a body is affected by the forces acting upon it, aeronautical engineers may be interested in finding out how automatic sensors can provide information to guide a spaceship, and biologists may be interested in how organisms react to chemical changes in their environment (for example, how much of a certain drug is required, on the average, to destroy a certain virus). Similarly, sociologists may be interested in the relationship of group size to problem-solving productivity, psychologists may be interested in the relationship of personality test scores to anxiety level, and educators may be interested in how high school grades can be used to predict college achievement. Indeed, virtually all scientific and technological disciplines require, as a part of their development, knowledge of the interrelationships among the variables which compose their empirical framework.

Mechanistic models (see Chapter 1) for natural phenomena express relationships between variables either in terms of the logic of causation (for example, "A causes B") or in terms of exact (or approximate) mathematical formulas (such as "force equals mass times acceleration"). Stochastic models, on the other hand, express such relationships in terms of statistical tendencies of joint occurrence. For example, stochastic models help quantify such statements as: "High values of X tend to be associated with high values of Y". The probabilistic framework for such stochastic models involves the notion of *composite experiments* (see Chapter 3). When the random experiments which form the components of such composite experiments give rise to more than one random variable, we say that the resulting probability model gives the *joint* (or *multivariate*) *distribution* of these variables.

To introduce the ideas embodied in the notion of multivariate distributions, we first discuss the joint distribution of two random variables (a *bivariate distribution*). Because the concepts that we discuss are more easily understood when both random variables are discrete, we begin in Sections 2 and 3 with an examination of discrete bivariate distributions. In Section 4, we consider continuous bivariate distributions (that is, probability models that describe the joint variation of two continuous random variables). The most famous and widely used continuous bivariate distribution, the bivariate normal distribution, is investigated in Section 5. Finally, in Section 6 we generalize our discussion to cover joint distributions of three random variables (*trivariate distributions*) and of more than three random variables.

2. DISCRETE BIVARIATE DISTRIBUTIONS

In Chapters 4 through 8, we have considered probability models for only one random variable. However, if we are observing more than one variable in a given (composite) random experiment, concentrating upon the distribution of each of these variables separately causes us to ignore any information about the frequency with which certain values of any one of these random variables, say X, can be expected to occur in connection with certain values of any other random variable, say Y. To obtain such information, we construct the *bivariate distribution* of X and Y. That is, we construct a probability distribution for the composite experiment whose outcomes are values of the *pair* (X,Y).

Suppose X and Y are both discrete random variables. Then their *joint probability mass function* (bivariate probability mass function) is the rule which assigns to each pair of numbers (x,y) the probability of the event $\{X = x \text{ and } Y = y\}$. Thus, the joint probability mass function $p_{X,Y}(x,y)$ is defined by the equation

(2.1) $\qquad p_{X,Y}(x,y) = P_{X,Y}\{X = x \text{ and } Y = y\}.$

Since the double subscript on $p_{X,Y}(x,y)$ is cumbersome, when no confusion is possible, we write $p(x,y)$ instead of $p_{X,Y}(x,y)$. Often, observations on the random variables X and Y are taken nearly simultaneously;

in such cases, $p_{X,Y}(x,y)$ gives the probability of the event that $X = x$ and simultaneously $Y = y$. On the other hand, one of the variables, say X, may be observed first; here $p_{X,Y}(x,y)$ tells us the probability of the event that $X = x$ and then $Y = y$.

Consider the case where the random variable X has three possible values, say x_1, x_2, x_3, and where the random variable Y has four possible values: y_1, y_2, y_3, and y_4. Then the possible outcomes of the random experiment in which X and Y are both observed are the pairs of numbers: (x_1,y_1), (x_1,y_2), (x_1,y_3), (x_1,y_4), (x_2,y_1), (x_2,y_2), (x_2,y_3), (x_2,y_4), (x_3,y_1), (x_3,y_2), (x_3,y_3), (x_3,y_4). There are 12 such outcomes; corresponding to these outcomes there are 12 simple events: $\{(x_1,y_1)\}$, $\{(x_1,y_2)\}$, and so on. To give the joint probability mass function $p_{X,Y}(x,y)$ for X and Y, it is not necessary to write down the probabilities of simple events corresponding to pairs of numbers (x,y) that cannot be observed since these probabilities are, of course, all equal to 0 [that is, $p_{X,Y}(x,y) = 0$ for $x \neq x_1, x_2, x_3$; $y \neq y_1, y_2, y_3, y_4$]. Nevertheless, we must still list 12 different numbers. A helpful way to list the probabilities $p_{X,Y}(x,y) = p(x,y)$ for $x = x_1, x_2, x_3$, $y = y_1, y_2, y_3, y_4$, is to give them in the form of a *contingency table*:

x \ y	y_1	y_2	y_3	y_4
x_1	$p(x_1,y_1)$	$p(x_1,y_2)$	$p(x_1,y_3)$	$p(x_1,y_4)$
x_2	$p(x_2,y_1)$	$p(x_2,y_2)$	$p(x_2,y_3)$	$p(x_2,y_4)$
x_3	$p(x_3,y_1)$	$p(x_3,y_2)$	$p(x_3,y_3)$	$p(x_3,y_4)$

Such a contingency table is both more compact than a vertical list of values $p_{X,Y}(x, y)$, and also more revealing. By looking across the first row of the table, we can see which values of the random variable Y are the most probable to occur with a value $X = x_1$ of the random variable X. Similarly, the third column of the contingency table tells us what values of the random variable X are the most probable to appear jointly with a value $Y = y_3$ of the random variable Y. Note that since 1 of the 12 outcomes (x_1, y_1), $(x_1,y_2), \ldots, (x_3, y_4)$ must occur, the sum of the entries $p_{X,Y}(x,y)$ in the contingency table is equal to 1.

As an alternative to a contingency table, we can graph the joint probability mass function $p_{X,Y}(x,y)$ (see Figure 2.1). We can view the possible pairs (x,y) as points in the plane. From these points (x,y), perpendicular dotted lines are extended to a height equal to $p_{X,Y}(x,y)$ from the plane; these lines represent the probabilities of the events $\{X = x_i$ and $Y = y_j\}$, $i = 1, 2, 3, j = 1, 2, 3, 4$. Thus, for example, if $p_{X,Y}(x_1,y_3) = 1/10$, the line rising from the point (x_1,y_3) has height 1/10. Since the sum of the values of the probability mass function $p_{X,Y}(x,y)$ over all possible points (x,y)

2. DISCRETE BIVARIATE DISTRIBUTIONS

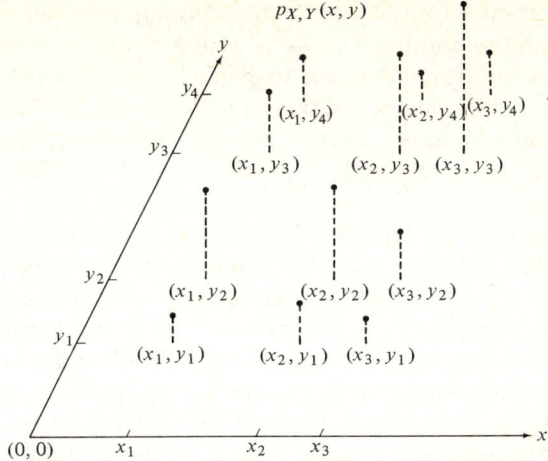

Figure 2.1 Graph of the joint probability mass function $p_{X,Y}(x,y)$.

is equal to 1, the sum of the heights of the dotted lines used to graph the probability mass function $p_{X,Y}(x,y)$ in Figure 2.1 is equal to 1. Note that Figure 2.1 and the contingency table exhibited above both give similar information about the joint variation of the random variables X and Y.

Example 2.1. Two professional wine tasters are asked to rate a sample of wine using a rating scale that allows them to give ratings of 1, 2, 3, 4, or 5 (a rating of 5 being equivalent to "top quality"). If standards of wine quality are more or less common to all wine tasters, and if the wine they are to taste is of a moderate quality, we might expect their responses to follow a table of probabilities such as Table 2.1. Here, wine taster I's rating is a random variable X, wine taster II's rating is a random variable

Table 2.1. A bivariate probability mass function for two wine tasters exhibited in the form of a contingency table.

Rating of Wine Taster I	II	\multicolumn{5}{c	}{y}	Totals
		1	2	3	4	5	
	1	0.03	0.02	0.01	0.00	0.00	0.06
	2	0.02	0.08	0.05	0.02	0.01	0.18
x	3	0.01	0.05	0.25	0.05	0.01	0.37
	4	0.00	0.02	0.05	0.20	0.02	0.29
	5	0.00	0.01	0.01	0.02	0.06	0.10
	Totals	0.06	0.18	0.37	0.29	0.10	1.00

Y, and we can think of both wine tasters announcing their ratings simultaneously. From the contingency table (Table 2.1), we note that most of the higher probabilities are given to pairs (x,y) for which x and y are nearly equal and in the middle of the range of ratings (ranks of 2, 3, or 4). The most probable simple event is that both wine tasters agree that the wine they are rating is of average quality, that is, the event $\{X = 3$ and $Y = 3\}$.

Notice that we have added one extra column and one extra row to the contingency table; this extra column and extra row give the row sums and column sums, respectively, of the contingency table. For example, in row 1 of the table (corresponding to the value $x = 1$), the last entry, 0.06, is the sum of the other entries 0.03, 0.02, 0.01, 0.00, 0.00 in that row. Similarly, in column 3 (corresponding to the value $y = 3$), the last entry 0.37 is the sum of the other entries 0.01, 0.05, 0.25, 0.05, 0.01, in that column. Because the row sums and column sums are placed at the margins of the contingency table, they are known as *marginal totals*, or (since they are the sum of probabilities) *marginal probabilities*. In fact, the sum (marginal total) of the entries in row 1 (excluding the entry in the "Total" column) of the contingency table gives the probability $P_X\{X=1\} = 0.06$. Similarly, the sum (marginal total) of the entries in column 3 of the contingency table (excluding the entry in the row marked "Total") gives $P_Y\{Y=3\} = 0.37$. We return to a discussion of marginal totals momentarily.

To show how different bivariate probability distributions can reveal different interrelationships between two random variables X and Y, consider Table 2.2, which gives another possible discrete joint probability model for the ratings of the two wine tasters. Here, as in Table 2.1, most of the probability is concentrated on simple events $\{(x,y)\}$ corresponding to situations where the wine being rated is considered to be average or nearly average by both wine tasters (ratings of 2, 3, or 4). However, in Table 2.2 there is less probability attached to simple events corresponding to cases where there is exact agreement between the raters [that is, events $\{(x,y)\}$ for which $x=y$], and more probability assigned to simple

Table 2.2. Another bivariate probability mass function for two wine tasters.

Rating of Wine Taster I	II	y					Totals
		1	2	3	4	5	
x	1	0.01	0.01	0.01	0.01	0.01	0.05
	2	0.01	0.09	0.09	0.09	0.01	0.29
	3	0.01	0.09	0.12	0.09	0.01	0.32
	4	0.01	0.09	0.09	0.09	0.01	0.29
	5	0.01	0.01	0.01	0.01	0.01	0.05
Totals		0.05	0.29	0.32	0.29	0.05	1.00

2. DISCRETE BIVARIATE DISTRIBUTIONS 437

events in which there is wide divergence of opinion between the wine tasters [that is, events such as {(1,5)}, {(5,1)}, {(1,4)}, {(4,1)}, and so on]. Table 2.2, in contrast to Table 2.1, may thus reflect a situation where standards of quality vary a great deal with the individual.

The two probability models given in Example 2.1 can also be compared visually through an inspection of the graphs of their respective bivariate probability mass functions. The graphs associated with Tables 2.1 and 2.2 are shown in Figure 2.2.

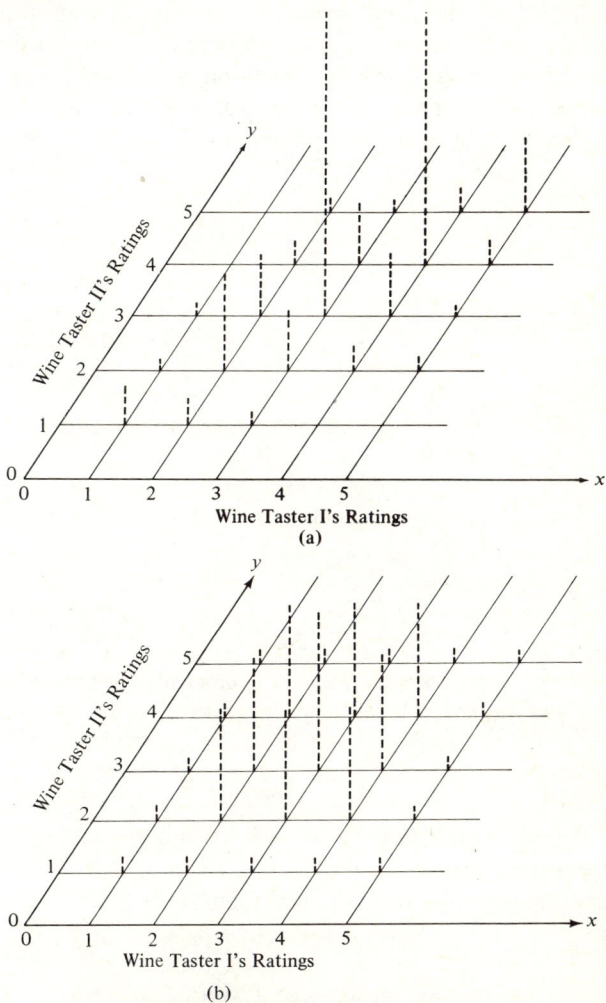

Figure 2.2 Graph of the bivariate probability mass function $p_{X,Y}(x, y)$ corresponding to the probability model for the joint ratings of Wine Tasters I and II: (a) associated with Table 2.1; (b) associated with Table 2.2.

The bivariate probability mass function $p_{X,Y}(x,y)$ contains all of the probabilistic information relevant for describing the simultaneous variation of the two discrete random variables X and Y. It is helpful to think of the pair (X,Y), resulting from the observation of X and Y, as a random point in the plane. That is, instead of the random experiment in which we actually observe X and Y, we can think of another random experiment in which X provides the first coordinate and Y provides the second coordinate of a point (X,Y) in the plane. On repeated trials of this last random experiment, different points will occur. Let R be any region in the plane (see Figure 2.3), and suppose we are interested in the probability that (X,Y) falls within the region R on any given performance of the experiment. In the case of discrete random variables, we obtain this probability by summing $p_{X,Y}(x,y)$ over all the possible points (x,y) contained within the region R.

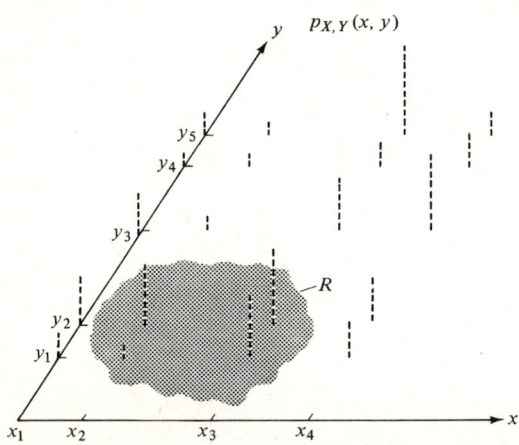

Figure 2.3 A region R in the plane. The probability that the random point (X,Y) falls within the region R is the sum of the heights $p_{X,Y}(x,y)$ of the dotted lines which correspond to points (x,y) in R.

Example 2.1 (Continued). Suppose we ask for the probability that the two wine tasters agree in their rankings (see Figure 2.4). Under the first probability model given for this problem (see Table 2.1),

$$P\{\text{tasters agree}\} = P\{(x,y): x = y\}$$
$$= P\{(1,1) \text{ or } (2,2) \text{ or } (3,3) \text{ or } (4,4) \text{ or } (5,5)\}$$
$$= 0.03 + 0.08 + 0.25 + 0.20 + 0.06 = 0.62.$$

Under the second probability model given for this problem (see Table 2.2),

2. DISCRETE BIVARIATE DISTRIBUTIONS

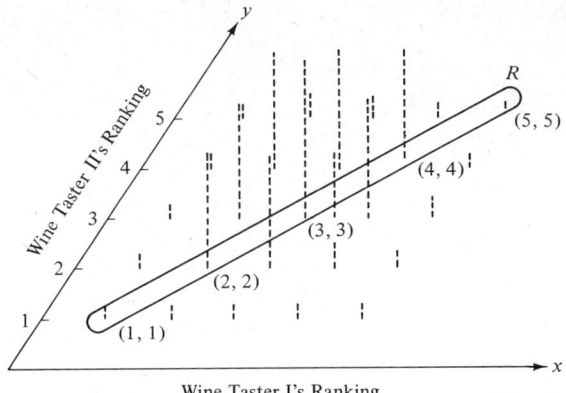

Figure 2.4 The region R corresponding to the event E that the Wine Tasters I and II agree in their rankings of a sample of wine. The sum of the probabilities of the points contained in R equals the probability of the event E.

we obtain

$$P\{\text{tasters agree}\} = 0.01 + 0.09 + 0.12 + 0.09 + 0.01 = 0.32.$$

This calculation partially verifies our earlier assertion that there is less agreement between the wine tasters under probability model of Table 2.2 than there is under the probability model of Table 2.1.

Marginal Distributions

Not only does the joint probability mass function $p_{X,Y}(x,y)$ contain all of the relevant statistical information concerning X and Y considered *simultaneously*, it also contains all the relevant information concerning X and Y considered *individually*. That is, from $p_{X,Y}(x,y)$ we can obtain the probability mass functions $p_X(x)$ and $p_Y(y)$ of X and Y, respectively. To see this, let R_1 be the region in the plane consisting of all possible points (see Figure 2.5) having x_0 for the first coordinate (no matter what the second coordinate y may be). The event $\{(X,Y) \text{ falls in } R_1\}$ is equivalent to the event $\{X = x_0\}$. Hence, $P_{X,Y}\{\text{point }(X,Y) \text{ is a member of } R_1\} = P_X\{X = x_0\} = p_X(x_0)$, and this probability can be obtained by summing $p_{X,Y}(x,y)$ over all points (x,y) in the region R_1. In an analogous way, $p_Y(y_0)$ can be obtained by summing $p_{X,Y}(x,y)$ over all points in the region R_2 in the plane consisting of points (x,y) whose second coordinate is equal to y_0 (see Figure 2.5). When the distributions of X and Y are obtained in this way from their joint probability function $p_{X,Y}(x,y)$, the distributions obtained are referred to as *marginal distributions*.

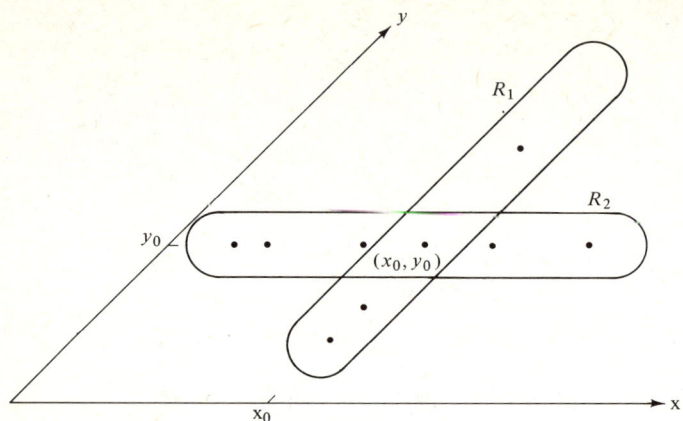

Figure 2.5 Region R_1 consists of all points (x,y) having $x = x_0$ (a fixed x-coordinate). Region R_2 consists of all points (x, y) having a fixed y-coordinate.

We have already considered marginal distributions in Example 2.1, where the probabilities $p_X(x_i)$ and $p_Y(y_j)$ appeared in the appended last column and last row, respectively, of the contingency tables, Table 2.1 and Table 2.2. There, these probabilities were called marginal probabilities. In general, suppose that we have summarized a discrete bivariate probability distribution in terms of a contingency table such as the following:

x \ y	y_1	y_2	y_3	y_4	Row Total
x_1	$p_{X,Y}(x_1,y_1)$	$p_{X,Y}(x_1,y_2)$	$p_{X,Y}(x_1,y_3)$	$p_{X,Y}(x_1,y_4)$	$p_X(x_1)$
x_2	$p_{X,Y}(x_2,y_1)$	$p_{X,Y}(x_2,y_2)$	$p_{X,Y}(x_2,y_3)$	$p_{X,Y}(x_2,y_4)$	$p_X(x_2)$
x_3	$p_{X,Y}(x_3,y_1)$	$p_{X,Y}(x_3,y_2)$	$p_{X,Y}(x_3,y_3)$	$p_{X,Y}(x_3,y_4)$	$p_X(x_3)$
Column Total	$p_Y(y_1)$	$p_Y(y_2)$	$p_Y(y_3)$	$p_Y(y_4)$	1.00

In such a contingency table, we can find the marginal distribution (marginal probability mass function) of X from the (appended) last column of the table (the column marked "Row Total"), and the marginal distribution of Y from the (appended) last row. Thus, in Table 2.1, the marginal distribution (probability mass function) of the rating X of wine taster I appears in the last column of the table. We reproduce that distribution in Table 2.3(a).

2. DISCRETE BIVARIATE DISTRIBUTIONS

Table 2.3(a). Probability mass function (given only for possible values of the random variable) of the rating X of wine taster I as taken from the last column of Table 2.1.

X	Probability $p_X(x) = P_X\{X = x\}$
1	0.06
2	0.18
3	0.37
4	0.29
5	0.10

Similarly, in Table 2.1, the marginal distribution of the rating Y of wine taster II appears in the last row of the table. We reproduce that distribution in Table 2.3(b).

Table 2.3(b). Probability mass function of the rating Y of wine taster II as taken from the last column of Table 2.1.

y	1	2	3	4	5
Probability $p_Y(y) = P_Y\{Y = y\}$	0.06	0.18	0.37	0.29	0.10

Comparing Tables 2.3(a) and (b), we note that X and Y have identical marginal distributions.

It should be emphasized that the marginal distributions $p_X(x)$ and $p_Y(y)$ obtained from the joint distribution $p_{X,Y}(x,y)$ of any pair (X,Y) of jointly distributed random variables must be the same as the distributions $p_X(x)$ and $p_Y(y)$ which we would have obtained if we had considered the variables X and Y individually.

Although $p_X(x)$ and $p_Y(y)$ can be obtained from $p_{X,Y}(x,y)$ in the form of marginal distributions, the converse is not true; that is, $p_{X,Y}(x,y)$ *cannot be determined from a knowledge of* $p_X(x)$ *and* $p_Y(y)$ *alone.* To illustrate this point, recall the two joint probability models for the sex of the first- and second-born children considered in Example 3.1 of Chapter 3.

Example 2.2. In Example 3.1 of Chapter 3, we observe the sex of the children in a family that has exactly 2 children. In that example, we represented the outcomes of this composite random experiment in terms of the pairs: (sex of first-born child, sex of second-born child). Let us now describe the outcomes of the experiment in terms of two random variables X and Y. Let $X = 1$ if the first-born child is a male, $X = 0$ if the first-born child is a female, $Y = 1$ if the second-born child is a male, and $Y = 0$ if the second-born child is a female. The possible outcomes in the experiment (which now consists of observing X and Y rather than the sexes of the children) are then: (1,1), (1,0), (0,1), (0,0). In Chapter 3 we considered two possible joint distributions for the pair (X,Y). Table 2.4(a) gives the first of these joint distributions (in the form of a contingency table).

Table 2.4(a). Joint probability distribution for the sex of the children in a family having 2 children. $X = 1$ when the first child is male, and 0 otherwise; $Y = 1$ when the second child is male, and 0 otherwise.

x \ y	1	0	Total
1	$\frac{1}{3}$	$\frac{1}{6}$	$\frac{1}{2}$
0	$\frac{1}{6}$	$\frac{1}{3}$	$\frac{1}{2}$
Total	$\frac{1}{2}$	$\frac{1}{2}$	1

In Chapter 3, we also considered the joint distribution given in Table 2.4(b).

Table 2.4(b). Another joint probability distribution for the sex of the children in a family having 2 children.

x \ y	1	0	Total
1	$\frac{1}{4}$	$\frac{1}{4}$	$\frac{1}{2}$
0	$\frac{1}{4}$	$\frac{1}{4}$	$\frac{1}{2}$
Total	$\frac{1}{2}$	$\frac{1}{2}$	1

The joint probability models given in Tables 2.4(a) and 2.4(b) both yield the same marginal probability distributions for X, namely,

X:

x	0	1
$p_X(x)$	$\frac{1}{2}$	$\frac{1}{2}$

Similarly, the joint probability models in Tables 2.4(a) and 2.4(b) have a common marginal distribution for Y, namely,

Y:

y	0	1
$p_Y(y)$	$\frac{1}{2}$	$\frac{1}{2}$

[*Note:* The fact that X and Y have the same marginal distribution is irrelevant to the point we are making here.] However, the probability model given in Table 2.4(a) is obviously not the same probability model as that given in Table 2.4(b). Hence, two different joint probability models can have the same marginal probability distributions. Knowing only the marginal probability distributions for X and for Y, as shown above, we cannot tell which joint probability model [Table 2.4(a) or Table 2.4(b)] is the appropriate one.

Example of the Construction of a Bivariate Probability Model

Example 2.2 illustrates an experimental situation where we have transformed a probability model for a composite random experiment consisting of nonquantitative outcomes into a bivariate probability model for two random variables X and Y. In the following example, we illustrate a somewhat more complicated composite random experiment in which two quantitative aspects X and Y are of interest, and show how the bivariate probability distribution for X and Y is obtained.

Example 2.3. Fifth-graders from a certain school district can be divided into three distinct categories: talented (T), average (A), and slow (S). Let p_T be the proportion of fifth-grade children who are talented, p_A be the proportion who are average, and p_S be the proportion who are slow; then, $p_T + p_A + p_S = 1$. A random sample of 4 fifth-graders is drawn *with* replacement. For each child drawn, we note only whether the child is talented, average, or slow. The 81 possible outcomes ω_i and the respective probabilities of the corresponding simple events $\{\omega_i\}$ are shown in Figure 2.6.

Let X denote the number of average children in the sample and Y denote the number of talented children. The list of possible outcomes (x,y) is

(0,0)   (1,0)   (2,0)   (3,0)   (4,0)
(0,1)   (1,1)   (2,1)   (3,1)
(0,2)   (1,2)   (2,2)
(0,3)   (1,3)
(0,4)

Note that any pair such as (1,4) cannot be an outcome because we only draw 4 children. It can be seen from the enumeration in Figure 2.6 that the joint distribution of the random variables X and Y can be given by a contingency table such as shown in Table 2.5.

It is evident from Figure 2.6 that a complete enumeration of the probabilities for some bivariate probability distributions can become a somewhat tedious task. Fortunately, in the present case, there is a convenient

Outcome ω	Probability of $\{\omega\}$	Outcome ω	Probability of $\{\omega\}$	Outcome ω	Probability of $\{\omega\}$
TTTT	p_T^4	AAAA	p_A^4		
TTTS, TTST, TSTT, STTT	$p_T^3 p_S$	AAAS, AASA, ASAA, SAAA	$p_A^3 p_S$	TTTA, TTAT, TATT, ATTT	$p_A p_T^3$
TTSS, TSTS, STTS, TSST, STST, SSTT	$p_T^2 p_S^2$	AASS, ASAS, SAAS, ASSA, SASA, SSAA	$p_A^2 p_S^2$	TTAA, TATA, ATTA, TAAT, ATAT, AATT	$p_A^2 p_T^2$
SSST, SSTS, STSS, TSSS	$p_T p_S^3$	SSSA, SSAS, SASS, ASSS	$p_A p_S^3$	AAAT, AATA, ATAA, TAAA	$p_A^3 p_T$
SSSS	p_S^4				
TTSA, TSTA, STTA, TTAS, TSAT, STAT, TATS, TAST, SATT, ATTS, ATST, ASTT	$p_T^2 p_S p_A$	SSTA, STSA, TSSA, SSAT, STAS, TSAS, SAST, SATS, TASS, ASST, ASTS, ATSS	$p_T p_A p_S^2$	AATS, ATAS, TAAS, AAST, ATSA, TASA, ASAT, ASTA, TSAA, SAAT, SATA, STAA	$p_T p_A^2 p_S$

Figure 2.6. Outcomes and corresponding probabilities for Example 2.3. The symbol TTTT for the first outcome in this list represents the outcome in which talented children are drawn on all four draws; the symbol STAA for the last outcome in the list represents the outcome in which a slow child is drawn on the first draw, a talented child is drawn on the second draw, and average children are drawn on the third and fourth draws. Outcomes ω for which the corresponding simple events $\{\omega\}$ have equal probabilities are grouped together.

2. DISCRETE BIVARIATE DISTRIBUTIONS

Table 2.5. A bivariate probability mass function for the number X of average children and the number Y of talented children observed in a random sample, with replacement, of 4 fifth-graders.

x \ y	0	1	2	3	4
0	p_S^4	$4p_T p_S^3$	$6p_T^2 p_S^2$	$4p_T^3 p_S$	p_T^4
1	$4p_A p_S^3$	$12p_T p_A p_S^2$	$12p_T^2 p_A p_S$	$4p_T^3 p_A$	0
2	$6p_A^2 p_S^2$	$12p_T p_A^2 p_S$	$6p_T^2 p_A^2$	0	0
3	$4p_A^3 p_S$	$4p_T p_A^3$	0	0	0
4	p_A^4	0	0	0	0

formula which enables us to write the bivariate probability mass function for X and Y in a compact form:

$$(2.2) \quad p_{X,Y}(x,y) = \begin{cases} \dfrac{4!}{x! y! (4-x-y)!} p_A^x p_T^y p_S^{4-x-y}, & \text{if } x, y \text{ are integers} \\ & \text{and } x+y \leq 4, \\ 0, & \text{otherwise.} \end{cases}$$

Thus, if $x = 2$ and $y = 1$, we have, by a direct substitution into Equation (2.1),

$$p_{X,Y}(2,1) = \frac{4!}{2! 1! 1!} p_A^2 p_T^1 p_S^1 = 12 p_A^2 p_T p_S.$$

If the proportion of average students in the given school district is $p_A = \frac{1}{2}$, the proportion of slow students is $p_S = \frac{1}{3}$, and the proportion of talented students is $p_T = \frac{1}{6}$, then the probability that in our sample of 4 students we observe 2 average students and 1 talented student is

$$p_{X,Y}(2,1) = P_{X,Y}\{X = 2 \text{ and } Y = 1\}$$
$$= 12 \left(\frac{1}{2}\right)^2 \left(\frac{1}{6}\right) \left(\frac{1}{3}\right) = \frac{1}{6}.$$

The probability that we observe 1 average student and 2 talented students is, from Equation (2.2),

$$p_{X,Y}(1,2) = \frac{4!}{1! 2! 1!} p_A^1 p_T^2 p_S^1 = 12 p_A p_T^2 p_S = 12 \left(\frac{1}{2}\right) \left(\frac{1}{6}\right)^2 \left(\frac{1}{3}\right) = \frac{1}{18}.$$

Conditional Distributions

We have mentioned earlier that in performing a composite random experiment, the value of one of two random variables may become known before the value of the other variable. Thus, we may know that the value of the random variable X is x before we know the value of Y. On the basis of the knowledge that $X = x$, we may want to reassess our calculations concerning probabilities for the various possible values of Y. Conditional probabilities (Chapter 3) now become relevant. The *conditional probability mass function* of Y given $X = x$, in symbols $p_{Y|X=x}(y)$, is defined to be

$$(2.3) \quad p_{Y|X=x}(y) = \frac{p_{X,Y}(x,y)}{p_X(x)},$$

assuming $p_X(x) > 0$. [The conditional probability mass function $p_{Y|X=x}(y)$ is defined only where $p_X(x) > 0$.] It can easily be verified that $p_{Y|X=x}(y)$ is, in fact, a probability mass function; that is, it is nonnegative and the sum over all possible values of y is 1. The probability mass function $p_{Y|X=x}(y)$ contains all the relevant probabilistic information concerning Y when we know that $X = x$.

Example 2.4. A political scientist has just completed a study of American voting habits. He has been studying how the tendency of people to vote in presidential elections is related to their tendency to vote in nonpresidential, congressional elections. Since, for purposes of machine tabulation, he has been marking a "1" if a person voted in a given election and "0" if the person did not vote, the political scientist decides to summarize his findings in terms of the joint distribution of the random variable X, which indicates whether or not a person voted in a presidential election, and the random variable Y, which indicates whether or not that person voted in the following nonpresidential, congressional election. That is, $X = 1$ if the person voted in the presidential election, and $X = 0$ if the person did not vote in that election. Similarly, $Y = 1$ if the person voted in the next nonpresidential, congressional election, and $Y = 0$ if he did not vote in that election. The political scientist's data led him to hypothesize the following joint discrete probability distribution for X and Y:

x \ y	0	1	Total
0	$\frac{1}{8}$	$\frac{1}{8}$	$\frac{1}{4}$
1	$\frac{1}{2}$	$\frac{1}{4}$	$\frac{3}{4}$
Total	$\frac{5}{8}$	$\frac{3}{8}$	1

2. DISCRETE BIVARIATE DISTRIBUTIONS

From this contingency table, we see that the marginal probability mass functions of X and Y are, respectively,

X:

x	0	1
$p_X(x)$	$\frac{1}{4}$	$\frac{3}{4}$

and

Y:

y	0	1
$p_Y(y)$	$\frac{5}{8}$	$\frac{3}{8}$

From these values we obtain

$$p_{Y|X=0}(0) = \frac{p_{X,Y}(0,0)}{p_X(0)} = \frac{\frac{1}{8}}{\frac{1}{4}} = \frac{1}{2},$$

$$p_{Y|X=0}(1) = \frac{p_{X,Y}(0,1)}{p_X(0)} = \frac{\frac{1}{8}}{\frac{1}{4}} = \frac{1}{2},$$

and

$$p_{Y|X=1}(0) = \frac{p_{X,Y}(1,0)}{p_X(1)} = \frac{\frac{1}{2}}{\frac{3}{4}} = \frac{2}{3},$$

$$p_{Y|X=1}(1) = \frac{p_{X,Y}(1,1)}{p_X(1)} = \frac{\frac{1}{4}}{\frac{3}{4}} = \frac{1}{3}.$$

For descriptive purposes, the political scientist records the conditional probability mass function of Y given $X=0$ and the conditional probability mass function of Y given $X=1$ in tables such as the following:

y	0	1
$p_{Y|X=0}(y)$	$\frac{1}{2}$	$\frac{1}{2}$

y	0	1
$p_{Y|X=1}(y)$	$\frac{2}{3}$	$\frac{1}{3}$

From an inspection of these tables, the political scientist concludes that if a person does not vote in a presidential election (that is, if $X=0$), then the conditional probabilities that he *will* vote in the next nonpresidential, congressional election and that he *will not* vote in that election are the same. On the other hand, if a person does vote in a presidential election (that is, if $X=1$), then the conditional probability that he will not vote in the next nonpresidential, congressional election is greater (twice as great) than the conditional probability that he will vote. Comparing the conditional probability mass function $p_{Y|X=1}(y)$ to the conditional probability mass function $p_{Y|X=0}(y)$, and each of these to the marginal probability mass function $p_Y(y)$, the political scientist suspects that when a person *does* vote in a presidential election, he has a tendency to believe that he has done his "civic duty," and thus does not feel the need to vote in the following nonpresidential election.

Example 2.1 (Continued). Recall that in Example 2.1 two professional wine tasters are asked to rate a sample of wine using a rating scale that allows them to give ratings of 1, 2, 3, 4, or 5 (a rating of 5 being equivalent to "top quality"). Table 2.1, reproduced below, gives a bivariate probability mass function for the rating X of wine taster I and the rating Y of wine taster II.

Wine Taster II

	x \ y	1	2	3	4	5	Total
Wine Taster I	1	0.03	0.02	0.01	0.00	0.00	0.06
	2	0.02	0.08	0.05	0.02	0.01	0.18
	3	0.01	0.05	0.25	0.05	0.01	0.37
	4	0.00	0.02	0.05	0.20	0.02	0.29
	5	0.00	0.01	0.01	0.02	0.06	0.10
	Total	0.06	0.18	0.37	0.29	0.10	1.00

We might be interested in seeing if we can predict the ratings of wine taster II if we are given the ratings of wine taster I. Thus, we are interested in finding the conditional probability mass functions for Y given $X = 1$, given $X = 2$, given $X = 3$, given $X = 4$, and given $X = 5$. We can construct such conditional probability mass functions directly from the contingency table given above. Note that the last column of the table (the marginal total) gives the values of $p_X(x)$, and that the entries in the contingency table in each row correspond to the values of the joint probability mass function $p_{X,Y}(x,y)$ for a given value x, and for $y = 1, 2, 3, 4, 5$. To calculate

$$p_{Y|X=x}(y) = \frac{p_{X,Y}(x,y)}{p_X(x)},$$

for $y = 1, 2, 3, 4, 5$, we need only divide every entry, $p_{X,Y}(x,y)$, in a given row by the last entry [the marginal total $p_X(x)$] in that row. The result is the following table (actually a collection of tables):

y	1	2	3	4	5
$p_{Y\|X=1}(y)$	$\dfrac{0.03}{0.06}$	$\dfrac{0.02}{0.06}$	$\dfrac{0.01}{0.06}$	0	0
$p_{Y\|X=2}(y)$	$\dfrac{0.02}{0.18}$	$\dfrac{0.08}{0.18}$	$\dfrac{0.05}{0.18}$	$\dfrac{0.02}{0.18}$	$\dfrac{0.01}{0.18}$
$p_{Y\|X=3}(y)$	$\dfrac{0.01}{0.37}$	$\dfrac{0.05}{0.37}$	$\dfrac{0.25}{0.37}$	$\dfrac{0.05}{0.37}$	$\dfrac{0.01}{0.37}$
$p_{Y\|X=4}(y)$	0	$\dfrac{0.02}{0.29}$	$\dfrac{0.05}{0.29}$	$\dfrac{0.20}{0.29}$	$\dfrac{0.02}{0.29}$
$p_{Y\|X=5}(y)$	0	$\dfrac{0.01}{0.10}$	$\dfrac{0.01}{0.10}$	$\dfrac{0.02}{0.10}$	$\dfrac{0.06}{0.10}$

Here, the first row of the table gives the conditional probability mass function $p_{Y|X=1}(y)$, the second row gives the conditional probability mass function $p_{Y|X=2}(y)$, and so on. Notice that the sum of the entries in each row is now equal to 1. Reducing all fractions, we obtain the following table(s):

y	1	2	3	4	5
$p_{Y\|X=1}(y)$	$\dfrac{1}{2}$	$\dfrac{1}{3}$	$\dfrac{1}{6}$	0	0
$p_{Y\|X=2}(y)$	$\dfrac{1}{9}$	$\dfrac{4}{9}$	$\dfrac{5}{18}$	$\dfrac{1}{9}$	$\dfrac{1}{18}$
$p_{Y\|X=3}(y)$	$\dfrac{1}{37}$	$\dfrac{5}{37}$	$\dfrac{25}{37}$	$\dfrac{5}{37}$	$\dfrac{1}{37}$
$p_{Y\|X=4}(y)$	0	$\dfrac{2}{29}$	$\dfrac{5}{29}$	$\dfrac{20}{29}$	$\dfrac{2}{29}$
$p_{Y\|X=5}(y)$	0	$\dfrac{1}{10}$	$\dfrac{1}{10}$	$\dfrac{1}{5}$	$\dfrac{3}{5}$

Note that we can also construct the conditional probability mass functions $p_{X|Y=1}(x)$, $p_{X|Y=2}(x)$, and so on. This task is easily accomplished by dividing every entry in a given column of the contingency table, say the column corresponding to $y = 1$, by the marginal total of that column. The result is the following table(s) of conditional probability mass functions:

x	$p_{X\|Y=1}(x)$	$p_{X\|Y=2}(x)$	$p_{X\|Y=3}(x)$	$p_{X\|Y=4}(x)$	$p_{X\|Y=5}(x)$
1	$\frac{1}{2}$	$\frac{1}{9}$	$\frac{1}{37}$	0	0
2	$\frac{1}{3}$	$\frac{4}{9}$	$\frac{5}{37}$	$\frac{2}{29}$	$\frac{1}{10}$
3	$\frac{1}{6}$	$\frac{5}{18}$	$\frac{25}{37}$	$\frac{5}{29}$	$\frac{1}{10}$
4	0	$\frac{1}{9}$	$\frac{5}{37}$	$\frac{20}{29}$	$\frac{1}{5}$
5	0	$\frac{1}{18}$	$\frac{1}{37}$	$\frac{2}{29}$	$\frac{3}{5}$

Observe that this time the *column* totals are each equal to 1.

It is worth remarking that in this example (Example 2.1) and in Example 2.4, the conditional probability mass functions $p_{Y|X=x}(y)$ of Y given $X = x$ are not the same as the marginal probability mass function $p_Y(y)$. Knowing information about the value x of X changes the probabilistic information we have concerning the variation of Y. For example, if we know nothing about the rank which wine taster I has given to a given sample of wine and are asked to guess what rank wine taster II has given (or will give) this sample of wine, we would be likely to say that wine taster II would give the wine a rank of $Y = 3$ since $p_Y(3)$ is larger than any other value of the marginal probability mass function $p_Y(y)$. Certainly, we would hesitate to say that wine taster II would give the wine a rank of $Y = 1$, since $p_Y(1) = 0.06$ is smaller than any other value of $p_Y(y)$. However, if we know that wine taster I has given the wine sample a rank of $X = 1$, then our guess of the rank which wine taster II gives this wine would be $Y = 1$ since the conditional probability $p_{Y|X=1}(1)$, of $Y = 1$ given that $X = 1$ is $\frac{1}{2}$, a probability that is larger than any other value of the conditional probability mass function $p_{Y|X=1}(y)$, and three times as large as the conditional probability, $p_{Y|X=1}(3) = \frac{1}{6}$. Thus, information about the value observed for the random variable X in this example leads us to change our prediction for the value of Y.

The Regression Function

Since, for each possible value x of X, the conditional probability mass function of Y given that $X = x$ is itself a probability mass function, we can consider the expected value of the random variable associated with this probability mass function. Formally, we define the *conditional expected value of Y given that X = x* (or, for short, the expected value of Y given that X = x) as the expected value associated with $p_{Y|X=x}(y)$. We denote the expected value of Y given $X = x$ by $E[Y|X = x]$.

Example 2.1 (Continued). In this example, the random variable X is the rank given a sample of wine by wine taster I, and the random variable Y is the rank given the same sample of wine by wine taster II. We have previously calculated (from Table 2.1) the conditional probability mass function $p_{Y|X=3}(y)$:

y	1	2	3	4	5
$p_{Y|X=3}(y)$	1/37	5/37	25/37	5/37	1/37

The expected value, $E[Y|X=3]$, of this distribution is

$$E[Y|X=3] = (1)p_{Y|X=3}(1) + (2)p_{Y|X=3}(2) + (3)p_{Y|X=3}(3)$$
$$+ (4)p_{Y|X=3}(4) + (5)p_{Y|X=3}(5)$$
$$= (1)\left(\frac{1}{37}\right) + (2)\left(\frac{5}{37}\right) + (3)\left(\frac{25}{37}\right) + (4)\left(\frac{5}{37}\right) + (5)\left(\frac{1}{37}\right)$$
$$= 3.$$

We have also calculated the conditional probability mass function $P_{Y|X=1}(y)$ for Y given that $X = 1$:

y	1	2	3	4	5
$P_{Y|X=1}(y)$	1/2	1/3	1/6	0	0

The expected value, $E[Y|X=1]$, of the distribution of the ranks wine taster II gives to samples of wine ranked $X = 1$ by wine taster I is then

$$E[Y|X=1] = (1)\left(\frac{1}{2}\right) + (2)\left(\frac{1}{3}\right) + (3)\left(\frac{1}{6}\right) = \frac{5}{3}.$$

The conditional expected value of Y given $X = x$ must be determined separately for each value of x. The function that assigns to every possible value x of X the value $E[Y|X = x]$ is called the *regression function of Y on X*. In practical applications, the regression function of one variable Y on another variable X is of considerable importance since it indicates, at least "on the average," the mathematical relationship existing between Y, as a variable to be predicted (*the dependent variable* in the mathematical relationship), and X, the predicting variable (*the independent variable* in the mathematical relationship).

Example 2.1 (Continued). Our interest in this example is, for the moment, centered on predicting the rank Y assigned to a sample of wine by wine taster II from knowledge of the rank X assigned to this wine sample by wine taster I. We have already found the conditional expectations, $E[Y|X=3]$ and $E[Y|X=1]$. The regression function, $E[Y|X=x]$, is given by the expression

$$E[Y|X=x] = \begin{cases} 5/3, & \text{if } x=1, \\ 23/9, & \text{if } x=2, \\ 3, & \text{if } x=3, \\ 109/29, & \text{if } x=4, \\ 43/10, & \text{if } x=5. \end{cases}$$

This result follows since $E[Y|X=3]=3$ and $E[Y|X=1]=\frac{5}{3}$, as we have already shown, and since

$$E[Y|X=2] = (1)\left(\frac{1}{9}\right) + (2)\left(\frac{4}{9}\right) + (3)\left(\frac{5}{18}\right) + (4)\left(\frac{1}{9}\right) + (5)\left(\frac{1}{18}\right) = \frac{23}{9},$$

$$E[Y|X=4] = (1)(0) + (2)\left(\frac{2}{29}\right) + (3)\left(\frac{5}{29}\right) + (4)\left(\frac{20}{29}\right) + (5)\left(\frac{2}{29}\right) = \frac{109}{29},$$

and

$$E[Y|X=5] = (1)(0) + (2)\left(\frac{1}{10}\right) + (3)\left(\frac{1}{10}\right) + (4)\left(\frac{1}{5}\right) + (5)\left(\frac{3}{5}\right) = \frac{43}{10}.$$

The regression function $E[Y|X=x]$ is graphed in Figure 2.7. Note that the values of $E[Y|X=x]$ increase as x increases.

*[*Remark:* We note that $E[Y|X=x]$ defines a number which is dependent on the observed value x of X. However, the value of X a priori is

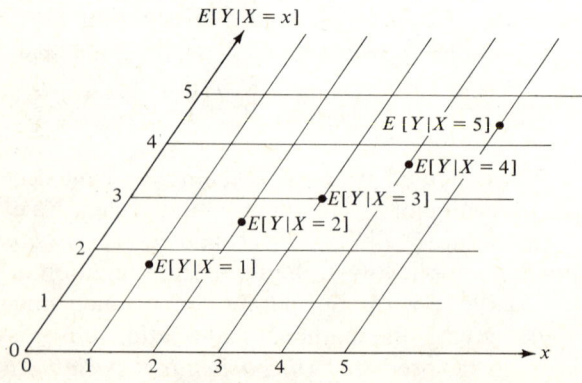

Figure 2.7 The regression function $E[Y|X=x]$ for the joint probability distribution given in Table 2.1 of Example 2.1.

random [it is not determined until a particular outcome $\omega = (x,y)$ occurs]. Thus, $E[Y|X = x]$ is, from this point of view, a random variable. To take note of this fact, $E[Y|X = x]$ is sometimes written $E[Y|X]$ and is called the *conditional expected value of Y given X*.]

3. DESCRIPTIVE PROPERTIES OF BIVARIATE DISTRIBUTIONS

We have seen that it is often the case that certain summarizing measures are sufficient to describe those aspects of univariate distributions which are of primary interest. A similar fact is true for bivariate distributions.

Suppose that we are considering two random variables X and Y. For the moment, assume that X and Y are both discrete random variables with the joint probability mass function $p_{X,Y}(x,y)$. We will discuss the case when X and Y are both continuous random variables in Section 4. To begin with, we can look for a measure for the bivariate distribution of X and Y which is analogous to a measure of location for a univariate distribution. The concept of a quantile (for example, the median) is not meaningful for a bivariate distribution, since the notion of the probability "to the left" and "to the right" of a point (x,y) in the plane does not have a unique definition. The *mode* of a bivariate distribution for discrete random variables is the pair (or pairs) of values (x_0, y_0) for which the value of joint probability mass function $p_{X,Y}(x,y)$ is largest. Since the mode of a bivariate distribution is subject to mathematical difficulties (and since even for univariate distributions it is rarely a good measure of location), the mode is seldom used as a measure of location for bivariate distributions.

The measure of location which is usually used for bivariate distributions is the pair (μ_X, μ_Y), where μ_X is the expected value of X [as calculated from the marginal probability mass function $p_X(x)$], and where μ_Y is the expected value of Y [calculated from $p_Y(y)$]. The pair (μ_X, μ_Y) is called the *joint expectation*, the *joint mean*, or the *mean vector* of the distribution of X and Y. Although μ_X and μ_Y can be computed from the marginal probability mass functions $p_X(x)$ and $p_Y(y)$, respectively, it is sometimes more convenient to compute these quantities directly from the joint probability mass function $p_{X,Y}(x,y)$. For example, we can calculate $E(X) = \mu_X$ using the formula

$$\mu_X = E(X) = x_1 p_X(x_1) + x_2 p_X(x_2) + x_3 p_X(x_3) + \cdots.$$

However, since

$$p_X(x_i) = p_{X,Y}(x_i, y_1) + p_{X,Y}(x_i, y_2) + p_{X,Y}(x_i, y_3) + \cdots,$$

for $i = 1, 2, 3, \ldots$, we can substitute this result into the formula for μ_X

given above and obtain

$$
\begin{aligned}
(3.1) \quad \mu_X &= x_1[p_{X,Y}(x_1,y_1) + p_{X,Y}(x_1,y_2) + p_{X,Y}(x_1,y_3) + \cdots] \\
&\quad + x_2[p_{X,Y}(x_2,y_1) + p_{X,Y}(x_2,y_2) + p_{X,Y}(x_2,y_3) + \cdots] + \cdots \\
&= x_1 p_{X,Y}(x_1,y_1) + x_1 p_{X,Y}(x_1,y_2) + x_1 p_{X,Y}(x_1,y_3) + \cdots \\
&\quad + x_2 p_{X,Y}(x_2,y_1) + x_2 p_{X,Y}(x_2,y_2) + x_2 p_{X,Y}(x_2,y_3) + \cdots.
\end{aligned}
$$

We can calculate μ_Y from the joint probability mass function $p_{X,Y}(x,y)$ in a similar fashion.

The joint expectation (μ_X, μ_Y), in analogy to the expected value in the univariate case, has the property that it is the center of gravity of the distribution of probability mass (in the plane) defined by $p_{X,Y}(x,y)$. Further, since μ_X and μ_Y are each univariate measures of location, we know from Section 2 of Chapter 5 that for any numbers a and b,

$$(3.2) \qquad \mu_{aX+b} = a\mu_X + b,$$

while for any numbers c and d,

$$(3.3) \qquad \mu_{cY+d} = c\mu_Y + d.$$

In the search for something analogous to a measure of dispersion for bivariate distributions, the variance σ_X^2 of X and the variance σ_Y^2 of Y come naturally to our attention. These measures can be computed from the respective marginal probability mass functions of X and Y; for example,

$$\sigma_X^2 = (x_1 - \mu_X)^2 p_X(x_1) + (x_2 - \mu_X)^2 p_X(x_2) + \cdots.$$

However, again it is sometimes more convenient to calculate the variance directly from the joint probability mass function $p_{X,Y}(x,y)$. For example,

$$
\begin{aligned}
(3.4) \quad \sigma_X^2 &= (x_1 - \mu_X)^2 p_{X,Y}(x_1,y_1) + (x_1 - \mu_X)^2 p_{X,Y}(x_1,y_2) + \cdots \\
&\quad + (x_2 - \mu_X)^2 p_{X,Y}(x_2,y_1) + (x_2 - \mu_X)^2 p_{X,Y}(x_2,y_2) + \cdots.
\end{aligned}
$$

A similar formula holds for σ_Y^2. Since σ_X^2 and σ_Y^2 are each univariate measures of dispersion [σ_X^2 is a measure of dispersion for the marginal probability mass function $p_X(x)$ and σ_Y^2 is a measure of dispersion for $p_Y(y)$], we know from Section 4 of Chapter 5 that for any two numbers a and b,

$$(3.5) \qquad \sigma_{aX+b}^2 = a^2 \sigma_X^2,$$

and that for any two numbers c and d,

$$(3.6) \qquad \sigma_{cY+d}^2 = c^2 \sigma_Y^2.$$

When the mass function $p_{X,Y}(x,y)$ is graphed in the plane, the indices σ_X^2 and σ_Y^2 indicate the dispersion of the joint probability mass function

3. DESCRIPTIVE PROPERTIES OF BIVARIATE DISTRIBUTIONS

$p_{X,Y}(x, y)$ around the lines $x = \mu_X$ and $y = \mu_Y$, respectively (see Figures 3.1 and 3.2). However, any line "$\alpha x + \beta y =$ constant" could define a center for a measure of dispersion; there is no obvious reason for choosing the lines $x = \mu_X$ or $y = \mu_Y$.

Further, although both σ_X^2 and σ_Y^2 can be defined solely in terms of the marginal distributions $p_X(x)$ and $p_Y(y)$, respectively, we have seen in Section 2 that the marginal distributions by themselves do not determine the joint distribution of X and Y. What we lack is a measure of the *interrelationship* or *covariation* of X and Y. That is, we lack a measure which reveals how X and Y vary jointly.

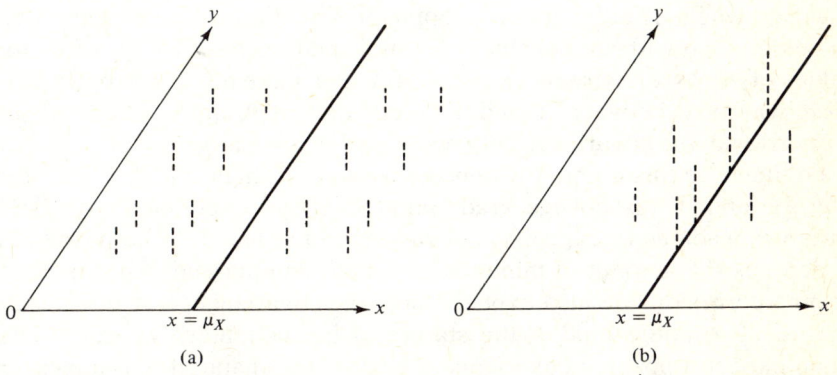

Figure 3.1 The variance σ_X^2 shows dispersion (or spread) of the joint probability mass function $p_{X,Y}(x, y)$ around the line $x = \mu_X$. The dispersion in (a) is larger than the dispersion in (b).

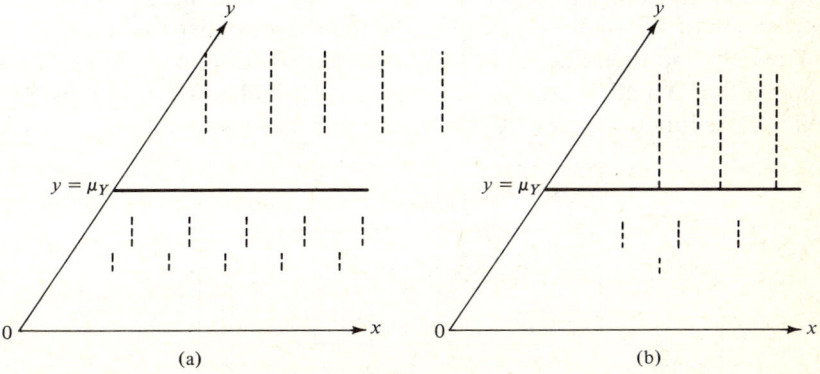

Figure 3.2 The variance σ_Y^2 shows dispersion (or spread) of the joint probability mass function $p_{X,Y}(x, y)$ around the line $y = \mu_Y$. The dispersion in (a) is larger than the dispersion in (b).

The formulas $\sigma_X^2 = E[X - \mu_X]^2$ and $\sigma_Y^2 = E[Y - \mu_Y]^2$ represent how X and Y vary individually. In a formal way, this suggests that $E[X - \mu_X][Y - \mu_Y]$ might represent how X and Y vary jointly. The quantity $E[X - \mu_X][Y - \mu_Y]$ is called the *covariance* between X and Y and is denoted by σ_{XY} or by $\text{Cov}(X,Y)$. Clearly $\text{Cov}(X,Y) = \text{Cov}(Y,X)$.

Suppose that $X = Y$. Then by substitution, $\text{Cov}(X,Y) = \text{Cov}(X,X) = E[X - \mu_X][X - \mu_X] = \sigma_X^2$. We conclude from this that $\sigma_X^2 = \sigma_{XX} = \text{Cov}(X,X)$ and $\sigma_Y^2 = \sigma_{YY} = \text{Cov}(Y,Y)$; that is, the notations σ_X^2 and σ_{XX}^2 are identical, as are σ_Y^2 and σ_{YY}.

The covariance can be positive or negative. It is positive when X and Y tend to move in the same direction; that is, when large values of X go with large values of Y and small values of X with small values of Y. It is negative when X and Y move in opposite directions; that is, when large values of X go with small values of Y and small values of X go with large values of Y. As an extreme example, if X and Y are directly proportional, the covariance between X and Y is positive; if X and Y are inversely proportional, the covariance between X and Y is negative.

To illustrate this point, if X denotes a student's high school grade-point average and Y his college grade-point average, experience with such variables leads us to expect the covariance of X and Y to be positive. If X denotes the amount of minerals in a water supply and Y is a measure of pipe corrosion, we also expect a positive covariance. If X denotes the stature of a mother and Y the stature of her daughter, we would once again expect a positive covariance. On the other hand, if X is a measure of motor coordination and Y denotes the amount of alcohol in the blood, the covariance between X and Y should be negative. Similarly, some experiments have shown that if X denotes age and Y denotes reading speed, the covariance between X and Y is negative.

In order to compute $\sigma_{XY} = \text{Cov}(X,Y)$, we could first determine the distribution of $W = (X - \mu_X)(Y - \mu_Y)$ from the joint distribution $p_{X,Y}(x,y)$ of X and Y, and then find $E[W]$ from the distribution of W. For example, suppose that X and Y are discrete random variables with the joint probability mass function given by the following contingency table:

x \ y	1	2	3	Total
1	$\frac{1}{4}$	$\frac{1}{12}$	$\frac{1}{6}$	$\frac{1}{2}$
2	$\frac{1}{12}$	$\frac{1}{4}$	$\frac{1}{6}$	$\frac{1}{2}$
Total	$\frac{1}{3}$	$\frac{1}{3}$	$\frac{1}{3}$	1

Then,

$$\mu_X = (1)[p_X(1)] + (2)[p_X(2)]$$
$$= (1)(\tfrac{1}{2}) + (2)(\tfrac{1}{2}) = \tfrac{3}{2},$$

3. DESCRIPTIVE PROPERTIES OF BIVARIATE DISTRIBUTIONS

$$\mu_Y = (1)[p_Y(1)] + (2)[p_Y(2)] + (3)[p_Y(3)]$$
$$= (1)(\tfrac{1}{3}) + (2)(\tfrac{1}{3}) + (3)(\tfrac{1}{3}) = 2.$$

Because

$$W = (X-\mu_X)(Y-\mu_Y) = \begin{cases} \tfrac{1}{2}, & \text{if } (X,Y) = (1,1) \text{ or } (2,3), \\ 0, & \text{if } (X,Y) = (1,2) \text{ or } (2,2), \\ -\tfrac{1}{2}, & \text{if } (X,Y) = (2,1) \text{ or } (1,3), \end{cases}$$

it follows that

$$p_W(-\tfrac{1}{2}) = p_{X,Y}(2,1) + p_{X,Y}(1,3) = (\tfrac{1}{12}) + (\tfrac{1}{6}) = \tfrac{1}{4},$$
$$p_W(0) = p_{X,Y}(1,2) + p_{X,Y}(2,2) = (\tfrac{1}{12}) + (\tfrac{1}{4}) = \tfrac{1}{3},$$
$$p_W(\tfrac{1}{2}) = p_{X,Y}(1,1) + p_{X,Y}(2,3) = (\tfrac{1}{4}) + (\tfrac{1}{6}) = \tfrac{5}{12},$$

and hence

$$\sigma_{XY} = E(X-\mu_X)(Y-\mu_Y) = EW = (-\tfrac{1}{2})(\tfrac{1}{4}) + (0)(\tfrac{1}{3}) + (\tfrac{1}{2})(\tfrac{5}{12}) = \tfrac{1}{12}.$$

It is often more convenient to calculate σ_{XY} directly from the joint probability mass function $p_{X,Y}(x,y)$ of X and Y. To do so, we can make use of the following formula for finding the expected value $E[g(X,Y)]$ of any function $g(X,Y)$ of X and Y:

$$(3.7) \quad E[g(X,Y)] = g(x_1,y_1)p_{X,Y}(x_1,y_1) + g(x_1,y_2)p_{X,Y}(x_1,y_2) + \cdots$$
$$+ g(x_2,y_1)p_{X,Y}(x_2,y_1) + g(x_2,y_2)p_{X,Y}(x_2,y_2) + \cdots.$$

That is, for each possible value (x,y) of the pair (X,Y), we multiply $g(x,y)$ by the probability $p_{X,Y}(x,y)$ that simultaneously $X = x$ and $Y = y$, and then sum such products over all possibilities (x,y). Note that Equation (3.7) is the analogy for bivariate distributions of the formula for the expectation of a function $g(X)$ of one random variable, X, given in Equation (4.2) of Chapter 5. Formula (3.7) for the expectation $E[g(X,Y)]$ of $g(X,Y)$ holds for any function $g(X,Y)$ of X and Y [for example, Formula (3.7) holds for $g(X,Y) = XY$, for $g(X,Y) = X+Y$, for $g(X,Y) = X/Y$, and so on].

In particular, in the case of the covariance, we require the expected value of $g(X,Y) = (X-\mu_X)(Y-\mu_Y)$. From Equation (3.7), we obtain

$$(3.8) \quad \sigma_{XY} = (x_1-\mu_X)(y_1-\mu_Y)p_{X,Y}(x_1,y_1) + (x_1-\mu_X)(y_2-\mu_Y)p_{XY}(x_1,y_2)$$
$$+ \cdots + (x_2-\mu_X)(y_1-\mu_Y)p_{X,Y}(x_2,y_1) + \cdots.$$

Applying this equation to the case of the contingency table

x \ y	1	2	3	Total
1	$\tfrac{1}{4}$	$\tfrac{1}{12}$	$\tfrac{1}{6}$	$\tfrac{1}{2}$
2	$\tfrac{1}{12}$	$\tfrac{1}{4}$	$\tfrac{1}{6}$	$\tfrac{1}{2}$
Total	$\tfrac{1}{3}$	$\tfrac{1}{3}$	$\tfrac{1}{3}$	1

given earlier (where we recall that $\mu_X = \frac{3}{2}$, $\mu_Y = 2$), we find that

$$\sigma_{XY} = (1-\mu_X)(1-\mu_Y)p_{X,Y}(1,1) + (1-\mu_X)(2-\mu_Y)p_{X,Y}(1,2)$$
$$+ (1-\mu_X)(3-\mu_Y)p_{X,Y}(1,3) + (2-\mu_X)(1-\mu_Y)p_{X,Y}(2,1)$$
$$+ (2-\mu_X)(2-\mu_Y)p_{X,Y}(2,2) + (2-\mu_X)(3-\mu_Y)p_{X,Y}(2,3)$$
$$- (-\tfrac{1}{2})(-1)(\tfrac{1}{4}) + (-\tfrac{1}{2})(0)(\tfrac{1}{12}) + (-\tfrac{1}{2})(1)(\tfrac{1}{6})$$
$$+ (\tfrac{1}{2})(-1)(\tfrac{1}{12}) + (\tfrac{1}{2})(0)(\tfrac{1}{4}) + (\tfrac{1}{2})(1)(\tfrac{1}{6})$$
$$= \tfrac{1}{12},$$

which is the result obtained earlier.

Another formula for $\sigma_{XY} = \text{Cov}(X,Y)$ which frequently is of help in calculations is

(3.9) $$\sigma_{XY} = E[XY] - \mu_X \mu_Y,$$

where from Equation (3.7) we know that we can calculate $E[XY]$ by forming the sum

(3.10) $$E[XY] = x_1 y_1 p_{X,Y}(x_1, y_1) + x_1 y_2 p_{X,Y}(x_1, y_2) + \cdots + x_2 y_1 p_{X,Y}(x_2, y_1)$$
$$+ x_2 y_2 p_{X,Y}(x_2, y_2) + \cdots.$$

Thus, we can also obtain the result $\sigma_{XY} = 1/12$ for the contingency table given above by the following calculations:

$$E[XY] = (1)(1)p_{X,Y}(1,1) + (1)(2)p_{X,Y}(1,2) + (1)(3)p_{X,Y}(1,3)$$
$$+ (2)(1)p_{X,Y}(2,1) + (2)(2)p_{X,Y}(2,2) + (2)(3)p_{X,Y}(2,3)$$
$$= (1)(1)\left(\frac{1}{4}\right) + (1)(2)\left(\frac{1}{12}\right) + (1)(3)\left(\frac{1}{6}\right) + (2)(1)\left(\frac{1}{12}\right)$$
$$+ (2)(2)\left(\frac{1}{4}\right) + (2)(3)\left(\frac{1}{6}\right) = \frac{37}{12},$$

and

$$\sigma_{XY} = E[XY] - \mu_X \mu_Y$$
$$= \left(\frac{37}{12}\right) - \left(\frac{3}{2}\right)(2) = \frac{1}{12}.$$

To see how Equation (3.9) follows from Equation (3.8), note that σ_{XY} in Equation (3.8) is the sum of terms of the form $(x-\mu_X)(y-\mu_Y)p_{X,Y}(x,y)$. However,

$$(x-\mu_X)(y-\mu_Y)p_{X,Y}(x,y) = xy p_{X,Y}(x,y) - \mu_X y p_{X,Y}(x,y)$$
$$- x \mu_Y p_{X,Y}(x,y) + \mu_X \mu_Y p_{X,Y}(x,y),$$

3. DESCRIPTIVE PROPERTIES OF BIVARIATE DISTRIBUTIONS

so that

(3.11) the sum of $(x-\mu_X)(y-\mu_Y)p_{X,Y}(x,y)$

$= $ [the sum of $xyp_{X,Y}(x,y)$] + [the sum of $(-\mu_X)yp_{X,Y}(x,y)$]

$+ $ [the sum of $(x)(-\mu_Y)p_{X,Y}(x,y)$] $+\mu_X\mu_Y$[the sum of $p_{X,Y}(x,y)$],

where the various sums are taken over all the possible pairs of values (x,y) for X and Y. From Equation (3.10), we have that

$$E[XY] = \text{the sum of } xyp_{X,Y}(x,y).$$

Since

$$[\text{the sum of } (-\mu_X)yp_{X,Y}(x,y)] = (-\mu_X)[\text{the sum of } yp_{X,Y}(x,y)]$$
$$= (-\mu_X)(\mu_Y),$$
$$[\text{the sum of } (x)(-\mu_Y)p_{X,Y}(x,y)] = (-\mu_Y)[\text{the sum of } xp_{X,Y}(x,y)]$$
$$= (-\mu_Y)(\mu_X),$$

and because the sum of $p_{X,Y}(x,y)$ over all possible pairs (x,y) is equal to 1, it follows by substitution into Equation (3.11) that

$$\sigma_{XY} = E[XY] + (-\mu_X)(\mu_Y) + (-\mu_Y)(\mu_X) + \mu_X\mu_Y(1)$$
$$= E[XY] - \mu_X\mu_Y.$$

We have thus verified Equation (3.9).

Because we have defined $\sigma_{XY} = \text{Cov}(X,Y)$ in a way analogous to the definitions of σ_X^2 and σ_Y^2, it is not surprising that $\text{Cov}(X,Y)$ has properties similar to these two measures of dispersion. One very important property of $\text{Cov}(X,Y)$ is given by the following: For any numbers a, b, c, and d,

(3.12) $\quad\quad\quad\quad \text{Cov}(aX+b, cY+d) = ac\,\text{Cov}(X,Y).$

Remark: If $a=c$, $b=d$, and $X=Y$, Equation (3.12) reduces to the equation

$$\text{Cov}(aX+b, aX+b) = a^2\,\text{Cov}(X,X).$$

Because $\text{Cov}(aX+b, aX+b) = \text{Var}(aX+b) = \sigma_{aX+b}^2$ and $\text{Cov}(X,X) = \text{Var}(X) = \sigma_X^2$, we obtain the result

$$\sigma_{aX+b}^2 = a^2\sigma_X^2,$$

which agrees with the result obtained in Chapter 5 [see Equation (4.9)].

To verify (3.12), note that from (3.2) and (3.3),

$$\mu_{aX+b} = a\mu_X + b, \quad\quad \mu_{cY+d} = c\mu_Y + d.$$

Thus,

$$\text{(3.13)} \quad \text{Cov}(aX+b, cY+d) = E[(aX+b-\mu_{aX+b})(cY+d-\mu_{cY+d})]$$
$$= E[(aX+b-a\mu_X-b)(cY+d-c\mu_Y-d)]$$
$$= E[a(X-\mu_X)c(Y-\mu_Y)]$$
$$= E[ac(X-\mu_X)(Y-\mu_Y)].$$

Let $W = (X-\mu_X)(Y-\mu_Y)$. Then since $E[W]$ is a measure of location for the distribution of W, we know from Section 2 of Chapter 5 that for any constants α and β,

$$E[\alpha W + \beta] = \alpha E[W] + \beta.$$

Letting $\alpha = ac, \beta = 0$, we obtain

$$\text{(3.14)} \quad E[ac(X-\mu_X)(Y-\mu_Y)] = E[(ac)W]$$
$$= acE[W] + 0 = acE[(X-\mu_X)(Y-\mu_Y)].$$

Combining (3.13) and (3.14) yields

$$\text{Cov}(aX+b, cY+d) = acE[(X-\mu_X)(Y-\mu_Y)]$$
$$= ac\text{Cov}(X,Y),$$

which verifies Equation (3.12).

Rescaling Techniques

Equations (3.2), (3.3), (3.5), (3.6), and (3.12) often provide an easier method for calculating the expected values μ_X and μ_Y, variances σ_X^2 and σ_Y^2, and covariance σ_{XY} of the jointly distributed random variables X and Y.

Example 3.1. Suppose that the random variable X is the score of an individual on a questionnaire designed to measure social status in a given community. If the questionnaire consists of 4 "yes-no" type questions, where each "yes" scores 5 points and each "no" scores 0 points, then the possible values of X are 0, 5, 10, 15, 20. Suppose that the random variable Y is the income (in dollars) of an individual. If individuals are selected from a community having a limited range of incomes, and if incomes have been rounded off to the nearest $2000, the possible values of Y might be $2000, $4000, $6000, $8000, and $10,000. Assume that the joint distribution of the variables X and Y is given by the contingency table, Table 3.1. From this contingency table, the reader is invited to calculate the values of $\mu_X, \mu_Y, \sigma_X^2, \sigma_Y^2$, and σ_{XY}. To illustrate the computational

3. DESCRIPTIVE PROPERTIES OF BIVARIATE DISTRIBUTIONS

problems encountered, note that

$$\mu_Y = (2000)(0.12) + (4000)(0.24) + (6000)(0.32) + (8000)(0.22)$$
$$+ (10{,}000)(0.10)$$
$$= 5880,$$

so that calculation of σ_Y^2 involves a sum of terms such as $(4000 - 5880)^2(0.24)$ and $(10{,}000 - 5880)^2(0.10)$.
Similarly,

$$\mu_X = (0)(0.12) + (5)(0.22) + (10)(0.30) + (15)(0.24) + (20)(0.12)$$
$$= 10.1,$$

so that computation of σ_X^2 involves a sum of terms such as $(5 - 10.1)^2 \times (0.22)$ and $(20 - 10.1)^2(0.30)$. In addition, the computation of σ_{XY} involves the sum of terms such as $(0 - 10.1)(4000 - 5880)(0.02)$ and $(15 - 10.1) \times (8000 - 5880)(0.10)$. From such considerations, it would appear that computation of σ_X^2, σ_Y^2, and σ_{XY} involves a considerable amount of laborious addition and multiplication.

Table 3.1. Joint probability mass function of the social status score X and income Y of a randomly chosen individual from a given community.

		\$2000	\$4000	\$6000	\$8000	\$10,000	Total
	0	0.05	0.02	0.03	0.01	0.01	0.12
Social	5	0.02	0.10	0.08	0.01	0.01	0.22
Status	10	0.03	0.08	0.10	0.08	0.01	0.30
Score	15	0.01	0.03	0.08	0.10	0.02	0.24
	20	0.01	0.01	0.03	0.02	0.05	0.12
	Total	0.12	0.24	0.32	0.22	0.10	1.00

However, let us *rescale* the variables X and Y. That is, let us construct new random variables $X^* = aX + b$, $Y^* = cY + d$, where a, b, c, and d are certain constants. Note that if (x,y) is a possible pair of values for the pair of random variables (X,Y), then $(ax+b, cy+d)$ is a possible pair of values for the pair of random variables (X^*, Y^*), and

$$p_{X^*, Y^*}(ax+b, cy+d) = p_{X,Y}(x,y).$$

It follows that the joint distribution of the random variables X^* and Y^* is given by the contingency table:

x^* \ y^*	$2000c+d$	$4000c+d$	$6000c+d$	$8000c+d$	$10{,}000c+d$	Total
b	0.05	0.02	0.03	0.01	0.01	0.12
$5a+b$	0.02	0.10	0.08	0.01	0.01	0.22
$10a+b$	0.03	0.08	0.10	0.08	0.01	0.30
$15a+b$	0.01	0.03	0.08	0.10	0.02	0.24
$20a+b$	0.01	0.01	0.03	0.02	0.05	0.12
Total	0.12	0.24	0.32	0.22	0.10	1.00

We now choose the numbers a, b, c, and d so as to make the possible values of X^* and Y^* convenient for computational purposes. Since small integers are perhaps easier to work with, we think of letting $a = 1/5$, $c = 1/2000$, and b and d be 0. Then the possible values of X^* are 0, 1, 2, 3, and 4; while the possible values of Y^* are 1, 2, 3, 4, and 5. (Other choices of a, b, c, and d may also be convenient; for example, we could let $d = -1$ so as to make the possible values of Y^* equal to 0, 1, 2, 3, and 4.) The joint distribution of X^* and Y^* for these choices of the constants a, b, c, and d is given by the contingency table:

x^* \ y^*	1	2	3	4	5	Total
0	0.05	0.02	0.03	0.01	0.01	0.12
1	0.02	0.10	0.08	0.01	0.01	0.22
2	0.03	0.08	0.10	0.08	0.01	0.30
3	0.01	0.03	0.08	0.10	0.02	0.24
4	0.01	0.01	0.03	0.02	0.05	0.12
Total	0.12	0.24	0.32	0.22	0.10	1.00

From this table, we quickly find that

$$\mu_{X^*} = (0)(0.12) + (1)(0.22) + (2)(0.30) + (3)(0.24) + (4)(0.12) = 2.02,$$

and

$$\mu_{Y^*} = (1)(0.12) + (2)(0.24) + (3)(0.32) + (4)(0.22) + (5)(0.10) = 2.94.$$

Using the formula (see Chapter 5, Section 5)

$$\sigma^2_{X^*} = E[X^{*2}] - \mu^2_{X^*}, \qquad \sigma^2_{Y^*} = E[Y^{*2}] - \mu^2_{Y^*},$$

we see that

$$\sigma^2_{X^*} = [(0)^2(0.12) + (1)^2(0.22) + (2)^2(0.30) + (3)^2(0.24) + (4)^2(0.12)] - (2.02)^2$$
$$= 5.50 - 4.0804 = 1.4196,$$

Table 7.3. Sizes of groups of pedestrians in Eugene, Oregon on a spring morning.

Size of Group = k	Observed Frequency of Groups of Size k	Theoretical Frequency of Groups of Size k (Truncated Poisson, $\lambda = 0.892$)
1	1486	1501
2	694	670
3	195	199
4	37	44
5	10	8
6	1	1
	2423	2423

Table 7.4. Size of shopping groups.

Size of Group = k	Observed Frequency of Groups of Size k	Theoretical Frequency of Groups of Size k (Truncated Poisson, $\lambda = 0.889$)
1	316	316
2	141	141
3	44	42
4	5	9
5	4	2
	510	510

Table 7.5. Size of playgroups.

Size of Group = k	Observed Frequency of Groups of Size k	Theoretical Frequency of Groups of Size k (Truncated Poisson, $\lambda = 1.362$)
1	570	599
2	435	408
3	203	185
4	57	63
5	11	17
6	1	4
7	0	1
	1277	1277

SOURCE: Tables 7.3, 7.4 and 7.5 from Chatfield, C. (1970). Distributions in market research, *Random Counts in Physical Science, Geo Science, and Business*, ed. G. P. Patil, pp. 163-181. Reprinted by permission of the Pennsylvania State University Press.

REFERENCES TO OTHER DISCRETE DISTRIBUTIONS

The following books describe a number of discrete distributions which have been used in practice.

1. *Classical and Contagious Distributions* (1965). Edited by G. P. Patil. Statistical Publishing Society, Calcutta.

   This book provides a bibliography of the statistical literature on discrete distributions.

2. Johnson, N. L., and Kotz, S. (1969). *Distributions in Statistics: Discrete Distributions*. Houghton Mifflin Company, Boston.

   This book provides a discussion of the theory and applications of a wide range of discrete distributions.

8. SUMMARY OF DISCRETE DISTRIBUTIONS

To summarize, the various discrete distributions discussed in this chapter are listed on p. 305 together with their parameters, expected values, and variances.

EXERCISES

1. An examination consists of 10 multiple-choice questions. Each question has 5 possible answers, of which only one is correct. Suppose that a student takes this examination and independently guesses the answer to each question.
   (a) What is the probability that the student obtains the correct answer to the first question on the examination?
   (b) What is the probability that the student obtains the correct answer to only one question on the examination (and answers 9 questions incorrectly)?
   (c) What is the probability that the student obtains correct answers to 3 questions on the examination (and answers 7 questions incorrectly)?
   (d) If Z denotes the number of correct answers obtained on the examination, what is the probability distribution of Z?
   (e) Find the expected value μ_Z and variance σ_Z^2 of Z.
   (f) Suppose 5 or more correct answers constitutes a passing grade. What is the probability that the student passes?

A2. Another student takes the examination described in Exercise 1, but this student has studied, and consequently has a probability of 0.8 of correctly answering a question. Complete parts (a) through (f) of Exercise 1 for this

Distribution	Probability Mass Function $p(k)^a$	Parameters	Expected Value	Variance
Bernoulli	$p^k(1-p)^{1-k}$ for $k=0,1$	$0 \leq p \leq 1$	p	$p(1-p)$
Binomial	$\binom{n}{k}p^k(1-p)^{n-k}$ for $k=0,1,2,\ldots,n$	$0 \leq p \leq 1$ $n=1,2,\ldots$	np	$np(1-p)$
Hypergeometric	$\dfrac{\binom{Np}{k}\binom{N-Np}{n-k}}{\binom{N}{n}}$ for $k=0,1,\ldots,n$	$0 \leq p \leq 1$ $q=1-p$ Np is an integer $n=1,2,\ldots$	np	$\left(\dfrac{N-n}{N-1}\right)np(1-p)$
Geometric	$p(1-p)^{k-1}$ for $k=1,2,\ldots$	$0 \leq p \leq 1$	$\dfrac{1}{p}$	$\dfrac{1-p}{p^2}$
Negative Binomial	$\binom{r+k-1}{k}p^k(1-p)^k$ for $k=0,1,2,\ldots$	$0 \leq p \leq 1$ $r=1,2,\ldots$	$\dfrac{r(1-p)}{p}$	$\dfrac{r(1-p)}{p^2}$
Generalized Negative Binomial	$H(k,r)p^r(1-p)^k$ for $k=0,1,2,\ldots$	$0 \leq p \leq 1$ $r>0$	$\dfrac{r(1-p)}{p}$	$\dfrac{r(1-p)}{p^2}$
Poisson	$\dfrac{e^{-\lambda}\lambda^k}{k!}$ for $k=0,1,2,\ldots$	$0 \leq \lambda$	λ	λ
Generalized Discrete Uniform	$\dfrac{1}{K}$ for $k=c+h,$ $c+2h,\ldots,c+Kh$	$k=1,2,\ldots$ $-\infty < c < \infty,$ $h>0$	$c+\dfrac{(K+1)h}{2}$	$\dfrac{h^2(K^2-1)}{12}$

^{a}Values of k not shown in the table have probability equal to zero.

student (that is, under the assumption that the student has probability 0.8 of getting the correct answer on each question). Suppose you are faced with this model; that is, suppose you believe that if you study, you have probability of 0.8 of answering a question correctly; however, if you do not study, you must guess the answer to each question. Do you think it is worth studying for the examination? Explain.

3. Five individuals are chosen from among the registered voters of a community and are asked if they favor a certain bond issue. Assume that the choice of these 5 individuals can be regarded as resulting from a simple random sample with replacement. If only 30 percent of the voters in the community favor the bond issue, what is the probability that the majority of the 5 individuals chosen will favor the proposal? [*Note:* According to *Webster's Seventh New Collegiate Dictionary*, a "majority" is "a number (strictly) greater than half of a total."]

4. The Internal Revenue Service has a special letter-opening machine which opens and removes the contents from an envelope. If the envelope is fed improperly into the machine, or if the machine malfunctions, the contents of the envelope may not be removed or may be damaged. In this case, we say that the machine has "failed." Assume that the fate of each envelope entering the machine is independent of the fate of all other envelopes handled by that machine.
   (a) If the machine has a probability of failure of 0.1, what is the probability of more than 3 failures occurring in a batch of 10 envelopes?
   (b) Suppose the probability of failure of the machine is 0.02. What is the probability that more than 3 failures of the machine will occur in a batch of 100 envelopes? [*Warning*: For this part and for part (c), you will need to refer to an existing set of tables, for example, Romig (1953).]
   (c) If the probability of failure of the machine is 0.01, what is the probability that more than 3 failures of the machine will occur in a batch of 100 envelopes? [See the warning in part (b) above.]

5. A large number (53,680) of German families with 8 children were contacted and, for each family, the number X of male children was noted. The results are given in Table E.1.

   Fit a binomial distribution to these data. [*Hint:* First determine the total number of children observed, and then the total number of boys observed.]

   [*Remark:* These data are taken from a study by Geissler, and are given by Fisher, R. A. (1950). *Statistical Methods for Research Workers* (11th ed.), Oliver & Boyd, Ltd., Edinburgh. Geissler's data are also discussed in Simpson, G. G., Roe, A. and Lewontin, R. C. (1960). *Quantitative Zoology*, Harcourt, Brace & World, Inc., New York.]

^A6. In the study of Mendelian laws of inheritance, litters of 5 mice were studied, and the number of "dominant" mice in each litter noted. The actual mating was a "back cross," that is, a mating of a parent having one dominant and one recessive allele with a parent having two recessive alleles. See Table E.2.

Table E.1. Results of observing the number X of male children in each of 53,680 German families.

Number x of Boys in German Families of 8 Children	Observed Number of German Families Having x Boys
0	215
1	1485
2	5331
3	10,649
4	14,959
5	11,929
6	6678
7	2092
8	342
	53,680

If the theoretical probability of a "dominant" character is 0.5 for each mouse, what are the theoretical frequencies and what are the theoretical relative frequencies that you would obtain using the binomial distribution? What assumptions are you making when you use the binomial distribution?

[*Remark:* The data in Table E.2 are taken from Detlefsen, J. A. (1918), Fluctuations of sampling in a Mendelian population. *Genetics*, vol. 3, pp. 599–607. A discussion of these data may also be found in Wright, S. (1968). *Genetic and Biometric Foundations*, vol. I. University of Chicago Press, Chicago.]

Table E.2. Results of observing the number X of mice with the dominant character in 330 litters of 5 mice.

Number x of Mice in Each Litter which Show the Dominant Character	Observed Number of Litters with x "Dominant" Mice
0	9
1	47
2	106
3	103
4	51
5	14
	330

7. In some military courts, 9 judges are appointed. Both the prosecution and the defense attorneys are entitled to a preemptory challenge of any judge, in which case that judge is removed from the case and is *not* replaced. Suppose that the prosecution attorney does not exercise his right to challenge the selection of judges, and the defense is limited to two challenges. A defendant is declared guilty only if there are a majority (see Exercise 3) of votes cast in favor of "guilty," and is declared innocent otherwise. Suppose each judge has a probability of 0.6 of voting "guilty," and that judges make their decisions independently.

If Z denotes the *number* of votes for a verdict of "guilty," find the values of the probability mass function for Z when there are:
(a) 9 judges, (b) 8 judges, (c) 7 judges.
What is the probability that the defendant is declared guilty when there are:
(d) 9 judges, (e) 8 judges, (f) 7 judges?
(g) If you were the defense attorney how many challenges would you exercise?

[*Optional:* If each attorney is entitled to exactly one challenge, and the prosecution has the first challenge, what would you do if you were (h) the prosecution, or (i) the defense?]

^A8. One of the classical examples of the application of statistical analysis is that of a tea-tasting experiment [discussed in detail by Neyman, J. (1950). *A First Course in Probability and Statistics*, Holt, Rinehart and Winston, Inc., New York]: "A lady declares that by tasting a cup of tea made with milk she can discriminate whether the milk or the tea infusion was first added to the cup." More exactly, the claim is "not that she could draw the distinction with invariable certainty, but that, though sometimes mistaken, she would be right more often than not." To test the claim, the lady is subjected to an experiment in which, while blindfolded, she will taste a number of pairs of cups of tea, each pair containing one cup of tea made by each of the two methods (that is, in one cup, tea is added before milk, while milk is added before tea in the other cup). Let us assume that every pair of cups of tea is tasted under identical circumstances, and that the outcome of tasting one pair of cups is independent of the outcome of tasting any other pair of cups.

Suppose the lady tastes 8 pairs of cups of tea.
(a) If for each pair of cups of tea, the lady only guesses which cup had tea added first, what is the probability distribution of the number Z of correct answers that she gives?
(b) On the other hand, if the lady really is skilled in detecting the difference, suppose that her probability of giving a correct answer is 0.7. Now, what is the probability distribution of the number Z of correct answers that she gives?
(c) In a particular performance of the experiment, the lady gives 6 correct answers. Would you feel that the lady has guessed and is particularly lucky, or would you conclude that she is skilled in detecting the difference between cups of tea in which tea is added before milk and cups of tea in which milk is added before tea? Justify your answer.

A9. The proportion of foreign-born in the city of New York is (for argument's sake) equal to 0.3. We are interested in interviewing only foreign-born people, but unfortunately until we talk to a person, we cannot determine whether or not he is foreign-born. Assume that people come to our office in such a way that: (i) only one person comes each day, and (ii) any New Yorker has an equal chance of showing up on *any given day*.
   (a) Let X be the number of foreign-born people among the first n people interviewed. Which of the following random variables has the same probability distribution as X?
      (i) The number of heads in n tosses of a biased coin where the probability of getting a head on 1 toss equals 0.3.
      (ii) The number of red balls in a random sample of size n with replacement from an urn containing 300 red balls and 700 black balls.
      (iii) The number of red balls in a random sample of n drawn without replacement from an urn containing 300 red balls and 700 black balls.
   (b) What is the probability that we find 3 or more foreign-born people among the first 10 people that we interview?
   (c) What would be the expected number of foreign-born people interviewed if we interviewed 1000 people?
   (d) What would be the variance of the number of foreign-born people interviewed if we interviewed 1000 people?
   (e) Suppose we had agreed to keep interviewing people until we found and interviewed the first foreign-born person. What is the probability that we would interview for exactly 5 days? (*Be Careful!*)

10. Suppose that the random variable Z has a binomial distribution with parameters n and p. In each of the following cases table and graph (i) the probability mass function $p_Z(k)$, and (ii) the cumulative distribution function $F_Z(\tau)$ of Z.
   (a) $n = 8$, $p = 0.3$,
   (b) $n = 8$, $p = 0.5$,
   (c) $n = 8$, $p = 0.7$,
   (d) $n = 10$, $p = 0.3$,
   (e) $n = 10$, $p = 0.5$,
   (f) $n = 10$, $p = 0.7$.
   For each part [(a) through (f)] of this exercise find (i) the mode of Z, and (ii) the median of Z. Which of these distributions are positively skewed? Which of these distributions are negatively skewed? Which distributions are symmetric?

A11. There is currently much discussion concerning minority representation in a venire (panel from which a jury is chosen). Complete the following table assuming that choice of one individual for the panel is statistically independent of the choice of any other individual for the panel. In a recent study, 30 venires were sampled and all had 5 or fewer non-whites. Is this result surprising? [*Note:* For further discussion of such problems, see Finkelstein (1966).]

Probability p of Selecting a Single Non-White Venireman	Probability P of Selecting a Single Venire of 30 with 5 or Fewer Non-Whites	Probability P^{30} of Selecting 30 Such Venires
0.20		
0.10		
0.05		

12. Suppose that the random variable Z has a binomial distribution with parameters n and p.
    (a) If you know that $p = 0.6$ and $\mu_Z = 6$, find n and σ_Z^2.
    (b) If you know that $\mu_Z = 6.0$ and $\sigma_Z^2 = 4.2$, find n and p.
    (c) If you know that $n = 25$ and $\mu_Z = 10$, find p and σ_Z^2.
    (d) Is it possible for $\mu_Z = 3$ and $\sigma_Z^2 = 5$ if Z has a binomial distribution?

A13. The following question is, in essence, the question which the Chevalier de Méré asked Pascal (see Chapter 1): Which of the following probabilities is the larger?
    (i) The probability that a "6" (the face with 6 dots on it) comes up *at least once* in 4 tosses of a balanced die.
    (ii) The probability that *at least once* in 24 tosses of 2 balanced dice, both dice simultaneously show a "6."
    Remember that a die is a cube with 6 faces, 1 of which has 6 dots upon it. (The Chevalier de Méré thought that he had given a theoretical argument which proved that these probabilities are equal, yet his observations suggested otherwise. He asked Pascal to explain the apparent contradiction.)

14. Answer the question posed in Exercise 3 under the assumption that the choice of the 5 individuals who are asked their opinion of the bond issue results from a simple random sample *without replacement* from among:
    (a) a total of 50 voters in the community,
    (b) a total of 100 voters in the community,
    (c) a total of 200 voters in the community.
    How do the answers that you obtain under assumptions (a), (b), and (c) compare with each other, and with the answer you obtained to Exercise 3? Explain.

15. Suppose that it is known that 20 percent of all patients who come to a clinic and exhibit certain symptoms have a certain disease. Final diagnosis (detection) of this disease depends on a blood test. Sixty individuals with symptoms of the disease come to the clinic. Individual blood tests are expensive, so the hematologist uses the following screening device: The blood of a number n of individuals is combined and tested. If none of the n persons has the disease, then the composite blood test is negative. However, if there is any contaminated blood (that is, if at least one person has the disease), the composite test will be positive. Assuming that the n persons whose blood is tested are chosen from among the 60 individuals by a simple random sample *without replacement*, what is the probability that the composite blood test

will be negative if
(a) $n = 2$,
(b) $n = 4$,
(c) $n = 6$,
(d) $n = 10$?

*^16. Consider a group of 10 individuals consisting of 7 males and 3 females. We will choose a committee of n people from among these 10 individuals by means of a simple random sample without replacement. We want the committee size n to be as small as possible, but we also want n to be large enough so that the probability that the committee will have at least one male and at least one female is no less than 0.9. How large should the committee size n be?

^17. A radio manufacturer intends to purchase 100 radio tubes from a supply house. He expects that some of these tubes will be defective, but is willing to tolerate only 4 defective tubes in the batch of 100 tubes. He decides to verify the quality of the batch by drawing a simple random sample of 3 tubes without replacement from among the 100 tubes, and testing these tubes with a tube tester.
   (a) Let X be the (random) number of defective tubes in the sample of 3 tubes. Determine the values of the probability mass function $p_X(k) = P_X\{X = k\}$ of X if there are exactly 4 defective tubes in the batch of 100 tubes.
   (b) What is the mode of the distribution of X? What is the median of X? What is the expected value of X?
   (c) The radio manufacturer decides to reject the batch of radio tubes if X exceeds 1. What is the probability that he will reject the batch of tubes if there are exactly 4 defective tubes in the batch? (That is, what is the probability that his test rejects a batch that just meets his specifications?)
   (d) Suppose there are actually 5 defective tubes in the batch of 100 tubes. Does the random variable X have the same probability mass function as the probability mass function you calculated in part (a)? What is the probability that the manufacturer will reject this batch of 100 tubes? That is, what is the probability that X exceeds 1, if there are 5 defective tubes in the batch of 100 tubes?

18. Let Z be a random variable which has a hypergeometric distribution with parameters $N = 10$, n, and p. Graph the probability mass function $p_Z(k) = P\{Z = k\}$ and the cumulative distribution function $F_Z(\tau)$ of Z when
   (a) $n = 8$, $p = 0.3$,
   (b) $n = 8$, $p = 0.5$,
   (c) $n = 8$, $p = 0.7$.
   How do your graphs compare with the graphs you obtained in Exercise 10, parts (a), (b), and (c)?

^19. Let Z be a random variable which has a hypergeometric distribution with parameters N, $n = 5$, and p.

(a) If $\mu_Z = 1$, $\sigma_Z^2 = 0.40$, what are the values of N and p?
(b) If $\mu_Z = 2$, $\sigma_Z^2 = 0.40$, what are the values of N and p?
(c) How would you find N if you know the values of μ_Z and σ_Z^2?
(d) Can you find the value of N if you only know the values of p and μ_Z?
Can you find the value of N if you only know the values of p and σ_Z^2?
(e) Can you find the value of N if you know the value of Np and of μ_Z? If so, show how. If not, why not?

20. There are an unknown number N of fish in a certain pond. Three fish are caught by net, tagged, and returned to the pond. Now, $n = 5$ fish are caught by net and the number Z of tagged fish in this catch is recorded. We assume that our second catch of 5 fish is a simple random sample without replacement from the total population of N fish in the pond. The value of Z that we observe is a single trial of the random experiment in which the variable Z is observed. Therefore, we may estimate μ_Z by the sample average $\hat{\mu}_Z$, which in this case is equal to the observed value of Z. Thus, if we observe 1 tagged fish in our sample of 5 fish, we estimate μ_Z by $\hat{\mu}_Z = 1$. We now know the value of n ($n = 5$), the value of Np ($Np =$ the number of tagged fish in the pond $= 3$), and we have an estimate of the value of μ_Z (namely, $\hat{\mu}_Z = 1$).

(a) Using your answers to Exercise 19, either suggest a way of estimating the number N of fish in the pond on the basis of the above information, or else indicate why you feel that such an estimate cannot be obtained.
(b) Suppose that in our catch of $n = 5$ fish, there are $Z = 0$ tagged fish. Can we estimate N now? If not, why not? If so, what estimate would you obtain? Is this a reasonable estimate? Why, or why not?

*^21. In studying the frequency of wars, Richardson (1945) tallied the number of wars X which began in each of the calendar years from 1500 A.D. to 1931 A.D. (Of course, it is not always clear as to the exact time when a war begins. In order to minimize any subjective bias, the tabulation was made by three different individuals.) The number of wars X in a calendar year was hypoth-

Table E.3. Results of observing the number X of wars in each of the calendar years from 1500 A.D. to 1931 A.D.

Number x of Outbreaks of War in a Particular Year	Observed Frequency of Outbreaks of x Wars in a Year
0	223
1	142
2	48
3	15
4	4
5 or more	0
	432

Sample Average $\hat{\mu}_X = 0.692$

esized to have a Poisson distribution. The 432 observed values of X are summarized in Table E.3.
(a) Verify that the sample average $\hat{\mu}_X$ obtained from the above data is indeed equal to 0.692. Show your work.
(b) Using Table T.3, fit a Poisson distribution to these data, assuming that $\hat{\mu}_X = 0.70$ (instead of the observed value of $\hat{\mu}_X = 0.692$). That is, find the values of the theoretical frequency of the outbreak of x wars in a given calendar year, for $x = 0, 1, 2, 3,$ and 4, and find the theoretical frequency of 5 or more outbreaks of war in a given year, when X has a Poisson distribution with parameter $\lambda = 0.70$.
(c) By interpolating in Table T.3, fit a Poisson distribution with parameter $\lambda = 0.69$ to these data.
(d) Does either the Poisson distribution with $\lambda = 0.70$ or the Poisson distribution with $\lambda = 0.69$ provide a good fit to these data? Explain your answer.

*22. The number of lost articles found in a large municipal office building and turned into the lost-and-found office was noted. Records for each day (except Sundays and holidays) were kept for the period November 1, 1923 to September 30, 1925 (excluding the summer months June, July, and August when there might be variations in the population of the building). The data so obtained are given in Table E.4 [Thorndike (1926)].

Table E.4. Results of observing the number X of lost articles turned into the lost-and-found office of a large municipal office building for each working day from November 1, 1923 to September 30, 1925.

Number x of Lost Articles in a Day	Observed Frequency of x Lost Articles in a Day
0	169
1	134
2	74
3	32
4	11
5	2
6	0
7	1
	423

Fit a Poisson distribution to these data. Is the fit a good one?

A23. In a geological study, Krumbein (1953) noted the number X of granite pebbles in each of 100 samples of beach pebbles (each sample of beach pebbles contained 10 pebbles of moderate size: 16–32 mm in diameter). The resulting observations are summarized in Table E.5.

Table E.5. Observations of the number X of granite pebbles in 100 samples of beach pebbles.

Number x of Granite Pebbles in a Sample	Observed Frequency of Samples with x Granite Pebbles
0	58
1	33
2	7
3	2
	100

Fit a Poisson distribution to these data. Does the Poisson distribution provide a good fit? Explain your answer.

*24. On each day for 3 consecutive years, the number of death notices in the *London Times* was counted, and the distribution of deaths of men over 85 years of age and of women over 80 years of age was noted. (See Table E.6.)

Table E.6. Observations of the number X of death notices of men over 85 years of age and the number Y of death notices for women over 80 years of age in 1096 daily issues of the *London Times*.

Men		Women	
Number x of Deaths in a Day	Observed Frequency of x Deaths in a Day	Number y of Deaths in a Day	Observed Frequency of y Deaths in a Day
0	484	0	162
1	391	1	267
2	164	2	271
3	45	3	185
4	11	4	111
5	1	5	61
	1096	6	27
		7	8
		8	3
		9	1
			1096

(a) Fit a Poisson distribution to the data giving the number X of deaths per day involving men over 85.

(b) Fit a Poisson distribution to the data giving the number Y deaths per day involving women over 80 years of age.

[*Note:* The above data are taken from Whitaker, L. (1914). On the Poisson law of small numbers. *Biometrika*, vol. 10, pp. 36–71.]

A25. Suppose that the number X of misprints on a page of a book has a Poisson distribution with parameter $\lambda = 1.5$.
  (a) What is the expected value of X? How many misprints "on the average" would you expect to find on a given page of this book?
  (b) Using Table T.3 graph the probability mass function $p_X(k)$ and the cumulative distribution $F_X(\tau)$ of X.
  (c) What is the median of the distribution of X? What is the mode of the distribution of X? What is the most probable number of misprints on a given page of the book?
  (d) What is the probability that a given page of the book will have no misprints? In a 600-page book, how many pages do you expect "on the average" to be misprint-free? (Assume that the number of mistakes on any one page of the book is statistically independent of the numbers of mistakes on all other pages.)

26. There has been much concern recently with the problem of abandoned cars on highways. Suppose that the number X of cars abandoned in a week on a particular highway has a Poisson distribution with parameter $\lambda = 2.0$.
  (a) Graph the probability mass function $p_X(k)$ and cumulative distribution function $F_X(\tau)$ of X.
  (b) It costs the state government \$100 per car to tow away and dispose of an abandoned car. What is the expected cost per week to the state government to dispose of cars abandoned on the given highway?
  (c) What is the most probable amount of money that the state government must pay to dispose of cars abandoned on the highway in a given week?
  (d) How probable is it that the state government will have to spend more than \$400 in a given week to dispose of cars abandoned on this highway?
  (e) At present the state government pays private tow truck operators \$25 a car to tow away abandoned automobiles. The cost of purchasing and running their own tow truck, amortized over the life of the tow truck, is \$50 a week. Would you recommend that the state purchase their own tow truck? Justify your answer.

A27. An archeologist has made a study of ancient trade routes, and in the process has come to the conclusion that along a certain stretch of coastal shelf, every acre of ocean bottom has a probability of 0.01 of containing a wrecked ancient ship which is in good enough condition to yield important archeological information. He chooses 100 acres of ocean bottom on the coastal shelf and asserts to a foundation that is interested in supporting his work that in these 100 acres there is at least a probability of 0.60 that one or more ancient ships in good condition will be found. Assuming that the event that a ship is found in any one acre of ocean bottom is statistically independent of the event that a ship is found in any other acre of ocean bottom, find the probability that one or more ancient ships (in good condition) will be found in these 100 acres. In the process of your work, answer the following questions:
  (a) What is the distribution of the number X of ships that will be found in

100 acres of ocean bottom? (Assume that at most one ship can be found in 1 acre of ocean bottom.)

(b) What is the probability that $X = 0$? Hence, what is the probability that X is greater than or equal to 1? Use the distribution of X to obtain your answers.

(c) Is there an approximation to the distribution of X that can simplify your calculations in part (b)? What is this approximation? Answer the questions in part (b) using this approximation. How good is the approximation?

(d) Is the archeologist's assertion to the foundation correct?

*A28. A medical researcher is interested in studying the etiology of a rare disease. This disease has symptoms similar to those of a more common disease. To distinguish the two diseases an expensive test must be run. Out of 500 individuals who have the symptoms of both diseases and who have visited clinics and hospitals to which the researcher has access, the researcher knows that about 10 have the rare disease. The researcher can only afford to test 50 individuals, and yet he needs at least 3 individuals with the rare disease for his studies. From among the 500 individuals with symptoms of both diseases, he chooses a simple random sample of 50 individuals *without replacement*.

(a) Let X be the number of individuals in the researcher's sample who have the rare disease. What distribution does X have?

(b) What is the value of the finite population factor for this experiment? Can you suggest a distribution to approximate the distribution of X? Explain.

(c) Note that a proportion of $10/500 = 0.02$ of the 500 individuals who have the symptoms of both diseases actually have the rare disease. Note also that the product of this proportion (0.02) and the sample size (50) is equal to $(50)(0.02) = 1.0$. Based on these facts, can you suggest a distribution to approximate the distribution that you named in part (b)? Explain.

(d) Putting together the approximations that you obtained in parts (b) and (c), give an approximate value for the probability that the medical researcher finds at least 3 people with the rare disease in his sample of 50 individuals.

Does the line of argument outlined above suggest a general principle for approximating the probability distribution of the number X of individuals who possess a rare characteristic in a random sample drawn without replacement from a large population of N individuals? What is this principle?

*A29. A man receives 5 letters per day on the average. On a certain day he receives no mail and wonders if it is a holiday. To decide this, he computes the probability of receiving no mail in a given day. He assumes that the number X of letters that he receives on a given day has a Poisson distribution.

(a) What is the probability of his receiving no mail on a given day? That is, what is the probability of the event $\{X = 0\}$?

(b) On second thought, the man realizes that this is the first mail delivery day in a year in which he has received no mail. Assuming 300 mail delivery days in a year, approximately what is the probability that in a year no delivery day will pass without the man's receiving at least one piece of mail? [*Hint:* Assume that the number of letters received in one day is statistically independent of the number of letters received in any or all other days. What is the distribution of the number Y of mail delivery days in a year that pass without the man's receiving at least one letter? What distribution is a good approximation to the distribution of Y?]

A30. How many raisins should be mixed into a bowl of cookie batter if 25 cookies of equal size are to be made from the batter, and if it is desired that with probability of 0.99 or greater a cookie sampled at random from among the 25 cookies should contain at least one raisin? [*Hint:* If there are R raisins in the batter, then on the average each cookie contains $R/25$ raisins. Let X be the number of raisins in the sampled cookie, and use a Poisson approximation to the distribution of X. Do you think this approximation is a good approximation? Why, or why not?]

31. In Table 4.1, with $p = 0.4$, find $P\{X \geq 4 | X \geq 3\}$ and $P\{X \geq 7 | X \geq 6\}$. Do your results suggest any general principle? Can you suggest a formula for $P\{X \geq k+1 | X \geq k\}$? Try it out for $p = 0.5$ and $p = 0.6$ on the following values of k: $k = 0, 1, 2, 3, 4, 5, 6$. [*Optional:* If you can, give a proof of your formula which holds for all values of p and all values of k.]

32. In Table 4.3 plot (a) the observed relative frequencies against k, and (b) the theoretical relative frequencies against k. Comment on these graphs.

A33. Let X have a geometric distribution with parameter p. Using Table 4.1, make graphs of the probability mass function $p_X(k)$ and the cumulative distribution function $F_X(\tau)$ of X when
 (a) $p = 0.4$,
 (b) $p = 0.6$,
 (c) $p = 0.8$.
In each of the cases (a), (b), and (c) above answer the following questions: What is the median of the distribution of X? What is the mode of X? Is the distribution of X positively skewed, symmetric, or negatively skewed?

34. Suppose that you are informed that a random quantity X has a geometric distribution with parameter $p = 0.5$. You are asked to predict the value of X that will occur on the next trial of the random experiment in which X is observed. What value would you predict for X? Why?

A35. In another traffic study, the number X of occupants carried in each of 1469 passenger cars was tabulated by Haight (1970). The cars were both eastbound and westbound, and were observed at the corner of Wilshire and Bundy (in Los Angeles) between 11:10 and 12:00 on the morning of March 23. The data obtained by Haight are tabulated in Table E.7.

Table E.7. Observations of the number X of occupants carried in each of 1469 passenger cars.

Number x of Occupants	Observed Frequency of x Occupants
1	902
2	403
3	106
4	38
5	16
6 or more	4
	1469

Assuming that the 4 observations corresponding to "6 or more occupants" were actually observations of "exactly 6 occupants," fit a geometric distribution to these data.

36. A sociologist is studying the publication behavior of researchers in his own field. He takes a sample of 363 active sociologists, and asks each individual to tell him the number of journals to which his most recent published paper was submitted before it was finally published. The data he obtains are summarized in Table E.8.

Table E.8. Observations of the number X of journals to which a sociologist's most recent paper was submitted before publication.

Number x of Journals	Observed Frequency of x Journals
1	261
2	75
3	24
4	3
5 and above	0
	363

Fit a geometric distribution to these data. Is the fit a good one? Would you anticipate that the fit is good? Why, or why not?

A37. A reporter is sent out to interview a "man in the street" for a human interest story. The reporter hypothesizes that any person whom he approaches will, with probability 0.6, agree to be interviewed. He also assumes that the event that any one individual will agree to be interviewed is statistically independent of the event that any other individual agrees to be interviewed.
  (a) What is the probability that the reporter will have to approach 4 or more people before he finds a person who agrees to be interviewed?
  (b) The reporter is aware that not all individuals whom he can interview are good subjects for a human interest story. He judges that the condi-

tional probability of getting a good interview from an individual, given that the individual has agreed to be interviewed, is 0.3. What is the expected number of individuals that the reporter must approach before he can interview an individual who will provide him with a good human interest story?

38. Two baseball teams, Team A and Team B, meet in the World Series. Assume that every game between these 2 teams is a Bernoulli trial in which "success" is a victory for Team A, and "failure" is a victory for Team B. Assume also that the result of every game between these 2 teams is statistically independent of the result of all other games between these teams, and that the teams are evenly matched. In the World Series, the first team to win 4 games wins the series.
   (a) What is the probability that Team A wins the series in 5 games?
   (b) What is the probability that the series lasts 5 games?
   (c) What is the probability that the series will end after exactly 7 games have been played?
   (d) What is the probability that Team A will win the series?
   (e) What is the conditional probability that Team A wins the series, given that the series lasts 7 games?

^{A}39. Answer parts (a) through (e) of Exercise 38 under the assumption that Team A has a probability of 0.6 of winning on any game played with Team B.

*40. What is the expected number of games that will be played in the World Series described in Exercise 38 if:
   (a) Team A and Team B are evenly matched,
   (b) Team A has probability 0.6 of winning any game played with Team B?

*^{A}41. The data summarized in Table E.9 list the number X of 1-day industrial

Table E.9. Observations of the number X of 1-day industrial absences during 1957 for each of 195 inspectors.

Number x of 1-Day Absences	Observed Frequency of x 1-Day Absences
0	27
1	35
2	35
3	37
4	19
5	20
6	10
7	6
8	3
9	3
10 or more	0
	195

absences (all causes) during 1957 of each of 195 nonsupervisory married male inspectors employed in one company throughout the year. Fit a negative binomial distribution to these data. How good is the "fit"?

42. The data summarized in Table E.10 list the number X of traffic accidents incurred during the period 1952–1955 by each of 708 public corporation bus drivers who drove regularly throughout the period. Fit a negative binomial distribution to these data. Does the "fit" appear to be a good one?

Table E.10. Observations of number X of traffic accidents incurred during 1952–1955 by each of 708 bus drivers.

Number x of Traffic Accidents During the Period	Observed Frequency of x Accidents
0	117
1	157
2	158
3	115
4	78
5	44
6	21
7	7
8	6
9	1
10	3
11	1
12 or more	0
	708

[*Note:* These data appear in a paper by Froggatt, P. (1970). Application of discrete distribution theory to the study of noncommunicable events in medical epidemiology, in *Random Counts in Biomedical and Social Sciences*, vol. 2, edited by G. P. Patil, Pennsylvania State University Press, University Park, Pennsylvania.]

43. Fit negative binomial distributions to the data summarized in:
    (a) Table 3.2,
    (b) Table 3.4.
    Does the negative binomial distribution give a better fit to the data in either of these tables than does the Poisson distribution? Explain.

*44. Let X have a negative binomial distribution with parameters r and p. Using Table T.5, graph the probability mass function $p_X(k)$ and cumulative distribution function $F_X(\tau)$ of X for
    (a) $r = 2$, $p = 0.6$,
    (b) $r = 2$, $p = 0.8$,

(c) $r = 3$, $p = 0.6$,
(d) $r = 3$, $p = 0.8$.

Compare the graphs that you have obtained in this exercise to the graphs that you obtained in Exercise 32. What conclusions do you reach about the dependence of the shapes of the graphs of the probability mass function of a negative binomial distribution on the parameters r and p? (Remember that if Y has a geometric distribution, then $Y - 1$ has a negative binomial distribution with parameter $r = 1$.)

^A45. In each of the cases (a)–(d) in Exercise 44, find the mode and the median of X.

46. Suppose that you are informed that a random quantity X has a negative binomial distribution with parameters $r = 2$, $p = 0.5$. You are asked to predict the value of X that will occur on the next trial of the random experiment in which X is observed. What value would you predict for X? Why? Suppose $r = 5$, $p = 0.2$. What value would you predict for X? Does it matter which value of X you choose? Explain.

*^A47. Suppose, in Example 5.2, there are $N = 1000$ fish in the pond and $M = 200$ of these are of the desired species. What is the probability that 10 or more fish that are not of the desired species are caught before 3 fish of the desired species are observed?

*^A48. Suppose, in Example 5.2, there are $N = 1000$ fish in the pond and $M = 200$ of these are of the desired species. However, in contrast to the experiment described in Example 5.2, when we catch a fish, we do not return it to the pond. We continue to catch fish until 3 fish of the desired species have been caught. Let Y equal the number of fish caught that are not of the desired species. Find the probability of each of the following events:
(a) $\{Y = 0\}$,
(b) $\{Y = 1\}$,
(c) $\{Y = 2\}$,
(d) $\{Y = 3\}$,
(e) $\{Y = 4\}$,
(f) $\{Y = 30\}$.

Compare these probabilities to the probabilities you would obtain if Y had a negative binomial distribution with $r = 3$, $p = 0.2$ (that is, if you threw the fish back after they were caught, instead of keeping them). Can we use the negative binomial distribution to calculate approximate probabilities for Y even when the fish are not replaced in the pond after they are caught? If not, are there at least some events of the form $\{Y = k\}$ for which the negative binomial approximation can be helpful? Explain your answers.

49. When an electric light bulb is purchased, the bulb can be tested to determine whether it is operating properly. If it is defective, the bulb is put aside to be returned to the manufacturer, and another bulb is tested. Bulbs are tested until a good one is found. Suppose we start with a box of 5 bulbs of which 2 are defective. The experiment consists of testing bulbs until 2 good ones are found. Let X denote the number of bulbs tested. Find the probability mass

function of X. Does $Y = X - 1$ have a negative binomial distribution? Is the negative binomial distribution a good approximation to the distribution of Y? Explain your answer.

A50. If we spin the roulette wheel shown in Figure 4.1 in this chapter, what is the probability that the wheel will stop on:
   (a) an even number (let 0 be even),
   (b) a number X which is strictly less than 10,
   (c) a number X which is divisible (without remainder) by 3?
   What is the distribution of the number X on which the wheel stops?

*51. Show by direct calculation that the expected value μ_X of a random variable X having a probability mass function of the form given in Equation (6.1) is equal to $\frac{1}{2}(K+1)$ when
   (a) $K = 5$,
   (b) $K = 10$,
   (c) $K = 15$.
   In cases (a)–(c) also show by direct calculation that the variance of X is given by the equation $\sigma_X^2 = (K^2 - 1)/12$.

*52. Fit a truncated Poisson distribution to the data in Table 7.1. Is the fit good? Is the fit better than the fit of the zeta distribution?

*53. Fit a zeta distribution to the data in Table 7.3. Is the fit good? Is the fit better than the fit of the truncated Poisson distribution?

54. The data summarized in Table E.11 give the number X of surgery consultations in the year 1962 of each of 2810 patients registered in a given clinic (group practice clinic). The sample average of these data is $\hat{\mu}_X = 2.768$ and the sample variance is $\hat{\sigma}_X^2 = 11.805$.

What type of discrete distribution do you think best fits the data? Demonstrate by fitting the type of discrete distribution which you think gives the best fit, and also one other type of discrete distribution which seems to be appropriate. If you can, give a theoretical justification for your choice in terms of some causal model which explains why a patient will have X surgery consultations in a year.

Table E.11. Observations of the number X of surgery consultations in 1962 by each of 2810 patients.

Number x of Surgery Consultations	Observed Frequency of x Consultations
0	820
1	535
2	369
3	283
4	201
5	149
6	106
7	76
8	77
9	54
10	32
11	31
12	27
13	14
14	3
15	6
16	8
17	3
18	2
19 or more	14
	2810

7

THE NORMAL DISTRIBUTION

> *I know of scarcely anything so apt to impress the imagination as the wonderful form of cosmic order expressed by the "Law of Frequency of Error." The law would have been personified by the Greeks and deified, if they had known it. It reigns with serenity and in complete self-effacement amidst the wildest confusion. The huger the mob and the greater the apparent anarchy, the more perfect its sway. It is the supreme law of Unreason. Whenever a large sample of chaotic elements are taken in hand and marshalled in the order of their magnitude, an unsuspected and most beautiful form of regularity proves to have been latent all along.*
>
> Sir Francis Galton.

1. INTRODUCTION

Of all distributions of continuous random variables, the most widely studied and frequently used is the *normal* distribution. The importance of the normal distribution is due in part to the success that investigators in many disciplines have had in using this distribution to describe quantitative phenomena of importance to their research, and in part to the fact that certain useful functions of nonnormally distributed random variables are approximately normally distributed.

The probability density function $f_X(\tau)$ of a normal distribution is defined by the equation

$$(1.1) \qquad f_X(\tau) = \frac{1}{\sqrt{2\pi\sigma^2}} e^{-(\tau-\mu)^2/2\sigma^2},$$

for all numbers τ. Here, the Greek symbol π (pi) is an absolute constant equal to 3.14159.... Although it relates the circumference of a circle to its diameter, this constant also frequently appears in expressions for probability distributions. The constants μ and σ^2 appearing in Equation (1.1) are parameters of the normal distribution; that is, when μ and σ^2 are known, the distribution is completely determined. The graph of $f_X(\tau)$ is shown in Figure 1.1. Note that this graph is symmetric about $\tau = \mu$, and that it extends over the entire horizontal axis.

Using the theory of integral calculus, it can be shown that μ is the expected value of that normal distribution whose probability density function $f_X(\tau)$ appears in Equation (1.1). On the other hand, because the

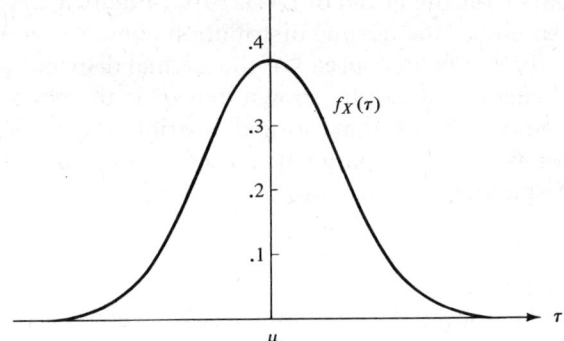

Figure 1.1 Graph of a normal density function $f_x(\tau)$. The graph is symmetric about the point μ.

graph of the density function $f_X(\tau)$ is symmetric about μ, the area under the graph of $f_X(\tau)$ to the right of $\tau = \mu$ is equal to the area under the graph of $f_X(\tau)$ to the left of $\tau = \mu$, and hence μ is a median of the distribution. Interestingly enough (see Figure 1.1), μ is also the mode of the distribution. Thus, a normal distribution has the property that its expected value, median, and mode all have the same value, μ. The parameter μ of a normal distribution tells us about the location of the distribution on the horizontal axis; μ can either be a positive number, a negative number, or zero.

The other parameter σ^2 of a normal distribution tells us something about the shape of the graph of the probability density function $f_X(\tau)$. In Figure 1.2, we show how different values of σ^2 ($\sigma^2 = 0.25, 1.0, 4.0$) correspond

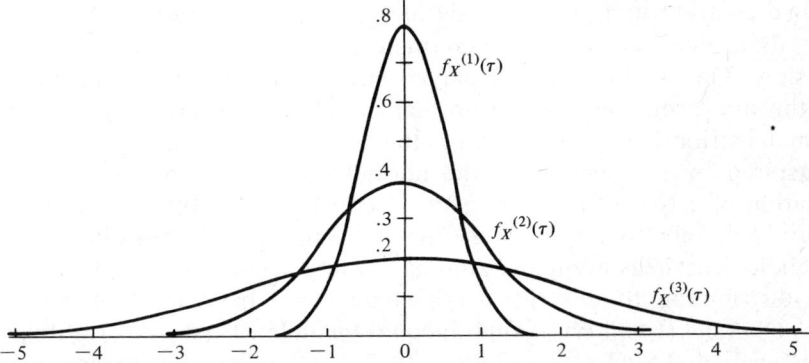

Figure 1.2 Graphs of the probability density functions of three normally distributed random variables having expected value $\mu = 0$. The densities $f_X^{(1)}(\tau)$, $f_X^{(2)}(\tau)$, $f_X^{(3)}(\tau)$ correspond to normal distributions with variances $\sigma^2 = 0.50$, $\sigma^2 = 1.00$, and $\sigma^2 = 2.00$, respectively.

to different shapes for the graph of $f_X(\tau)$. From Figure 1.2, we see that σ^2 describes the spread of the normal distribution about $\tau = \mu$; the larger the value of σ^2 is, the larger the spread of the normal distribution is. Indeed, using integral calculus, it can be shown that σ^2 is the variance (and σ is the standard deviation) of that normal distribution whose probability density function is $f_X(\tau)$. The parameter σ^2 of a normal distribution is thus a measure of dispersion for the distribution.

History

The form of the normal distribution was discovered early in the history of probability theory. Originally studied by de Moivre, Laplace, and others because it arose as the answer to an important theoretical problem in the emerging theory of probability, the normal distribution soon became of interest to scientists in a variety of disciplines because it seemed to describe the variation from observation to observation of many important quantitative phenomena. One of the very early applications of the normal distribution occurred in astronomy, where Laplace and Gauss used it to describe errors of measurement in the observation of the motions of planetoids. Gauss' very persuasive theoretical arguments justifying the applicability of the normal distribution to errors of measurement, and the success of that distribution in describing the variability of measurement errors in many different contexts, led to widespread acceptance by the scientific community of the "Law of Errors" (which states that errors of measurement have a normal distribution). Although it has been recognized that this "law" has its limitations, and that not all errors of measurement have normal probability distributions, the normal distribution still plays a central role in the analysis of measurement errors in the experimental sciences. Because the German mathematician, astronomer, and physicist Gauss (1777-1855) played such a prominent role in demonstrating the usefulness of the normal distribution in error analysis, the normal distribution is also known as the *Gaussian* distribution.

Inspired by the success of the normal distribution in describing the variation of errors of measurement, scientists began applying this distribution not only to physical phenomena, but also to physiological and psychological measurements. Among the first scientists to apply the normal distribution to describe physiological and behavioral phenomena were Quetelet (Lambert Adolph Jacques Quetelet, 1796-1874), a Belgian statistician, and Sir Francis Galton (1822-1911), a British pioneer in many scientific fields. Quetelet was concerned with the development of standard methods of statistical data analysis (data collection, tabulation, and presentation). Using an analogy with the concept of the center of gravity

1. INTRODUCTION

in physical mechanics, Quetelet advocated the use of the concept of "the average man" (l'homme moyen) to summarize and describe the extensive statistics on anthropometric measurements which he had collected. Through a further physical analogy, this time with the "Law of Errors," he also advocated a general "Law of Accidental Causes" to justify modeling the variation of his measurements about the measurements of "the average man" by means of a normal distribution. Since "the average man" in Quetelet's theory represented the typical, or normal, individual, the adjective "normal" gradually began to be used to describe the distribution of variations about "the average man," thus giving the distribution defined by Equation (1.1) its present name.

Galton's work, apparently carried out independently, was very similar to that of Quetelet. He used the normal curve to describe and explain anthropometric measurements. His main results appear in *Natural Inheritance* (1889), and in *Finger Prints* (1892).

Both Quetelet and Galton had passionate disciples, and in the early 1900s there appeared numerous studies of anthropometric data in which the normal distribution was asserted to "fit" (or describe) the variation in the obtained data. Although we have not yet discussed how to "fit" a normal distribution to data (this is done in Section 3), at this point it may be of interest to mention some of these early studies, and note how well the observed results in these experiments agreed with the particular normal distribution "fit" to the data.

Example 1.1 (Study of Stature). In a study of the inheritance of physical characteristics in human beings, Pearson and Lee (1903) examined family data. As examples of their theoretical results, they compared the stature of father and son, the span of the arms of mother and daughter, and the length of the son's forearm and the daughter's arm span. They also went into considerable detail to try to justify their assertions that such physical measurements as stature, arm span, and length of forearm have normal distributions. In Figure 1.3, we show the distribution of the mother's stature in inches.

Example 1.2 (Cranial Studies). The detection of differences between racial characteristics was another problem that motivated many studies. In particular, various craniological measurements were taken. Table 1.1 and Figure 1.4 give two examples of such craniological measurements: one in which the greatest forehead breadth was measured, and one in which the greatest skull breadth was measured. Both measurements were made on 2000 male Hungarian skulls, and are given to the nearest millimeter.

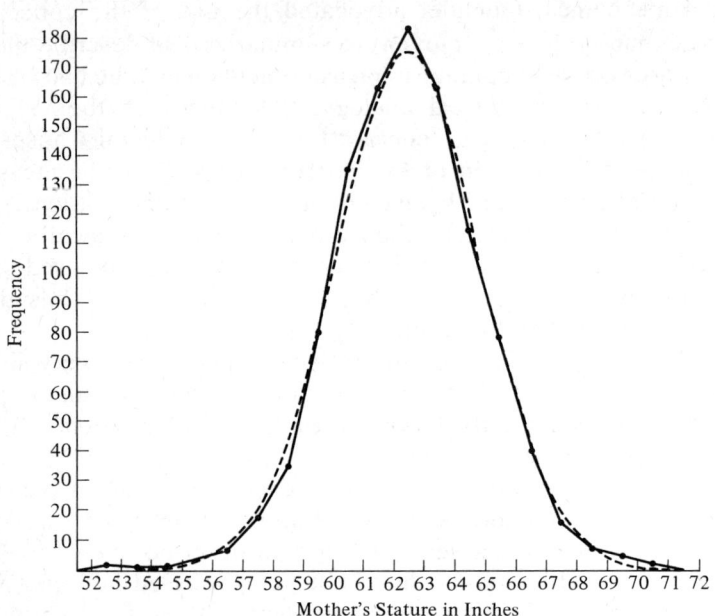

Stature in Mothers. 1052 Cases. Estimated $\mu = 62''.484$, estimated $\sigma^2 = 5''.7140$.

Stature in Inches	52–53	53–54	54–55	55–56	56–57	57–58	58–59	59–60	60–61	61–62	62–63	63–64	64–65	65–66	66–67	67–68	68–69	69–70	70–71
Observed Frequency	1.5	.5	1	2	6.5	18	34.5	79.5	135.5	163	183	163	114.5	78.5	41	16	7.5	4.5	2
Normal Frequency			.9		2.6	7.9	20.9	44.5	80.8	124.1	160.3	174.3	159.4	122.8	79.5	43.2	20.1	7.7	2.5	.8

Figure 1.3 Distribution of mother's stature.

SOURCE: Pearson, K., and Lee, A. (1903). On the laws of inheritance in man. I. Inheritance of physical characters. *Biometrika*, vol. 2, pp. 357–462. Reprinted by permission.

Table 1.1. Distribution of observed and theoretical frequencies of greatest forehead breadth and greatest skull breadth of 2000 Hungarian skulls.

mm.	Greatest Forehead Breadth Frequency		mm.	Greatest Skull Breadth Frequency	
	OBSERVED	THEORETICAL		OBSERVED	THEORETICAL
Below *103*	1	2.5	Below *127*	2	5.5
103	4	1.9	*127*	6	3.7
104	5	3.1	*128*	1	5.8
105	5	4.8	*129*	9	8.7

Table 1.1. *(continued)*

mm.	Greatest Forehead Breadth Frequency		mm.	Greatest Skull Breadth Frequency	
	OBSERVED	THEORETICAL		OBSERVED	THEORETICAL
106	3	7.4	*130*	13	12.9
107	14	11.1	*131*	20	18.3
108	13	16.0	*132*	28	25.5
109	21	22.7	*133*	36	34.3
110	33	30.9	*134*	36	45.0
111	50	40.9	*135*	58	57.2
112	43	52.7	*136*	60	70.6
113	49	65.8	*137*	88	84.7
114	89	79.9	*138*	108	98.7
115	99	94.2	*139*	118	111.6
116	108	107.7	*140*	119	122.6
117	124	119.5	*141*	143	130.8
118	114	128.9	*142*	138	135.5
119	151	134.9	*143*	127	136.0
120	148	137.1	*144*	140	133.5
121	141	135.2	*145*	128	126.4
122	130	129.6	*146*	118	116.5
123	122	120.6	*147*	115	104.3
124	109	108.8	*148*	84	90.6
125	80	95.5	*149*	78	76.5
126	81	81.2	*150*	52	62.7
127	60	67.2	*151*	46	49.9
128	45	53.9	*152*	29	38.6
129	46	42.0	*153*	30	29.0
130	33	31.8	*154*	23	21.1
131	26	23.3	*155*	19	15.0
132	17	16.6	*156*	8	10.3
133	19	11.5	*157*	5	6.6
134	4	7.8	*158*	3	4.7
135	7	5.1	*159*	5	2.8
136	1	3.2	*160*	4	1.7
137	2	2.0	*161*	0	1.1
138	0	1.2	*162*	1	0.6
139	1	0.7	*163*	1	0.3
above *139*	2	0.8	above *163*	1	0.4

SOURCE: (Table 1.1 and Figure 1.4) Pearson, K. (1903). Craniological notes. *Biometrika*, vol. 2, pp. 338–347. Reprinted by permission.

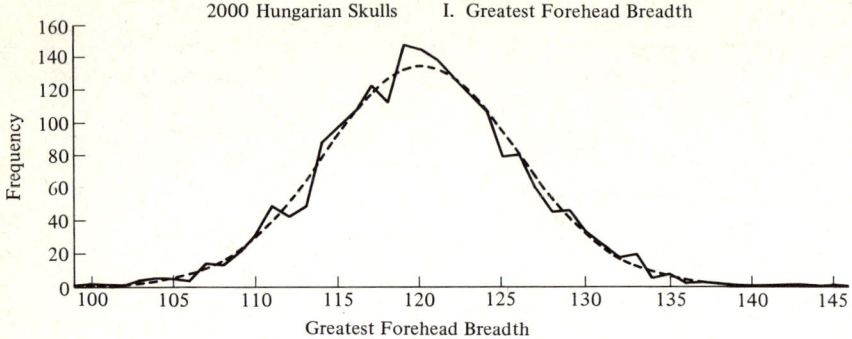

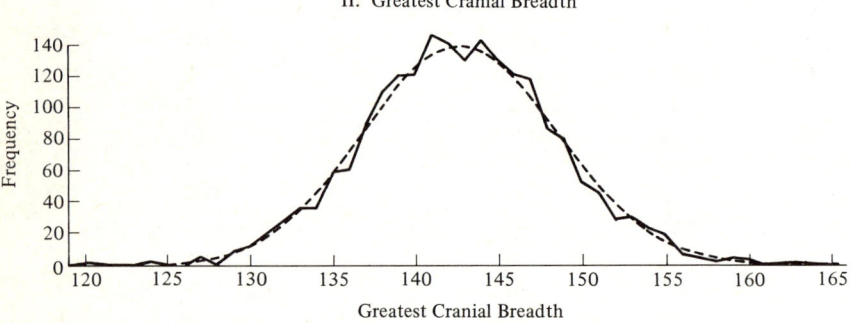

Figure 1.4 Distribution of greatest forehead breadth and of greatest cranial breadth.

Example 1.3 (Brain-Weight). In this study, variations in brain-weight with respect to sex, other skull characteristics, and national origin were examined. The data are one of a series, and are based on measurements taken on 416 Swedish males between the ages of 20 and 80. The weights are given in grams. See Table 1.2 and Figure 1.5.

Many other examples of these kinds of studies could be given: measurements on birds' eggs [Latter (1901)], anthropometric measurements taken on criminals and noncriminals to see if criminality could be predicted from physiological dimensions [Macdonell (1901)], weights of human viscera taken to see if healthy and diseased hearts could be distinguished by weight [Greenwood (1901)], and so on. Indeed, the early issues of the journal *Biometrika*, founded by Galton, Karl Pearson (1857–1936), and W. F. R. Weldon (1860–1906), are fascinating to read, both because of the interesting studies that were reported and because of the atmosphere of controversy and discovery that can be found in many of the articles.

1. INTRODUCTION

Table 1.2. Distribution of observed and theoretical frequencies of brain-weight of 416 adult Swedish males.

Grams of Brain-Weight	Observed	Theoretical
Under 1100	0	0.981
1100–1150	1	2.9
1150–1200	10	8.5
1200–1250	21	20.3
1250–1300	44	39.0
1300–1350	53	60.4
1350–1400	86	75.2
1400–1450	72	75.3
1450–1500	60	60.8
1500–1550	28	39.4
1550–1600	25	20.6
1600–1650	12	8.7
1650–1700	3	2.9
1700–1750	1	0.8
1750 and over	0	0.036
Totals	416	415.817

Estimated $\mu = 1400.481$, estimated $\sigma^2 = 4.5224$

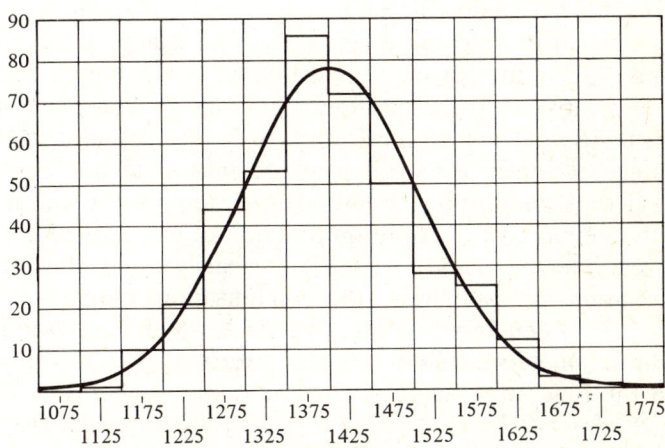

Figure 1.5 Distribution of brain-weight of adult Swedish Males.

SOURCE: (Table 1.2 and Figure 1.5) Pearl, R. (1905). Biometrical studies on man. I. Variation and correlation in brain-weight. *Biometrika*, vol. 4, pp. 13–104. Reprinted by permission.

2. PROBABILITY CALCULATIONS FOR THE NORMAL DISTRIBUTION

An educator is planning to make use of a certain reading comprehension test to classify students in the primary grades in his school district according to their reading readiness. The manual for the test asserts that the test score X of a randomly chosen student is normally distributed with parameters $\mu = 80$ and $\sigma^2 = 16$. During the course of his analysis of the scores of the students on the reading test, the educator knows that he is going to want to ask such questions as: "What is the probability of obtaining a score X no greater than a certain number a?" and "What is the probability that a student will obtain a score X between the numbers a and b?" The answers to such questions will allow him to compare the scores of his students on the reading test against national norms. Thus, the educator needs to be able to calculate probabilities of events of the form $\{X \leq a\}$, $\{a \leq X \leq b\}$, $\{X > b\}$, and so on, where X has a normal distribution with parameters μ and σ^2.

From Chapter 4, Section 3, we know that the educator only needs a table (or graph) of the values of the cumulative distribution function $F_X(\tau)$ of X for all (or selected) values of τ in order to perform the necessary probability calculations.

However, the task of tabulating the cumulative distribution functions for *all* possible normal distributions appears to be quite formidable. This is because the dependence of the form of a normal distribution upon the two parameters μ and σ^2 seems to necessitate a separate table for each pair (μ, σ^2) of values of these parameters. If such an infinite number of tables were required, the educator might find that no one had ever bothered to table the cumulative distribution function of a normal distribution having parameters $\mu = 80$ and $\sigma^2 = 16$, forcing him to do the tabulating task himself.

Fortunately, all normal distributions are related to one another in a simple way; thus, only one standard table is required. From this table, tables of the cumulative distribution function of any normal distribution (with any parameter values μ and σ^2) are easily obtained. Let us now study the relationship that permits this condensation of tables.

Let $\mathcal{N}(X; \mu, \sigma^2)$ denote a random variable X which has a normal distribution with parameters μ and σ^2. By the assertion "X is $\mathcal{N}(X; \mu, \sigma^2)$" we mean that the random variable X has a normal distribution with parameters μ and σ^2, while by $P\{\mathcal{N}(X; \mu, \sigma^2) \leq a\}$, for example, we represent the probability that a normally distributed random variable with parameters μ and σ^2 does not exceed the number a. Even though this notation does not distinguish a random variable from its distribution, we find it convenient, and use it in the present context with the hope that no misunderstanding results.

2. PROBABILITY CALCULATIONS FOR THE NORMAL DISTRIBUTION

In the example given above of the educator who is planning to analyze the scores X of his students on a reading comprehension test, we are given that X is $\mathscr{N}(X; 80,16)$. To evaluate the cumulative distribution function

$$F_X(\tau) = P_X\{X \leq \tau\} = P\{\mathscr{N}(X; 80,16) \leq \tau\}$$

for $\tau = a$, we can find the area of the shaded region shown in Figure 2.1. Because a normally distributed random variable is a continuous random variable, $P_X\{X < a\} = P_X\{X \leq a\}$, and thus the area of the shaded region in Figure 2.1 also equals $P_X\{X < a\} = P\{\mathscr{N}(X; 80,16) < a\}$.

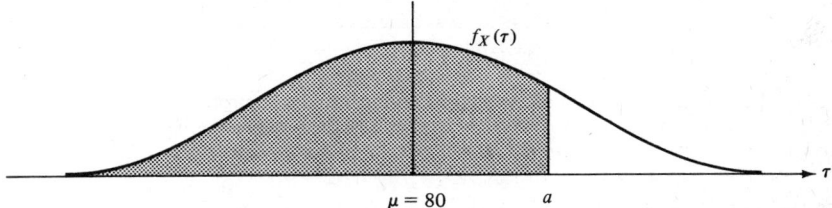

Figure 2.1 The area of the shaded region equals the probability that the score X on a reading comprehension test is no greater than a.

The area of the shaded region in Figure 2.1 cannot be given in terms of a simple formula, and hence a table of such areas (probabilities) is needed. As we have already remarked, this task would be tremendous if we had to form a separate table for each different pair of parameter values (μ,σ^2). Fortunately, the fundamental identity

$$(2.1) \qquad P\{\mathscr{N}(X; \mu,\sigma^2) \leq \tau\} = P\left\{\mathscr{N}(Z; 0,1) \leq \frac{\tau-\mu}{\sigma}\right\},$$

where $\sigma = \sqrt{\sigma^2}$, allows us to obtain the cumulative distribution function of any normally distributed random variable X from the cumulative distribution function $F_Z(\tau)$ of the *standard normal* random variable Z, where Z has a normal distribution with expected value 0 and variance 1. Pictorially, the identity (2.1) means that the area of the shaded region in Figure 2.1 equals the area of the shaded region shown in Figure 2.2.

To partially justify the identity (2.1), we note that the events $\{X \leq \tau\}$ and $\{(X-\mu)/\sigma \leq (\tau-\mu)/\sigma\}$ are the same event. Thus,

$$P_X\{X \leq \tau\} = P_X\left\{\frac{X-\mu}{\sigma} \leq \frac{\tau-\mu}{\sigma}\right\}.$$

Let us write $Z = (X-\mu)/\sigma$, in which case $P_Z\{Z \leq (\tau-\mu)/\sigma\} =$

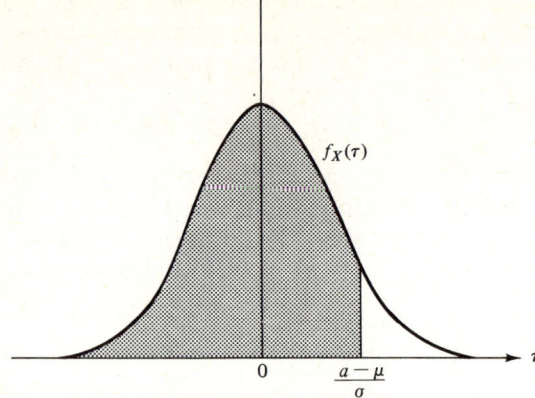

Figure 2.2 The area of the shaded region equals the probability that a standard normal random variable Z is no greater than $(a-\mu)/\sigma$ [that is, $P\{\mathcal{N}(Z, 0, 1) \leq (a-\mu)/\sigma\}$].

$P_X\{(X-\mu)/\sigma \leq (\tau-\mu)/\sigma\}$, so that

$$P_Z\left\{Z \leq \frac{\tau-\mu}{\sigma}\right\} = P_X\{X \leq \tau\}.$$

To verify Equation (2.1), we now only need to verify that if X is $\mathcal{N}(X; \mu, \sigma^2)$, then $Z = (X-\mu)/\sigma$ must be $\mathcal{N}(Z; 0,1)$. From Chapter 5, Equation (2.7), and because $E(X) = \mu$,

$$E(Z) = E\left(\frac{X-\mu}{\sigma}\right) = E\left(\frac{1}{\sigma}X - \frac{\mu}{\sigma}\right) = \frac{1}{\sigma}E(X) - \frac{\mu}{\sigma} = \frac{\mu}{\sigma} - \frac{\mu}{\sigma} = 0.$$

Thus, Z has expected value 0. Also, from Chapter 5, Equation (4.9), and because the variance of X is σ^2,

$$\text{Var}(Z) = \text{Var}\left(\frac{1}{\sigma}X - \frac{\mu}{\sigma}\right) = \left(\frac{1}{\sigma}\right)^2 \text{Var}(X) = 1.$$

Therefore, Z has variance equal to 1. Hence, we know that *if* Z has a normal distribution, *then* that normal distribution must have parameters 0 and 1 [that is, Z must be $\mathcal{N}(X; 0,1)$]. Verification of the assertion that $Z = (X-\mu)/\sigma$ is normally distributed follows from the fact that X is normally distributed, and is accomplished by using the theory of differential calculus.

Table T.6 gives values of the cumulative distribution function

$$F_Z(k) = P\{\mathcal{N}(Z; 0,1) \leq k\}$$

for values $k = 0(0.01)4$. Because the graph of the probability density function of the normal distribution with parameters $\mu = 0$ and $\sigma = 1$ is

2. PROBABILITY CALCULATIONS FOR THE NORMAL DISTRIBUTION

symmetric about the vertical axis (see Figure 2.2), the area under this graph to the left of $-k$ equals the area under the graph to the right of k; that is,

(2.2) $$P\{\mathcal{N}(Z;0,1) \leq -k\} = P\{\mathcal{N}(Z;0,1) \geq k\},$$

for all positive numbers k. Further, since normally distributed random variables are continuous variables,

$$P\{\mathcal{N}(Z;0,1) \geq k\} = P\{\mathcal{N}(Z;0,1) > k\}.$$

Therefore, it follows from the Law of Complementation (Rule 4.2 of Chapter 2) that $P\{\mathcal{N}(Z;0,1) > k\} = 1 - P\{\mathcal{N}(Z;0,1) \leq k\}$, and we conclude from Equation (2.2) that

(2.3) $$P\{\mathcal{N}(Z;0,1) \leq -k\} = 1 - P\{\mathcal{N}(Z;0,1) \leq k\},$$

for all positive numbers k. Hence, if we want to calculate $F_Z(-k) = P\{\mathcal{N}(Z;0,1) \leq -k\}$, where k is a positive number, we look up $F_Z(k) = P\{\mathcal{N}(Z;0,1) \leq k\}$ in Table T.6 and subtract this number from 1. For example, Table T.6 tells us that $F_Z(1.96) = P\{\mathcal{N}(Z;0,1) \leq 1.96\} = 0.975$. Thus, if we want $F_X(-1.96) = P\{\mathcal{N}(Z;0,1) \leq -1.96\}$, we subtract 0.975 from 1, yielding

$$F_Z(-1.96) = P\{\mathcal{N}(Z;0,1) \leq -1.96\} = 1 - 0.975 = 0.025.$$

From the identity (2.1) and Table T.6, we now have a way of calculating the cumulative density function. $F_X(a) = P\{\mathcal{N}(X;\mu,\sigma^2) \leq a\}$ for any values of μ, σ^2, and a. To obtain $P\{\mathcal{N}(X;\mu,\sigma^2) \leq a\}$, we proceed as follows:

(i) Compute $k = (a-\mu)/\sigma$.
(ii) If $k \geq 0$, look up the entry in Table T.6 corresponding to this value of k. The number that is obtained is the desired probability $P\{\mathcal{N}(X;\mu,\sigma) \leq a\}$.
(iii) If $k < 0$, look up the entry in Table T.6 corresponding to $-k$ and subtract this entry from 1. The number so obtained is the desired probability $P\{\mathcal{N}(X;\mu,\sigma^2) \leq a\}$.

For example, the educator mentioned at the beginning of this section has just given the reading comprehension test to two students. One student has made a score of 85 on the test and the other student has made a score of 70. The educator would like to know the quantile ranks of these two students relative to the national distribution of test scores; that is, he wants to find the probabilities of the events $\{X \leq 85\}$ and $\{X \leq 70\}$, where X is the score on the reading test of a randomly chosen student. Because the educator knows from the test manual that X has a normal distribution with expected value 80 and variance 16, he calculates

$$P_X\{X \leq 85\} = P\{\mathcal{N}(X; 80,16) \leq 85\}$$
$$= P\left\{\mathcal{N}(Z;0,1) \leq \frac{85-80}{\sqrt{16}}\right\}$$
$$= P\{\mathcal{N}(Z;0,1) \leq 1.25\} = 0.8944,$$

since from Table T.6, $P\{\mathcal{N}(Z; 0,1) \leq 1.25\} = F_Z(1.25) = 0.8944$. Similarly,

$$P_X\{X \leq 70\} = P\{\mathcal{N}(X; 80,16) \leq 70\}$$
$$= P\left\{\mathcal{N}(Z;0,1) \leq \frac{70-80}{\sqrt{16}}\right\}$$
$$= P\{\mathcal{N}(Z;0,1) \leq -2.5\}.$$

Table T.6 is entered with $k = 2.5$, and a value of $F_Z(2.5) = 0.9938$ is obtained. Thus,

$$P_X\{X \leq 70\} = P\{\mathcal{N}(Z;0,1) \leq -2.5\} = 1 - F_Z(2.5) = 1 - 0.9938$$
$$= 0.0062.$$

The educator concludes that the student with a score of 85 is at the (0.8944)th quantile of the national distribution of scores on the reading test (that is, his score is at least as high as 89.44 percent of the scores in the country), while the student with a score of 70 is at the (0.0062)th quantile.

Identity (2.1), Table T.6, and the steps of calculation illustrated above can be used to obtain the cumulative distribution functions

$$F_X(\tau) = P\{\mathcal{N}(X; \mu, \sigma^2) \leq \tau\}$$

for any values of the parameters μ and σ^2 that we wish to choose. In Figure 2.3, graphs of the cumulative distribution function $P\{\mathcal{N}(X; \mu, \sigma^2) \leq \tau\}$ are shown for several different values of the parameters μ and σ^2. Notice that the graphs of all of these functions have an S-shaped form. The graph of the cumulative distribution function of a normally distributed random variable is often called a *normal ogive*.

Tail Probabilities

Probabilities of the form $P_X\{X > b\}$ are often called *tail probabilities*. The reason for such a terminology can be seen in Figure 2.4, where the area of the shaded region equals $P_X\{X > b\} = P\{\mathcal{N}(X; 80,16) > b\}$ and the shaded region is in the "tail" of the graph of the density function of X. Since the area under the normal curve is unity, $P_X\{X > b\} = 1 - P_X\{X \leq b\} = F_X(b)$. Hence we can obtain tail probabilities for the

2. PROBABILITY CALCULATIONS FOR THE NORMAL DISTRIBUTION

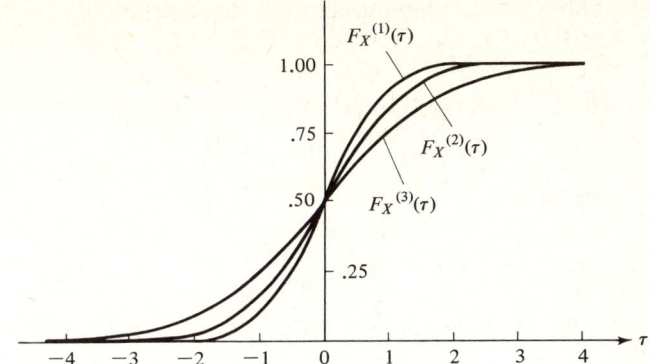

Figure 2.3 Graphs of the cumulative distribution functions of three normally distributed random variables. The cumulative distribution functions $F_X^{(1)}(\tau)$, $F_X^{(2)}(\tau)$, $F_X^{(3)}(\tau)$ correspond to normally distributed random variables with expected value $\mu = 0$ and variances $\sigma^2 = 0.5$, 1, and 2, respectively.

normal distribution by finding the value of the cumulative distribution function $F_X(b)$ by the methods illustrated above, and then subtracting $F_X(b)$ from 1.

Suppose the educator who is evaluating scores of students on a reading comprehension test finds that one of the students who took the test in his district has scored a 90 on the test. Since this is a high score, the educator would like to know the probability that this score would be exceeded by the score X of a randomly chosen student taken from the national population of students. Since X is $\mathcal{N}(X; 80, 16)$,

$$F_X(90) = P\{\mathcal{N}(X; 80,16) \leq 90\}$$
$$= P\{\mathcal{N}(Z; 0,1) \leq 2.5\} = 0.9938,$$

and thus

$$P_X\{X > 90\} = 1 - F_X(90) = 1 - 0.9938 = 0.0062.$$

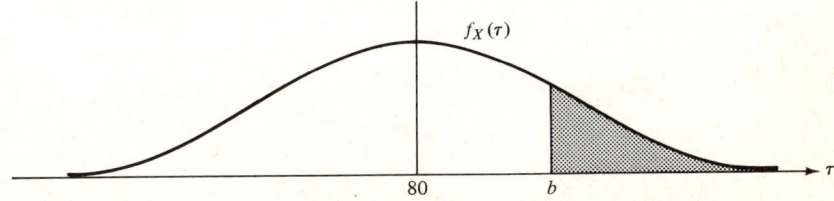

Figure 2.4 The area of the shaded region equals the probability that the score X on the reading test exceeds the number b.

Since the probability of achieving a score larger than 90 on the reading test is less than 0.01, the educator concludes that a score of 90 must be quite exceptional.

Although we can make a similar computation every time we want to compute a probability of the form $P\{\mathcal{N}(X; \mu,\sigma^2) > b\}$, it is often more convenient to provide a direct method. From Identity (2.1) and the Law of Complementation,

(2.4) $$P\{\mathcal{N}(X; \mu,\sigma^2) > b\} = P\left\{\mathcal{N}(Z; 0,1) > \frac{b-\mu}{\sigma}\right\},$$

which suggests the need for a table of $P\{\mathcal{N}(Z; 0,1) > k\}$. Table T.7 gives values of $P\{\mathcal{N}(Z; 0,1) > k\}$ for $k = 0(0.01)4$. For reasons already noted, we do not need to table $P\{\mathcal{N}(Z; 0,1) > k\}$ for negative values of k since

(2.5) $$P\{\mathcal{N}(Z; 0,1) > -\tau\} = 1 - P\{\mathcal{N}(Z; 0,1) > \tau\}$$

for all numbers τ.

To obtain the value of $P\{\mathcal{N}(X; \mu,\sigma^2) > b\}$, we thus proceed as follows:

(i) Compute $k = (b-\mu)/\sigma$, where $\sigma = \sqrt{\sigma^2}$.
(ii) If $k \geq 0$, look up the entry in Table T.7 corresponding to that value of k. This entry gives the desired value of $P\{\mathcal{N}(X; \mu,\sigma^2) > b\}$.
(iii) If $k < 0$, look up the entry in Table T.7 corresponding to $-k$ and subtract this entry from 1. The resulting number gives the desired value of $P\{\mathcal{N}(X; \mu,\sigma^2) > b\}$.

Using Table T.7, the educator could have directly determined the probability that the score X of a randomly chosen student would exceed 90; that is,

$$P\{\mathcal{N}(X; 80,16) > 90\} = P\left\{\mathcal{N}(Z; 0,1) > \frac{90-80}{4} = 2.5\right\}$$
$$= 0.0062.$$

If he found that another of his students had scored 75 on the test, he could find the probability of the event $\{X > 75\}$ as follows:

$$P_X\{X > 75\} = P\{\mathcal{N}(X; 80,16) > 75\}$$
$$= P\left\{\mathcal{N}(Z; 0,1) > \frac{75-80}{4} = -1.25\right\}$$
$$= 1 - P\{\mathcal{N}(Z; 0,1) > 1.25\},$$

and since from Table T.7, $P\{\mathcal{N}(Z; 0,1) > 1.25\} = 0.1056$,

$$P_X\{X > 75\} = 1 - 0.1056 = 0.8944.$$

2. PROBABILITY CALCULATIONS FOR THE NORMAL DISTRIBUTION

[*Note:* If $\tau = -k$, where k is a positive number, then because $1 - P\{\mathcal{N}(Z;0,1) > \tau\} = P\{\mathcal{N}(Z;0,1) \leq \tau\}$, it follows from (2.5) that Table T.7 can be used to compute the value of the cumulative distribution function $F_Z(-k)$ of a standard normal random variable Z.]

Probabilities of Intervals of Numbers

One other type of probability calculation that is of interest concerns probabilities of events of the form: $\{a \leq X \leq b\}, \{a < X \leq b\}, \{a \leq X < b\}$, or $\{a < X < b\}$. When X is $\mathcal{N}(X; \mu, \sigma^2)$, all of these events have the same probability because X is a continuous random variable. That is,

(2.6) $\quad P\{a \leq \mathcal{N}(X; \mu, \sigma^2) \leq b\} = P\{a < \mathcal{N}(X; \mu, \sigma^2) \leq b\}$
$$= P\{a \leq \mathcal{N}(X; \mu, \sigma^2) < b\}$$
$$= P\{a < \mathcal{N}(X; \mu, \sigma^2) < b\}.$$

Recalling from Chapter 4, Table 3.2, that

(2.7) $\quad P_X\{a < X \leq b\} = F_X(b) - F_X(a),$

it follows that each of the probabilities shown in Equation (2.6) can be computed as a difference between two entries obtained from Table T.6. Thus, for example,

(2.8) $\quad P\{a \leq \mathcal{N}(X; \mu, \sigma^2) \leq b\}$
$$= P\{\mathcal{N}(X; \mu, \sigma^2) \leq b\} - P\{\mathcal{N}(X; \mu, \sigma^2) \leq a\}$$
$$= P\left\{\mathcal{N}(Z; 0,1) \leq \frac{b-\mu}{\sigma}\right\} - P\left\{\mathcal{N}(Z; 0,1) \leq \frac{a-\mu}{\sigma}\right\}$$
$$= F_Z\left(\frac{b-\mu}{\sigma}\right) - F_Z\left(\frac{a-\mu}{\sigma}\right).$$

When X is $\mathcal{N}(X; 80, 16)$, $a = 75$, $b = 90$, we obtain from Equation (2.8) that

$$P_X\{75 \leq X \leq 90\} = P\{75 \leq \mathcal{N}(X; 80, 16) \leq 90\}$$
$$= F_Z\left(\frac{90-80}{4}\right) - F_Z\left(\frac{75-80}{4}\right)$$
$$= F_Z(2.5) - F_Z(-1.25)$$
$$= 0.9938 - 0.1056 = 0.8882.$$

Similarly,
$$P_X\{80 < X \leq 90\} = P\{80 < \mathcal{N}(X; 80, 16) \leq 90\}$$
$$= F_Z\left(\frac{90-80}{4}\right) - F_Z\left(\frac{80-80}{4}\right)$$
$$= F_Z(2.5) - F_Z(0)$$
$$= 0.9938 - 0.5000 = 0.4938.$$

A related computation for normally distributed random variables which is often of interest concerns the probability that the random variable X falls within plus or minus k standard deviations of the expected value μ_X. Suppose that X is $\mathcal{N}(X;\mu,\sigma^2)$, so that the expected value of X is μ and the variance of X is σ^2. The probability that we are interested in is

(2.9) $\quad P_X\{-k\sigma \leq X-\mu \leq k\sigma\} = P_X\{\mu-k\sigma \leq X \leq \mu+k\sigma\}$
$$= P\{\mu-k\sigma \leq \mathcal{N}(X;\mu,\sigma^2) \leq \mu+k\sigma\}$$
$$= P\{-k \leq \mathcal{N}(Z;0,1) \leq k\}$$
$$= F_Z(k) - F_Z(-k).$$

From Equation (2.3) we know that $F_Z(-k) = 1 - F_Z(k)$. Hence,

(2.10) $\quad P_X\{-k\sigma \leq X-\mu \leq k\sigma\} = F_Z(k) - [1 - F_Z(k)]$
$$= 2F_Z(k) - 1.$$

For $k = 0(0.5)3$ we obtain, from (2.10) and Table T.6, a table of values for $P_X\{-k\sigma \leq X-\mu \leq k\sigma\}$, as given in Table 2.1.

Table 2.1. Probability that a normally distributed random variable is within $\pm k$ standard deviations of its expected value.

k	0	0.5	1	1.5	2	2.5	3
$P_X\{-k\sigma \leq X-\mu \leq k\sigma\}$	0.0000	0.3829	0.6827	0.8664	0.9545	0.9876	0.9973

These results can also be interpreted pictorially (Figure 2.5). It is of interest to compare these probabilities with the lower limits for such probabilities which are valid for random variables X which may have any distribution whatsoever (see Chapter 5, Section 6, the Bienaymé–Chebychev Inequality).

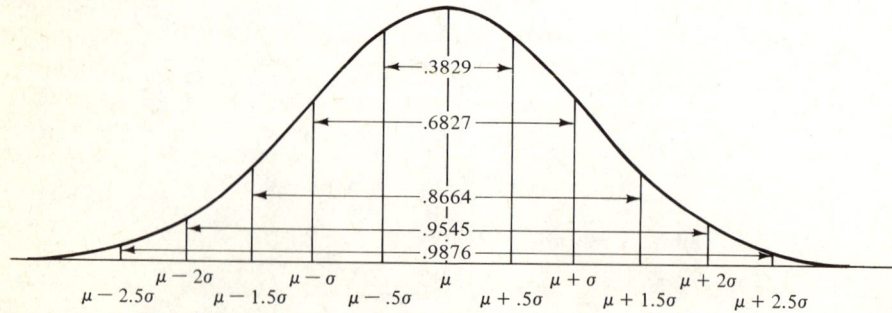

Figure 2.5 The areas indicated by the arrows are equal to the probability that a normally distributed random variable is with $\pm k$ standard deviations σ of its expected value.

As an example of the use of the calculations in Table 2.1, consider once again the educator who is giving a reading comprehension test to the students in his school district. Suppose there are 1000 students in his district who will take the test. Since the score X on this test has a normal distribution with expected value $\mu = 80$ and variance $\sigma^2 = 16$, it follows from Table 2.1 that among the 1000 students tested the educator can expect $1000\,P\{76 \leq \mathcal{N}(X; 80, 16) \leq 84\} = 1000(0.6826)$, or approximately 683 students to receive a score on the test between 76 and 84 [that is, between plus or minus 1 standard deviation from the expected value $E(X) = 80$]. Similarly, the educator can expect approximately 954 students to receive scores between 72 and 88 (that is, plus or minus 2 standard deviations from the expected value), and approximately 997 students to have scores between 68 and 92. Such calculations give the educator some feeling for what he can expect if the distribution of the levels of reading comprehension for students in his school district is similar to the national distribution of levels of reading comprehension.

3. FITTING A NORMAL DISTRIBUTION

We now consider a classical example of the applicability of the normal distribution: L. M. Terman's study of the intelligence quotients (I.Q.) of school children. Using this example, we discuss how to fit a theoretical normal distribution.

Example 3.1 (Intelligence Quotients). In his well-known book, *The Intelligence of School Children*, L. M. Terman describes several studies on school children which utilized the Stanford Revision of the Binet-Simon Intelligence Scale. In one of these studies, 112 children (65 boys and 47 girls) attending 5 kindergarten classes in San Jose and San Mateo, California were measured on this intelligence scale. In Table 3.1, the original ungrouped data are given. In Table 3.2, we have grouped the original data to form a frequency distribution preparatory to giving the frequency distribution in graphical form as a histogram (see Chapter 4, Section 3) in Figure 3.1.

It appears from the histogram in Figure 3.1 that a normal distribution might serve as an approximation to the actual probability distribution of I.Q.'s. However, this suggestion may seem unrealistic since we earlier stated that a normally distributed random variable has all real numbers as possible values, whereas in Figure 3.1 the distribution of I.Q. appears to assign nonzero probability only to values ranging from a low of 60 to a high of 160. Nevertheless, this discrepancy is not serious. The assumption that a random variable has a normal distribution is a mathematical model

Table 3.1. Original I.Q. data in the Terman study of 112 children attending kindergarten classes in San Jose and San Mateo, California.

152	120	111	106	98	88
146	119	111	106	97	86
142	119	110	105	97	86
136	118	110	103	96	85
130	117	110	103	96	85
130	117	110	103	96	85
129	114	109	102	96	84
126	114	109	102	94	82
126	114	109	102	94	81
125	114	109	102	93	80
124	114	109	102	93	80
124	114	108	101	93	80
123	113	108	101	92	80
122	113	107	100	91	79
121	113	107	100	91	77
121	113	107	100	90	76
121	112	107	99	90	75
121	112	106	98	90	72
121			98	90	61

Table 3.2. Observed frequency distribution of I.Q. of 112 children in Terman study.

Interval of I.Q. Values	Midpoint of I.Q. Intervals	Frequency
60–69	65	1
70–79	75	5
80–89	85	13
90–99	95	22
100–109	105	28
110–119	115	23
120–129	125	14
130–139	135	3
140–149	145	2
150–159	155	1
		112

SOURCE: Tables 3.1 and 3.2 from Terman, L. M., *The Intelligence of School Children.* Houghton Mifflin Company, 1919. Reprinted by permission of the publisher.

3. FITTING A NORMAL DISTRIBUTION

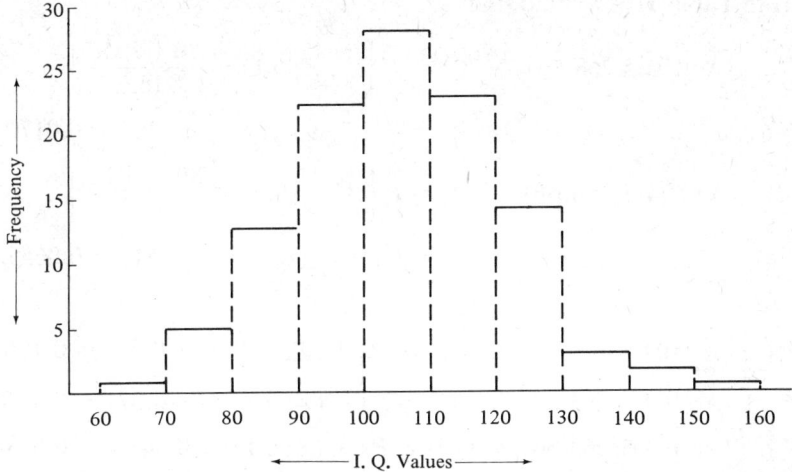

Figure 3.1 Histogram corresponding to grouped data (Table 3.2) in Terman study.

of a real phenomenon. The model is useful if it gives a sufficiently close approximation to "reality." In the present example, a normal distribution provides a close approximation to the actual variation observed in the I.Q.'s of children, if I.Q.'s which are less than 60 and greater than 160 are very rarely observed.

As a first step in determining a normal distribution to "fit" the observed data, we need to determine values for the parameters μ and σ^2 (since μ and σ^2 determine the normal distribution). From the original data in Table 3.1, we find that (see Chapter 5, Sections 2 and 4)

$$\hat{\mu} = \text{estimate of } \mu = \text{average of observed values} = 104.5,$$

$$\hat{\sigma}^2 = \text{estimate of } \sigma^2 = \frac{1}{112}[(152-\hat{\mu})^2 + (146-\hat{\mu})^2 + \cdots$$
$$+ (72-\hat{\mu})^2 + (61-\hat{\mu})^2] = 263.66,$$

and

$$\hat{\sigma} = \sqrt{\hat{\sigma}^2} = \sqrt{263.66} = 16.2.$$

Thus, we wish to fit a theoretical normal distribution with $\mu = 104.4$ and $\sigma^2 = 263.66$. If we were to use the grouped data shown in Table 3.2 to calculate $\hat{\mu}$ and $\hat{\sigma}^2$, the values we would obtain would differ slightly from the values 104.5 and 263.66, respectively, that we have obtained from the ungrouped data in Table 3.1.

From Table T.6, we find that

$$P\{\mathcal{N}(X; 104.5, 263.66) \leq 70\} = P\left\{\mathcal{N}(Z; 0,1) \leq \frac{70 - 104.5}{16.2}\right\}$$

$$= P\{\mathcal{N}(Z; 0,1) < -2.12\} = 0.0170,$$

$$P\{\mathcal{N}(X; 104.5, 263.66) \leq 80\} = P\left\{\mathcal{N}(Z; 0,1) \leq \frac{80 - 104.5}{16.2}\right\}$$

$$= P\{\mathcal{N}(Z; 0,1) \leq -1.51\} = 0.0655,$$

and also

$$P\{\mathcal{N}(X; 104.5, 263.66) \leq 90\} = P\{\mathcal{N}(Z; 0,1) \leq -0.89\} = 0.1867,$$
$$P\{\mathcal{N}(X; 104.5, 263.66) \leq 100\} = P\{\mathcal{N}(Z; 0,1) \leq -0.27\} = 0.3936,$$
$$P\{\mathcal{N}(X; 104.5, 263.66) \leq 110\} = P\{\mathcal{N}(Z; 0,1) \leq 0.34\} = 0.6331,$$
$$P\{\mathcal{N}(X; 104.5, 263.66) \leq 120\} = P\{\mathcal{N}(Z; 0,1) \leq 0.96\} = 0.8315,$$
$$P\{\mathcal{N}(X; 104.5, 263.66) \leq 130\} = P\{\mathcal{N}(Z; 0,1) \leq 1.57\} = 0.9418,$$
$$P\{\mathcal{N}(X; 104.5, 263.66) \leq 140\} = P\{\mathcal{N}(Z; 0,1) \leq 2.19\} = 0.9857,$$
$$P\{\mathcal{N}(X; 104.5, 263.66) \leq 150\} = P\{\mathcal{N}(Z; 0,1) \leq 2.80\} = 0.9974,$$
$$P\{\mathcal{N}(X; 104.5, 263.66) \leq 160\} = P\{\mathcal{N}(Z; 0,1) \leq 3.42\} = 0.9997.$$

Table 3.3. Comparison of observed and theoretical frequency distribution for the I.Q.'s of 112 children in the Terman study.

Interval of I.Q. Scores	Observed Frequency of Interval of I.Q. Scores	Theoretical Frequency from Normal Distribution with Expected Value 104.5 and Variance 263.66
60–69	1	1.9
70–79	5	5.4
80–89	13	13.6
90–99	22	23.2
100–109	28	26.8
110–119	23	22.2
120–129	14	12.4
130–139	3	4.9
140–149	2	1.3
150–159	1	0.3
	112	112.0

Since there are 112 children, the expected number of children with scores under 70 is $(112)(0.0170) = 1.9$. The expected number of children with scores under 80 is $(112)(0.0655) = 7.3$, so that the expected number of children between 70 and 80 is $7.3 - 1.9 = 5.4$. Continuing in this way, we obtain Table 3.3. The normal distribution with parameters $\mu = 104.5$ and $\sigma^2 = 263.66$ appears to provide a reasonably close approximation to the distribution of I.Q. scores.

4. THE CENTRAL LIMIT THEOREM

We have seen that the normal distribution can be used to describe the variation of many random variables which are directly of interest in scientific and practical investigations. However, the normal distribution is also useful in other ways. In the present section, we describe one way, not already mentioned, in which the normal distribution is used in experimental investigations. The theoretical basis for the usefulness of the normal distribution in such applications, the so-called *Central Limit Theorem*, has also been used by Gauss and others to explain why the normal distribution so frequently describes the variation from observation to observation of quantitative phenomena.

In Chapter 5, it was suggested that if we are interested in estimating the expected value μ_X of a random variable X, and if we have n independent observations on X (that is, n independent repetitions of a random experiment $\mathscr{E}$ which, in its most simplified form, can be described as the experiment in which X is observed), then we may estimate μ_X by means of the average $\hat{\mu}_X$ of the n observations made on X. Because each observation on X results from a trial of the random experiment $\mathscr{E}$, the observations $X_1, X_2, \ldots, X_n$ made in the n trials of experiment $\mathscr{E}$ are all random variables. Hence, the average

(4.1) $$\hat{\mu}_X = \frac{1}{n}(X_1 + X_2 + \cdots + X_n)$$

of the random variables $X_1, X_2, \ldots, X_n$ is itself a random variable. If we use $\hat{\mu}_X$ to estimate the expected value μ_X of X, we cannot expect $\hat{\mu}_X$ always to be exactly equal to μ_X. Every time we repeat the *composite* random experiment (see Chapter 3) which consists of making n independent trials of the random experiment $\mathscr{E}$, we observe a new value of the average $\hat{\mu}_X$. We can measure how closely $\hat{\mu}_X$ varies about μ_X by calculating the difference $\hat{\mu}_X - \mu_X$ between these quantities. An implication of the Central Limit Theorem is that *when the number n of statistically independent observations $X_1, X_2, \ldots, X_n$ is sufficiently large, we can calculate the probabilities of events such as* $\{\hat{\mu}_X - \mu_X \leq a\}$, $\{a \leq \hat{\mu}_X - \mu_X \leq b\}$, *and so on*, under the assumption that the random variable $\hat{\mu}_X - \mu_X$ has a normal distribution with expected value 0 and variance σ_X^2/n. *The resulting probabilities are then approximately equal to the true probabilities of these events*, as calculated using the exact distribution of the random variable $\hat{\mu}_X$. Thus, speaking loosely, for sufficiently large values of n, the sample average $\hat{\mu}_X = (X_1 + \cdots + X_n)/n$ is approximately normally distributed. (A more detailed discussion of the Central Limit Theorem may be found in Chapter 10, Section 4.)

As an example of the application of this consequence of the Central Limit Theorem, suppose we wish to know the probability that $\hat{\mu}_X$ differs from μ_X by no more than $\pm d$ units. That is, suppose we want to know the probability of the event $\{-d \leq \hat{\mu}_X - \mu_X \leq d\}$. Then, from the Central Limit Theorem,

(4.2) $\quad P_{\hat{\mu}_X}\{-d \leq \hat{\mu}_X - \mu_X \leq d\} \cong P\{-d \leq \mathcal{N}(\hat{\mu}_X - \mu_X; 0, \sigma_X^2/n) \leq d\},$

where "$\cong$" means "approximately equal to." However, letting $d = k\sqrt{\sigma_X^2/n}$, we have, from Equations (2.9) and (2.10), that

$$P\{-d \leq \mathcal{N}(\hat{\mu}_X - \mu_X; 0, \sigma_X^2/n) \leq d\} = P\{-k \leq \mathcal{N}(Z; 0,1) \leq k\}$$
(4.3) $\hspace{4cm} = 2F_Z(k) - 1,$

where $F_Z(k)$ is the cumulative distribution function of a standard normal random variable [that is, $F_Z(k) = P\{\mathcal{N}(Z; 0,1) \leq k\}$] and is tabled in Table T.6. Hence, from Equations (4.2) and (4.3), and noting that $k = d/\sqrt{\sigma_X^2/n} = \sqrt{n}d/\sqrt{\sigma_X^2} = \sqrt{n}d/\sigma_X$, we find that

(4.4) $\quad P_{\hat{\mu}_X}\{-d \leq \hat{\mu}_X - \mu_X \leq d\} \cong 2F_Z\left(\dfrac{\sqrt{n}d}{\sigma_X}\right) - 1.$

Example 4.1. An astronomer is interested in measuring the distance D in light years between a certain star and his observatory. He has a measuring device that will provide a measurement X of this distance, but even if the star remains always at the same distance D from his observatory, the astronomer knows that repeated trials of the experiment which consists of measuring the desired distance will yield observations (measurements) $X_1, X_2, \ldots$ which vary (due to atmospheric disturbances, minute variations in operating characteristics of the measuring instrument, and so on). He assumes that the measurements $X_1, X_2, \ldots$ all are random variables having the same distribution, that the common expected value of $X_1, X_2, \ldots$ is $\mu_X = D$, and that the measurements are statistically independent of one another. He also has enough knowledge of the characteristics of his measuring instrument to know how widely the common distribution of $X_1, X_2, \ldots$ is dispersed around the common expected value D; that is, he knows the common variance σ_X^2 of his measurements $X_1, X_2, \ldots$. For the sake of discussion, assume that $\sigma_X^2 = 4$. The astronomer decides to take 100 measurements, $X_1, X_2, \ldots, X_{100}$, and decides to estimate the true distance D by the sample average $\hat{\mu}_X = (X_1 + X_2 + \cdots + X_{100})/100$ of his measurements. He would like his estimate of D to be accurate to within ± 0.25 light years. Because he questions whether the "Law of Errors" (see Section 1) applies to the errors of measurement in his experiment, the astronomer is unwilling to assume that each of his observations $X_1, X_2, \ldots, X_{100}$ has a normal distribution. He hopes, however, that his sample size of $n = 100$ is "sufficiently large" to allow the Central Limit Theorem to provide him with a good approximation to the probability of the event $\{-0.25 \leq \hat{\mu}_X - \mu_X \leq 0.25\}$ that his estimate of the distance has the desired accuracy. Using Equation (4.4), he finds that

$$P_{\hat{\mu}_X}\{-0.25 \leq \hat{\mu}_X - D \leq 0.25\} \cong 2F_Z\left(\frac{(\sqrt{100}\,(0.25))}{\sqrt{4}}\right) - 1$$

$$= 2F_Z(1.25) - 1 = 2(0.8944) - 1$$

$$= 0.7888.$$

Because he would like a higher probability of achieving the accuracy of estimation that he desires, the astronomer casts around for a way to improve the accuracy of his estimation technique. One way, of course, is to construct a better instrument for measuring distances. However, the astronomer remembers that from the frequency interpretation of probability, it follows that the sample average and the expected value $\mu_X = D$ agree more and more closely as the number n of trials is increased (see Chapter 5, Section 2). Thus, he decides to try taking $n = 400$ observations. Again applying Equation (4.4) he finds that

$$P_{\hat{\mu}_X}\{-0.25 \leq \hat{\mu}_X - D \leq 0.25\} \cong 2F_Z\left(\frac{(\sqrt{400})\,(0.25)}{\sqrt{4}}\right) - 1$$

$$= 2F_Z(2.5) - 1 = 0.9876.$$

Although this is a more satisfactory probability of achieving the desired accuracy than the probability 0.7888, the astronomer must balance this improvement against the cost of making an additional 300 measurements.

It should be emphasized that *the Central Limit Theorem makes no assumptions about the common distribution of the independent random variables* $X_1, X_2, \ldots, X_n$. As a result, the Central Limit Theorem can only state (speaking loosely again) that for n large enough, $\hat{\mu}_X - \mu_X$ has *approximately* a normal distribution. On the other hand, if it is known that the common distribution of $X_1, X_2, \ldots, X_n$ *is* a normal distribution, then we do not need the Central Limit Theorem, for in this case it can be shown that the distribution of $\hat{\mu}_X - \mu_X$ is *exactly* a normal distribution with expected value equal to 0 and variance equal to σ_X^2/n. Thus, in Example 4.1, if the astronomer is willing to assert that his measurements are each normally distributed with expected value $\mu_X = D$ and variance $\sigma_X^2 = 4$, then he knows that in Equation (4.4) he can replace the symbol "≅" by an equality sign (that is, "=") *regardless of the number n of observations* $X_1, X_2, \ldots, X_n$ *which he has taken.*

In Example 4.1, the astronomer was not sure what (common) distribution his observations had. His observations could have been normally distributed (in which case, his probability calculations give exact answers), or they might have had some other distribution (in which case, his probability calculations give approximate answers). In the following example, we describe a situation in which the investigator is certain his observations are *not* normally distributed.

Example 4.2. An actuary is attempting to fit the Poisson distribution (Chapter 6, Section 3) to accident data. He has $n = 900$ observations $X_1, X_2, \ldots, X_{900}$, where each X_i is an independent observation of the number X of accidents in a given time period. To estimate the parameter λ of the Poisson distribution, he takes the advice given in Chapter 6, Section 3, and estimates $\lambda = \mu_X$ by $\hat{\mu}_X = (X_1 + X_2 + \cdots + X_{900})/900$. Since he wants his estimate of λ not to be in error by more than ± 0.1, he wants to know the probability of the event $\{-d \leq \hat{\mu}_X - \mu_X \leq d\} = \{-d \leq \hat{\mu}_X - \lambda \leq d\}$, where $d = 0.1$. Because the Central Limit Theorem holds no matter what distribution the random variables $X_1, X_2, \ldots, X_{900}$ follow (just so long as the number n of such variables is large; here, $n = 900$), the fact that $X_1, X_2, \ldots, X_{900}$ all have a Poisson distribution does not prevent the actuary from applying Equation (4.4). Applying this equation, he finds that

$$P_{\hat{\mu}_X}\{-0.1 \leq \hat{\mu}_X - \lambda \leq 0.1\} \cong 2F_Z\left(\frac{\sqrt{900}(0.1)}{\sqrt{\lambda}}\right) - 1,$$

$$= 2F_Z\left(\frac{3}{\sqrt{\lambda}}\right) - 1,$$

where we recall that the variance σ_X^2 of a Poisson distribution equals λ (and thus $\sigma_X = \sqrt{\sigma_X^2} = \sqrt{\lambda}$). The actuary notes that the probability he desires depends on the true value of the parameter λ. Since he has a general idea of how large λ should be (he believes that λ is between 1 and 9), the actuary tries a few values of λ. For $\lambda = 1$, he finds from Table T.6 that

$$P_{\hat{\mu}_X}\{-0.1 \leq \hat{\mu}_X - \lambda \leq 0.1\} \cong 2F_Z\left(\frac{3}{\sqrt{1}}\right) - 1$$

$$= 0.9973.$$

4. THE CENTRAL LIMIT THEOREM

For $\lambda = 4$, he finds that

$$P_{\hat{\mu}_X}\{0.1 \leq \hat{\mu}_X - \lambda \leq 0.1\} \cong 2F_Z\left(\frac{3}{\sqrt{4}}\right) - 1 = 0.8664,$$

while for $\lambda = 9$,

$$P_{\hat{\mu}_X}\{-0.1 \leq \hat{\mu}_X - \lambda \leq 0.1\} \cong 2F_Z\left(\frac{3}{\sqrt{9}}\right) - 1 = 0.6826.$$

This actuary is surprised that he cannot get better accuracy with 900 observations. However, he notes that the probabilities $P_{\hat{\mu}_X}\{-d \leq \hat{\mu}_X - \lambda \leq d\}$ decrease with increasing λ, and that when $d = 0.2$, $\lambda = 9$, Equation (4.4) yields him the approximation

$$P_{\hat{\mu}_X}\{-0.2 \leq \hat{\mu}_X - \lambda \leq 0.2\} \cong 2F_Z\left(\frac{\sqrt{900}(0.2)}{\sqrt{9}}\right) - 1$$

$$= 2F_Z(2) - 1 = 0.9545.$$

Thus, the actuary decides that with high probability he can trust his estimate of λ to be accurate to at least within an error of ± 0.2.

*A Justification of the Normal Distribution by the Central Limit Theorem

We can invoke the Central Limit Theorem to explain why so many quantitative phenomena have (at least approximately) a normal distribution. Suppose, for example, that we are interested in the height Y of an individual. We can think of this height as being the result of the sum of individual growth spurts, each spurt taking place over some short interval of time. For purposes of discussion, we can conceive of each growth spurt as being a random variable X_i with the same distribution as any other growth spurt X_j. This assumption may not appear to be very realistic when we consider that different influences affect growth at different stages of growth (in time). However, over very short time spans (say, minutes or seconds), this assumption is more nearly realistic (in any case, a certain amount of difference in the distributions of the growth spurts does not affect our final conclusions; see Chapter 10, Section 4). Also, if the time spans are very short, it is reasonable to assume that the growth spurts at different times are statistically independent, and that Y is the sum of a great many growth spurts. Let the growth spurts at the various times be represented by $X_1, X_2, \ldots, X_n$. The random growth spurts $X_1, X_2, \ldots, X_n$ are then, by our assumptions, statistically independent and identically distributed random variables with a common expected value μ_X and a common variance σ_X^2. Also, $Y = X_1 + X_2 + \cdots + X_n$, and n is very large. We have mentioned that it is a consequence of the Central Limit Theorem that in such circumstances, probabilities of events of the form

$\{a \leq [(1/n)Y - \mu_X] \leq b\}$, $\{[(1/n)Y - \mu_X] \leq a\}$, and so on, can be approximated by corresponding probabilities for the normal distribution with expected value 0 and variance σ_X^2/n. Thus, for any number b,

$$(4.5) \qquad P_Y\{Y \leq b\} = P_Y\left\{\frac{Y}{n} - \mu_X \leq \frac{b}{n} - \mu_X\right\}$$

$$\cong P\left\{\mathcal{N}\left(W; 0, \frac{\sigma_X^2}{n}\right) \leq \frac{b}{n} - \mu_X\right\},$$

where W represents a random variable which has a normal distribution with expected value 0 and variance σ_X^2/n. From Equation (2.1),

$$(4.6) \qquad P\left\{\mathcal{N}\left(W; 0, \frac{\sigma_X^2}{n}\right) \leq \frac{b}{n} - \mu_X\right\} = P\left\{\mathcal{N}(Z; 0,1) < \frac{b/n - \mu_X}{\sqrt{\sigma_X^2/n}}\right\}$$

$$= P\left\{\mathcal{N}(Z; 0,1) \leq \frac{b - n\mu_X}{\sqrt{n\sigma_X^2}}\right\}.$$

Let U be a random variable having a normal distribution with expected value $n\mu_X$ and variance $n\sigma_X^2$. From Equation (2.1),

$$P_U\{U \leq b\} = P\{\mathcal{N}(U; n\mu_X, n\sigma_X^2) \leq b\} = P\left\{\mathcal{N}(Z; 0,1) \leq \frac{b - n\mu_X}{\sqrt{n\sigma_X^2}}\right\};$$

combining this result with Equations (4.5) and (4.6) yields

$$(4.7) \qquad P_Y\{Y \leq b\} \cong P_U\{U \leq b\} = P\{\mathcal{N}(U; n\mu_X, n\sigma_X^2) \leq b\}.$$

Thus, the cumulative distribution function of Y can be approximated by the cumulative distribution function of a normally distributed random variable U with expected value $n\mu_X$ and variance $n\sigma_X^2$. Since the cumulative distribution function of a random variable determines the distribution of that random variable (see Chapter 4, Section 3), we conclude that the random variable Y has approximately a normal distribution.

The above argument gives an indication of how the Central Limit Theorem can be used to argue (at least in a heuristic fashion) why many quantitative phenomena appear to follow a normal distribution. Whenever a measurement Y can be thought of as resulting from the summation of a great many independent and (nearly) identically distributed random influences, we can assume that the measurement Y has (approximately) a normal distribution.

The Central Limit Theorem asserts that when the sum or average of a large number n of independent and identically distributed random variables $X_1, X_2, ..., X_n$ is being considered, the distribution of this sum (or average) is approximately the normal distribution. The question often arises of determining how large n should be before we can assert that

$Y = X_1 + X_2 + \cdots + X_n$ or $\hat{\mu}_X - \mu_X = (1/n)Y - \mu_X$ is approximately normally distributed. The answer depends on the common distribution of $X_1, X_2, X_3, \ldots, X_n$; the more this distribution resembles the normal distribution, the smaller n can be. It is frequently the case in practice that an n of from 5 to 8 may suffice; even for a moderately skewed distribution (see Chapter 5, Section 5), an n of from 20 to 40 is often sufficient. The question of how large n should be before we can apply the Central Limit Theorem is investigated in one special case in Chapter 10, where we further discuss the Central Limit Theorem and its applications.

EXERCISES

1. Suppose that the random variable X has a normal distribution with an expected value $\mu_X = 3$ and a variance $\sigma_X^2 = 4$. Find the probabilities of the following events:
   (a) $\{X > 5\}$,
   (b) $\{X \geq 5\}$,
   (c) $\{X \leq 5\}$,
   (d) $\{3 < X \leq 5\}$,
   (e) $\{2 \leq X \leq 5\}$,
   (f) $\{2 < X \leq 5\}$.

A2. Suppose that the random variable Z has a standard normal distribution. Let A be the event $\{-0.3 \leq Z \leq 0.7\}$, and let B be the event $\{0.1 \leq Z \leq 3.0\}$. What are the values of:
   (a) $P(A)$,
   (b) $P(B)$,
   (c) $P(A \cup B)$,
   (d) $P(A \cap B)$,
   (e) $P(A|B)$,
   (f) $P(B|A)$?

A3. Consider an I.Q. test for which the scores of adult Americans are known to have a normal distribution which expected value 100 and variance 324, and a second I.Q. test for which the scores of adult Americans are known to have a normal distribution with expected value 50 and variance 100. Under the assumption that both tests measure the same phenomenon ("intelligence"), what score on the second test is comparable to a score of 127 on the first test? Explain your answer.

A4. Suppose it is known that the reaction time of cats to a certain stimulus is normally distributed with expected value 0.1 seconds and variance 0.000169 (seconds)^2. Three cats are chosen independently and subjected to this stimulus.
   (a) What is the probability that all 3 cats respond in less than 0.126 seconds to the stimulus?
   (b) What is the probability that at least 1 of the cats takes more than 0.113 seconds to respond?
   (c) What is the probability that the average of the response times of the 3 cats is strictly greater than 0.110 seconds?
   (d) If the stimulus is given to 100 cats, what is the probability that the average of the response times of these 100 cats is strictly greater than 0.110 seconds?

5. The College Entrance Examination Board Advanced Placement Examination in English was administered on May 18, 1964 to 11,329 secondary-school students seeking advanced placement in college. The test consists of three parts: (i) analysis of a poem, (ii) literature, and (iii) composition. A detailed distributional analysis of 370 of the composition scores appears in Table E.1.

Table E.1. Distribution of scores X on the composition portion of the College Entrance Examination Board Advanced Placement Examination in English.

Interval of Scores	Observed Frequency
68–71	2
64–67	6
60–63	13
56–59	21
52–55	35
48–51	41
44–47	58
40–43	63
36–39	46
32–35	34
28–31	28
24–27	17
20–23	3
16–19	1
12–15	1
8–11	1
	370

For these data the sample average $\hat{\mu}_X$ is 42.96 and the sample variance $\hat{\sigma}_X^2 = 101.2036$. Thus, the sample standard deviation is $\hat{\sigma}_X = \sqrt{\hat{\sigma}_X^2} = 10.06$. To simplify the numerical computations that follow, however, we will use the approximate values $\hat{\mu}_X = 43.0$ and $\hat{\sigma}_X = 10.0$. Do the following:
(a) Plot a frequency histogram for these data.
(b) Plot a relative frequency histogram for these data.
(c) Plot a sample cumulative distribution function for these data.
(d) Fit a theoretical normal curve to the data, and compare the graph of the density function of this distribution to the histogram obtained in part (a) by superimposing the graphs.
(e) Compare the cumulative distribution function of the theoretical normal

distribution fitted to the data in part (d) with the sample cumulative distribution function obtained in part (c).

Based on your answers to parts (d) and (e) of this question, does the normal distribution provide a good "fit" to the data? Explain.

A6. Let X denote the score of a student on the composition test of Exercise 5. Assuming that X has an expected value of 43 and a variance of 100, and that X has a normal distribution, find:
  (a) the probability of the event $\{X \geq 63\}$,
  (b) the 90th, 95th, and 99th percentiles [the (0.90)th, (0.95)th, and (0.99)th quantiles] of the distribution of X,
  (c) the median and mode of the distribution of X,
  (d) the semi-interquartile range of the distribution of X.

7. Men's short sleeve shirts are frequently classified according to size as S, M, L, XL for short, medium, large, and extra large neck sizes. These classifications correspond respectively to neck circumferences of under 15 inches (for S), between 15 and 16 inches (for M), between 16 and 17 inches (for L), and over 17 inches (for XL). Suppose that neck circumferences for adult males have a normal distribution with expected value 15.75 inches and variance 0.49 (inches)^2.
  (a) How many shirts should a manufacturer manufacture in each category of shirt?
  (b) If you wanted to define categories S, M, L, XL so that each category contained 25 percent of the total population of adult males, what neck sizes would you assign to each of these categories?

A8. Suppose we know that the scores X of individuals on a certain test of personality are normally distributed, but the publisher of the test keeps the expected value μ and variance σ^2 of the scores on the test confidential. However, the publisher does announce the values of the 90th percentile and the 95th percentile of the distribution of scores on this test. Suppose that the 90th percentile is 87 and the 95th percentile is 96. Are the values of the expected value μ and variance σ^2 of the test scores really confidential? If you believe that μ and σ^2 are still unknown, explain why. If you believe that μ and σ^2 can be obtained from the given information, then find the values of μ and σ^2.

A9. In data collected at the Lick observatory on 80 bright stars in a certain celestial area, the radial velocity X is measured. The data [reported by Trumpler and Weaver (1953)] are shown in Table E.2.
  (a) Plot a frequency histogram and a relative frequency histogram for these data.
  (b) Fit a normal distribution to these data.
  (c) Graph the density function of the fitted normal distribution and compare this graph to the relative frequency distribution plotted in part (a).
  (d) Assuming that the radial velocity X has the normal distribution obtained in part (b), find the probability of obtaining a positive radial velocity.

Table E.2. Observed values of radial velocity X for 80 bright stars.

Interval of Velocities	Midpoint of Interval	Observed Frequency
-80 to -70	-75	1
-70 to -60	65	2
-60 to -50	-55	2
-50 to -40	-45	2
-40 to -30	-35	8
-30 to -20	-25	24
-20 to -10	-15	26
-10 to 0	-5	11
0 to 10	5	2
10 to 20	15	1
20 to 30	25	1
		$\overline{80}$

10. Simple flashlights generally contain two or three batteries. Suppose that the life of each battery is normally distributed with an expected value of 30 hours and a variance of 25 (hours)^2. The flashlight will cease to function if one or more of its batteries fails (we ignore failure due to storage, and assume batteries have statistically independent lifetimes). Find the probability that a flashlight will operate for at least 30 hours if the flashlight has
    (a) 2 batteries,
    (b) 3 batteries.
    Find the probability that the flashlight will operate no more than 40 hours if the flashlight has
    (c) 2 batteries,
    (d) 3 batteries.
    If you take a 2-battery flashlight and a 3-battery flashlight on a trip, what is the probability that at least one of these flashlights will last the trip, if the trip requires use of each of the flashlights for 40 hours?

A11. Two populations of guinea pigs are examined for length. Guinea pigs from one population have a length X which is normally distributed with an expected value μ_X of 6.9 inches and a variance of 2.56 (inches)^2. Guinea pigs from the second population have a length Y which is normally distributed with an expected value μ_Y of 7.0 inches and a variance σ_Y^2 of 2.89 (inches)^2.
    (a) Find the probability that a guinea pig from the second population has a length Y which exceeds the median length of the first population.
    (b) Find the probability that a guinea pig from the first population has a length X which exceeds the median length of the second population.
    (c) Find the value of the (0.90)th quantile $Q_{0.90}(X)$ of the distribution of the lengths X of guinea pigs in the first population. To what quantile in the second population does $Q_{0.90}(X)$ correspond?

(d) Find the value of the (0.90)th quantile $Q_{0.90}(Y)$ of the distribution of the lengths Y of guinea pigs in the second population. To what quantile in the first population does $Q_{0.90}(Y)$ correspond?

12. Repeat parts (a) through (d) of Exercise 11, but under the assumption that $\sigma_Y^2 = 2.25$. What difference does this new assumption about the variance of the lengths Y in the second guinea pig population make in your answers? Can you explain this difference?

13. An expert witness in a paternity suit testifies that the length (in days) of pregnancy in Caucasian females (from time of impregnation to delivery of the child) is approximately normally distributed with an expected value of 270 days and a standard deviation of 10 days. The male defendant of the paternity suit is able to prove that he was out of the country during a period which started from 293 days before the birth of the child whose paternity is in question, and which lasted until 240 days before the birth of the child. If it were the case that the defendant is the father of the child, what is the probability that the mother of the child could have the very long (or the very short) pregnancy indicated by the testimony? Based on the evidence presented and this probability, if you were on the jury trying this case, would you vote to convict the defendant? Explain your answer.

A14. An airfreight company has 5000 cubic feet of cargo space available for a given flight. The company accepts packages for shipment on the basis of weight alone, since it is the weight of the cargo that will determine the margin of safety for the plane (that is, the plane's ability to take off, fly, and land with safety). However, there is the question of storing the cargo. Previous experience has shown that the cargo space X required (in cubic feet) to store a shipment of Y pounds in weight is a random variable having approximately a normal distribution with expected value $0.34\,Y$ and variance $0.01\,Y^2$. The company figures that their plane can fly and land safely with a cargo of up to 16,000 pounds. However, if they cannot store packages in the cargo space, they have to delay sending some of these packages and pay a penalty. Thus, they want to determine a weight limit for acceptance of packages that with probability no less than 0.99 will allow them to store all packages accepted for shipment. Can the company continue to accept packages for shipment up to the weight limit for flight safety of the plane (that is, 16,000 pounds), or must they stop accepting packages when the total weight of shipment is a lower value than 16,000 pounds? Explain your answer. Show your calculations. If the company must stop accepting packages when the total shipment weight is less than 16,000 pounds, what is the largest weight of shipment that they can accept?

15. Recall (see Chapter 6, Section 1) that the number Y of successes in n statistically independent Bernoulli trials has a binomial distribution with parameters n and p, where p is the probability of a success on any single Bernoulli trial. Let X_i be the Bernoulli random variable resulting from the ith Bernoulli trial. That is, let X_i equal 1 if a "success" is observed on the ith Bernoulli trial, and let X_i equal 0 if a "failure" is observed on the ith Bernoulli trial.

(a) Show that $Y = X_1 + X_2 + X_3 + \cdots + X_n$.
(b) What does the Central Limit Theorem tell you about the distribution of Y when the number n of Bernoulli trials is very large?
(c) Although the probability of obtaining a male infant on a given birth is usually taken to be 0.50, the *actual* value of this probability for human beings is closer to 0.51. If we look at 1000 births, what is the approximate probability of obtaining more than 500 males? If we look at 10,000 births, what is the approximate probability of obtaining more than 5000 males? Use the actual probability of obtaining a male infant in making your calculations.

A16. A chemistry professor has given his class a sample of acid to titrate. The sample of acid actually had a pH of 4.90, but the average of the pH values obtained by the students is 5.25. The professor wonders if his sample has been diluted or contaminated, or if the equipment used by the class members was not rinsed properly. From experience, he knows that students who use their equipment properly obtain a pH value X which is "on the average" correct, but which is a random variable with a variance of 0.98. He also knows from experience that the distribution of X is always slightly positively skewed (when the actual solution to be titrated is acidic), so that X is not normally distributed. However, since there are 50 students in the class, he decides that he can use the Central Limit Theorem to approximate the distribution of the average of the 50 pH values obtained by the class.
(a) Using the Central Limit Theorem, what is the (approximate) probability that the average of the 50 pH values obtained by the class is 5.25 *or larger*, if the true pH of the solution is 4.90 and if the members of the class used their equipment properly?
(b) Do you believe that the "approximate" probability which you obtained in part (a) is an under-approximation, an over-approximation, or the exact value of the actual probability which would be obtained if the precise distribution of the average of 50 pH values were known? Explain your answer.
(c) Use the Chebyshev inequality in Table 6.4 of Chapter 5 to approximate the probability that the average pH value obtained by the 50 students is 5.25 or greater. Assume, as in part (a), that the true pH of the solution is 4.90 and that the members of the class used their equipment properly. How does this approximation compare to the approximation obtained in part (a)? Which approximation would you use? Why?
(d) Based on your calculations in part (a), or in part (c), do you believe that the students obtained such a large average pH value because of the random measurement error inherent even when the titration equipment is properly used, or do you think that some other explanation (a contaminated solution, poor measurement techniques, and so on) is needed? Explain your answer.

17. Suppose that in Example 4.1, the common variance of the astronomer's measurements is later found to be $\sigma_X^2 = 9$. If the astronomer uses $n = 400$

observations to estimate the distance D between the star and his observatory, what is the probability that his estimate will be within 0.25 light years of the true distance? If he wants his estimate of the distance to be within 0.25 light years of the true distance D with probability no less than 0.9876, what is the smallest number n of measurements that he can use?

8
SPECIAL DISTRIBUTIONS: CONTINUOUS CASE

The only certainty is that nothing is certain.
Pliny the Elder (Historia Naturalis)

False facts are highly injurious to the progress of science, for they often endure long; but false views, if supported by some evidence, do little harm, for every one takes a salutary pleasure in proving their falseness.
Charles Darwin (Descent of Man)

In Chapter 6, we presented some discrete distributions which frequently arise in applications. When one of these discrete probability distributions is chosen to model the random phenomenon under observation, the choice is usually made either for theoretical reasons, or for empirical ones; that is, either because the probability distribution appears to be a logical consequence of certain properties of the random phenomenon which are known to the experimenter, or because the probabilities derived from the distribution agree with the relative frequencies obtained from repeated observation of the phenomenon. Sometimes a probability distribution is chosen for both theoretical and empirical reasons.

In the present chapter we discuss some continuous distributions other than the normal distribution (see Chapter 7) which are used repeatedly in practice. As in the case of the discrete probability distributions discussed in Chapter 6, each continuous distribution is presented in a parametric form. That is, the general shape of the distribution is given in a mathematical form in which certain constants (parameters) are left unspecified. (As, for example, p was left unspecified in the case of the Bernoulli distribution, and μ and σ^2 were left unspecified in the case of the normal distribution.) When these constants *are* specified, then the probability distribution (and thus the probability of any event) is completely determined. We remark that it is a considerable achievement when we arrive at the conclusion that the probability distribution of a given random variable has a certain parametric form. When such a parametric form is found, not only is the number of possible distributions which we must consider greatly reduced, but insights beyond the range of past experience (that is, extrapolations and interpolations) are obtainable.

1. THE EXPONENTIAL DISTRIBUTION

The continuous distributions that are discussed in this chapter are the exponential distribution (Section 1), the gamma distribution (Section 2), the Weibull distribution (Section 3), the uniform distribution (Section 4), and the beta distribution (Section 5). In Section 6, we briefly sketch some facts about two other continuous distributions that are of some interest to probabilists: the lognormal distribution and the Cauchy distribution.

1. THE EXPONENTIAL DISTRIBUTION

A light bulb is turned on and left to burn until it burns out. How long will it last? A student is given a learning task to master. How long will it take him to complete the task? An alpha particle is emitted from a sample of radioactive material. How long will it be before the next alpha particle is emitted?

Many experiments of interest to science and industry involve the measurement of the duration of time X between an initial point of time and the occurrence of some phenomenon of interest. Frequently, more than one such measurement is made; in particular, as in the case of measurements of time between the successive emissions of alpha particles from radioactive material, measurements are made of the durations of time X between successive observations of a given phenomenon of interest. For example, psychologists have studied the length of time X between successive presses of a bar by a rat placed in a Skinner box, when a certain reinforcement schedule for operant conditioning has just been applied to modify the rat's behavior [see Mueller (1950)]. Industrial engineers have investigated the period of time X between successive failures of certain kinds of components in complex machines (for example, rockets, automobiles, electric appliances, and so on). Telephone companies are interested in the duration X of a telephone call, since knowledge of the statistical properties of this random variable is important for planning the provision of telephone service to their customers. As a final example, both psychologists and biophysicists have been interested in the time X between successive electrical impulses received by measuring devices implanted in the spinal cords of various mammals (mice, cats, monkeys, and so on).

If we make repeated measurements of the duration of time X of interest in such experiments, the histogram (see Chapter 4, Section 5) which we obtain from such data often starts on the left with very tall rectangles (corresponding to large probabilities assigned to small values of the random variable X). As we move to the right, the heights of the rectangles that form the histogram decrease rapidly in magnitude. A typical such histogram is shown in Figure 1.1.

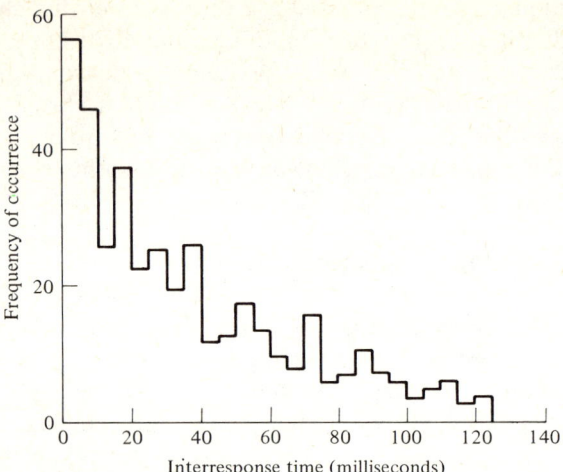

Figure 1.1 Frequency distribution of interresponse times in the spontaneous activity of a single spinal interneurone. The distribution consists of 391 intervals recorded by a micropipette inserted in the spinal cord of a cat.

SOURCE: McGill, William J., Stochastic latency mechanisms, *Handbook of Mathematical Psychology*, eds. R. D. Luce, R. R. Bush, and E. Galanter, © 1963, John Wiley & Sons, Inc., by permission of John Wiley & Sons, Inc.

One probability density function whose graph resembles the histogram shown in Figure 1.1 (in the sense that it falls off sharply in height as we look from left to right along the horizontal axis) is a density function $f_X(x)$ of the form

(1.1) $$f_X(x) = \begin{cases} \theta e^{-\theta x}, & \text{if } x \geqslant 0, \\ 0, & \text{if } x < 0, \end{cases}$$

where the constant θ can be any positive number. A random variable X having a density function $f_X(x)$ of the form shown in Equation (1.1) is said to have an *exponential distribution*; the density function (1.1) is called the *exponential density function*.

The density function of the exponential distribution is not difficult to graph because of many readily available tables. Figure 1.2 provides graphs of the exponential density function for the following values of the constant θ: $\theta = 0.1, 0.5, 1.0$, and 2.0.

The positive constant θ is the parameter of the exponential distribution. Once we have established the value of θ, we have determined which exponential distribution governs the variability of the random variable X of interest to us. Notice from Figure 1.2 that the steepness of the decrease in height of the graph of the exponential density function $f_X(x)$, as

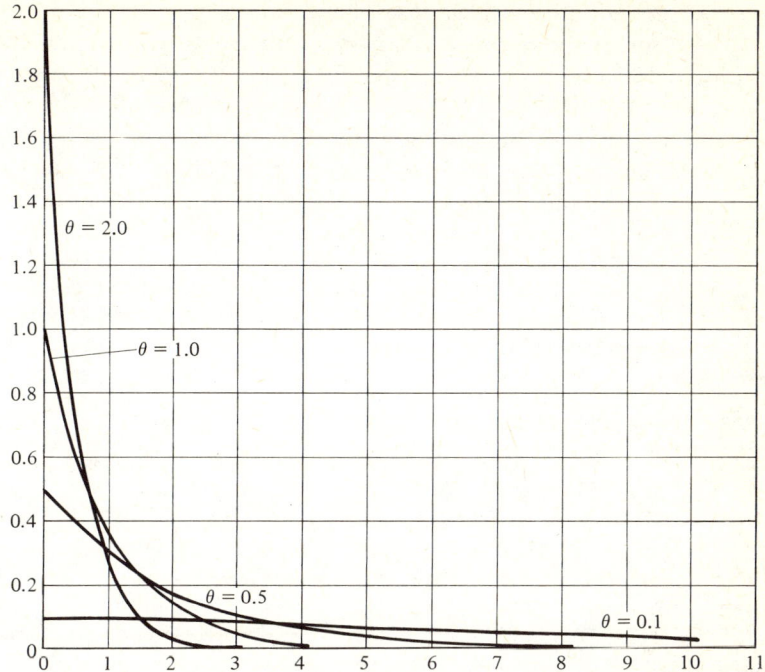

Figure 1.2 Graphs of the exponential density function for $\theta = 0.1$, 0.5, 1.0, and 2.0.

we move from small to large values of x, depends on the value of θ; the larger the value of θ, the steeper the decrease in height of the graph of the exponential density function $f_X(x)$.

It can be shown that if X has an exponential distribution, then the expected value μ_X and the variance σ_X^2 of X are given by

$$(1.2) \qquad \mu_X = \frac{1}{\theta}, \qquad \sigma_X^2 = \frac{1}{\theta^2}.$$

Thus, the expected value of X is equal to the standard deviation σ_X of X (that is, the ratio μ_X/σ_X is equal to 1). Note from Equation (1.2) that if we know μ_X, then $\theta = 1/\mu_X$ is directly obtained. Hence, we may determine the particular exponential distribution which describes the variation of a random variable X of interest to us either by finding the value of the parameter θ that describes the distribution of X, or else, equivalently, by finding the expected value μ_X of X.

Fitting the Exponential Distribution to Data

Proschan (1963) provides records giving the durations of time between successive failures of the air conditioning systems of each member of a fleet of 13 Boeing 720 jet airplanes. These durations of time are listed in Table 1.1. Thus, plane number 7907 had a failure of its air conditioning system after 194 hours of service, a second failure 15 hours of service after the first failure, a third failure 41 hours of service after the second failure, and so on. After roughly 2000 hours of service, each plane received a major overhaul; if a time interval between two failures of the air conditioning system included the major overhaul, the length of this time interval

Table 1.1. Table of the durations of time between successive failures of the air conditioning systems of each member of a fleet of 13 Boeing 720 jet airplanes (213 observations in all).

						Plane						
7907	7908	7909	7910	7911	7912	7913	7914	7915	7916	7917	8044	8045
194	413	90	74	55	23	97	50	359	50	130	487	102
15	14	10	57	320	261	51	44	9	254	493	18	209
41	58	60	48	56	87	11	102	12	5		100	14
29	37	186	29	104	7	4	72	270	283		7	57
33	100	61	502	220	120	141	22	603	35		98	54
181	65	49	12	239	14	18	39	3	12		5	32
	9	14	70	47	62	142	3	104			85	67
	169	24	21	246	47	68	15	2			91	59
	447	56	29	176	225	77	197	438			43	134
	184	20	386	182	71	80	188				230	152
	36	79	59	33	246	1	79				3	27
	201	84	27	**	21	16	88				130	14
	118	44	**	15	42	106	46					230
	**	59	153	104	20	206	5					66
	34	29	26	35	5	82	5					61
	31	118	326		12	54	36					34
	18	25			120	31	22					
	18	156			11	216	139					
	67	310			3	46	210					
	57	76			14	111	97					
	62	26			71	39	30					
	7	44			11	63	23					
	22	23			14	18	13					
	34	62			11	191	14					
		**			16	18						
		130			90	163						
		208			1	24						
		70			16							
		101			52							
		208			95							

**Indicates major overhaul

1. THE EXPONENTIAL DISTRIBUTION

is not listed in Table 1.1, since its magnitude may have been affected by the fact that repairs were made on the plane.

A frequency histogram obtained from the data given in Table 1.1 is shown in Figure 1.3.

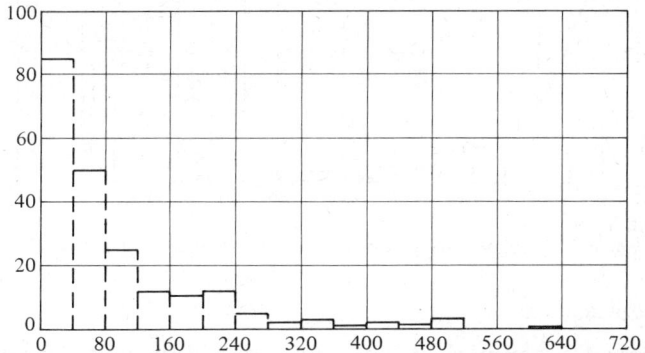

Figure 1.3 Frequency histogram for the data of Table 1.1 representing durations of time between successive failures of the air conditioning systems of each member of a fleet of 13 Boeing 720 jet airplanes.

To fit an exponential distribution to these data, we must obtain a value for the parameter θ. Recall that the expected value μ_X of an exponentially distributed random variable X is related to θ by the formula $\mu_X = 1/\theta$. Consequently, we may estimate θ by estimating μ_X by the sample average $\hat{\mu}_X$ obtained from the data given in Table 1.1 (see the Appendix to Chapter 5), and then solving the equation $\hat{\mu}_X = 1/\theta$ for θ. From the data in Table 1.1,

$$\hat{\mu}_X = \frac{1}{213}(194 + 15 + 41 + 29 + 33 + 181 + 413 + 14 + \cdots$$

$$+ 66 + 61 + 34)$$

$$= 93.14 \text{ hours},$$

so that the estimated θ equals $1/\hat{\mu}_X = 1/93.14$, or $\theta = 0.0107$. Thus, the fitted exponential probability density function of the duration of time X between successive failures of the air conditioning system of a Boeing 720 jet airplane is

$$f_X(x) = (0.0107) \, e^{-(0.0107)x}.$$

A graph of this density function appears in Figure 1.4. In Figure 1.4, we have also graphed the relative frequency histogram (see Chapter 4, Section 5) of the data in Table 1.1 in order to permit comparison of this histogram to the graph of the fitted probability density function.

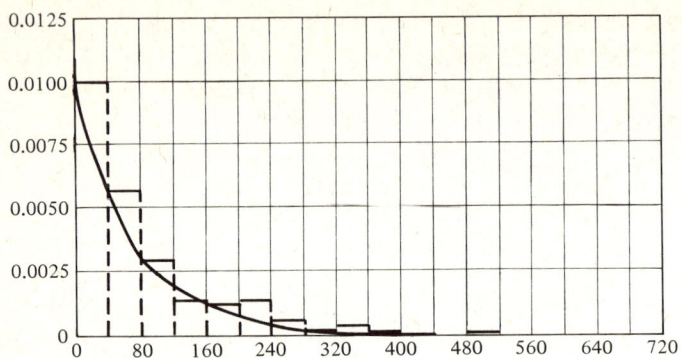

Figure 1.4 Relative frequency histogram for the data of Table 1.1.

Probability Calculations

In Chapter 4, we remarked that knowledge of the cumulative distribution function $F_X(x)$ of a continuous random variable X allows us to calculate probabilities of events of the form $\{X \leq a\}, \{b < X\}, \{a \leq X \leq b\}$, and so on. The cumulative distribution function $F_X(x)$ of a random variable X, which has an exponential distribution with parameter θ, is given by the expression

(1.3) $$F_X(x) = \begin{cases} 0, & \text{if } x < 0, \\ 1 - e^{-\theta x}, & \text{if } x \geq 0. \end{cases}$$

From Equation (1.3), from Table 3.2 of Chapter 4, and from a table which gives values of $e^{-\tau}$ for various values of τ, we can compute probabilities of events of the form indicated above. Table T.8 provides values of $e^{-\tau}$ for $\tau = 0.00(0.01)7.99$; values of $e^{-\tau}$ for larger values of τ are smaller than 0.0004.

Example 1.1. In the example of the duration of time X between failures of the air conditioning system of a Boeing 720 jet airplane, which we used to demonstate fitting of the exponential distribution to data, we found that X (at least approximately) has an exponential distribution with parameter θ equal to 0.0107. Suppose that this is indeed exactly the case. In order to determine whether his maintenance costs could become excessive, a purchaser of a Boeing 720 jet airplane would want to know the probabilities of events of the form "the time X between failures of the air conditioning system exceeds b" (that is, events of the form $\{X > b\}$).

It follows from Table 3.2 of Chapter 4 that the probability, $P_X\{X > b\}$, of such an event is equal to $1 - F_X(b)$. Hence,

$$P_X\{X > b\} = 1 - F_X(b) = e^{-(0.0107)b},$$

and the value of $e^{-(0.0107)b}$ may be obtained from Table T.8. The probability that the air conditioning system of a Boeing 720 jet airplane will last more than $b = 100$ hours of operating time is thus equal to $e^{-(0.0107)(100)} = 0.343$.

Example 1.2. We have mentioned earlier an experiment described by Mueller (1950) in which white rats were periodically reconditioned. The duration of time X in seconds between presses of the rat on a certain bar has (approximately) an exponential distribution with parameter $\theta = 0.20$. Suppose we wish to obtain the probability that the rat will press the bar any time between a seconds and b seconds, $a < b$, after the last time it pressed the bar (that is, we wish to find the probability of the event $\{a \leq X \leq b\}$). From Table 3.2 of Chapter 4 (and the fact that X is a continuous random variable), it follows that

$$P_X\{a \leq X \leq b\} = F_X(b) - F_X(a)$$
$$= (1 - e^{-\theta b}) - (1 - e^{-\theta a})$$
$$= e^{-\theta a} - e^{-\theta b}$$
$$= e^{-(0.20)a} - e^{-(0.20)b}.$$

Thus if, for example, we want the probability that the length of time X between bar-presses by the rat is between 1 and 3 seconds ($a = 1, b = 3$), then using Table T.8,

$$P_X\{1 \leq X \leq 3\} = e^{-(0.20)(1)} - e^{-(0.20)(3)}$$
$$= e^{-0.20} - e^{-0.60}$$
$$= 0.819 - 0.549 = 0.270.$$

The Survival Function

In a number of experiments, the lifetime X of a certain system, phenomenon, item, individual, and so on, is observed. In such cases, the cumulative distribution function $F_X(x)$ of the length of life X gives the probability that this lifetime does not exceed x units of time; or, equivalently, $F_X(x)$ gives the probability that the "system" will "die" before x units of time have passed. The quantity $1 - F_X(x)$, on the other hand, equals the probability of the event $\{X > x\}$, and thus gives the probability that the "system" *survives* more than x units of time. For this reason,

$$1 - F_X(x) = P_X\{X > x\}$$

is sometimes called the *survival function* (or survival cumulative distribution function) of the random lifetime X.

366 SPECIAL DISTRIBUTIONS: CONTINUOUS CASE 8

For the exponential distribution with parameter θ, the survival function $1 - F_X(x)$ is given by the equation

$$1 - F_X(x) = \begin{cases} 1, & \text{if } x < 0, \\ e^{-\theta x}, & \text{if } x \geq 0, \end{cases}$$

so that the value of the survival function for a given number x can be obtained by looking up $\tau = \theta x$ in Table T.8. Figure 1.5 provides graphs of the survival functions $1 - F_X(x)$ corresponding to exponentially distributed random variables X with parameter values θ equal to 0.1, 0.5, 1.0, 2.0, 5.0, and 10.0.

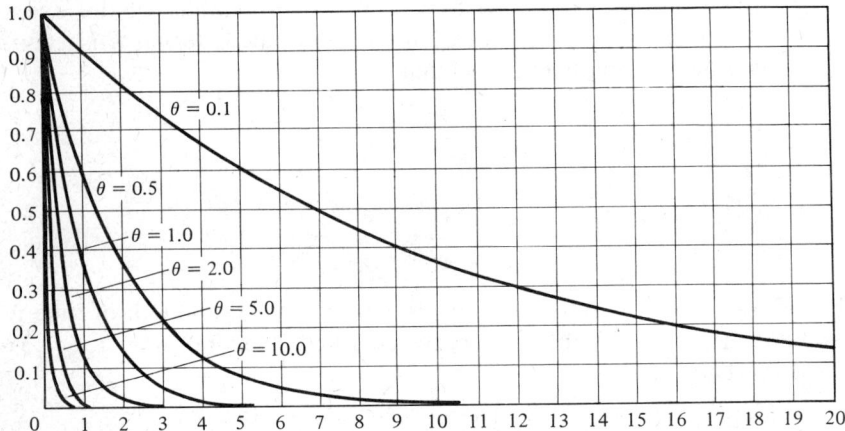

Figure 1.5 Graphs of the survival function $1 - F_X(x)$ for the exponential distribution with values of the parameter $\theta = 0.1, 0.5, 1.0, 2.0, 5.0,$ and 10.0.

The length of life (survival time) of a "system" is often of importance. For this reason, and because knowledge of the survival function is equivalent to knowledge of the cumulative distribution function $F_X(x)$ of the distribution of the lifetime X, investigators often check the fit of a given theoretical distribution (like the exponential distribution) to observed data by comparing the graph of the theoretical survival function, $1 - F_X(x)$, calculated from the theoretical distribution, to the graph of the *sample survival function*,

$$1 - F_n(x) = \frac{\text{number of observations which exceed } x}{n},$$

which is calculated from n observations $X_1, X_2, ..., X_n$ of X. (See Chapter 4, Exercise 14 for a discussion of the computation of $F_n(x)$ from grouped data. Computation of $1 - F_n(x)$ proceeds similarly.)

1. THE EXPONENTIAL DISTRIBUTION

In the case of the data of Table 1.1 (where the lifetime X of interest is the length of time during which the air conditioning system of the jet plane is operational), a comparison of the graphs of the theoretical survival function and the sample survival function (see Figure 1.6) bears out our earlier finding that there is a close fit between the observed data and the exponential distribution with parameter $\theta = 0.0107$.

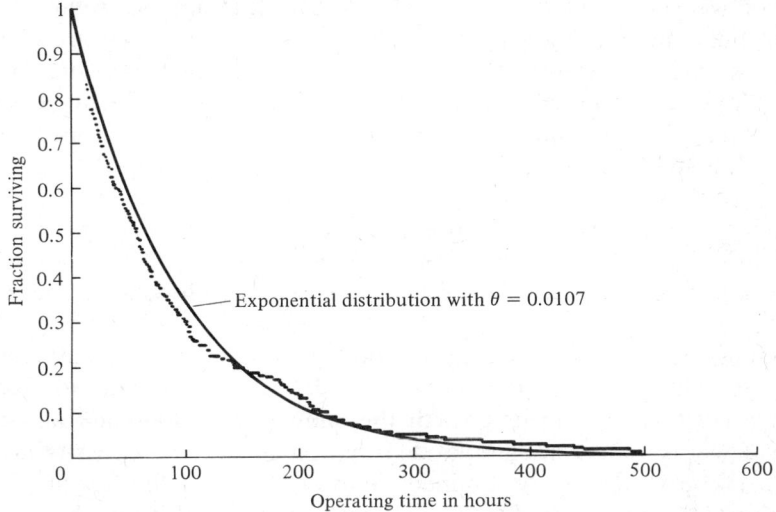

Figure 1.6 Comparison between theoretical survival function $1 - F_X(x) = e^{-(0.0107)x}$, and sample survival function $1 - F_n(x)$ calculated from the data in Table 1.1. The smooth line gives the graph of $1 - F_X(x)$.

An Interesting Property of the Exponential Distribution

The exponential distribution has the following interesting property. Suppose we know that a random variable X, having an exponential distribution with parameter θ, exceeds a number x^*. For example, suppose we know that the duration of time X between an initial time point t_0 and the time of occurrence of a given phenomenon is larger than x^*. That is, the time is now $t_0 + x^*$, and the phenomenon of interest has not yet occurred. What is the conditional distribution of the length, $X - x^*$, of additional time that must pass until the phenomenon finally does occur, *given* that the length of time that has already passed is equal to x^*? It can be shown that this conditional distribution is the *same* exponential distribution as the unconditional distribution of X. That is, given that X exceeds x^*, the conditional distribution of $Y = X - x^*$ is the same as the unconditional distribution of X. When we ask, "How much longer must we wait for the phenomenon to occur" *given* that x^* units of time has already passed, we can "forget" that we started the experiment x^* units of time in the past, and act as if we are just now starting the experiment. Thus, in this sense, the exponential distribution "has no memory" of the past.

In terms of the length of a telephone conversation, the above property of the exponential distribution has the following implication. If a conversation between two individuals is still in progress at any given time t_0, the time remaining until the conversation is completed is a random variable, and this random variable has the same statistical properties as the random variable representing the length of time until completion of a conversation just begun. Consequently, if the exponential distribution is in operation, it makes no sense to get in line in front of the phone booth that has been occupied for the longest time!

Among all distributions of nonnegative continuous random variables, only the exponential distributions "have no memory." The distribution of a random variable X will "have no memory" if the forces that act to influence the magnitude of X have a certain uniformity. If X, for example, is the diameter of a snowflake, and if the forces that act to make the snowflake grow in size continue to act probabilistically at the same strength no matter what dimension of growth the snowflake has already achieved, then X will have an exponential distribution since the process of growth "forgets" growth already attained. Similarly, if X is the length of time required for an individual to finish a certain task, and if the abilities and handicaps of the individual probabilistically have the same influence on the additional time needed to finish the task, regardless of the time which the individual has already spent on the task, then X has an exponential distribution. In general, if we believe that a random variable X is the resultant of a process of growth, duration, or increase for which the strength of the forces that act (probabilistically) to control the process are at all stages of the process independent of the growth, duration, or increase already achieved, then X has an exponential distribution.

NOTES AND REFERENCES

The exponential and Poisson distributions (see Section 3 of Chapter 6) are related in an interesting way. Suppose we have a batch of light bulbs whose lifetimes are each exponentially distributed with parameter θ. We start at a given time (say $t = 0$) and burn 1 bulb until extinction. We replace that bulb instantly with another bulb, wait until that bulb burns out, replace it with another, wait until this third bulb burns out, and so on. At a given time $t = T$, we stop this process and count the number Y of light bulbs that we have burned out. It then follows that Y is a random variable having the Poisson distribution with parameter $\lambda = \theta T$.

2. THE GAMMA DISTRIBUTION

Visualize a system (such as a radio, an assembly line, an airplane, and so on) in which the proper functioning of a certain component is essential for the proper functioning of the system as a whole. In order to increase the reliability of the system (increase the time between failures of the system), the system may be designed to carry $(r-1)$ spare components to be used in case the given component fails. When the original component fails, one of the $(r-1)$ spare components is activated to take its place. As this component fails, one of the $(r-2)$ remaining spare components takes over. This process continues until all r components have failed [the original component and the $(r-1)$ spares]; at this point, the entire system suffers failure. Assuming that the entire system can only fail if the single essential component fails, the lifetime (time until failure) of the entire system is the sum of the times until failure, $X_1, X_2, \ldots, X_r$ of the r components. Let Y be the time until the entire system fails. Then $Y = X_1 + X_2 + \cdots + X_r$. If each of the times until failure, $X_1, X_2, \ldots, X_r$, has the same exponential distribution with parameter θ, and if the times $X_1, X_2, \ldots, X_r$ are statistically independent (this is a plausible assumption since each component is "active" over a different period of time), then Y has the probability density function

$$(2.1) \qquad f_Y(y) = \begin{cases} \dfrac{\theta^r y^{r-1} e^{-\theta y}}{\Gamma(r)}, & \text{if } y \geq 0, \\ 0, & \text{if } y < 0, \end{cases}$$

where θ and r are positive numbers, and $\Gamma(r)$ is the gamma function (see Chapter 6, Section 5).

Any random variable Y having the probability density function $f_Y(y)$ given by Equation (2.1) is said to have a *gamma distribution*; any density function of the form (2.1) is said to be a *gamma density function*. The constants θ and r in Equation (2.1) are the parameters of the gamma distribution; given the values of θ and r, the distribution (2.1) is completely determined. Although in the example used above to motivate the gamma distribution (the example of the system with r spare parts) the parameter r is an integer, this need not always be the case. In many applications of the gamma distribution, r is not an integer.

Example 2.1 (Radioactivity). Slack and Krumbein (1955) describe an experiment in which the average mean value Y of radioactivity (alphas per minute) within a sample of Pennsylvania black shale is measured. A gamma distribution with parameters $\theta = 0.9083$ and $r = 2.493$ is found to provide a reasonably close fit to the observed frequency distribution of Y.

The Standard Gamma Density Function

As in the case of the normal distribution, if we had to table the gamma distribution for all values of the pair of parameters θ and r, our task would indeed be difficult. Although we cannot, as in the case of the normal distribution, reduce this tabulating problem to that of constructing a single table, some simplification is possible. Consider the new random variable

(2.2) $$V = aY.$$

If Y has a gamma distribution with parameters $\theta = a$ and r, then V has a gamma distribution with parameters $\theta = 1$ and r. That is, V has the probability density function

(2.3) $$f_V(v) = \begin{cases} \dfrac{v^{r-1}e^{-v}}{\Gamma(r)}, & \text{if } v \geq 0, \\ 0, & \text{if } v < 0. \end{cases}$$

This density function is called the *standardized gamma density function* with parameter r. To gain some insight into the dependence of the form of the gamma distribution upon the parameter r, we can graph the standardized gamma density function (in which the influence of the parameter θ has been removed) for various values of r. In Figure 2.1(a), graphs of the standardized gamma density function for $r = 0.5, 2, 3, 4$, and 5 are given. Note that for $r < 1$, the graph of the corresponding standardized gamma density function drops off steeply (just like the graph of the exponential density function). When r exceeds 1 ($r = 2, 3, 4, 5$), the graph of the standardized gamma density function takes on a "humped" form. The density function is positively skewed (that is, it has a long right-hand "tail"; see Chapter 5, Section 4), but this skewness becomes less pronounced as r increases. For r large enough (say, $r = 50$), the standard gamma distribution closely resembles a normal distribution with mean and variance both approximately equal to r [see Figure 2.1(b)].

The influence of the parameter θ upon the shape of the graph of the density function of the gamma distribution with parameters θ and r can be illustrated by the case $r = 1$. When $r = 1$, *the gamma distribution with parameters θ and $r = 1$ is the same as the exponential distribution with parameter θ.* Thus, in this case, the influence of θ on the shape of the graph of the gamma density function can be seen by inspecting Figure 1.2 in

2. THE GAMMA DISTRIBUTION

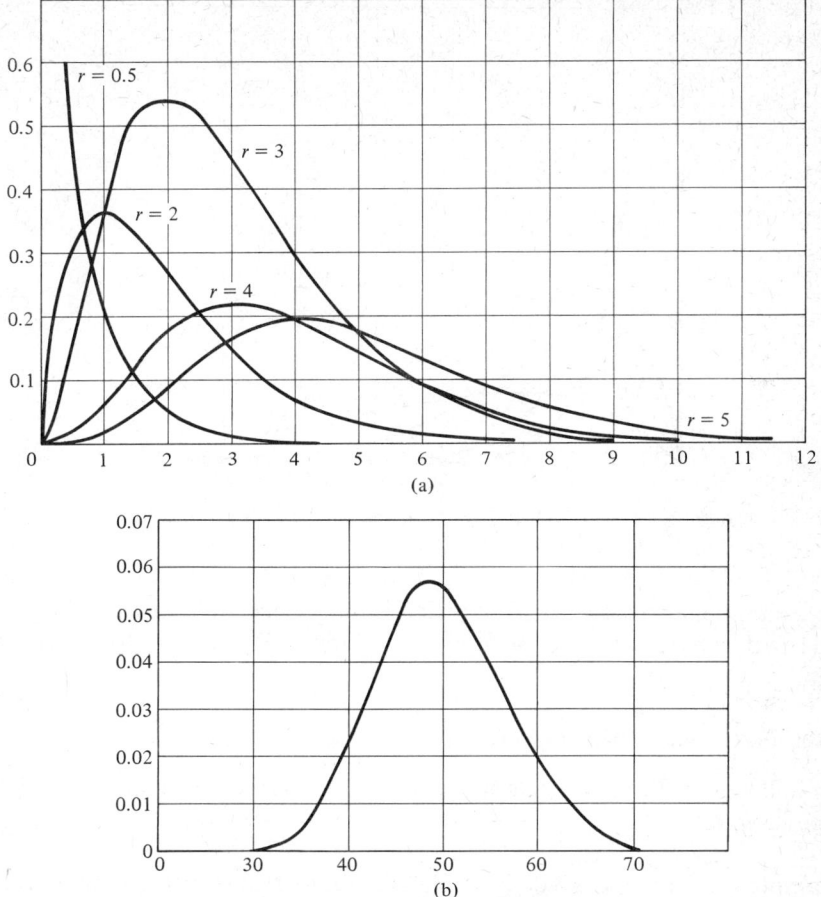

Figure 2.1 Graphs of the standardized gamma density function for
(a) $r = 0.5, 2, 3, 4$, and 5; (b) $r = 50$.

Section 1. Figure 2.2 illustrates the influence of the parameter θ on the shape of the graph of the gamma density function when $r = 2$.

When Y has the gamma distribution with parameters θ and r, the expected value μ_Y and variance σ_Y^2 of Y are given by

$$(2.4) \qquad \mu_Y = \frac{r}{\theta}, \qquad \sigma_Y^2 = \frac{r}{\theta^2}.$$

Since r and θ can be determined by knowing μ_Y and σ_Y^2 through the equations

$$(2.5) \qquad \theta = \frac{\mu_Y}{\sigma_Y^2}, \qquad r = \frac{\mu_Y^2}{\sigma_Y^2},$$

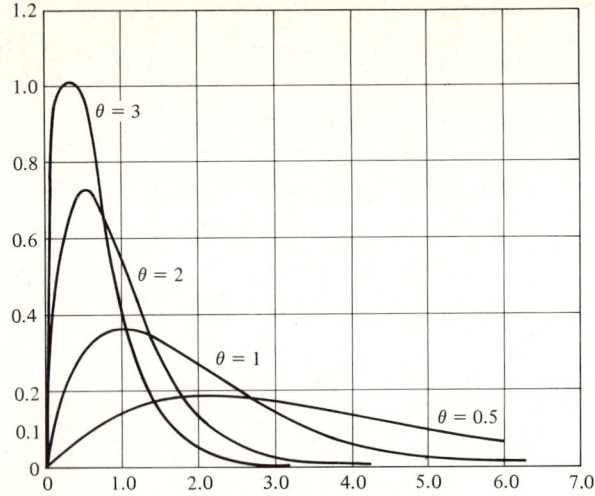

Figure 2.2 Changes in the graph of the gamma density function for $r = 2$ and $\theta = 0.5, 1, 2,$ and 3.

the gamma distribution function is determined once we know μ_Y and σ_Y^2. This fact provides a way of fitting the gamma distribution to observed data.

Fitting the Gamma Distribution to Data

To illustrate fitting the gamma distribution to data, consider the following example.

Example 2.2. The data given in Table 2.1 are the failure times Y (in weeks) of 31 transistors in an accelerated life test (34 transistors were actually under observation, but 3 survived the entire time period of observation). These observations are reported by Wilk, Gnanadesikan, and Huyett (1962), who indicate that from past experience there is reason to believe that the gamma distribution might reasonably approximate the distribution of failure times.

Table 2.1. Times until failure of 34 transistors undergoing an accelerated life test (only the 31 transistors that failed are listed).

3	4	5	6	6	7	8	8
9	9	9	10	10	11	11	11
13	13	13	13	13	17	17	19
19	25	29	33	42	42	52	

2. THE GAMMA DISTRIBUTION

If we wish to fit a gamma distribution to these data, we need to estimate the values of the parameters θ and r. As mentioned above, we can first estimate μ_Y and σ_Y^2 by the methods described in Chapter 5, and then find θ and r by means of Equation (2.5). From Table 2.1,

$$\hat{\mu}_Y = \frac{1}{31}(3+4+5+6+7+\cdots+42+42+52)$$
$$= 15.71,$$

and

$$\hat{\sigma}_Y^2 = \frac{1}{31}[(3-\hat{\mu}_Y)^2+(4-\hat{\mu}_Y)^2+\cdots+(52-\hat{\mu}_Y)^2]$$
$$= 141.50.$$

Substituting the values of $\hat{\mu}_Y$ and $\hat{\sigma}_Y^2$ for μ_Y and σ_Y^2, respectively, in Equation (2.5), we obtain the following estimates for θ and r:

$$\hat{\theta} = \frac{\hat{\mu}_Y}{\hat{\sigma}_Y^2} = \frac{15.71}{141.50} = 0.11,$$

$$\hat{r} = \frac{(\hat{\mu}_Y)^2}{\hat{\sigma}_Y^2} = \frac{(15.71)^2}{141.50} = 1.74.$$

Using these values of θ and r and Equation (2.1), we may prepare a graph of the (fitted) probability density function of the failure times Y of the transistors; this graph appears in Figure 2.3.

[*Note:* In the article by Wilk, Gnanadesikan, and Huyett from which the data in Table 2.1 are taken, different estimates of the parameters θ and r are obtained, because these authors use a different method to estimate these parameters. As we have remarked in Chapter 6, there are many ways in which we can fit a given distributional form, such as the gamma distribution, to data. A comparison of alternative methods of fitting is a subject dealt with in texts on statistical inference, and is beyond the scope of this book.]

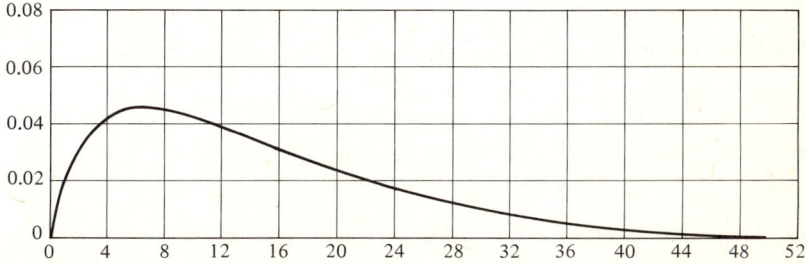

Figure 2.3 Graph of the gamma density function with parameters $\theta = 0.11$ and $r = 1.74$.

Probability Calculations

Suppose we want to calculate probabilities of events of the form $\{Y \leq a\}$, $\{b < Y\}$, $\{a \leq Y \leq b\}$, and so on, when Y has a gamma distribution with parameters θ and r. As we have already noted in the case of the exponential distribution (which we recall is a gamma distribution with parameter $r = 1$), such probabilities can be calculated once we have tables of the cumulative distribution function $F_Y(y)$ of Y. Because the gamma distribution depends on the two parameters θ and r, so does the cumulative distribution function $F_Y(y)$. Thus, it would seem that we may need a separate table of the cumulative distribution function $F_Y(y)$ for every pair of values (θ, r) for the parameters θ and r. However, as already noted, we can transform Y to a random variable $V = \theta Y$ having a standard gamma distribution with the single parameter r. Further,

$$(2.6) \qquad F_Y(y) = P_Y\{Y \leq y\} = P_Y\{\theta Y \leq \theta y\} = F_V(\theta y).$$

The cumulative distribution function $F_V(\tau)$ of a random variable V having a standard gamma distribution with parameter r is extensively tabulated, and is known as the *incomplete gamma function*. A special notation $I_r(\tau)$ is often used for this function. The symbol $I_r(\tau)$ represents the probability $P_V\{V \leq \tau\}$, where V has a standard gamma distribution with parameter r. For example, $I_3(a)$ equals the shaded area shown in Figure 2.4. Tables of $I_r(\tau)$ for $r = 1(1)5$, $\tau = 0.2(0.2)8.0(0.5)15.0$, and for $r = 6(1)10$, $\tau = 1.0(0.2)8.0(0.5)17.0$ are given in Table T.9. From Equation (2.6) and the definition of $I_r(\tau)$, it follows that if Y has a gamma

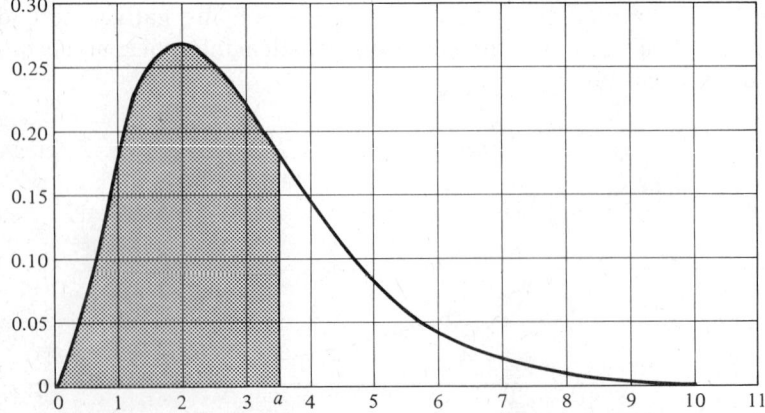

Figure 2.4 The shaded area represents the probability that a variate having a gamma distribution is less than or equal to a.

distribution with parameters θ and r, then

(2.7) $$F_Y(y) = I_r(\theta y).$$

Using Equation (2.7), Table 3.2 of Chapter 4, and Table T.9, we can find probabilities of events of the form $\{Y \leq a\}$, $\{b < Y\}$, $\{a \leq Y \leq b\}$, and so on.

Example 2.3. A psychologist is studying the solution of a complex learning task by young children. For each child who is given the learning task (say, putting together a geometrical puzzle), the psychologist measures the length Y of time (in minutes) that it takes the child to complete the task. Because the psychologist has a theory that each child solves this complex learning task by independently and sequentially solving r simple tasks, each of approximately the same difficulty, and because he is willing to assume that the time X_i taken by a child to solve the ith such simple task is exponentially distributed with parameter θ for $i = 1, 2, 3, ..., r$, he concludes that the total length of time $Y = X_1 + X_2 + \cdots + X_r$ required by a child to finish the entire learning task has a gamma distribution with parameters θ and r. Using the methods already described for fitting the gamma distribution to data, the psychologist gives a large number of children this learning task, and finds that the distribution of Y is approximately a gamma distribution with parameters $\theta = 0.40 = 1/2.50$ and $r = 5.01$. Since he believes that the number r of simple tasks solved sequentially by a child in order to complete the learning task must be an integer (and because $r = 5.01$ is so close to 5), the psychologist decides that Y has a gamma distribution with parameters $\theta = 1/2.50$ and $r = 5$.

The psychologist is planning a new experiment in which children in a certain school will be given this learning task under various environmental conditions (alone, in small groups of children, in a large classroom of children, and so on). Because he has only a limited time (30 minutes) in which to observe the children, and because he wants to make sure that most (say 95 percent) of the children finish the task, the psychologist asks for the probability, $P_Y\{Y > 30\}$, that a child cannot finish the learning task. Under the assumption that the different experimental conditions have no effect on a child's performance (that is, assuming that Y has a gamma distribution with $\theta = 1/2.50$ and $r = 5$), it follows from Table 3.2 of Chapter 4, from Equation (2.7), and from Table T.9 that

$$\begin{aligned} P_Y\{Y > 30\} &= 1 - F_Y(30) \\ &= 1 - I_5\left(\frac{30}{2.50}\right) \\ &= 1 - I_5(12) \\ &= 1 - 0.9924 = 0.0076. \end{aligned}$$

Thus, the psychologist concludes that there is enough time available to him (provided that there are no experimental effects) to ensure that most of the children will finish the learning task before his 30 minutes is up.

If, for some reason, the psychologist wants to know the probability that a child takes between 15 and 20 minutes to finish the learning task, then from Table 3.2 of Chapter 4, Equation (2.7), and Table T.9, he can calculate

$$P_Y\{15 \leq Y \leq 20\} = F_Y(20) - F_Y(15)$$
$$= I_5\left(\frac{20}{2.50}\right) - I_5\left(\frac{15}{2.50}\right)$$
$$= I_5(8) - I_5(6)$$
$$= 0.90037 - 0.71494$$
$$= 0.18543.$$

Notice that $P_Y\{15 \leq Y \leq 20\} = P_Y\{15 < Y \leq 20\}$ since Y is a continuous random variable.

Equation (2.7) and Table T.9 allow us to graph the cumulative distribution function $F_Y(y)$ of random variables Y having a gamma distribution. Since the survival function (see Section 1) is equal to $1 - F_Y(y)$, we can also graph that function. Graphs of the cumulative distribution functions (Figure 2.5) and survival functions (Figure 2.6) for random variables having gamma distributions with various pairs of parameters (θ, r) are given in Figures 2.5 and 2.6.

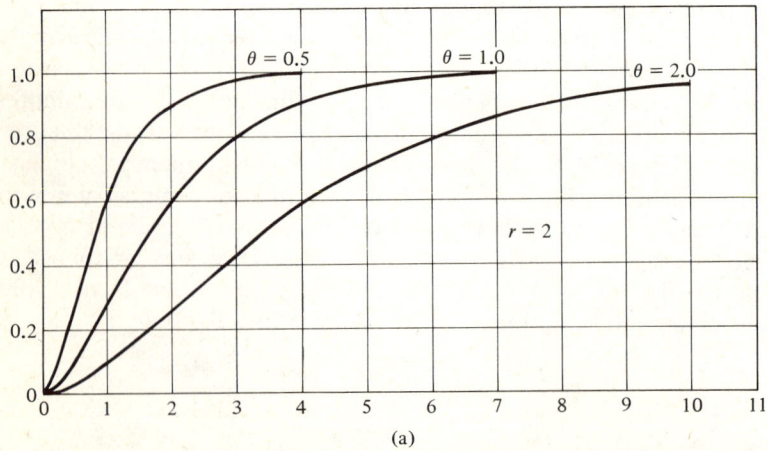

Figure 2.5 Graphs of the cumulative gamma distribution functions with (a) $r = 2$ and $\theta = 0.5, 1.0$, and 2.0; (b) $r = 5$ and $\theta = 0.25, 0.5$, and 1.0.

2. THE GAMMA DISTRIBUTION

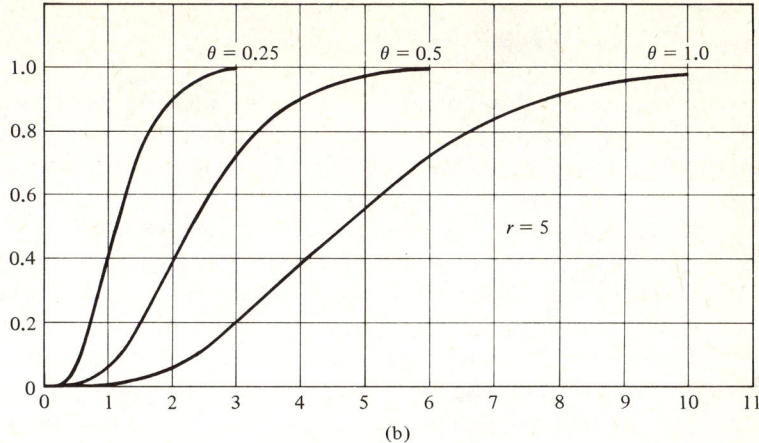

(b)

Figure 2.5 *(continued)*

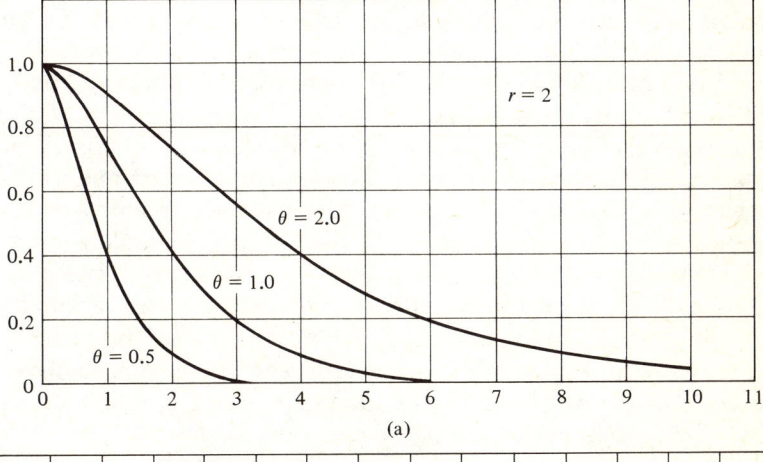

(a)

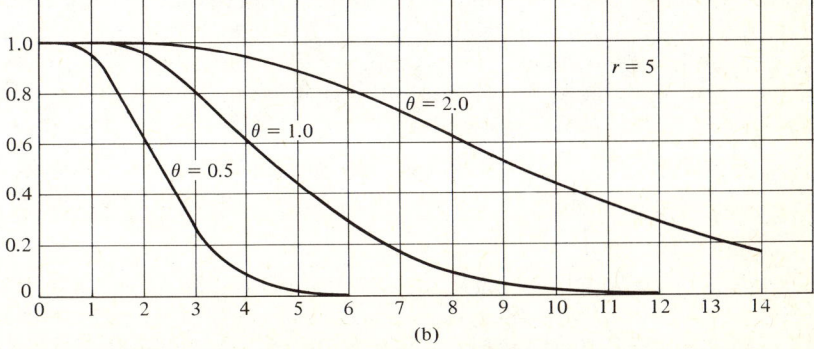

(b)

Figure 2.6 Graphs of the survival function for the gamma distribution with parameters (a) $r = 2$ and $\theta = 0.5$, 1.0, and 2.0; (b) $r = 5$ and $\theta = 0.5$, 1.0, and 2.0.

*3. THE WEIBULL DISTRIBUTION

The exponential distributions and gamma distributions are two large classes of continuous distributions that investigators have used in their attempts to find distributions which explain or describe the variation of nonnegative random variables. Important examples of such nonnegative random variables are lifetimes of individuals, lifetimes of mechanical, electrical, biological, or social systems, waiting times for service, learning times for animals or individuals, durations of epidemics, and traveling times for buses, trains, or planes. Nontemporal quantities that are nonnegative random variables include the strengths of various structural materials, the geometric dimensions of solid particles (for example, rocks), the physical dimensions of biological organisms, the intensity of the average level of radioactivity in rocks, the amount of rainfall in a given period of time, and the cost (in dollars) of industrial accidents. Although exponential or gamma distributions provide a reasonable "fit" to the frequency distributions of some of these random variables (and to the distributions of many other, equally important, nonnegative continuous random variables not mentioned above), in some cases the "fit" is not always as close as is desired, and in other cases the "fit" is unsatisfactory. Hence, scientists have continued to look for a class (or parametric family) of distributions that is large enough to explain the variability of as many nonnegative phenomena as possible, while still being simple enough in its mathematical form so that in any given experimental context we can select one distribution from this family to describe the variability of a given nonnegative random variable without the necessity of determining the values of a large number of parameters. The gamma distributions comprise one such parametric family of distributions. Another parametric family of distributions which has been used successfully to model the variability of nonnegative quantitative random phenomena is the family of *Weibull distributions*, named after the Swedish physicist, Waloddi Weibull, who in 1939 suggested this family of distributions.

If a random variable X has a Weibull distribution, then the probability density function $f_X(x)$ of X has the form

$$(3.1) \quad f_X(x) = \begin{cases} \left(\dfrac{\beta}{\alpha}\right)\left(\dfrac{x-\nu}{\alpha}\right)^{\beta-1} e^{-[(x-\nu)/\alpha]^\beta}, & \text{if } x \geq \nu, \\ 0, & \text{if } x < \nu. \end{cases}$$

Here, α and β are positive constants, while ν is either positive or zero (that is, ν is nonnegative). The three constants α, β, and ν are the parameters of the Weibull distribution; when α, β, and ν are known, the distribution whose probability density function is given by Equation (3.1) is completely determined.

3. THE WEIBULL DISTRIBUTION

Notice that the parameter ν tells us the smallest possible value of the random variable X [that is, the smallest number x for which the density function $f_X(x)$ is positive]. If we know the value of this parameter, we can subtract ν from X and obtain a new continuous random variable, $W = X - \nu$, which can be shown (see Section 1 of Chapter 10) to have the probability density function

$$(3.2) \qquad f_W(w) = \begin{cases} \left(\dfrac{\beta}{\alpha}\right)\left(\dfrac{w}{\alpha}\right)^{\beta-1} e^{-(w/\alpha)^\beta}, & \text{if } w \geq 0, \\ 0, & \text{if } w < 0. \end{cases}$$

Hence, W has a Weibull distribution with the same values of α and β as X, but with the value of the third parameter ν equal to zero. The smallest possible value of W is $w = 0$. Figure 3.1 gives the graphs of some Weibull

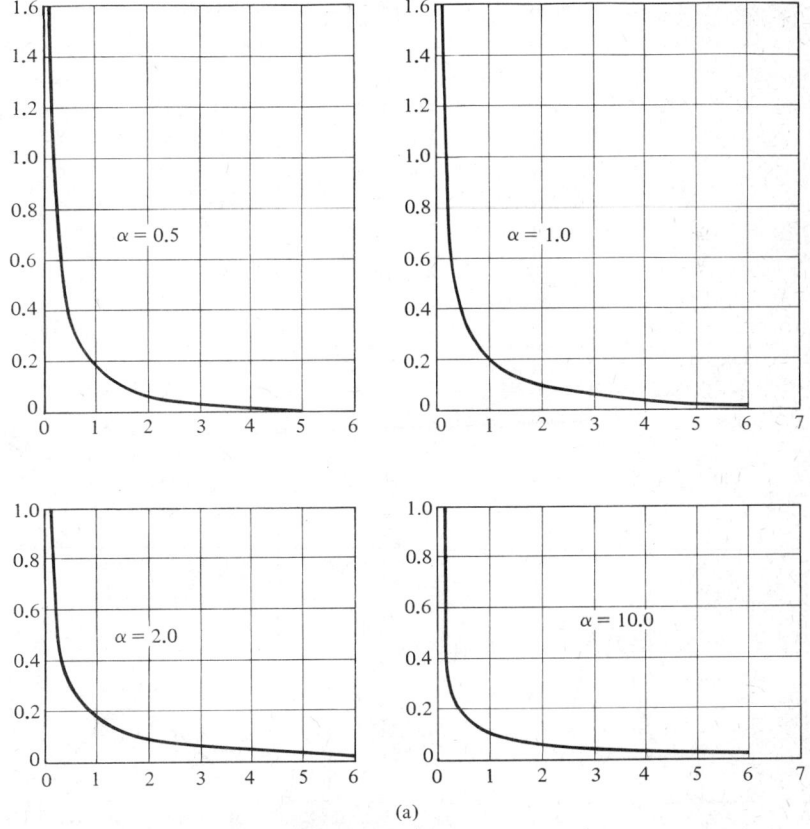

(a)

Figure 3.1 Graphs of the Weibull density function for (a) $\nu = 0$, $\beta = 1.0$, and $\alpha = 0.5, 1.0, 2.0,$ and 10.0; (b) $\nu = 0$, $\beta = 2.0$, and $\alpha = 0.5, 1.0, 2.0,$ and 10.0; (c) $\beta = 10.0$, $\nu = 0$ and $\alpha = 0.5, 1.0, 2.0,$ and 10.0.

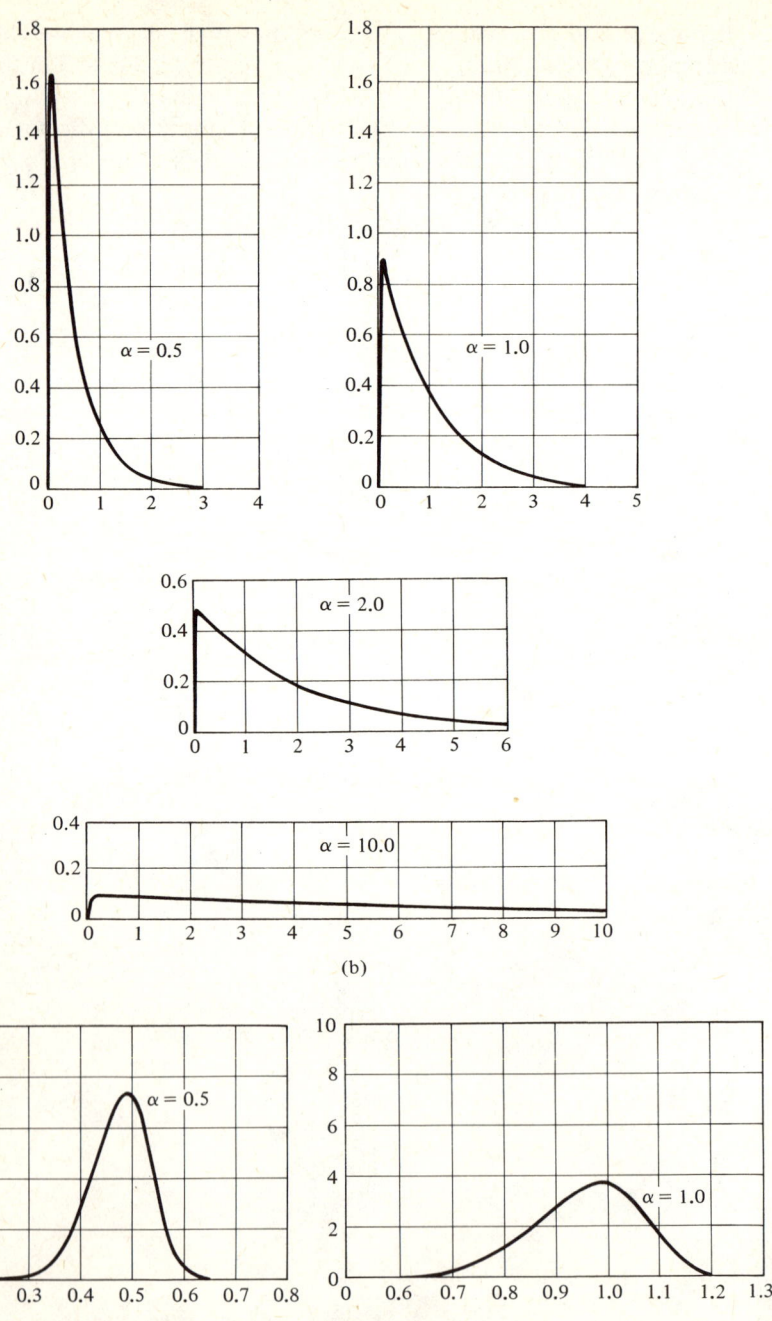

Figure 3.1 (*continued*)

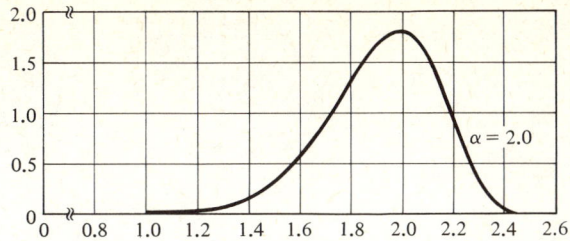

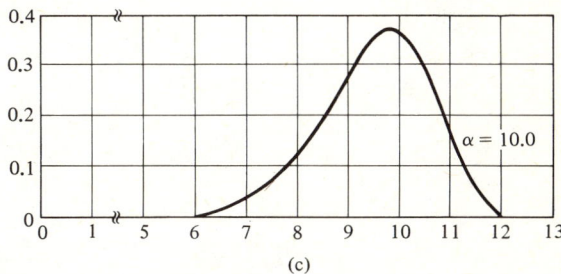

(c)

Figure 3.1 (*continued*)

density functions $f_W(w)$ for $\nu = 0$ and various values of the parameters α and β.

To show the effect of the parameter ν upon the density function $f_X(x)$ of the general Weibull distribution, Figure 3.2 gives the graphs of Weibull density functions $f_X(x)$ for $\nu = 0.5$ and the same values of α and β used in Figure 3.1. A comparison of Figures 3.1(a) and 3.2(a), for example, shows that the parameter ν acts as a measure of location for the general Weibull distribution (3.1) when the other two parameters α and β are fixed. However, when α, β, and ν are all allowed to vary, the location of the Weibull distribution is determined by all three parameters. Thus, for example, it can be shown that if X has a Weibull distribution with parameters α, β, and ν [that is, if X has the density function $f_X(x)$ given by Equation (3.1)], then the expected value μ_X of X equals

$$(3.3) \qquad \mu_X = \nu + \alpha \Gamma\left(1 + \frac{1}{\beta}\right),$$

where $\Gamma(1+a)$ can be found from tables of the gamma function (see Table 5.2 of Chapter 6).

The dispersion of a random variable X having a Weibull distribution with parameters α, β, and ν, depends on α and β, but not on ν. Indeed,

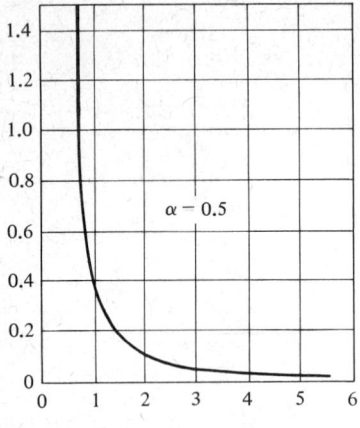
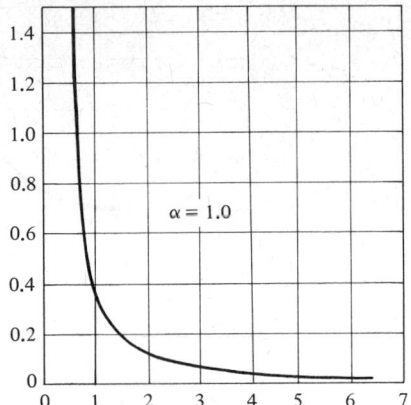
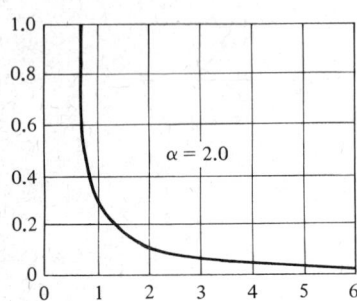
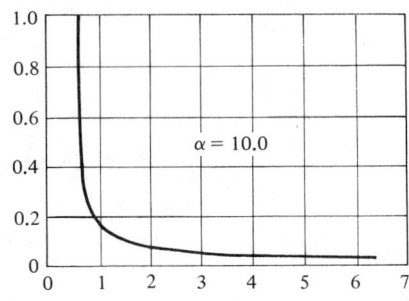

(a)

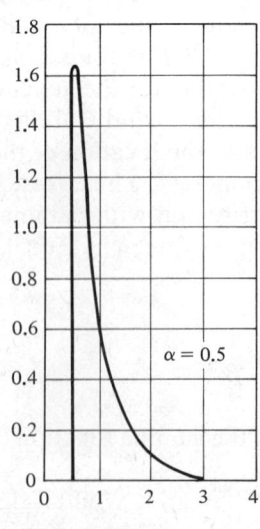
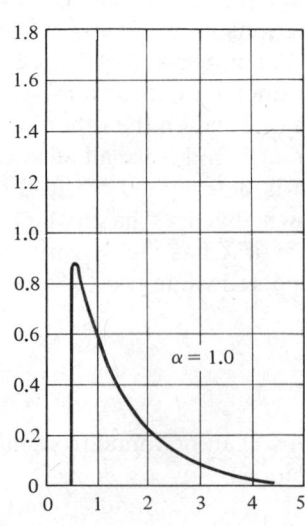

(b)

3. THE WEIBULL DISTRIBUTION

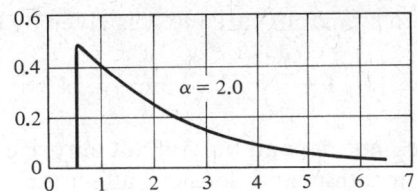

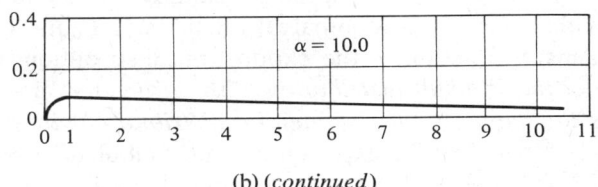

(b) *(continued)*

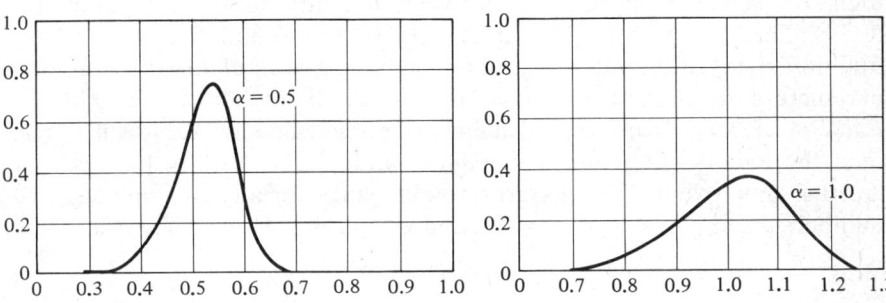

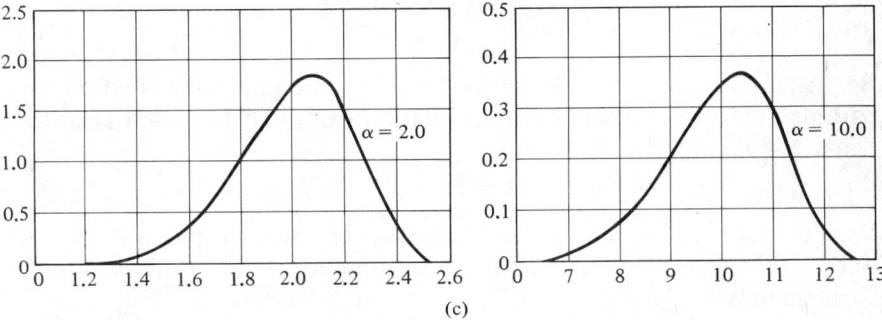

(c)

Figure 3.2 Graphs of the Weibull density function for (a) $\nu = 0.5$, $\beta = 1.0$, and $\alpha = 0.5, 1.0, 2.0,$ and 10.0; (b) $\nu = 0.5, \beta = 2.0$, and $\alpha = 0.5, 1.0, 2.0,$ and 10.0; (c) $\nu = 0.5, \beta = 10, \alpha = 0.5$, $1.0, 2.0,$ and 10.0.

the variance σ_X^2 of such a random variable X is given by the formula

(3.4) $$\sigma_X^2 = \alpha^2 \left[\Gamma\left(1 + \frac{2}{\beta}\right) - \Gamma\left(1 + \frac{1}{\beta}\right) \Gamma\left(1 + \frac{1}{\beta}\right) \right].$$

The fact that σ_X^2 does not depend on ν is not surprising since we know that ν is a measure of location, and does not affect the *shape* of the graph of the density function (compare Figures 3.1 and 3.2).

When $\alpha = 1/\theta$, $\beta = 1$, and $\nu = 0$, the probability density function $f_X(x)$ in (3.1) is of the same form [compare Equations (1.1) and (3.2)] as the probability density function of the exponential distribution with parameter θ. Hence, *the Weibull distribution with parameters $\alpha = 1/\theta$, $\beta = 1$, and $\nu = 0$ is the same as the exponential distribution with parameter θ.* Another relation between the exponential and Weibull distributions is the following: If $Y = [(X - \nu)/\alpha]^\beta$ has an exponential distribution with parameter $\theta = 1$, then X has a Weibull distribution with parameters α, β, and ν. Alternatively, if X has a Weibull distribution with parameters α, β, and ν, then $Y = [(X - \nu)/\alpha]^\beta$ has an exponential distribution with parameter $\theta = 1$. This last fact enables us to obtain the cumulative distribution function $F_X(x)$ of a random variable X having a Weibull distribution with parameters α, β, and ν. To do this, note that when $x < \nu$, $F_X(x) = P_X\{X \leq x\} = 0$, since ν is the smallest possible value for X. Now if $x \geq \nu$, then the events $\{(X - \nu)/\alpha \leq (x - \nu)/\alpha\}$ and $\{[(X - \nu)/\alpha]^\beta \leq [(x - \nu)/\alpha]^\beta\}$ are the same event. (This assertion holds since for any two nonnegative numbers a and b, $a \leq b$, if $a^\beta \leq b^\beta$, and vice versa.) Hence, if $x \geq \nu$,

(3.5) $$F_X(x) = P_X\{X \leq x\}$$
$$= P_X\left\{\frac{X - \nu}{\alpha} \leq \frac{x - \nu}{\alpha}\right\}$$
$$= P_X\left\{\left(\frac{X - \nu}{\alpha}\right)^\beta \leq \left(\frac{x - \nu}{\alpha}\right)^\beta\right\}.$$

Because we know that $[(X - \nu)/\alpha]^\beta$ has an exponential distribution with parameter $\theta = 1$, it follows from Equation (3.5) and Equation (1.3) that when $x \geq \nu$,

$$F_X(x) = 1 - e^{-[(x-\nu)/\alpha]^\beta}.$$

We conclude that if X has a Weibull distribution with parameters α, β, and ν, then the cumulative distribution function $F_X(x)$ of X is given by the formula

(3.6) $$F_X(x) = \begin{cases} 0, & \text{if } x < \nu, \\ 1 - e^{-[(x-\nu)/\alpha]^\beta}, & \text{if } x \geq \nu. \end{cases}$$

Both the gamma family of distributions and the Weibull family of distributions generalize the exponential family of distributions in the sense

3. THE WEIBULL DISTRIBUTION

that an exponentially distributed random variable X with parameter θ can also be said to have a gamma distribution with parameters θ and $r = 1$, and a Weibull distribution with parameters $\alpha = 1/\theta$, $\beta = 1$, and $\nu = 0$. However, there are gamma distributions that are not also Weibull distributions (for example, any gamma distribution with parameter $r \neq 1$) and Weibull distributions that are not gamma distributions (for example, any Weibull distributions with $\beta \neq 1$ or $\nu \neq 0$). Thus, the gamma and Weibull families of distributions are two different families of distributions, each of which generalizes the exponential family of distributions.

A Theoretical Explanation

The experience of many investigators has shown that the Weibull distributions provide good probability models for describing "length of life" and other endurance data. One explanation for the success of the Weibull distributions in describing such data is related to the following "weakest link" interpretation of endurance. Suppose an object is put under stress. Think of the object as being composed of a large number of separate parts, each of which has its own statistically independent random endurance time (lifetime). If any one of these parts fails (or breaks) under stress, the whole object experiences failure (breakdown). For example, the object could be a metal chain composed of a large number of links. If any link ("the weakest link") breaks, so does the chain. Thus, the lifetime of the object (the chain) is equal to the *minimum* lifetime of any of its parts (links). If the lifetimes of any class of objects (chains, rockets, transistors, and so on) have this property, then it can be shown that a Weibull distribution provides a close approximation to the distribution of these lifetimes. If instead of lifetimes we consider waiting times until the occurrence of some phenomenon, and if the given phenomenon can instantaneously occur if any one of a large number of statistically independent causes comes into effect (so that the waiting time until the given phenomenon occurs is the minimum of the times which elapse until one of the causes comes into effect), the Weibull distribution again provides a close fit to the distribution of such waiting times.

Fitting the Weibull Distribution to Data

Suppose that we have n statistically independent observations $X_1, X_2, \ldots, X_n$ of a random variable X which we believe has a Weibull distribution, and we want to find the values of the parameters α, β, and ν. Since ν is the smallest possible value of X, it seems reasonable to estimate ν by the smallest observed value $X_{\min}$ of X. That is,

(3.7) $\quad$ estimate of $\nu = $ minimum of $X_1, X_2, \ldots, X_n$
$\qquad\qquad\qquad = X_{\min}.$

[*Note:* In many cases, we know for theoretical reasons that ν has some value ν_0. In this case, we do not need to estimate ν from the data.] The remaining two parameters, α and β, can now be obtained through the use of Equations (3.3) and (3.4). First note that

$$\mu_X - \nu = \alpha \Gamma\left(1 + \frac{1}{\beta}\right), \qquad \sigma_X^2 + (\mu_X - \nu)^2 = \alpha^2 \Gamma\left(1 + \frac{2}{\beta}\right).$$

Thus, β can be obtained by dividing $(\mu_X - \nu)^2$ by $\sigma_X^2 + (\mu_X - \nu)^2$. This operation yields

$$(3.8) \qquad \frac{(\mu_X - \nu)^2}{\sigma_X^2 + (\mu_X - \nu)^2} = \frac{\Gamma\left(1 + \frac{1}{\beta}\right)\Gamma\left(1 + \frac{1}{\beta}\right)}{\Gamma\left(1 + \frac{2}{\beta}\right)}.$$

Once we have estimated values of μ_X and σ_X^2 (these can be obtained from the data in the usual manner; see the Appendix of Chapter 5), we can use the estimated value $\nu = X_{\min}$ for ν and then solve Equation (3.8) for β. In order to facilitate this computation, Table 3.1 provides values of $\Gamma(1+z)\Gamma(1+z)/\Gamma(1+2z)$ for values of z between 0 and 1. [*Note:* If z exceeds 1, Table 3.1 can still be used by recalling that $\Gamma(z+1) = z\Gamma(z)$.] Once an estimate of β has been determined from Equation (3.8), we can obtain an estimate of α through the formula

$$(3.9) \qquad \alpha = \frac{\mu_X - \nu}{\Gamma\left(1 + \frac{1}{\beta}\right)},$$

which is straightforwardly obtained from (3.3) by algebraic manipulation.

In a paper published in 1951, Weibull provided three examples in which a distribution of the form (3.1) was fit to data. Since that time, there have been many other applications of the Weibull distribution.

Table 3.1. Values of $\Gamma(1+z)\Gamma(1+z)/\Gamma(1+2z)$.

	.00	.01	.02	.03	.04	.05	.06	.07	.08	.09
0.00	1.0000	.9998	.9993	.9985	.9975	.9961	.9945	.9928	.9906	.9884
0.10	.9858	.9830	.9801	.9768	.9735	.9699	.9664	.9625	.9585	.9545
0.20	.9502	.9458	.9412	.9367	.9319	.9271	.9221	.9170	.9119	.9067
0.30	.9015	.8961	.8906	.8852	.8796	.8741	.8683	.8626	.8568	.8512
0.40	.8453	.8395	.8336	.8274	.8215	.8156	.8095	.8035	.7975	.7914
0.50	.7854	.7793	.7732	.7672	.7611	.7550	.7489	.7428	.7367	.7307
0.60	.7246	.7186	.7125	.7064	.7004	.6944	.6885	.6825	.6765	.6706
0.70	.6647	.6588	.6529	.6470	.6412	.6354	.6296	.6239	.6182	.6125
0.80	.6068	.6012	.5955	.5899	.5844	.5789	.5734	.5679	.5625	.5571
0.90	.5518	.5464	.5411	.5359	.5306	.5254	.5203	.5152	.5101	.5050

3. THE WEIBULL DISTRIBUTION

Example 3.1 (Size Distribution of Particles). The term fly ash refers to fine solid particles of noncombustible ash that are carried out of a bed of solid fuel by the draft of combustion. Table 3.2 shows a frequency distribution of the size (particle diameter) of fly ash, based on $n = 211$ observations.

For these data, Weibull notes that the smallest observed value of X is equal to 30 microns or 1.5 20-micron units. Thus, the estimated value for the parameter ν based on these data is

$$\nu = 1.5 \quad \text{(in 20-micron units).}$$

Notice that we do not obtain this estimate from Table 3.2, since from that table we can only determine the *interval* of numbers 1.5–2.5, 2.5–3.5, and so on, in which the smallest observed value falls.

Weibull does not supply us with estimated values of μ_X and σ_X^2. However, from Table 3.2 an estimate of μ_X is

$$\hat{\mu}_X = 2\left(\frac{3}{211}\right) + 3\left(\frac{11}{211}\right) + \cdots + 13\left(\frac{6}{211}\right) + 14\left(\frac{3}{211}\right) = 7.18,$$

and an estimate of σ_X^2 is

$$\hat{\sigma}_X^2 = (2.00 - 7.18)^2\left(\frac{3}{211}\right) + (3.00 - 7.18)^2\left(\frac{11}{211}\right)$$
$$+ \cdots + (14.00 - 7.18)^2\left(\frac{3}{211}\right) = 6.70.$$

Table 3.2. Observed frequency distribution of the size X of fly ash (in 20-micron units).

Midpoint x of Interval of Sizes	Observed Frequency of the Event $\{x - 0.5 \leq X \leq x + 0.5\}$
2	3
3	11
4	20
5	22
6	29
7	41
8	24
9	25
10	13
11	9
12	5
13	6
14	3
	211

Thus,

$$\frac{(\hat{\mu}_X - \nu)^2}{\hat{\sigma}_X^2 + (\hat{\mu}_X - \nu)^2} = \frac{(7.18 - 1.50)^2}{6.70 + (7.18 - 1.50)^2} = 0.828.$$

Solving the equation

$$0.828 = \frac{\Gamma\left(1 + \frac{1}{\beta}\right)\Gamma\left(1 + \frac{1}{\beta}\right)}{\Gamma\left(1 + \frac{2}{\beta}\right)}$$

for β [see Equation (3.8)], we find from Table 3.1 that $z = 1/\beta$ lies between 0.42 and 0.43. By linear interpolation we obtain $z = 0.429$, so that $\beta = 1/0.429 = 2.331$. Finally, from Equation (3.9) and Table 5.1 of Chapter 6,

$$\alpha = \frac{\hat{\mu}_X - \nu}{\Gamma\left(1 + \frac{1}{\beta}\right)} = \frac{7.18 - 1.5}{\Gamma(1 + 0.429)} = \frac{5.68}{0.8860} = 6.41.$$

Thus, based on the data in Table 3.2, we have determined that the distribution of the size X (in 20-micron units) of fly ash seems to fit a Weibull distribution with parameters $\alpha = 6.41$, $\beta = 2.331$, and $\nu = 1.50$.

Example 3.2 (Strength of Steel). Weibull (1951) also discusses an example in which the yield strength X of a Bofors steel was measured. The resulting frequency distribution is given in Table 3.3.

The minimum value of the yield strength observed was 38.57 kilograms per millimeter squared (kg/mm²) or 30.25 in units of 1.275 kg/mm². Thus, to fit a Weibull distribution to the data in Table 3.3, we take $\nu = 30.25$.

From Table 3.3, we find that

$$\hat{\mu}_X = (32)\left(\frac{10}{389}\right) + (33)\left(\frac{23}{389}\right) + \cdots + (40)\left(\frac{14}{389}\right) + (41)\left(\frac{6}{389}\right) = 36.152,$$

and that

$$\hat{\sigma}_X^2 = (32.000 - 36.152)^2 \left(\frac{10}{389}\right) + (33.000 - 36.152)^2 \left(\frac{23}{389}\right) + \cdots$$

$$+ (40.000 - 36.152)^2 \left(\frac{14}{389}\right) + (41.000 - 36.152)^2 \left(\frac{6}{389}\right)$$

$$= 3.995.$$

Thus,

$$\frac{(\hat{\mu}_X - \nu)^2}{\hat{\sigma}_X^2 + (\hat{\mu}_X - \nu)^2} = \frac{(36.152 - 30.250)^2}{3.995 + (36.152 - 30.250)^2} = 0.8971,$$

so that from Equation (3.8) and by interpolation in Table 3.1,

$$\text{estimated } \beta = \frac{1}{0.308} = 3.247.$$

3. THE WEIBULL DISTRIBUTION

Table 3.3. Observed frequency distribution of the yield strength X in 1.275 kg/mm² units of a Bofors steel.

Midpoint x of Interval of Yield Strengths from $x-0.5$ to $x+0.5$	Observed Frequency of the Event $\{x-0.5 \leq X \leq x+0.5\}$
32	10
33	23
34	48
35	80
36	63
37	65
38	47
39	33
40	14
41[a]	6
	389

[a] In the published version of Weibull's paper (1951), this number is given as 42, not 41. However, we suspect that this is a typographical error.

Finally, from Equation (3.9) and Table 5.1 of Chapter 6,

$$\text{estimated } \alpha = \frac{\hat{\mu}_X - \nu}{\Gamma\left(1+\frac{1}{\beta}\right)} = \frac{36.152 - 30.250}{\Gamma(1+0.308)} = \frac{5.902}{0.8963} = 6.58.$$

Therefore, we have determined that the distribution of the yield strength X (in units of 1.275 kg/mm²) of a Bofors steel seems to fit a Weibull distribution with parameters $\alpha = 6.58$, $\beta = 3.247$, and $\nu = 30.25$.

The Survival Function

As in the case of the exponential distribution, the survival function of a Weibull distribution can be expressed in a convenient form. If X has a Weibull distribution with parameters α, β, and ν, then the survival function $1 - F_X(x)$ of X is given by the equation

$$(3.10) \qquad 1 - F_X(x) = \begin{cases} 1, & \text{if } x < \nu, \\ e^{-[(x-\nu)/\alpha]^\beta}, & \text{if } x \geq \nu. \end{cases}$$

Recall that if X denotes the length of life of an item, individual, or system, the survival function gives the probability that the item, individual, or system will "live" or "survive" more than x units of time. We can use the survival function (3.10) to test the fit of the Weibull distribution to data (see Section 1) by comparing the graphs of $1 - F_X(x)$, where α, β, and ν are given values determined by the data, and the sample survival function

$$1 - F_n(x) = \frac{\text{number of observations which exceed } x}{n},$$

where n is the total number of observations $X_1, X_2, \ldots, X_n$ in the data. In Figures 3.3 and 3.4, we compare the theoretical survival function and sample survival function based on the data in Examples 3.1 and 3.2, respectively. These comparisons support our earlier assertion that the variability of each of the random variables discussed in these two examples is adequately described by a Weibull distribution.

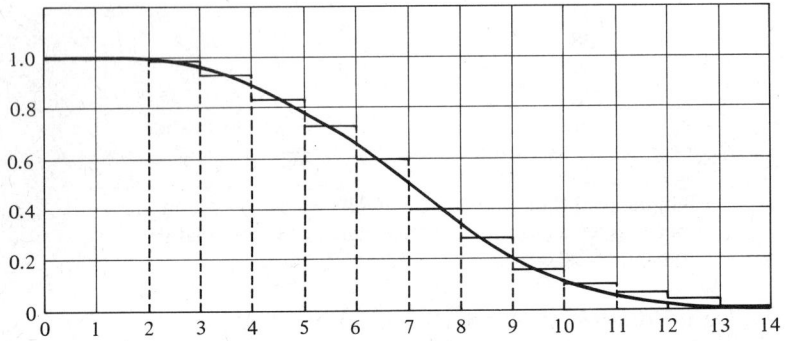

Figure 3.3 Comparison of the observed and theoretical survival functions for the data of Example 3.1. The theoretical survival function is based on the Weibull distribution with parameters $\alpha = 6.41$, $\beta = 2.298$, and $\nu = 1.5$.

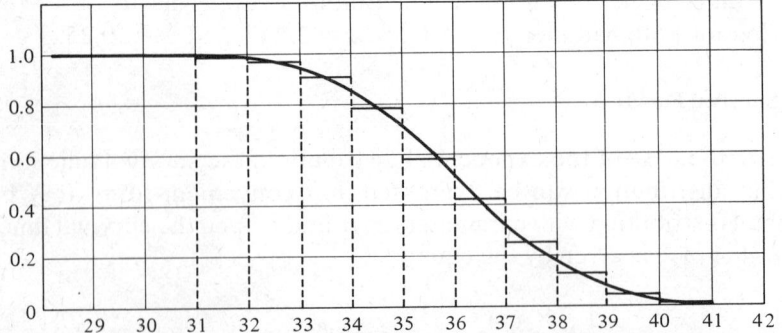

Figure 3.4 Comparison of the observed and theoretical survival functions for the data of Example 3.2. The theoretical survival function is based on the Weibull distribution with parameters $\alpha = 6.58$, $\beta = 3.247$, and $\nu = 1.5$.

4. THE UNIFORM DISTRIBUTION

A continuous random variable X has a *uniform distribution* over the interval of numbers from a to b if the probability of the event $\{x_1 \leq X \leq x_2\}$ is the same as the probability of the event $\{x_3 \leq X \leq x_4\}$ for all numbers x_1, x_2, x_3, x_4 between a and b for which $x_2 - x_1 = x_4 - x_3$. That is, X has a uniform distribution if intervals of possible values which have equal length have equal probability of containing an observed value of X.

The probability density function $f_X(x)$ of a random variable X having a uniform distribution over the interval of numbers from a to b is given by the equation

(4.1) $$f_X(x) = \begin{cases} \dfrac{1}{b-a}, & \text{if } a \leq x \leq b, \\ 0, & \text{if } x < a \text{ or } x > b. \end{cases}$$

The graph of the density function $f_X(x)$ for $a = 0$, $b = 1$, and the graph of the density function $f_X(x)$ for $a = 1$, $b = 3$, are shown in Figure 4.1. From Figure 4.1, we can see why the uniform distribution is also called the *rectangular distribution*.

When X has the uniform distribution over the range of numbers from a to b, knowledge of a and b completely determines the density function (4.1) of X. Thus, a and b are the parameters of the uniform distribution.

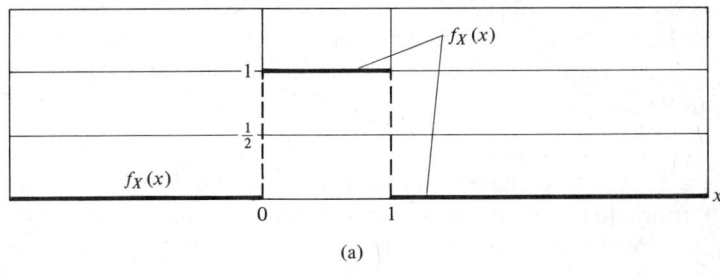

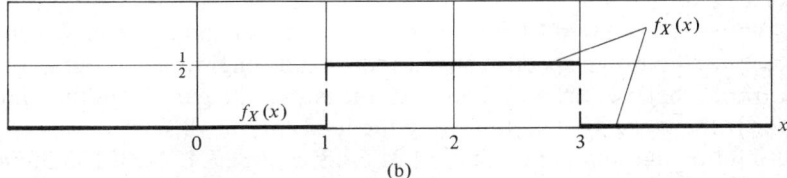

Figure 4.1 Graph of the probability density function $f_X(x)$ of the uniform distribution for (a) $a = 0$, $b = 1$; (b) $a = 1$, $b = 3$.

Note from Equation (4.1) that a is the smallest possible value of X and that b is the largest possible value of X. That is,

$$P_X\{X < a\} = P_X\{X > b\} = 0,$$

or equivalently, $P_X\{a \leq X \leq b\} = 1$. Because it is awkward to keep saying "X has a uniform distribution over the interval of numbers between a and b," we adopt a shorthand notation for this statement, and simply say "X is $U[a,b]$." Thus "X is $U[1,3]$" means that X has a uniform distribution over the interval of numbers between $a = 1$ and $b = 3$, or equivalently that X has a uniform distribution with parameters $a = 1$ and $b = 3$.

It can be shown, using the theory of integral calculus, that if X is $U[a,b]$, then the expected value μ_X and variance σ_X^2 of X are given by

$$(4.2) \qquad \mu_X = \frac{a+b}{2}, \qquad \sigma_X^2 = \frac{(b-a)^2}{12}.$$

Thus, knowledge of μ_X and σ_X^2 is equivalent to knowledge of the parameters a and b. However, in most scientific applications of the uniform distribution the parameters a and b are known (usually on theoretical grounds) and do not need to be estimated from data. Also, when a and b are not known, estimation of a and b through the use of estimates of μ_X and σ_X^2 can lead to unreasonable values of a and b. (See Chapter 6, Section 6 for a demonstration of a similar assertion concerning the discrete uniform distribution.) If statistically independent observations $X_1, X_2, \ldots, X_n$ are made on X, and if the parameters a and b are not known, then since a is the smallest possible value and b is the largest possible value of X, a can be estimated by the smallest value among the observations $X_1, X_2, \ldots, X_n$, and b can be estimated by the largest value among the observations $X_1, X_2, \ldots, X_n$.

Example 4.1. A traffic engineer is studying the length X of gaps (in feet) between following cars on a two-lane highway. He assumes that X is $U[a,b]$, where the parameter b is the largest gap that a given individual will allow between his car and the next car, and a is the smallest such gap. Over a period of time in which the lead car in a group of cars holds to a constant speed of 60 miles per hour, and the cars in a line of cars seem to be keeping their own speeds constant, the traffic engineer photographs the cars from a helicopter, and then later measures the gaps from the photograph. If there are 10 cars following the lead car, the observed gaps might be the following numbers: 15.1, 31.3, 64.2, 51.6, 87.4, 25.0, 103.3, 74.9, 98.5, and 118.7. Since the smallest observed gap is 15.1 feet and the largest observed gap is 118.7 feet, the traffic engineer estimates the parameter a to be 15.1 and the parameter b to be 118.7.

Probability Calculations

Probability calculations for the uniform distribution are quite easy to perform. Suppose, for example, that X is $U[a,b]$ and that we want the probability of the event $\{X \leq x\}$. Calculation of the probability of events of this form for all numbers x allows us to determine the cumulative distribution function $F_X(x)$ of X. If the number x is strictly less than a (that is, $x < a$), then $P\{X < x\} = 0$. This assertion follows since no numbers less than a are possible values of X. If the number x is strictly greater than b, then the event $\{X \leq x\}$ contains all possible values of X, so that $P_X\{X \leq x\} = 1$. Finally, if the number x is between a and b, then $F_X(x)$ is given by the shaded area shown in Figure 4.2. Note that this area is the area of a

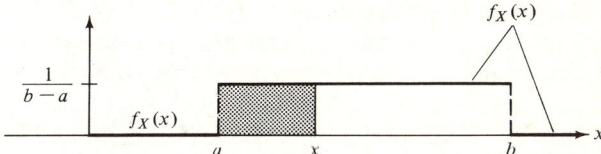

Figure 4.2 Graph of the probability density function $f_X(x)$ of a uniform distribution. The area of the shaded rectangle equals the probability $P_X\{X \leq x\}$ of the event $\{X \leq x\}$.

rectangle with base of length $x - a$ and height equal to $1/(b-a)$. Thus,

$$F_X(x) = \text{area of shaded rectangle} = (x-a)\left(\frac{1}{b-a}\right).$$

Putting these results together, we obtain the following formula for $F_X(x)$:

(4.3) $$F_X(x) = \begin{cases} 0, & \text{if } x < a, \\ \dfrac{x-a}{b-a}, & \text{if } a \leq x \leq b, \\ 1, & \text{if } x > b. \end{cases}$$

The graph of $F_X(x)$ for $a = 1$, $b = 3$, is shown in Figure 4.3.

Using Formula (4.3) for the cumulative distribution function $F_X(x)$ of X, and using Table 3.2 of Chapter 4, we can find the probabilities of events of the form: $\{x < X\}$, $\{x_1 \leq X \leq x_2\}$, $\{x_1 < X \leq x_2\}$, $\{x_1 \leq X < x_2\}$, $\{x_1 < X < x_2\}$, and so on.

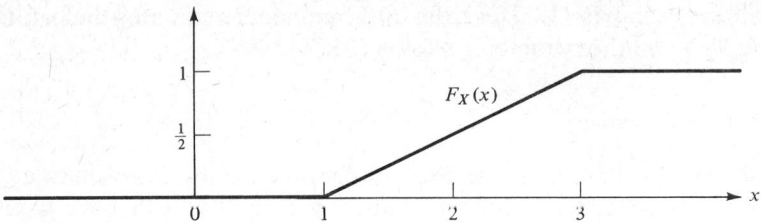

Figure 4.3 Graph of the cumulative distribution function $F_X(x)$ of the uniform distribution with parameters $a = 1, b = 3$.

Example 4.2. The last commuter train to a certain suburb of a large city leaves the railroad station at 6:00 p.m. A businessman who wants to catch that train is late leaving his office and does not catch a cab until 5:40 p.m. Under ideal driving conditions, it takes 10 minutes to drive from the businessman's office to the train station, but in rush hours it has taken as much as 50 minutes in the past. The businessman assumes that it will take him 3 minutes to get on his train once he arrives at the train station. Let X be the number of minutes needed to drive from office to train station. The businessman assumes that it is equally likely that X will be any number from $a = 10$ to $b = 50$; that is, he assumes that X is $U[10,50]$. The businessman will miss his train if $X > 17$ minutes (since if X is greater than 17, it will take the businessman more than 20 minutes to get on the train, and he has only 20 minutes, from 5:40 p.m. to 6:00 p.m., to make the train). Using Equation (4.3) and Table 3.2 of Chapter 4, we can calculate that

$$P_X\{X > 17\} = 1 - F_X(17)$$
$$= 1 - \frac{17 - a}{b - a}$$
$$= 1 - \frac{17 - 10}{50 - 10} = \frac{33}{40}.$$

Thus, the businessman has a probability of 33/40 of missing his train.

Example 4.3. A biologist is studying the homing behavior of pigeons. He has taken the pigeons from their cote and kept them for 2 or 3 days in a room which is artificially lit in a fashion that simulates natural lighting (that is, light provided by the sun), but in which dawn is made to appear 12 hours out of phase. If the pigeons orient themselves only by the position of the sun, then when they are brought back into the open air, their

4. THE UNIFORM DISTRIBUTION

time sense will be confused and they will misorient themselves. As a working hypothesis, the biologist assumes that once released, the confused pigeon will initially pick a direction in which to fly according to a uniform distribution. More precisely, if the bird is released at point A (see Figure 4.4), then relative to line L his line of flight will make an angle of X degrees with line L, where X is uniformly distributed over the interval of numbers from $a = 0$ to $b = 360$ (that is, X is $U[0,360]$). It has been observed that the lines of flight taken by pigeons who were not confused in the manner described above (and who had come from the same cote as the pigeons who were confused) varied from an angle of 210 degrees from line L to an angle of 220 degrees from line L (see Figure 4.4). The biologist decides that the range of values for X between 210 degrees and 220 degrees corresponds to "accurate orientation." Given that the biologist's working hypothesis is correct and that the angle X of line of flight for a confused pigeon is $U[0,360]$, the probability that a confused pigeon chooses a line of flight corresponding to "accurate orientation" is equal to

$$P_X\{210 \leq X \leq 220\} = F_X(220) - F_X(210)$$

$$= \frac{220-a}{b-a} - \frac{210-a}{b-a}$$

$$= \frac{220-0}{360-0} - \frac{210-0}{360-0}$$

$$= \frac{1}{36}.$$

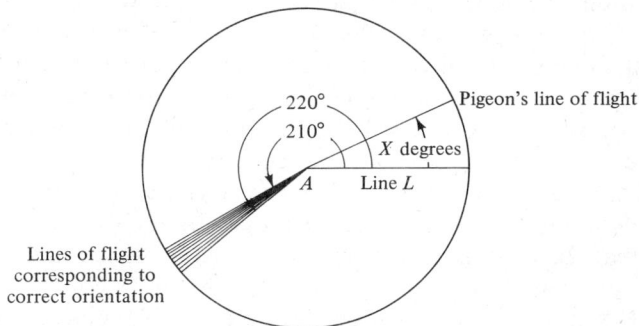

Figure 4.4 Diagram illustrating the experiment described in Example 4.3. The pigeon is released at point A. Its line of flight makes an angle of X degrees (measured counterclockwise) with the fixed line L. Lines of flight making angles of from 210 to 220 degrees correspond to orientations ordinarily taken by pigeons returning home.

Thus, if the hypothesis of the biologist is correct, it is rather improbable that a confused pigeon will choose a line of flight corresponding to "correct orientation."

Applications

Examples 4.1 and 4.3 illustrate two areas of scientific research into random phenomena in which the uniform distribution serves as a reasonable model. In these applications, and in other scientific applications, the uniform distribution is often adopted as a working hypothesis.

In measurement theory, the uniform distribution with $a = 0$, $b = (0.5)10^{-k}$ is often used to represent the distribution of "rounding-off" errors in values tabulated to the nearest k decimal places. That is, the unsigned difference, $X = |Y - Y_R|$, between the actual observed value Y of a random variable and the rounded-off value Y_R is assumed to have a uniform distribution with parameters $a = 0$ and $b = (0.5)10^{-k}$. The assumption that $|Y - Y_R|$ has a uniform distribution can be used to motivate the general method of estimation for probability density functions (that is, the construction of histograms) described in Chapter 4.

The uniform distribution is mathematically perhaps the simplest possible continuous distribution [see Equation (4.1)]. For this reason, general probabilistic results (holding true for large classes of continuous distributions) are often checked by "trying them out" for the uniform distribution. The particular uniform distribution for which $a = 0$, $b = 1$, is of theoretical importance for another reason. Computer programs (and tables) exist which can provide a succession of values that resemble, in their variation, the successive observations which would be obtained by observing a random variable X having a uniform distribution over the interval of numbers from $a = 0$ to $b = 1$. Such computer programs are called *random number generators*. Using such computer programs, we can generate successions of values that resemble, in their variation, the successive (and statistically independent) observations made upon a random variable Y having *any* continuous distribution that we desire. This *simulation* of the random variable Y can be accomplished as follows. Suppose that the cumulative distribution function of the random variable Y which we are simulating is $F_Y(y)$. Recall that for a continuous random variable, $F_Y(y)$ is always nondecreasing and has no jumps (see Chapter 4). We use our random number generator to generate a value x of a random variable X having a uniform distribution over the interval of numbers from $a = 0$ to $b = 1$. Then, from a graph (or table) of $F_Y(y)$, we find the smallest value y^* of y for which $F_Y(y) = x$. This value y^* is then the observed value of the (simulated) random variable Y. If we want yet another simulated observation on Y, we generate another value x of X using our random number generator and repeat the above steps.

Example 4.4 (Simulation of an Exponentially Distributed Random Variable). Suppose we desire to simulate a random variable Y having an exponential distribution with parameter $\theta = 1$ (see Section 1). The cumulative distribution function, $F_Y(y)$, of this distribution is (for $y > 0$)

$$F_Y(y) = 1 - e^{-y}.$$

Using our random number generator, we generate a value x of the uniformly distributed random variable X. Suppose the value so obtained is $x = 0.93$. Then, to obtain the observed value y^* of the simulated random variable Y, we solve

$$F_Y(y) = 1 - e^{-y} = 0.93$$

for y. Thus, $e^{-y^*} = 1 - 0.93 = 0.07$, or $y^* = -\log(0.07)$, where $\log u$ is the *natural logarithm* of u. Since $-\log(0.07)$ is equal to $2.65926...$, the observed value of the simulated random variable Y is equal to $2.65926...$.

The procedures described above for simulating the observations which we might observe if we observed a random variable Y having a given distribution are often called *Monte Carlo simulation methods* in the literature of probability and statistics. Monte Carlo simulations provide us with a way of mimicking observations that might be made on natural phenomena, and thus allow us to establish and verify results that are difficult to obtain by straightforward mathematical calculations, or whose verification would involve expensive experimentation. As an example of the verification of results that are difficult to obtain mathematically, Monte Carlo simulation might provide us with a way of answering the question (raised in Chapter 7) as to the number n of statistically independent observations $Y_1, Y_2, ..., Y_n$ on a given random variable Y which is sufficient to permit us to approximate the distribution of $S_n = Y_1 + Y_2 + \cdots + Y_n$ by a normal distribution with specified parameters μ and σ^2. For example, if we wanted to see if $n = 10$ is large enough, we could use Monte Carlo simulation to generate, say, 10,000 observations of the random variable Y. Then, adding these 10,000 observations 10 at a time, we would obtain 1000 statistically independent observations of S_{10}. From these 1000 observations of S_{10}, we could construct a sample frequency distribution. The closeness of the agreement between the observed and theoretical frequencies would then give us some idea of whether or not $n = 10$ is a sufficiently large number of observations to permit us to say that S_{10} is approximately normally distributed.

Monte Carlo simulations are also frequently used to simulate probabilistic learning models (where the task to be learned involves spending periods of time which are prohibitively long), the lifetime of extremely complex electrical systems (which are too expensive to construct to permit them to be used in experimentation), or the lifetime of human beings exposed to extreme stress (for example, disease, heat, low gravity, and so on).

*5. THE BETA DISTRIBUTION

There are many continuous quantitative phenomena of interest to science and industry which take on values that are bounded above and below by known numbers a and b. We have given three examples of such phenomena in Section 4 (Examples 4.1, 4.2, and 4.3). Other examples include: (i) the distance from one end of a steel bar of known length to the point at which failure (breakage) occurs when the bar is placed under stress, (ii) the proportion of total farm acreage spoiled by a certain kind of fungus, (iii) the percentage of defective items in a given shipment of items, (iv) the ratio of the length of the femur to the total length of the arm of a given individual, and (v) the fraction of individuals in a given population who are able to answer a given question correctly at a given time. Of these last examples, notice that examples (ii), (iv), and (v) all involve random variables that are fractions; thus, these variables have possible values between $a = 0$ and $b = 1$. The random variable described in example (iii), on the other hand, is a percentage, and thus has possible values between $a = 0$ and $b = 100$. Finally, the random variable mentioned in example (i) has possible values between $a = 0$ and $b =$ length of the steel bar.

Out of all of the continuous probability models which we have discussed so far (both in Sections 1 to 4 of this chapter and in Chapter 7), only the uniform distribution is truly appropriate for modeling a quantitative random phenomenon whose possible values lie in a restricted interval of numbers (that is, an interval of numbers lying between two numbers a and b). It is true that the normal, exponential, gamma, and Weibull distributions can provide close approximations to the distributions of such random variables in certain cases; namely, cases in which the probability of the event that the value of the random variable X lies outside of the interval of numbers from a to b (that is, the event $\{X < a \text{ or } X > b\}$) is very small when calculated under one of these distributions (normal, exponential, gamma, or Weibull). Strictly speaking, however, these families of distributions are not applicable for describing the variation of a random variable whose possible values lie in the restricted interval of numbers from a to b, because each distribution in these families of distributions (normal, exponential, gamma, and Weibull) assigns positive probability density to values outside of this restricted interval of numbers.

If we know that a random variable X of interest to us has possible values between the *known* numbers a and b, then (as we have seen in Section 4)

5. THE BETA DISTRIBUTION

there is only *one* uniform distribution that can be used to model the variability of X. Since the shape of the distributions (probability density functions) of such random variables are surely not *all* of rectangular form, the class of uniform distributions is not by itself adequate to model the variability of all random variables X whose possible values lie in a restricted interval of numbers.

A class of distributions which includes the uniform distribution, and which is rich enough to provide models for most random variables having a restricted range of possible values, is the class of *beta distributions*. Because the class of beta distributions is usually defined only for random variables Y having possible values in the interval of numbers between 0 and 1, to use this class of distributions to model the variability of a random variable X having possible values in the interval of numbers between a and b, we must first consider the transformed random variable

$$Y = \frac{X-a}{b-a},$$

which has possible values between 0 and 1. Once we have found a beta distribution which describes the variability of Y, it is not difficult to obtain the form of the distribution which describes the variability of X (see Section 1 of Chapter 10).

Example 5.1. Steel bars 12 inches in length are held in a vise at one end and loaded with a heavy weight at the other end. At very low temperatures these bars will often break in one or more places. Assuming that the bar breaks in only one place, let X be the distance, in inches, from the end of the bar which is held in the vise to the point of breakage. Then X is a random variable with possible values between $a = 0$ and $b = 12$. To find a distribution for X from among the class of beta distributions, we let $Y = (X-0)/(12-0) = X/12$, and find a beta distribution which describes the variation of Y. Since $Y = (X-a)/(b-a)$, the events $\{X \leq x\}$ and $\{Y \leq (x-a)/(b-a)\}$ are the same event. Thus, from the distribution of Y, we can find the cumulative distribution function, $F_X(x) = P_X\{X \leq x\} = F_Y((x-a)/(b-a))$ of X. Knowledge of the cumulative distribution function of X tells us the distribution of X [in that we can compute the probability of any event concerning X from knowledge of $F_X(x)$; see Chapter 4]. Therefore, to obtain the distribution of X, it is sufficient to find the distribution of Y.

Example 5.2. A meteorolgist is studying the proportion X of the area of a certain geographical region that recieves rain when a certain type of cloud formation is observed over that region. Since X is a proportion, the possible values of X are between $a=0$ and $b=1$. In this case, $Y = (X-0)/(1-0) = X$, so that no transformation is needed.

Assume now that we have transformed our random variable X into the random variable Y, where Y has possible values between 0 and 1. If Y has a beta distribution, then the probability density function $f_Y(y)$ of Y has the form

(5.1) $$f_Y(y) = \begin{cases} \dfrac{y^{r-1}(1-y)^{s-1}}{B(r,s)}, & \text{if } 0 \leq y \leq 1, \\ 0, & \text{if } y < 0 \text{ or } y > 1. \end{cases}$$

Here r and s are positive numbers, and are the parameters of the beta distribution; when the values of r and s are known, the distribution (5.1) is completely determined. The constant $B(r,s)$, which depends only on the values of r and s, is called the *beta function* by mathematicians; the value of $B(r,s)$ can be calculated from the formula

(5.2) $$B(r,s) = \frac{\Gamma(r)\Gamma(s)}{\Gamma(r+s)},$$

where $\Gamma(u)$ is the gamma function (see Table 5.1 of Chapter 6). To illustrate the diversity of shapes taken on by the graph of the probability density function $f_Y(y)$ of the beta distribution, Figure 5.1 gives graphs of $f_Y(y)$ for

$$r = \tfrac{1}{2}, 2, 3, 5, \qquad s = \tfrac{1}{2}, 1, 2, 3, 5, 10.$$

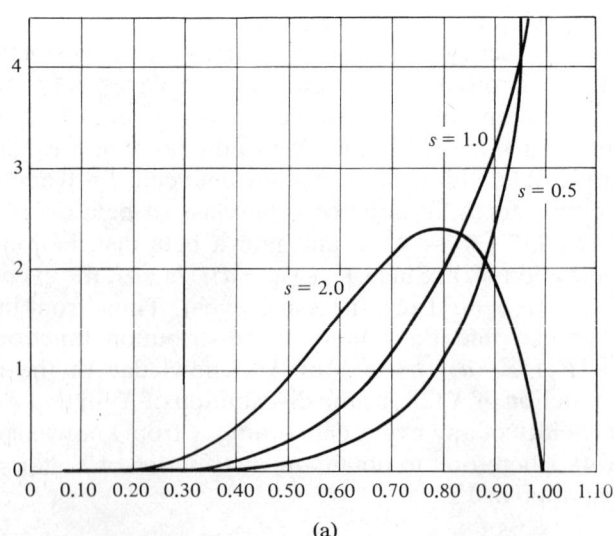

(a)

Figure 5.1 Graphs of the beta density function for (a) $r=5$ and $s=0.5, 1.0, 2.0, 3.0, 5.0,$ and 10.0; (b) $r=2$ and $s=0.5, 1.0, 2.0, 3.0, 5.0,$ and 10.0; (c) $r=3$ and $s=0.5, 1.0, 2.0, 3.0, 5.0,$ and 10.0; (d) $r=0.5$ and $s=0.5, 1.0, 2.0, 3.0, 5.0,$ and 10.0.

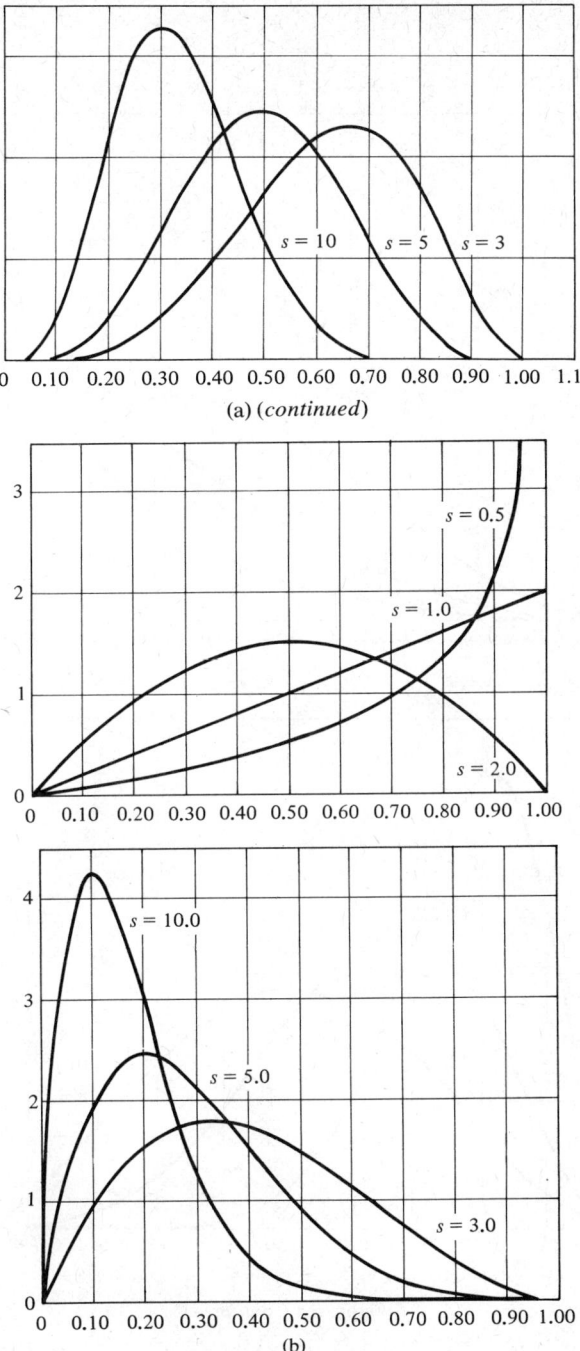

Figure 5.1 (continued)

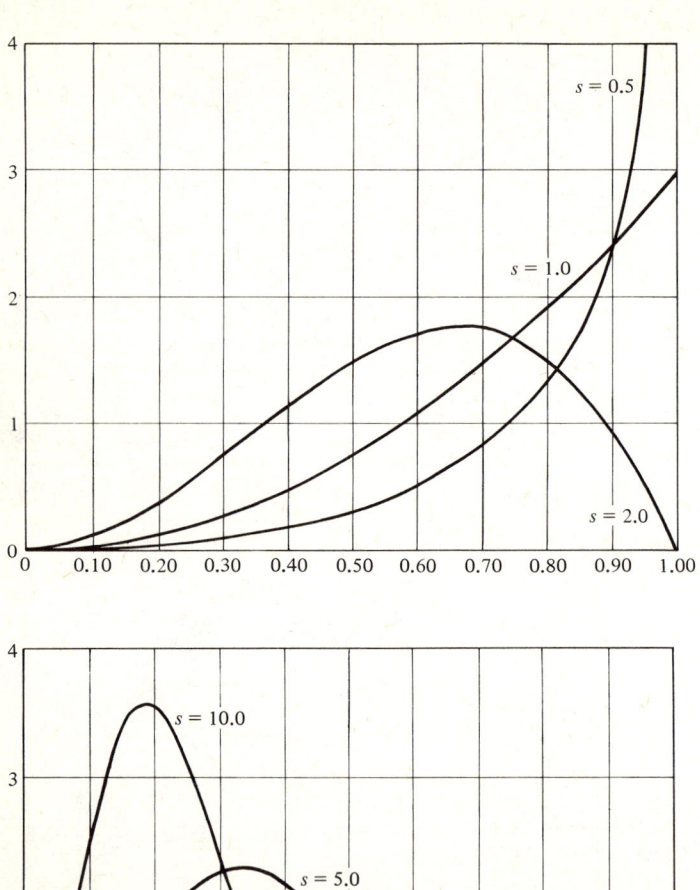

(c)

Figure 5.1 (*continued*)

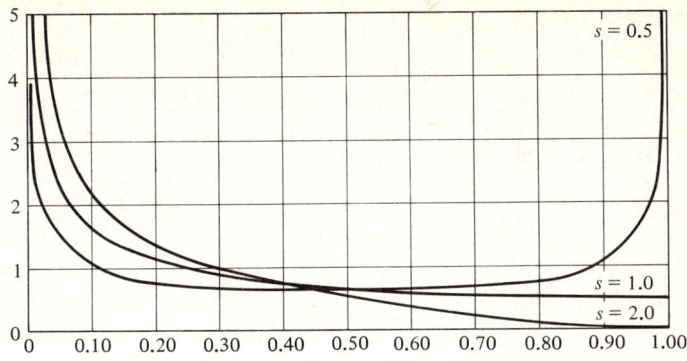

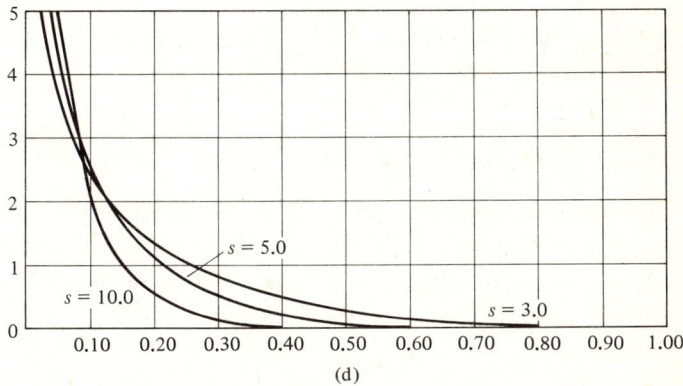

Figure 5.1 (*continued*)

Note from Figure 5.1 that the beta distribution for $r = 1$, $s = 1$ is the same as the uniform distribution $U[0,1]$.

If Y has a beta distribution with parameters r and s, then the expected value μ_Y and the variance σ_Y^2 of Y are given by

(5.3) $$\mu_Y = \frac{r}{r+s}, \qquad \sigma_Y^2 = \frac{rs}{(r+s)^2(r+s+1)}.$$

If we know the values of μ_Y and σ_Y^2, we may determine r and s by noting that

$$\sigma_Y^2 = \frac{rs}{(r+s)^2(r+s+1)} = \mu_Y(1-\mu_Y)\frac{1}{r+s+1}.$$

Thus,

$$r+s = \frac{\mu_Y(1-\mu_Y)}{\sigma_Y^2} - 1,$$

and it follows from Equation (5.3) that

$$(5.4) \quad r = \mu_Y \left[\frac{(\mu_Y)(1-\mu_Y)}{\sigma_Y^2} - 1 \right], \quad s = (1-\mu_Y)\left[\frac{(\mu_Y)(1-\mu_Y)}{\sigma_Y^2} - 1 \right].$$

Fitting the Beta Distribution to Data

Suppose that we are interested in a random variable Y having possible values between 0 and 1, and that we believe that Y has a beta distribution. If we have n statistically independent observations $Y_1, Y_2, \ldots, Y_n$ upon Y and want to "fit" a beta distribution to these observations, we can estimate μ_Y and σ_Y^2 from the data (either from the original data or from a frequency distribution of these data) by the methods described in Chapter 5. Thus, we estimate μ_Y by the sample average $\hat{\mu}_Y$, and we estimate σ_Y^2 by the sample variance $\hat{\sigma}_Y^2$. The values $\hat{\mu}_Y$ and $\hat{\sigma}_Y^2$ are then substituted for μ_Y and σ_Y^2, respectively, in Equation (5.4) to obtain estimates for the parameters r and s. We obtain as a result the formulas

$$(5.5) \quad \hat{r} = \hat{\mu}_Y \left[\frac{(\hat{\mu}_Y)(1-\hat{\mu}_Y)}{\hat{\sigma}_Y^2} - 1 \right], \quad \hat{s} = (1-\hat{\mu}_Y)\left[\frac{(\hat{\mu}_Y)(1-\hat{\mu}_Y)}{\hat{\sigma}_Y^2} - 1 \right].$$

Example 5.3. A study has been made of the proportion Y of change in wholesale prices of commodities from one year to the next when such a change is in the downward direction (that is, when the prices fall). That is, if in the first year the price of a commodity was 1 unit and in the second year it was 0.57 units, then the price fell 0.43 units from the first year to the second year, and this fall of 0.43 units represented a proportion of change of 0.43 of the price of the commodity for the first year (that is, $Y = 0.43$). In general, if the first-year price is d_1 and the second-year price is $d_2 \le d_1$, then $Y = (d_1 - d_2)/d_1$. Note that if $d_2 > d_1$ (that is, prices rose), no value of Y is recorded. A total of 2314 cases of falling commodity prices were observed. These data are summarized in the form of a frequency distribution in Table 5.1. It is decided to try to fit a beta distribution to the data in Table 5.1. From Table 5.1, we find that

$$\hat{\mu}_Y = (0.02)\left(\frac{780}{2413}\right) + (0.06)\left(\frac{567}{2413}\right) + \cdots + (0.50)\left(\frac{2}{2314}\right) + (0.54)\left(\frac{1}{2314}\right) + 0$$
$$= 0.089,$$

and

$$\hat{\sigma}_Y^2 = (0.2 - \hat{\mu}_Y)^2 \left(\frac{780}{2314}\right) + (0.06 - \hat{\mu}_Y)^2 \left(\frac{567}{2314}\right)$$
$$+ \cdots + (0.05 - \hat{\mu}_Y)^2 \left(\frac{2}{2314}\right) + (0.54 - \hat{\mu}_Y)^2 \left(\frac{1}{2314}\right)$$
$$= 0.0064.$$

5. THE BETA DISTRIBUTION

Table 5.1. Frequency distribution of the proportion Y of change in the wholesale prices of commodities from one year to the next (only falling prices are considered).

y	Observed Frequency of the Event $\{y-0.02 \leq Y \leq y+0.02\}$
0.02	780
0.06	567
0.10	373
0.14	227
0.18	147
0.22	84
0.26	49
0.30	43
0.34	17
0.38	12
0.42	9
0.46	3
0.50	2
0.54	1
	2314

Thus, using Equation (5.5), the estimates of the parameters r and s are

$$\hat{r} = (0.089)\left[\frac{(0.089)(1-0.089)}{0.0064} - 1\right] = 1.038,$$

$$\hat{s} = (1-0.089)\left[\frac{(0.089)(1-0.089)}{0.0064} - 1\right] = 10.63.$$

Thus, the observed variability of the proportion of change Y has been "fit" by a beta distribution with parameters $r = 1.038$ and $s = 10.63$. In Figure 5.2, a relative frequency histogram based on Table 5.1 is graphed along with the probability density function of a beta distribution with parameters $r = 1.038$ and $s = 10.63$. From this figure, it appears that the beta distribution provides a reasonable "fit" to the observed variation of Y.

Probability Calculations

Suppose that the random variable Y has a beta distribution with parameters r and s. If we desire to find probabilities of events of the form $\{Y \leq y_1\}$, $\{Y > y_2\}$, $\{y_1 \leq Y \leq y_2\}$, and so on, it is sufficient (see Section 3, Chapter 4) to have available tables of the cumulative distribution function $F_Y(y)$ of Y. Unfortunately, unlike some of the two-parameter families

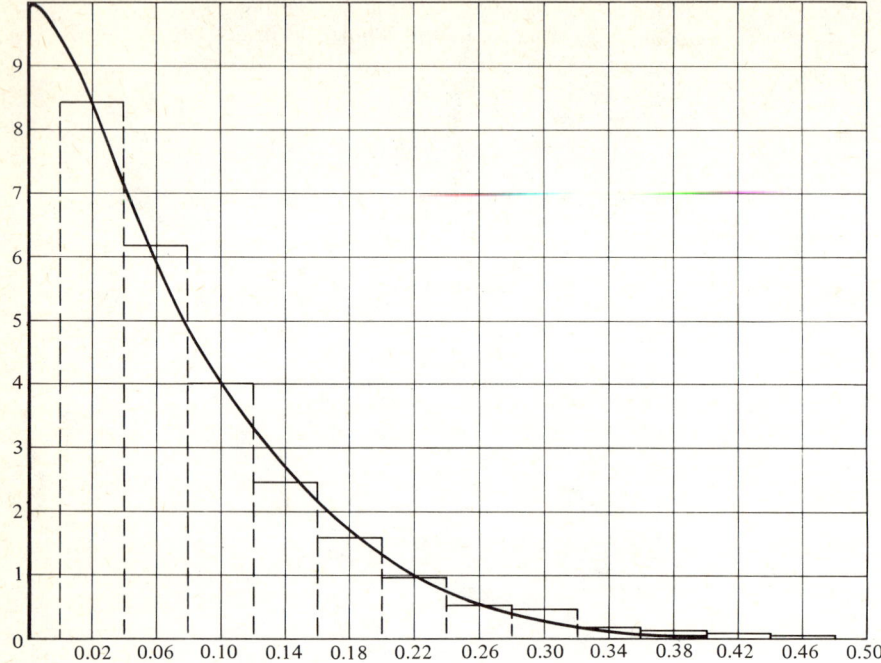

Figure 5.2 Fitting a beta distribution with parameters $r = 1.038$ and $s = 10.61$ to the data of Table 5.1. (Note that to make the comparison, the relative frequencies in Table 5.1 should be divided by 0.04, which is the length of the interval.)

of continuous distributions which we have considered earlier (for example, the normal, gamma, and Weibull distributions), to table the cumulative distribution $F_Y(y)$ of Y, we must construct a separate table for every pair (r,s) of parameter values. Thus, to have a complete collection of tables for all possible pairs of parameter values (r,s) which might be met in practice, a complete book of tables is required. Because such a book of tables *is* available (see Notes and References at the end of this section), we give only two such tables here: Table 5.2(a) gives values of $F_Y(y)$ for $r = 3$, $s = 2$, while Table 5.2(b) gives values of $F_Y(y)$ for $r = 12$, $s = 48$.

For any pair of values (r,s), the cumulative distribution function $F_Y(y)$ is known to mathematicians as the *incomplete beta function*. Note that since any random variable Y which has a beta distribution has possible values between 0 and 1, it is always the case, for any pair (r,s) of parameter values, that $F_Y(y) = 0$ whenever $y \leq 0$, and that $F_Y(y) = 1$ whenever $y \geq 1$. However, the values of $F_Y(y)$ for $0 < y < 1$ depend on the values of r and s (and on y).

Table 5.2 Values of the incomplete beta function.

(a) $r = 3$, $s = 2$

	0.00	0.01	0.02	0.03	0.04	0.05	0.06	0.07	0.08	0.09
0.0	0.00000	0.00000	0.00003	0.00011	0.00025	0.00048	0.00083	0.00130	0.00193	0.00272
0.1	0.00370	0.00488	0.00629	0.00793	0.00982	0.01198	0.01442	0.01715	0.02018	0.02353
0.2	0.02720	0.03121	0.03556	0.04027	0.04534	0.05078	0.05659	0.06279	0.06937	0.07634
0.3	0.08370	0.09146	0.09961	0.10817	0.11713	0.12648	0.13624	0.14639	0.15693	0.16787
0.4	0.17920	0.19091	0.20300	0.21546	0.22829	0.24148	0.25502	0.26890	0.28312	0.29765
0.5	0.31250	0.32765	0.34308	0.35879	0.37476	0.39098	0.40743	0.42409	0.44095	0.45800
0.6	0.47520	0.49255	0.51002	0.52760	0.54526	0.56298	0.58074	0.59852	0.61629	0.63402
0.7	0.65170	0.66929	0.68678	0.70412	0.72130	0.73828	0.75504	0.77154	0.78776	0.80365
0.8	0.81920	0.83436	0.84911	0.86340	0.87720	0.89048	0.90320	0.91532	0.92680	0.93761
0.9	0.94770	0.95704	0.96557	0.97327	0.98009	0.98598	0.99090	0.99481	0.99766	0.99941

(b) $r = 12$, $s = 48$.

	0.00	0.01	0.02	0.03	0.04	0.05	0.06	0.07	0.08	0.09
0.0	0.00000	0.00000	0.00000	0.00000	0.00000	0.00003	0.00017	0.00069	0.00219	0.00571
0.1	0.01280	0.02539	0.04552	0.07499	0.11496	0.16571	0.22646	0.29551	0.37037	0.44813
0.2	0.52582	0.60067	0.67041	0.73337	0.78857	0.83562	0.87468	0.90630	0.93127	0.95053
0.3	0.96507	0.97579	0.98353	0.98900	0.99279	0.99536	0.99707	0.99818	0.99889	0.99934
0.4	0.99961	0.99978	0.99987	0.99993	0.99996	0.99998	0.99999	0.99999	1.00000	1.00000

Example 5.4. An environmental engineer is studying the percent X of saturation of dissolved oxygen in a major American river at a given point on the river. The variable X is an important index of water quality. From some past data, the engineer has hypothesized that if the quality of the water in the river does not change, then the fraction $Y = X/100$ of saturation of dissolved oxygen in the river has a beta distribution with parameters $r = 3$, $s = 2$. It has been proposed to set up a water quality warning system that will monitor the saturation of dissolved oxygen. If the fraction $Y = X/100$ of saturation of dissolved oxygen goes below a certain level y^*, the state or federal government will be required by law to force neighboring towns and industries to improve their sewage treatment plants. A colleague of the engineer has proposed setting $y^* = 0.40$. The engineer decides that he had better calculate the probability that $Y \leq 0.40$ when the quality of the water has actually not changed, since if water quality remains unchanged, a "false-alarm" would be unnecessarily expensive to taxpayers and industry. Since when the water quality of the river is unchanged, Y has a beta distribution with parameters $r = 3$, $s = 2$, the engineer looks up $F_Y(0.40) = P_Y\{Y \leq 0.40\}$ in a table of the incomplete beta function (for $r = 3$, $s = 2$) and finds [see Table 5.2(a)] that

$$F_Y(0.40) = 0.1792.$$

This probability of a false alarm is much too high, considering the expenses involved in improving sewage treatment if Y falls below the warning level; thus, a search for a lower value of the warning level y^* is indicated. Since a probability of false alarm of about 0.03 seems tolerable to all parties concerned, the engineer looks in Table 5.2(a) and notices that $F_Y(0.20) = 0.027$. Thus, he proposes that $y^* = 0.20$ be used as the warning level for water quality. If the fraction Y of saturation of dissolved oxygen in the river at the given point is less than or equal to $y^* = 0.20$, improvement of new sewage treatment plants in the neighboring area will be required.

Example 5.5. A psychologist is studying sleeping states in adult subjects. By means of instruments such as the EEG, and by means of direct observation of the subject's eye movements, he is able to ascertain when a subject is in a certain stage of sleep (called "REM sleep") which has been shown to be highly associated with dreaming. Each subject goes to sleep at his normal bedtime and is observed from the time he begins actually sleeping until he is fully awake. The psycholgist is interested in the fraction (or proportion) Y of the total sleeping time which is spent in "REM sleep" by the subject. His previous studies have suggested that the distribution of Y is (approximately) a beta distribution with parameters $r = 12$,

$s = 48$. In order to compare normal sleepers with sleepers who are under the influence of stress (for example, lack of previous sleep), alcohol, or drugs, the psychologist would like to define a "normal" or typical range of values for the proportion Y of total sleeping time spent in REM sleep. Thus, the psychologist looks for two values y_1, y_2 such that $P_Y\{y_1 \le Y \le y_2\} = 0.95$, say. That is, he looks for two numbers y_1 and y_2 such that approximately 95 percent of all healthy adult sleepers have a proportion between y_1 and y_2 of total sleeping time spent in "REM sleep." From Table 3.2 of Chapter 4, the psychologist knows that

$$P_Y\{y_1 \le Y \le y_2\} = F_Y(y_2) - F_Y(y_1).$$

Since the expected value of Y is $\mu_Y = r/(r+s) = 0.2$, the psychologist looks for y_1 among possible values of Y less than $\mu_Y = 0.20$, and he looks for y_2 among values of y_2 greater than $\mu_Y = 0.20$. In fact, for the sake of convenience, the psychologist decides to find y_1 and y_2 such that $(y_1 - \mu_Y) = (y_2 - \mu_Y)$ (that is, such that y_1 is as much less than μ_Y as y_2 is greater than μ_Y). He tries $y_1 = 0.10$ and $y_2 = 0.30$ and finds from Table 5.2(b) that

$$P_Y\{0.10 \le Y \le 0.30\} = F_Y(0.30) - F_Y(0.10)$$
$$= 0.96507 - 0.01280$$
$$= 0.952.$$

Thus, the psychologist decides that a "normal" range of values for the proportion Y of total sleeping time spent in "REM sleep" by adult subjects is the interval of numbers between $y_1 = 0.10$ and $y_2 = 0.30$.

NOTES AND REFERENCES

Values of $F_Y(y)$ are presented in Table T.10 for $y = 0.05(0.5)0.95$, $r = 1(1)5$, and $s = 2(1)5$.

Extensive tables of the incomplete beta function appear in:

Tables of the Incomplete Beta-function (1934), edited by Karl Pearson. The "Biometrika" office, University College, London.

In these tables a different set of symbols is used: in place of our "r," they use "p"; in place of our "s," they use "q"; finally, in place of $F_Y(y)$, they use $I_y(p,q)$ to symbolize $P_Y\{Y \le y\}$.

*6. OTHER CONTINUOUS DISTRIBUTIONS

Even though the normal, exponential, gamma, Weibull, uniform, and beta distributions are frequently used to model the variability of continuous random phenomena, many other distributions (and classes of distributions) have proved to be useful in certain contexts. Indeed, in a recent survey of continuous distributions, Johnson and Kotz (1970) list over 50 different families of such distributions. Thus, no short list of distributions (or classes of distributions) can be expected to serve as models for all quantitative random phenomena; while a complete list and description of all continuous distributions is the subject for one (or more) additional books. The families of distributions that we have discussed in Chapter 7 and in Sections 1 to 5 of this chapter are varied enough to provide adequate models for a wide range of random phenomena. When a continuous quantitative random phenomenon does not appear to follow any of the continuous distributions discussed in Chapters 7 and 8, there are several source books (see Notes and References) that may be of help in suggesting more appropriate models.

In this section, therefore, we have chosen to mention (and briefly describe) two additional classes of continuous distributions: the lognormal distributions and the Cauchy distributions, not solely because of their importance as models of the variation of quantitative random phenomena (although they are of importance in this respect), but rather because these distributions illustrate facts and concepts about continuous probability distributions that are worthy of some attention.

The Lognormal Distribution

The applicability of the normal distribution to many quantitative random phenomena can be theoretically justified (see Chapter 7, Section 4) by assuming that such quantitative random phenomena arise from the summation of many statistically independent and (nearly) identically distributed random causes. That is, we can anticipate that a random variable Y has approximately a normal distribution if we can conceive of Y being equal to the *sum* $Y = X_1 + X_2 + \cdots + X_n$ of a large number n of statistically independent realizations of the same random variable X.

The *lognormal distribution* arises from the *product* of many independent and (nearly) identically distributed random variables. A random variable Z has approximately a lognormal distribution if we can conceive of Z as being equal to the product $(W_1)(W_3)(W_3)\ldots(W_n)$, where $W_1, W_2, \ldots, W_n$ are statistically independent realizations of the same *nonnegative* random variable W. The name "lognormal" given to such a distribution comes from noting that

(6.1)
$$\log Z = \log[(W_1)(W_2)\cdots(W_{n-1})(W_n)]$$
$$= \log W_1 + \log W_2 + \cdots + \log W_{n-1} + \log W_n.$$

This result holds true since the logarithm (natural logarithm) of the product of n numbers equals the sum of the logarithms of these numbers. Letting $Y = \log Z$, $X_1 = \log W_1$, $X_2 = \log W_2, \ldots, X_n = \log W_n$, we see that $Y = \log Z$ is the sum of a large number n of statistically independent, identically distributed random variables, and hence by the Central Limit Theorem, $Y = \log Z$, is (approximately) normally distributed. These remarks motivate the definition: *A nonnegative random variable Z has a lognormal distribution whenever $Y = \log Z$ has a normal distribution.*

Suppose that the random variable Z does have a lognormal distribution, and that $Y = \log Z$ is normally distributed with expected value $\mu_Y = \xi$ and variance $\sigma_Y^2 = \delta^2$. Then, using the theory of differential calculus, it can be shown that the probability density function $f_Z(Z)$ of Z is given by the following expression:

(6.2)
$$f_Z(z) = \begin{cases} \dfrac{1}{\sqrt{2\pi z^2 \delta^2}} e^{-(1/2\delta^2)(\log z - \xi)^2}, & \text{if } z \geq 0, \\ 0, & \text{if } z < 0. \end{cases}$$

The constants ξ and δ^2 are parameters of the lognormal distribution (6.2); knowledge of the values of ξ and δ^2 determines the distribution.

In contrast to the graph of the density function of the normal distribution (see Chapter 7), the graph of the probability density function $f_Z(z)$ of the lognormal distribution is not a symmetric, bell-shaped curve. Rather, the lognormal distribution is positively skewed (see Chapter 5). In Figure 6.1, graphs of the probability density function (6.2) of the lognormal distribution are given for various values of ξ and δ^2.

If Z has a lognormal distribution with parameters ξ and δ^2, it can be shown that

(6.3)
$$\mu_Z = e^{\xi + (1/2)\delta^2}, \qquad \sigma_Z^2 = e^{2\xi + \delta^2}(e^{\delta^2} - 1).$$

Probabilities of events of the form $\{Z \leq a\}$, $\{b < Z\}$, $\{a \leq Z \leq b\}$, and so on (for $a \geq 0$, $b \geq 0$) can be calculated by noting that $Y = \log Z$ has a normal distribution with expected value ξ and variance δ^2, and that

$$\{Z \leq a\} = \{Y = \log Z \leq \log a\},$$
$$\{b < Z\} = \{\log b < \log Z = Y\},$$
$$\{a \leq Z \leq b\} = \{\log a \leq \log Z = Y \leq \log b\},$$

and so on.

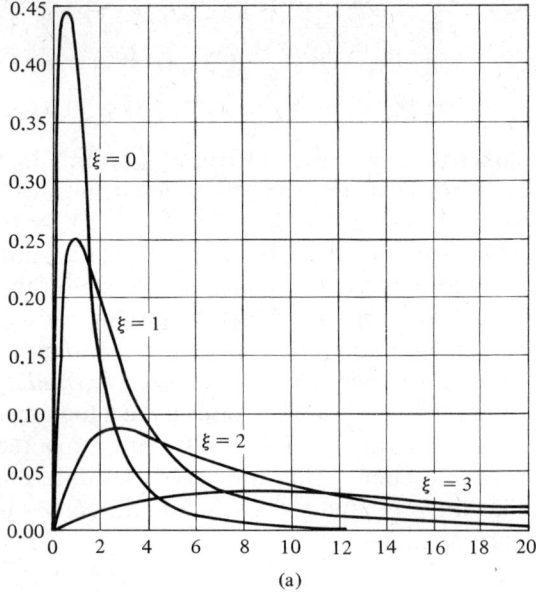

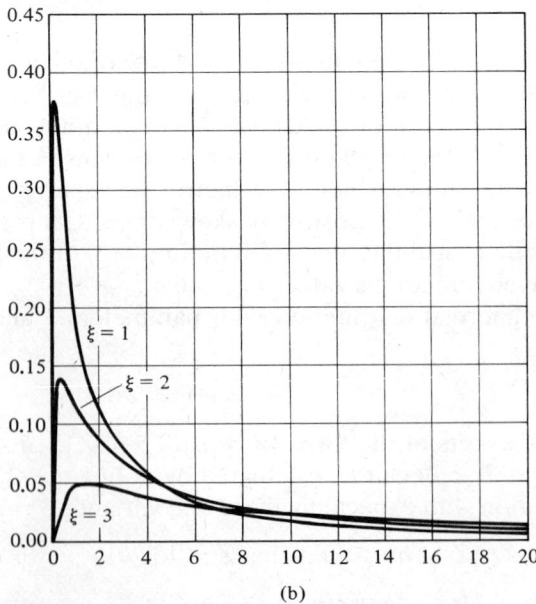

Figure 6.1 Graphs of the probability density function of the lognormal distribution for (a) $\delta^2 = 1$ and $\xi = 0, 1, 2,$ and 3; (b) $\delta^2 = 3$ and $\xi = 1, 2,$ and 3.

Thus,

(6.4) $$P_Z\{Z \leq a\} = P_Y\{Y \leq \log a\} = P\{\mathcal{N}(Y;\xi,\delta^2) \leq \log a\},$$

$$P_Z\{b < Z\} = P_Y\{\log b < Y\} = P\{\log b < \mathcal{N}(Y;\xi,\delta^2)\},$$

$$P_Z\{a \leq Z \leq b\} = P_Y\{\log a \leq Y \leq \log b\}$$
$$= P\{\log a \leq \mathcal{N}(Y;\xi,\delta^2) \leq \log b\},$$

where the notation $\mathcal{N}(Y;\xi,\delta^2)$ and ways of calculating the quantities $P\{\mathcal{N}(Y;\xi,\delta^2) \leq \tau\}$, $P\{\tau < \mathcal{N}(Y;\xi,\delta^2)\}$, and $P\{\tau_1 \leq \mathcal{N}(Y;\xi,\delta^2) \leq \tau_2\}$ in Equation (6.4) are discussed in Chapter 7.

The lognormal distribution is used and known under a variety of alternative names. For example, in some fields the lognormal distribution is known as the antilognormal distribution [since the antilogarithm of a normally distributed random variable has the distribution (6.2)]. Because Galton (1879) and McAlister (1879) were perhaps the first scientists to use the lognormal distribution (and to motivate the applicability of this distribution by means of arguments based on the Central Limit Theorem), the lognormal distribution is sometimes known as the Galton-McAlister distribution. In econometric theory, the lognormal distribution is often applied to production data, and in such contexts it is called the Cobb-Douglas distribution [for example, see Dhrymes (1962)]. In psychophysical studies, the lognormal distribution has been mentioned by Fechner (1897), while Gaddum (1945) and Bliss (1934) have applied this distribution to the study of critical doses of drugs in human beings and animals. Among other nonnegative quantitative random phenomena whose variation has been modeled by the lognormal distribution are:

(i) particle sizes in naturally occurring aggregates [Hatch and Choute (1929); Krumbein (1936); Herdan (1960)],
(ii) lengths of words [Herdan (1958)], and sentences [Williams (1940)],
(iii) concentrations of the chemical elements in geological materials [Ahrens (1954a, b; 1957); Chayes (1954)],
(iv) lifetimes of mechanical and electrical systems [Epstein (1947; 1948)] and other survival data [Feinlieb (1960); Goldthwaite (1961); Adams (1962)],
(v) the abundance of species of animals [Grundy (1951)],
(vi) the incubation periods of infectious diseases [Sartwell (1950)].

From this list, it is apparent that the lognormal distributions are important competitors to the exponential, gamma, or Weibull distributions as models for the variation of the nonnegative random variables encountered in many scientific and practical disciplines.

Because methods of "fitting" the normal distribution to observed data have been discussed previously (Chapter 7, Section 3), and since we know that the logarithm of a random variable Z, which has the lognormal distribution, is normally distributed, a detailed discussion of methods for "fitting" the lognormal distribution is not necessary here. Suppose we have observed statistically independent observations $Z_1, Z_2, \ldots, Z_n$ upon a random variable Z, which we believe has a lognormal distribution. To estimate the parameters ξ and δ^2 from these data, we can *transform* the data $Z_1, Z_2, \ldots, Z_n$ by taking the logarithms $Y_1 = \log Z_1$, $Y_2 = \log Z_2$, $\ldots$, $Y_n = \log Z_n$. Since $Y_1, Y_2, \ldots, Y_n$ are then statistically independent observations upon a random variable $Y = \log Z$ which has a normal distribution with expected value ξ and variance δ^2, estimation of ξ and δ^2 can now be accomplished using the transformed data $Y_1, Y_2, \ldots, Y_n$ and the methods of Chapter 7, Section 3.

Because of the demonstrated applicability of the lognormal distribution to nonnegative quantitative phenomena observed in many scientific and practical contexts, the logarithmic transformation of observations taken upon such phenomena (so as to obtain normally distributed observations) has become a standard technique of data analysis in these disciplines. Once we can conceive of logarithmic transformations of data, other transformations (exponential, quadratic, square root, and so on) also suggest themselves. Hence, many scientists start the analysis of their data by trying various transformations in an attempt to transform the random variable originally observed into a new (but equivalent) random variable which has a convenient distribution (such as the normal distribution). The lognormal distributions illustrate and motivate this technique, and it is for this reason that we have included this class of distributions in the present section.

The Cauchy Distribution

There is a common view among scientists that mathematical difficulties (problems of mathematical rigor, paradoxes, and so on) which arise in connection with models of real-world phenomena are theoretical curiosities which have no relevance to experimental practice. While there may be some truth in this notion, it is not necessarily the case. In science, today's paradoxes often become the stimulus for tomorrow's theories. Nonintuitive mathematical consequences (such as paradoxes) arising from a seemingly reasonable set of assumptions indicate areas not fully understood, and can stimulate research aimed at exposing the causes for such "paradoxical" results. When this research is successful, the scientist gains a greater understanding of the phenomena he is studying, and becomes aware of the limitations of the model (or theory) he is using to describe that phenomena.

Within the context of probability theory, the *Cauchy distribution* [named after the French mathematician Augustin Cauchy (1789–1857)] exhibits certain nonintuitive behavior; behavior that seems to be inconsistent with certain consequences of the frequency interpretation of probability theory.

The probability density function $f_X(x)$ of a random variable X having a Cauchy distribution has the form

$$(6.5) \qquad f_X(x) = \frac{1}{\pi} \frac{1}{[1 + (x - \theta)^2]}$$

for all real numbers x. The graph of this density function for $\theta = -1, 0, 1$, appears in Figure 6.2. Notice that the graph of a Cauchy probability density function, like that of the probability density function of a normal distribution, is symmetric about a central value (in this case, θ). This central value θ is both the unique median and the unique mode of the Cauchy distribution. Thus, the parameter θ is a measure of location for the Cauchy distribution.

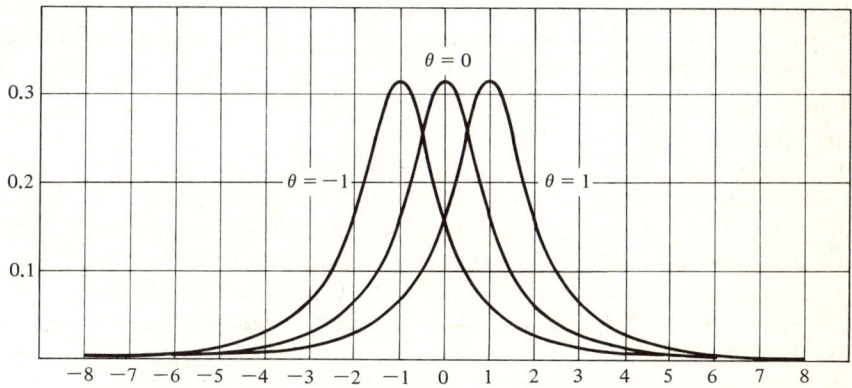

Figure 6.2 Graphs of the Cauchy density function for $\theta = -1, 0,$ and 1.

Apparently, then, the Cauchy distribution shares many properties with the normal distribution: (i) both distributions assign positive probability density to all numbers, (ii) both distributions are symmetric about a central value (μ for the normal distribution, θ for the Cauchy distribution) which is both the unique median and the unique mode of the distribution, and (iii) both distributions are unimodal (see Chapter 5). From Figure 6.3 (in which the density functions of a normal distribution with parameters $\mu = 0$ and $\sigma^2 = 1$ and a Cauchy distribution with parameter $\theta = 0$ are graphed together), we see that the normal and Cauchy distributions do seem to differ in the shape of the graphs of their density functions, at least

416 SPECIAL DISTRIBUTIONS: CONTINUOUS CASE 8

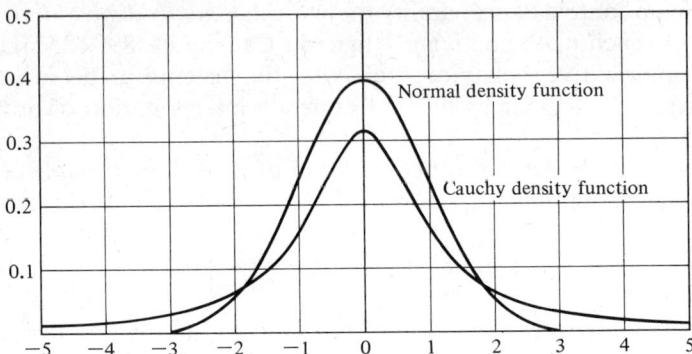

Figure 6.3 Comparison of the standard normal density function and the Cauchy density function with $\theta = 0$.

in the sense that the Cauchy distribution is flatter and more "spread out" than the normal distribution. However, if we compare graphs of the density functions of a normal distribution with parameters $\mu = 0$ and $\sigma^2 = 2$ and a Cauchy distribution with parameter $\theta = 0$ (see Figure 6.4), this difference in shape is less pronounced, and may even vanish if a large enough value of σ^2 is chosen. Thus, the Cauchy distribution seems to be well enough behaved. What, then, is nonintuitive about the Cauchy distribution?

To begin with, if X has a Cauchy distribution, its expected value $\mu_X = E(X)$ is not well-defined. Recall that the expected value of X is the area (taken algebraically) between the graph of $xf_X(x)$ and the horizontal axis (Section 2, Chapter 5). However, the area between the horizontal axis and the graph of $xf_X(x)$ to the right of $x = 0$ is assigned a positive sign and is infinitely large, while the area between the horizontal axis and the

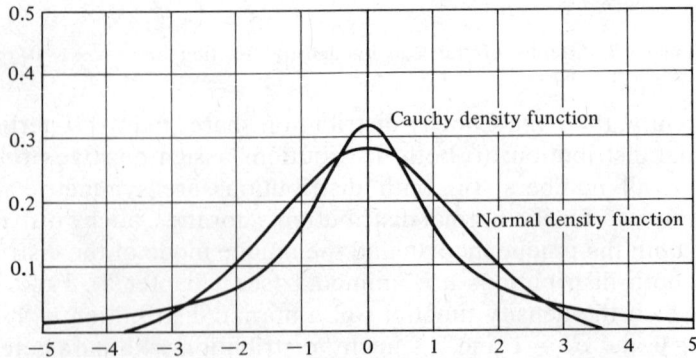

Figure 6.4 Comparison of the normal density function with $\sigma^2 = 2$ and the Cauchy distribution with $\theta = 0$.

graph of $xf_X(x)$ to the left of $x = 0$ is infinitely large but is assigned a negative sign. The total area between the graph of $xf_X(x)$ and the horizontal axis is thus equal to the difference of two infinitely large numbers; we have no unique way of defining what that difference should be (in a way which is consistent with our notion of the addition of areas), and thus we must say that this difference [and, therefore, $E(X)$] is undefined. A similar argument, based on the definition, can show that the variance σ_X^2 is *infinite* even if we, somewhat arbitrarily, define $E(X)$ to be equal to θ.

The fact that the expected value of a Cauchy distribution is not well-defined, and that the variance of a Cauchy distribution is infinite, may perhaps be assigned by the reader to the perversity of mathematics. In any case, the expected value and variance of a distribution are not the only indices of the location and dispersion, respectively, of a distribution. We can use the median (or the mode) as a measure of location for the Cauchy distribution, and we can use the semi-interquartile range (see Chapter 5, Section 4) as a measure of dispersion. Since both of these indices are well-defined for the Cauchy distribution, why should we fuss over a mathematical technicality?

If this last argument appears satisfactory, then consider the following, quite nonintuitive, property of the Cauchy distribution: Suppose that X is a random variable having a Cauchy distribution with parameter θ, and suppose that we obtain n statistically independent observations, $X_1, X_2, ..., X_n$, of X. Then it can be shown that *the sample average* $\hat{\mu}_X = (X_1 + X_2 + \cdots + X_n)/n$ *of these n observations has the same distribution as X* (that is, a Cauchy distribution with parameter θ) *no matter how many observations we obtain upon X*. The frequency interpretation of probability would lead us to believe that the variability of $\hat{\mu}_X$ about some central value, say θ, decreases as the number n of observations upon X increases (see Chapter 5, Section 2), and that for very large n, $\hat{\mu}_X$ and θ become approximately equal to one another. When $X_1, X_2, ..., X_n$ have the (same) Cauchy distribution, however, the distribution of $\hat{\mu}_X$ is always the same, regardless of how large n is, and thus the variability of $\hat{\mu}_X$ about a central value does not decrease as n increases. This apparent contradiction to the frequency interpretation of probabilities (actually, the frequency interpretation of sample averages) is explained by the fact that the Cauchy distribution does not have a well-defined expected value. The frequency interpretation of sample averages tells us that the variability of $\hat{\mu}_X$ around the expected value μ_X of X decreases as the number n of observations $X_1, X_2, ..., X_n$ upon X increases. However, since the expected value of X is *not* well-defined when X has a Cauchy distribution, there is no obvious central point around which the variability of $\hat{\mu}_X$ can decrease. It is now known that a sufficient condition for the frequency interpretation of sample averages to be valid for a given probability distribution is that the expected value of that distribution be well-defined (and finite).

Faced with the apparent contradiction between the frequency interpretation of sample averages and the above-mentioned characteristic of the Cauchy distribution, it is tempting to dismiss the Cauchy distribution as a mathematical invention having "no practical importance." However, the Cauchy distribution is useful in certain scientific contexts: in mechanics and electrical theory, in psychophysics [Urban (1909)], in physical anthropology [Fieller (1932)], and in measurement and calibration problems in many fields of science. Thus, the warnings which the Cauchy distribution provides about the interpretation and estimation of the expected value of a distribution must be given serious attention in scientific practice.

NOTES AND REFERENCES

A survey of classes of continuous distributions which have proved useful in scientific and practical problems appear in the following books:

1. Johnson, N. L., and Kotz, S. (1970). *Continuous Univariate Distributions*, vols. 1 and 2, Houghton Mifflin Company, Boston.

The lognormal distribution is discussed at some length in:

2. Aitchison, J., and Brown, J. A. C. (1963). *The Lognormal Distribution*. Cambridge University Press, Cambridge, England.

EXERCISES

A1. The magnitude M of an earthquake, as measured on the Richter scale, is a random variable of considerable practical importance. As a consequence of studies of earthquakes with large magnitudes (say, magnitudes exceeding 3.25), it has been suggested that the "excess," $X = M - 3.25$, of the magnitude of these earthquakes over the threshold value of 3.25 follows an exponential distribution. Over the period January 1934 to May 1943, the magnitudes of shock (earthquakes) were recorded. The resulting data are summarized by Gutenberg and Richter (1944). From these data, the following table of observed frequencies (Table E.1) for the excess magnitude, $X = M - 3.25$, has been obtained.
   (a) From the given data, estimate the parameter θ of the exponential distribution which is followed by the "excess" $X = M - 3.25$.
   (b) Does the exponential distribution provide a good "fit" to these data? Support your answer by calculating the appropriate theoretical fre-

Table E.1. Observed frequency distribution of earthquakes with large magnitudes, January 1934 to May 1943.

Midpoint m of Interval of Magnitudes	Midpoint x of Interval of Excesses	Observed Frequency of Earthquakes Whose "Excess" $X = M - 3.25$ Is between $x - 0.25$ and $x + 0.25$ (and whose magnitude is between $m - 0.25$ and $m + 0.25$)
3.5	0.25	579
4.0	0.75	311
4.5	1.25	108
5.0	1.75	32
5.5	2.25	13
6.0	2.75	5
6.5	3.25	2
		1050

quencies, and by comparing the graph of the theoretical survival function to the graph of the sample survival function.

(c) What is the probability that the magnitude M of an earthquake shock will exceed 6 on the Richter scale? [Regardless of your answer to part (b), assume that $X = M - 3.25$ has an exponential distribution with a parameter θ equal to the value you obtained in part (a).]

(d) Suppose we define large earthquakes to be those with magnitudes greater than 3.75, and define the "excess" to be $Y = M - 3.75$. Use the data of Gutenberg and Richter to fit an exponential distribution for Y. [*Note:* There are now only $1050 - 579 = 471$ observations available on Y, since earthquakes with magnitudes between 3.25 and 3.75 are not "large" under our new definition of "large."] Compare the graph of the theoretical survival function for Y to the graph of the sample survival function. Does the exponential distribution better "fit" the observed distribution of X, or the observed distribution of Y?

A2. Rasch (1960) considers a stochastic model for reading time in which the time X in minutes required for an individual to read a given passage is a random variable following an exponential distribution with parameter θ. Experimental evidence indicates that for a certain reading exercise the value of θ is 2.0.

(a) Find the probability of the event that it takes an individual more than 1 minute to read the given reading exercise.

(b) Find the median and 90th percentile of the distribution of the time X required to read the given reading exercise.

(c) For a different reading exercise, the value of the parameter θ describing the (exponential) distribution of the time Y required to read this exercise is unknown. However, it is known that the probability of the event

$\{Y < 1.0\}$ is equal to 0.75. (That is, 75 percent of the population takes less than 1 minute to read this exercise.) Find the value of θ.

3. The distribution of the duration of pauses (and the duration of vocalizations) that occur in a monologue has an exponential distribution [Jaffe and Feldstein (1970)]. If the expected value of the duration of pauses is 0.70 seconds, what is the variance of the duration of pauses? What is the 90th percentile?

4. Suppose that a random variable X has an exponential distribution with parameter $\theta = 1.0$. Which is larger:
   (a) The mode of the distribution of X or the median of the distribution of X?
   (b) The median of the distribution of X or the expected value of X?
   (c) The expected value of X or the mode of X?
   (d) The median of X or the variance of X?
   Support your assertions. Does the value of θ matter? Would your answer be the same if $\theta = 10.0$? If $\theta = 100.0$? Explain.

A5. In the context of Example 1.2, find the following probabilities:
   (a) The probability of the event that the rat presses the bar again any time between 2 and 4 seconds after the last time it pressed the bar.
   (b) The probability of the event that the length of time between 2 bar presses of the rat exceeds 1.5 seconds.
   (c) The probability of the event that the length of time between 2 bar presses of the rat is less than 3.0 seconds.

6. In the context of Example 1.1, assume that each time X between failures of the air conditioning system of a Boeing 720 jet airplane has an exponential distribution with parameter $\theta = 0.0107$. Assume also that when the air conditioning system fails, it is instantaneously repaired, and that the lifetime of the repaired system is statistically independent of all previous lengths of life of this system. What is the distribution of the number Y of separate repairs which must be made on the air conditioning system in order to keep it running for 200 hours of flight time? What are the expected value and variance of Y?

A7. A trucking company knows that the length of time X in hours required for one of its trucks to complete a typical haul (round trip) is exponentially distributed with parameter $\theta = 1/48$. Given that a truck has already been gone for 48 hours, what is the conditional probability that this truck will arrive within the next hour? What is the expected value of the additional length of time which the company must wait until the truck completes its route, given that the truck has already been gone for 48 hours?

A8. Suppose that X and Y are statistically independent random variables, and that both X and Y have the same exponential distribution with parameter $\theta = 0.10$.
   (a) Find the probability that the minimum value of X and Y exceeds 5.0. That is, find the probability of the event $\{X > 5.0 \text{ and } Y > 5.0\}$.
   (b) Find a general formula for the probability that the minimum value of X and Y exceeds a given value, say, τ.

(c) Use your answer to part (b) to find the probability that the minimum value of X and Y is less than or equal to τ. Let Z be the (random) minimum value of X and Y. Your answer gives the probability of the event $\{Z \leq \tau\}$. In other words, your answer gives the form of the cumulative distribution function $F_Z(\tau)$ of Z. Based on your answer, what is the distribution of Z?

(d) Complex electrical and mechanical systems often are composed of several identical components. If one of these components fails, the system fails. Suppose that an electrical or mechanical system is composed of 2 identical components, and each such component has a lifetime (in hours) which has an exponential distribution with parameter $\theta = 0.10$. Assume further that the lifetimes of the 2 components are statistically independent. What is the expected length of life of the entire system?

9. Consider again the reading model described in Exercise 2. Suppose that a student is asked to read N similar exercises (similar, but not necessarily identical). Assume that the student reads these exercises one right after the other, and that the length of time, in minutes, required for the student to read one such exercise is statistically independent of the lengths of times required for the student to read the other exercises. If the reading model described in Exercise 2 holds, and the reading time for each exercise is exponentially distributed with parameter θ, then the total time Y required for the student to read all N exercises has a gamma distribution with parameters θ and N.

(a) Suppose the student reads 5 similar exercises and $\theta = 2.0$. Graph the theoretical density function and the theoretical cumulative distribution function of the total reading time Y.

(b) What is the probability that the student reads all 5 exercises in 2.5 minutes or less?

(c) What are the median and 90th percentiles of the distribution of total reading time?

(d) Which is largest: the median of Y, the mode of Y, or the expected value of Y? Which of these quantities is the smallest?

10. The gamma distribution has been used in a medical application by Masuyama and Kuroiwa (1952). They considered the sedimentation rate X at various stages during normal pregnancy. In the 20th week after commencement of pregnancy, the sedimentation rate had a gamma distribution with (estimated) parameter values of $r = 3.73$ and $\theta = 1/10.18$. In the 30th week after commencement of pregnancy, the sedimentation rate had a gamma distribution with (estimated) parameter values of $r = 5.07$ and $\theta = 1/9.98$. Graph (roughly) the density functions of both theoretical distributions. Compare the medians and 90th percentiles of these two distributions.

*A11. The gamma distribution has been used as a model for the lifetimes of metals subjected to stress. In one experiment, 101 rectangular strips of aluminum of standardized dimensions were submitted to repeated alternating stresses (at a frequency of 18 cycles per second). The lifetimes of these strips of aluminum (expressed in thousands of cycles) are summarized in Table E.2.

Table E.2. Lifetimes under periodic loading, maximum stress 21,000 PSI, 18 cycles per second, 6061-T6 aluminum coupon cut parallel to the direction of rolling. Observations listed in increasing order.

370	1055	1270	1502	1763
706	1085	1290	1505	1768
716	1102	1293	1513	1781
746	1102	1300	1522	1782
785	1108	1310	1522	1792
797	1115	1313	1530	1820
844	1120	1315	1540	1868
855	1134	1330	1560	1881
858	1140	1355	1567	1890
886	1199	1390	1578	1893
886	1200	1416	1594	1895
930	1200	1419	1602	1910
960	1203	1420	1604	1923
988	1222	1420	1608	1940
990	1235	1450	1630	1945
1000	1238	1452	1642	2023
1010	1252	1475	1674	2100
1016	1258	1478	1730	2130
1018	1262	1481	1750	2215
1020	1269	1485	1750	2268
				2440

SOURCE: Z. W. Birnbaum and S. C. Saunders (1958). A statistical model for the life-length of materials. *Journal of the American Statistical Association*, 53, pp. 151–160. Reprinted by permission.

(a) Graph the sample cumulative distribution function or the sample survival function obtained from these data.

(b) Verify that the sample mean and sample variance of these data are 1400.91 and 151,618.36, respectively.

(c) Fit a gamma distribution to these data. Either compare the graph of the theoretical cumulative distribution function to the graph of the sample cumulative distribution function, or else compare the graph of the theoretical survival function to the graph of the sample survival function. [*Hint:* Interpolate in Table T.9] Does the gamma distribution provide a good model for the lifetimes of the aluminum strips? Why, or why not?

12. Let X have an exponential distribution with parameter $\theta = 1$ and let Y have a gamma distribution with parameters $r = 2$ and $\theta = 2$.

(a) Which random variable has the larger expected value? Support your assertion.

(b) Which random variable has the larger variance? Support your assertion.

(c) Which event has the larger probability:
 (i) the event that X exceeds its expected value by at least 2 standard deviations,
 (ii) the event that Y exceeds its expected value by at least 2 standard deviations?
(d) Which random variable has the larger median? Support your assertion.

A13. Suppose that the random variable X has a gamma distribution with parameters r and θ. We are told that the expected value of X is equal to 4.0 and that the variance of X is equal to 8.0.
(a) Find the values of r and θ.
(b) Find the probability of the event $\{1.0 \leq X \leq 4.0\}$.
(c) What is the median of the distribution of X?

*14. Fit a Weibull distribution to the lifetimes of the strips of aluminum described in Exercise 11. Does the Weibull distribution provide a better fit to the observed data? Support your answer by either comparing the theoretical survival functions of the fitted Weibull and gamma distributions to the sample survival function, or by comparing the theoretical cumulative distribution functions of the fitted Weibull and gamma distributions to the sample cumulative distribution function.

A15. It is known that the duration X, in days, of an epidemic of a certain infectious disease has a Weibull distribution with parameters $\alpha = 10$, $\beta = 2$, and $\nu = 2$.
(a) What is the probability that an epidemic of the disease will last at least 2 weeks (14 days)?
(b) What is the probability that an epidemic of the disease will last for less than 6 days?
(c) What is the expected length of duration (in days) of an epidemic of the disease?
(d) What is the median length of duration (in days) of an epidemic of the disease?
(e) An epidemic of the disease has just been reported. What prediction would you give as to the length of time (in days) that the epidemic will last? Explain your answer.

16. If the random variable X has a Weibull distribution with parameters $\alpha = 5$, $\beta = 3$, and $\nu = 1$, what is the distribution of
(a) $Y = X - 1$?
(b) $Z = (X - 1)/5$?
(c) $W = (X - 1)^3/125$?
(d) $V = (X - 1)^2/25$?
(e) $Q = (X - 1)^2$?

A17. In the context of Example 3.1, what is the probability that:
(a) A particle of fly ash has a diameter greater than 8 twenty-micron units?
(b) A particle of fly ash has a diameter between 6 and 10 twenty-micron units?
(c) A particle of fly ash has a diameter less than 4 twenty-micron units?
What is the median particle diameter of fly ash?

424   SPECIAL DISTRIBUTIONS: CONTINUOUS CASE   8

18. In the context of Example 3.2, find:
    (a) The 90th percentile of the distribution of the yield strength of a Bofors steel.
    (b) The median yield strength of a Bofors steel.
    (c) The probability that the yield strength of a Bofors steel is less than or equal to 40 kg/mm².
    (d) The probability that the yield strength of a Bofors steel is between 45 and 50 kg/mm².
    If you had to use steel in a structure which was subjected to a strain of 40 kg/mm², would you feel confident using a Bofors steel? Why, or why not?

*A19. The Weibull distribution has been used by Indow (1971) as a model for determining the effect of advertising. Let X be the length of time (in days) after the end of a certain advertising campaign that an individual is able to remember the name (or description) of the brand of merchandise being advertised. Indow assumes that X has a Weibull distribution. In a particular experiment, an advertising campaign for a brand of chocolate was initiated, after which a series of questions were asked to randomly selected individuals at each of several times. Using the proportions of individuals who could recognize the brand of chocolate at these several time points, Indow determined that in this advertising context X has a Weibull distribution with parameters $\nu = 1.0$, $\beta = 0.98$, and $\alpha = 7360.0$.
    (a) Graph the cumulative distribution function of X.
    (b) Find the quartiles $Q_{0.25}(X)$ and $Q_{0.75}(X)$ of the distribution of X.
    (c) If we interviewed a random sample of individuals 1 week (7 days) after the advertising campaign for the brand of chocolate was ended, what percentage of these individuals would you expect to remember the name of the brand of chocolate that was advertised? Explain your answer.

20. Suppose that X and Y are statistically independent random variables. Further, suppose that X and Y each have a uniform distribution with parameters $a = 1$ and $b = 2$.
    (a) Find the probability of the event $\{X > 1.5 \text{ and } Y > 1.5\}$.
    (b) Find the probability of the event $\{X \leq 1.5 \text{ and } Y \leq 1.5\}$.
    (c) Find the probability of the event $\{X \leq 1.5 \text{ and } Y > 1.5\}$.
    (d) What is the expected value of X? What is the expected value of Y?

A21. A bus travels between two cities, A and B, which are 100 miles apart. If the bus has a breakdown, the distance X of the point of the breakdown from city A has a uniform distribution with parameters $a = 0$ and $b = 100$.
    (a) If there is a service garage in city A, a service garage in city B, and a service garage in the center of the route between the 2 cities, and if a tow truck is dispatched to help the bus from the service garage closest to the point at which the bus breaks down, what is the probability that the tow truck has to travel more than 10 miles to reach the bus?
    (b) An operations research expert suggests that travel distances for tow trucks would be less if the 3 service garages were placed 25, 50, and 75 miles, respectively, from city A. Do you agree? Justify your answer.

22. The monthly fire incidence in buildings in England and Wales in 1961 is given in Table E.3.

Table E.3. Monthly fire incidence in buildings in England and Wales in 1961.

Month	Incidence of Fires
January	0.0936
February	0.0739
March	0.0917
April	0.0704
May	0.0861
June	0.0809
July	0.0771
August	0.0699
September	0.0705
October	0.0842
November	0.0942
December	0.1075

The incidence of fires in a given month is equal to the number of fires observed during that month divided by the total number of fires observed during the year. Let X equal the exact time in days (after 12:00 a.m. on the morning of January 1, 1961) that a given fire occurs in a building in England and Wales. (We assume that this fire is chosen at random from among all building fires observed in England and Wales in 1961.) Thus, if a given fire occurs at 6:00 a.m. on February 5, 1961, the value of X for this fire is $X = 35.25$ days. It has been asserted that X has a uniform distribution with parameters $a = 0$ and $b = 365$.

(a) Assume that every one of the 12 months of the year has the *same* number of days. Using this assumption and Table E.3, construct a relative frequency histogram for X. Compare this histogram to a graph of the probability density function of the uniform distribution on the interval from $a = 0$ to $b = 365$. Is the uniform distribution a good "fit" to the distribution of the times of fires? Explain.

(b) Actually, of course, the months of the year do not have the same number of days. How would you adjust your relative frequency histogram to take account of this fact? Make this adjustment. Does this adjustment improve the "fit"? Explain.

23. If buses run on the half hour, and on a given day a commuter arrives at a time X in the morning which is uniformly distributed between 8:15 a.m. and 8:45 a.m., what is the probability that the commuter will have to wait for more than 15 minutes for a bus? What is the probability that the commuter will have to wait for more than 20 minutes for a bus? What is the expected length of time that the commuter must wait for a bus?

*A24. Suppose that you wish to simulate a random variable W having a Weibull distribution with parameters $v = 1.6$, $\alpha = 3.0$, and $\beta = 4.0$. You have a random number generator which simulates a random variable X having a uniform distribution with parameters $a = 0$ and $b = 1$. How would you simulate W? Suppose that your random number generator yields the value $x = 0.77$ for X. What value w would be simulated for W?

25. A clock radio is set to turn on at 7:00 a.m. and is tuned to a news station so that the sleeper can wake up to the news headlines and the weather report.
   (a) Because of random delays in the broadcasting schedule, and because the clock radio is not exactly synchronous with the clock used by the radio station, the radio station may actually start broadcasting the news at a time X on the clock radio which is uniformly distributed over the interval of times between $a = 6:58$ a.m. and $b = 7:02$ a.m. The news starts with a reading of the headlines. It takes the announcer 2 minutes to read the headlines. What is the probability that the clock radio turns on too late to hear any of the headlines?
   (b) The weather forecast is presented immediately after the headlines and lasts 1 minutes. What is the probability that the clock radio turns on in time to catch all of the weather report?
   (c) What is the probability that the clock radio turns on in time for the sleeper to hear all of the weather report and all of the headlines?

*A26. Ore samples from the transvaal vein of level 10 of the Frisco mine were assayed for metal content. Among the metals studied were copper (Cu) and lead (Pb). For each of 1000 ore samples, the ratio X of the weight of Cu to the sum of the weights of Cu and Pb [that is, the ratio Cu/(Cu + Pb)] were measured. Table E.4 (p. 427) gives a frequency distribution (in the 1000 ore samples) of the ratio X.

The sample average $\hat{\mu}_X$ of these data is equal to 2.87 and the sample variance $\hat{\sigma}_X^2$ is equal to 13.82. Using these values, fit a beta distribution to these data. Then graph the theoretical cumulative distribution function of X by making use of tables of the incomplete beta function. Compare this graph to a graph of the sample cumulative distribution function. Is the beta distribution a good model for the variation of the ratio X? Assuming that X does have the beta distribution which you have fit using the data described in Table E.4, calculate the probability that in a given ore sample there will be more than twice as much lead (Pb) by weight as there is copper (Cu).

27. A certain shirt manufacturing machine makes collars whose circumferences X are supposed to be 16 inches, but actually vary around that quantity according to a continuous distribution with a probability density function $f_X(\tau)$ which is equal to $(960)(\tau - 15.75)^2(16.25 - \tau)^2$ when $15.75 < \tau < 16.25$, and is equal to 0 otherwise. It can then be shown that $(X - 15.75)/(16.25 - 15.75) = 2X - 31.50$ has a beta distribution with parameters $r = 3$ and $s = 3$.
   (a) Using this information, find $P\{X > 16.00\}$ and $P\{X < 16.00\}$.
   (b) Also find the expected value and median of the distribution of X.

A28. Let X have a beta distribution with parameters r and s.
   (a) Assume that X has expected value μ_X equal to 0.6 and variance σ_X^2 equal to 0.048. What are the values of r and s?

Table E.4. Frequency distribution of the ratio Cu/(Cu + Pb) in 1000 samples of ore.

Midpoint x of Interval of Values	Observed Frequency of the Event $\{x-0.2 \leq X < x+0.2\}$
0.02	209
0.06	264
0.10	170
0.14	105
0.18	58
0.22	53
0.26	33
0.30	22
0.34	12
0.38	11
0.42	11
0.46	10
0.50	4
0.54	5
0.58	4
0.62	8
0.66	1
0.70	5
0.74	4
0.78	5
0.82	2
0.86	1
0.90	1
0.94	0
0.98	2

SOURCE: Based on G. S. Koch, Jr. and R. F. Link, *Statistical Analysis of Geological Data*, vol 2, © 1971, John Wiley & Sons, Inc., by permission of John Wiley & Sons, Inc.

(b) Find the probability that X is within 2 standard deviations of its expected value. That is, find $P\{0.6-2\sqrt{0.048} \leq X \leq 0.6+2\sqrt{0.048}\}$. Compare this probability with the lower bound provided by Table 6.2 of Chapter 5, and with the probability that a normally distributed random variable is within 2 standard deviations of its expected value.

(c) Find the median of the distribution of X.

29. In the context of Example 5.3, find the following quantities:
   (a) The expected proportion Y of change in the wholesale prices of commodities from one year to the next, when prices fall.

(b) The probability that the wholesale price of a commodity falls to less than 50 percent (0.50) of its value of the preceding year.

(c) If you know that the wholesale price of a commodity will fall from the previous year, and you know that the price of the commodity during the previous year was 100, what price would you predict for this commodity this year? Explain your answer.

*30. The gamma and lognormal distributions have both been used as models for the distribution of hourly median power and instantaneous power of received radio signals. Siddiqui and Weiss (1963) present data for values of the hourly median power D (expressed in decibels) received over a particular path from Detroit, Michigan to Hudson, Ohio during the period from May 1950 to October 1950. These data, expressed in terms of $Q = 10 \log_{10} D$, and rounded to the nearest integer, are summarized in Table E.5.

Table E.5. Observed frequency distribution of $Q = 10 \log_{10} D$.

Midpoint q	Observed Frequency of $\{q-0.5 \leq Q \leq q+0.5\}$	Midpoint q	Observed Frequency
1	0	25	15
2	26	26	13
3	5	27	7
4	8	28	17
5	9	29	5
6	22	30	12
7	27	31	9
8	36	32	10
9	27	33	4
10	46	34	6
11	68	35	5
12	48	36	2
13	67	37	3
14	38	38	2
15	24	39	6
16	58	40	2
17	33	41	1
18	54	42	3
19	38	43	1
20	41	44	1
21	25	45	2
22	34	46	0
23	20	47	0
24	21	48	1

(a) Fit a normal probability distribution to the observed data on the random variable Q. Compare the theoretical cumulative distribution function obtained from the fitted normal distribution to the sample cumulative distribution of Q obtained from the data. Is the fit a good one? What does your answer imply about the distribution of the random variable D?

(b) Note that $D = 10^{Q/10}$. Convert the above table of observations on Q to a table of observations on
$$X = 1/D = 1/10^{Q/10} = 10^{-Q/10}.$$
Do this by calculating the value of X corresponding to the midpoint q of each interval of Q values, and by assigning to this value of X the observed frequency of the interval of Q values whose midpoint is q.

(c) Using the table of observations on X obtained in part (b) above, fit a gamma probability distribution. Is the fit a good one? Explain.

(d) You wish to find the probability that the hourly median power D exceeds 50 decibels. Show how your results in parts (a) and (c) can each be used to assign probability values to this event. Are the probability values assigned the same? If not, which probability value do you think is a more accurate evaluation of the true probability of the event $\{D > 50\}$?

*A31. The lognormal distribution has been used as a model for some meteorological data by Pretorius (1930). Measurements of the height of the barometer at Greenwich were taken on alternate days during the period 1848–1926. A

Table E.6. Observed frequency distribution of barometric heights at Greenwich.

Barometric Height	Observed Frequency
30.75	1
30.65	1
30.55	10
30.45	32
30.35	111
30.25	214
30.15	386
30.05	365
29.95	288
29.85	199
29.75	129
29.65	86
29.55	62
29.45	26
29.35	17
29.25	10
29.15	9
29.05	2
28.95	2
28.85	1
	1951

distribution was made for the readings on the first day when the reading on the third day was between 30.1 and 30.2 inches. These data appear in Table E.6.

Fit a lognormal distribution to these data. How good is the fit?

32. Ages of the bride and bridegroom were noted for 301,785 marriages contracted in Australia during the period 1907–1914. (Ages of brides over 85 and bridegrooms over 90 were not included.) It was also noted that mis-

Table E.7. Distribution of ages of brides and bridegrooms, Australian marriages.

	Brides			Bridegrooms	
AGE OF BRIDE	OBSERVED FREQUENCY	THEOR. FREQ. LOG. NORMAL	AGE OF BRIDEGROOM	OBSERVED FREQUENCY	THEOR. FREQ. LOG. NORMAL
12.5	5	...	16.5	294	259
15.5	2,975	2,207	19.5	10,995	19,453
18.5	38,291	44,776	22.5	61,001	56,819
21.5	80,847	77,051	25.5	73,054	64,141
24.5	71,010	64,894	28.5	56,501	51,989
27.5	44,541	43,568	31.5	33,478	36,894
30.5	24,261	27,008	34.5	20,569	24,735
33.5	13,752	16,318	37.5	14,281	16,203
36.5	8,883	9,831	40.5	9,320	10,542
39.5	6,062	5,966	43.5	6,236	6,869
42.5	3,478	3,664	46.5	4,770	4,503
45.5	2,605	2,281	49.5	3,620	2,977
48.5	1,805	1,442	52.5	2,190	1,987
51.5	1,139	924	55.5	1,655	1,339
54.5	645	601	58.5	1,100	912
57.5	513	396	61.5	810	627
60.5	291	264	64.5	649	435
63.5	242	179	67.5	487	305
66.5	206	122	70.5	326	216
69.5	130	84	73.5	211	154
72.5	56	59	76.5	119	111
75.5	25	41	79.5	73	80
78.5	16	29	82.5	27	59
81.5	6	21	85.5	14	43
84.5	1	15	88.5	5	32
...	...	$\Big\{$ 44	...	...	$\Big\{$ 101
...	...		...	...	
Totals	301,785	301,785		301,785	301,785

Expected value: 25.72   29.38
Variance: 5.016 (in 3-year units)   6.974 (in 3-year units)

statement of ages by persons under 21 years is quite common in marriage statistics; however, no adjustment of the data was made. The results of the fit are shown in Table E.7 [Pretorius (1930)]. Comment on the use of the lognormal model for these data. Does the lognormal distribution provide a good fit? Do you think any of the other families of continuous distributions which have been discussed in Chapter 8 (or in Chapter 7) would provide a better fit? Explain your answer.

9
MULTIVARIATE DISTRIBUTIONS

Order and simplification are the first steps toward the mastery of a subject—the actual enemy is the unknown.

Thomas Mann (The Magic Mountain)

Happy the man, who, studying Nature's laws, through known effects can trace the secret cause.

Virgil (Georgics)

1. INTRODUCTION

Many of the most important problems in science and technology are concerned with the question of how the values of one measurable quantity (or several measurable quantities) can be used to explain, predict, or control another quantity (or quantities). Thus, physicists may be interested in how the motion of a body is affected by the forces acting upon it, aeronautical engineers may be interested in finding out how automatic sensors can provide information to guide a spaceship, and biologists may be interested in how organisms react to chemical changes in their environment (for example, how much of a certain drug is required, on the average, to destroy a certain virus). Similarly, sociologists may be interested in the relationship of group size to problem-solving productivity, psychologists may be interested in the relationship of personality test scores to anxiety level, and educators may be interested in how high school grades can be used to predict college achievement. Indeed, virtually all scientific and technological disciplines require, as a part of their development, knowledge of the interrelationships among the variables which compose their empirical framework.

Mechanistic models (see Chapter 1) for natural phenomena express relationships between variables either in terms of the logic of causation (for example, "*A* causes *B*") or in terms of exact (or approximate) mathematical formulas (such as "force equals mass times acceleration"). Stochastic models, on the other hand, express such relationships in terms of statistical tendencies of joint occurrence. For example, stochastic models help quantify such statements as: "High values of X tend to be associated with high values of Y". The probabilistic framework for such stochastic models involves the notion of *composite experiments* (see Chapter 3). When the random experiments which form the components of such composite experiments give rise to more than one random variable, we say that the resulting probability model gives the *joint* (or *multivariate*) *distribution* of these variables.

To introduce the ideas embodied in the notion of multivariate distributions, we first discuss the joint distribution of two random variables (a *bivariate distribution*). Because the concepts that we discuss are more easily understood when both random variables are discrete, we begin in Sections 2 and 3 with an examination of discrete bivariate distributions. In Section 4, we consider continuous bivariate distributions (that is, probability models that describe the joint variation of two continuous random variables). The most famous and widely used continuous bivariate distribution, the bivariate normal distribution, is investigated in Section 5. Finally, in Section 6 we generalize our discussion to cover joint distributions of three random variables (*trivariate distributions*) and of more than three random variables.

2. DISCRETE BIVARIATE DISTRIBUTIONS

In Chapters 4 through 8, we have considered probability models for only one random variable. However, if we are observing more than one variable in a given (composite) random experiment, concentrating upon the distribution of each of these variables separately causes us to ignore any information about the frequency with which certain values of any one of these random variables, say X, can be expected to occur in connection with certain values of any other random variable, say Y. To obtain such information, we construct the *bivariate distribution* of X and Y. That is, we construct a probability distribution for the composite experiment whose outcomes are values of the *pair* (X, Y).

Suppose X and Y are both discrete random variables. Then their *joint probability mass function* (bivariate probability mass function) is the rule which assigns to each pair of numbers (x, y) the probability of the event $\{X = x \text{ and } Y = y\}$. Thus, the joint probability mass function $p_{X,Y}(x, y)$ is defined by the equation

(2.1) $$p_{X,Y}(x, y) = P_{X,Y}\{X = x \text{ and } Y = y\}.$$

Since the double subscript on $p_{X,Y}(x, y)$ is cumbersome, when no confusion is possible, we write $p(x, y)$ instead of $p_{X,Y}(x, y)$. Often, observations on the random variables X and Y are taken nearly simultaneously;

in such cases, $p_{X,Y}(x,y)$ gives the probability of the event that $X = x$ and simultaneously $Y = y$. On the other hand, one of the variables, say X, may be observed first; here $p_{X,Y}(x,y)$ tells us the probability of the event that $X = x$ and then $Y = y$.

Consider the case where the random variable X has three possible values, say x_1, x_2, x_3, and where the random variable Y has four possible values: y_1, y_2, y_3, and y_4. Then the possible outcomes of the random experiment in which X and Y are both observed are the pairs of numbers: (x_1,y_1), (x_1,y_2), (x_1,y_3), (x_1,y_4), (x_2,y_1), (x_2,y_2), (x_2,y_3), (x_2,y_4), (x_3,y_1), (x_3,y_2), (x_3,y_3), (x_3,y_4). There are 12 such outcomes; corresponding to these outcomes there are 12 simple events: $\{(x_1,y_1)\}$, $\{(x_1,y_2)\}$, and so on. To give the joint probability mass function $p_{X,Y}(x,y)$ for X and Y, it is not necessary to write down the probabilities of simple events corresponding to pairs of numbers (x,y) that cannot be observed since these probabilities are, of course, all equal to 0 [that is, $p_{X,Y}(x,y) = 0$ for $x \neq x_1, x_2, x_3$; $y \neq y_1, y_2, y_3, y_4$]. Nevertheless, we must still list 12 different numbers. A helpful way to list the probabilities $p_{X,Y}(x,y) = p(x,y)$ for $x = x_1, x_2, x_3$, $y = y_1, y_2, y_3, y_4$, is to give them in the form of a *contingency table*:

x \ y	y_1	y_2	y_3	y_4
x_1	$p(x_1,y_1)$	$p(x_1,y_2)$	$p(x_1,y_3)$	$p(x_1,y_4)$
x_2	$p(x_2,y_1)$	$p(x_2,y_2)$	$p(x_2,y_3)$	$p(x_2,y_4)$
x_3	$p(x_3,y_1)$	$p(x_3,y_2)$	$p(x_3,y_3)$	$p(x_3,y_4)$

Such a contingency table is both more compact than a vertical list of values $p_{X,Y}(x, y)$, and also more revealing. By looking across the first row of the table, we can see which values of the random variable Y are the most probable to occur with a value $X = x_1$ of the random variable X. Similarly, the third column of the contingency table tells us what values of the random variable X are the most probable to appear jointly with a value $Y = y_3$ of the random variable Y. Note that since 1 of the 12 outcomes (x_1, y_1), $(x_1,y_2), \ldots, (x_3,y_4)$ must occur, the sum of the entries $p_{X,Y}(x,y)$ in the contingency table is equal to 1.

As an alternative to a contingency table, we can graph the joint probability mass function $p_{X,Y}(x,y)$ (see Figure 2.1). We can view the possible pairs (x,y) as points in the plane. From these points (x,y), perpendicular dotted lines are extended to a height equal to $p_{X,Y}(x,y)$ from the plane; these lines represent the probabilities of the events $\{X = x_i \text{ and } Y = y_j\}$, $i = 1, 2, 3, j = 1, 2, 3, 4$. Thus, for example, if $p_{X,Y}(x_1,y_3) = 1/10$, the line rising from the point (x_1,y_3) has height $1/10$. Since the sum of the values of the probability mass function $p_{X,Y}(x,y)$ over all possible points (x,y)

2. DISCRETE BIVARIATE DISTRIBUTIONS 435

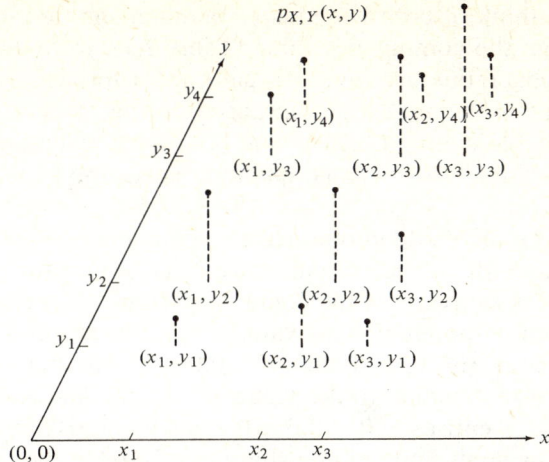

Figure 2.1 Graph of the joint probability mass function $p_{X,Y}(x,y)$.

is equal to 1, the sum of the heights of the dotted lines used to graph the probability mass function $p_{X,Y}(x,y)$ in Figure 2.1 is equal to 1. Note that Figure 2.1 and the contingency table exhibited above both give similar information about the joint variation of the random variables X and Y.

Example 2.1. Two professional wine tasters are asked to rate a sample of wine using a rating scale that allows them to give ratings of 1, 2, 3, 4, or 5 (a rating of 5 being equivalent to "top quality"). If standards of wine quality are more or less common to all wine tasters, and if the wine they are to taste is of a moderate quality, we might expect their responses to follow a table of probabilities such as Table 2.1. Here, wine taster I's rating is a random variable X, wine taster II's rating is a random variable

Table 2.1. A bivariate probability mass function for two wine tasters exhibited in the form of a contingency table.

Rating of Wine Taster I	II			y			
		1	2	3	4	5	Totals
	1	0.03	0.02	0.01	0.00	0.00	0.06
	2	0.02	0.08	0.05	0.02	0.01	0.18
x	3	0.01	0.05	0.25	0.05	0.01	0.37
	4	0.00	0.02	0.05	0.20	0.02	0.29
	5	0.00	0.01	0.01	0.02	0.06	0.10
	Totals	0.06	0.18	0.37	0.29	0.10	1.00

Y, and we can think of both wine tasters announcing their ratings simultaneously. From the contingency table (Table 2.1), we note that most of the higher probabilities are given to pairs (x,y) for which x and y are nearly equal and in the middle of the range of ratings (ranks of 2, 3, or 4). The most probable simple event is that both wine tasters agree that the wine they are rating is of average quality, that is, the event $\{X = 3$ and $Y = 3\}$.

Notice that we have added one extra column and one extra row to the contingency table; this extra column and extra row give the row sums and column sums, respectively, of the contingency table. For example, in row 1 of the table (corresponding to the value $x = 1$), the last entry, 0.06, is the sum of the other entries 0.03, 0.02, 0.01, 0.00, 0.00 in that row. Similarly, in column 3 (corresponding to the value $y = 3$), the last entry 0.37 is the sum of the other entries 0.01, 0.05, 0.25, 0.05, 0.01, in that column. Because the row sums and column sums are placed at the margins of the contingency table, they are known as *marginal totals*, or (since they are the sum of probabilities) *marginal probabilities*. In fact, the sum (marginal total) of the entries in row 1 (excluding the entry in the "Total" column) of the contingency table gives the probability $P_X\{X=1\} = 0.06$. Similarly, the sum (marginal total) of the entries in column 3 of the contingency table (excluding the entry in the row marked "Total") gives $P_Y\{Y=3\} = 0.37$. We return to a discussion of marginal totals momentarily.

To show how different bivariate probability distributions can reveal different interrelationships between two random variables X and Y, consider Table 2.2, which gives another possible discrete joint probability model for the ratings of the two wine tasters. Here, as in Table 2.1, most of the probability is concentrated on simple events $\{(x,y)\}$ corresponding to situations where the wine being rated is considered to be average or nearly average by both wine tasters (ratings of 2, 3, or 4). However, in Table 2.2 there is less probability attached to simple events corresponding to cases where there is exact agreement between the raters [that is, events $\{(x,y)\}$ for which $x = y$], and more probability assigned to simple

Table 2.2. Another bivariate probability mass function for two wine tasters.

Rating of Wine Taster I	II	y 1	2	3	4	5	Totals
x 1		0.01	0.01	0.01	0.01	0.01	0.05
2		0.01	0.09	0.09	0.09	0.01	0.29
3		0.01	0.09	0.12	0.09	0.01	0.32
4		0.01	0.09	0.09	0.09	0.01	0.29
5		0.01	0.01	0.01	0.01	0.01	0.05
Totals		0.05	0.29	0.32	0.29	0.05	1.00

2. DISCRETE BIVARIATE DISTRIBUTIONS

events in which there is wide divergence of opinion between the wine tasters [that is, events such as {(1,5)}, {(5,1)}, {(1,4)}, {(4,1)}, and so on]. Table 2.2, in contrast to Table 2.1, may thus reflect a situation where standards of quality vary a great deal with the individual.

The two probability models given in Example 2.1 can also be compared visually through an inspection of the graphs of their respective bivariate probability mass functions. The graphs associated with Tables 2.1 and 2.2 are shown in Figure 2.2.

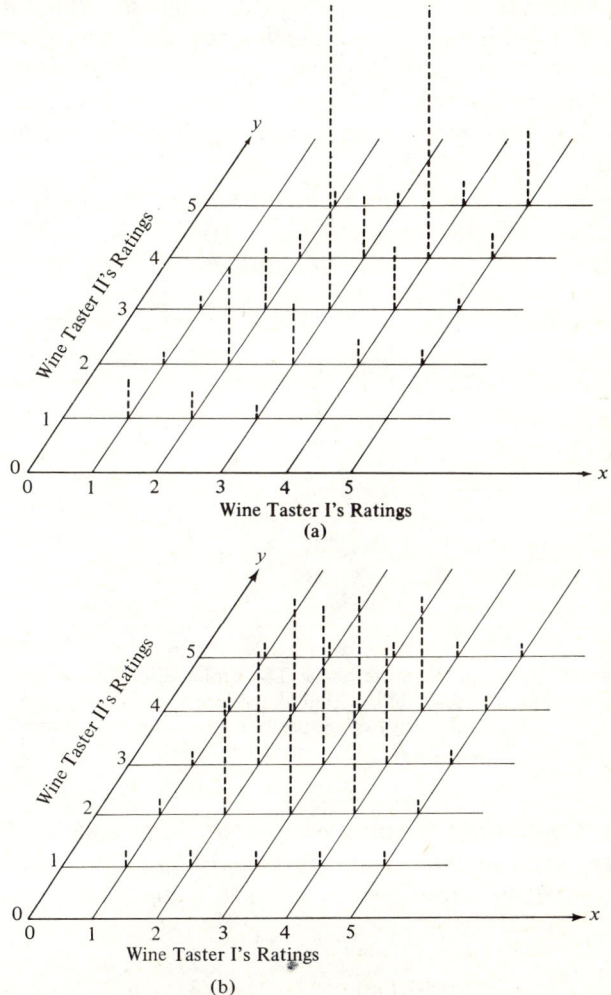

Figure 2.2 Graph of the bivariate probability mass function $p_{X,Y}(x, y)$ corresponding to the probability model for the joint ratings of Wine Tasters I and II: (a) associated with Table 2.1; (b) associated with Table 2.2.

The bivariate probability mass function $p_{X,Y}(x,y)$ contains all of the probabilistic information relevant for describing the simultaneous variation of the two discrete random variables X and Y. It is helpful to think of the pair (X,Y), resulting from the observation of X and Y, as a random point in the plane. That is, instead of the random experiment in which we actually observe X and Y, we can think of another random experiment in which X provides the first coordinate and Y provides the second coordinate of a point (X,Y) in the plane. On repeated trials of this last random experiment, different points will occur. Let R be any region in the plane (see Figure 2.3), and suppose we are interested in the probability that (X,Y) falls within the region R on any given performance of the experiment. In the case of discrete random variables, we obtain this probability by summing $p_{X,Y}(x,y)$ over all the possible points (x,y) contained within the region R.

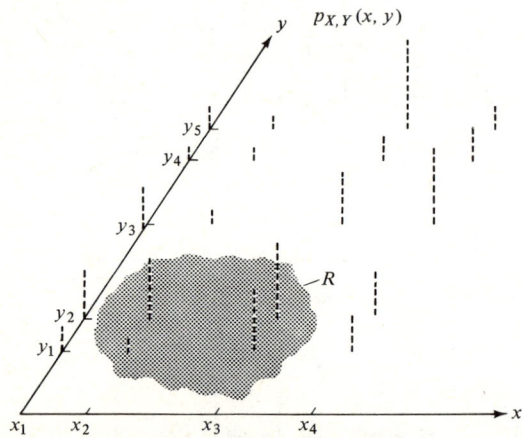

Figure 2.3 A region R in the plane. The probability that the random point (X,Y) falls within the region R is the sum of the heights $p_{X,Y}(x,y)$ of the dotted lines which correspond to points (x,y) in R.

Example 2.1 (Continued). Suppose we ask for the probability that the two wine tasters agree in their rankings (see Figure 2.4). Under the first probability model given for this problem (see Table 2.1),

$$P\{\text{tasters agree}\} = P\{(x,y): x = y\}$$
$$= P\{(1,1) \text{ or } (2,2) \text{ or } (3,3) \text{ or } (4,4) \text{ or } (5,5)\}$$
$$= 0.03 + 0.08 + 0.25 + 0.20 + 0.06 = 0.62.$$

Under the second probability model given for this problem (see Table 2.2),

2. DISCRETE BIVARIATE DISTRIBUTIONS 439

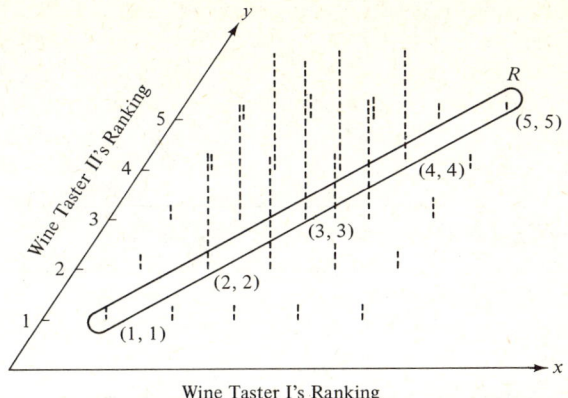

Figure 2.4 The region R corresponding to the event E that the Wine Tasters I and II agree in their rankings of a sample of wine. The sum of the probabilities of the points contained in R equals the probability of the event E.

we obtain

$$P\{\text{tasters agree}\} = 0.01 + 0.09 + 0.12 + 0.09 + 0.01 = 0.32.$$

This calculation partially verifies our earlier assertion that there is less agreement between the wine tasters under probability model of Table 2.2 than there is under the probability model of Table 2.1.

Marginal Distributions

Not only does the joint probability mass function $p_{X,Y}(x,y)$ contain all of the relevant statistical information concerning X and Y considered *simultaneously*, it also contains all the relevant information concerning X and Y considered *individually*. That is, from $p_{X,Y}(x,y)$ we can obtain the probability mass functions $p_X(x)$ and $p_Y(y)$ of X and Y, respectively. To see this, let R_1 be the region in the plane consisting of all possible points (see Figure 2.5) having x_0 for the first coordinate (no matter what the second coordinate y may be). The event $\{(X,Y) \text{ falls in } R_1\}$ is equivalent to the event $\{X = x_0\}$. Hence, $P_{X,Y}\{\text{point } (X,Y) \text{ is a member of } R_1\} = P_X\{X = x_0\} = p_X(x_0)$, and this probability can be obtained by summing $p_{X,Y}(x,y)$ over all points (x,y) in the region R_1. In an analogous way, $p_Y(y_0)$ can be obtained by summing $p_{X,Y}(x,y)$ over all points in the region R_2 in the plane consisting of points (x,y) whose second coordinate is equal to y_0 (see Figure 2.5). When the distributions of X and Y are obtained in this way from their joint probability function $p_{X,Y}(x,y)$, the distributions obtained are referred to as *marginal distributions*.

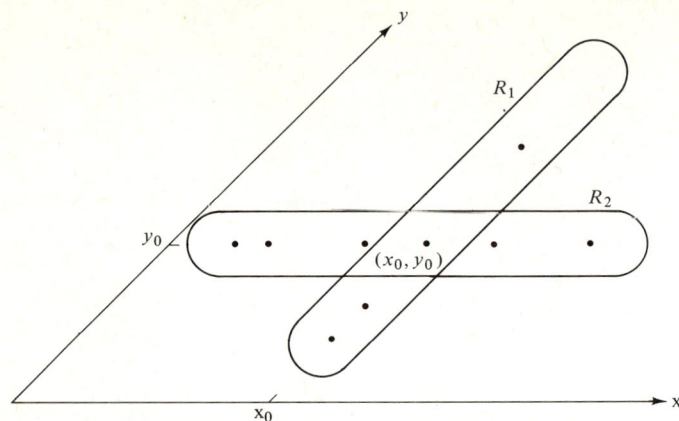

Figure 2.5 Region R_1 consists of all points (x,y) having $x = x_0$ (a fixed x-coordinate). Region R_2 consists of all points (x, y) having a fixed y-coordinate.

We have already considered marginal distributions in Example 2.1, where the probabilities $p_X(x_i)$ and $p_Y(y_j)$ appeared in the appended last column and last row, respectively, of the contingency tables, Table 2.1 and Table 2.2. There, these probabilities were called marginal probabilities. In general, suppose that we have summarized a discrete bivariate probability distribution in terms of a contingency table such as the following:

x \ y	y_1	y_2	y_3	y_4	Row Total
x_1	$p_{X,Y}(x_1,y_1)$	$p_{X,Y}(x_1,y_2)$	$p_{X,Y}(x_1,y_3)$	$p_{X,Y}(x_1,y_4)$	$p_X(x_1)$
x_2	$p_{X,Y}(x_2,y_1)$	$p_{X,Y}(x_2,y_2)$	$p_{X,Y}(x_2,y_3)$	$p_{X,Y}(x_2,y_4)$	$p_X(x_2)$
x_3	$p_{X,Y}(x_3,y_1)$	$p_{X,Y}(x_3,y_2)$	$p_{X,Y}(x_3,y_3)$	$p_{X,Y}(x_3,y_4)$	$p_X(x_3)$
Column Total	$p_Y(y_1)$	$p_Y(y_2)$	$p_Y(y_3)$	$p_Y(y_4)$	1.00

In such a contingency table, we can find the marginal distribution (marginal probability mass function) of X from the (appended) last column of the table (the column marked "Row Total"), and the marginal distribution of Y from the (appended) last row. Thus, in Table 2.1, the marginal distribution (probability mass function) of the rating X of wine taster I appears in the last column of the table. We reproduce that distribution in Table 2.3(a).

2. DISCRETE BIVARIATE DISTRIBUTIONS

Table 2.3(a). Probability mass function (given only for possible values of the random variable) of the rating X of wine taster I as taken from the last column of Table 2.1.

X	Probability $p_X(x) = P_X\{X = x\}$
1	0.06
2	0.18
3	0.37
4	0.29
5	0.10

Similarly, in Table 2.1, the marginal distribution of the rating Y of wine taster II appears in the last row of the table. We reproduce that distribution in Table 2.3(b).

Table 2.3(b). Probability mass function of the rating Y of wine taster II as taken from the last column of Table 2.1.

y	1	2	3	4	5
Probability $p_Y(y) = P_Y\{Y = y\}$	0.06	0.18	0.37	0.29	0.10

Comparing Tables 2.3(a) and (b), we note that X and Y have identical marginal distributions.

It should be emphasized that the marginal distributions $p_X(x)$ and $p_Y(y)$ obtained from the joint distribution $p_{X,Y}(x,y)$ of any pair (X,Y) of jointly distributed random variables must be the same as the distributions $p_X(x)$ and $p_Y(y)$ which we would have obtained if we had considered the variables X and Y individually.

Although $p_X(x)$ and $p_Y(y)$ can be obtained from $p_{X,Y}(x,y)$ in the form of marginal distributions, the converse is not true; that is, $p_{X,Y}(x,y)$ *cannot be determined from a knowledge of $p_X(x)$ and $p_Y(y)$ alone*. To illustrate this point, recall the two joint probability models for the sex of the first- and second-born children considered in Example 3.1 of Chapter 3.

Example 2.2. In Example 3.1 of Chapter 3, we observe the sex of the children in a family that has exactly 2 children. In that example, we represented the outcomes of this composite random experiment in terms of the pairs: (sex of first-born child, sex of second-born child). Let us now describe the outcomes of the experiment in terms of two random variables X and Y. Let $X = 1$ if the first-born child is a male, $X = 0$ if the first-born child is a female, $Y = 1$ if the second-born child is a male, and $Y = 0$ if the second-born child is a female. The possible outcomes in the experiment (which now consists of observing X and Y rather than the sexes of the children) are then: (1,1), (1,0), (0,1), (0,0). In Chapter 3 we considered two possible joint distributions for the pair (X,Y). Table 2.4(a) gives the first of these joint distributions (in the form of a contingency table).

Table 2.4(a). Joint probability distribution for the sex of the children in a family having 2 children. $X = 1$ when the first child is male, and 0 otherwise; $Y = 1$ when the second child is male, and 0 otherwise.

x \ y	1	0	Total
1	$\frac{1}{3}$	$\frac{1}{6}$	$\frac{1}{2}$
0	$\frac{1}{6}$	$\frac{1}{3}$	$\frac{1}{2}$
Total	$\frac{1}{2}$	$\frac{1}{2}$	1

In Chapter 3, we also considered the joint distribution given in Table 2.4(b).

Table 2.4(b). Another joint probability distribution for the sex of the children in a family having 2 children.

x \ y	1	0	Total
1	$\frac{1}{4}$	$\frac{1}{4}$	$\frac{1}{2}$
0	$\frac{1}{4}$	$\frac{1}{4}$	$\frac{1}{2}$
Total	$\frac{1}{2}$	$\frac{1}{2}$	1

The joint probability models given in Tables 2.4(a) and 2.4(b) both yield the same marginal probability distributions for X, namely,

X:

x	0	1
$p_X(x)$	$\frac{1}{2}$	$\frac{1}{2}$

Similarly, the joint probability models in Tables 2.4(a) and 2.4(b) have a common marginal distribution for Y, namely,

Y:

y	0	1
$p_Y(y)$	$\frac{1}{2}$	$\frac{1}{2}$

[*Note:* The fact that X and Y have the same marginal distribution is irrelevant to the point we are making here.] However, the probability model given in Table 2.4(a) is obviously not the same probability model as that given in Table 2.4(b). Hence, two different joint probability models can have the same marginal probability distributions. Knowing only the marginal probability distributions for X and for Y, as shown above, we cannot tell which joint probability model [Table 2.4(a) or Table 2.4(b)] is the appropriate one.

Example of the Construction of a Bivariate Probability Model

Example 2.2 illustrates an experimental situation where we have transformed a probability model for a composite random experiment consisting of nonquantitative outcomes into a bivariate probability model for two random variables X and Y. In the following example, we illustrate a somewhat more complicated composite random experiment in which two quantitative aspects X and Y are of interest, and show how the bivariate probability distribution for X and Y is obtained.

Example 2.3. Fifth-graders from a certain school district can be divided into three distinct categories: talented (T), average (A), and slow (S). Let p_T be the proportion of fifth-grade children who are talented, p_A be the proportion who are average, and p_S be the proportion who are slow; then, $p_T + p_A + p_S = 1$. A random sample of 4 fifth-graders is drawn *with* replacement. For each child drawn, we note only whether the child is talented, average, or slow. The 81 possible outcomes ω_i and the respective probabilities of the corresponding simple events $\{\omega_i\}$ are shown in Figure 2.6.

Let X denote the number of average children in the sample and Y denote the number of talented children. The list of possible outcomes (x,y) is

(0,0)	(1,0)	(2,0)	(3,0)	(4,0)
(0,1)	(1,1)	(2,1)	(3,1)	
(0,2)	(1,2)	(2,2)		
(0,3)	(1,3)			
(0,4)				

Note that any pair such as (1,4) cannot be an outcome because we only draw 4 children. It can be seen from the enumeration in Figure 2.6 that the joint distribution of the random variables X and Y can be given by a contingency table such as shown in Table 2.5.

It is evident from Figure 2.6 that a complete enumeration of the probabilities for some bivariate probability distributions can become a somewhat tedious task. Fortunately, in the present case, there is a convenient

Outcome ω	Probability of $\{\omega\}$	Outcome ω	Probability of $\{\omega\}$	Outcome ω	Probability of $\{\omega\}$
TTTT	p_T^4	AAAA	p_A^4		
TTTS TTST TSTT STTT	$p_T^3 p_S$	AAAS AASA ASAA SAAA	$p_A^3 p_S$	TTTA TTAT TATT ATTT	$p_A p_T^3$
TTSS TSTS STTS TSST STST SSTT	$p_T^2 p_S^2$	AASS ASAS SAAS ASSA SASA SSAA	$p_A^2 p_S^2$	TTAA TATA ATTA TAAT ATAT AATT	$p_A^2 p_T^2$
SSST SSTS STSS TSSS	$p_T p_S^3$	SSSA SSAS SASS ASSS	$p_A p_S^3$	AAAT AATA ATAA TAAA	$p_A^3 p_T$
SSSS	p_S^4				
TTSA TSTA STTA TTAS TSAT STAT TATS TAST SATT ATTS ATST ASTT	$p_T^2 p_S p_A$	SSTA STSA TSSA SSAT STAS TSAS SAST SATS TASS ASST ASTS ATSS	$p_T p_A p_S^2$	AATS ATAS TAAS AAST ATSA TASA ASAT ASTA TSAA SAAT SATA STAA	$p_T p_A^2 p_S$

Figure 2.6. Outcomes and corresponding probabilities for Example 2.3. The symbol TTTT for the first outcome in this list represents the outcome in which talented children are drawn on all four draws; the symbol STAA for the last outcome in the list represents the outcome in which a slow child is drawn on the first draw, a talented child is drawn on the second draw, and average children are drawn on the third and fourth draws. Outcomes ω for which the corresponding simple events $\{\omega\}$ have equal probabilities are grouped together.

2. DISCRETE BIVARIATE DISTRIBUTIONS

Table 2.5. A bivariate probability mass function for the number X of average children and the number Y of talented children observed in a random sample, with replacement, of 4 fifth-graders.

x \ y	0	1	2	3	4
0	p_S^4	$4p_T p_S^3$	$6p_T^2 p_S^2$	$4p_T^3 p_S$	p_T^4
1	$4p_A p_S^3$	$12p_T p_A p_S^2$	$12p_T^2 p_A p_S$	$4p_T^3 p_A$	0
2	$6p_A^2 p_S^2$	$12p_T p_A^2 p_S$	$6p_T^2 p_A^2$	0	0
3	$4p_A^3 p_S$	$4p_T p_A^3$	0	0	0
4	p_A^4	0	0	0	0

formula which enables us to write the bivariate probability mass function for X and Y in a compact form:

$$(2.2) \quad p_{X,Y}(x,y) = \begin{cases} \dfrac{4!}{x!\,y!\,(4-x-y)!} p_A^x p_T^y p_S^{4-x-y}, & \text{if } x, y \text{ are integers} \\ & \text{and } x+y \leq 4, \\ 0, & \text{otherwise.} \end{cases}$$

Thus, if $x = 2$ and $y = 1$, we have, by a direct substitution into Equation (2.1),

$$p_{X,Y}(2,1) = \frac{4!}{2!\,1!\,1!} p_A^2 p_T^1 p_S^1 = 12 p_A^2 p_T p_S.$$

If the proportion of average students in the given school district is $p_A = \tfrac{1}{2}$, the proportion of slow students is $p_S = \tfrac{1}{3}$, and the proportion of talented students is $p_T = \tfrac{1}{6}$, then the probability that in our sample of 4 students we observe 2 average students and 1 talented student is

$$p_{X,Y}(2,1) = P_{X,Y}\{X = 2 \text{ and } Y = 1\}$$

$$= 12 \left(\frac{1}{2}\right)^2 \left(\frac{1}{6}\right)\left(\frac{1}{3}\right) = \frac{1}{6}.$$

The probability that we observe 1 average student and 2 talented students is, from Equation (2.2),

$$p_{X,Y}(1,2) = \frac{4!}{1!\,2!\,1!} p_A^1 p_T^2 p_S^1 = 12 p_A p_T^2 p_S = 12\left(\frac{1}{2}\right)\left(\frac{1}{6}\right)^2\left(\frac{1}{3}\right) = \frac{1}{18}.$$

Conditional Distributions

We have mentioned earlier that in performing a composite random experiment, the value of one of two random variables may become known before the value of the other variable. Thus, we may know that the value of the random variable X is x before we know the value of Y. On the basis of the knowledge that $X = x$, we may want to reassess our calculations concerning probabilities for the various possible values of Y. Conditional probabilities (Chapter 3) now become relevant. The *conditional probability mass function* of Y given $X = x$, in symbols $p_{Y|X=x}(y)$, is defined to be

$$(2.3) \qquad p_{Y|X=x}(y) = \frac{p_{X,Y}(x,y)}{p_X(x)},$$

assuming $p_X(x) > 0$. [The conditional probability mass function $p_{Y|X=x}(y)$ is defined only where $p_X(x) > 0$.] It can easily be verified that $p_{Y|X=x}(y)$ is, in fact, a probability mass function; that is, it is nonnegative and the sum over all possible values of y is 1. The probability mass function $p_{Y|X=x}(y)$ contains all the relevant probabilistic information concerning Y when we know that $X = x$.

Example 2.4. A political scientist has just completed a study of American voting habits. He has been studying how the tendency of people to vote in presidential elections is related to their tendency to vote in nonpresidential, congressional elections. Since, for purposes of machine tabulation, he has been marking a "1" if a person voted in a given election and "0" if the person did not vote, the political scientist decides to summarize his findings in terms of the joint distribution of the random variable X, which indicates whether or not a person voted in a presidential election, and the random variable Y, which indicates whether or not that person voted in the following nonpresidential, congressional election. That is, $X = 1$ if the person voted in the presidential election, and $X = 0$ if the person did not vote in that election. Similarly, $Y = 1$ if the person voted in the next nonpresidential, congressional election, and $Y = 0$ if he did not vote in that election. The political scientist's data led him to hypothesize the following joint discrete probability distribution for X and Y:

x \ y	0	1	Total
0	$\frac{1}{8}$	$\frac{1}{8}$	$\frac{1}{4}$
1	$\frac{1}{2}$	$\frac{1}{4}$	$\frac{3}{4}$
Total	$\frac{5}{8}$	$\frac{3}{8}$	1

From this contingency table, we see that the marginal probability mass functions of X and Y are, respectively,

X:

x	0	1
$p_X(x)$	$\frac{1}{4}$	$\frac{3}{4}$

and

Y:

y	0	1
$p_Y(y)$	$\frac{5}{8}$	$\frac{3}{8}$

From these values we obtain

$$p_{Y|X=0}(0) = \frac{p_{X,Y}(0,0)}{p_X(0)} = \frac{\frac{1}{8}}{\frac{1}{4}} = \frac{1}{2},$$

$$p_{Y|X=0}(1) = \frac{p_{X,Y}(0,1)}{p_X(0)} = \frac{\frac{1}{8}}{\frac{1}{4}} = \frac{1}{2},$$

and

$$p_{Y|X=1}(0) = \frac{p_{X,Y}(1,0)}{p_X(1)} = \frac{\frac{1}{2}}{\frac{3}{4}} = \frac{2}{3},$$

$$p_{Y|X=1}(1) = \frac{p_{X,Y}(1,1)}{p_X(1)} = \frac{\frac{1}{4}}{\frac{3}{4}} = \frac{1}{3}.$$

For descriptive purposes, the political scientist records the conditional probability mass function of Y given $X = 0$ and the conditional probability mass function of Y given $X = 1$ in tables such as the following:

y	0	1
$p_{Y|X=0}(y)$	$\frac{1}{2}$	$\frac{1}{2}$

y	0	1
$p_{Y|X=1}(y)$	$\frac{2}{3}$	$\frac{1}{3}$

From an inspection of these tables, the political scientist concludes that if a person does not vote in a presidential election (that is, if $X = 0$), then the conditional probabilities that he *will* vote in the next nonpresidential, congressional election and that he *will not* vote in that election are the same. On the other hand, if a person does vote in a presidential election (that is, if $X = 1$), then the conditional probability that he will not vote in the next nonpresidential, congressional election is greater (twice as great) than the conditional probability that he will vote. Comparing the conditional probability mass function $p_{Y|X=1}(y)$ to the conditional probability mass function $p_{Y|X=0}(y)$, and each of these to the marginal probability mass function $p_Y(y)$, the political scientist suspects that when a person *does* vote in a presidential election, he has a tendency to believe that he has done his "civic duty," and thus does not feel the need to vote in the following nonpresidential election.

Example 2.1 (Continued). Recall that in Example 2.1 two professional wine tasters are asked to rate a sample of wine using a rating scale that allows them to give ratings of 1, 2, 3, 4, or 5 (a rating of 5 being equivalent to "top quality"). Table 2.1, reproduced below, gives a bivariate probability mass function for the rating X of wine taster I and the rating Y of wine taster II.

Wine Taster II

	x \ y	1	2	3	4	5	Total
	1	0.03	0.02	0.01	0.00	0.00	0.06
	2	0.02	0.08	0.05	0.02	0.01	0.18
Wine Taster I	3	0.01	0.05	0.25	0.05	0.01	0.37
	4	0.00	0.02	0.05	0.20	0.02	0.29
	5	0.00	0.01	0.01	0.02	0.06	0.10
	Total	0.06	0.18	0.37	0.29	0.10	1.00

We might be interested in seeing if we can predict the ratings of wine taster II if we are given the ratings of wine taster I. Thus, we are interested in finding the conditional probability mass functions for Y given $X = 1$, given $X = 2$, given $X = 3$, given $X = 4$, and given $X = 5$. We can construct such conditional probability mass functions directly from the contingency table given above. Note that the last column of the table (the marginal total) gives the values of $p_X(x)$, and that the entries in the contingency table in each row correspond to the values of the joint probability mass function $p_{X,Y}(x,y)$ for a given value x, and for $y = 1, 2, 3, 4, 5$. To calculate

$$p_{Y|X=x}(y) = \frac{p_{X,Y}(x,y)}{p_X(x)},$$

for $y = 1, 2, 3, 4, 5$, we need only divide every entry, $p_{X,Y}(x,y)$, in a given row by the last entry [the marginal total $p_X(x)$] in that row. The result is the following table (actually a collection of tables):

y	1	2	3	4	5
$p_{Y\mid X=1}(y)$	$\dfrac{0.03}{0.06}$	$\dfrac{0.02}{0.06}$	$\dfrac{0.01}{0.06}$	0	0
$p_{Y\mid X=2}(y)$	$\dfrac{0.02}{0.18}$	$\dfrac{0.08}{0.18}$	$\dfrac{0.05}{0.18}$	$\dfrac{0.02}{0.18}$	$\dfrac{0.01}{0.18}$
$p_{Y\mid X=3}(y)$	$\dfrac{0.01}{0.37}$	$\dfrac{0.05}{0.37}$	$\dfrac{0.25}{0.37}$	$\dfrac{0.05}{0.37}$	$\dfrac{0.01}{0.37}$
$p_{Y\mid X=4}(y)$	0	$\dfrac{0.02}{0.29}$	$\dfrac{0.05}{0.29}$	$\dfrac{0.20}{0.29}$	$\dfrac{0.02}{0.29}$
$p_{Y\mid X=5}(y)$	0	$\dfrac{0.01}{0.10}$	$\dfrac{0.01}{0.10}$	$\dfrac{0.02}{0.10}$	$\dfrac{0.06}{0.10}$

Here, the first row of the table gives the conditional probability mass function $p_{Y\mid X=1}(y)$, the second row gives the conditional probability mass function $p_{Y\mid X=2}(y)$, and so on. Notice that the sum of the entries in each row is now equal to 1. Reducing all fractions, we obtain the following table(s):

y	1	2	3	4	5
$p_{Y\mid X=1}(y)$	$\dfrac{1}{2}$	$\dfrac{1}{3}$	$\dfrac{1}{6}$	0	0
$p_{Y\mid X=2}(y)$	$\dfrac{1}{9}$	$\dfrac{4}{9}$	$\dfrac{5}{18}$	$\dfrac{1}{9}$	$\dfrac{1}{18}$
$p_{Y\mid X=3}(y)$	$\dfrac{1}{37}$	$\dfrac{5}{37}$	$\dfrac{25}{37}$	$\dfrac{5}{37}$	$\dfrac{1}{37}$
$p_{Y\mid X=4}(y)$	0	$\dfrac{2}{29}$	$\dfrac{5}{29}$	$\dfrac{20}{29}$	$\dfrac{2}{29}$
$p_{Y\mid X=5}(y)$	0	$\dfrac{1}{10}$	$\dfrac{1}{10}$	$\dfrac{1}{5}$	$\dfrac{3}{5}$

Note that we can also construct the conditional probability mass functions $p_{X\mid Y=1}(x)$, $p_{X\mid Y=2}(x)$, and so on. This task is easily accomplished by dividing every entry in a given column of the contingency table, say the column corresponding to $y = 1$, by the marginal total of that column. The result is the following table(s) of conditional probability mass functions:

x	$p_{X\|Y=1}(x)$	$p_{X\|Y=2}(x)$	$p_{X\|Y=3}(x)$	$p_{X\|Y=4}(x)$	$p_{X\|Y=5}(x)$
1	$\frac{1}{2}$	$\frac{1}{9}$	$\frac{1}{37}$	0	0
2	$\frac{1}{3}$	$\frac{4}{9}$	$\frac{5}{37}$	$\frac{2}{29}$	$\frac{1}{10}$
3	$\frac{1}{6}$	$\frac{5}{18}$	$\frac{25}{37}$	$\frac{5}{29}$	$\frac{1}{10}$
4	0	$\frac{1}{9}$	$\frac{5}{37}$	$\frac{20}{29}$	$\frac{1}{5}$
5	0	$\frac{1}{18}$	$\frac{1}{37}$	$\frac{2}{29}$	$\frac{3}{5}$

Observe that this time the *column* totals are each equal to 1.

It is worth remarking that in this example (Example 2.1) and in Example 2.4, the conditional probability mass functions $p_{Y|X=x}(y)$ of Y given $X = x$ are not the same as the marginal probability mass function $p_Y(y)$. Knowing information about the value x of X changes the probabilistic information we have concerning the variation of Y. For example, if we know nothing about the rank which wine taster I has given to a given sample of wine and are asked to guess what rank wine taster II has given (or will give) this sample of wine, we would be likely to say that wine taster II would give the wine a rank of $Y = 3$ since $p_Y(3)$ is larger than any other value of the marginal probability mass function $p_Y(y)$. Certainly, we would hesitate to say that wine taster II would give the wine a rank of $Y = 1$, since $p_Y(1) = 0.06$ is smaller than any other value of $p_Y(y)$. However, if we know that wine taster I has given the wine sample a rank of $X = 1$, then our guess of the rank which wine taster II gives this wine would be $Y = 1$ since the conditional probability $p_{Y|X=1}(1)$, of $Y = 1$ given that $X = 1$ is $\frac{1}{2}$, a probability that is larger than any other value of the conditional probability mass function $p_{Y|X=1}(y)$, and three times as large as the conditional probability, $p_{Y|X=1}(3) = \frac{1}{6}$. Thus, information about the value observed for the random variable X in this example leads us to change our prediction for the value of Y.

The Regression Function

Since, for each possible value x of X, the conditional probability mass function of Y given that $X = x$ is itself a probability mass function, we can consider the expected value of the random variable associated with this probability mass function. Formally, we define the *conditional expected value of Y given that X = x* (or, for short, the expected value of Y given that $X = x$) as the expected value associated with $p_{Y|X=x}(y)$. We denote the expected value of Y given $X = x$ by $E[Y|X = x]$.

Example 2.1 (Continued). In this example, the random variable X is the rank given a sample of wine by wine taster I, and the random variable Y is the rank given the same sample of wine by wine taster II. We have previously calculated (from Table 2.1) the conditional probability mass function $p_{Y|X=3}(y)$:

y	1	2	3	4	5
$p_{Y|X=3}(y)$	1/37	5/37	25/37	5/37	1/37

The expected value, $E[Y|X = 3]$, of this distribution is

$$E[Y|X=3] = (1)p_{Y|X=3}(1) + (2)p_{Y|X=3}(2) + (3)p_{Y|X=3}(3)$$
$$+ (4)p_{Y|X=3}(4) + (5)p_{Y|X=3}(5)$$
$$= (1)\left(\frac{1}{37}\right) + (2)\left(\frac{5}{37}\right) + (3)\left(\frac{25}{37}\right) + (4)\left(\frac{5}{37}\right) + (5)\left(\frac{1}{37}\right)$$
$$= 3.$$

We have also calculated the conditional probability mass function $P_{Y|X=1}(y)$ for *Y given* that $X = 1$:

y	1	2	3	4	5
$P_{Y|X=1}(y)$	1/2	1/3	1/6	0	0

The expected value, $E[Y|X = 1]$, of the distribution of the ranks wine taster II gives to samples of wine ranked $X = 1$ by wine taster I is then

$$E[Y|X=1] = (1)\left(\frac{1}{2}\right) + (2)\left(\frac{1}{3}\right) + (3)\left(\frac{1}{6}\right) = \frac{5}{3}.$$

The conditional expected value of Y given $X = x$ must be determined separately for each value of x. The function that assigns to every possible value x of X the value $E[Y|X = x]$ is called the *regression function of Y on X*. In practical applications, the regression function of one variable Y on another variable X is of considerable importance since it indicates, at least "on the average," the mathematical relationship existing between Y, as a variable to be predicted (*the dependent variable* in the mathematical relationship), and X, the predicting variable (*the independent variable* in the mathematical relationship).

Example 2.1 (Continued). Our interest in this example is, for the moment, centered on predicting the rank Y assigned to a sample of wine by wine taster II from knowledge of the rank X assigned to this wine sample by wine taster I. We have already found the conditional expectations, $E[Y|X=3]$ and $E[Y|X=1]$. The regression function, $E[Y|X=x]$, is given by the expression

$$E[Y|X=x] = \begin{cases} 5/3, & \text{if } x = 1, \\ 23/9, & \text{if } x = 2, \\ 3, & \text{if } x = 3, \\ 109/29, & \text{if } x = 4, \\ 43/10, & \text{if } x = 5. \end{cases}$$

This result follows since $E[Y|X=3]=3$ and $E[Y|X=1]=\frac{5}{3}$, as we have already shown, and since

$$E[Y|X=2] = (1)\left(\frac{1}{9}\right) + (2)\left(\frac{4}{9}\right) + (3)\left(\frac{5}{18}\right) + (4)\left(\frac{1}{9}\right) + (5)\left(\frac{1}{18}\right) = \frac{23}{9},$$

$$E[Y|X=4] = (1)(0) + (2)\left(\frac{2}{29}\right) + (3)\left(\frac{5}{29}\right) + (4)\left(\frac{20}{29}\right) + (5)\left(\frac{2}{29}\right) = \frac{109}{29},$$

and

$$E[Y|X=5] = (1)(0) + (2)\left(\frac{1}{10}\right) + (3)\left(\frac{1}{10}\right) + (4)\left(\frac{1}{5}\right) + (5)\left(\frac{3}{5}\right) = \frac{43}{10}.$$

The regression function $E[Y|X=x]$ is graphed in Figure 2.7. Note that the values of $E[Y|X=x]$ increase as x increases.

*[*Remark:* We note that $E[Y|X=x]$ defines a number which is dependent on the observed value x of X. However, the value of X a priori is

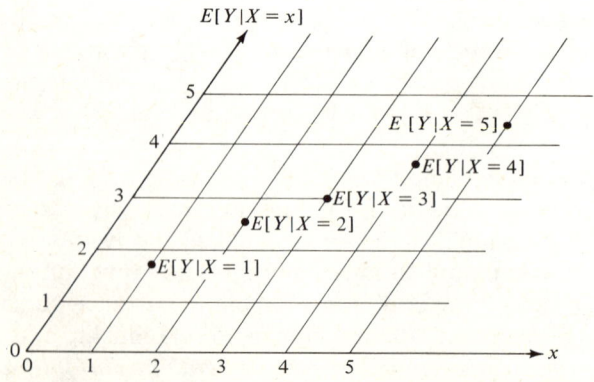

Figure 2.7 The regression function $E[Y|X=x]$ for the joint probability distribution given in Table 2.1 of Example 2.1.

random [it is not determined until a particular outcome $\omega = (x,y)$ occurs]. Thus, $E[Y|X = x]$ is, from this point of view, a random variable. To take note of this fact, $E[Y|X = x]$ is sometimes written $E[Y|X]$ and is called the *conditional expected value of Y given X*.]

3. DESCRIPTIVE PROPERTIES OF BIVARIATE DISTRIBUTIONS

We have seen that it is often the case that certain summarizing measures are sufficient to describe those aspects of univariate distributions which are of primary interest. A similar fact is true for bivariate distributions.

Suppose that we are considering two random variables X and Y. For the moment, assume that X and Y are both discrete random variables with the joint probability mass function $p_{X,Y}(x,y)$. We will discuss the case when X and Y are both continuous random variables in Section 4. To begin with, we can look for a measure for the bivariate distribution of X and Y which is analogous to a measure of location for a univariate distribution. The concept of a quantile (for example, the median) is not meaningful for a bivariate distribution, since the notion of the probability "to the left" and "to the right" of a point (x,y) in the plane does not have a unique definition. The *mode* of a bivariate distribution for discrete random variables is the pair (or pairs) of values (x_0, y_0) for which the value of joint probability mass function $p_{X,Y}(x,y)$ is largest. Since the mode of a bivariate distribution is subject to mathematical difficulties (and since even for univariate distributions it is rarely a good measure of location), the mode is seldom used as a measure of location for bivariate distributions.

The measure of location which is usually used for bivariate distributions is the pair (μ_X, μ_Y), where μ_X is the expected value of X [as calculated from the marginal probability mass function $p_X(x)$], and where μ_Y is the expected value of Y [calculated from $p_Y(y)$]. The pair (μ_X, μ_Y) is called the *joint expectation*, the *joint mean*, or the *mean vector* of the distribution of X and Y. Although μ_X and μ_Y can be computed from the marginal probability mass functions $p_X(x)$ and $p_Y(y)$, respectively, it is sometimes more convenient to compute these quantities directly from the joint probability mass function $p_{X,Y}(x,y)$. For example, we can calculate $E(X) = \mu_X$ using the formula

$$\mu_X = E(X) = x_1 p_X(x_1) + x_2 p_X(x_2) + x_3 p_X(x_3) + \cdots.$$

However, since

$$p_X(x_i) = p_{X,Y}(x_i, y_1) + p_{X,Y}(x_i, y_2) + p_{X,Y}(x_i, y_3) + \cdots,$$

for $i = 1, 2, 3, \ldots$, we can substitute this result into the formula for μ_X

given above and obtain

$$(3.1) \quad \mu_X = x_1[p_{X,Y}(x_1,y_1) + p_{X,Y}(x_1,y_2) + p_{X,Y}(x_1,y_3) + \cdots]$$
$$+ x_2[p_{X,Y}(x_2,y_1) + p_{X,Y}(x_2,y_2) + p_{X,Y}(x_2,y_3) + \cdots] + \cdots$$
$$= x_1 p_{X,Y}(x_1,y_1) + x_1 p_{X,Y}(x_1,y_2) + x_1 p_{X,Y}(x_1,y_3) + \cdots$$
$$+ x_2 p_{X,Y}(x_2,y_1) + x_2 p_{X,Y}(x_2,y_2) + x_2 p_{X,Y}(x_2,y_3) + \cdots.$$

We can calculate μ_Y from the joint probability mass function $p_{X,Y}(x,y)$ in a similar fashion.

The joint expectation (μ_X, μ_Y), in analogy to the expected value in the univariate case, has the property that it is the center of gravity of the distribution of probability mass (in the plane) defined by $p_{X,Y}(x,y)$. Further, since μ_X and μ_Y are each univariate measures of location, we know from Section 2 of Chapter 5 that for any numbers a and b,

$$(3.2) \quad \mu_{aX+b} = a\mu_X + b,$$

while for any numbers c and d,

$$(3.3) \quad \mu_{cY+d} = c\mu_Y + d.$$

In the search for something analogous to a measure of dispersion for bivariate distributions, the variance σ_X^2 of X and the variance σ_Y^2 of Y come naturally to our attention. These measures can be computed from the respective marginal probability mass functions of X and Y; for example,

$$\sigma_X^2 = (x_1 - \mu_X)^2 p_X(x_1) + (x_2 - \mu_X)^2 p_X(x_2) + \cdots.$$

However, again it is sometimes more convenient to calculate the variance directly from the joint probability mass function $p_{X,Y}(x,y)$. For example,

$$(3.4) \quad \sigma_X^2 = (x_1 - \mu_X)^2 p_{X,Y}(x_1,y_1) + (x_1 - \mu_X)^2 p_{X,Y}(x_1,y_2) + \cdots$$
$$+ (x_2 - \mu_X)^2 p_{X,Y}(x_2,y_1) + (x_2 - \mu_X)^2 p_{X,Y}(x_2,y_2) + \cdots.$$

A similar formula holds for σ_Y^2. Since σ_X^2 and σ_Y^2 are each univariate measures of dispersion [σ_X^2 is a measure of dispersion for the marginal probability mass function $p_X(x)$ and σ_Y^2 is a measure of dispersion for $p_Y(y)$], we know from Section 4 of Chapter 5 that for any two numbers a and b,

$$(3.5) \quad \sigma_{aX+b}^2 = a^2 \sigma_X^2,$$

and that for any two numbers c and d,

$$(3.6) \quad \sigma_{cY+d}^2 = c^2 \sigma_Y^2.$$

When the mass function $p_{X,Y}(x,y)$ is graphed in the plane, the indices σ_X^2 and σ_Y^2 indicate the dispersion of the joint probability mass function

3. DESCRIPTIVE PROPERTIES OF BIVARIATE DISTRIBUTIONS

$p_{X,Y}(x, y)$ around the lines $x = \mu_X$ and $y = \mu_Y$, respectively (see Figures 3.1 and 3.2). However, any line "$\alpha x + \beta y = $ constant" could define a center for a measure of dispersion; there is no obvious reason for choosing the lines $x = \mu_X$ or $y = \mu_Y$.

Further, although both σ_X^2 and σ_Y^2 can be defined solely in terms of the marginal distributions $p_X(x)$ and $p_Y(y)$, respectively, we have seen in Section 2 that the marginal distributions by themselves do not determine the joint distribution of X and Y. What we lack is a measure of the *interrelationship* or *covariation* of X and Y. That is, we lack a measure which reveals how X and Y vary jointly.

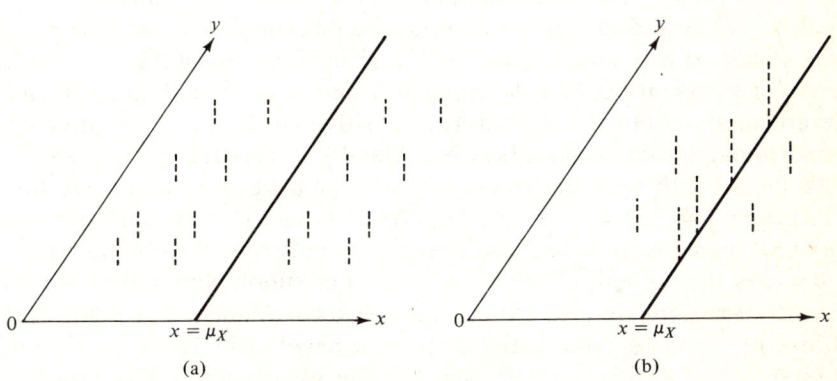

Figure 3.1 The variance σ_X^2 shows dispersion (or spread) of the joint probability mass function $p_{X,Y}(x, y)$ around the line $x = \mu_X$. The dispersion in (a) is larger than the dispersion in (b).

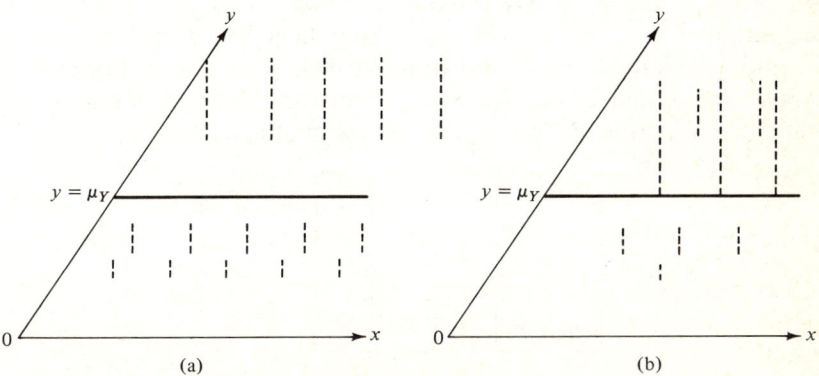

Figure 3.2 The variance σ_Y^2 shows dispersion (or spread) of the joint probability mass function $p_{X,Y}(x, y)$ around the line $y = \mu_Y$. The dispersion in (a) is larger than the dispersion in (b).

The formulas $\sigma_X^2 = E[X-\mu_X]^2$ and $\sigma_Y^2 = E[Y-\mu_Y]^2$ represent how X and Y vary individually. In a formal way, this suggests that $E[X-\mu_X][Y-\mu_Y]$ might represent how X and Y vary jointly. The quantity $E[X-\mu_X][Y-\mu_Y]$ is called the *covariance* between X and Y and is denoted by σ_{XY} or by $\text{Cov}(X,Y)$. Clearly $\text{Cov}(X,Y) = \text{Cov}(Y,X)$.

Suppose that $X = Y$. Then by substitution, $\text{Cov}(X,Y) = \text{Cov}(X,X) = E[X-\mu_X][X-\mu_X] = \sigma_X^2$. We conclude from this that $\sigma_X^2 = \sigma_{XX} = \text{Cov}(X,X)$ and $\sigma_Y^2 = \sigma_{YY} = \text{Cov}(Y,Y)$; that is, the notations σ_X^2 and σ_{XX}^2 are identical, as are σ_Y^2 and σ_{YY}.

The covariance can be positive or negative. It is positive when X and Y tend to move in the same direction; that is, when large values of X go with large values of Y and small values of X with small values of Y. It is negative when X and Y move in opposite directions; that is, when large values of X go with small values of Y and small values of X go with large values of Y. As an extreme example, if X and Y are directly proportional, the covariance between X and Y is positive; if X and Y are inversely proportional, the covariance between X and Y is negative.

To illustrate this point, if X denotes a student's high school grade-point average and Y his college grade-point average, experience with such variables leads us to expect the covariance of X and Y to be positive. If X denotes the amount of minerals in a water supply and Y is a measure of pipe corrosion, we also expect a positive covariance. If X denotes the stature of a mother and Y the stature of her daughter, we would once again expect a positive covariance. On the other hand, if X is a measure of motor coordination and Y denotes the amount of alcohol in the blood, the covariance between X and Y should be negative. Similarly, some experiments have shown that if X denotes age and Y denotes reading speed, the covariance between X and Y is negative.

In order to compute $\sigma_{XY} = \text{Cov}(X,Y)$, we could first determine the distribution of $W = (X-\mu_X)(Y-\mu_Y)$ from the joint distribution $p_{X,Y}(x,y)$ of X and Y, and then find $E[W]$ from the distribution of W. For example, suppose that X and Y are discrete random variables with the joint probability mass function given by the following contingency table:

x \ y	1	2	3	Total
1	$\frac{1}{4}$	$\frac{1}{12}$	$\frac{1}{6}$	$\frac{1}{2}$
2	$\frac{1}{12}$	$\frac{1}{4}$	$\frac{1}{6}$	$\frac{1}{2}$
Total	$\frac{1}{3}$	$\frac{1}{3}$	$\frac{1}{3}$	1

Then,
$$\mu_X = (1)[p_X(1)] + (2)[p_X(2)]$$
$$= (1)(\tfrac{1}{2}) + (2)(\tfrac{1}{2}) = \tfrac{3}{2},$$

3. DESCRIPTIVE PROPERTIES OF BIVARIATE DISTRIBUTIONS

$$\mu_Y = (1)[p_Y(1)] + (2)[p_Y(2)] + (3)[p_Y(3)]$$
$$= (1)(\tfrac{1}{3}) + (2)(\tfrac{1}{3}) + (3)(\tfrac{1}{3}) = 2.$$

Because

$$W = (X - \mu_X)(Y - \mu_Y) = \begin{cases} \tfrac{1}{2}, & \text{if } (X,Y) = (1,1) \text{ or } (2,3), \\ 0, & \text{if } (X,Y) = (1,2) \text{ or } (2,2), \\ -\tfrac{1}{2}, & \text{if } (X,Y) = (2,1) \text{ or } (1,3), \end{cases}$$

it follows that

$$p_W(-\tfrac{1}{2}) = p_{X,Y}(2,1) + p_{X,Y}(1,3) = (\tfrac{1}{12}) + (\tfrac{1}{6}) = \tfrac{1}{4},$$
$$p_W(0) = p_{X,Y}(1,2) + p_{X,Y}(2,2) = (\tfrac{1}{12}) + (\tfrac{1}{4}) = \tfrac{1}{3},$$
$$p_W(\tfrac{1}{2}) = p_{X,Y}(1,1) + p_{X,Y}(2,3) = (\tfrac{1}{4}) + (\tfrac{1}{6}) = \tfrac{5}{12},$$

and hence

$$\sigma_{XY} = E(X-\mu_X)(Y-\mu_Y) = EW = (-\tfrac{1}{2})(\tfrac{1}{4}) + (0)(\tfrac{1}{3}) + (\tfrac{1}{2})(\tfrac{5}{12}) = \tfrac{1}{12}.$$

It is often more convenient to calculate σ_{XY} directly from the joint probability mass function $p_{X,Y}(x,y)$ of X and Y. To do so, we can make use of the following formula for finding the expected value $E[g(X,Y)]$ of any function $g(X,Y)$ of X and Y:

$$(3.7) \quad E[g(X,Y)] = g(x_1,y_1)p_{X,Y}(x_1,y_1) + g(x_1,y_2)p_{X,Y}(x_1,y_2) + \cdots$$
$$+ g(x_2,y_1)p_{X,Y}(x_2,y_1) + g(x_2,y_2)p_{X,Y}(x_2,y_2) + \cdots.$$

That is, for each possible value (x,y) of the pair (X,Y), we multiply $g(x,y)$ by the probability $p_{X,Y}(x,y)$ that simultaneously $X = x$ and $Y = y$, and then sum such products over all possibilities (x,y). Note that Equation (3.7) is the analogy for bivariate distributions of the formula for the expectation of a function $g(X)$ of one random variable, X, given in Equation (4.2) of Chapter 5. Formula (3.7) for the expectation $E[g(X,Y)]$ of $g(X,Y)$ holds for any function $g(X,Y)$ of X and Y [for example, Formula (3.7) holds for $g(X,Y) = XY$, for $g(X,Y) = X + Y$, for $g(X,Y) = X/Y$, and so on].

In particular, in the case of the covariance, we require the expected value of $g(X,Y) = (X-\mu_X)(Y-\mu_Y)$. From Equation (3.7), we obtain

$$(3.8) \quad \sigma_{XY} = (x_1 - \mu_X)(y_1 - \mu_Y)p_{X,Y}(x_1,y_1) + (x_1 - \mu_X)(y_2 - \mu_Y)p_{XY}(x_1,y_2)$$
$$+ \cdots + (x_2 - \mu_X)(y_1 - \mu_Y)p_{X,Y}(x_2,y_1) + \cdots.$$

Applying this equation to the case of the contingency table

x \ y	1	2	3	Total
1	$\tfrac{1}{4}$	$\tfrac{1}{12}$	$\tfrac{1}{6}$	$\tfrac{1}{2}$
2	$\tfrac{1}{12}$	$\tfrac{1}{4}$	$\tfrac{1}{6}$	$\tfrac{1}{2}$
Total	$\tfrac{1}{3}$	$\tfrac{1}{3}$	$\tfrac{1}{3}$	1

given earlier (where we recall that $\mu_X = \frac{3}{2}$, $\mu_Y = 2$), we find that

$$\begin{aligned}\sigma_{XY} &= (1-\mu_X)(1-\mu_Y)p_{X,Y}(1,1) + (1-\mu_X)(2-\mu_Y)p_{X,Y}(1,2) \\ &+ (1-\mu_X)(3-\mu_Y)p_{X,Y}(1,3) + (2-\mu_X)(1-\mu_Y)p_{X,Y}(2,1) \\ &+ (2-\mu_X)(2-\mu_Y)p_{X,Y}(2,2) + (2-\mu_X)(3-\mu_Y)p_{X,Y}(2,3) \\ &= (-\tfrac{1}{2})(-1)(\tfrac{1}{4}) + (-\tfrac{1}{2})(0)(\tfrac{1}{12}) + (-\tfrac{1}{2})(1)(\tfrac{1}{6}) \\ &+ (\tfrac{1}{2})(-1)(\tfrac{1}{12}) + (\tfrac{1}{2})(0)(\tfrac{1}{4}) + (\tfrac{1}{2})(1)(\tfrac{1}{6}) \\ &= \tfrac{1}{12},\end{aligned}$$

which is the result obtained earlier.

Another formula for $\sigma_{XY} = \text{Cov}(X,Y)$ which frequently is of help in calculations is

(3.9) $$\sigma_{XY} = E[XY] - \mu_X \mu_Y,$$

where from Equation (3.7) we know that we can calculate $E[XY]$ by forming the sum

(3.10) $$E[XY] = x_1 y_1 p_{X,Y}(x_1, y_1) + x_1 y_2 p_{X,Y}(x_1, y_2) + \cdots + x_2 y_1 p_{X,Y}(x_2, y_1)$$
$$+ x_2 y_2 p_{X,Y}(x_2, y_2) + \cdots.$$

Thus, we can also obtain the result $\sigma_{XY} = 1/12$ for the contingency table given above by the following calculations:

$$\begin{aligned}E[XY] &= (1)(1)p_{X,Y}(1,1) + (1)(2)p_{X,Y}(1,2) + (1)(3)p_{X,Y}(1,3) \\ &+ (2)(1)p_{X,Y}(2,1) + (2)(2)p_{X,Y}(2,2) + (2)(3)p_{X,Y}(2,3) \\ &= (1)(1)\left(\frac{1}{4}\right) + (1)(2)\left(\frac{1}{12}\right) + (1)(3)\left(\frac{1}{6}\right) + (2)(1)\left(\frac{1}{12}\right) \\ &+ (2)(2)\left(\frac{1}{4}\right) + (2)(3)\left(\frac{1}{6}\right) = \frac{37}{12},\end{aligned}$$

and

$$\sigma_{XY} = E[XY] - \mu_X \mu_Y$$
$$= \left(\frac{37}{12}\right) - \left(\frac{3}{2}\right)(2) = \frac{1}{12}.$$

To see how Equation (3.9) follows from Equation (3.8), note that σ_{XY} in Equation (3.8) is the sum of terms of the form $(x-\mu_X)(y-\mu_Y)p_{X,Y}(x,y)$. However,

$$(x-\mu_X)(y-\mu_Y)p_{X,Y}(x,y) = xyp_{X,Y}(x,y) - \mu_X y p_{X,Y}(x,y)$$
$$- x\mu_Y p_{X,Y}(x,y) + \mu_X \mu_Y p_{X,Y}(x,y),$$

3. DESCRIPTIVE PROPERTIES OF BIVARIATE DISTRIBUTIONS

so that

(3.11) $\quad$ the sum of $(x-\mu_X)(y-\mu_Y)p_{X,Y}(x,y)$

$= $ [the sum of $xyp_{X,Y}(x,y)$] + [the sum of $(-\mu_X)yp_{X,Y}(x,y)$]

$\quad +$ [the sum of $(x)(-\mu_Y)p_{X,Y}(x,y)$] + $\mu_X\mu_Y$[the sum of $p_{X,Y}(x,y)$],

where the various sums are taken over all the possible pairs of values (x,y) for X and Y. From Equation (3.10), we have that

$$E[XY] = \text{the sum of } xyp_{X,Y}(x,y).$$

Since

$$[\text{the sum of } (-\mu_X)yp_{X,Y}(x,y)] = (-\mu_X)[\text{the sum of } yp_{X,Y}(x,y)]$$
$$= (-\mu_X)(\mu_Y),$$
$$[\text{the sum of } (x)(-\mu_Y)p_{X,Y}(x,y)] = (-\mu_Y)[\text{the sum of } xp_{X,Y}(x,y)]$$
$$= (-\mu_Y)(\mu_X),$$

and because the sum of $p_{X,Y}(x,y)$ over all possible pairs (x,y) is equal to 1, it follows by substitution into Equation (3.11) that

$$\sigma_{XY} = E[XY] + (-\mu_X)(\mu_Y) + (-\mu_Y)(\mu_X) + \mu_X\mu_Y(1)$$
$$= E[XY] - \mu_X\mu_Y.$$

We have thus verified Equation (3.9).

Because we have defined $\sigma_{XY} = \text{Cov}(X,Y)$ in a way analogous to the definitions of σ_X^2 and σ_Y^2, it is not surprising that $\text{Cov}(X,Y)$ has properties similar to these two measures of dispersion. One very important property of $\text{Cov}(X,Y)$ is given by the following: For any numbers a, b, c, and d,

(3.12) $\quad\quad\quad \text{Cov}(aX+b, cY+d) = ac\,\text{Cov}(X,Y).$

Remark: If $a=c$, $b=d$, and $X=Y$, Equation (3.12) reduces to the equation

$$\text{Cov}(aX+b, aX+b) = a^2\,\text{Cov}(X,X).$$

Because $\text{Cov}(aX+b, aX+b) = \text{Var}(aX+b) = \sigma_{aX+b}^2$ and $\text{Cov}(X,X) = \text{Var}(X) = \sigma_X^2$, we obtain the result

$$\sigma_{aX+b}^2 = a^2\sigma_X^2,$$

which agrees with the result obtained in Chapter 5 [see Equation (4.9)].

To verify (3.12), note that from (3.2) and (3.3),

$$\mu_{aX+b} = a\mu_X + b, \quad\quad \mu_{cY+d} = c\mu_Y + d.$$

Thus,

$$\text{(3.13)} \quad \text{Cov}(aX+b, cY+d) = E[(aX+b-\mu_{aX+b})(cY+d-\mu_{cY+d})]$$
$$= E[(aX+b-a\mu_X-b)(cY+d-c\mu_Y-d)]$$
$$= E[a(X-\mu_X)c(Y-\mu_Y)]$$
$$= E[ac(X-\mu_X)(Y-\mu_Y)].$$

Let $W = (X-\mu_X)(Y-\mu_Y)$. Then since $E[W]$ is a measure of location for the distribution of W, we know from Section 2 of Chapter 5 that for any constants α and β,

$$E[\alpha W + \beta] = \alpha E[W] + \beta.$$

Letting $\alpha = ac$, $\beta = 0$, we obtain

$$\text{(3.14)} \quad E[ac(X-\mu_X)(Y-\mu_Y)] = E[(ac)W]$$
$$= acE[W]+0 = acE[(X-\mu_X)(Y-\mu_Y)].$$

Combining (3.13) and (3.14) yields

$$\text{Cov}(aX+b, cY+d) = acE[(X-\mu_X)(Y-\mu_Y)]$$
$$= ac\text{Cov}(X,Y),$$

which verifies Equation (3.12).

Rescaling Techniques

Equations (3.2), (3.3), (3.5), (3.6), and (3.12) often provide an easier method for calculating the expected values μ_X and μ_Y, variances σ_X^2 and σ_Y^2, and covariance σ_{XY} of the jointly distributed random variables X and Y.

Example 3.1. Suppose that the random variable X is the score of an individual on a questionnaire designed to measure social status in a given community. If the questionnaire consists of 4 "yes-no" type questions, where each "yes" scores 5 points and each "no" scores 0 points, then the possible values of X are 0, 5, 10, 15, 20. Suppose that the random variable Y is the income (in dollars) of an individual. If individuals are selected from a community having a limited range of incomes, and if incomes have been rounded off to the nearest $2000, the possible values of Y might be $2000, $4000, $6000, $8000, and $10,000. Assume that the joint distribution of the variables X and Y is given by the contingency table, Table 3.1. From this contingency table, the reader is invited to calculate the values of μ_X, μ_Y, σ_X^2, σ_Y^2, and σ_{XY}. To illustrate the computational

3. DESCRIPTIVE PROPERTIES OF BIVARIATE DISTRIBUTIONS

problems encountered, note that

$$\mu_Y = (2000)(0.12) + (4000)(0.24) + (6000)(0.32) + (8000)(0.22)$$
$$+ (10{,}000)(0.10)$$
$$= 5880,$$

so that calculation of σ_Y^2 involves a sum of terms such as $(4000 - 5880)^2 (0.24)$ and $(10{,}000 - 5880)^2(0.10)$.
Similarly,

$$\mu_X = (0)(0.12) + (5)(0.22) + (10)(0.30) + (15)(0.24) + (20)(0.12)$$
$$= 10.1,$$

so that computation of σ_X^2 involves a sum of terms such as $(5 - 10.1)^2 \times (0.22)$ and $(20 - 10.1)^2(0.30)$. In addition, the computation of σ_{XY} involves the sum of terms such as $(0 - 10.1)(4000 - 5880)(0.02)$ and $(15 - 10.1) \times (8000 - 5880)(0.10)$. From such considerations, it would appear that computation of σ_X^2, σ_Y^2, and σ_{XY} involves a considerable amount of laborious addition and multiplication.

Table 3.1. Joint probability mass function of the social status score X and income Y of a randomly chosen individual from a given community.

		\multicolumn{6}{c}{Income}					
	y x	$2000	$4000	$6000	$8000	$10,000	Total
	0	0.05	0.02	0.03	0.01	0.01	0.12
Social	5	0.02	0.10	0.08	0.01	0.01	0.22
Status	10	0.03	0.08	0.10	0.08	0.01	0.30
Score	15	0.01	0.03	0.08	0.10	0.02	0.24
	20	0.01	0.01	0.03	0.02	0.05	0.12
	Total	0.12	0.24	0.32	0.22	0.10	1.00

However, let us *rescale* the variables X and Y. That is, let us construct new random variables $X^* = aX + b$, $Y^* = cY + d$, where a, b, c, and d are certain constants. Note that if (x,y) is a possible pair of values for the pair of random variables (X,Y), then $(ax+b, cy+d)$ is a possible pair of values for the pair of random variables (X^*, Y^*), and

$$p_{X^*, Y^*}(ax + b, cy + d) = p_{X,Y}(x, y).$$

It follows that the joint distribution of the random variables X^* and Y^* is given by the contingency table:

x^* \ y^*	$2000c+d$	$4000c+d$	$6000c+d$	$8000c+d$	$10{,}000c+d$	Total
b	0.05	0.02	0.03	0.01	0.01	0.12
$5a+b$	0.02	0.10	0.08	0.01	0.01	0.22
$10a+b$	0.03	0.08	0.10	0.08	0.01	0.30
$15a+b$	0.01	0.03	0.08	0.10	0.02	0.24
$20a+b$	0.01	0.01	0.03	0.02	0.05	0.12
Total	0.12	0.24	0.32	0.22	0.10	1.00

We now choose the numbers a, b, c, and d so as to make the possible values of X^* and Y^* convenient for computational purposes. Since small integers are perhaps easier to work with, we think of letting $a = 1/5$, $c = 1/2000$, and b and d be 0. Then the possible values of X^* are 0, 1, 2, 3, and 4; while the possible values of Y^* are 1, 2, 3, 4, and 5. (Other choices of a, b, c, and d may also be convenient; for example, we could let $d = -1$ so as to make the possible values of Y^* equal to 0, 1, 2, 3, and 4.) The joint distribution of X^* and Y^* for these choices of the constants a, b, c, and d is given by the contingency table:

x^* \ y^*	1	2	3	4	5	Total
0	0.05	0.02	0.03	0.01	0.01	0.12
1	0.02	0.10	0.08	0.01	0.01	0.22
2	0.03	0.08	0.10	0.08	0.01	0.30
3	0.01	0.03	0.08	0.10	0.02	0.24
4	0.01	0.01	0.03	0.02	0.05	0.12
Total	0.12	0.24	0.32	0.22	0.10	1.00

From this table, we quickly find that

$$\mu_{X^*} = (0)(0.12) + (1)(0.22) + (2)(0.30) + (3)(0.24) + (4)(0.12) = 2.02,$$

and

$$\mu_{Y^*} = (1)(0.12) + (2)(0.24) + (3)(0.32) + (4)(0.22) + (5)(0.10) = 2.94.$$

Using the formula (see Chapter 5, Section 5)

$$\sigma^2_{X^*} = E[X^{*2}] - \mu^2_{X^*}, \qquad \sigma^2_{Y^*} = E[Y^{*2}] - \mu^2_{Y^*},$$

we see that

$$\sigma^2_{X^*} = [(0)^2(0.12) + (1)^2(0.22) + (2)^2(0.30) + (3)^2(0.24) + (4)^2(0.12)] - (2.02)^2$$
$$= 5.50 - 4.0804 = 1.4196,$$

3. DESCRIPTIVE PROPERTIES OF BIVARIATE DISTRIBUTIONS

and
$$\sigma^2_{Y*} = [(1)^2(0.12) + (2)^2(0.24) + (3)^2(0.32) + (4)^2(0.22)$$
$$+ (5)^2(0.10)] - (2.94)^2$$
$$= 9.98 - 8.6436$$
$$= 1.3364.$$

Finally, using Equation (3.9),
$$\sigma_{X*Y*} = \text{Cov}(X*, Y*) = E[X*Y*] - \mu_{X*}\mu_{Y*}$$
$$= [(0)(1)(0.05) + (0)(2)(0.02) + \cdots + (4)(4)(0.02)$$
$$+ (4)(5)(0.01)] - (2.02)(2.94)$$
$$= 6.50 - 5.9388 = 0.5612.$$

From these computations and Equations (3.2), (3.3), (3.5), (3.6), and (3.12), we know that
$$2.02 = \mu_{X*} = a\mu_X + b = (1/5)\mu_X + 0,$$
$$2.94 = \mu_{Y*} = c\mu_Y + d = (1/2000)\mu_Y + 0,$$
$$1.4196 = \sigma^2_{X*} = a^2\sigma_X^2 = (1/5)^2\sigma_X^2,$$
$$1.3364 = \sigma^2_{Y*} = c^2\sigma_Y^2 = (1/2000)^2\sigma_Y^2,$$

and
$$0.5612 = \text{Cov}(X*, Y*) = ac\text{Cov}(X, Y)$$
$$= (1/5)(1/2000)\text{Cov}(X, Y) = (1/10,000)\sigma^2_{XY}.$$

Solving these five equations for μ_X, μ_Y, σ_X^2, σ_Y^2, and σ_{XY}, we obtain
$$\mu_X = 5(2.02) = 10.1,$$
$$\mu_Y = (2000)(2.94) = 5880,$$
$$\sigma_X^2 = (25)(1.4196) = 35.49,$$
$$\sigma_Y^2 = (4,000,000)(1.3364) = 5,345,600,$$
$$\sigma_{XY} = (10,000)(0.5612) = 5612.0.$$

Note that the values of μ_X and μ_Y agree with the values which we have earlier obtained by direct methods. The superiority of the rescaling method of computation illustrated above over direct methods of computation lies in the fact that in the rescaling method of computation we do not have to deal with sums and products of large numbers until the very last step (when we convert from the values of μ_{X*}, μ_{Y*}, σ^2_{X*}, σ^2_{Y*}, σ_{X*Y*} to the desired values of μ_X, μ_Y, σ_X^2, σ_Y^2, and σ_{XY}).

The Correlation Coefficient

As a measure of how closely X and Y move together, the covariance possesses one main defect: it is sensitive to the scales of measurement used for X and Y. To illustrate, suppose X denotes the length (measured in feet) of a randomly selected rod and Y denotes the length (measured in feet) of that rod as estimated by a given individual. Suppose that the covariance is found to be 0.01, a value which is reasonably small and might serve to indicate that the true length X and the perceived length Y do not move together at all. Now suppose we measure the lengths in inches. Thus, we have new random variables $X^* = 12X$ and $Y^* = 12Y$ that result from giving the measurements in inches (that is, rescaling X and Y). But from Equation (3.12), with $a = c = 12$ and $b = d = 0$,

$$\text{Cov}(X^*, Y^*) = 12(12)\,\text{Cov}(X, Y)$$
$$= 144(0.01)$$
$$= 1.44,$$

which is a respectably large number and might lead us to believe that the true length X and the perceived length Y are related. If we measure in millimeters, we obtain an even larger number for the covariance. Therefore, if we want an interpretable measure of how X and Y interrelate, we need a measure that does not change with scale.

Note that in the above example, application of Equations (3.5) and (3.6) gives

$$\sigma_{X^*} = \sqrt{\sigma^2_{X^*}} = \sqrt{(12)^2 \sigma_X^2} = 12\sigma_X,$$

and $\sigma_{Y^*} = 12\sigma_Y$. Thus,

$$\frac{\sigma_{X^*Y^*}}{\sigma_{X^*}\sigma_{Y^*}} = \frac{\sigma_{XY}}{\sigma_X\sigma_Y}.$$

That is, the quotient resulting from dividing the covariance σ_{XY} by the product $\sigma_X\sigma_Y$ of the standard deviations remains constant, no matter what scale of measurement we use for the true length X and the perceived length Y.

This leads us to use the *correlation coefficient*

$$(3.15) \qquad \rho_{X,Y} = \frac{\sigma_{XY}}{\sigma_X \sigma_Y} = \frac{\text{Cov}(X,Y)}{\sqrt{\text{Var}(X)\,\text{Var}(Y)}},$$

between X and Y as a measure of the co-relationship between X and Y. The correlation coefficient is also sometimes denoted by $\text{Corr}(X,Y)$.

Since the correlation coefficient between X and Y is formed by dividing $\text{Cov}(X,Y)$ by σ_X and σ_Y, it is without dimension. $\text{Corr}(aX+b, cY+d)$ is equal to $\text{Corr}(X,Y)$ provided a and c are either both positive or both negative.

3. DESCRIPTIVE PROPERTIES OF BIVARIATE DISTRIBUTIONS

Example 3.1 (Continued). In Example 3.1, we considered a situation where the social status score X and income Y of individuals in a given community were of interest. A social scientist might be interested in determining how closely (in a statistical sense) these variables are related to one another. Looking at the covariance $\sigma_{XY} = 5612.0$ between X and Y, he cannot really tell whether or not this extremely large value of σ_{XY} indicates a high interrelationship between social status score and income. The value he has found for σ_{XY} may be large because of the scales of measurement he is using for status and income. Thus, the social scientist decides to compute the correlation coefficient between X and Y. Since

$$\sigma_X^2 = 35.49, \qquad \sigma_Y^2 = 5{,}345{,}600,$$

it follows that

$$\sigma_X = \sqrt{35.49} = 5.957, \qquad \sigma_Y = \sqrt{5{,}345{,}600} = 2312.06,$$

and

$$\rho_{X,Y} = \frac{\sigma_{XY}}{\sigma_X \sigma_Y} = \frac{5612.0}{(5.957)(2312.06)} = 0.41.$$

The correlation coefficient ρ_{XY} assumes values only between -1 and $+1$. The extreme values of -1 and $+1$ are assumed by ρ_{XY} only when Y is a linear function of X, that is, when $Y = aX + b$ for some nonzero constant a and some constant b. The correlation between X and Y is $+1$ when X and Y are linear functions of one another having positive slope (X and Y increase together); the correlation is -1 when X and Y are linear functions of one another having negative slope (X increases when Y decreases).

Example 3.2. Suppose that the random variable X has the following probability mass function $p_X(x)$:

x	-2	-1	0	1	2
$p_X(x)$	$\frac{1}{5}$	$\frac{1}{5}$	$\frac{1}{5}$	$\frac{1}{5}$	$\frac{1}{5}$

If $W = 3X + 2$, then X and W have a joint probability mass function given by the contingency table

x \ w	-4	-1	2	5	8	Total
2	0	0	0	0	$\frac{1}{5}$	$\frac{1}{5}$
1	0	0	0	$\frac{1}{5}$	0	$\frac{1}{5}$
0	0	0	$\frac{1}{5}$	0	0	$\frac{1}{5}$
-1	0	$\frac{1}{5}$	0	0	0	$\frac{1}{5}$
-2	$\frac{1}{5}$	0	0	0	0	$\frac{1}{5}$
Total	$\frac{1}{5}$	$\frac{1}{5}$	$\frac{1}{5}$	$\frac{1}{5}$	$\frac{1}{5}$	1

From the marginal probability mass function $p_X(x)$ of X we find that

$$\mu_X = 0, \qquad \sigma_X^2 = 2.$$

We could similarly calculate μ_W and σ_W^2 from the marginal probability mass function $p_W(w)$. However, since we know that $W = 3X+2$, we have, from Equation (3.2),

$$\mu_W = \mu_{3X+2} = 3\mu_X + 2 = 2,$$

and from Equation (3.5),

$$\sigma_W^2 = \sigma_{3X+2}^2 = (3)^2 \sigma_X^2 = 18.$$

Also, from Equation (3.12),

$$\sigma_{XW} = \text{Cov}(X,W) = \text{Cov}(X, 3X+2) = 3\,\text{Cov}(X,X) = 3\sigma_X^2 = 6.$$

Thus,

$$\rho_{X,W} = \frac{\sigma_{XW}}{\sigma_X \sigma_W} = \frac{6}{\sqrt{2}\sqrt{18}} = 1.$$

The joint probability mass function $p_{X,W}(x,w)$ of X and $W = 3X+2$ is graphed in Figure 3.3.

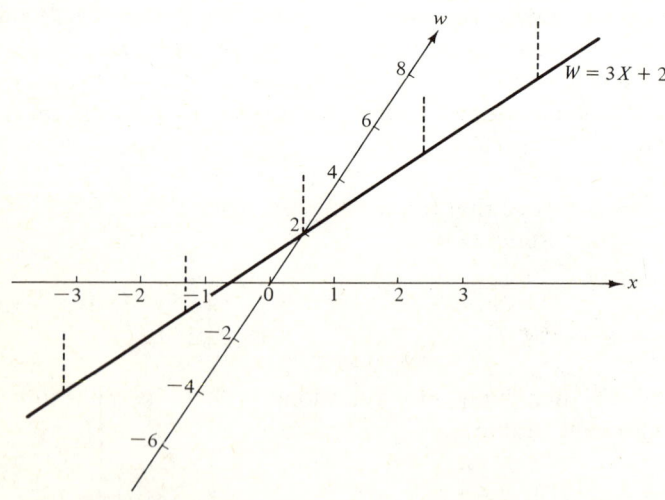

Figure 3.3 Graph of the joint probability mass function $p_{X,W}(x,w)$ of X and $W = 3X+2$.

If instead, $W = (-3)X+2$, then X and W have a joint probability mass function given by the contingency table

3. DESCRIPTIVE PROPERTIES OF BIVARIATE DISTRIBUTIONS

x \ w	-4	-1	2	5	8	Total
2	$\frac{1}{5}$	0	0	0	0	$\frac{1}{5}$
1	0	$\frac{1}{5}$	0	0	0	$\frac{1}{5}$
0	0	0	$\frac{1}{5}$	0	0	$\frac{1}{5}$
-1	0	0	0	$\frac{1}{5}$	0	$\frac{1}{5}$
-2	0	0	0	0	$\frac{1}{5}$	$\frac{1}{5}$
Total	$\frac{1}{5}$	$\frac{1}{5}$	$\frac{1}{5}$	$\frac{1}{5}$	$\frac{1}{5}$	1

and

$$\mu_W = -3\mu_X + 2 = 2, \qquad \sigma_W^2 = (-3)^2 \sigma_X^2 = 18,$$
$$\sigma_{XW} = (-3)\mathrm{Cov}(X,X) = -3\sigma_X^2 = -6,$$

so that

$$\rho_{X,W} = \frac{-6}{\sqrt{2}\sqrt{18}} = -1.$$

The joint probability mass function of X and $W = (-3)X + 2$ is graphed in Figure 3.4.

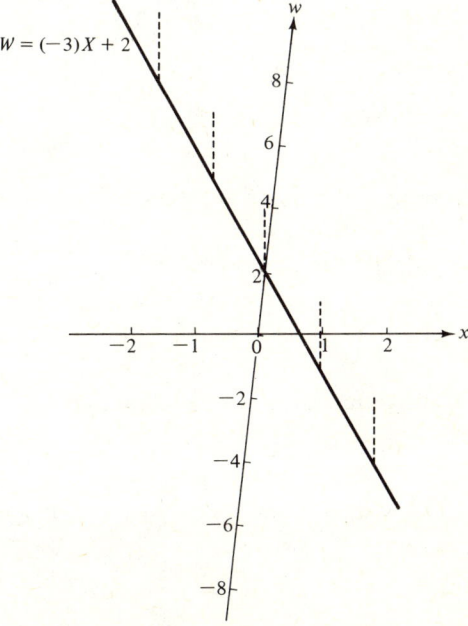

Figure 3.4 Graph of the joint probability mass function $p_{X,W}(x,w)$ of X and $W = (-3)X + 2$.

The reader may wonder whether the correlation coefficient is sensitive to *all* kinds of mathematical relationships between X and Y. Does it, for example, assume large positive values when $Y = X^2$? The answer, in general, is *no*; the correlation coefficient is sensitive *only to linear relationships* (that is, relationships of the form $Y = aX + b$).

Example 3.2 (Continued). Recall that in this example, the random variable X has the probability mass function $p_X(x)$ given below:

x	-2	-1	0	1	2
$p_X(x)$	$\frac{1}{5}$	$\frac{1}{5}$	$\frac{1}{5}$	$\frac{1}{5}$	$\frac{1}{5}$

and that $\mu_X = 0$, $\sigma_X^2 = 2$. Suppose now that $Z = X^2$. Then X and Z have a bivariate probability mass function $p_{X,Z}(x,z)$ determined by the following contingency table:

x \ z	0	1	4	Total
2	0	0	$\frac{1}{5}$	$\frac{1}{5}$
1	0	$\frac{1}{5}$	0	$\frac{1}{5}$
0	$\frac{1}{5}$	0	0	$\frac{1}{5}$
-1	0	$\frac{1}{5}$	0	$\frac{1}{5}$
-2	0	0	$\frac{1}{5}$	$\frac{1}{5}$
Total	$\frac{1}{5}$	$\frac{2}{5}$	$\frac{2}{5}$	1

From the marginal probability mass function of Z, we find that

$$\mu_Z = (0)\left(\frac{1}{5}\right) + (1)\left(\frac{2}{5}\right) + (4)\left(\frac{2}{5}\right) = 2,$$

$$\sigma_Z^2 = (0-2)^2\left(\frac{1}{5}\right) + (1-2)^2\left(\frac{2}{5}\right) + (4-2)^2\left(\frac{2}{5}\right) = \frac{14}{5}.$$

From the above contingency table, we also find that

$$\sigma_{XZ} = E(X - \mu_X)(Z - \mu_Z) = E(X - 0)(Z - 2)$$

$$= (-2)(0)(0) + (-2)(1)(0) + (-2)(4)\left(\frac{1}{5}\right) + \cdots + (2)(1)(0) + (2)(4)\left(\frac{1}{5}\right)$$

$$= 0.$$

Thus,

$$\rho_{X,Z} = \frac{\sigma_{XZ}}{\sigma_X \sigma_Y} = \frac{0}{\sqrt{2}\sqrt{14/5}} = 0.$$

3. DESCRIPTIVE PROPERTIES OF BIVARIATE DISTRIBUTIONS

Thus, we have here an example in which the correlation $\rho_{X,Z}$ between X and $Z = X^2$ is equal to 0, yet Z *is completely determined from X*.

Keeping this warning in mind, statisticians, in practice, nonetheless tend to regard a large positive correlation between X and Y as evidence that X and Y "on the average" increase together (large values of X tending to occur with large values of Y and small values of X tending to occur with small values of Y) and a large negative correlation as evidence that X and Y vary inversely. When the magnitude of $\rho_{X,Y}$ is very small, one feels that one cannot predict X from Y or Y from X. This is particularly true when $\rho_{X,Y} = 0$; we say in such a case that X and Y are *uncorrelated*.

Correlation and Statistical Independence

The notion of statistical independence as applied to events (and as applied to components of composite random experiments) was discussed in Chapter 3. We may extend this concept to define the independence of two discrete random variables. Two discrete random variables X and Y possessing a joint probability mass function $p_{X,Y}(x,y)$ are said to be *statistically independent* if $p_{Y|X=x}(y) = p_Y(y)$ for *all* values of x and y; that is, X and Y are statistically independent if the conditional distribution of Y given $X = x$ is the same as the marginal distribution of Y, whatever the value of x is. We also can say that X and Y are statistically independent if $p_{X|Y=y}(x) = p_X(x)$, for all x and y; indeed, these two definitions are equivalent. The essence of statistical independence is that knowledge of the value of one of two statistically independent random variables provides no new probabilistic knowledge whatsoever about the other variable.

Recall the definition (2.3) of the conditional probability mass function

$$p_{Y|X=x}(y) = \frac{p_{X,Y}(x,y)}{p_X(x)}$$

and suppose that X and Y are statistically independent, so that

$$p_{Y|X=x}(y) = p_Y(y).$$

By combining both of the above formulas, we arrive at the following alternative and equivalent definition of statistical independence: *Two discrete random variables X and Y are statistically independent if*

(3.16) $$p_{X,Y}(x,y) = p_X(x)p_Y(y)$$

for all possible pairs of values (x,y).

Note that when the two discrete random variables X and Y are statistically independent, their joint probability mass function $p_{X,Y}(x,y)$ can be obtained from the individual marginal distributions, $p_X(x)$ and $p_Y(y)$, by use of Equation (3.16). This is in contrast to the general case, mentioned in Section 2, where the marginal distributions $p_X(x)$ and $p_Y(y)$ cannot, by themselves, determine the joint probability mass function $p_{X,Y}(x,y)$.

An example of statistically independent random variables occurs when a random experiment is performed in physically independent stages and the random variable X depends for its numerical value only on what takes place at the first stage, while the random variable Y depends only on what takes place at the second stage. For example, assume two individuals are selected, *with replacement* between selections, at random from a population. Let X denote the age of the first individual and Y denote the yearly income of the second individual. Then it can be reasonably assumed that the two random variables are statistically independent.

However, X and Y may be statistically independent even when the experiment is not performed in physically independent stages. For example, suppose one individual is selected at random from a given population. Let X denote the yearly income of the individual selected and let Y denote the length of time since the last birthday of the individual selected. It would seem reasonable to expect that these two random variables would be statistically independent.

It can be shown that when the random variables X and Y are statistically independent, the correlation coefficient $\rho_{X,Y}$ is always equal to 0. For example, suppose that the random variable X has two possible values, x_1 and x_2, and suppose that the random variable Y has two possible values, y_1 and y_2. If the random variables X and Y are statistically independent, then the joint probability mass function $p_{X,Y}(x,y)$ equals the product $p_X(x)p_Y(y)$ of the marginal probability functions. Since $\rho_{X,Y} = \sigma_{XY}/\sigma_X\sigma_Y$ and since we know, from Equation (3.9), that $\sigma_{XY} = E[XY] - \mu_X\mu_Y$, let us start by calculating $E[XY]$. Because $p_{X,Y}(x,y) = p_X(x)p_Y(y)$, it follows that

$$E[XY] = x_1y_1 p_{X,Y}(x_1,y_1) + x_1y_2 p_{X,Y}(x_1,y_2) + x_2y_1 p_{X,Y}(x_2,y_1)$$
$$+ x_2y_2 p_{X,Y}(x_2,y_2),$$
$$= x_1y_1 p_X(x_1)p_Y(y_1) + x_1y_2 p_X(x_1)p_Y(y_2) + x_2y_1 p_X(x_2)p_Y(y_1)$$
$$+ x_2y_2 p_X(x_2)p_Y(y_2)$$
$$= [x_1 p_X(x_1) + x_2 p_X(x_2)][y_1 p_Y(y_1) + y_2 p_Y(y_2)].$$

However,
$$\mu_X = [x_1 p_X(x_1) + x_2 p_X(x_2)],$$
$$\mu_Y = [y_1 p_Y(y_1) + y_2 p_Y(y_2)],$$

so that
$$E[XY] = \mu_X\mu_Y.$$

Hence,
$$\sigma_{XY} = E[XY] - \mu_X\mu_Y = \mu_X\mu_Y - \mu_X\mu_Y = 0,$$

3. DESCRIPTIVE PROPERTIES OF BIVARIATE DISTRIBUTIONS

and thus

$$\rho_{X,Y} = \frac{\sigma_{XY}}{\sigma_X \sigma_Y} = 0.$$

We have therefore shown that when X and Y are statistically independent (and when X and Y each have only two possible values), the correlation coefficient $\rho_{X,Y}$ equals 0. When either X, or Y, or both X and Y have more than two possible values, a similar approach [using the relation $p_{X,Y}(x,y) = p_X(x)p_Y(y)$ to show that $E[XY] = \mu_X \mu_Y$, and then using this result to prove that $\rho_{X,Y} = 0$] can be used to demonstrate the same fact, namely, that *statistically independent random variables always have a correlation coefficient which is equal to zero.*

What about the converse of this last statement? Are uncorrelated random variables X and Y necessarily statistically independent? Notwithstanding the common usage of the words "independent" and "uncorrelated" as synonyms, *uncorrelated random variables are not necessarily statistically independent.*

We have given an example (Example 3.2) earlier in this section which illustrates this fact. There, we considered the uncorrelated random variables X and Z for which $Z = X^2$ (so that given X, we can always *exactly* predict the value of Z that will occur). The joint probability mass function of X and Z was given by the contingency table:

x \ z	0	1	4	Total
2	0	0	$\tfrac{1}{5}$	$\tfrac{1}{5}$
1	0	$\tfrac{1}{5}$	0	$\tfrac{1}{5}$
0	$\tfrac{1}{5}$	0	0	$\tfrac{1}{5}$
-1	0	$\tfrac{1}{5}$	0	$\tfrac{1}{5}$
-2	0	0	$\tfrac{1}{5}$	$\tfrac{1}{5}$
Total	$\tfrac{1}{5}$	$\tfrac{2}{5}$	$\tfrac{2}{5}$	1

From this distribution we earlier calculated that $\mu_X = 0$, $\mu_Z = 2$, $\sigma_X^2 = 2$, $\sigma_Z^2 = 14/5$, and $\rho_{X,Y} = \sigma_{XY} = 0$. To show that X and Z are *not* statistically independent, note that $p_{X,Y}(-2,0) = 0$, $p_X(-2) = \tfrac{1}{5}$, and $p_Z(0) = \tfrac{1}{5}$, so that

$$p_{X,Z}(-2,0) \neq p_X(-2)p_Z(0).$$

Thus, from the definition (3.16), X and Z are not statistically independent. This is what we anticipated, since we know that $Z = X^2$.

Although the value of the correlation coefficient $\rho_{X,Y}$ is not always an unequivocal indicator of the presence of statistical independence between the variables X and Y, calculation of this coefficient often serves as a convenient check for statistical independence. If $\rho_{X,Y}$ is *not* 0, we can be *sure* that X and Y are not statistically independent (since if X and Y are statistically independent, $\rho_{X,Y}$ must be 0). Thus, a nonzero value of $\rho_{X,Y}$ indicates lack of statistical independence, or *statistical dependence*, between the random variables X and Y, and suggests that one of these variables can be used to help predict the values of the other variable (through the use of the concepts of conditional probability and conditional expectation; see Section 2). If $\rho_{X,Y}$ is equal to 0, this may be because X and Y are statistically independent. On the other hand, we have just seen an example of two random variables X and Z for which $\rho_{X,Z} = 0$ and yet $Z = X^2$. Thus, even in the case where two random variables are uncorrelated, it may still be possible to predict the values of one random variable from a knowledge of the values of the other variable.

Correlation and Statistical Dependence

Because the value of the correlation coefficient $\rho_{X,Y}$ is used so frequently to measure the "strength" of the statistical relationship between two random variables X and Y, it may be helpful to describe a simple experimental situation in which the "strength" of the statistical relationship between two discrete random variables is of interest. By giving various possible joint probability models for such an experimental situation and by computing $\rho_{X,Y}$ for each such model, we can see how the value of the correlation coefficient $\rho_{X,Y}$ relates to the closeness of the statistical relationship between the random variables X and Y.

Example 3.3. A very common experimental technique in the behavioral sciences makes use of panels of judges to "rate" various products, actions, personalities, individuals, and so on, as for example in the wine-tasting experiment mentioned in Example 2.1 of this chapter. Of frequent interest in the context of such experiments is the question of how closely the "raters" agree with one another. If, at one extreme, one judge's ratings are completely predictable from the ratings of another judge, then there is good reason to believe that both judges use the same criteria and perceptions in making judgments, and thus a true psychological or behavioral dimension is being measured by the ratings of the panel of judges. If, at the other extreme, the ratings of the various judges are statistically independent of one another, then the judgments of the panel do not have a common behavioral dimension.

The following contingency tables give some possible bivariate probability distributions for the ranks X and Y given by two judges to a certain series of alternatives. The ranking scale used by each judge can assign ranks of 1 (corresponding to "least favored"), 2, or 3 ("most favored") to any given alternative. Along with each contingency table we have

3. DESCRIPTIVE PROPERTIES OF BIVARIATE DISTRIBUTIONS

Table 3.2. Joint probability model for the ratings of two judges.

(a) $\rho_{X,Y} = 0$

x \ y	1	2	3	Total
3	1/9	1/9	1/9	1/3
2	1/9	1/9	1/9	1/3
1	1/9	1/9	1/9	1/3
Total	1/3	1/3	1/3	1

(b) $\rho_{X,Y} = 1/2$

x \ y	1	2	3	Total
3	1/18	1/18	4/18	1/3
2	1/18	4/18	1/18	1/3
1	4/18	1/18	1/18	1/3
Total	1/3	1/3	1/3	1

(c) $\rho_{X,Y} = -1/2$

x \ y	1	2	3	Total
3	4/18	1/18	1/18	1/3
2	1/18	4/18	1/18	1/3
1	1/18	1/18	4/18	1/3
Total	1/3	1/3	1/3	1

(d) $\rho_{X,Y} = 5/9$

x \ y	1	2	3	Total
3	1/27	2/27	6/27	1/3
2	2/27	5/27	2/27	1/3
1	6/27	2/27	1/27	1/3
Total	1/3	1/3	1/3	1

(e) $\rho_{X,Y} = -5/9$

x \ y	1	2	3	Total
3	6/27	2/27	1/27	1/3
2	2/27	5/27	2/27	1/3
1	1/27	2/27	6/27	1/3
Total	1/3	1/3	1/3	1

(f) $\rho_{X,Y} = 2/3$

x \ y	1	2	3	Total
3	1/36	2/36	9/36	1/3
2	2/36	8/36	2/36	1/3
1	9/36	2/36	1/36	1/3
Total	1/3	1/3	1/3	1

(g) $\rho_{X,Y} = -2/3$

x \ y	1	2	3	Total
3	9/36	2/36	1/36	1/3
2	2/36	8/36	2/36	1/3
1	1/36	2/36	9/36	1/3
Total	1/3	1/3	1/3	1

provided the value of the correlation coefficient $\rho_{X,Y}$. Graphs of some of the joint probability mass functions $p_{X,Y}(x,y)$ are shown in Figure 3.5. Notice that the marginal probability mass functions $p_X(x)$ and $p_Y(y)$ are kept constant from contingency table to contingency table; the differences between the various contingency tables thus reflect differences in the (statistical) interrelationships between the variables X and Y.

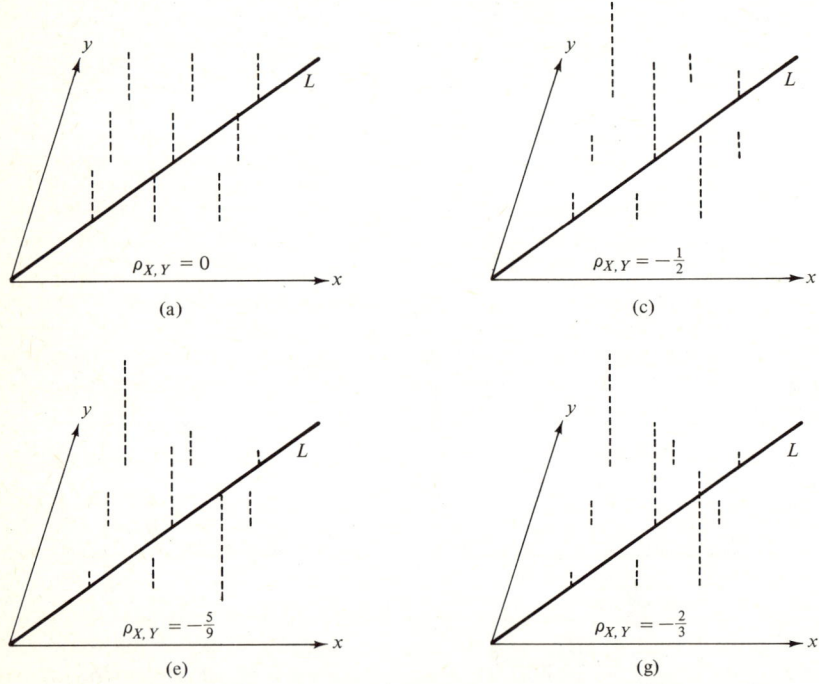

Figure 3.5 Graph of the mass functions corresponding to Tables 3.2(a), 3.2(c), 3.2(e), and 3.2(g).

Comparing the graphs in Figure 3.5 to the corresponding Tables 3.2(a), 3.2(c), 3.2(e), and 3.2(g), we see that as more and more probability mass is taken away from points (x,y) which are distant from the rising line L shown in each graph in Figure 3.5 [the line formed of points (x,y) for which $x = y$] and assigned to points (x,y) which fall nearer that line, the correlation coefficient $\rho_{X,Y}$ becomes more and more positive. From the probability mass functions and the corresponding graphs, we see that $\rho_{X,Y}$ becomes more and more negative as more and more probability mass is assigned to points (x,y) near the falling line L' shown in Figure 3.6 [the line formed of points (x,y) for which $y = 4 - x$].

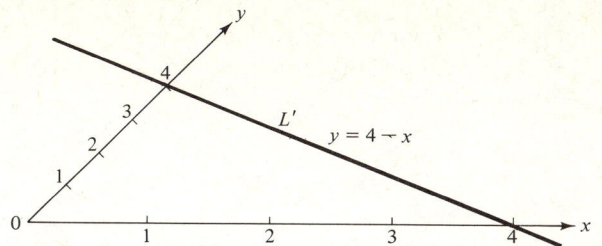

Figure 3.6 Graph of the line $y = 4-x$. The graph is the slanting line L'.

4. CONTINUOUS BIVARIATE DISTRIBUTIONS

In Chapters 4 and 5 we noted that the probability concepts and mathematical formulas used to describe the distributions of discrete random variables have a direct correspondence with analogous concepts and formulas that are used to describe continuous probability distributions. Thus, for discrete distributions, we can represent the assignment of probability to the possible values of a random variable X by a probability mass function, and can find probabilities of events by adding the heights of lines which graphically represent probability masses. Correspondingly, for continuous distributions, we can represent the assignment of probability to the possible values of a random variable by a probability density function, and can find probabilities of events by computing certain areas under the graph of the probability density function. Also, for both discrete and continuous random variables, the various indices of distributions (for example, the expected value and variance) obey similar mathematical equations, even though they are not computed in the same fashion in the discrete case as they are in the continuous case.

A similar correspondence holds between the concepts and formulas used to describe bivariate discrete distributions (see Sections 2 and 3) and the concepts and formulas used to describe bivariate continuous distributions. In particular, the joint probability mass function $p_{X,Y}(x,y)$, which assigns probability mass to the possible pairs (x,y) of values for the discrete random variables X and Y, corresponds to the *joint probability density function* (or *bivariate density function*), $f_{X,Y}(x,y)$, which assigns probability to the possible pairs (x,y) of values for two continuous random variables X and Y. In the case of discrete bivariate probability distributions, we have seen (Section 2) that we can find probabilities of events by adding the heights of lines which graphically represent probability mass assigned to the points (x,y) in the plane. To show how to obtain probabilities of events in the case of bivariate continuous probability distributions, we first need to provide a geometrical (graphical) description of the joint probability density function $f_{X,Y}(x,y)$.

The graph of a bivariate density function $f_{X,Y}(x,y)$ is a surface above the (x,y) plane (see Figure 4.1).

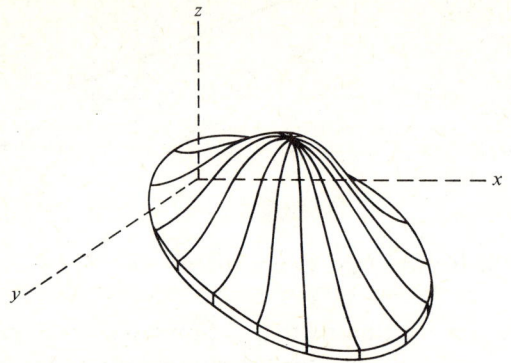

Figure 4.1 The joint density function $f_{X,Y}(x,y)$.

Any event E concerning the random variables X and Y can be represented as a region R in the (x,y) plane; this region R contains all points in the plane corresponding to those pairs of values (outcomes) (x,y) which are contained in the event E. For example, in Figure 4.2 the event E represented by the region R is the event that the random variable X is between 1 and 2 *and* the random variable Y is between 3 and 4; that is,

$$E = \{1 \leq X \leq 2 \text{ and } 3 \leq Y \leq 4\}.$$

Recall that in the case of a single random variable X, the area under the graph of the density function $f_X(x)$ over a certain region represents the probability of the occurrence of a given event (see Chapter 4, Section 5). For more than one variable, the natural extension of area is to volume.

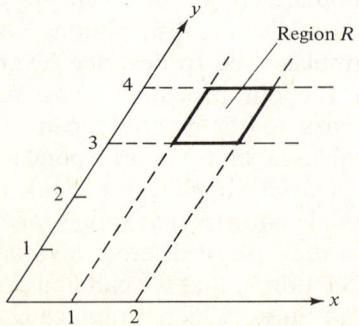

Figure 4.2 The event E represented by region R in the plane is the event $\{1 \leq X \leq 2 \text{ and } 3 \leq Y \leq 4\}$.

4. CONTINUOUS BIVARIATE DISTRIBUTIONS

Thus, if we are interested in the probability of an event E which is represented graphically by a region R in the (x,y) plane, then *the probability of E is calculated by finding the volume over the region R between the (x,y) plane and the surface determined by the graph of the density function $f_{X,Y}(x,y)$* (see Figure 4.2).

Example 4.1. For simplicity, let us consider the bivariate density function

(4.1) $$f_{X,Y}(x,y) = \begin{cases} 1, & \text{if } 0 \leq x \leq 1, \ 0 \leq y \leq 1, \\ 0, & \text{otherwise.} \end{cases}$$

The graph of this density function $f_{X,Y}(x,y)$ is shown in Figure 4.3.

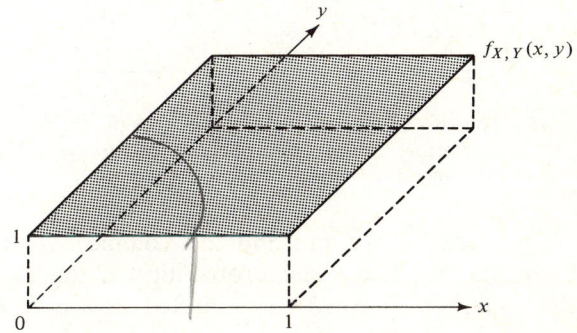

Figure 4.3 Graph of the bivariate density function $f_{X,Y}(x,y)$ given by Equation (4.1).

If we wish to evaluate $P_{X,Y}\{0 \leq X \leq \frac{1}{4} \text{ and } \frac{1}{2} \leq Y \leq \frac{3}{4}\}$, then we must determine the volume of the shaded portion in Figure 4.4.

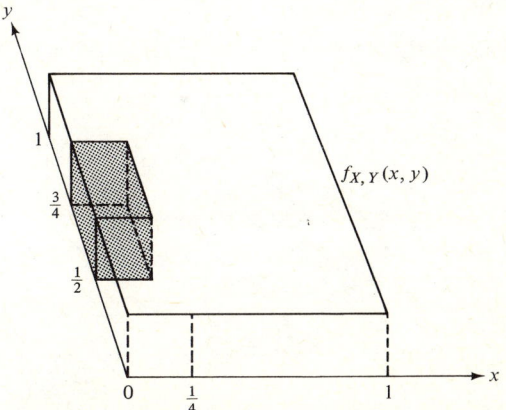

Figure 4.4 The probability $P_{X,Y}\{0 \leq X \leq \frac{1}{4} \text{ and } \frac{1}{2} \leq Y \leq \frac{3}{4}\}$ equals the volume of the shaded three-dimensional rectangle.

Of course, the region R need not be a rectangle, but may be of any shape (as in Figure 4.5). However, it may be more difficult to actually determine the volume over an irregularly shaped region, although, conceptually,

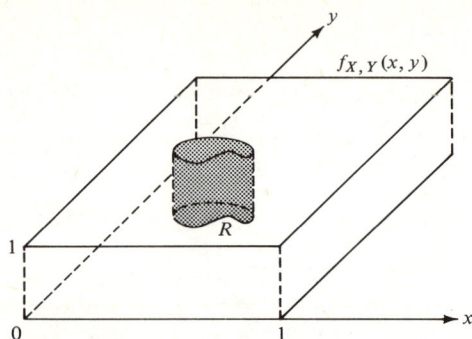

Figure 4.5 The shaded volume [the volume over the region R in the (x,y) plane] equals the probability of the event E which corresponds to R.

the problem is solvable. If we can find the volume over the region R shown in Figure 4.5, we know the probability of the event E corresponding to the region R. Thus, if R is a circle (see Figure 4.6) centered at the point $(\frac{1}{2},\frac{1}{2})$ and with radius $\frac{1}{2}$, then the volume over R between the (x,y) plane and the graph of the bivariate density function (4.1) is equal to the volume of a cylinder with a height of 1 and a base of area $\pi(\frac{1}{2})^2$. Hence, the desired volume equals the height of the cylinder times the area of its base, or $\pi(\frac{1}{2})^2 = 0.78539\ldots$. The probability of the event $E = \{(X-\frac{1}{2})^2 + (Y-\frac{1}{2})^2 \leq \frac{1}{4}\}$ corresponding to the region R shown in Figure 4.6 is thus

$$P_{X,Y}(E) = P_{X,Y}\{(X-\tfrac{1}{2})^2 + (Y-\tfrac{1}{2})^2 \leq \tfrac{1}{4}\} = 0.78539\ldots.$$

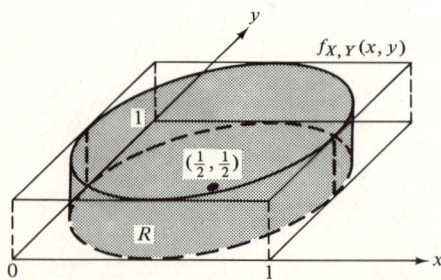

Figure 4.6 The probability of the event $E = \{(X-\frac{1}{2})^2 + (Y-\frac{1}{2})^2 \leq \frac{1}{4}\}$ equals the volume of the shaded cylinder whose base is the circle R.

4. CONTINUOUS BIVARIATE DISTRIBUTIONS

Marginal Probability Density Functions

Suppose that X and Y are continuous random variables with a joint distribution determined by the bivariate density function $f_{X,Y}(x,y)$. Since X is a continuous random variable, if we consider the distribution of X taken alone (without taking account of the fact that Y has also been observed), then we know that X has a probability density function $f_X(x)$. Similarly, if we consider the distribution of Y taken alone, then we know that Y has a probability density function $f_Y(y)$. The density functions $f_X(x)$ and $f_Y(y)$, in analogy to the discrete case (see Section 2), are called *marginal probability density functions*.

The marginal probability density functions $f_X(x)$ and $f_Y(y)$ can be obtained from the joint probability density function $f_{X,Y}(x,y)$ in any one of a variety of methods. A graphical method, which may help us conceptualize why $f_X(x)$ and $f_Y(y)$ are called "marginal" density functions, can be described in the following metaphoric terms: Think of the surface over the (x,y) plane defined by the graph of $f_{X,Y}(x,y)$ [and all of the volume contained between this graph and the (x,y) plane] as being a mountain rising out of the (x,y) plane (see Figure 4.7). To obtain the marginal probability density function $f_Y(y)$, think of a plane $\mathscr{P}$ rising perpendicular to the (x,y) plane from a line parallel to the y-axis (see Figure 4.8). Assume that this plane $\mathscr{P}$ is at the (left) margin of the mountain (see Figure 4.8) and there

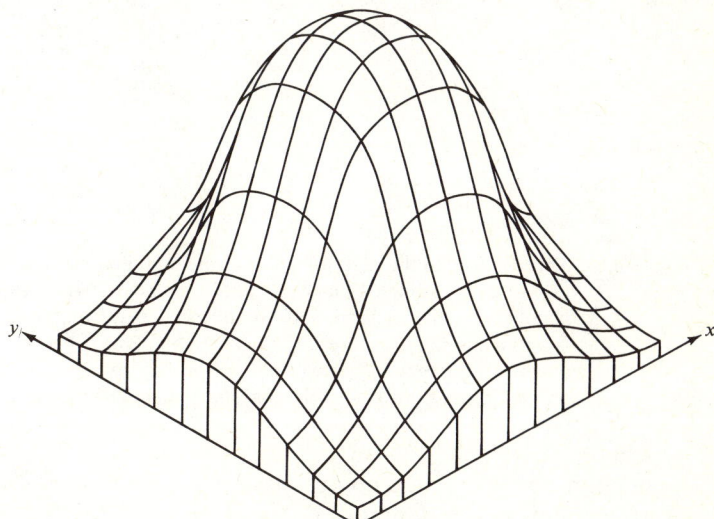

Figure 4.7 The surface over the (x,y) plane defined by the graph of $f_{X,Y}(x,y)$ can be visualized as a mountain rising from the (x,y) plane.

is another plane parallel to plane $\mathscr{P}^*$ at the opposite (right) edge of the mountain. As this last plane is moved toward the marginal plane $\mathscr{P}$, the mountain is *accumulated* on $\mathscr{P}$, until at the end (when the planes meet) the mountain has been massed entirely on $\mathscr{P}$, and the impression left by the mountain on the plane $\mathscr{P}$ is a certain geometric shape (see Figure 4.8). The curve that forms the top of this geometric figure (see Figure 4.8) is the graph of $f_Y(y)$. A similar operation can be conceptualized for the construction of the graph of $f_X(x)$.

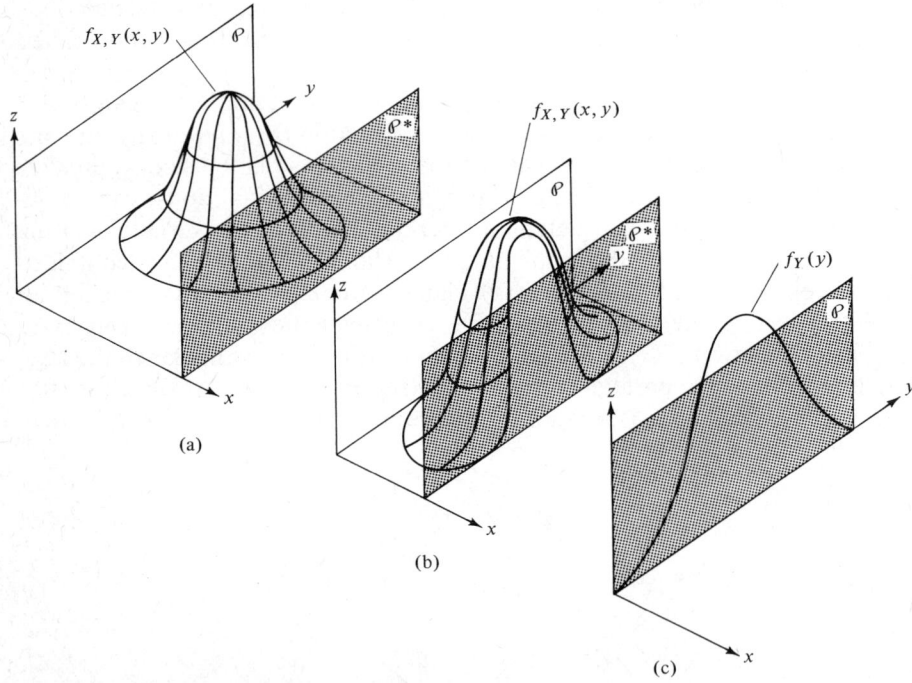

(a)

(b)

(c)

Figure 4.8 Three stages in the determination of the marginal distribution of Y by moving a plane $\mathscr{P}^*$ until it coincides with the plane $\mathscr{P}$. The first figure represents the density function $f_{X,Y}$ as a mountain (before any movement of plane $\mathscr{P}^*$ takes place). In the second figure, plane $\mathscr{P}^*$ moves toward plane $\mathscr{P}$, and in the third figure, the mass is accumulated on the plane $\mathscr{P}$.

The above geometrical method could possibly be carried out if we were to use clay to model the "mountain" $f_{X,Y}(x,y)$, but it is hardly practical. One of the actual methods used by probabilists to find the marginal probability density functions $f_X(x)$ and $f_Y(y)$ from knowledge of the bivariate density function $f_{X,Y}(x,y)$ is the following: To obtain the marginal prob-

4. CONTINUOUS BIVARIATE DISTRIBUTIONS

ability density function $f_X(\tau)$ of X, draw a plane perpendicular to the (x,y) plane along the line $x = \tau$ (see Figure 4.9). This plane cuts the surface determined by $f_{X,Y}(x,y)$ in a curve $f_{X,Y}(\tau,y) = g(y)$. The area between this curve $g(y)$ and the line $x = \tau$ is $f_X(\tau)$. If we do this computation for every τ, we obtain $f_X(x)$. A similar computation gives us $f_Y(y)$.

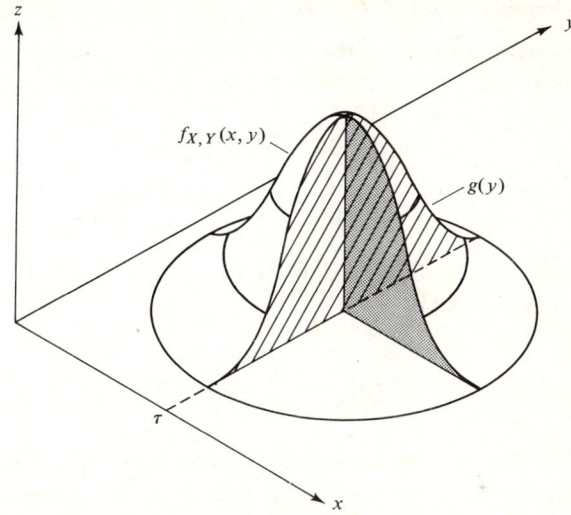

Figure 4.9 The striped area under the graph of $g(y)$ equals $f_X(\tau)$.

Example 4.1 (Continued). We return to the simple example of a bivariate density function given in Example 4.1. If we wish to determine the marginal distribution of X, we make a cut at τ (see Figure 4.10). The area of the rectangle formed by the curve $f_{X,Y}(\tau,y) = g(y)$ and the line $x = \tau$ is equal

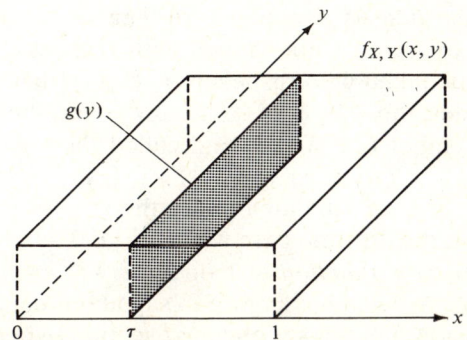

Figure 4.10 The shaded area under the graph of $g(y)$ equals 1, so that $f_X(\tau) = 1$.

to 1 for all τ, $0 \leq \tau \leq 1$. Thus, the marginal distribution of X is the uniform distribution with parameters $a = 0$ and $b = 1$ (Chapter 8, Section 4). The marginal distribution of Y is also the uniform distribution with parameters $a = 0$ and $b = 1$.

Conditional Probability Density Functions

Continuing our analogies to the discrete bivariate case, we may define the conditional probability density function of Y given $X = x$ [denoted by $f_{Y|X=x}(y)$] as

(4.2) $$f_{Y|X=x}(y) = \frac{f_{X,Y}(x,y)}{f_X(x)}, \quad \text{when } f_X(x) > 0.$$

In a similar way, the conditional density function of X given $Y = y$ [denoted by $f_{X|Y=y}(x)$] is defined by

(4.3) $$f_{X|Y=y}(x) = \frac{f_{X,Y}(x,y)}{f_Y(y)}, \quad \text{when } f_Y(y) > 0.$$

It can be shown that $f_{Y|X=x}(y)$ and $f_{X|Y=y}(x)$ are, indeed, probability density functions (that is, they are nonnegative, and the total areas under their graphs are unity). An idea of the *shape* of the graph of $f_{Y|X=x^*}(y)$ can be obtained by drawing a plane perpendicular to the (x,y) plane along the line $x = x^*$ and observing the curves $g(y) = f_{X,Y}(x^*,y)$ cut by the plane on the surface formed by the graph of $f_{X,Y}(x,y)$. Similarly, the shape of the graph of $f_{X|Y=y^*}(x)$ can be obtained by drawing a plane perpendicular to the (x,y) plane along the line $y = y^*$ and observing the curve $h(x) = f_{X,Y}(x,y^*)$ cut by the plane on the surface formed by the graph of $f_{X,Y}(x,y)$ (see Figure 4.11).

Since conditional density functions are themselves density functions, we can consider expected values taken with respect to these densities. The expected value taken with respect to $f_{Y|X=x}(y)$ [that is, the area, taken algebraically, under the graph of $y f_{Y|X=x}(y)$ over the entire y-axis] is denoted by the symbol $E[Y|X = x]$ and called the *conditional expected value of Y given X = x*. Similarly, $E[X|Y = y]$, the conditional expected value of X given $Y = y$, is the area under the graph of $f_{X|Y=y}(x)$ over the entire x-axis. As in the discrete case we call $E[Y|X = x]$ (which is a function of x) the *regression function* of Y on X; $E[X|Y = y]$ is the regression function of X on Y. We will have more to say about conditional expected values and regression functions in the next section, where these concepts are discussed for the case where X and Y have a bivariate normal distribution.

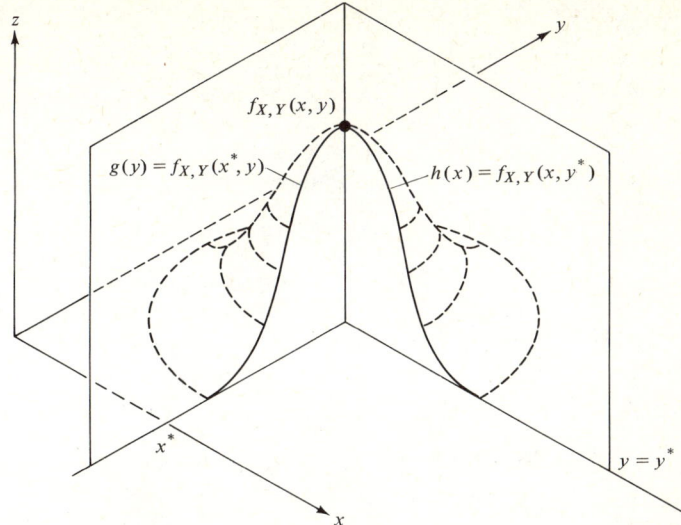

Figure 4.11 Shapes of the graphs of $f_{Y|X=x^*}(y)$ and $f_{X|Y=y^*}(x)$. The shape of the graph of $f_{Y|X=x^*}(y)$ is given by the graph of $g(y)$. The shape of the graph of $f_{X|Y=y^*}(x)$ is given by the graph of $h(x)$.

Statistical Independence

The notion of the statistical independence of discrete variables X and Y is extendable to the continuous case. We say that the continuous random variables X and Y are *statistically independent* if

(4.4) $\qquad f_{Y|X=x}(y) = f_Y(y) \quad \text{or} \quad f_{X|Y=y}(x) = f_X(x),$

for all x and y. That is, random variables are statistically independent when their conditional density functions do not differ from their unconditional density functions. An equivalent definition of statistical independence is the following: X and Y are statistically independent if

(4.5) $\qquad f_{X,Y}(x,y) = f_X(x)f_Y(y)$

for all x and y; that is, the joint density factors into the product of the two marginal densities.

Example 4.1 (Continued). In this example, the bivariate density function $f_{X,Y}(x,y)$ is given by the equation

$$f_{X,Y}(x,y) = \begin{cases} 1, & \text{if } 0 \leq x \leq 1 \text{ and } 0 \leq y \leq 1, \\ 0, & \text{otherwise.} \end{cases}$$

We have determined that the marginal distributions obtained from $f_{X,Y}(x,y)$ are

$$f_X(x) = \begin{cases} 1, & \text{if } 0 \leq x \leq 1, \\ 0, & \text{if } x < 0 \text{ or } x > 1, \end{cases}$$

$$f_Y(y) = \begin{cases} 1, & \text{if } 0 \leq y \leq 1, \\ 0, & \text{if } y < 0 \text{ or } y > 1. \end{cases}$$

Thus, the product of the marginal probability densities is equal to

$$f_X(x)f_Y(y) = \begin{cases} (1)(1), & \text{if } 0 \leq x \leq 1 \text{ and } 0 \leq y \leq 1, \\ (1)(0), & \text{if } 0 \leq x \leq 1 \text{ and } y < 0 \text{ or } y > 1, \\ (0)(1), & \text{if } x < 0 \text{ or } x > 1 \text{ and } 0 \leq y \leq 1, \\ (0)(0), & \text{if } x < 0 \text{ or } x > 1 \text{ and } y < 0 \text{ or } y > 1. \end{cases}$$

$$= \begin{cases} 1, & \text{if } 0 \leq x \leq 1 \text{ and } 0 \leq y \leq 1, \\ 0, & \text{otherwise,} \end{cases}$$

$$= f_{X,Y}(x,y),$$

so that X and Y are independent random variables.

Two random variables X and Y can be expected to be independent when the variables X and Y are, in reality, unrelated (such as when X and Y result from physically independent experiments). However, the mathematical definition of statistical independence is broader than this interpretation, in that even physically related random variables may, in fact, be statistically independent (that is, their joint density function may factor into the product of their marginal density functions).

Descriptive Properties of Bivariate Continuous Distributions

As in the case of jointly distributed discrete random variables, the indices most commonly used to describe aspects of bivariate continuous distributions are the joint expected value or mean vector (μ_X, μ_Y), the variances σ_X^2 and σ_Y^2, the covariance σ_{XY}, and the correlation coefficient $\rho_{X,Y}$. Each of these quantities may be obtained from the joint probability density function $f_{X,Y}(x,y)$ of X and Y through use of the following general formula for finding the expected value $E[g(X,Y)]$ of any function $g(X,Y)$ of X and Y:

(4.6) $E[g(X,Y)] =$ "The volume (taken algebraically) between the surface defined by the graph of the function $g(x,y)f_{X,Y}(x,y)$ and the (x,y) plane."

Note that Equation (4.6) is the analogy for bivariate distributions of the formula for the expectation of a function $g(X)$ of one continuous random variable X given in Equation (4.3) of Chapter 5. Formula (4.6) for the

4. CONTINUOUS BIVARIATE DISTRIBUTIONS

expectation $E[g(X,Y)]$ of $g(X,Y)$ holds for any function $g(X,Y)$ of X and Y [for example, Formula (4.6) holds when $g(X,Y) = XY$, when $g(X,Y) = X+Y$, and so on]. In particular, when $g(X,Y) = X$, Formula (4.6) yields the result

(4.7) $\quad \mu_X = E[X] =$ "The volume (taken algebraically) between the surface defined by the graph of the function $xf_{X,Y}(x,y)$ and the (x,y) plane,"

and when $g(x,y) = (x-\mu_X)^2$, Formula (4.6) yields the result

(4.8) $\quad \sigma_X^2 = E(X-\mu_X)^2 =$ "The volume between the surface defined by the graph of the function $(x-\mu_X)^2 f_{X,Y}(x,y)$ and the (x,y) plane."

Similar formulas can be used to compute μ_Y and σ_Y^2. However, it is usually more convenient to compute μ_X, μ_Y, σ_X^2, and σ_Y^2 from the marginal probability density functions $f_X(x)$ and $f_Y(y)$; see Chapter 5.

We can compute the covariance, $\sigma_{XY} = \text{Cov}(X,Y)$, between x and y either directly by the formula

(4.9) $\quad \sigma_{XY} = E[(X-\mu_X)(Y-\mu_Y)]$

$\quad\quad\quad =$ "The volume (considered algebraically) between the surface defined by the graph of the function $(x-\mu_X)(y-\mu_Y)f_{X,Y}(x,y)$ and the (x,y) plane,"

or indirectly, using the relationship

(4.10) $\quad\quad\quad \sigma_{XY} = E[XY] - \mu_X\mu_Y,$

which was demonstrated in Section 3 for the case of discrete random variables, and which is also true for continuous random variables.

Finally, the correlation coefficient $\rho_{X,Y} = \text{Cov}(X,Y)$ between X and Y is defined by

(4.11) $\quad\quad\quad \rho_{X,Y} = \dfrac{\sigma_{XY}}{\sigma_X \sigma_Y}.$

In direct analogy to the discrete case (Section 3), the indices μ_X, μ_Y, σ_X^2, σ_Y^2, σ_{XY}, and $\rho_{X,Y}$ have the following properties:

(i) The joint expected value (μ_X, μ_Y) is a measure of location for the joint distribution of X and Y. It gives the center of gravity for the distribution of probability over the plane. The expected values μ_X and μ_Y obey the equations

$$\mu_{aX+b} = a\mu_X + b,$$
$$\mu_{cY+d} = c\mu_Y + d,$$

for any numbers a, b, c, and d.

(ii) The variances σ_X^2 and σ_Y^2 are measures of the dispersion of the joint probability distribution of X and Y around the lines $x = \mu_X$ and $y = \mu_Y$, respectively. The variances σ_X^2 and σ_Y^2 obey the equations

$$\sigma_{aX+b}^2 = a^2 \sigma_X^2,$$
$$\sigma_{cY+d}^2 = c^2 \sigma_Y^2,$$

for any numbers a, b, c, and d.

(iii) The covariance σ_{XY} is a measure of the interrelationship between X and Y. The covariance is scale dependent; that is, for any numbers a, b, c, and d,

$$\sigma_{(aX+b)(cY+d)} = ac\, \sigma_{XY}.$$

Also, the covariance of X with itself is $\sigma_{XX} = \sigma_X^2$, and the covariance of Y with itself is $\sigma_{YY} = \sigma_Y^2$.

(iv) The correlation $\rho_{X,Y}$ is a scale-free measure of the interrelationship between X and Y. When X and Y are statistically independent, $\rho_{X,Y} = 0$. However, it is possible that $\rho_{X,Y} = 0$, and yet X and Y are *not* statistically independent. The correlation coefficient is a convenient check for the statistical dependence of two random variables. If $\rho_{X,Y}$ is not zero, the random variables X and Y cannot be statistically independent. The correlation coefficient $\rho_{X,Y}$ is sensitive to linear dependence. The correlation $\rho_{X,Y}$ reaches its maximum possible value of 1 when X and Y are directly linearly related (that is, $Y = aX + b$, where $a > 0$, so that X and Y increase together), and attains its minimum value of -1 when X and Y are inversely linearly related (that is, $Y = aX + b$, where $a < 0$, so that Y decreases as X increases). The more concentrated is the joint probability distribution of X and Y about a line $Y = aX + b$ in the (x,y) plane, the closer $\rho_{X,Y}$ becomes to its maximum possible value of 1 (if $a > 0$) or minimum value of -1 (if $a < 0$).

A variety of bivariate continuous distributions $f_{X,Y}(x,y)$ in which X and Y are statistically independent have been used to model random phenomena. However, among those bivariate continuous distributions for which X and Y are statistically dependent, only the bivariate normal distribution is used widely enough to justify our devoting detailed attention to its properties. We discuss the bivariate normal distribution in the next section.

5. THE BIVARIATE NORMAL DISTRIBUTION

In Chapter 7, we emphasized that the normal distribution is extensively used to model the variation of individual quantitative phenomena. In particular, we noted that various physiological and behavioral measurements obeyed normal distributions. Similarly, the bivariate normal distribution is frequently used to model the joint variation of *pairs* of quantitative phenomena.

Two continuous random variables X and Y have a bivariate normal distribution if the graph of their joint probability density function $f_{X,Y}(x,y)$ has a shape similar to that shown in Figure 5.1.

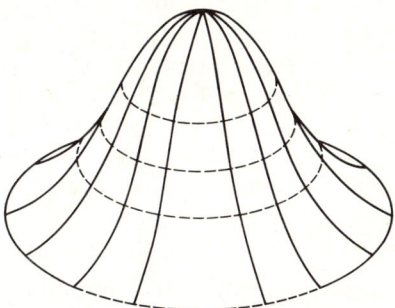

Figure 5.1 The bivariate normal probability density function.

The mathematical formula for the joint probability density function of a bivariate normal distribution is somewhat complicated. Thus, although we present it here for the benefit of those readers who might find it useful, we will try in this section to make as little use of this expression as possible, depending instead on graphs of $f_{X,Y}(x,y)$ to illustrate our assertions.

The joint probability density function $f_{X,Y}(x,y)$ of a bivariate normal distribution is defined by the equation

(5.1) $$f_{X,Y}(x,y) = \frac{1}{\sqrt{(2\pi)^2 \sigma_X^2 \sigma_Y^2 (1-\rho_{X,Y}^2)}} e^{-Q(x,y)/2},$$

where

$$Q(x,y) = \frac{1}{1-\rho_{X,Y}^2}\left[\frac{(x-\mu_X)^2}{\sigma_X^2} - 2\rho_{X,Y}\frac{(x-\mu_X)(y-\mu_Y)}{\sigma_X \sigma_Y} + \frac{(y-\mu_Y)^2}{\sigma_Y^2}\right].$$

The quantities μ_X, μ_Y, σ_X^2, σ_Y^2, and $\rho_{X,Y}$ are the *parameters* of the bivariate normal distribution; the bivariate normal distribution is completely determined when the values of these five parameters are specified. The parameters μ_X and μ_Y are the expected values of X and of Y, respectively, and tell us where the center of the "mountain" shown in Figure 5.1 is located. The parameters σ_X^2 and σ_Y^2 are the variances of X and Y, respectively, and measure the spread of the "mountain" in the x- and y-directions. Finally, the parameter $\rho_{X,Y}$ is the correlation coefficient between X and Y. As we shall see, $\rho_{X,Y}$ determines the shape and orientation on the (x,y) plane of the "mountain" shown in Figure 5.1.

If X and Y have a bivariate normal distribution with parameters μ_X, μ_Y, σ_X^2, σ_Y^2, and $\rho_{X,Y}$, then it can be shown that the marginal distribution of X is the normal distribution with parameters μ_X and σ_X^2, and that the marginal distribution of Y is the normal distribution with parameters μ_Y and σ_Y^2. Indeed, not only are X and Y each normally distributed, but also *any linear combination $aX+bY+c$ is normally distributed*. The bivariate normal distribution is the *only* joint distribution for two random variables, X and Y, that has the property that *all* linear combinations $aX+bY+c$ of X and Y are normally distributed. This property *characterizes* the normal distribution. On the other hand, many bivariate distributions possess normal marginal distributions. Thus, *the property of having normal marginal distributions does not characterize the bivariate normal distribution*. Even though the marginals of a joint distribution may be normal distributions, this is no reason to jump to the conclusion that the joint distribution is a bivariate normal distribution!

If X and Y have a bivariate normal distribution with parameters μ_X, μ_Y, σ_X^2, σ_Y^2, and $\rho_{X,Y}$, then by using the definition (4.2) of the conditional probability density function $f_{Y|X=x}(y)$ of Y given $X=x$, we can show (after some algebra) that $f_{Y|X=x}(y)$ is the probability density function of a (univariate) normal distribution having an expected value equal to

$$(5.2) \qquad \mu_{Y|X=x} = E[Y|X=x] = \mu_Y + \rho_{X,Y}\frac{\sigma_Y}{\sigma_X}(x-\mu_X),$$

and conditional variance

$$(5.3) \qquad \sigma^2_{Y|X=x} = \sigma_Y^2(1-\rho^2_{X,Y}).$$

Similarly, it can be shown that the conditional probability density function of X given $Y=y$, is also a univariate normal density function with an expected value

$$(5.4) \qquad \mu_{X|Y=y} = E[X|Y=y] = \mu_X + \rho_{X,Y}\frac{\sigma_X}{\sigma_Y}(y-\mu_Y),$$

and conditional variance

$$(5.5) \qquad \sigma^2_{X|Y=y} = \sigma_X^2(1-\rho^2_{X,Y}).$$

From Equations (5.2) and (5.4), we see that the regressions of Y on X (see Section 2) and of X on Y are *linear functions*. That is, the regression $E[Y|X]$ of Y on X is a linear function $aX+b$ of X with intercept

$$b = \mu_Y - \rho_{X,Y}\frac{\sigma_Y}{\sigma_X}\mu_X,$$

and slope

$$a = \rho_{X,Y}\frac{\sigma_Y}{\sigma_X}.$$

5. THE BIVARIATE NORMAL DISTRIBUTION

Similarly, the regression of X on Y is a linear function $cY + d$ of Y with intercept

$$d = \mu_X - \rho_{X,Y} \frac{\sigma_X}{\sigma_Y} \mu_Y,$$

and slope

$$c = \rho_{X,Y} \frac{\sigma_X}{\sigma_Y}.$$

When $\rho_{X,Y} = 0$, the slopes a and c are both equal to 0 and the regression lines are constant (flat).

It should be noted that the conditional variance $\sigma^2_{Y|X=x}$ of the distribution of Y given that $X = x$ does not depend for its value on the value of x. Also, $\sigma^2_{Y|X=x}$ is equal to the product of the variance σ_Y^2 of the marginal distribution of Y and the factor $(1 - \rho^2_{X,Y})$. Since $\rho_{X,Y}$ is a number between -1 and 1, $1 - \rho^2_{X,Y}$ is a number no greater than 1. Therefore, except for the case when X and Y are uncorrelated (that is, when $\rho_{X,Y} = 0$), the conditional variance of Y given $X = x$ is always *smaller* than the marginal variance of Y. It follows that when X and Y have a bivariate normal distribution with nonzero correlation $\rho_{X,Y}$, knowledge of the value x of X yields a conditional probability density for Y given $X = x$ which has *smaller dispersion* than the probability density for Y considered alone (that is, the marginal density of Y). Thus, if we know the value of X, we obtain a more precise type of predictive statement about the unknown random variable Y than could be obtained when we do not know the value of X. By interchanging the roles of X and Y in the above argument, we see that $\sigma^2_{X|Y=y} = \sigma_X^2(1 - \rho^2_{X,Y})$ is less than σ_X^2, unless $\rho_{X,Y} = 0$. Thus, if we know the value of Y, we obtain a more precise type of predictive statement about X than could be obtained when we do not know Y and have to obtain predictive statements about the variable X from its marginal distribution, $f_X(x)$.

Note that when X and Y have a bivariate normal distribution and the parameter $\rho_{X,Y}$ is 0 (that is, X and Y are uncorrelated), then the conditional distribution of Y given $X = x$ is the normal distribution with expected value $E[Y|X = x] = \mu_Y$ and variance $\sigma^2_{Y|X=x} = \sigma_Y^2$. Thus, the conditional probability density function $f_{Y|X=x}(y)$ of Y given $X = x$ is equal to the marginal density function $f_Y(y)$, since both of these density functions are those of a normally distributed random variable having expected value μ_Y and variance σ_Y^2. Hence, using the characterization of the statistical independence of X and Y given in Equation (4.4), it follows that X and Y are statistically independent. In general, we have seen that uncorrelated random variables are not necessarily statistically independent, but *if we know that the random variables X and Y have a bivariate normal distribution, then if X and Y are uncorrelated, they are also statistically* *independent.* Since any two random variables that are statistically independent are also uncorrelated, it follows that *when X and Y have a bivariate normal distribution, lack of correlation* (that is, $\rho_{X,Y} = 0$) *between X and Y is equivalent to statistical independence.*

The Probability Contours of a Bivariate Normal Distribution

If we slice the surface formed by the graph of the normal probability density function (see Figure 5.2) by a plane parallel to the (x,y) plane, then the cross section (contour) created by the intersection of the plane and the surface is ellipsoidal in shape. No matter what horizontal plane we use to slice the surface in Figure 5.1 (as long as this plane intersects the surface), the resulting cross section has the same shape (but not the same area). Knowledge of this shape [and its orientation in the (x,y) plane] is equivalent to knowledge of how probability is distributed in the plane by the bivariate normal distribution. Thus, we call this characteristic cross-sectional shape the *probability contour* of the bivariate normal distribution.

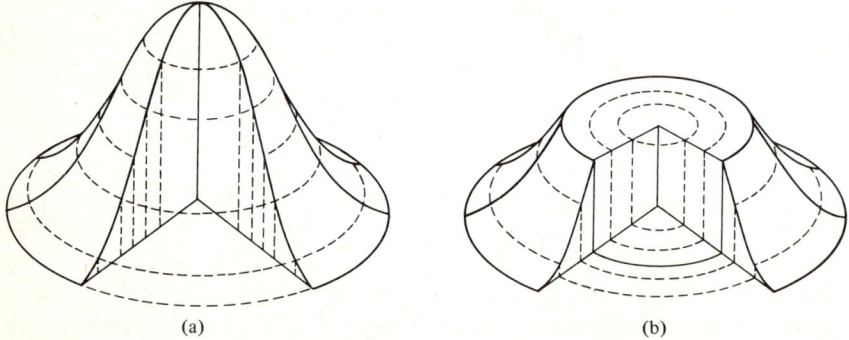

Figure 5.2 Probability contours of the bivariate normal distribution, given in full view and in cross section.

Figures 5.3, 5.4, and 5.5 represent the different elliptical probability contours that can define the bivariate normal distribution. They illustrate how the correlation coefficient $\rho_{X,Y}$ determines the orientation and shape of the surface formed by a graph of the bivariate normal probability density function.

First consider Figure 5.3. These three graphs illustrate the probability contours of the bivariate normal distribution when X and Y are uncorrelated ($\rho_{X,Y} = 0$). The probability contour is a circle when the variances of X and Y are the same ($\sigma_X^2 = \sigma_Y^2$), is an ellipse stretched in the y-direction in the plane if $\sigma_Y^2 > \sigma_X^2$, and is an ellipse stretched in the x-direction if $\sigma_X^2 > \sigma_Y^2$. In all three cases, the probability contours are

5. THE BIVARIATE NORMAL DISTRIBUTION

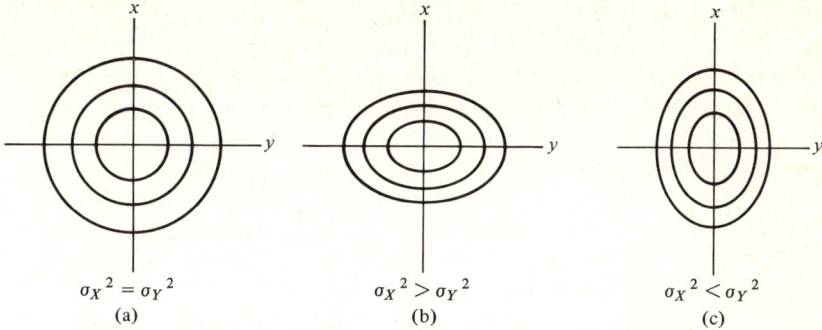

$\sigma_X^2 = \sigma_Y^2$
(a)

$\sigma_X^2 > \sigma_Y^2$
(b)

$\sigma_X^2 < \sigma_Y^2$
(c)

Figure 5.3 Contours of the bivariate normal distribution when X and Y are uncorrelated.

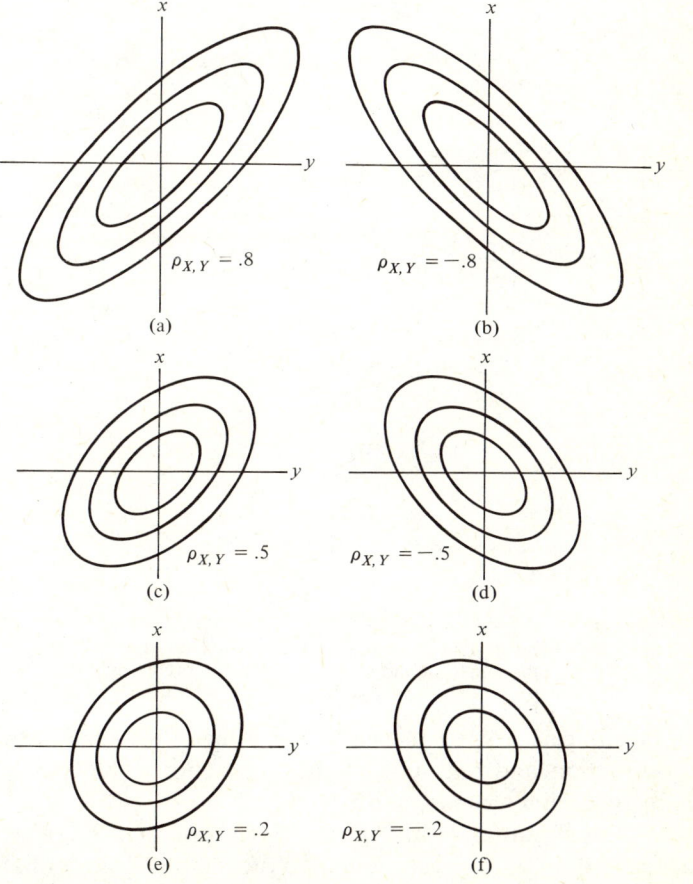

$\rho_{X,Y} = .8$ (a)
$\rho_{X,Y} = -.8$ (b)
$\rho_{X,Y} = .5$ (c)
$\rho_{X,Y} = -.5$ (d)
$\rho_{X,Y} = .2$ (e)
$\rho_{X,Y} = -.2$ (f)

Figure 5.4 Contours of the bivariate normal distribution for different values of the correlation coefficient $\rho_{X,Y}$ when $\sigma_X^2 = \sigma_Y^2 = 1$.

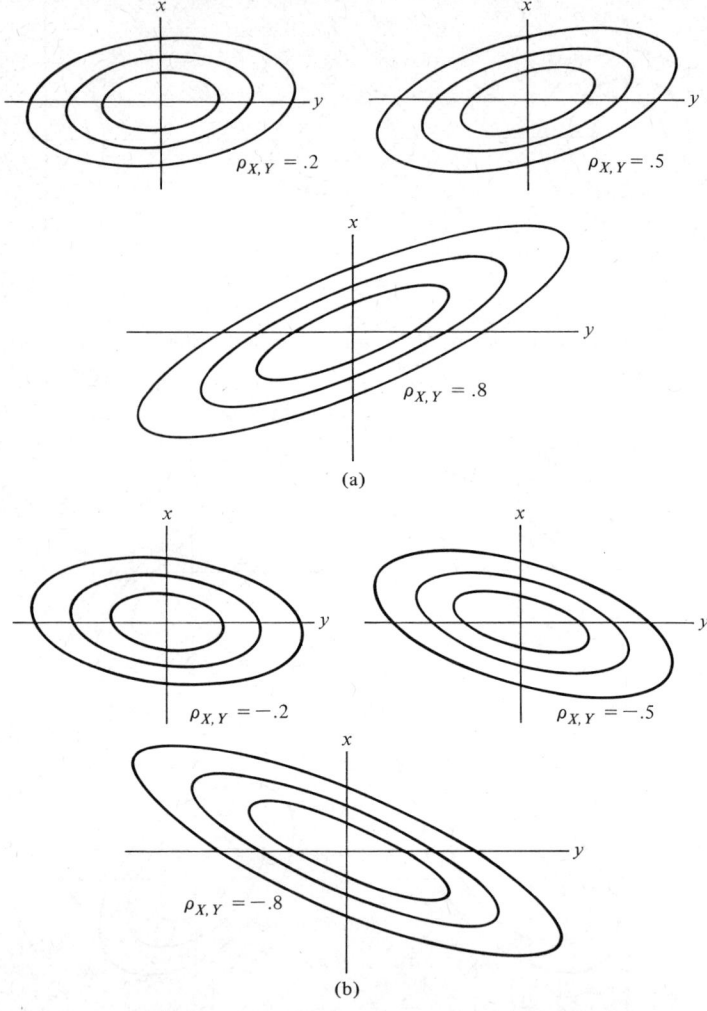

Figure 5.5 (a) Contours of the bivariate normal distribution for different positive values of the correlation coefficient $\rho_{X,Y}$ when $\sigma_X^2 = 4$ and $\sigma_Y^2 = 1$; (b) for different negative values.

ellipses (a circle is a special case of an ellipse) oriented with one axis in the x-direction and one axis in the y-direction. The two axes of the ellipse are on the lines $y = \mu_Y$ and $x = \mu_X$.

Now look at Figure 5.4. In Figure 5.3 we saw that the relative size of the variances σ_X^2 and σ_Y^2 can influence the shape of the probability contours; here, we eliminate this influence by setting $\sigma_X^2 = \sigma_Y^2$. The six graphs shown thus illustrate the influence of $\rho_{X,Y}$ on the shape and orientation of the probability contours. In all six cases, the probability contours are ellipses with one axis directed along a rising straight line (that is, a line having a positive slope) and one axis directed along a falling straight line (a line having a negative slope). If $\rho_{X,Y}$ is positive, the long axis of the ellipse lies on a rising straight line. As $\rho_{X,Y}$ approaches 1, the probability contour elongates along this longest axis (that is, the ellipse becomes longer and thinner), until when $\rho_{X,Y} = 1$, the contour and the rising straight line are identical. Similarly, if $\rho_{X,Y}$ is negative, the long axis of the ellipse lies on a falling straight line. As $\rho_{X,Y}$ approaches -1, the probability contour elongates along this longest axis until, when $\rho_{X,Y} = -1$, the contour and the falling straight line are identical.

Figure 5.5 shows how inequality of the variances σ_X^2 and σ_Y^2 can either modify or accentuate the influence of the correlation $\rho_{X,Y}$ on the shape and orientation of the probability contours. Each graph in Figure 5.5 should be compared with the graph in Figure 5.4 corresponding to the same value of $\rho_{X,Y}$. Even though inequality of the variances σ_X^2 and σ_Y^2 can speed up or slow down the change of the shape of the probability contours with changing values of the correlation coefficient $\rho_{X,Y}$, the nature of the change is the same: as $\rho_{X,Y}$ approaches $+1$, the probability mass of a bivariate normal distribution becomes concentrated on a straight line having positive slope; as $\rho_{X,Y}$ approaches -1, the probability mass of a bivariate normal distribution concentrates on a straight line having a negative slope.

The Standard Bivariate Normal Distribution

Recall our notation for a univariate normal distribution, namely, $\mathcal{N}(X; \mu, \sigma^2)$. To denote a bivariate normal distribution we use

$$\mathcal{N}(X,Y; (\mu_X, \mu_Y), (\sigma_X^2, \sigma_Y^2, \rho_{X,Y})).$$

Although it is a somewhat cumbersome notation, it has the advantage of providing all the necessary ingredients in one expression. For the univariate normal distribution, recall the result

$$P\{\mathcal{N}(X; \mu, \sigma) \leq a\} = P\left\{\mathcal{N}(Z; 0, 1) \leq \frac{a - \mu}{\sigma}\right\}.$$

It was this result that permitted us to compute the probabilities for any normal distribution from only one table, that of the cumulative distribution function of the standard normal distribution (see Chapter 7).

In the bivariate case we have a similar key result, namely, if X and Y have a $\mathcal{N}(X,Y; (\mu_X, \mu_Y), (\sigma_X^2, \sigma_Y^2, \rho_{X,Y}))$ distribution, then

(5.6) $\quad P_{X,Y}\{X \leq a \text{ and } Y \leq b\} = P\left\{X^* \leq \frac{a - \mu_X}{\sigma_X} \text{ and } Y^* \leq \frac{b - \mu_Y}{\sigma_Y}\right\},$

where now X^* and Y^* have a $\mathcal{N}(X^*,Y^*; (0,0), (1,1,\rho_{X,Y}))$ distribution. Put another way, every pair of bivariate normal random variables X and Y can be transformed to standardized bivariate normal random variables

$$X^* = \frac{X - \mu_X}{\sigma_X}, \qquad Y^* = \frac{Y - \mu_Y}{\sigma_Y},$$

so that the expected values of X^* and Y^* are 0, the variances of X^* and Y^* are 1, and the correlation ρ_{X^*,Y^*} between X^* and Y^* is equal to the correlation $\rho_{X,Y}$ between X and Y. Notice that although there is only one standard univariate normal distribution, there are many standard bivariate normal distributions, one for each value of the correlation coefficient $\rho_{X,Y}$.

Tables of probabilities for standard bivariate normal distributions are not as easy to construct as the tables for the standard univariate normal distribution given in Chapter 7. Tables of probabilities of events of the form $\{X^* > h \text{ and } Y^* > k\}$ are provided in a publication of the National Bureau of Standards ("Tables of the Bivariate Normal Distribution Function and Related Functions," PB 176 520, 15 June, 1969). Here X^* and Y^* are $\mathcal{N}(X^*,Y^*; (0,0), (1,1,\rho_{X^*,Y^*}))$, and the tables are given for $h, k = 0(0.1)4$ and for values of ρ_{X^*,Y^*} (called r in the tables) ranging in steps of 0.05 from 0.00 to 0.95, and then in steps of 0.01 from 0.95 to 1.00 [that is, $\rho_{X^*,Y^*} = 0.00\ (0.05)\ 0.95\ (0.01)\ 1.00$]. Similar tables are also given for $-\rho_{X^*,Y^*} = 0.00\ (0.05)\ 0.95\ (0.01)\ 1.00$.

As examples of the National Bureau of Standards tables, we give, in Table T.11, values of $P\{X^* > h \text{ and } Y^* > k\}$ for $k, h = 0.00\ (0.1)\ 4.00$, when X^* and Y^* are $\mathcal{N}(X^*,Y^*; (0,0), (1,1,\rho_{X^*,Y^*}))$ for $\rho_{X^*,Y^*} = 0.5, -0.5, 0.85, -0.85$.

Since the tables of $P\{X^* > h \text{ and } Y^* > k\}$ given in T.11 (and the National Bureau of Standards tables) cover only nonnegative values of h and k, we need to have ways of calculating $P\{X^* > h \text{ and } Y^* > k\}$ when either h, or k, or both h and k are negative numbers:

If h is nonnegative and k is negative. Look up $P\{\mathcal{N}(Z; 0,1) > h\}$ in Table T.7 of Chapter 7. Then

$$P\{X^* > h \text{ and } Y^* > k\} = P\{\mathcal{N}(Z; 0,1) > h\} - P\{V > h \text{ and } W > -k\},$$

where $P\{V > h \text{ and } W > -k\}$ is found by looking in the table giving probability values for $\mathcal{N}(V,W; (0,0),(1,1,-\rho_{X^*,Y^*}))$; that is, in the table giving probability values for standard bivariate normal random variables with correlation equal to $-\rho_{X^*,Y^*}$.

If h is negative and k is nonnegative. Look up $P\{\mathcal{N}(Z;0,1) > k\}$ in

5. THE BIVARIATE NORMAL DISTRIBUTION

Table T.7. Then

$$P\{X^* > h \text{ and } Y^* > k\} = P\{\mathcal{N}(Z; 0,1) > k\} - P\{V > -h \text{ and } W > k\},$$

where $P\{V > -h \text{ and } W > k\}$ is found by looking in the table giving probabilities for $\mathcal{N}(V, W; (0,0), (1,1, -\rho_{X^*,Y^*}))$.

If both h and k are negative. Look up $P\{\mathcal{N}(Z; 0,1) \leq -k\}$ and $P\{\mathcal{N}(Z; 0,1) \leq -h\}$ in Table T.6. Then

$$P\{X^* > h \text{ and } Y^* > k\} = P\{\mathcal{N}(Z; 0,1) \leq -k\} + P\{\mathcal{N}(Z; 0,1) \leq -h\} + P\{X^* > -h \text{ and } Y^* > -k\} - 1.$$

[*Note:* In the case of the standard normal distribution there is a relationship between probabilities of the form $P\{\mathcal{N}(Z; 0,1) \leq a\}$ and the tail probabilities $P\{\mathcal{N}(Z; 0,1) > b\}$; namely,

$$P\{\mathcal{N}(Z; 0,1) \leq a\} = P\{\mathcal{N}(Z; 0,1) > -a\}.]$$

Similarly, for standard bivariate normal distributions we have the relationship

(5.7) $\qquad P\{X^* \leq \tau \text{ and } Y^* \leq \nu\} = P\{V > -\tau \text{ and } W > -\nu\},$

where X^* and Y^* are $\mathcal{N}(X^*, Y^*; (0,0), (1,1, \rho_{X^*,Y^*}))$, and V and W are $\mathcal{N}(V, W; (0,0), (1,1, -\rho_{X^*,Y^*}))$. Equation (5.7) allows us to compute probabilities of the form shown in Equation (5.6) through the use of the National Bureau of Standards tables mentioned above.

We now apply Table T.11 in checking the "fit" of the bivariate normal distribution to data arising from the work of Pearson and Lee (1903) on the inheritance of family characteristics (see Chapter 7, Section 1).

Example 5.1. Pearson and Lee (1903) measured various physical dimensions of fathers, mothers, sons, and daughters in order to determine how inheritable such physical characteristics are. In Table 5.1 we provide data that they collected on the heights (in inches) of mothers and of their sons. The range of heights Y for mothers is broken up into 19 intervals of equal length (just as if we were going to construct a histogram of the distribution of the mothers' heights). Similarly, the range of heights X for the sons is broken up into 21 intervals of equal length. Table 5.1 can be read very much like a contingency table, except that the numbers appearing in the table are frequencies (not probabilities). Thus, the number 30.5 appearing in the row corresponding to a son's stature of from 68 to 69 inches and in the column corresponding to a mother's stature of from 62 to 63 inches is the observed frequency (in 1054 mother-son pairs) with which a mother of height from 62 to 63 inches had a son whose height at maturity was from

68 to 69 inches. [*Note:* Pearson and Lee followed a practice that is not recommended by most statisticians today. If an observation coincided with interval limits common to a number of cells in the contingency table, equal fractional frequencies adding to unity were allocated to each one of these cells. This explains the occurrence of fractional frequencies in the table.]

To attempt to fit a bivariate normal distribution to these data, we need to estimate the values of the parameters μ_X, μ_Y, σ_X^2, σ_Y^2, and $\rho_{X,Y}$. If we do this from the original data, we compute the estimates $\hat{\mu}_X$ and $\hat{\sigma}_X^2$ from the relative frequency distribution of the son's height X considered by itself (see Chapter 5), and the estimates $\hat{\mu}_Y$ and $\hat{\sigma}_Y^2$ from the relative frequency distribution of the mother's height Y considered by itself. Thus,

$$\hat{\mu}_X = t_1 \, \text{r.f.}\{X = t_1\} + t_2 \, \text{r.f.}\{X = t_2\} + \ldots,$$

$$\hat{\sigma}_X^2 = (t_1 - \hat{\mu}_X)^2 \, \text{r.f.}\{X = t_1\} + (t_2 - \hat{\mu}_X)^2 \, \text{r.f.}\{X = t_2\} + \ldots,$$

where $t_1, t_2, \ldots$ are the distinct observed values of the son's height in Pearson and Lee's (1903) ungrouped data. Similar formulas allow us to estimate $\hat{\mu}_Y$ and $\hat{\sigma}_Y^2$. The covariance σ_{XY} is estimated from the relative frequency distribution of the pairs (X,Y); thus,

$$\hat{\sigma}_{XY} = (t_1 - \hat{\mu}_X)(w_1 - \hat{\mu}_Y) \text{r.f.}\{X = t_1 \text{ and } Y = w_1\}$$
$$+ (t_1 - \hat{\mu}_X)(w_2 - \hat{\mu}_Y) \text{r.f.}\{X = t_1 \text{ and } Y = w_2\} + \cdots,$$

where $t_1, t_2, \ldots$ are the distinct observed values of X and $w_1, w_2, \ldots$ are the distinct observed values of Y. Finally, the correlation $\rho_{X,Y}$ is estimated by using the formula

$$\hat{\rho}_{X,Y} = \frac{\hat{\sigma}_{XY}}{\sqrt{\hat{\sigma}_X^2 \hat{\sigma}_Y^2}}.$$

[*Note:* The parameters μ_X, μ_Y, σ_X^2, σ_Y^2, and $\rho_{X,Y}$ can also be estimated from the grouped data in Table 5.1. To do this we assume that whenever the pair (X,Y) is counted in a particular cell of the table, the observed values of X and Y are the values of the respective midpoints of the intervals of X values and Y values which define that cell. For example, if an observed pair (X,Y) falls into the cell of Table 5.1 corresponding to a son's height X of between 68 and 69 inches and to a mother's height Y of between 62 and 63 inches, we assume that the actual observed value of (X,Y) is $(68.5, 62.5)$. Using this assumption, the parameters are estimated as indicated above. The estimated values for the parameters obtained from the ungrouped, original data will usually differ slightly from the estimated values for the parameters obtained from Table 5.1.]

Pearson and Lee used the original data to form estimates of the parameters μ_X, μ_Y, σ_X^2, σ_Y^2, and $\rho_{X,Y}$ of the bivariate normal distribution and

Table 5.1 Mother's stature and son's stature.

Mother's Stature.

Son's Stature	52–53	53–54	54–55	55–56	56–57	57–58	58–59	59–60	60–61	61–62	62–63	63–64	64–65	65–66	66–67	67–68	68–69	69–70	70–71	Totals
59–60	—	—	—	—	—	—	—	0.25	0.25	0.25	0.25	—	—	—	—	—	—	—	—	1
60–61	—	—	—	—	—	—	—	0.75	0.25	0.25	0.25	—	—	—	—	—	—	—	—	1.5
61–62	—	—	—	—	—	—	1	0.5	—	—	—	—	—	—	—	—	—	—	—	1.5
62–63	—	—	—	—	—	—	—	3.5	1.5	2	0.5	—	—	—	—	—	—	—	—	8
63–64	0.5	0.5	—	—	0.25	1.25	2.5	6.75	9	5	3.75	1.25	1.75	0.5	—	—	—	—	—	30
64–65	—	—	0.25	0.25	1	3.25	5.75	13	7.75	7	6.75	6.75	1.75	0.5	1	0.25	—	—	—	49
65–66	—	—	0.5	1	3.75	5.5	4.5	9	14.25	18.25	7.25	10	7.75	2	1.25	0.75	0.5	—	—	74
66–67	—	—	0.25	0.75	1.75	2.75	8.5	15	19.5	20	22.25	27	12.75	4	2.75	1.5	0.5	—	—	114.5
67–68	—	—	—	—	0.25	4.25	5	11.25	27.25	31	36	26.75	23.25	12.75	2.25	1.75	0.75	—	—	163
68–69	—	—	—	—	—	—	6.75	12.25	27.5	27.25	31.5	20.5	20	13.25	5	1	0.75	0.5	—	175.5
69–70	—	—	—	—	—	—	—	6	15.25	17	23.25	23.5	15.75	14.25	4.5	2.75	0.75	0.5	—	124
70–71	—	—	—	—	—	2	0.5	2.75	9.5	22	21.5	16	13.25	12.75	7.5	1	0.5	—	—	122
71–72	—	—	—	—	—	—	0.5	1	2.75	10.25	13.5	11	7.5	4	6	2.5	1	—	—	78
72–73	—	—	—	—	—	1	0.5	1	5	4.25	6	6.25	3.75	7	4.75	1.75	1.25	1.5	—	47.5
73–74	—	—	—	—	—	—	—	—	1.25	4	6.75	2.5	1	3	1.5	0.75	1.25	1.5	1	36
74–75	—	—	—	—	—	—	—	—	—	2.5	2.5	0.5	—	1	2	1	1.5	—	—	17
75–76	—	—	—	—	—	—	—	—	—	0.5	—	—	—	—	1.5	—	—	—	—	6.5
76–77	—	—	—	—	—	—	—	—	—	—	—	1	—	—	0.75	0.25	—	—	0.5	3.5
77–78	—	—	—	—	—	—	—	—	—	—	—	—	—	0.5	0.25	0.75	—	—	0.5	1.5
78–79	—	—	—	—	—	—	—	—	—	0.5	—	—	—	—	0.5	0.5	—	—	—	2
79–80	—	—	—	—	—	—	—	—	—	0.5	—	—	—	—	—	1	—	—	—	1
Totals	0.5	0.5	1	2	7	20	35.5	83	141	172.5	182	153	108.5	76.5	41.5	17.5	9	4	2	1057

obtained the following estimates:

$$\hat{\mu}_X = 68.65 \text{ inches}, \quad \hat{\mu}_Y = 62.48 \text{ inches},$$
$$\hat{\sigma}_X^2 = 7.344, \quad \hat{\sigma}_Y^2 = 5.712,$$
$$\hat{\rho}_{X,Y} = 0.494.$$

To check the "fit" to the observed data of the bivariate normal distribution having these values of the parameters, we can calculate theoretical frequencies for each of the cells in Table 5.1 and compare these to the observed frequencies. To obtain the theoretical frequency for the cell of Table 5.1 corresponding to a son's height X of between 68 and 69 inches and a mother's height Y of between 62 and 63 inches, we must find the theoretical probability

(5.8) $\quad P\{68 \leq X \leq 69 \text{ and } 62 \leq Y \leq 63\}$

$$= P\left\{\frac{68-68.65}{\sqrt{7.344}} \leq X^* \leq \frac{69-68.65}{\sqrt{7.344}} \text{ and } \frac{62-62.48}{\sqrt{5.712}} \leq Y^* \leq \frac{63-62.48}{\sqrt{5.712}}\right\},$$

where X^* and Y^* are $\mathcal{N}(X^*, Y^*; (0,0), (1,1,0.494))$. This theoretical probability (5.8) is then multiplied by the number 1057 of observed mother-son pairs to obtain the desired theoretical frequency.

We can obtain the theoretical probability

$$P\left\{\frac{68-68.65}{\sqrt{7.344}} \leq X^* \leq \frac{69-68.65}{\sqrt{7.344}} \text{ and } \frac{62-62.48}{\sqrt{5.712}} \leq Y^* \leq \frac{63-62.48}{\sqrt{5.712}}\right\}$$

(5.9) $\quad = P\{-0.24 \leq X^* \leq 0.13 \text{ and } -0.20 \leq Y^* \leq 0.22\}$

from Table T.11 if we approximate the estimated correlation coefficient $\hat{\rho}_{X,Y} = 0.494$ by 0.50. Thus, we compute the probability in (5.9) assuming that (X^*, Y^*) are $\mathcal{N}(X^*, Y^*; (0,0), (1,1,0.50))$.

Note from Figure 5.6 that we wish to compute the probability of the event A. From Table 5.1, we can find probabilities of the events C, $B \cup C$, $C \cup D$, and $A \cup B \cup C \cup D$, where A, B, C, and D are disjoint events (see Figure 5.6). Since

$$P(B \cup C) = P(B) + P(C),$$
$$P(C \cup D) = P(C) + P(D),$$
$$P(A \cup B \cup C \cup D) = P(A) + P(B) + P(C) + P(D),$$

it follows that

(5.10) $\quad P(A) = P(A \cup B \cup C \cup D) + P(C)$
$\quad\quad\quad - P(B \cup C) - P(C \cup D).$

5. THE BIVARIATE NORMAL DISTRIBUTION

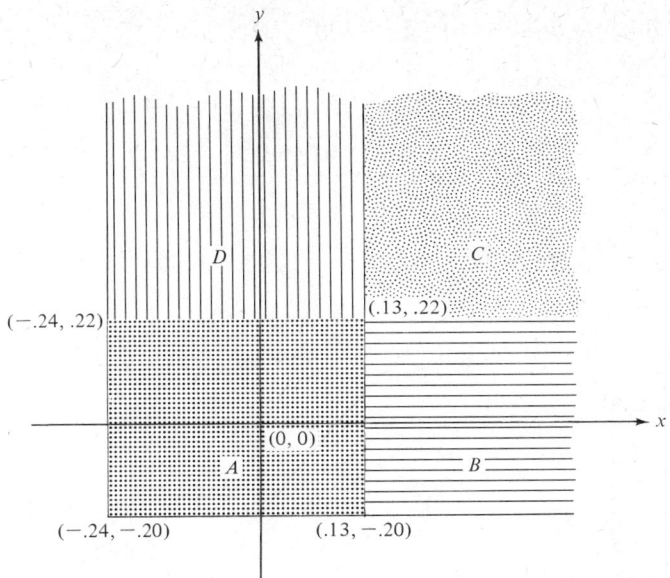

Figure 5.6 Events A, B, C and D required for the calculation of $P\{A\}$.

Thus, from Table T.11, we need to find the following probabilities:

(5.11) $P(A \cup B \cup C \cup D) = P\{X^* > -0.24 \text{ and } Y^* > -0.20\}$,

$P(B \cup C) = P\{X^* > 0.13 \text{ and } Y^* > -0.20\}$,

$P(C \cup D) = P\{X^* > -0.24 \text{ and } Y^* > 0.22\}$,

$P(C) = P\{X^* > 0.13 \text{ and } Y^* > 0.22\}$.

The probability $P(C)$ is directly computed from Table T.11. However, in order to use Table T.11 to compute $P(C)$, we must do a certain amount of interpolation. Table T.11 gives values

k \ h	0.1	0.2
0.2	0.275161	0.257709
0.3	0.255392	0.239718

for $P\{X^* > h, Y^* > k\}$. We may first linearly interpolate by rows to obtain

k \ h	0.1	0.2
0.2	0.275161	0.257709
0.22	0.271202	0.254111
0.3	0.255392	0.239718

and then linearly interpolate by columns to obtain

k \ h	0.1	0.13	0.2
0.22	0.271202	0.266075	0.254111

Hence,
$$P(C) = P\{X^* > 0.13 \text{ and } Y^* > 0.22\} = 0.266075 \cong 0.2661.$$

To obtain
$$P(B \cup C) = P\{X^* > 0.13 \text{ and } Y^* > -0.20\},$$

we first note that

$$P\{X^* > 0.13 \text{ and } Y^* > -0.20\}$$
$$= P\{\mathcal{N}(Z;0,1) > 0.13\} - P\{V > 0.13 \text{ and } W > 0.20\},$$

where (V,W) are $\mathcal{N}(V,W;(0,0),(1,1,-0.5))$. From Table T.7 we find that $P\{\mathcal{N}(Z;0,1) > 0.13\} = 0.4483$, while from Table T.11, we find that $P\{V > 0.13 \text{ and } W > 0.20\} = 0.1086$. Thus,

$$P(B \cup C) = 0.4483 - 0.1086 = 0.3397.$$

Similarly,

$$P(C \cup D) = P\{\mathcal{N}(Z;0,1) > 0.22\} - P\{V > 0.24 \text{ and } W > 0.22\}$$
$$= 0.4129 - 0.0901 = 0.3228.$$

Finally,

$$P(A \cup B \cup C \cup D) = P\{\mathcal{N}(Z;0,1) \leq 0.24\} + P\{\mathcal{N}(Z;0,1) \leq 0.20\}$$
$$+ P\{X^* > 0.24 \text{ and } Y^* > 0.20\} - 1$$
$$= 0.5948 + 0.5793 + 0.2541 - 1$$
$$= 0.4282.$$

Hence from Equation (5.10),

$$P(A) = 0.4282 + 0.2661 - 0.3397 - 0.3228$$
$$= 0.0318,$$

so that the theoretical frequency of the event $\{68 \leq X \leq 69 \text{ and } 62 \leq Y \leq 63\}$ is equal to $(1057)(0.0318) = 33.61$. The observed frequency of this event is, from Table 5.1, equal to 30.5. Thus, at least for this cell of Table 5.1, the observed and theoretical frequencies are reasonably close to one another.

6. SOME MULTIVARIATE GENERALIZATIONS

The procedure illustrated above is tedious if we wish to fit a complete distribution, but is not too troublesome if we need only a few cells. If a complete distribution is needed, then we may have to resort to a computer program. Since we have achieved our twin goals of demonstrating how to check "fit" of a bivariate normal distribution and how to use tables such as Table T.11 to compute bivariate normal probabilities, we leave this task to the interested reader.

*6. SOME MULTIVARIATE GENERALIZATIONS: TRIVARIATE DISCRETE DISTRIBUTIONS

In the previous sections, the notion of a bivariate distribution was introduced, the essential idea being that if two (random) variables are of interest in an experiment, the probability model should reflect how they jointly vary. When several random variables $X_1, X_2, X_3, \ldots, X_p$ are considered, we again require knowledge of their joint variation. This knowledge is presented in terms of a *multivariate distribution*, usually given either by a multivariate probability mass function or a multivariate probability density function.

The discussion in this section concentrates upon *trivariate discrete distributions*. From this discussion, it should be apparent how to extend the concepts involved to quadrivariate, 5-variate, 6-variate, or p-variate distributions. We have chosen to study trivariate distributions in preference to a general discussion of p-variate distributions only because of notational difficulties involved in the discussion of the general case. The extension of the concepts involved in bivariate continuous distributions to the concepts involved with trivariate and p-variate continuous distributions closely parallels the discussion we give here for discrete distributions. However, the graphical methods that we have used in Sections 4 and 5 to discuss bivariate continuous distributions would require three-dimensional paper if we were to use them to illustrate trivariate continuous distributions. Alternative methods for discussing continuous multivariate distributions either make use of mathematical techniques such as differential and integral calculus and matrix algebra, or else become smothered in awkward notation. Hence, we do not further discuss continuous multivariate distributions in this chapter.

As an example of a trivariate problem, consider selecting an individual at random from the population of American college seniors. We measure his college grade-point average X, his high school grade-point average Y, and his score Z on a prognostic test (say, an I.Q. test). We might be interested in predicting X given Y and Z, or in deciding whether such a prediction is possible.

For three discrete random variables X, Y, and Z, a *trivariate probability mass function* is a rule assigning to every triple (x,y,z) of possible values for the three variables the probability $p_{X,Y,Z}(x,y,z)$ of its occurrence (that is, the probability that $X = x$ and $Y = y$ and $Z = z$). It would be nice to graph such a probability mass function, but unfortunately this requires a graph in four dimensions. However, even though we cannot specifically visualize some of the concepts, we can understand through tables how the bivariate case extends to the trivariate case.

Example 6.1 (Wine-Tasting Experiment). Recall Example 2.1, in which two professional wine tasters are asked to rate a sample of wine using a rating scale which allows them to give ratings of 1, 2, 3, 4, or 5. If we have three wine tasters, we might have the following table (Table 6.1) of probabilities, where X, Y, and Z denote the ratings given by wine tasters I, II, and III, respectively.

Because we require four dimensions to graph the probability mass function for (X,Y,Z), we resort to alternative forms of display. For example, we could use Table 6.2 which is a *three-way contingency table*.

Table 6.1. Probability model for the joint ratings (x,y,z) given by three wine tasters.

Outcome (x,y,z)	$p_{X,Y,Z}(X,Y,Z)$	Outcome (x,y,z)	$p_{X,Y,Z}(X,Y,Z)$	Outcome (x,y,z)	$p_{X,Y,Z}(X,Y,Z)$
111	0.010	151	0.000	241	0.000
112	0.008	152	0.000	242	0.001
113	0.003	153	0.000	243	0.002
114	0.002	154	0.000	244	0.002
115	0.001	155	0.000	245	0.000
121	0.007	211	0.008	251	0.000
122	0.006	212	0.002	252	0.000
123	0.001	213	0.004	253	0.000
124	0.001	214	0.000	254	0.001
125	0.000	215	0.000	255	0.001
131	0.002	221	0.006	311	0.005
132	0.001	222	0.045	312	0.010
133	0.001	223	0.010	313	0.011
134	0.000	224	0.005	314	0.002
135	0.000	225	0.001	315	0.008
141	0.001	231	0.001	321	0.015
142	0.000	232	0.002	322	0.020
143	0.000	233	0.002	323	0.020
144	0.000	234	0.000	324	0.015
145	0.000	235	0.000	325	0.015

Table 6.1 (*continued*)

Outcome (x,y,z)	$p_{X,Y,Z}(X,Y,Z)$	Outcome (x,y,z)	$p_{X,Y,Z}(X,Y,Z)$	Outcome (x,y,z)	$p_{X,Y,Z}(X,Y,Z)$
331	0.030	421	0.000	511	0.000
332	0.060	422	0.001	512	0.000
333	0.100	423	0.006	513	0.000
334	0.080	424	0.005	514	0.001
335	0.050	425	0.004	515	0.001
341	0.010	431	0.000	521	0.000
342	0.008	432	0.004	522	0.000
343	0.060	433	0.005	523	0.002
344	0.060	434	0.008	524	0.002
345	0.030	435	0.004	525	0.001
351	0.008	441	0.002	531	0.000
352	0.007	442	0.002	532	0.000
353	0.010	443	0.008	533	0.002
354	0.015	444	0.028	534	0.004
355	0.015	445	0.012	535	0.004
411	0.003	451	0.001	541	0.000
412	0.002	452	0.002	542	0.000
413	0.004	453	0.004	543	0.003
414	0.003	454	0.012	544	0.009
415	0.000	455	0.010	545	0.009
				551	0.000
				552	0.002
				553	0.005
				554	0.009
				555	0.015

Table 6.2. A three-way contingency table giving the probability mass function for (X,Y,Z).

(a)

$z = 1$

x \ y	1	2	3	4	5
5	0.000	0.000	0.000	0.000	0.000
4	0.003	0.000	0.000	0.002	0.001
3	0.005	0.015	0.030	0.010	0.008
2	0.008	0.006	0.001	0.000	0.000
1	0.010	0.007	0.002	0.001	0.000

Table 6.2 (*continued*)

(b)
$z = 2$

x \ y	1	2	3	4	5
5	0.000	0.000	0.000	0.000	0.002
4	0.002	0.001	0.004	0.002	0.002
3	0.010	0.020	0.060	0.008	0.007
2	0.002	0.045	0.002	0.001	0.000
1	0.008	0.006	0.001	0.000	0.000

(c)
$z = 3$

x \ y	1	2	3	4	5
5	0.000	0.002	0.002	0.003	0.005
4	0.004	0.006	0.005	0.008	0.004
3	0.011	0.020	0.100	0.060	0.010
2	0.004	0.010	0.002	0.002	0.000
1	0.003	0.001	0.001	0.000	0.000

(d)
$z = 4$

x \ y	1	2	3	4	5
5	0.001	0.002	0.004	0.009	0.009
4	0.003	0.005	0.008	0.028	0.012
3	0.002	0.015	0.080	0.060	0.015
2	0.000	0.005	0.000	0.002	0.001
1	0.002	0.001	0.000	0.000	0.000

(e)
$z = 5$

x \ y	1	2	3	4	5
5	0.001	0.001	0.004	0.009	0.015
4	0.000	0.004	0.004	0.012	0.010
3	0.008	0.015	0.050	0.030	0.015
2	0.000	0.001	0.000	0.000	0.001
1	0.001	0.000	0.000	0.000	0.000

Certain notions already mentioned for bivariate distributions can be generalized to trivariate distributions. Although we discuss these notions in terms of trivariate probability mass functions, the same ideas carry over directly to probability mass functions for four or more variables.

6. SOME MULTIVARIATE GENERALIZATIONS

If $p_{X,Y,Z}(x,y,z)$ is a trivariate probability mass function, then the *marginal bivariate probability mass function* $p_{X,Y}(x,y)$ can be obtained by summing $p_{X,Y,Z}(x,y,z)$ over all possible values of z, holding the x and y values fixed. We can evaluate $p_{X,Z}(x,z)$ and $p_{Y,Z}(y,z)$ in a similar fashion. For our wine-tasting example we obtain the following marginal contingency tables (Table 6.3) corresponding respectively to the marginal bivariate distribution of X and Y, the marginal bivariate distribution of X and Z, and the marginal bivariate distribution of Y and Z.

Table 6.3. Marginal bivariate probability mass functions.

(a) $p_{X,Y}(x,y)$

x \ y	1	2	3	4	5
5	0.002	0.005	0.010	0.021	0.031
4	0.012	0.016	0.021	0.052	0.029
3	0.036	0.085	0.320	0.168	0.055
2	0.014	0.067	0.005	0.005	0.002
1	0.024	0.015	0.004	0.001	0.000

(b) $p_{X,Z}(x,z)$

x \ z	1	2	3	4	5
5	0.000	0.002	0.012	0.025	0.030
4	0.006	0.011	0.027	0.056	0.030
3	0.068	0.105	0.201	0.172	0.118
2	0.015	0.050	0.018	0.008	0.002
1	0.020	0.015	0.005	0.003	0.001

(c) $p_{Y,Z}(y,z)$

y \ z	1	2	3	4	5
5	0.009	0.011	0.019	0.037	0.041
4	0.013	0.011	0.073	0.099	0.051
3	0.033	0.067	0.110	0.092	0.058
2	0.028	0.072	0.039	0.028	0.021
1	0.026	0.022	0.022	0.008	0.010

The *univariate marginal probability mass function* $p_X(x)$ is obtained either by summing $p_{X,Y}(x,y)$ over all y values (for fixed x) or by summing $p_{X,Z}(x,z)$ over all z values (for fixed x). Similarly, we can find $p_Y(y)$ and $p_Z(z)$. These are given in Table 6.4.

From the joint probability mass function (or more easily from the marginal univariate and bivariate probability mass functions), we can

Table 6.4. Marginal univariate probability mass functions.

(a)

x	1	2	3	4	5
$p_X(x)$	0.044	0.093	0.664	0.130	0.069

(b)

y	1	2	3	4	5
$p_Y(y)$	0.088	0.188	0.360	0.247	0.117

(c)

z	1	2	3	4	5
$p_Z(z)$	0.109	0.183	0.263	0.264	0.181

compute the expected values μ_X, μ_Y, μ_Z, the variances $\sigma_X^2, \sigma_Y^2, \sigma_Z^2$, the covariances $\sigma_{XY}, \sigma_{XZ}, \sigma_{YZ}$, and the correlations $\rho_{X,Y}, \rho_{X,Z}, \rho_{Y,Z}$, of the random variables X, Y, and Z.

Conditional Distributions and Partial Correlation Coefficients

Although we might consider such measures as

$$E(X - \mu_X)(Y - \mu_Y)(Z - \mu_Z)$$

to measure the joint variation of X, Y, and Z, this is rarely done in practice. Joint variation is usually studied either by considering all of the different marginal bivariate distributions, or by considering all of the different *conditional bivariate distributions*. In the former case attention is paid to the correlation coefficients $\rho_{X,Y}, \rho_{X,Z}$, and $\rho_{Y,Z}$; in the latter case, we study the *partial correlation coefficients*.

As an example of a conditional bivariate distribution, we can define $p_{X,Y|Z=z}(x,y)$ to be

(6.1) $$p_{X,Y|Z=z}(x,y) = \frac{p_{X,Y,Z}(x,y,z)}{p_Z(z)},$$

for $p_Z(z) > 0$. We call $p_{X,Y|Z=z}(x,y)$ the *conditional bivariate probability mass function* given $Z = z$, or more briefly the conditional bivariate distribution of X and Y given $Z = z$.

Returning to our wine-tasting example, the conditional bivariate probability mass functions of X and Y given $Z = 1$ and of X and Y given $Z = 2$ appear in Table 6.5. Note that for each value of z the conditional bivariate mass function is actually a bivariate probability mass function. Consequently, we can consider the conditional correlation coefficient $\rho_{X,Y|Z=z}$ of X and Y given $Z = z$.

This conditional correlation coefficient $\rho_{X,Y|Z=z}$ is called the *partial*

6. SOME MULTIVARIATE GENERALIZATIONS

Table 6.5. Conditional bivariate distribution of X and Y, given $Z = 1$ and of X and Y, given $Z = 2$.

(a) $p_{X,Y|Z=1}(x,y)$

x \ y	1	2	3	4	5
5	0.000	0.000	0.000	0.000	0.000
4	0.028	0.000	0.000	0.018	0.009
3	0.046	0.138	0.275	0.092	0.074
2	0.073	0.055	0.009	0.000	0.000
1	0.092	0.064	0.018	0.009	0.000

(b) $p_{X,Y|Z=2}(x,y)$

x \ y	1	2	3	4	5
5	0.000	0.000	0.000	0.000	0.011
4	0.011	0.005	0.022	0.011	0.011
3	0.055	0.109	0.328	0.044	0.038
2	0.011	0.246	0.011	0.005	0.000
1	0.044	0.033	0.005	0.000	0.000

correlation coefficient of X and Y given $Z = z$; it reflects the covariation between X and Y for fixed given values of Z. Usually the value of $\rho_{X,Y|Z=z}$ depends on the value of z.

For example, in the wine-tasting experiment, from Table 6.5, we have

$$\sigma_{XY|Z=1} = 0.457, \quad \sigma^2_{X|Z=1} = 0.723, \quad \sigma^2_{Y|Z=1} = 1.462,$$

so that

$$\rho_{X,Y|Z=1} = 0.445;$$

similarly,

$$\sigma_{XY|Z=2} = 0.393, \quad \sigma^2_{X|Z=2} = 0.579, \quad \sigma^2_{Y|Z=2} = 0.970,$$

so that

$$\rho_{X,Y|Z=2} = 0.525.$$

We can also define the conditional univariate distribution of X given that $Y = y$ and $Z = z$:

(6.2) $$p_{X|Y=y,Z=z}(x) = \frac{p_{X,Y,Z}(x,y,z)}{p_{Y,Z}(y,z)},$$

for $p_{Y,Z}(y,z) > 0$.

For example, in the wine-tasting experiment, the conditional mass function of X given $Y = 1$ and $Z = 1$ and of X given $Y = 1$ and $Z = 2$ are given in Table 6.6. For each pair (y,z) the conditional probability mass function $p_{X|Y=y,Z=z}(x)$ is a true probability mass function. Thus, it is

Table 6.6. Conditional mass functions of X, given $Y = y$ and $Z = z$.

(a)

x	1	2	3	4	5
$p_{X \mid Y=1 \, and \, Z=1}(x)$	0.385	0.308	0.192	0.115	0.000

(b)

x	1	2	3	4	5
$p_{X \mid Y=1 \, and \, Z=2}(x)$	0.364	0.091	0.454	0.091	0.000

reasonable to consider the expected value $E[X \mid Y = y \text{ and } Z = z]$ taken with respect to this mass function. The function of y and z given by $E[X \mid Y = y \text{ and } Z = z]$ is called the *regression of X on Y and Z*.

In our example, we have

$$E[X \mid Y = 1 \text{ and } Z = 1] = 2.037,$$
$$E[X \mid Y = 1 \text{ and } Z = 2] = 2.272,$$

whereas

$$E[X \mid Y = 1] = 2.477.$$

We can also define other conditional distributions, for example,

$$p_{X,Z \mid Y=y}(x,z), \qquad p_{Y,Z \mid X=x}(y,z), \qquad p_{Z \mid X=x,\, Y=y}(z),$$

and so on, and the corresponding conditional expected values, variances, and correlations.

Statistical Independence

The discrete random variables X, Y, and Z are said to be *mutually statistically independent* if, for every value of x, y, and z,

(6.3) $$p_{X,Y,Z}(x,y,z) = p_X(x) p_Y(y) p_Z(z).$$

Note that in this case, the joint multivariate probability mass function is determined by the marginal univariate distributions.

Note that (6.3) defines *mutual* independence. We have already defined *pairwise* statistical independence in Section 2. X and Y are pairwise statistically independent if

$$p_{X,Y}(x,y) = p_X(x) p_Y(y);$$

similarly, X and Z are pairwise statistically independent if

$$P_{X,Z}(x,z) = p_X(x) p_Z(z),$$

6. SOME MULTIVARIATE GENERALIZATIONS

and Y and Z are pairwise statistically independent if

$$p_{Y,Z}(y,z) = p_Y(y)p_Z(z).$$

It is a fact that if X, Y, and Z are mutually independent, then X and Y, X and Z, and Y and Z are all pairwise independent. However, the converse is not true; that is, pairwise statistical independence of X and Y, X and Z, and Y and Z does not necessarily imply mutual statistical independence of X, Y, and Z (see Exercise 1).

It is often of interest to consider the *conditional statistical independence* of X and Y given $Z = z$. Following the definition used for pairwise independence, we say that X and Y are statistically independent given $Z = z$ if for every value of x and y,

(6.4) $$P_{X,Y|Z=z}(x,y) = p_{X|Z=z}(x) p_{Y|Z=z}(y).$$

There are similar definitions of the conditional statistical independence of X and Z given $Y = y$, or of Y and Z given $X = x$.

The Multinomial Distribution

One of the most important discrete multivariate probability mass functions is that belonging to the *multinomial distribution*. This distribution is a generalization of the binomial distribution.

The binomial distribution resulted from n statistically independent repetitions of a random experiment having two possible outcomes: success or failure. There are many situations, however, where the number of possible outcomes of such an experiment can be larger than two. For example, when a numerical scale is reduced to nominal categories, we may have categories such as "high," "medium," or "low." In questionnaires it is common to use the categories "strongly disagree," "mildly disagree," "disagree," "indifferent," "agree," "mildly agree," and "strongly agree."

The multinomial distribution arises from a random experiment in which there are r possible outcomes. If this experiment is independently repeated n times, we can let X_1 denote the number of occurrences of category 1, X_2 denote the number of occurrences of category 2, ... X_r denote the number of occurrences of category r. Since the categories are assumed to be distinct, we have

(6.5) $$X_1 + X_2 + \cdots + X_r = n.$$

Furthermore, suppose p_1 is the probability of observing category 1, p_2 is the probability of observing category 2, ..., p_r is the probability of observing category r. Thus,

$$p_1 + p_2 + \cdots + p_r = 1.$$

The joint probability mass function of $(X_1, X_2, \ldots, X_r)$ is called the multinomial distribution and is given by

(6.6) $$p_{X_1,X_2,\ldots,X_r}(x_1,x_2,\ldots,x_r) = \frac{n!}{x_1! \, x_2! \ldots x_r!} p_1^{x_1} p_2^{x_2} \ldots p_r^{x_r},$$

where the $x_1, x_2, \ldots, x_r$ are integers satisfying $x_1 + x_2 + \cdots + x_r = n$.
For $r = 3$,

(6.7) $$p_{X_1,X_2,X_3}(x_1,x_2,x_3) = \frac{n!}{x_1! \, x_2! \, x_3!} p_1^{x_1} p_2^{x_2} p_3^{x_3},$$

where $x_1 + x_2 + x_3 = n$.

As an example, suppose 100 persons chosen randomly with replacement from a given school district are interviewed and asked their opinion concerning a certain bond issue, the permissible responses being "favor," "indifferent," and "opposed." Let X_1 denote the number in favor of the bond issue, X_2 the number indifferent, and X_3 the number opposed. Then $X_1 + X_2 + X_3 = 100$. Suppose that experience with a similar bond issue in an adjacent school district recorded the following proportions: proportion in favor is 0.52; proportion indifferent is 0.04; proportion opposed is 0.44. If similar proportions hold in the given school district, then

$$p_{X_1,X_2,X_3}(40,15,45) = \frac{100!}{40! \, 15! \, 45!} (0.52)^{40}(0.04)^{15}(0.44)^{45} = 0.309 \times 10^{-6}.$$

Similarly,

$$p_{X_1,X_2,X_3}(52,4,44) = \frac{100!}{52! \, 4! \, 44!} (0.52)^{52}(0.04)^{4}(0.44)^{44} = 0.016.$$

To compute the above, no convenient tables are available. However, the use of logarithms in this situation is usually the most direct method for making the computations.

EXERCISES

A1. The following is a joint probability mass function for the three random variables X, Y, and Z.

(x,y,z)	(1,0,0)	(0,1,0)	(0,0,1)	(1,1,1)
Probability	$\frac{1}{4}$	$\frac{1}{4}$	$\frac{1}{4}$	$\frac{1}{4}$

(a) Show that the joint probability mass functions of the pairs (X,Y), (X,Z), and (Y,Z) are the same.
(b) Show that the marginal probability mass functions of X, of Y, and of Z are the same.

(c) Are the random variables X and Y statistically independent? Are the random variables Y and Z statistically independent? Are X and Z statistically independent?
(d) Find the correlation between X and Y. Does this answer conflict with the answer you gave in part (c)? Explain.
(e) Are the random variables X, Y, and Z mutually statistically independent?
(f) Find the conditional probability mass function of X, given that $Y = 1$, directly from the definition of a conditional probability mass function.
(g) Does your answer in part (c) provide a quicker way to obtain the distribution of X given that $Y = 1$? Explain.
(h) Find the conditional joint distribution function of X and Y given that $Z = 1$.

2. Suppose that you wanted to rescale X, Y, and Z in Exercise 1 so that the expected values are 0 and the variances are 1. That is, suppose that instead of considering X, Y, and Z, you wish to consider $X^* = a_1 X + b_1$, $Y^* = a_2 Y + b_2$, and $Z^* = a_3 Z + b_3$, where the numbers a_1, a_2, a_3, b_1, b_2, and b_3 are chosen in such a way that the expected values of X^*, Y^*, and Z^* are equal to 0 and their variances are equal to 1. Find a_1, a_2, a_3, b_1, b_2, and b_3, and determine the joint probability mass function of X^*, Y^*, and Z^*.

A3. A sample of 455 married couples were asked to (i) rate the happiness of their marriage, and (ii) indicate the extent to which they agree on ways of dealing with their in-laws, Burgess and Cottrell (1955). Let X be the rating of the happiness of the marriage of a given couple. The possible values of X are -2 (very unhappy), -1, 0, 1, 2 (very happy). Let Y be a rating of the extent of agreement between the marriage partners on ways of dealing with in-laws. The possible values of Y are 0 (always disagree), 1, 2, 3, 4, 5 (always agree). The joint distribution of X and Y is summarized in Table E.1.

Table E.1. Comparison of ratings of happiness in marriage with extent of agreement on ways of dealing with in-laws. Observed relative frequencies for 455 couples.

	x	Rating of Happiness				
	y	−2	−1	0	1	2
	5	0.015	0.011	0.046	0.077	0.251
	4	0.011	0.015	0.042	0.073	0.103
Extent of	3	0.011	0.028	0.033	0.029	0.042
Agreement	2	0.018	0.024	0.022	0.015	0.015
	1	0.009	0.022	0.009	0.002	0.002
	0	0.028	0.029	0.009	0.007	0.002

Suppose that the relative frequencies shown in Table E.1 are the actual probabilities for the joint distribution of X and Y.
(a) Find the marginal distributions of X and of Y.

(b) Find the conditional distributions of Y given that $X = x$, for $x = -2, -1, 0, 1,$ and 2.
(c) Find $E[Y|X = x]$ for $x = -2, -1, 0, 1,$ and 2.
(d) Find $\rho_{X,Y}$
(e) Do you think that the extent Y that married couples agree on ways of dealing with their in-laws and the rating X of marital happiness are statistically related? Explain.

4. In the context of Exercise 3, suppose that we regard the observed values of X and of Y as values of continuous variables which have been rounded off to the nearest integer. A colleague suggests that these continuous variables have a bivariate normal distribution.
   (a) Refer to Table E.1 of Exercise 3. Based on these data, do you think the bivariate normal distribution provides a good fit? Support your answer, pro or con, without carrying out any computations.
   (b) *Optional*: Support your answer to part (a) by carrying out whatever computations that you feel will establish your assertion.

^5. Suppose we perform 5 statistically independent and identical Bernoulli trials (see Chapter 6, Section 1), where on each trial there is a probability $p = 0.6$ of a "success." Let X be the number of "successes" in these 5 trials, and let Y be the length of the longest "run" in the 5 trials. A "run" is is a consecutive sequence of 1 or more identical outcomes; either a sequence of 1 or more consecutive "successes," or a sequence of 1 or more consecutive "failures." Thus, if the 5 Bernoulli trials resulted in 1 "success," then 2 "failures," 1 "success," and finally 1 "failure" (that is, SFFSF), then these trials would result in runs of length 1, 2, 1, and 1 (the 1 "success," the 2 "failures," the 1 "success," and the 1 "failure"). If the 5 Bernoulli trials resulted in the outcome FSSSF, then there are runs of length 1, 3, and 1, respectively. In the first example, $Y = 2$, while in the second example, $Y = 3$.
   (a) Find the joint distribution (joint probability mass function) of X and Y.
   (b) Find the marginal distribution, expected value, and variance of Y.
   (c) Find the conditional probability mass function of X given that $Y = 3$.
   (d) Find the conditional probability mass function of Y given that $X = 3$.
   (e) Calculate $P\{X > 3 | Y \leq 2\}$.
   (f) Calculate $P\{Y \geq 3 | X = 3\}$.

6. At a certain university, two university fellowships are available which will pay tuition and a stipend to the students who hold them. After an initial screening for ability and need, the choice must be made among 10 candidates of whom 5 are male whites, 2 are male blacks, 2 are female whites, and 1 is a female black. The university decides that the only fair method of selection is to choose the 2 fellowship winners by a simple random sample *without* replacement from among the 10 finalists. Let X denote the number of women who win a university fellowship, and let Y denote the number of blacks who win a fellowship.
   (a) Find the joint probability mass function of X and Y.

(b) Find the conditional probability mass function of X given that $Y = 1$.
(c) Find $E[X|Y = 1]$.
(d) Find $\rho_{X,Y}$.

A7. Two random variables X and Y have a joint probability distribution, and the covariance σ_{XY} between X and Y is equal to 3.0. The expected value of X is 0.5 and the variance of X is 4.0. The expected value of Y is 1.5 and the variance of Y is 9.0.
(a) What is the correlation $\rho_{X,Y}$ between X and Y?
(b) What is the covariance σ_{ZW} between $Z = \frac{1}{2}X - \frac{1}{4}$ and $W = \frac{1}{3}Y - \frac{1}{2}$?
(c) What is the correlation $\rho_{Z,W}$ between Z and W?
(d) Which correlation is greater: $\rho_{X,Y}$ or $\rho_{Z,W}$?
(e) Find σ_{ZY} and σ_{XW}. Which is the larger covariance?
(f) Find $\rho_{Z,Y}$ and $\rho_{X,W}$. Which is the larger correlation?

8. The following bivariate data might have been obtained in a study of the dispersal of population in an urban environment. In this study, the population density X and the distance Y from a downtown landmark (say, city hall) was observed for each of 30 tracts of land. (A tract is an area of land defined by law for zoning purposes.) The resulting data are listed in Table E.2.

Table E.2. Observations of the population density X and the distance Y for each of 30 tracts of land in an urban environment.

Tract Numbers	Population Density x (thousands of people per square mile)	Distance y from the Downtown Landmark (in thousands of feet)
1	27.5	5
2	21.0	5
3	14.5	5
4	20.8	10
5	20.0	10
6	16.6	10
7	11.0	10
8	19.1	15
9	17.2	15
10	10.7	15
11	7.8	15
12	15.2	20
13	11.6	20
14	9.8	20
15	13.2	25
16	10.6	25
17	7.9	25
18	6.5	25
19	9.0	30
20	6.7	30

Table E.2 (continued)

Tract Numbers	Population Density x (thousands of people per square mile)	Distance y from the Downtown Landmark (in thousands of feet)
21	6.2	30
22	3.9	30
23	7.9	35
24	5.3	35
25	1.9	35
26	5.3	40
27	3.4	40
28	3.0	40
29	1.7	45
30	1.0	45

(a) Construct a sample joint relative frequency distribution for these data by filling in the entries of Table E.3.

Table E.3. Relative frequencies of the events. $\{x-3.75 < X < x+3.75 \text{ and } y-6.0 \leq Y < y+6.0\}$

Midpoint of: y-interval \ x-interval	3.75	11.25	18.75	26.25
6.0		$\frac{2}{30}$		
18.0				
30.0				
42.0				

Note that one of the entries of this table has already been filled in.

(b) Find the (sample) correlation $\hat{\rho}_{X,Y}$ of X and Y. (To do this, assume that the relative frequencies in the table of relative frequencies you constructed in part (a) are concentrated at the midpoints of the intervals.)
(c) Construct a relative frequency histogram for the conditional distribution of Y given that $7.5 \leq X < 15.0$.
(d) Construct a relative frequency histogram for the conditional distribution of Y given that $15.0 \leq X < 22.5$.
(e) Using the given data, how would you estimate the conditional expected value of Y given that $7.5 \leq X < 15.0$? Describe your method and give the estimate.

A9. A park guide at Yellowstone National Park asserts that the longer the geyser "Old Faithful" is in eruption, the longer will be the wait until the basin that feeds the geyser refills, and thus the longer will be the wait until the next eruption. The National Park Service records the length of time X of each eruption, and the time Y that is required until the next eruption. The following data were taken from the records of the National Park Service. Here, X is measured to the nearest $\frac{1}{4}$ minute, and Y is measured to the nearest minute. There are 40 pairs of observations.

x	y	x	y	x	y	x	y
1.75	48	2.00	53	2.00	52	2.25	51
4.00	79	4.25	70	4.00	71	4.00	76
1.75	54	3.00	73	3.50	68	4.00	63
3.50	72	4.25	74	3.75	69	3.75	71
2.00	45	2.00	50	4.00	68	3.75	65
3.50	74	4.50	79	4.00	77	4.00	70
2.00	61	2.00	51	1.75	47	3.50	65
4.50	70	4.50	72	4.25	77	4.00	77
2.00	51	3.75	72	2.50	56	4.00	63
3.75	83	4.00	70	4.00	71	3.50	73

Even though both X and Y are continuous random variables, treat them as if they are discrete random variables. Under this assumption:
(a) Determine the sample joint probability mass function for X and Y from the above data.
  Assume that the distribution obtained in part (a) is, in fact, the true joint distribution of X and Y.
(b) Find the marginal probability mass functions of X and of Y.
(c) Find the correlation $\rho_{X,Y}$ between X and Y. Do you agree with the park guide that the length of time X that an eruption lasts helps us to predict the duration of time Y until the next eruption? Explain your answer.
(d) Suppose we observe an eruption that lasts 4 minutes. What is the conditional probability mass function $p_{Y|X=4}(\tau)$ of the duration of time until the next eruption?
(e) What value would you predict for Y if you are told that $X = 4$? Explain your answer. Is this value of Y different from the value you would predict for Y if you are told that $X = 2$? Explain.

*A10. The relationship, in German words, between the number X of syllables and the number Y of phonemes (letters) was studied for 20,453 headwords in Vietor-Meyer's *Deutsches Aussprachewörterbuch* [see Herdan (1966), p. 301]. The data are given in Table E.4. The estimates of the expected values of X and of Y obtained from these data are 2.780 and 7.296, respectively. The estimated variances of X and of Y are 1.2122 and 6.6793, respectively, while the estimated covariance between X and Y is 2.4040.
Even though X and Y are both discrete random variables, it is convenient, for distributional approximations, to regard each of these variables as if they are continuous random variables which have been rounded off to the nearest integer. Table E.4 then provides information about the joint distribution of X and Y similar to the information that Table 5.1 (Example 5.1) provided about the joint distribution of the heights of mothers and their sons. Keeping

Table E.4. Number of syllables and number of letters in 20,453 head-words in Vietor-Meyer's *Deutsches Aussprachewörterbuch*.

Number of Syllables X

	1	2	3	4	5	6	7	8	9
y=22	—	—	—	—	—	—	—	—	3
21	—	—	—	—	—	—	2	1	3
20	—	—	—	—	—	2	3	3	0
19	—	—	—	—	—	2	3	0	1
18	—	—	—	—	—	5	2	2	1
17	—	—	—	—	3	10	11	1	—
16	—	—	—	4	14	20	8	4	—
15	—	—	—	6	44	29	8	—	—
14	—	—	1	43	81	50	4	—	—
13	—	—	4	99	154	52	1	—	—
12	—	—	28	268	200	31	—	—	—
11	—	—	157	572	201	10	—	—	—
10	—	25	566	896	167	3	—	—	—
9	—	91	1125	883	43	—	—	—	—
8	—	430	1893	643	12	—	—	—	—
7	2	1394	1840	204	1	—	—	—	—
6	69	1603	1105	22	—	—	—	—	—
5	444	1492	253	—	—	—	—	—	—
4	962	1256	7	—	—	—	—	—	—
3	645	101	—	—	—	—	—	—	—
2	114	4	—	—	—	—	—	—	—
1	9	—	—	—	—	—	—	—	—

Number of Letters Y

this background in mind, do the following:
(a) Transform (rescale) the variables X and Y to new variables X^* and Y^* so that the expected values of these new variables are both equal to 0 and the variances of these new variables are both equal to 1.
(b) Suppose that we use a standard bivariate normal distribution to describe the joint distribution of X^* and Y^*. What value of ρ_{X^*,Y^*} would we use? Which probability contour in Figure 5.4 would most closely resemble the probability contour of the fitted joint distribution of X^* and Y^*?
(c) Using the bivariate normal approximation to the joint distribution of X and Y, calculate $P_{X,Y}\{X > 5.5 \text{ and } Y > 16.5\}$.
(d) Using the bivariate normal approximation to the joint distribution of X and Y, calculate $P_{X,Y}\{X \leq 3.5 \text{ and } Y \leq 5.5\}$.
(e) Compare the probabilities which you calculated in parts (c) and (d) with the observed relative frequencies of the events $\{X > 5.5 \text{ and } Y > 16.5\}$ and $\{X \leq 3.5 \text{ and } Y \leq 5.5\}$. Based on your comparisons, do you think that the bivariate normal distribution provides a helpful

approximation to the true joint distribution of X and Y? Explain your answer.

11. Continuing with the experiment described in Exercise 10, and under the (possibly false) assumption that X and Y have a bivariate normal distribution, do the following:
    (a) Find the marginal distributions of X and of Y.
    (b) Find the conditional distribution of Y given that $X = 5$.
    (c) Find $E[Y|X=5]$.
    (d) Find the variance of the conditional distribution of Y given that $X = 5$.
    If you are told that a German word has 5 syllables, and are asked to guess the number of letters it has, what number Y of letters would you guess? What number of letters would you guess if you were told that $X = 3$? Explain your answers.

12. The College Board English Composition Test score X and the Scholastic Aptitude Test (SAT) Verbal score Y have (approximately) a bivariate normal distribution with parameters $\mu_X = \mu_Y = 650$, $\sigma_X = 81$, $\sigma_Y = 64$, and $\rho_{X,Y} = 0.6$.
    (a) Find $P_{X,Y}\{X > 750 \text{ and } Y > 750\}$.
    (b) Find the (0.90)th quantile $Q_{0.90}(X)$ of X.
    (c) Find the (0.90)th quantile $Q_{0.90}(Y)$ of Y.
    (d) Find $P\{X > Q_{0.90}(X) \text{ and } Y > Q_{0.90}(Y)\}$. Does the result surprise you? Explain.
    (e) Compare the answer in (d) with the answer you would get if X and Y are statistically independent.
    (f) If you know a student's SAT Verbal score, $Y = y$, what value would you predict (or guess) for his English Composition Test score X? Would your prediction be an improvement over the prediction of the student's English Composition Test score which you would give if you were *not* told his SAT Verbal score Y? Explain your answer.

^A 13. A certain industrial process produces two interlocking steel pipes at a time. The lengths X and Y of the pipes have a bivariate normal distribution with parameters $\mu_X = \mu_Y = 3.25$ inches, $\sigma_X = \sigma_Y = 0.05$ inch, and $\rho_{X,Y} = \rho$. After the pipes are produced, they are fitted together. Assuming that the pipes are joined without their combined length being lessened, find the probability that the combined length of the two joined pipes is less than 6.60 inches when:
    (a) $\rho = 0.0$,
    (b) $\rho = 0.5$,
    (c) $\rho = -0.5$,
    (d) $\rho = 1.0$,
    (e) $\rho = -1.0$.

*14. A "random walk" is the name given to a certain sequence of movements (steps) of a particle, where at each step the particle either moves one unit forward with probability p, or one unit backward with probability $1 - p$. The results of the various steps of the particle are mutually statistically independent. Let X_k represent the result of the kth step of the particle, $X_k = 1$ if the particle moves one step forward, and $X_k = -1$ if the particle moves one step backward. Thus, $P_{X_k}\{X_k = 1\} = p$ and $P_{X_k}\{X_k = -1\} = 1 - p$, and the variables $X_1, X_2, X_3, \ldots$, are mutually statistically independent.

(a) Find the joint probability mass function of X_1, X_2, and X_3.
(b) Find the joint probability mass function of X_1, X_2, X_3, and X_4.
(c) Of great interest in a random walk is the position (with respect to the starting point) of the particle after it has taken k steps. Let U_k be the position of the particle after k steps so that

$$U_k = X_1 + X_2 + \cdots + X_k.$$

Find the distribution of U_3. Find the distribution of U_4. [*Hint:* Use the joint distributions obtained in parts (a) and (b).]

(d) Find the joint distribution of U_1 and U_2.
(e) Find the joint distribution of U_1, U_2, and U_3.
(f) Find the joint distribution of U_1, U_2, U_3, and U_4.

[*Note:* If possible, try to answer parts (d), (e), and (f) for all values of p. However, if you find this too difficult a task, try to answer parts (d), (e), and (f) for the case when $p = \frac{1}{2}$ and $p = \frac{3}{4}$.]

15. The discrete random variables X, Y, and Z have the following joint probability mass function:

(x,y,z)	(1,1,1)	(1,1,0)	(1,0,1)	(0,1,1)
$P_{X,Y,Z}(x,y,z)$	$\frac{1}{4}$	$\frac{1}{4}$	$\frac{1}{4}$	$\frac{1}{4}$

Find $P_{X,Y,Z}\{X \geq Y\}$, $P_{X,Y,Z}\{Y \geq Z\}$, and $P_{X,Y,Z}\{Z \geq X\}$. Do the results which you obtain surprise you? Why, or why not?

10
WEIGHTED SUMS OF RANDOM VARIABLES

What I tell you three times is true.
Lewis Carroll (Hunting of the Snark)

As far as the propositions of mathematics refer to reality, they are not certain; and as far as they are certain, they do not refer to reality.
Albert Einstein ("Geometry and Experience," lecture before the Prussian Academy of Sciences, January 27, 1921)

The random variables that we use to summarize the results of a random experiment are not always the obtained measurements themselves. Frequently, functions (or transformations) of these measurements are used. For example, if an angle θ is measured, the tangent, $\tan \theta$, of that angle may be reported. Or, after observing n independent and identically distributed observations $X_1, X_2, \ldots, X_n$ upon a random variable X, we may summarize our findings by an estimate $\hat{\mu}_X = (X_1 + X_2 + \cdots + X_n)/n$ of the expected value μ_X of the distribution of X (see Chapter 5). Again, the n independent and identically distributed observations $X_1, X_2, \ldots, X_n$ upon the random variable X may be used to estimate one or more parameters of some hypothesized type of distribution for X (see Chapters 6, 7, and 8), and these estimates (which, since they depend for their values on the values of $X_1, X_2, \ldots, X_n$, are functions of $X_1, X_2, \ldots, X_n$) are often reported instead of the observations themselves. In the present chapter, we concentrate on some simple functions of random variables, namely, location and scale transformations of a random variable, and weighted sums of random variables. These transformations are used so frequently in practice that they deserve special attention. Further, in contrast to other functions of random variables, methods of obtaining distributions and properties of such random variables can be discussed in terms of elementary mathematics.

1. STANDARDIZATION OF RANDOM VARIABLES

One of the most commonly used transformations of a random variable X is the transformation which changes the *scale* in which X is measured and which redefines the origin of this scale (a change of *location*). This transformation changes X into a new random variable $Y = aX + b$, where a is a positive number and b is any number. The factor a converts the units of measurement (for example, from yards to inches), whereas the quantity b changes the point of origin of the scale (for example, instead of writing 10.1, 10.2, 10.5, 10.6, we may write 1, 2, 5, 6; here, $a = 10$, $b = -100$).

For example, suppose X is a measurement of temperature given in degrees Fahrenheit. We might decide instead to give the measurement in degrees Centigrade. In such a case $Y = (5/9)X - 160/9$ gives the random temperature X in terms of the Centigrade scale.

When X is a discrete random variable with probability mass function $p_X(\tau)$, then we have already verified (for example, in Chapter 5, Section 2) that the probability mass function $p_Y(y)$ of $Y = aX + b$ is given by the formula

$$(1.1) \qquad p_Y(y) = p_X\left(\frac{y-b}{a}\right).$$

Equation (1.1) holds because when $X = (Y-b)/a$, the events $\{Y = y\}$ and $\{X = (y-b)/a\}$ are the same event.

When X is a continuous random variable, it can be shown (by means of the differential calculus) that the probability density function $f_Y(y)$ of the continuous random variable $Y = aX + b$ is related to the probability density function $f_X(x)$ of X by the formula

$$(1.2) \qquad f_Y(y) = \frac{1}{|a|} f_X\left(\frac{y-b}{a}\right),$$

where $|a|$ is the absolute (unsigned) value of a.

An important application of Equation (1.2) is a result which we used in Chapter 7, Section 2 to simplify the tabulating of probabilities for the normal distribution. There, we needed to show that if X is normally distributed with expected value μ_X and variance σ_X^2, then $Z = (1/\sigma_X)X - (\mu_X/\sigma_X)$ has a (standard) normal distribution with expected value 0 and variance 1. The probability density function of X is

$$f_X(x) = \frac{1}{\sqrt{2\pi}\sigma_X} e^{-(x-\mu_X)^2/2\sigma_X^2}$$

for all values of x. If we let $a = 1/\sigma_X > 0$ and $b = -\mu_X/\sigma_X$, then from Equation (1.2) we can conclude that $Z = aX + b = (X - \mu_X)/\sigma_X$ has the probability density function

$$(1.3) \qquad f_Z(z) = \frac{1}{\sqrt{2\pi}} e^{-z^2/2}, \qquad -\infty \leqslant z \leqslant \infty,$$

which we know (see Chapter 7, Section 1) is the probability density func-

1. STANDARDIZATION OF RANDOM VARIABLES

tion of a normal distribution with expected value 0 and variance 1. Thus, $Z = (X - \mu_X)/\sigma_X$ has a standard normal distribution.

For both discrete and continuous random variables X, we can obtain the cumulative distribution function $F_Y(y)$ of $Y = aX + b$ from the cumulative distribution function $F_X(x)$ of X by using the formula

(1.4) $$F_Y(y) = F_X\left(\frac{y-b}{a}\right).$$

In Chapter 5 we showed that the expected value μ_Y and variance σ_Y^2 of $y = aX + b$ are related to the expected value μ_X and variance σ_X^2 of X by the equations

(1.5) $$\mu_Y = a\mu_X + b,$$
$$\sigma_Y^2 = a^2\sigma_X^2.$$

Formulas (1.1), (1.2), and (1.4) allow us to find the distribution of the rescaled quantity $Y = aX + b$ from the distribution of the original observed quantity X. Formula (1.5) allows us to obtain the new measures of location and dispersion for the distribution of Y from the original measures of location and dispersion for the distribution of X. In Section 3 of Chapter 9, we showed how the computations are simplified by the technique of rescaling.

The particular transformation of location and scale, $Y = aX + b$, which transforms the distribution of X into a new distribution which has expected value (location) equal to 0 and variance equal to 1 is of special importance in between-experimental comparisons. This transformation is the one for which $a = 1/\sigma_X$ and $b = -(\mu_X/\sigma_X)$; that is,

(1.6) $$Y = \frac{1}{\sigma_X}X + \left(-\frac{\mu_X}{\sigma_X}\right) = \frac{X - \mu_X}{\sigma_X}.$$

Such a transformation is called a *standardization* of the random variable X. We have already mentioned the standardization of a normally distributed random variable. There, the standardization permitted probability calculations for every normal distribution to be converted to probability calculations for a standard normal distribution. However, even when X is not normally distributed, such a transformation can be useful *because it converts all measurements to the same scale of measurement*, and thus permits comparisons between random observations uninfluenced by the scales of measurements that have been used.

Example 1.1. There are many different kinds of mathematical aptitude tests. A certain preparatory school is considering applications from two students. One student, student A, has taken a mathematical aptitude test on which, nationally, the scores X are known to have expected value $\mu_X =$

50 and variance $\sigma_X^2 = 25$. The other student, B, has taken a different and much harder mathematical aptitude test on which, nationally, the scores Y are known to have expected value $\mu_Y = 10$ and variance $\sigma_Y^2 = 36$. Student A reports a score of $X = 56$ on the mathematical aptitude test he has taken, while student B, on the mathematical aptitude test he took, reports a score of $Y = 22$. If we compare the scores of student A and student B directly (that is, compare $X = 56$ and $Y = 22$), without taking account of the different levels of difficulty of the two tests, we would conclude that student A has higher mathematical aptitude than student B. On the other hand, student A's standardized score is

$$Z_A = \frac{56 - \mu_X}{\sigma_X} = \frac{56 - 50}{\sqrt{25}} = 1.2,$$

while student B's standardized score is

$$Z_B = \frac{22 - \mu_Y}{\sigma_Y} = \frac{22 - 10}{\sqrt{36}} = 2.0,$$

so that student B's standardized score is higher than student A's standardized score. Student A's score of $Z_A = 1.2$ means that his score on the mathematical aptitude test which he took was 1.2 standard deviations above the "expected" value on that test. If scores X on the mathematical aptitude test taken by student A are normally distributed, the fact that $Z_A = 1.2$ means that student A's score is at the (0.8849)th quantile of the national distribution of scores on that test. If scores Y on the mathematical aptitude test taken by student B are normally distributed, the fact that $Z_B = 2.0$ means that student B's score is at the (0.9773)th quantile of the national distribution of scores on *that* test. Even if neither X nor Y have normal distributions, the probability inequalities given in Chapter 5, Section 6 argue strongly that student A and student B should be compared on the basis of their standardized scores Z_A and Z_B, since these standardized scores show how well student A and student B have done relative to national norms. For this reason, the preparatory school decides that student B has a higher mathematical aptitude than student A.

In Section 4 of this chapter, we will see that the Central Limit Theorem (see Chapter 7, Section 4) is really an assertion concerning the standardization Z_n of the average or of the sum of a large number n of statistically independent observations on a random variable X. Because the Central Limit Theorem permits many otherwise difficult probability calculations to be made conveniently (up to a reasonably close approximation) even when the number of repetitions n of X is small, investigators often calculate such a standardized average or sum as a routine part of the analysis of a given random experiment.

2. WEIGHTED SUMS OF RANDOM VARIABLES

One of the problems that must often be met in scientific practice is that of finding measures which summarize the findings of an experiment. If we have r random numerical observations $X_1, X_2, ..., X_r$, one summary measure which is frequently used is a weighted sum

$$Y = a_1 X_1 + a_2 Y_2 + \cdots + a_r X_r$$

of these observations. Although usually the weights $a_1, a_2, ..., a_r$ are positive numbers, this is not always the case.

Example 2.1. Suppose $X_1, X_2, ..., X_r$ are r independent repeated measurements on the same random variable X. Then $\hat{\mu}_X = (X_1 + X_2 + \cdots + X_r)/r$, the average of the observations, is an index of location for the data values and frequently serves as a good estimate of the expected value μ_X of X (see Chapter 5).

Example 2.2. Suppose an experiment is performed in two parts. In the first part, drug A is given to a randomly drawn individual and his basal metabolism X_1 is recorded. In the second part of the experiment (1 month later), drug B is given to the same individual and his basal metabolism X_2 is recorded. The random variable $Y = X_1 - X_2$ (which is in the form of a weighted sum with $a_1 = 1$, $a_2 = -1$, and $r = 2$) is of interest in comparing the effects of the two drugs upon the basal metabolism of human beings.

Example 2.3. Suppose we wish to determine a final grade for a course and are given 3 weekly examination grades X_1, X_2, X_3, a midterm grade X_4, and a final examination grade X_5. In order to arrive at a summary score we might weight the weekly examinations with a weight of 1, the midterm examinations by a weight of 3, and the final examination by a weight of 5, thus yielding the summary score

$$X_1 + X_2 + X_3 + 3X_4 + 5X_5.$$

If each grade X_1, X_2, X_3, X_4, X_5 is a number between 0 and 100, to make our summary score have a range of possible values between 0 and 100, we can divide the above score by the total weight, $1 + 1 + 1 + 3 + 5 = 11$, thus obtaining

$$\frac{1}{11} X_1 + \frac{1}{11} X_2 + \frac{1}{11} X_3 + \frac{3}{11} X_4 + \frac{5}{11} X_5.$$

Both the quantity $X_1 + X_2 + X_3 + 3X_4 + 5X_5$ and the quantity $(1/11)X_1 +$

$(1/11)X_2+(1/11)X_3+(3/11)X_4+(5/11)X_5$ give equivalent rankings of the students.

Because the weighted sum $Y = a_1X_1 + a_2X_2 + \cdots + a_rX_r$ is a sum of random variables, it is itself a random variable. In theory, we can always obtain the distribution of Y from a knowledge of the joint distribution of $X_1, X_2, \ldots, X_r$; in practice, however, such a task often is very complicated. (Of course, if the random variables $X_1, X_2, \ldots, X_r$ are statistically independent, the task is simplified.) The following are two examples of how the distribution of a weighted sum is found in simple discrete cases.

Example 2.4. Assume that the joint distribution of the random variables X_1, X_2, and X_3 is given by the following joint probability mass function:

(x_1,x_2,x_3)	(0,0,0)	(1,0,0)	(1,1,0)	(1,0,1)	(0,1,0)	(0,1,1)	(0,0,1)	(1,1,1)
$p_{X_1,X_2,X_3}(x_1,x_2,x_3)$	$\frac{1}{8}$	$\frac{1}{4}$	$\frac{1}{16}$	$\frac{1}{8}$	$\frac{1}{16}$	$\frac{1}{16}$	$\frac{1}{16}$	$\frac{1}{4}$

Let $Y = \frac{1}{3}(X_1 + X_2 + X_3)$ be the average of X_1, X_2, and X_3. Then Y can take on any one of the values $0, \frac{1}{3}, \frac{2}{3}, 1$. The probability that $Y = 0$ is the probability of $(0,0,0)$; that is, $P_Y\{Y=0\} = p_{X_1,X_2,X_3}(0,0,0) = \frac{1}{8}$. Similarly,

$$P_Y\left\{Y=\frac{1}{3}\right\} = p_{X_1,X_2,X_3}(1,0,0) + p_{X_1,X_2,X_3}(0,1,0) + p_{X_1,X_2,X_3}(0,0,1)$$

$$= \frac{1}{4} + \frac{1}{16} + \frac{1}{16} = \frac{3}{8},$$

$$P_Y\left\{Y=\frac{2}{3}\right\} = p_{X_1,X_2,X_3}(1,1,0) + p_{X_1,X_2,X_3}(1,0,1) + p_{X_1,X_2,X_3}(0,1,1)$$

$$= \frac{1}{16} + \frac{1}{8} + \frac{1}{16} = \frac{2}{8},$$

$$P_Y\{Y=1\} = p_{X_1,X_2,X_3}(1,1,1) = \frac{2}{8}.$$

The distribution of Y can thus be given by means of the following table of probabilities:

y	0	$\frac{1}{3}$	$\frac{2}{3}$	1
$P_Y\{Y=y\}$	$\frac{1}{8}$	$\frac{3}{8}$	$\frac{2}{8}$	$\frac{2}{8}$

Example 2.5. Assume that X_1 and X_2 are statistically independent, each

having the distribution whose probability mass function is

x	-1	0	1
$p_X(x)$	$\frac{1}{4}$	$\frac{1}{2}$	$\frac{1}{4}$

To find the distribution of $Y = 2X_1 + X_2$, we proceed as follows. We first note that the possible values of Y are $-3, -2, -1, 0, 1, 2, 3$. Next, we find that

$$P_Y\{Y = -3\} = P_{X_1, X_2}\{X_1 = -1 \text{ and } X_2 = -1\}$$
$$= P_{X_1}\{X_1 = -1\} P_{X_2}\{X_2 = -1\}$$
$$= \left(\frac{1}{4}\right)\left(\frac{1}{4}\right) = \frac{1}{16},$$

$$P_Y\{Y = -2\} = P_{X_1, X_2}\{X_1 = -1 \text{ and } X_2 = 0\} = \left(\frac{1}{4}\right)\left(\frac{1}{2}\right) = \frac{1}{8},$$

$$P_Y\{Y = -1\} = P_{X_1, X_2}[\{X_1 = -1 \text{ and } X_2 = 1\} \cup \{X_1 = 0 \text{ and } X_2 = -1\}]$$
$$= P_{X_1, X_2}\{X_1 = -1 \text{ and } X_2 = 1\} + P_{X_1, X_2}\{X_1 = 0 \text{ and } X_2 = -1\}$$
$$= \left(\frac{1}{4}\right)\left(\frac{1}{4}\right) + \left(\frac{1}{2}\right)\left(\frac{1}{4}\right) = \frac{3}{16},$$

with similar calculations yielding

$$P_Y\{Y = 0\} = \tfrac{4}{16}, \qquad P_Y\{Y = 1\} = \tfrac{3}{16}, \qquad P_Y\{Y = 2\} = \tfrac{2}{16},$$
$$P_Y\{Y = 3\} = \tfrac{1}{16}.$$

Thus, the probability mass function $p_Y(y)$ of Y is given by the table

y	-3	-2	-1	0	1	2	3
$p_Y(y)$	$\frac{1}{16}$	$\frac{2}{16}$	$\frac{3}{16}$	$\frac{4}{16}$	$\frac{3}{16}$	$\frac{2}{16}$	$\frac{1}{16}$

As the probability models for the joint distribution of $X_1, X_2, X_3, \ldots$ become more complicated than the models of Examples 2.4 and 2.5, a direct enumeration is no longer a feasible way of obtaining the distributions of weighted sums. (However, it should be borne in mind that with sufficient energy and time, a direct enumeration would yield the result.) In such cases, algebraic techniques or analytic techniques (such as differential and integral calculus) are used. With the general availability of computers, we may soon use the computer to help with the enumeration. One technique which is extremely useful, and which may be carried out with rather elementary mathematics, is that of moment-generating functions. Because a discussion of moment-generating functions at this point would serve to distract us from the main topic of our discussion, we postpone consideration of this topic to an Appendix to this chapter.

Some Special Cases

The following are special cases where the distributions of certain weighted sums are known:

(1) *The Normal Distribution.* If $X_1, X_2, \ldots, X_r$ are mutually statistically independent, and each random variable X_i has a normal distribution with expected value μ_{X_i} and variance $\sigma^2_{X_i}$ ($i = 1, 2, 3, \ldots, r$), then $Y = a_1 X_1 + a_2 X_2 + \cdots + a_r X_r$ also has a normal distribution with expected value

$$\mu_Y = a_1 \mu_{X_1} + a_2 \mu_{X_2} + \cdots + a_r \mu_{X_r},$$

and variance

$$\sigma_Y^2 = a_1^2 \sigma^2_{X_1} + a_2^2 \sigma^2_{X_2} + \cdots + a_r^2 \sigma^2_{X_r}.$$

In the special case when $X_1, X_2, \ldots, X_r$ all have the *same* normal distribution with expected value μ and variance σ^2, then the sample average

$$\hat{\mu}_X = \frac{(X_1 + X_2 + \cdots + X_r)}{r}$$

has a normal distribution with expected value μ and variance σ^2/r.

(2) *The Binomial Distribution.* If $X_1, X_2, \ldots, X_r$ are mutually statistically independent, and if each X_i has a binomial distribution with parameters n_i and p, where p is the probability of "success," then

$$Y = X_1 + X_2 + \cdots + X_r$$

also has a binomial distribution with parameters $n = n_1 + n_2 + \cdots + n_r$ and p.

(3) *The Poisson Distribution.* If $X_1, X_2, \ldots, X_r$ are mutually statistically independent, and if each X_i has a Poisson distribution with parameter λ_i, then

$$Y = X_1 + X_2 + \cdots + X_r$$

also has a Poisson distribution with parameter $\lambda = \lambda_1 + \lambda_2 + \cdots + \lambda_r$.

(4) *The Exponential Distribution.* If $X_1, X_2, \ldots, X_r$ are mutually statistically independent, and if each X_i has an exponential distribution

with parameter θ, then
$$Y = X_1 + X_2 + \cdots + X_r$$
has a gamma distribution with parameters θ and r.

(5) *The Gamma Distribution.* If $X_1, X_2, \ldots, X_r$ are mutually statistically independent, and if each X_i has a gamma distribution with parameters θ and d_i, then
$$Y = X_1 + X_2 + \cdots + X_r$$
has a gamma distribution with parameters θ and $d = d_1 + d_2 + \cdots + d_r$.

(6) *The Cauchy Distribution.* If $X_1, X_2, \ldots, X_r$ are mutually statistically independent, and if each X_i has a Cauchy distribution with parameter θ, then
$$Y = \frac{X_1 + X_2 + \cdots + X_r}{r}$$
has a Cauchy distribution with the same parameter θ.

All of the special cases (1) through (6) listed above deal with weighted sums of statistically independent random variables. In Chapter 9, Section 5, we gave one example where the weighted sum of statistically *dependent* random variables is known:

(7) *Bivariate Normal Distribution.* If X and Y have a bivariate normal distribution with parameters μ_X, μ_Y, σ_X^2, σ_Y^2, and $\rho_{X,Y}$, then the weighted sum $aX + bY + c$ has a normal distribution with expected value
$$\mu_{aX+bY+c} = a\mu_X + b\mu_Y + c,$$
and variance
$$\sigma^2_{aX+bY+c} = a^2\sigma_X^2 + b^2\sigma_Y^2 + 2ab\sigma_X\sigma_Y\rho_{X,Y}.$$

Expected Values and Variances

In some practical cases, it is sufficient to determine the expected value and variance of $Y = a_1X_1 + a_2X_2 + \cdots + a_rX_r$ for the purpose of summarizing the probability distribution of Y (see Chapter 5). The following rules provide a convenient way of calculating μ_Y and σ_Y^2 without knowledge of the exact form of the distribution of Y.

Rule 2.1. Let $X_1, X_2, \ldots, X_r$ have any joint distribution, and let $E(X_1) = \mu_1$, $E(X_2) = \mu_2, \ldots, E(X_r) = \mu_r$. Then the expected value $\mu_Y = E(Y)$ of $Y = a_1X_1 + a_2X_2 + \cdots + a_rX_r$ is found from the formula

(2.1) $$\mu_Y = a_1\mu_1 + a_2\mu_2 + \cdots + a_r\mu_r.$$

In the special case where $Y = X_1 + X_2 + \cdots + X_r$ is the sum of the X_i's (that is, in the case where $a_1 = a_2 = \cdots = a_r = 1$),

$$\mu_Y = \mu_1 + \mu_2 + \cdots + \mu_r.$$

Hence, *the expected value of the sum of the random variables $X_1, X_2, \ldots, X_r$ is equal to the sum $\mu_1 + \mu_2 + \cdots + \mu_r$ of their expected values.*

We can demonstrate the validity of Rule 2.1 in the case where $r = 2$, the random variables $V = X_1$ and $W = X_2$ have a discrete bivariate distribution, and V and W each has two possible values. Let v_1 and v_2 be the possible values of V, and w_1 and w_2 be the possible values of W. Further, let $p_{V,W}(v,w)$ be the joint probability mass function of V and W. From Equation (3.7) in Chapter 9, we know that if $Y = g(V,W) = aV + bW$, then

(2.2) $\mu_Y = E[aV + bW]$
$= (av_1 + bw_1)p_{V,W}(v_1,w_1) + (av_1 + bw_2)p_{V,W}(v_1,w_2)$
$\quad + (av_2 + bw_1)p_{V,W}(v_2,w_1) + (av_2 + bw_2)p_{V,W}(v_2,w_2)$

$= a[v_1 p_{V,W}(v_1,w_1) + v_1 p_{V,W}(v_1,w_2) + v_2 p_{V,W}(v_2,w_1)$
$\quad + v_2 p_{V,W}(v_2,w_2)] + b[w_1 p_{V,W}(v_1,w_1) + w_2 p_{V,W}(v_1,w_2)$
$\quad + w_1 p_{V,W}(v_2,w_1) + w_2 p_{V,W}(v_2,w_2)].$

Since (see Chapter 9, Section 3)

$\mu_V = [v_1 p_{V,W}(v_1,w_1) + v_1 p_{V,W}(v_1,w_2) + v_2 p_{V,W}(v_2,w_1) + v_2 p_{V,W}(v_2,w_2)],$
$\mu_W = [w_1 p_{V,W}(v_1,w_1) + w_2 p_{V,W}(v_1,w_2) + w_1 p_{V,W}(v_2,w_1) + w_2 p_{V,W}(v_2,w_2)],$

we conclude from (2.2) that

$$\mu_Y = a\mu_V + b\mu_W,$$

thus proving Rule 2.1 in this special case. Rule 2.1 can be proven in the case of two discrete random variables (or more than two discrete random variables), each with any number of possible values, by a similar manner of breaking apart and recombining the various terms that make up the sum defining $E[a_1 X_1 + a_2 X_2 + \cdots + a_r X_r]$. The proof of Rule 2.1 for continuous random variables involves either use of facts about volumes in many dimensions or, equivalently, use of the theory of integral calculus.

Turning now to the variance of $Y = a_1 X_1 + a_2 X_2 + \cdots + a_r X_r$, the following formula holds for any jointly distributed random variables $X_1, X_2, \ldots, X_r$.

Rule 2.2. Let $X_1, X_2, \ldots, X_r$ have a joint distribution. Let $E(X_1) = \mu_1$, $E(X_2) = \mu_2, \ldots, E(X_r) = \mu_r$, and let $\text{Var}(X_1) = \sigma_1^2$, $\text{Var}(X_2) = \sigma_2^2, \ldots$, $\text{Var}(X_r) = \sigma_r^2$. Finally, let $\text{Cov}(X_1, X_2) = \sigma_{12}$, $\text{Cov}(X_1, X_3) = \sigma_{13}, \ldots$,

2. WEIGHTED SUMS OF RANDOM VARIABLES

$\text{Cov}(X_i, X_j) = \sigma_{ij}, \ldots, \text{Cov}(X_{r-1}, X_r) = \sigma_{r-1,r}$. Then if $Y = a_1 X_1 + a_2 X_2 + \cdots + a_r X_r$,

(2.3) $\quad \sigma_Y^2 = a_1^2 \sigma_1^2 + a_2^2 \sigma_2^2 + \cdots + a_r^2 \sigma_r^2 + 2 a_1 a_2 \sigma_{12} + 2 a_1 a_3 \sigma_{13}$
$\qquad + \cdots + 2 a_i a_j \sigma_{ij} + \cdots + 2 a_{r-1} a_r \sigma_{r-1,r}.$

In the special case where $Y = X_1 + X_2 + \cdots + X_r$, Rule 2.2 yields

$$\sigma_Y^2 = [\sigma_1^2 + \sigma_2^2 + \cdots + \sigma_r^2] + 2[\sigma_{12} + \sigma_{13} + \cdots + \sigma_{ij} + \cdots + \sigma_{r-1,r}];$$

that is, the variance of the sum is equal to the sum of the variances plus twice the sum of the covariances.

We verify Rule 2.2 from Rule 2.1 in the case of $r = 2$ random variables. Note that when $Y = a_1 X_1 + a_2 X_2$,

(2.4) $\quad \sigma_Y^2 = E[(a_1 X_1 + a_2 X_2) - \mu_Y]^2$
$\qquad = E[(a_1 X_1 + a_2 X_2) - (a_1 \mu_1 + a_2 \mu_2)]^2$
$\qquad = E[a_1(X_1 - \mu_1) + a_2(X_2 - \mu_2)]^2$
$\qquad = E[a_1^2(X_1 - \mu_1)^2 + 2 a_1 a_2 (X_1 - \mu_1)(X_2 - \mu_2)$
$\qquad \quad + a_2^2 (X_2 - \mu_2)^2].$

Let $U = (X_1 - \mu_1)^2$, $V = (X_1 - \mu_1)(X_2 - \mu_2)$, and $W = (X_2 - \mu_2)^2$. From Equation (2.4) and Rule 2.1,

$\sigma_Y^2 = E[a_1^2 U + 2 a_1 a_2 V + a_2^2 W]$
$\quad = a_1^2 E[U] + 2 a_1 a_2 E[V] + a_2^2 E[W]$
$\quad = a_1^2 E(X_1 - \mu_1)^2 + 2 a_1 a_2 E(X_1 - \mu_1)(X_2 - \mu_2) + a_2^2 E(X_2 - \mu_2)^2,$

so that

$$\sigma_Y^2 = a_1^2 \sigma_1^2 + a_2^2 \sigma_2^2 + 2 a_1 a_2 \sigma_{12},$$

thus verifying Rule 2.1 in the case where $r = 2$.

When the random variables $X_1, X_2, \ldots, X_r$ are statistically independent, then (Chapter 9, Section 3) we know that

$$\sigma_{ij} = \text{Cov}(X_i, X_j) = 0,$$

for $i \neq j$. That is, when $X_1, X_2, \ldots, X_r$ are pairwise statistically independent, the covariance between each pair of variables X_i and X_j is equal to 0. Substituting this result into Equation (2.3), we obtain the following rule.

Rule 2.3. If $X_1, X_2, \ldots, X_r$ are pairwise statistically independent random variables with $\text{Var}(X_1) = \sigma_1^2$, $\text{Var}(X_2) = \sigma_2^2$, ..., $\text{Var}(X_r) = \sigma_r^2$, and if $Y = a_1 X_1 + a_2 X_2 + \cdots + a_r X_r$, then

(2.5) $\quad \sigma_Y^2 = a_1^2 \sigma_1^2 + a_2^2 \sigma_2^2 + \cdots + a_r^2 \sigma_r^2.$

In the special case where $Y = X_1 + X_2 + \cdots + X_r$, Equation (2.5) yields

$$\sigma_Y^2 = \sigma_1^2 + \sigma_2^2 + \cdots + \sigma_r^2.$$

That is, *the variance of the sum of pairwise statistically independent random variables is the sum of the variances of these random variables.*

Notice that if X_1 and X_2 are statistically independent, then from Rule 2.3,

$$\sigma_{X_1+X_2}^2 = \sigma_1^2 + \sigma_2^2,$$

and

$$\sigma_{X_1-X_2}^2 = \sigma_1^2 + \sigma_2^2,$$

where $\text{Var}(X_1) = \sigma_1^2$, $\text{Var}(X_2) = \sigma_2^2$. Thus, *the variance of the sum of two statistically independent random variables is equal to the variance of the difference of these two random variables.* On the other hand,

$$\mu_{X_1+X_2} = \mu_1 + \mu_2 \neq \mu_1 - \mu_2 = \mu_{X_1-X_2},$$

unless $\mu_2 = E(X_2) = 0$.

3. THE LAW OF LARGE NUMBERS

Perhaps the most important weighted sum in terms of its practical use is the sample average. Suppose $X_1, X_2, \ldots, X_n$ are mutually statistically independent observations of a random variable X. By the *sample average* $\hat{\mu}_X$, we mean the average

(3.1) $$\hat{\mu}_X = \frac{1}{n}(X_1 + X_2 + \cdots + X_n)$$

of the random variables $X_1, X_2, \ldots, X_n$. Further, suppose that X (and thus each of the random variables $X_1, X_2, \ldots, X_n$) has expected value $E(X) = \mu$ and variance $\text{Var}(X) = \sigma^2$. Since $X_1, X_2, \ldots, X_n$ are mutually statistically independent, they are also pairwise statistically independent (see Chapter 9, Section 6). Hence, applying Rule 2.3 with $a_1 = 1/n$, $a_2 = 1/n$, ..., $a_n = 1/n$, we find that

(3.2) $$\sigma_{\hat{\mu}_X}^2 = \left(\frac{1}{n}\right)^2 \sigma^2 + \left(\frac{1}{n}\right)^2 \sigma^2 + \cdots + \left(\frac{1}{n}\right)^2 \sigma^2$$

$$= n\left[\left(\frac{1}{n}\right)^2 \sigma^2\right] = \frac{\sigma^2}{n}.$$

Because the variance of the sample average $\hat{\mu}_X$ is equal to the variance σ^2 of X divided by the number n of observations, it follows that *the variance of the sample average $\hat{\mu}_X$ decreases as the sample size n increases.*

3. THE LAW OF LARGE NUMBERS

From Rule 2.1, with $a_1 = a_2 = \cdots = a_n = 1/n$,

(3.3)
$$E(\hat{\mu}_X) = \frac{1}{n}\mu + \frac{1}{n}\mu + \cdots + \frac{1}{n}\mu$$

$$= n\left(\frac{1}{n}\mu\right) = \mu.$$

Thus, *the expected value of the sample average equals the expected value of X* (or of any of the observations $X_1, X_2, \ldots, X_n$). Although the sample average has the same expected value μ as any single observation X_j, it has a smaller variance (σ^2/n as compared to σ^2). As the number of observations forming the sample average increases, the variance σ^2/n of the sample average decreases, and *for large enough n, most of the probability mass of the distribution of $\hat{\mu}_X$ is concentrated around the expected value μ*. Indeed from the Chebychev Inequality (see Chapter 5, Section 6), we know that for any positive number c,

(3.4)
$$P_{\hat{\mu}_X}\left\{\mu - c\sqrt{\frac{\sigma^2}{n}} \leq \hat{\mu}_X \leq \mu + c\sqrt{\frac{\sigma^2}{n}}\right\} \geq 1 - \frac{1}{c^2},$$

or equivalently,

(3.5)
$$P_{\hat{\mu}_X}\left\{-c\frac{\sigma}{\sqrt{n}} \leq \hat{\mu}_X - \mu \leq c\frac{\sigma}{\sqrt{n}}\right\} \geq 1 - \frac{1}{c^2}.$$

Thus, as n grows large (that is, if we take a larger number of observations), the factor $c\sigma/\sqrt{n}$ becomes smaller, so that *the probability distribution of the sample average concentrates more and more closely around its expected value μ*. Indeed, if $\sigma^2 = 1$ and $n = 10{,}000$, Equation (3.5) tells us that the probability is at least 0.99 that $\hat{\mu}_X$ is within 0.10 units of its expected value. This computation is made as follows. In (3.5) we insert the values for σ^2 and n:

$$P_{\hat{\mu}_X}\left\{-\frac{c}{100} \leq \hat{\mu}_X - \mu \leq \frac{c}{100}\right\} \geq 1 - \frac{1}{c^2}.$$

Now we wish the difference of $\hat{\mu}_X$ and μ to be within 0.10 units; hence, c is chosen to be 10, and thus

$$P\{-0.10 \leq \hat{\mu}_X - \mu \leq 0.10\} \geq 1 - \frac{1}{100} = 0.99.$$

If $\sigma^2 = 1$ and $n = 1{,}000{,}000$, the probability is at least 0.99 that $\hat{\mu}_X$ is within 0.01 units of its expected value. [To obtain this result insert $c = 10$ in Equation (3.5).] Table 3.1 gives values of the sample size n required to guarantee with probability at least P (for $P = 0.95, 0.99$) that $\hat{\mu}_X$ is within various distances from its expected value μ when $\sigma^2 = 1$.

Table 3.1. Values of the sample size n required to assure at least a probability of P that $\hat{\mu}_X$ falls within a specified number c of units from its expected value μ when $\sigma^2 = 1$.

Number of Units c	$P = 0.95$ n	$P = 0.99$ n
0.01	200,000	1,000,000
0.10	2000	10,000
0.50	80	400
1.00	20	100
1.50	9	45
2.00	5	25
2.50	4	16
3.00	3	12

The conclusion we can draw from this analysis is that by making n large enough, we ensure a high probability of having $\hat{\mu}_X$ fall as close as we please to its expected value μ. This assertion is a somewhat informal statement of the *Law of Large Numbers*. We point out that this assertion is a mathematical fact derived from the axioms of probability theory; it corresponds to the frequency interpretation of the expected value which we gave in Chapter 5, Section 2.

A popular misconception concerning the law of large numbers is that it is a statement about the sample total $X_1 + X_2 + \cdots + X_n$, rather than about the sample average $(1/n)(X_1 + X_2 + \cdots + X_n)$. For example, in tossing a fair coin (where the probability of "heads" at any toss equals $\frac{1}{2}$), the so-called "man in the street" feels that the number of "heads" observed should equal the number of "tails." Thus, if he observes 50 "heads" in a row he tends to believe that a "tail" will appear on the 51st toss, claiming that the "law of averages" says that the total number of heads must equal the total number of tails, and therefore, "tails" is "due" to come up. But if the coin tosses are statistically independent, as assumed, observing 50 heads on the first 50 tosses tells us *nothing new* about the probability of the 51st toss resulting in a tail; the probability of "tail" is still $\frac{1}{2}$. The law of large numbers does not say (in this case) that the total number of "heads" must eventually equal the total number of "tails"; rather it says that the *proportion* of heads eventually equals the *proportion* of tails. The difference between these two statements can be illustrated by assuming that the first 10 tosses are "heads" and that after the first 10 tosses, the number of "heads" does equal the number of "tails." Thus, in 1000 tosses we have 505 "heads" and 495 "tails," in 10,000 tosses we have 5005 "heads" and 4995 "tails," and so on. Always, the number of "heads" is 10 more than the number of "tails," but the proportions are 0.505 as against 0.495 (after 1000 tosses), 0.5005 as against 0.4995 (after 10,000 tosses), 0.50005 as against 0.49995 (after 100,000 tosses), and so on. That is, when we compute proportions rather than totals, the excess of "heads" is overwhelmed by the total number of tosses; the difference between the total number of "heads" observed and the total number of "tails" observed becomes a smaller and smaller fraction of the total number of tosses as the number of tosses gets large. [*Note:* The sample average being discussed here is the proportion of heads; this proportion can be given as the average of the observations $X_1, X_2, ..., X_n$, where X_i is 1 if the ith toss is "heads" and 0 otherwise.]

To obtain some meaningful distributional statements about the sample total $X_1 + X_2 + \cdots + X_n = S_n$, we can either use the distributions derived for special cases mentioned in Section 2 (provided the assumptions concerning the distribution of the X_i's fit one of the special cases listed), search the literature to see if someone else has discussed our probability model and has derived the distribution of S_n, or attempt to derive the distribution ourselves. This last task (as we have mentioned earlier) can be quite difficult. If the sample size, however, is reasonably large, we can use the *Central Limit Theorem* to obtain approximate probability statements about S_n which are of great assistance in practice.

4. FURTHER COMMENTS ON THE CENTRAL LIMIT THEOREM

Let $X_1, X_2, ..., X_n$ be independent and identically distributed observations of a random variable X, where X (and thus each X_i) has expected value μ and variance σ^2. We are interested in the sample total $S_n = X_1 + X_2 + \cdots + X_n$ and in the sample average $\hat{\mu}_X = S_n/n$, since (as we have remarked previously) these random variables are frequently used to summarize the results of a random experiment. We have already seen that the determination of the distribution of S_n (or of $\hat{\mu}_X$) when n is small may be a tedious or difficult task. For moderately large values of n, however, we can use the famous *Central Limit Theorem* to find an approximation to the distribution of S_n.

A formal statement of the Central Limit Theorem (rather than the somewhat vague version of this theorem given in Section 4 of Chapter 7) concerns the cumulative distribution function of the standardized random variable:

(4.1) $$Z_n = \frac{S_n - n\mu}{\sqrt{n\sigma^2}} = \frac{(X_1 + X_2 + \cdots + X_n) - n\mu}{\sqrt{n\sigma^2}}.$$

The Central Limit Theorem asserts that by choosing the number n of observations to be large enough, we can make the (unsigned) difference between the value of the cumulative distribution function $F_{Z_n}(z)$ at any number z and the value of the cumulative distribution function $F_Z(z) = P\{\mathcal{N}(Z;0,1) \leq z\}$ at z to be as small as we please. That is, by choosing a large enough number n of observations, we can make the difference $|F_{Z_n}(z) - F_Z(z)|$ as small as we wish. For example, suppose we want to find the probability $F_{Z_n}(1.96) = P_{Z_n}\{Z_n \leq 1.96\}$ correct to within two decimal places of accuracy (that is, correct up to an error of ± 0.005). The Central Limit Theorem states that there is a value of the sample size n large enough so that the approximation of $F_{Z_n}(1.96)$ by the value $F_Z(1.96) = P\{\mathcal{N}(Z;0,1) \leq 1.96\} = 0.975$ results in an error of no more than ± 0.005. This enables us to obtain the accuracy that we require. For many applications, a sample size of 25 (or more) is sufficient to make such an approximation to $F_{Z_n}(z)$ accurate to the second decimal place.

The remarkable aspect of the Central Limit Theorem is that Z_n has approximately a normal distribution *regardless of the (common) distribution* of the independent random variables $X_1, \ldots, X_n$. The only requirement is that σ^2 must be finite. In fact, the Central Limit Theorem sometimes holds under conditions where the random variables $X_1, \ldots, X_n$ do not necessarily have the same distribution. For this reason, in applications a random variable is often *assumed* to have a normal distribution, since quite frequently the random variable of interest may be conceptualized as being composed of the sum of many independent random variables (see also Chapter 7, Section 4).

A more concrete indication of what the Central Limit Theorem says can perhaps be obtained by considering statistically independent repetitions $X_1, X_2, \ldots, X_n$ of a Bernoulli random variable X having parameter $p = 0.4$. In this case X has only two possible values, 0 and 1, and

$$p_X(1) = 1 - p_X(0) = p = 0.4.$$

From Chapter 6, Section 1, we know that

$$\mu = E(X) = p = 0.4,$$
$$\sigma^2 = \text{Var}(X) = p(1-p) = 0.24.$$

For various values of n ($n = 10, 12, 15, 20, 30,$ and 50), we have calculated and graphed the cumulative distribution function $F_{Z_n}(z)$ of

$$Z_n = \frac{(X_1 + X_2 + \cdots + X_n) - n\mu}{\sqrt{n\sigma^2}}$$
$$= \frac{(X_1 + X_2 + \cdots + X_n) - n(0.4)}{\sqrt{n(0.24)}}.$$

4. FURTHER COMMENTS ON THE CENTRAL LIMIT THEOREM

Since

$$F_{Z_n}(z) = P_{Z_n}\{Z_n \leq z\}$$
$$= P\left\{\frac{X_1 + X_2 + \cdots + X_n - n(0.4)}{\sqrt{n(0.24)}} \leq z\right\}$$
$$= P\{X_1 + X_2 + \cdots + X_n \leq n(0.4) + z(\sqrt{0.24})(\sqrt{n})\},$$

and since we know from Chapter 6, Section 1, that $X_1 + X_2 + \cdots + X_n$ has a binomial distribution with parameters $p = 0.4$ and n, calculation of $F_{Z_n}(z)$ can be accomplished either by using tables of the cumulative distribution function of a binomial random variable (see Chapter 6, Section 1) or by using the computer. For comparison to the graph of $F_{Z_n}(z)$, we have provided a graph of the c.d.f. $F_Z(z) = P\{\mathcal{N}(Z;0,1) \leq z\}$ of the standard normal distribution. The graph of $F_{Z_{10}}(z)$ and $F_Z(z)$ appears in Figure 4.1(a), the graph of $F_{Z_{12}}(z)$ and $F_Z(z)$ appears in Figure 4.1(b), ..., the graph of $F_{Z_{50}}(z)$ and $F_Z(z)$ appears in Figure 4.1(f). Note how the graphs of $F_{Z_n}(z)$ and $F_Z(z)$ more and more closely resemble one another as n increases. For $n = 50$ [see Figure 4.1(f)], the graph of $F_{Z_{50}}(z)$ is nearly identical to the graph of $F_Z(z)$.

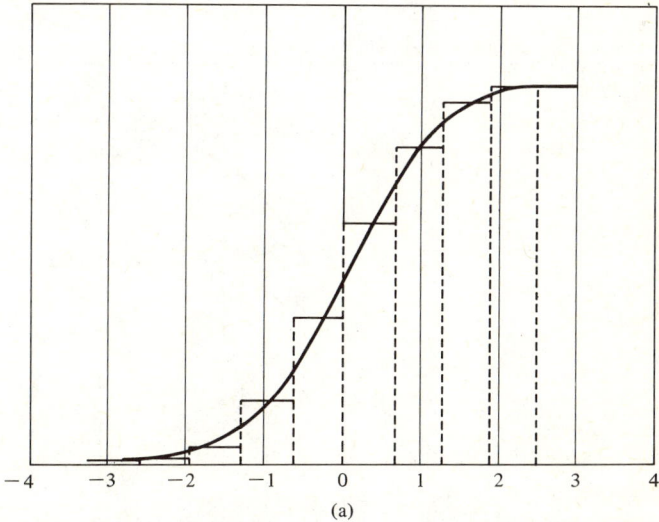

Figure 4.1 (a) Comparison of the graph of the cumulative distribution function of a standardized binomial random variable with $n = 10$ and $p = 0.4$ and the standard normal random variable; (b) with $n = 12$; (c) with $n = 15$; (d) with $n = 20$; (e) with $n = 30$; (f) with $n = 50$.

536 WEIGHTED SUMS OF RANDOM VARIABLES 10

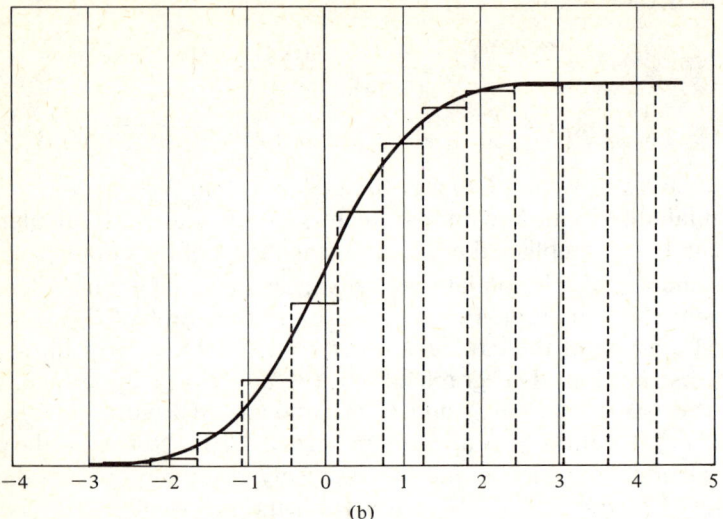

(b)

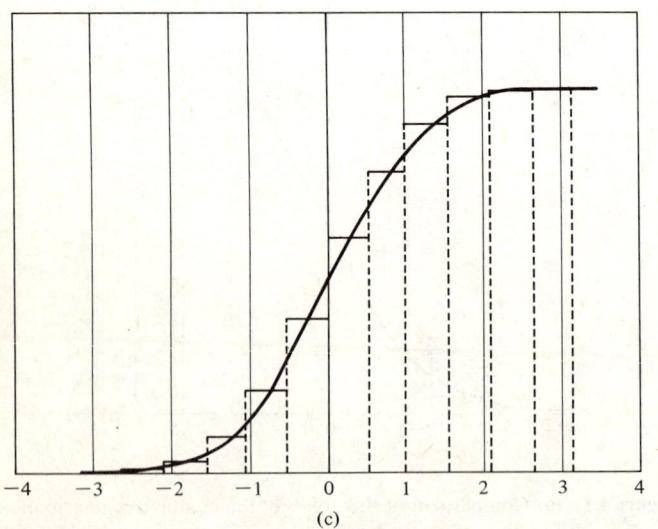

(c)

Figure 4.1 (*continued*)

4. FURTHER COMMENTS ON THE CENTRAL LIMIT THEOREM

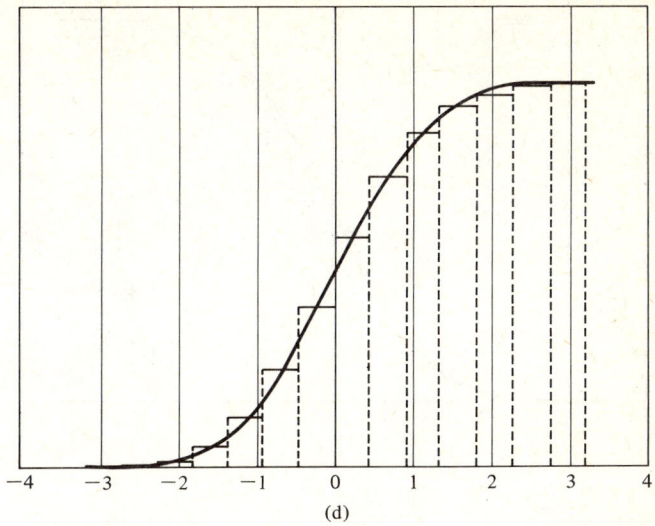

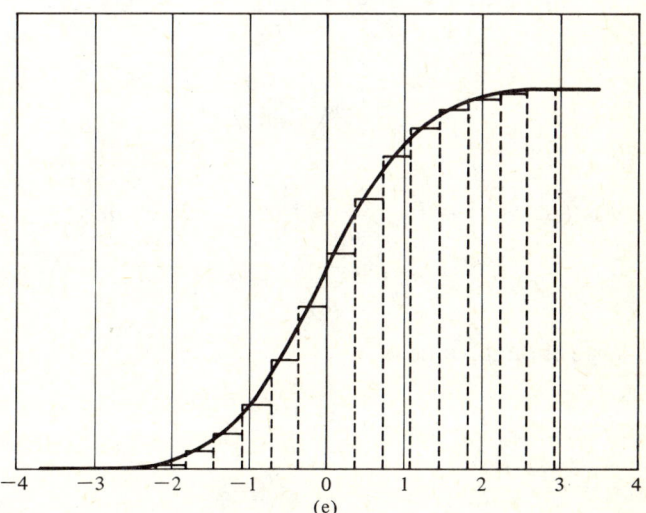

Figure 4.1 (*continued*)

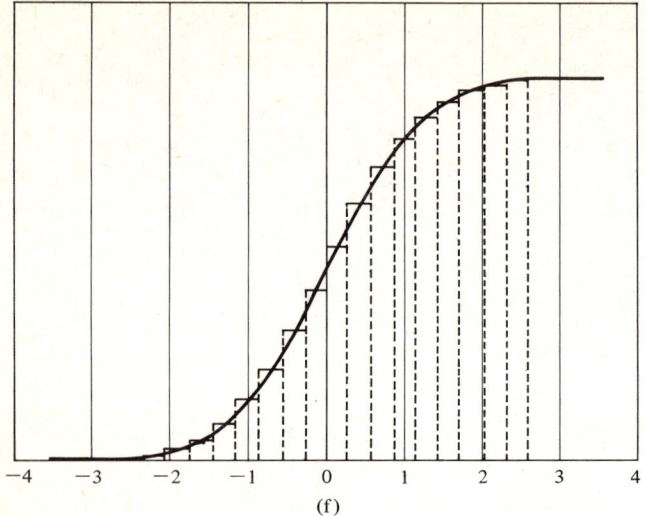
(f)

Figure 4.1 (*continued*)

In general, to use the Central Limit Theorem to approximate the distributions of $S_n = X_1 + X_2 + \cdots + X_n$ and of $\hat{\mu}_X = S_n/n$, we note that

(4.2) $$P_{S_n}\{S_n \leq s\} = P_{S_n}\left\{\frac{S_n - n\mu}{\sqrt{n\sigma^2}} \leq \frac{s - n\mu}{\sqrt{n\sigma^2}}\right\}$$
$$= F_{Z_n}\left(\frac{s - n\mu}{\sqrt{n\sigma^2}}\right),$$

and that

(4.3) $$P_{\hat{\mu}_X}(\hat{\mu}_X \leq t) = P_{S_n}\left\{\frac{S_n}{n} \leq t\right\} = P_{S_n}\{S_n \leq nt\}$$
$$= F_{Z_n}\left(\frac{nt - n\mu}{\sqrt{n\sigma^2}}\right) = F_{Z_n}\left(\sqrt{n}\left(\frac{t - \mu}{\sigma}\right)\right).$$

Thus, for a large enough value of n,

(4.4) $$P_{S_n}\{S_n \leq s\} \cong F_Z\left(\frac{s - n\mu}{\sqrt{n\sigma^2}}\right),$$

and

(4.5) $$P_{\hat{\mu}_X}\{\hat{\mu}_X \leq t\} \cong F_Z\left(\sqrt{n}\left(\frac{t - \mu}{\sigma}\right)\right),$$

where $F_Z(z)$ is the cumulative distribution function of the standard normal distribution (see Table T.6), and "$\cong$" means "approximately equal to."

4. FURTHER COMMENTS ON THE CENTRAL LIMIT THEOREM

The Continuity Correction

If S_n or $\hat{\mu}_X$ is computed from random variables $X_1, X_2, \ldots, X_n$, having discrete distributions whose possible values are equally spaced at intervals of length 2Δ, then, for moderate sample sizes n a *continuity correction* helps to improve the accuracy of the Central Limit Theorem approximation. (That is, it helps correct for the error we make in approximating a discrete distribution by the normal distribution, which is continuous.) It is assumed in these cases that the numbers s and t in the expressions $P\{S_n \leq s\}$ and $P\{\hat{\mu}_X \leq t\}$ are possible values of the random variable S_n and $\hat{\mu}_X$, respectively. If so, the continuity correction consists in replacing s by $s+\Delta$ and t by $t+\Delta/n$ in computing the approximation. Thus, we approximate $P_{S_n}\{S_n \leq s\}$ by $F_Z((s+\Delta-n\mu)/\sqrt{n}\sigma)$, and we approximate $P\{\hat{\mu}_X \leq t\}$ by $F_Z(\sqrt{n}(t-\mu+\Delta/n)/\sigma)$ when the random variables $X_1, X_2, \ldots, X_n$ are discrete with possible values 2Δ apart.

Example 4.1 (Continuous Case). We wish to estimate the weight of a minute quantity of blood serum to within 1 milligram. It is known that any measurement X of this weight has mean μ equal to the true weight (that is, the weighing scale is "unbiased") and a standard deviation σ of 4 milligrams. The experimenter decides to take 100 measurements $X_1, X_2, \ldots, X_{100}$ of the weight and to use $\hat{\mu}_X = (X_1 + X_2 + \cdots + X_{100})/100$ as his estimate of the true weight. What is the probability that his estimate $\hat{\mu}_X$ has the desired accuracy? We must find $P\{\mu - 1 \leq \hat{\mu}_X \leq \mu + 1\}$ which is equal to $F_{\hat{\mu}_X}(\mu+1) - F_{\hat{\mu}_X}(\mu-1)$, and which can be approximated by

$$F_Z\left(\sqrt{100}\left[\frac{(\mu+1)-\mu}{4}\right]\right) - F_Z\left(\sqrt{100}\left[\frac{(\mu-1)-\mu}{4}\right]\right),$$

or $F_Z(2.50) - F_Z(-2.50)$. Using Table T.6, we find this probability to be $0.9938 - 0.0062$ or 0.9876. (A similar approximation was given in Example 4.1 of Chapter 7.)

Example 4.2 (Discrete Case). A gambler is playing blackjack at a casino. He decides to play 100 times and to bet $10 each time. On each play, his probability of winning $10 is 0.49; the probability of losing $10 is 0.51. Let $X_i (i = 1, 2, \ldots, 100)$ denote his winnings on the ith play. We assume that the deck is shuffled after each hand so that the play of each hand of blackjack is statistically independent from the play of all other hands. In this case, we find that

$$\mu = EX_i = \$10(0.49) + (-\$10)(0.51) = -\$0.20,$$

and that
$$\sigma^2 = \text{Var } X_i = (\$10 + \$0.20)^2(0.49) + (-\$10 + \$0.20)^2(0.51)$$
$$= 99.96 \text{ dollars squared,}$$
from which we obtain the result
$$\sigma = \sqrt{\sigma^2} = \sqrt{99.96} = \$10.00.$$

The total winnings of the gambler are $S_{100} = X_1 + X_2 + \cdots + X_{100}$. Note that each X_i is a discrete random variable whose values are spaced $20 apart. Thus, to use the continuity correction, we let $\Delta = 10$. Let us calculate the probability, $P_{S_{100}}\{S_{100} > 0\}$, that the gambler finishes the evening with a profit. This probability is $1 - P_{S_{100}}\{S_{100} \leq 0\}$, which by the Central Limit Theorem (using the continuity correction) is approximately equal to

$$1 - F_Z\left(\frac{S + \Delta - n\mu}{\sqrt{n}\sigma}\right) = 1 - F_Z\left(\frac{0 + \$10 - 100(-\$0.20)}{\sqrt{100}(\$10.00)}\right)$$

or $1 - F_Z(0.30)$. From Table T.6, we find that this probability is 0.3821. It is interesting to calculate the probability that the gambler makes a profit if he plays 1000 times instead of 100. In this case, S_{1000}, his total winnings, has a probability approximately $1 - F_Z(0.66)$ or 0.2546 of being greater than 0. We see that the probability that the gambler finishes the evening with a profit decreases with the number of times he plays.

One of the important special uses of the Central Limit Theorem is to provide an approximation for the probabilities of a binomial distribution. If S_n has the binomial distribution (Chapter 6, Section 2) with parameters n and p, then we have already remarked that $S_n = X_1 + \cdots + X_n$, where the X_i are independent Bernoulli trials each having parameter p. For n even moderately large [say such that $np \geq 5$ and $n(1-p) \geq 5$], the cumulative distribution function of a standard normal random variable provides a close approximation to the cumulative distribution function of $(S_n - np)/\sqrt{np(1-p)}$. [We have already demonstrated this fact when $p = 0.4$; see Figures 4.1(a) through 4.1(f).] This approximation is improved by a continuity correction with $\Delta = \frac{1}{2}$ because each X_i is a discrete random variable with values a distance of 1 apart. Thus, when S_n has a binomial distribution with parameters n and p,

$$P_{S_n}\{S_n \leq s\} \cong F_Z\left(\frac{s + \frac{1}{2} - np}{\sqrt{np(1-p)}}\right).$$

Example 4.3. Suppose that a random sample of 25 is taken *with* replacement from the population of registered voters in a given city. We ask every individual drawn whether he will vote for Candidate A in the coming election. Let S_{25} be the number of people in our sample who say they will vote for Candidate A. If the proportion of registered voters in the population who favor Candidate A is p, then an estimate of p can be obtained from the proportion $S_{25}/25$ of favorable responses to Candidate A in our sample. What is the probability that our estimate $S_{25}/25$ of p will be within 0.04 of the correct value of p? From Chapter 6, Section 2, we know that S_{25} has a binomial distribution with parameters $n = 25$ and p. Thus we want to find

$$P_{S_{25}}\left\{p - 0.04 \leq \frac{S_{25}}{25} \leq p + 0.04\right\}$$
$$= P_{S_{25}}\{25(p-0.04) \leq S_{25} \leq 25(p+0.04)\}.$$

The trouble is that this probability depends explicitly on the value of p, and we do not know the value of p! Suppose we try several values of p to see what effect p has. Let us first try $p = 0.40$. Then we want the probability that S_{25} is strictly greater than 8 and less than or equal to 11. But this is approximately equal to (using the continuity correction) the probability that a random variable Z having the standard normal distribution is between $[8 + \frac{1}{2} - 25(0.4)]/\sqrt{25(0.4)(0.6)} = -0.61$ and $[11 + \frac{1}{2} - 25(0.4)]/\sqrt{25(0.4)(0.6)} = 0.61$. Thus when $p = 0.4$,

$$P_{S_{25}}\left\{p - 0.04 \leq \frac{S_{25}}{25} \leq p + 0.04\right\} = P_{S_{25}}\{9 \leq S_{25} \leq 11\}$$
$$= P_{S_{25}}\{S_{25} \leq 11\} - P_{S_{25}}\{S_{25} \leq 8\}$$
$$\cong F_Z(0.61) - F_Z(-0.61) = 0.4582.$$

If $p = 0.36$, we can proceed in a similar manner to that for $p = 0.40$, and find that the desired probability is 0.4741. If $p = 0.8$, the desired probability is 0.5468. The probability computed in this way is, in fact, always greater than 0.4, no matter what the value of p may be.

*5. APPENDIX. MOMENT-GENERATING FUNCTIONS

In Section 2, we mentioned that it is often difficult to obtain the distribution of a sum of random variables. One method of approaching this problem is through the concept of the moment-generating function of a random variable. Actually, the moment-generating function of a random variable has a much wider application. Indeed, it can be used in place of the cumulative distribution function, probability mass function, or probability density function as a summary of all of the information concerning the probability distribution of X.

The moment-generating function of X [in symbols $m_X(s)$] is defined by assigning to every number s the corresponding expected value of e^{sX}, provided that this expected value exists (Chapter 5, Section 2). Thus,

(5.1) $$m_X(s) = E(e^{sX}).$$

There are two principal facts about moment-generating functions that make them useful in finding the distribution of (weighted) sums of random variables. First, under quite general assumptions, there is a unique correspondence between a distribution function and its moment-generating function: *to every moment-generating function there corresponds one and only one probability distribution.* The second fact is that *if $X_1, X_2, \ldots, X_r$ are mutually statistically independent random variables and if $Y = X_1 + X_2 + \cdots + X_r$, then*

(5.2) $$m_Y(s) = m_{X_1}(s) m_{X_2}(s) \cdots m_{X_r}(s).$$

That is, the moment-generating function of the sum of statistically independent random variables is the product of the individual moment-generating functions of these variables.

From these two facts, we can replace the difficult computations often involved in finding the distribution of $Y = a_1 X_1 + a_2 X_2 + \cdots + a_r X_r$, by two simple steps:

(i) Find $m_Y(s) = [m_{a_1 X_1}(s)][m_{a_2 X_2}(s)] \cdots [m_{a_r X_r}(s)]$
(ii) Use a table of correspondences between generating functions and distributions to find the distribution corresponding to $m_Y(s)$. If no such table is available, or if $m_Y(s)$ is not listed in the table, use an inversion formula to invert $m_Y(s)$ to its corresponding distribution (this usually requires a knowledge of the integral calculus). The distribution corresponding to $m_Y(s)$ is then the probability distribution of Y.

In a sense, the procedure described in the above paragraph is analogous to the procedure for multiplying numbers using a table of logarithms. The multiplication of r numbers is a relatively more complex operation than the addition of r numbers. However, since the logarithm of the product of r numbers is the sum of their respective logarithms, by adding the respective logarithms, looking up the resulting number in a table of logarithms and finding the antilogarithm, we replace a complicated task of multiplication by a simple job of addition, table lookup, and conversion.

Table 5.1 gives the moment-generating functions of certain discrete distributions; Table 5.2 gives the moment-generating functions of some continuous distributions.

5. APPENDIX. MOMENT-GENERATING FUNCTIONS

Table 5.1. Moment-generating functions of some discrete probability functions.

Distribution	Probability Mass Function[a]	Moment-Generating Function
Binomial	$p_Z(k) = \dfrac{n!}{k!(n-k)!} p^k (1-p)^{n-k},$ $k = 0, 1, \ldots, n$	$(pe^s + 1 - p)^n$
Geometric	$p_Z(k) = p(1-p)^{k-1},$ $k = 1, 2, \ldots$	$\dfrac{pe^s}{1-(1-p)e^s}$
Negative Binomial	$p_Z(k) = \dfrac{(r+k-1)!}{k!(r-1)!} p^r (1-p)^k,$ $k = 0, 1, \ldots$	$\left(\dfrac{p}{1-(1-p)e^s}\right)^r$
Poisson	$p_Z(k) = \dfrac{e^{-\lambda} \lambda^k}{k!},$ $k = 0, 1, \ldots$	$e^{\lambda(e^s - 1)}$
Discrete Uniform	$p_Z(k) = \dfrac{1}{K},$ $k = c+h, \ldots, c+Kh$	$\dfrac{e^{s(c+h)}[1 - e^{sK}]}{K(1-e^s)}$

[a] In each case, the probability mass function is equal to 0 for values of k not indicated.

Table 5.2. Moment-generating functions of some continuous distributions.

Type of Distribution	Probability Density Function[a]	Moment-Generating Function
Uniform	$\dfrac{1}{b-a},$ for $a \leq x \leq b$	$\dfrac{e^{sb} - e^{sa}}{s(b-a)}$
Exponential	$\theta e^{-\theta},$ for $x > 0$	$\dfrac{\theta}{\theta - s}$
Gamma	$\dfrac{\theta}{\Gamma(r)} (\theta x)^{r-1} e^{-\theta x},$ for $x > 0$	$\left(\dfrac{\theta}{\theta - s}\right)^r$
Normal	$\dfrac{1}{\sqrt{2\pi\sigma^2}} e^{-\frac{1}{2}[(x-\mu)/\sigma]^2},$ for all x	$e^{s\mu + \frac{1}{2} s^2 \sigma^2}$

[a] In each case, the probability density function is equal to 0 for values of x not indicated.

An important relationship between a random variable and its corresponding moment-generating function is

(5.3) $$m_{aX+b}(s) = e^{sb}m_X(as).$$

Thus when $Y = aX + b$, the moment-generating function of Y is easily obtained in terms of the moment-generating function of X.

Example 5.1. Suppose X has a normal distribution with parameters μ and σ. From Table 6.2 we find that

$$m_X(s) = e^{s\mu + \frac{1}{2}s^2\sigma^2}.$$

Let $Z = (X - \mu)/\sigma$. Hence, from Equation (6.3),

$$m_Z(s) = e^{-s\mu/\sigma} m_X\left(\frac{s}{\sigma}\right) = e^{-s\mu/\sigma}(e^{(s/\sigma)\mu + \frac{1}{2}(s^2/\sigma^2)\sigma^2})$$
$$= e^{\frac{1}{2}s^2}.$$

Looking up $m_Z(s)$ in Table 6.2, we see that Z has the standard normal distribution.

Example 5.2. Suppose $X_1, X_2, ..., X_r$ are mutually statistically independent, each having a Poisson distribution with parameter λ. We wish to find the distribution of

$$Y = X_1 + X_2 + \cdots + X_r.$$

From Table 5.1, the moment-generating function of each X_i is $e^{\lambda(e^s - 1)}$. Using Equation (5.2), the product of these moment-generating functions is

$$m_Y(s) = e^{r\lambda(e^s - 1)}.$$

Using Table 5.1, we see that $m_Y(s)$ is the moment-generating function of a Poisson distribution with parameter $\lambda^* = r\lambda$. Hence, Y has a Poisson distribution with parameter $\lambda^* = r\lambda$.

EXERCISES

1. Let the random variable X have an exponential distribution with parameter θ.
    (a) What is the form of the density function $f_Y(y)$ of $Y = 5X$?
    (b) What is the expected value of Y?
    (c) What is the variance of Y?
    (d) What is the distribution of Y? Does Y have an exponential distribution? Explain your answer.
    (e) Find $P_Y\{Y \leq 3\}$ when $\theta = 2$.

A2. Let the random variable X have a Poisson distribution with parameter λ.
  (a) What is the form of the probability mass function $P_Y(y)$ of $Y = 5X$?
  (b) What is the expected value of Y?
  (c) What is the variance of Y?
  (d) Does Y have a Poisson distribution? Support your assertion.
  (e) Find $P_Y\{Y \leq 7\}$ when $\lambda = 1$.

3. Suppose X is the score of a randomly chosen student on a midterm examination and that Y is the score of the student on a final examination in a probability course. The midterm examination is difficult, and the scores X on this examination have a normal distribution with expected value $\mu_X = 60$ and variance $\sigma_X^2 = 16$. The final examination is easier; the scores Y on this examination are normally distributed with expected value $\mu_Y = 80$ and variance $\sigma_Y^2 = 9$.
  (a) Suppose that the instructor assigns the student a grade on the basis of his average score $(X+Y)/2$ on the two examinations. What is the distribution of $(X+Y)/2$? Assume that the two test scores are statistically independent.
  (b) Suppose that the instructor standardizes the scores on the two tests: he rescales X and Y so that the new scores $X^* = (X - \mu_X)/\sigma_X$ and $Y^* = (Y - \mu_Y)/\sigma_Y$ have expected values of 0 and variances of 1. A student is now assigned a grade on the basis of the average $(X^* + Y^*)/2$ of the standardized scores. What is the distribution of $(X^* + Y^*)/2$?

The instructor awards an "A" to students whose "average" score is at least 2 standard deviations over the expected value. A certain student scores 76 on both examinations.
  (c) If the "average score" used by the instructor is $(X+Y)/2$, does the student receive an "A" for the course? Support your assertion.
  (d) If the "average score" used by the instructor is $(X^* + Y^*)/2$, does the student receive an "A" for the course?
  (e) Which method of grading the students do you think is the fairer one? Why?

A4. Suppose X and Y are scores on two parts of an examination, and these scores have a bivariate normal distribution with respective expected values 30 and 40, with respective variances 9 and 16, and with a correlation of 0.7. Find the distribution of the total score $X + Y$ on the examination.

5. Suppose X and Y are statistically independent scores on an examination, each having a Cauchy distribution (with parameter $\theta = 0$).
  (a) What is the distribution of the average grade $(X+Y)/2$?
  (b) A student says he did not feel well when he took the second exam, and asks his instructor to base his grade for the course on his first score only. Should the instructor agree? Explain.

A6. Two blood banks are in search of people with a certain type of blood. The proportion p of people in the population at large who have this type of blood is equal to p. Blood donors arrive at the blood banks as if they were chosen by a simple random sample with replacement from the population at large. At

the end of a certain week, one blood bank has received blood from 100 donors of whom X had the desired type of blood, while the other blood bank had received blood from 300 donors of whom Y had the desired blood type.
- (a) What is the distribution of X?
- (b) What is the distribution of Y?
- (c) What is the distribution of $X+Y$?
- (d) If every donor gives 1 pint of blood, what is the probability that the blood bank has at least 50 pints of the desired type of blood at the end of the week if $p=0.1$? [*Hint:* Does the Central Limit Theorem give you any help?]

7. The number of automobile accidents occurring on any given non-holiday weekend at a particular intersection has a Poisson distribution with parameter $\lambda = 2$. The number of accidents occurring on any one weekend is statistically independent of the number of accidents occurring on all other weekends.
- (a) What is the distribution of the total number of accidents that occur on weekends in a month (4 weekends) which has no holiday weekends?
- (b) In a year there are 52 weekends of which 6 are holiday weekends and 46 are non-holiday weekends. On any holiday weekend, the number of accidents that occur has a Poisson distribution with parameter $\lambda = 3.5$. What is the distribution of the total number of weekend accidents Y that occur in a year?
- (c) Suppose that $X_1, X_2, \ldots, X_{52}$ are mutually statistically independent random variables each of which has a Poisson distribution with parameter $\lambda = 113/52$. Show that $Z = X_1 + X_2 + \cdots + X_{52}$ has the same distribution as the random variable Y described in part (b). Thus, show that $P_Y\{Y \geq 200\}$ and $P_Z\{Z \geq 200\}$ are equal.
- (d) Use the result you obtained in part (c) and the Central Limit Theorem to find the probability that at least 200 automobile accidents occur on weekends at the given intersection in a year.

A8. In Exercise 1 of Chapter 8, it was noted that the excess magnitudes of large earthquakes have been assumed to have an exponential distribution with a parameter θ which depends on what we mean by a "large" earthquake. Suppose that our definition of a "large" earthquake results in a value 0.50 for θ. If 5 "large" earthquakes occur in a given year, and if the excess magnitudes of these earthquakes are mutually statistically independent, what is the probability distribution of the average excess magnitude (the total excess magnitude of the 5 earthquakes divided by 5) of these earthquakes? What is the expected value and variance of this distribution? What is the probability that the average excess magnitude of these 5 earthquakes exceeds 2.5?

9. The probability of obtaining a response to a certain interviewing procedure is 0.80. We attempt to interview 400 independently selected people.
- (a) What is the probability of getting exactly 320 responses?
- (b) What is the probability of getting strictly more than 300 responses?

[*Hint:* Use the Central Limit Theorem approximation to the probabilities of a binomial distribution. Do you need to use the continuity correction here?]

^{A}10. The Internal Revenue Service uses a letter opening machine which removes checks from the envelopes. Suppose that the probability that the machine fails to remove a check from any given envelope is 0.03, and that the actions of the machine on the various envelopes are mutually statistically independent trials. Suppose 50 envelopes are opened in a given day.
   (a) Use the Central Limit Theorem (without the continuity correction) to find the (approximate) probability that the machine fails to remove checks from 5 or more envelopes in that day.
   (b) Use the Central Limit Theorem, together *with* the continuity correction, to find the (approximate) probability that the machine fails to remove checks from 5 or more envelopes.
   (c) Compare your answers in parts (a) and (b) to the correct probability obtained from tables of the binomial distribution. What differences do you note?

11. In the context of Exercise 10 suppose 5000 envelopes are opened.
   (a) Find the (approximate) probability that the machine fails to remove checks from 175 or more envelopes. Use the Central Limit Theorem without the correction for continuity.
   (b) Answer the question in part (a), but now use the continuity correction.
   (c) What is the discrepancy between your answers in part (a) and part (b)? Is the difference more or less than that which you found in part (c) of Exercise 10? Can you explain these results?

^{A}12. Inspectors checking canned goods keep records of defective cans. Canned peaches are reported to be defective if part of a pit is not removed. If the probability that a can is defective is 0.01, what is the (approximate) probability that 3 or more cans out of 100 will be defective? What assumptions do you make in order to obtain an answer? In this chapter we have discussed approximations to the desired probability using the Central Limit Theorem (with or without the continuity approximation). What other approximation could we use to obtain this probability (see Chapter 6)? How do the answers provided by these methods of approximation compare? Which method of approximation would you prefer? Why?

*^{A}13. Use the Central Limit Theorem to answer the following question. Suppose we want to know the actual probability p that a randomly chosen human birth (excluding multiple births, such as twins) will result in a male child. (This probability p is not 0.50, as we remarked earlier.) Assuming that births are statistically independent of one another, how many births must we observe in order that the probability is at least 0.90 that the proportion of male births observed will be within 0.005 of the true probability p of a male birth? That is, what is the smallest number n of human births that must be observed in order that the proportion $\hat{p}$ = (total number of male births)/n is, with probability at least 0.90, within 0.005 of p? [*Hint:* Let $X_k = 1$ if the kth

birth results in a male child, and let $X_k = 0$ if the kth birth is a female child. Then, $\hat{p} = (X_1 + X_2 + \cdots + X_n)/n$, and we want to choose n sufficiently large so that $P\{-0.005 \leq \hat{p} - p \leq 0.005\} \geq 0.90$.]

*14. Use Inequality (3.5) to answer the question posed in Exercise 13. Are the answers obtained by the two methods very different? Which answer do you trust more? Why?

*A15. Two randomly selected groups of individuals are given a reaction time test. The individuals in one group take this test after having taken a certain drug (in pill form). The individuals in the second group take this test after having taken a pill which resembles the drug in size, shape, and taste, but which is not potent (a so-called "placebo"). The reaction times $X_1, X_2, \ldots, X_n$ of the individuals who took the drug are assumed to be mutually statistically independent, and each X_i has expected value μ and variance σ^2. The reaction times $Y_1, Y_2, \ldots, Y_m$ of the individuals who took the placebo are assumed to be mutually statistically independent, and each Y_j has expected value ν and variance σ^2.

(a) To evaluate the effect on reaction time of taking the drug, we desire to estimate $\mu - \nu$. Let $\Delta = \mu - \nu$. We know that $\hat{\mu}_X = (X_1 + X_2 + \cdots + X_n)/n$ is a reasonable approximation to the expected value μ of any X_i when n is large. We know that $\hat{\mu}_Y = (Y_1 + Y_2 + \cdots + Y_m)/m$ is a reasonable approximation to the expected value ν of any Y_j when m is large. This suggests estimating Δ by $\hat{\mu}_X - \hat{\mu}_Y$. Find the expected value and variance of $\hat{\mu}_X - \hat{\mu}_Y$.

(b) Use the Bienaymé-Chebychev Inequality (see Chapter 5, Section 6) and the answer you obtained to part (a) to show that the probability distribution of $\hat{\mu}_X - \hat{\mu}_Y$ concentrates more and more probability around the value Δ as the sample sizes n and m become large.

(c) If the X_i's and Y_j's are all normally distributed, find the distribution of $\hat{\mu}_X - \hat{\mu}_Y$.

(d) Under the assumptions in part (c), find the probability that $\hat{\mu}_X - \hat{\mu}_Y$ is within 1 unit of Δ when $n = m = 100$, $\sigma^2 = 4$.

(e) Assume that $n = m$ and that $\sigma^2 = 4$. How many individuals must take the drug (and how many individuals must take the placebo) in order that $\hat{\mu}_X - \hat{\mu}_Y$ is within 1 unit of Δ with probability no less than 0.95? That is, how large must n be in order that

$$P\{-1 \leq \hat{\mu}_X - \hat{\mu}_Y - \Delta \leq 1\} \geq 0.95?$$

16. In Exercise 14 of Chapter 9, a so-called "random walk" model was discussed. In this model, the mutually statistically independent motions $X_1, X_2, X_3, \ldots$ determine the path of a particle with reference to the starting point of that particle. The motion X_k of the particle at the kth step has the probability model

x	-1	1
Probability	$1-p$	p

The position U_k of the particle after k steps is equal to $X_1+X_2+\cdots+X_k$. You found the joint distribution of U_1, U_2, U_3, and U_4 in Exercise 14 of Chapter 9.
(a) Using this joint distribution, find the expected values of U_1, U_2, U_3, and U_4.
(b) Find the expected values of U_1, U_2, U_3, and U_4 directly from the expected values of X_1, X_2, X_3, and X_4.
(c) Find the variance of U_3 and the variance of U_4 from the joint distribution of U_1, U_2, U_3, and U_4. (To do this, first obtain the marginal joint distribution of U_3 and U_4.)
(d) Find the variance of U_3 and the variance of U_4 directly from the variances and covariances of X_1, X_2, X_3, and X_4.
(e) Find the correlation between U_3 and U_4 from the joint marginal distribution of U_3 and U_4.
(f) Find the correlation between U_3 and U_4 directly from the variances, covariances, and correlations of X_1, X_2, X_3, and X_4.
(g) Are U_3 and U_4 statistically independent? Justify your assertion.

11
MARKOV CHAINS

People make the mistake of talking about "natural laws." There are no natural laws. There are only temporary habits of nature.
Alfred North Whitehead

It is impossible to trap modern physics into predicting anything with perfect determinism because it deals with probabilities from the outset.
Sir Arthur Stanley Eddington

1. INTRODUCTION

In previous chapters, we have discussed ways of modeling the variation of a single random phenomenon, and ways of modeling the joint variation of two or more random phenomena. In particular, we have studied possible probability models for individual random variables, and probability models for two or more random variables considered together.

These models, when applied to physical, biological, or social processes observed over time, only offer a static picture of the variational behavior of these processes, in the sense that these models focus on the variability of the "happenings" of a process at a specified and finite number of instants of time, rather than upon the variability of the process regarded as a whole. To reveal this latter information, it is necessary to take account of the dynamic nature of most natural processes. Over an extended period of time, "happenings" at any particular time are influenced by "happenings" at previous times, and in turn influence "happenings" in the future. Although we can certainly analyze a natural process through a description of the joint variability of the qualitative or quantitative characteristics of the process at specified points in time, this method of analysis obscures the dynamics that determine how such a process evolves over time.

For example, suppose the natural process of interest to us is the size (number of individuals) of a population of, say, seals. At an initial point in time, time 0, there may be X_0 such seals (of the same species). Between this point in time and the next point in time, time 1, at which we observe the population, fatal accidents may by chance occur to some of the seals, but other seals in the population may mate and produce a random number

1. INTRODUCTION

of new offspring. If the same random forces influence all of the X_0 seals between time 0 and time 1, the size X_1 of the population of seals at time 1 will be a random variable (random because of the random forces which act to increase or reduce the seal population), but, intuitively, the magnitude of X_1 will depend, in a statistical sense (expressed in terms of conditional probabilities), on the number X_0 of seals that were alive at time 0. This assertion can be at least partially justified by considering extreme cases. For example, if X_0 is less than 2 ($X_0 = 0$ or 1), there are not enough seals available for mating to occur, and the size of the population can only stay the same (if no seals die) or decrease. If X_0 is larger, then more matings of seals can occur (assuming that nearly equal numbers of fertile male and female seals exist among the X_0 seals), and thus there is a greater likelihood that new seals will be born to take the place of those seals that die. From this argument, we see that the size X_1 of the population of seals at time 1 depends in a probabilistic sense on the size X_0 of the population of seals at time 0. Similarly, the size X_2 of the population of seals at the next time, time 2, of observation depends probabilistically on the sizes X_0, X_1 of the population of seals at times 0 and 1; the size X_3 of the population of seals at time 3 depends probabilistically on the sizes X_0, X_1, and X_2, and so on. Although we may not be able to deterministically isolate all of the causes that influence the growth or decrease of the population of seals over time, we can construct a probabilistic model for this process (using the concepts of conditional probability; see Chapter 3) in which the dynamics of growth (or decrease) of the population of seals between any two time points, time i and time j (where time i is previous to time j; that is $i < j$) is dependent on the sizes X_0, X_1, X_2, ..., X_i at time points previous to (and including) time point i. From such a model, and a knowledge of the size of the population of seals at the initial time point, time 0, we can determine the probability model which describes the variability of the size X_i of the population of seals at any time i, the probability model which describes the joint variability of the sizes X_i and X_j of the population of seals at any pair of time points time i and time j, and so on. Thus, such a dynamic probabilistic model for the process of growth (or decrease) of the population of seals offers a powerful scientific tool for the description and prediction of that process. In general, when a probabilistic model is determined for a given process, and is studied in the light of the theoretical knowledge which exists within a given scientific discipline about the process in question, new theoretical insights may be obtained concerning the causal factors which act to generate or moderate the process.

Scientists are becoming increasingly interested in modeling and understanding dynamic random processes. Biologists, ecologists, sociologists, and many concerned laymen are interested in the growth and decline of animal and human populations. Traffic engineers devote considerable attention to the dynamics of vehicular traffic with an eye toward controlling traffic flow. Physicists study the movement and collisions of small particles (molecules, atoms, and so on) in an enclosed space (such as a reactor chamber); while on a larger scale, astronomers observe the evolution of the universe. Chemists investigate the dynamics of chemical reactions, while biochemists and biophysicists explore the internal dynamics of living organisms. Finally, psychologists have embarked upon the study of the processes of learning and understanding in the individual. Although science has dealt with dynamic *mechanistic* processes since Newton's time (or before), only in this century (and, indeed, only in the past 25 years) have scientists been actively concerned with dynamic random, or stochastic, processes.

Stochastic (random) processes can be classified into two categories. In one category, that of *continuous-time stochastic processes*, the process is considered to be observable at any time t after some initial time. That is, for any number t, it is assumed that we can wait until t units of time have passed, and then instantaneously observe the "happenings" of the process at that time. Because we must consider the outcomes of a continuous-time stochastic process at all possible times t, the modeling and analysis of such processes is quite complicated. *Discrete-time stochastic processes*, on the other hand, are only assumed to be observable (or observed) at specified times. For example, the voting behavior of Americans is observable only at election times, the growth of the population of a given city is observed only at census times, the transmission of genetic material is observable in mice only when new mice are born, and the extent of learning in a child is (objectively) observed only when the child is tested. It follows from the definition of discrete-time stochastic processes that the observations of such processes occur at only a denumerable number of time points (time 0, time 1, time 2, and so on; see the footnote to Example 2.8 of Chapter 2 for comments on "denumerable"). Thus, the task of modeling and analyzing discrete-time processes is not as complex as the corresponding task for continuous-time processes. Nevertheless, discrete-time stochastic processes are of interest and importance in a wide diversity of disciplines, as can be seen by some of the examples discussed in Section 3. In the remainder of this chapter, we discuss only discrete-time stochastic processes.

2. MARKOV CHAINS

In their most general form, the "happenings" of a discrete-time stochastic process can be either qualitative or quantitative in nature. In previous chapters, we distinguished between qualitative and quantitative "happenings." However, for the purpose of discussing many stochastic processes which are of scientific interest, this distinction need not be emphasized. Indeed, it is often possible (and convenient) to describe the "happening" of a stochastic process at any time (say, time j) as being one of a denumerable collection of possible *states* for the process. For example, the process under observation may be the political affiliation of a randomly chosen individual. In this case the "states" of the process are the various political parties, plus the "state" of no party affiliation at all. Note that an observation (outcome) of this process can be described by listing the states (political affiliations) of the individual at all times (time 0, time 1, time 2, and so on) at which the individual is observed.

As another example, consider the population of seals mentioned in Section 1. At any given time (say, time j), we can describe the "state" of the seal population by counting the number X_j of seals in the population. The possible "states" of this process are then any of the nonnegative integers 0, 1, 2, 3, and so on. The total process that we observe is the change in the size (number of seals) of the seal population over time. An outcome of the random experiment which consists of observing the seal population process (for example, the growth or decline of that population) is a list of the numbers $X_0, X_1, X_2, \ldots$ of seals in the seal population at each of the various times of observation: time 0, time 1, time 2, and so on.

Even though political affiliation is a qualitative description of the "happening" observed at any time in the first process described above (the process which observes the change of political affiliation of an individual over time), and the size of the seal population is a quantitative description of the "happening" observed at any time in the second process described, the notion of the "state" of a stochastic process allows us to study the process of change of the "happenings" in both kinds of process without need for distinguishing between qualitative and quantitative "happenings."

Suppose, then, that we are interested in constructing a probability model for a discrete-time stochastic process, where at a given time t we can describe the "happenings" of the process as being one of a (denumerable) collection of states: State S_1, State S_2, State S_3, and so on. Let X_n be the index of the observed state of the process at time n. That is, $X_n = i$ if at time n the process is in State S_i. Using this notion, the outcomes of the stochastic process under consideration can now be described by listing the indices $X_0, X_1, X_2, \ldots$ of the states observed for the process at each time n, $n = 0, 1, 2, \ldots$. Thus, the outcomes of the process are sequences $(X_0, X_1, X_2, \ldots)$ of integers; knowing that the outcome $(4, 3, 6, 1, \ldots)$ has occurred, for example, means that we know that the process is in State S_4 at time 0, in State S_3 at time 1, in State S_6 at time 2, in State S_1 at time 3, and so on. Alternatively, we can describe the outcomes in the process simply in terms of the sequence $(S_4, S_3, S_6, S_1, \ldots)$ of states which are observed over time.

Example 2.1. A randomly selected individual is observed yearly by his doctor, starting when the individual is 30 years of age. At each medical observation, the cholesterol level in the individual's blood is measured. Rather than record the actual quantitative measurement of the cholesterol level in the individual, the doctor records whether the level is (1) very low, (2) low, (3) normal, (4) high, or (5) very high. Thus, the states of this stochastic process (which consist of observing the blood cholesterol level of the individual over time) are

State S_1: very low cholesterol level,
State S_2: low cholesterol level,
State S_3: normal cholesterol level,
State S_4: high cholesterol level,
State S_5: very high cholesterol level.

The times of observation are

time 0 = when the individual is 30 years of age,
time 1 = when the individual is 31 years of age,
time 2 = when the individual is 32 years of age,
.... .

Thus, if the cholesterol level in the individual's blood is "low" at age 30, "low" at age 31, "normal" at age 32, "very low" at age 33, "high" at age 34, ..., then $X_0 = 2, X_1 = 2, X_2 = 3, X_3 = 1, X_4 = 4$, and so on.

Example 2.2. An economist is studying the sale of seats on the New York Stock Exchange. Every month, starting with January 1965, he takes published reports of the Securities Exchange Commission and from these he obtains the number of Exchange seats that have been sold during that month. In this process, the states are

State S_1: 0 seats sold,
State S_2: 1 seat sold,
State S_3: 2 seats sold,
State S_4: 3 seats sold,

and so on, while the times of observations are

time 0 = January 1965,
time 1 = February 1965,
time 2 = March 1965,

and so on. Thus, if 3 seats were sold in January 1965, 1 seat was sold in February 1965, 0 seats were sold in March 1965, and so on, then $X_0 = 3$, $X_1 = 1, X_2 = 0$, and so on.

To construct probability models for discrete-time stochastic processes with a denumerable number of states, we need to assign probabilities to the events of the process, for example, the (simple) events consisting of one and only one outcome of the process. Thus, we need to determine probabilities for events of the form

$$\{X_0 = 7, X_1 = 3, X_2 = 5, X_3 = 6, \ldots\},$$

or, in general, of the form

(2.1) $$\{X_0 = i_0, X_1 = i_1, X_2 = i_2, X_3 = i_3, \ldots\},$$

where i_0, i_1, i_2, i_3, ... are integers indexing those states of the process which are observed at times 0, 1, 2, 3, and so on. However, even in the case where there are only two possible states, State S_1 and State S_2, for the process, it is not hard to see that the number of possible events of the form (2.1) is infinitely large. Thus, in any finite number of repeated trials of the random experiment which consists of observing the given stochastic process over all times (time 0, time 1, and so on), most of the possible events of the form (2.1) would not be observed at all. We faced a similar problem when we considered constructing probability models for continuous random variables (see Chapter 4). In the present situation, however, we cannot fall back on the devices (histograms, cumulative distribution functions) which were used in Chapter 4 because we are not observing a single number, but rather sequences of states. However, even assuming that we can overcome this problem and obtain probabilities of events of the form (2.1), these probabilities do not directly reveal the probabilistic dynamics of the observed process (for example, dynamics such as described for the seal population process mentioned in Section 1). More directly revealing are the conditional probabilities $P\{X_1 = i_1 | X_0 = i_0\}$, $P\{X_2 = i_2 | X_1 = i_1, X_0 = i_0\}$, $P\{X_3 = i_3 | X_2 = i_2, X_1 = i_1, X_0 = i_0\}$, and so on, which tell how the state of the process observed at a given time depends probabilistically on the states observed for the process at previous times. Note that if we are interested in the probabilities of events of the form (2.1), we can always calculate such probabilities from the conditional probabilities:

(2.2) $$P\{X_1 = i_1 | X_0 = i_0\},$$
$$P\{X_2 = i_2 | X_1 = i_1, X_0 = i_0\},$$
$$P\{X_3 = i_3 | X_2 = i_2, X_1 = i_1, X_0 = i_0\},$$

and so on, and the unconditional probability $P\{X_0 = i_0\}$ by making use of the Law of Multiplication for conditional probabilities (Chapter 3, Section 4). Knowledge of all of the conditional probabilities (2.2), plus knowledge of the unconditional probabilities $P\{X_0 = i\}$ that the process initially (at time 0) is in State S_i, for $i = 1, 2, \ldots$, is thus equivalent to knowledge of the probability model for the entire stochastic process. Because of this fact, and because the conditional probabilities (2.2) reveal how the state of the process at a given time depends on the states of the process at previous times, we say that the conditional probabilities (2.2) give the *dynamic probabilistic structure* of the given stochastic process.

In general, if we want to construct a probability model for a given stochastic process, there are just as many probabilities of the form (2.2) to be specified as there are probabilities of the form (2.1). However, there is a large and very important class of discrete-time stochastic processes in which the state of the process at any time, time k, does not depend on *all* previous states of the process (at time $k-1$, time $k-2$, ..., time 1, and time 0) but only on the state of the process observed at the immediately previous time, time $k-1$. That is, for such processes,

(2.3) $\quad P\{X_k = i_k | X_{k-1} = i_{k-1}, X_{k-2} = i_{k-2}, \ldots, X_1 = i_1, X_0 = i_0\}$
$= P\{X_k = i_k | X_{k-1} = i_{k-1}\},$

for all times k, $k = 1, 2, 3, \ldots$, and for all state indices i_0, i_1, i_2, and so on. Put into words, the present state of the process depends probabilistically on past states of the process only through the state observed in the most immediate past. This type of probabilistic dependence is called a *Markov structure*, and a discrete-time stochastic process with conditional probabilities of the form (2.3) is called a *Markov chain*, in honor of the Russian probabilist A. A. Markov (1856–1922) who developed the probability theory for such processes.

To see what a Markov structure means for the probabilistic structure of a stochastic process, suppose that such a process has only four states: State S_1, State S_2, State S_3, and State S_4. If we wish to determine the probability of the event that the process is in State S_3 at time 0, State S_4 at time 1, State S_1 at time 2, and State S_2 at time 3 (that is, if we wish to find $P\{X_0 = 3, X_1 = 4, X_2 = 1, X_3 = 2\}$), then the Law of Multiplication for conditional probabilities allows us to find this probability as follows:

$P\{X_0 = 3, X_1 = 4, X_2 = 1, \text{ and } X_3 = 2\}$
$= P\{X_3 = 2 | X_2 = 1, X_1 = 4, X_0 = 3\} P\{X_2 = 1 | X_1 = 4, X_0 = 3\}$
$\times P\{X_1 = 4 | X_0 = 3\} P\{X_0 = 3\}.$

However, the Markov structure of the process tells us that

$P\{X_3 = 2 | X_2 = 1, X_1 = 4, X_0 = 3\} = P\{X_3 = 2 | X_2 = 1\},$

and

$P\{X_2 = 1 | X_1 = 4, X_0 = 3\} = P\{X_2 = 1 | X_1 = 4\},$

so that

$P\{X_3 = 2, X_2 = 1, X_1 = 4, \text{ and } X_0 = 3\}$
$= P\{X_3 = 2 | X_2 = 1\} P\{X_2 = 1 | X_1 = 4\} P\{X_1 = 4 | X_0 = 3\} P\{X_0 = 3\}.$

2. MARKOV CHAINS

Thus, a joint probability of the form $P\{X_n = i_n, X_{n-1} = i_{n-1}, \ldots, X_2 = i_2, X_1 = i_1, X_0 = i_0\}$ can be determined from knowledge of the conditional probabilities

$$P\{X_k = i_k | X_{k-1} = i_{k-1}\}, \qquad k = 1, 2, 3, \ldots, n,$$

and the initial probability $P\{X_0 = i_0\}$. Indeed, it can be shown that *the probability of any event concerning the outcome of a Markov chain can be determined through knowledge of the initial probabilities* $P\{X_0 = i_0\}$, *and the conditional probabilities*

(2.4) $\quad P\{X_1 = i_1 | X_0 = i_0\}, \quad P\{X_2 = i_2 | X_1 = i_1\}, \quad P\{X_3 = i_3 | X_2 = i_2\}, \ldots,$

for all integers $i_0, i_1, i_2, i_3, \ldots$. For example, if there are only two possible states, State S_1 and State S_2, for a stochastic process which is a Markov chain, we need to know only the following conditional and unconditional probabilities in order to determine the probability model for the process:

Initial (unconditional) probabilities: $P\{X_0 = 1\}, P\{X_0 = 2\}$.

Transition probabilities: $P\{X_k = 1 | X_{k-1} = 1\}, \qquad P\{X_k = 2 | X_{k-1} = 1\},$

$P\{X_k = 1 | X_{k-1} = 2\}, \qquad P\{X_k = 2 | X_{k-1} = 2\}, \qquad \text{for } k = 1, 2, 3, \ldots.$

The transition probability $P\{X_k = j | X_{k-1} = i\}$ gives the conditional probability that the process will be in State S_j at time k *given* that the process is known to have been in State S_i at time $k-1$. That is, speaking nonrigorously, the transition probability $P\{X_k = j | X_{k-1} = i\}$ gives the probability that the state of the process will change (undergo a transition) from State S_i to State S_j during the time from time $k-1$ to time k. For a general Markov chain, the transition probabilities $P\{X_k = j | X_{k-1} = i\}$ have values depending on the state indices i and j *and* on the time index k. Thus, when $k > 1$, $P\{X_k = 1 | X_{k-1} = 2\}$ and $P\{X_1 = 1 | X_0 = 2\}$ are not the same. When the transition probabilities $P\{X_k = j | X_{k-1} = i\}$ of a Markov chain are independent of the time index k, so that

(2.5) $\qquad P\{X_k = j | X_{k-1} = i\} = P\{X_1 = j | X_0 = i\}$

for all $k = 1, 2, 3, \ldots$, and all States S_i and S_j, then we say that these transition probabilities are *stationary* (over time). When a Markov chain has stationary transition probabilities, then the transition probabilities $P\{X_k = j | X_{k-1} = i\}$ depend only on the prior state State S_i and the new state State S_j and *not* on the time, time k, of transition.

If a Markov chain has stationary transition probabilities, then in comparison to more general kinds of Markov chains, the task of determining a probability model for such a process is greatly simplified. For example, if there are two possible states, State S_1 and State S_2, for the process, then the probability model for the process is completely determined once we know the values of the transition probabilities:

$$p_{11} = P\{X_1 = 1 | X_0 = 1\} = P\{X_k = 1 | X_{k-1} = 1\},$$
$$p_{12} = P\{X_1 = 2 | X_0 = 1\} = P\{X_k = 2 | X_{k-1} = 1\},$$
$$p_{21} = P\{X_1 = 1 | X_0 = 2\} = P\{X_k = 1 | X_{k-1} = 2\},$$
$$p_{22} = P\{X_1 = 2 | X_0 = 2\} = P\{X_k = 2 | X_{k-1} = 2\},$$

and the initial probabilities $a_1 = P\{X_0 = 1\}$, $a_2 = P\{X_0 = 2\}$. In our discussion in this chapter, we always assume that the transition probabilities of whatever Markov chain is under discussion are stationary.

The probability model for a Markov chain having N possible states and having stationary transition probabilities can be completely determined in terms of the transition probabilities

$$p_{ij} = P\{X_1 = j | X_0 = i\} = P\{X_k = j | X_{k-1} = i\}$$

for $i, j = 1, 2, 3, \ldots, N$, and the initial probabilities

$$a_i = P\{X_0 = i\},$$

for $i = 1, 2, \ldots, N$. Once the initial probabilities $a_1, a_2, \ldots a_N$ are known, the process (Markov chain) may be represented by the array of transition probabilities given in Table 2.1, since knowledge of these transition probabilities (and the initial probabilities) tells us the probability model for the process.

Table 2.1. Array of transition probabilities.

		State at time 1				
		S_1	S_2	S_3	$\cdots$	S_N
State at Time 0	S_1	p_{11}	p_{12}	p_{13}	$\cdots$	p_{1N}
	S_2	p_{21}	p_{22}	p_{23}	$\cdots$	p_{2N}
	S_3	p_{31}	p_{32}	p_{33}	$\cdots$	p_{3N}
	.	.	.	.		.
	.	.	.	.		.
	.	.	.	.		.
	S_N	p_{N1}	p_{N2}	p_{N3}	$\cdots$	p_{NN}

The array in Table 2.1 is known as a *transition* (or *stochastic*) *matrix*. Note that since the process must be in *one* of the N states at time 1, then for any State S_i the sum (over j) of the conditional probabilities that the process moves from State S_i at time 0 to State S_j at time 1 is equal to 1. That is,

(2.6) $\quad 1 = P\{X_1 = 1, 2, 3, \ldots, \text{or } N | X_0 = i\}$
$\quad\quad = P\{X_1 = 1 | X_0 = i\} + P\{X_1 = 2 | X_0 = i\} + \cdots + P\{X_1 = N | X_0 = i\}$
$\quad\quad = p_{i1} + p_{i2} + p_{i3} + \cdots + p_{iN}.$

It follows that the row sums of Table 2.1 are all equal to 1. *The row sums of any transition matrix* (table of transition probabilities) *are always equal to 1.*

As we have progressed in this section, we have gradually reduced the scope of our discussion from general discrete-time stochastic processes to Markov chains, and then from Markov chains to Markov chains with a finite number N of states and with stationary transition probabilities. Despite these restrictions, such models have an astonishing breadth of application. In the next section, we illustrate, by means of three examples, the wide applicability of Markov chains having stationary transition probabilities.

3. APPLICATIONS

We have indicated that the theory of Markov chains is rich in applications. The following three examples serve to illustrate the breadth of use and the importance of Markov chain models.

An Example from Psychology

Example 3.1 (Mother-Infant Vocalization). The capabilities and personality of a human infant develop as a consequence of a continuous interaction with its immediate environment. As the infant develops, he influences his environment and in turn is influenced by that environment. Since the infant's mother is an important and fairly constant component of his immediate environment, the joint behavior of an infant and his mother has been of particular interest to psychologists studying infant development. One study of joint infant-mother behavior has been conducted by Freedle and Lewis (1971), who concentrate their attention on mother-infant vocalization. In deciding what to observe in the study, Freedle and Lewis define 6 distinct (and exhaustive) states of mother-infant vocalization:

(3.1) S_1: neither mother nor infant vocalize,
S_2: infant vocalizes alone,
S_3: mother vocalizes alone to infant,
S_4: mother vocalizes alone to some other person,
S_5: mother and infant both vocalize (the mother vocalizes to infant),
S_6: mother vocalizes to another person and the infant vocalizes.

In the Freedle-Lewis experiments, observations were made at 10-second intervals. This time interval was brief enough so that one and only one of the above 6 states was in effect (could occur) during any given interval of time. To obtain an approximate model for the process underlying the variation of the states of mother-infant vocalization over time, we can assume that the state of mother-infant vocalization at a given time, time $t+1$, is influenced by past states of mother-infant vocalization only through the state of vocalization at the immediately previous time, time t, of observation. If we further simplify our model by assuming that identical psychological mechanisms cause the state of vocalization at time t to influence the state of vocalization at time $t+1$ for all times $t = 0, 1, 2, \ldots$, then the process of mother-infant vocalization can be modeled probabilistically as a Markov chain having stationary transition probabilities.

Although the two assumptions mentioned above lead us to conclude that we are observing a Markov chain, they do not allow us to specify the transition probabilities of this chain. One way in which these transition probabilities can be estimated is to take one or more mother-infant pairs and observe their vocalizations (or nonvocalizations) over many successive times: time 0, time 1, time 2, time 3, and so on. Suppose we observe only one mother-infant pair. For every pair of states S_i and S_j, we can count the number n_{ij} of times in which the process is in State S_i at one time, say time t, and in State S_j at the very next time, time $t+1$. Note that i can equal j; in that case, n_{ii} gives the number of times in which the process is in State S_i at a given time, time t, and remains in that state at the next time, time $t+1$. Once $n_{11}, n_{12}, \ldots, n_{16}, n_{21}, n_{22}, \ldots, n_{26}$, and so on have been calculated, we can form the sums

(3.2)
$$n_{1.} = n_{11} + n_{12} + n_{13} + n_{14} + n_{15} + n_{16},$$
$$n_{2.} = n_{21} + n_{22} + n_{23} + n_{24} + n_{25} + n_{26},$$
$$\vdots \qquad \vdots \qquad \vdots \qquad \vdots \qquad \vdots$$
$$n_{6.} = n_{61} + n_{62} + n_{63} + n_{64} + n_{65} + n_{66}.$$

The quantity $n_{i.}$ equals the number of times (not counting the very last time observed) at which the process is in State S_i, for $i = 1, 2, 3, 4, 5, 6$. Thus, the ratio

(3.3) $$\hat{p}_{ij} = \frac{n_{ij}}{n_{i.}}$$

gives the proportion of instances in which the process changes from State S_i to State S_j *given* that the process is initially in State S_i. [*Note:* If the process is not in State S_i at *any* time during the period of observation, then $n_{i.} = 0$ and we define $\hat{p}_{ij} = n_{ij}/n_{i.}$ to be $\frac{1}{6}$ for all $j = 1, 2, 3, 4, 5, 6$.] Because $\hat{p}_{ij}$ measures the proportion of transitions from State S_i to State S_j (in one time period) among all transitions in which State S_i is the initial state of the transition, it is intuitively reasonable to use $\hat{p}_{ij}$ to estimate p_{ij}, the probability of a transition from State S_i to State S_j (in one time period) given that the transition started at State S_i. As an example of the calculation of $\hat{p}_{ij}$, suppose that a given mother-infant pair is observed at $M+1 = 41$ time points; that is, the mother and infant are observed at times 0, 1, 2, 3, ..., 40. Suppose that the observations obtained are those given in Table 3.1. Then, direct enumeration shows that the values of the

Table 3.1. Hypothetical sequence of observed states of vocalization for a given mother-infant pair.

Time	0	1	2	3	4	5	6	7	8	9	10	11	12	13	14
State	S_2	S_1	S_1	S_3	S_4	S_5	S_3	S_6	S_4	S_2	S_5	S_1	S_2	S_2	S_4

Time	15	16	17	18	19	20	21	22	23	24	25	26	27
State	S_3	S_1	S_4	S_1	S_6	S_2	S_3	S_3	S_5	S_2	S_4	S_4	S_6

Time	28	29	30	31	32	33	34	35	36	37	38	39	40
State	S_6	S_5	S_3	S_2	S_2	S_6	S_1	S_5	S_4	S_1	S_6	S_6	S_3

n_{ij}'s are given by the entries in Table 3.2. Thus, for example, $n_{11} = 1$, $n_{24} = 2$, and $n_{56} = 0$. The row sum of the ith row in Table 3.2 is equal to $n_{i1} + n_{i2} + n_{i3} + n_{i4} + n_{i5} + n_{i6} = n_{i\cdot}$. Thus, for example, $n_{3\cdot} = 6$ and $n_{5\cdot} = 5$. To obtain the estimate $\hat{p}_{ij}$ of the transition probabilities p_{ij}, we simply divide every entry in the ith row of Table 3.2 by the row sum $n_{i\cdot}$, $i = 1, 2$,

Table 3.2. Table of values of n_{ij}, $i,j = 1, 2, 3, 4, 5, 6$, as calculated from the data in Table 3.1. (The entry in the ith row and jth column is n_{ij}.)

i \ j	1	2	3	4	5	6	Row Sum
1	1	1	1	1	1	2	7
2	1	2	1	2	1	1	8
3	1	1	1	1	1	1	6
4	2	1	1	1	1	1	7
5	1	1	2	1	0	0	5
6	1	1	1	1	1	2	7
							40

3, 4, 5, 6. The result is Table 3.3. Hence, for example, based on the hypothetical data of Table 3.1, our estimate $\hat{p}_{12}$ of the conditional probability p_{12} that the state of vocalization of the mother-infant pair is State S_2 at time $t+1$ *given* that the state of vocalization at time t is State S_1 is equal to 1/7, the entry in the first row and second column of Table 3.3.

Table 3.3. Estimated transition probabilities for the mother-infant vocalization process as calculated from the hypothetical data of Table 3.1.

		State at Time $t+1$					
		S_1	S_2	S_3	S_4	S_5	S_6
State at Time t	S_1	$\frac{1}{7}$	$\frac{1}{7}$	$\frac{1}{7}$	$\frac{1}{7}$	$\frac{1}{7}$	$\frac{2}{7}$
	S_2	$\frac{1}{8}$	$\frac{2}{8}$	$\frac{1}{8}$	$\frac{2}{8}$	$\frac{1}{8}$	$\frac{1}{8}$
	S_3	$\frac{1}{6}$	$\frac{1}{6}$	$\frac{1}{6}$	$\frac{1}{6}$	$\frac{1}{6}$	$\frac{1}{6}$
	S_4	$\frac{2}{7}$	$\frac{1}{7}$	$\frac{1}{7}$	$\frac{1}{7}$	$\frac{1}{7}$	$\frac{1}{7}$
	S_5	$\frac{1}{5}$	$\frac{1}{5}$	$\frac{2}{5}$	$\frac{1}{5}$	0	0
	S_6	$\frac{1}{7}$	$\frac{1}{7}$	$\frac{1}{7}$	$\frac{1}{7}$	$\frac{1}{7}$	$\frac{2}{7}$

In one of the actual experiments performed by Freedle and Lewis (1971), a mother-infant pair was observed for a total of 720 successive 10-second time periods. The resulting estimated transition probabilities appear in Table 3.4.

Table 3.4. Estimated transition probabilities for vocalization states for highly vocal participants.

		State at Time $t+1$					
		S_1	S_2	S_3	S_4	S_5	S_6
State at Time t	S_1	0.42	0.09	0.13	0.22	0.02	0.12
	S_2	0.22	0.46	0.00	0.08	0.02	0.22
	S_3	0.18	0.04	0.51	0.12	0.05	0.10
	S_4	0.05	0.01	0.05	0.71	0.01	0.17
	S_5	0.27	0.13	0.20	0.07	0.07	0.26
	S_6	0.05	0.06	0.01	0.33	0.02	0.53

Because the properties of the Markov chain are stationary in time, we are able to estimate the transition probabilities of the vocalization process between any two successive time points t and $t+1$ from the observed transitions of the process between *all* observed successive time points (0 to 1, 1 to 2, 2 to 3, and so on). The stationarity of the process in time means that we can regard each transition of the process from one state at one time to another state at the very next time as being a repetition (but not necessarily a statistically independent repetition) of the experiment in which the transition in state of the process is observed from time t to time $t+1$. It is this point of view which suggests that we estimate the transition probabilities p_{ij} of the process from observation of the process over many times, using the method of calculation described above.

To estimate the initial (unconditional) probabilities a_1, a_2, a_3, a_4, a_5, and a_6 of the states S_1, S_2, S_3, S_4, S_5, and S_6, several repetitions of the same mother-infant vocalization process could be utilized. If m similar mother-infant pairs are observed, and if m_1 of these pairs are initially (at time 0) in State S_1, m_2 of these pairs are initially in State S_2, ..., and m_6 of these pairs are initially in State S_6, where $m = m_1 + m_2 + m_3 + m_4 + m_5 + m_6$, then we estimate the initial probability a_1 by $\hat{a}_1 = m_1/m$, the initial probability a_2 by $\hat{a}_2 = m_2/m$, and so on. Thus, if 10 similar mother-infant pairs were observed, and of these 10 pairs, 2 pairs began in State S_1 (no vocalization), then we would estimte the initial (unconditional) probability of nonvocalization (State S_1) to be $\hat{a}_1 = 2/10 = 0.20$.

Actually, however, the estimates of the initial probabilities in the Freedle-Lewis experiment described above were obtained by a quite different method. Their estimates of the initial probabilities are

$$\hat{a}_1 = 0.13, \quad \hat{a}_2 = 0.07, \quad \hat{a}_3 = 0.09, \quad \hat{a}_4 = 0.44, \quad \hat{a}_5 = 0.02, \quad \hat{a}_6 = 0.25.$$

In the previous example taken from the studies of Freedle and Lewis, the infant was female, and both mother and infant were the most vocal of all mother-infant pairs studied. In a second study, a male infant and mother who were low in vocalization behavior were chosen. The estimated transition probabilities are given in Table 3.5.

Table 3.5. Estimated transition probabilities for vocalization states for low-vocal participants.

		State at Time $t+1$					
		S_1	S_2	S_3	S_4	S_5	S_6
State at Time t	S_1	0.77	0.05	0.07	0.08	0.01	0.02
	S_2	0.71	0.12	0.07	0.05	0.00	0.05
	S_3	0.43	0.05	0.42	0.06	0.03	0.01
	S_4	0.57	0.06	0.01	0.28	0.01	0.07
	S_5	0.56	0.11	0.11	0.11	0.00	0.11
	S_6	0.41	0.11	0.07	0.15	0.00	0.26

The estimated initial probabilities that Freedle and Lewis obtain are

$$\hat{a}_1 = 0.70, \quad \hat{a}_2 = 0.06, \quad \hat{a}_3 = 0.10, \quad \hat{a}_4 = 0.09, \quad \hat{a}_5 = 0.01, \quad \hat{a}_6 = 0.04.$$

Social Processes

Example 3.2 (Conformity). The following experiment was first performed by Asch (1952). A subject is seated at the end of a row of pretrained confederates of the experimenter. Certain instructions are read aloud to all participants, so that the subject believes that he and the confederates of the experimenter are *all* part of an experiment in which certain perceptual judgments are to be made. A series of questions, each with only one clearly correct response, are put to the group. Responses to each question are given by one participant at a time, with the subject always responding last. Furthermore, the responses of the confederates of the experimenter are always identical and incorrect. The subject (who answers last) is thus faced with a choice between giving that answer to the question which he knows is correct, or conforming to the unanimous and incorrect answer of the group. Cohen (1958) carried out several experiments using this same experimental context, and compared results with those predicted by a Markov chain model in which at each trial t of the experiment, one of the following four states of mind for the subject could be in effect:

S_1: The subject is motivated to answer correctly, independent of the answers of the group, at all trials.

S_2: The subject is motivated to answer correctly, but is still indecisive as to whether or not to yield to group pressure and conform in the future.

S_3: The subject is motivated to conform to the incorrect answer of the group, but is still indecisive as to whether or not to yield to group pressure and conform in the future.

S_4: The subject is motivated to conform to the unanimous answer of the group at all trials.

States S_1 and S_4 represent "resolution" states. If the subject is in one of these states at trial t, he will continue to be in that state at future trials $t+1$, $t+2$, and so on. If in State S_1, the subject rejects the pressure to conform to the group choice and always answers according to his perception; whereas, if the subject is in State S_4, he yields to group pressure and accepts their choice. States S_2 and S_3 represent "conflict" states—the conflict being whether to maintain one's integrity or to maintain one's status in the group. If the subject is in one of these two states at the tth trial of the experiment, he or she may respond one way at trial t, but may later respond differently. For example, if the subject is in State S_2 at trial t, he or she will respond correctly on the tth trial, but may or may not respond correctly on subsequent trials.

Using the definitions of the four states S_1, S_2, S_3, S_4 of the above Markov chain, we may argue that the conditional probability that the subject remains in State S_1 at the $(t+1)$st trial *given* that the subject is in State S_1 at the tth trial is equal to 1, and thus that the conditional probability that the subject is in states $S_2, S_3,$ or S_4 at the $(t+1)$st trial given that the subject is in State S_1 at the tth trial is equal to 0. That is, $p_{11} = 1$ and $p_{12} = p_{13} = p_{14} = 0$. Similarly, $p_{44} = 1$ and $p_{41} = p_{42} = p_{43} = 0$. In constructing a Markov chain model for the type of experiment described above, Cohen

(1958) also assumes that it is not possible for a subject to make the abrupt change from an indecisively conforming state of mind (that is, State S_3) at a trial t to a resolutely independent state of mind (State S_1) at the very next trial (trial $t+1$). Thus, Cohen assumes that $p_{31} = 0$. Similarly, he assumes that a subject cannot abruptly change from a state of irresolute nonconformity (S_2) at a given trial t to a state of resolute conformity (S_4) at the very next trial (trial $t+1$), so that $p_{24} = 0$. The other transition probabilities ($p_{21}, p_{22}, p_{23}, p_{32}, p_{33}, p_{34}$) are not specified. Thus, it is hypothesized that the transition probabilities for a Markov chain which describes the process of individual reactions to group pressures has the form shown in Table 3.6.

Table 3.6. Transition probabilities for a conformity model.

		State at Trial $t+1$			
		S_1	S_2	S_3	S_4
State at Trial t	S_1	1	0	0	0
	S_2	p_{21}	p_{22}	p_{23}	0
	S_3	0	p_{32}	p_{33}	p_{34}
	S_4	0	0	0	1

In this type of experiment, the states of the process are not directly observable (we cannot read the subject's mind). However, indirect inferences about the subject's state of mind at the tth trial can be made from his responses to that and all future trials. The method by which we can estimate the transition probabilities and initial probabilities of the process through observations on the responses of the subject at each trial of the experiment is rather complex. Details of the method can be found in the article by Cohen (1958) mentioned above, or in the book by Bartos (1967).

In one experiment reported by Cohen (1958), 33 subjects were used, and the deliberate errors made by the confederates did not deviate greatly from the correct answer. The transition probabilities obtained were

		State at Trial $t+1$			
		S_1	S_2	S_3	S_4
State at Trial t	S_1	1	0	0	0
	S_2	0.06	0.76	0.18	0
	S_3	0	0.27	0.69	0.04
	S_4	0	0	0	1

In a second experiment with 27 subjects, the errors made by the confederates deviated from the correct answer as widely as possible. The transition probabilities obtained were

		\multicolumn{4}{c}{State at Trial $t+1$}
		S_1	S_2	S_3	S_4
State at Trial t	S_1	1	0	0	0
	S_2	0.13	0.48	0.39	0
	S_3	0	0.39	0.595	0.015
	S_4	0	0	0	1

Genetics

Example 3.3 (Breeding Experiments). The theory of inheritance [originated by the Austrian botanist Gregor Johann Mendel (1822–1884)] provides numerous examples of Markov chain models. The applicability of such models to genetic inheritance processes follows from the biological fact that the genetic characteristics of a child are transmitted directly from his parents. Once we know the exact genetic makeup of a child's parents, no further information about the child's potential genetic characteristics can be gained from knowledge of the genetic makeup of previous generations (for example, grandparents).

In the simplest form of an inheritance model, each individual has a pair of genes, either

$$AA, \quad Aa \quad (\text{or } aA), \quad \text{or} \quad aa,$$

which govern his or her inheritable characteristics. These three pairs of genes are called *genotypes*. For example, the A gene may carry brown eye color, and the a gene may carry blue eye color. In this case, the three genotypes are distinguishable as brown, hazel, or blue eyes.

In the mating of two parents, the offspring inherits one gene from each of its parents. Suppose that one of the parents has genotype Aa. Then this parent may donate gene A or gene a to its child; the usual inheritance model assumes that the child has equal probability $\frac{1}{2}$ of getting either one of its parent's genes. Thus, for example, a single offspring of the mating of one parent with genotype AA and one parent with genotype Aa (represented symbolically by the notation $AA \times Aa$) has genotype AA with probability $\frac{1}{2}$ and genotype Aa with probability $\frac{1}{2}$. This fact follows since such an offspring always obtains gene A from the AA parent, but may receive either gene A (resulting in genotype AA) or gene a (resulting in genotype Aa) from the Aa parent, each gene having probability $\frac{1}{2}$ of being transmitted. Similarly, a given offspring of an $Aa \times Aa$ mating receives genotype AA from its parents with probability $\frac{1}{4}$, since the event that the offspring receives gene A from its father has probability $\frac{1}{2}$, and the event that the offspring receives gene A from its mother has probability $\frac{1}{2}$, and we assume that these events are statistically independent. An offspring of the mating $Aa \times Aa$ can also receive genotype aa with probability $\frac{1}{4}$, or genotype Aa with probability $1 - \frac{1}{4} - \frac{1}{4} = \frac{1}{2}$. The probability distributions of the genotypes of a given offspring resulting from the matings of various genotypes (for the parents) are shown in Table 3.7.

Table 3.7. Probability distribution for the genotypes of an offspring resulting from the matings of various genotypes for the parents.

Mating	Probability Assigned to Offspring's Genotype		
	AA	Aa	aa
$AA \times AA$	1	0	0
$AA \times Aa$	$\frac{1}{2}$	$\frac{1}{2}$	0
$AA \times aa$	0	1	0
$Aa \times AA$	$\frac{1}{2}$	$\frac{1}{2}$	0
$Aa \times Aa$	$\frac{1}{4}$	$\frac{1}{2}$	$\frac{1}{4}$
$Aa \times aa$	0	$\frac{1}{2}$	$\frac{1}{2}$
$aa \times AA$	0	1	0
$aa \times Aa$	0	$\frac{1}{2}$	$\frac{1}{2}$
$aa \times aa$	0	0	1

A very simple type of genetic mating experiment is the following: One parent, parent I, is required to possess a specified genotype, say AA. The other parent, parent II, is selected at random from a population in which all genotypes AA, Aa, and aa are represented. From the offspring of the mating of parents I and II, one individual is selected at random and mated with a parent with the same genotype as parent I (here, AA). One offspring of that mating is selected at random, and again this offspring is mated with another parent having the same genotype (AA) as parent I. Such a process is continued over many generations. The state of this process at a given stage of the experiment is the genotype of the offspring selected at random for the next mating (with the exception of the initial stage, time 0, when the state of the process is the genotype of parent II). This process is a Markov chain with states $S_1 =$ genotype AA, $S_2 =$ genotype Aa, and $S_3 =$ genotype aa.

At time 0, the state of the process described above is any of the genotypes AA, Aa, aa. From Table 3.7 (the first 3 rows) we see that the transition probabilities for this process are as given in Table 3.8. The "time t" mentioned in Table 3.8 refers to the matings of the experiment ($t = 0$ refers to the initial mating, $t = 1$ to the next mating, and so on); the transition probabilities are stationary over time. If at state $t = 0$, parent II is selected from a population of individuals in which equal numbers of indi-

Table 3.8. Transition probabilities for genotypes arising from a mating where one parent always has genotype AA, the other parent is randomly selected from the offspring of a previous such mating, and one offspring is selected at random from their children.

		Genotype of Randomly Selected Offspring (at Time $t+1$)		
		State S_1	State S_2	State S_3
Genotype of Parent with Unspecified Genotype (at Time t)	State S_1 (AA)	1	0	0
	State S_2 (Aa)	$\frac{1}{2}$	$\frac{1}{2}$	0
	State S_3 (aa)	0	1	0

viduals have each genotype (AA, Aa, or aa), then the initial probabilities

$$a_1 = P\{\text{parent II has genotype } AA\},$$
$$a_2 = P\{\text{parent II has genotype } Aa\},$$
$$a_3 = P\{\text{parent II has genotype } aa\},$$

are all equal to $\frac{1}{3}$.

If in this same kind of experiment, the parent with specified genotype (parent I) has genotype Aa, then from rows 4 to 6 of Table 3.7 we find that the transition probability matrix of this process is the one given in Table 3.9. [*Note:* One implicit assumption used in constructing Tables 3.8 and 3.9 is that in multiple births from a given mating, each offspring's genotype is statistically independent of all of the other offsprings' genotypes.]

Table 3.9. Transition probabilities for genotypes arising from a mating where one parent always has genotype Aa, the other parent is randomly selected from the offspring of a previous such mating, and one offspring is selected at random from their children.

		Genotype of Randomly Selected Offspring (at Time $t+1$)		
		State S_1	State S_2	State S_3
Genotype of Parent with Unspecified Genotype (at Time t)	S_1: (AA)	$\frac{1}{2}$	$\frac{1}{2}$	0
	S_2: (Aa)	$\frac{1}{4}$	$\frac{1}{2}$	$\frac{1}{4}$
	S_3: (aa)	0	$\frac{1}{2}$	$\frac{1}{2}$

In a slightly more complicated experiment, the so-called "brother-sister mating," two individuals are mated. From among their descendents

3. APPLICATIONS 569

(assumed to be a large enough population to contain offspring of both sexes), two individuals of opposite sex are chosen at random (each from a population consisting of offspring of their own sex) and mated. From the descendents of this new mating, two individuals of opposite sex are chosen at random and mated. The process continues in this manner indefinitely. For this process, the states refer to the pairs of genotypes of the brother and sister selected for mating. The complete list of states is as follows:

State S_1: $AA \times AA$, State S_2: $AA \times Aa$, State S_3: $AA \times aa$,
State S_4: $Aa \times Aa$, State S_5: $Aa \times aa$, State S_6: $aa \times aa$.

In our list of states we do not distinguish between (i) a mating in which the brother has genotype AA and the sister has genotype Aa, and (ii) a mating in which the brother has genotype Aa and the sister has genotype AA. Assuming that there is no genetic (or biological) connection between the sex of an individual and the genes A and a, either such mating results in the same probability distribution for the genotypes of the offspring.

Under all of the above assumptions, the process resulting from a brother-sister mating experiment is a Markov chain with stationary transition probabilities. The probability transition matrix of the Markov chain is given in Table 3.10.

Table 3.10. Transition probabilities for a brother-sister mating model.

		State of Offspring					
		S_1	S_2	S_3	S_4	S_5	S_6
State of Parents	S_1	1	0	0	0	0	0
	S_2	$\frac{1}{4}$	$\frac{1}{2}$	0	$\frac{1}{4}$	0	0
	S_3	0	0	0	1	0	0
	S_4	$\frac{1}{16}$	$\frac{1}{4}$	$\frac{1}{8}$	$\frac{1}{4}$	$\frac{1}{4}$	$\frac{1}{16}$
	S_5	0	0	0	$\frac{1}{4}$	$\frac{1}{2}$	$\frac{1}{4}$
	S_6	0	0	0	0	0	1

To see how the transition probabilities in Table 3.10 are obtained, let us compute, say, p_{43}. Recall that p_{43} is the probability of going from State S_4 at mating t to State S_3 at mating $(t+1)$. Since State S_4 is the state in which both of the individuals mated have genotype Aa, we know from row 5 of Table 3.7 that any given offspring of this mating has probability $\frac{1}{4}$ of having genotype AA, probability $\frac{1}{2}$ of having genotype Aa, and probability $\frac{1}{4}$ of having genotype aa. There may be k males and l females in the total offspring of the mating $Aa \times Aa$; each individual offspring has the above-given probabilities for having genotypes AA, Aa, and aa, respectively. From the k males, we choose one male at random, and from the l females, we choose one female at random. In order for State S_3: $AA \times aa$ to be observed, either (i) the male chosen has genotype AA and the female chosen has genotype aa, or (ii) the male chosen has genotype aa, and the female chosen has genotype AA. Let us first compute the probability of case (i), keeping in mind that all probabilities are computed *given* that the tth mating is $Aa \times Aa$. Among the k male offspring of the mating $Aa \times Aa$, a random number, say X, have genotype AA. Since the genotypes of the offspring are mutally statistically independent, the arguments of Section 1 of Chapter 6 can be used to show that X has a binomial distribution with parameters k and $p = P\{\text{genotype } AA\} = \frac{1}{4}$. Thus,

$$P_X\{X = x\} = \binom{k}{x} p^x (1-p)^{k-x} = \binom{k}{x} \left(\frac{1}{4}\right)^x \left(\frac{3}{4}\right)^{k-x},$$

for $x = 0, 1, 2, \ldots, k$. Given that there are $X = x$ males with genotype AA among the k males in the offspring, when we choose one male at random, the probability that we obtain a male with genotype AA is equal to x/k. That is,

$$P\{\text{male chosen has genotype } AA \mid X = x\} = \frac{x}{k}.$$

Using Rule 4.6 of Chapter 3, we find that

$P\{\text{male chosen for } (t+1)\text{st mating has genotype } AA \mid t\text{th mating is } Aa \times Aa\}$

$$= \frac{0}{k} P_X\{X = 0\} + \frac{1}{k} P_X\{X = 1\} + \cdots + \frac{k}{k} P_X\{X = k\}$$

$$= \frac{1}{k} [0 P_X\{X = 0\} + 1 P_X\{X = 1\} + \cdots + k P_X\{X = k\}]$$

$$= \frac{1}{k} E(X).$$

Since X has a binomial distribution with parameters k and $p = \frac{1}{4}$, we know from Section 1 of Chapter 6 that $E(X) = kp = k(\frac{1}{4})$. Thus,

$P\{\text{male chosen for } (t+1)\text{st mating has genotype } AA \mid t\text{th}$
$\text{mating is } Aa \times Aa\}$

$$= \frac{1}{k} E(X)$$

$$= \frac{1}{k}\left(k \frac{1}{4}\right) = \frac{1}{4}.$$

A similar argument shows that

$P\{\text{female chosen for } (t+1)\text{st mating has genotype } aa \mid t\text{th mating}$
$\text{is } Aa \times Aa\}$

$$= \frac{1}{4}.$$

4. CALCULATION OF k-STEP TRANSITION PROBABILITIES

Since the choices of the male and female for the $(t+1)$st mating are statistically independent,

$P\{$male chosen has genotype AA and female chosen has genotype $aa\,|\,t$th mating is $Aa\times Aa\}$

$$= \left(\frac{1}{4}\right)\left(\frac{1}{4}\right) = \frac{1}{16}.$$

In an exactly parallel fashion we find that the probability of case (ii) is

$P\{$male chosen for $(t+1)$st mating has genotype aa and female chosen has genotype $AA\,|\,t$th mating is $Aa\times Aa\}$

$$= \frac{1}{16}.$$

Since case (i) and case (ii) are mutually exclusive events,

$p_{43} = P\{$the $(t+1)$st mating is $AA\times aa\,|\,t$th mating is $Aa\times Aa\}$

$= P\{$male chosen for $(t+1)$st mating has genotype AA *and* female chosen has genotype $aa\,|\,t$th mating is $Aa\times Aa\}$
$+ P\{$male chosen for $(t+1)$st mating has genotype aa *and* female chosen has genotype $AA\,|\,t$th mating is $Aa\times Aa\}$

$$= \frac{1}{16} + \frac{1}{16} = \frac{1}{8}.$$

Using the above form of argument, the other transition probabilities in Table 3.10 can be obtained.

Various other breeding experiments are used in genetics. Many of these experiments are describable by Markov chain models. If a Markov chain model *does* describe the generation-to-generation variation in genotypes, such a model can be used to predict the genotypes of offspring several generations in the future of a given mating. If the model of a given such breeding experiment leads to the prediction that *all* of the offspring of a future mating with high probability will be of a given desirable genotype, such a breeding experiment can then be undertaken with the expectation of producing such desirable offspring in the future. For this reason, Markov chain models can serve as a guide to agriculturists interested in breeding methods designed to improve crops and herds.

4. CALCULATION OF k-STEP TRANSITION PROBABILITIES

In Section 2, we defined what is meant by a Markov chain with stationary transition probabilities. There, we indicated that the probabilistic structure of any Markov chain with stationary transition probabilities is completely determined once we know the values of the initial probabilities, $a_i = P\{X_0 = i\}$, for all states S_i and the values of the transition probabilities

$$p_{ij} = P\{X_{t+1} = j | X_t = i\},$$

for all states S_i and S_j. In Section 3, we described three processes which have been modeled as Markov chains with stationary transition probabilities, and for which the transition probabilities have been determined. In this section, in Section 5, and in Chapter 12, we concern ourselves with certain probabilistic calculations based on the initial probabilities and the transition probabilities of a Markov chain. These calculations allow us to describe, in both quantitative and qualitative terms, various probabilistic dynamic properties of a Markov chain. In the present section, we begin discussion of such calculations by showing how to compute the conditional probability that a Markov chain will undergo a transition from State S_i to State S_j in k time units (k *time steps*) given that the process (Markov chain) begins this transition in State S_i. Such a conditional probability is called a *k-step transition probability*, and is denoted by the symbol $p_{ij}^{(k)}$. Thus,

(4.1) $$p_{ij}^{(k)} = P\{X_{t+k} = j | X_t = i\}.$$

As we will see, the conditional probability $P\{X_{t+k} = j | X_t = i\}$ does not depend for its value on the time, time t, at which the transition from State S_i to State S_j begins. This fact explains why, in Equation (4.1), we have not taken account of t in our notation for the k-step transition probability $p_{ij}^{(k)}$. Thus, $p_{ij}^{(k)} = P\{X_k = j | X_0 = i\}$, $p_{ij}^{(k)} = P\{X_{k+1} = j | X_1 = i\}$, $p_{ij}^{(k)} = P\{X_{k+2} = j | X_2 = i\}$, and so on.

Note that the transition probability p_{ij} is a special case of a k-step transition probability (namely, the case $k = 1$). Consistent with the notation in Equation (4.1), we could denote p_{ij} by $p_{ij}^{(1)}$. However, since the (1-step) transition probabilities p_{ij} are more fundamental to the definition of a Markov chain than are the k-step transition probabilities $p_{ij}^{(k)}$ for $k \neq 1$, and since no confusion between the 1-step and k-step transition probabilities should arise if we drop the superscript "(1)" when symbolizing 1-step transition probabilities, we use the notation p_{ij} for the transition probabilities $P\{X_{t+1} = j | X_t = i\}$. As we will see, knowledge of all of the transition probabilities p_{uv} (that is, knowledge of the transition matrix of the Markov chain; see Section 2) allows us to calculate any k-step transition probability $p_{ij}^{(k)}$.

2-Step Transition Probabilities

Suppose that we are interested in a Markov chain in which only two states, S_1 and S_2, are possible. Assume that at a given time, time t, the process is in State S_i. We might wish to know the conditional probability

4. CALCULATION OF k-STEP TRANSITION PROBABILITIES

$p_{ij}^{(2)}$ that the process will be in State S_j *two* time units in the future (that is, at time $t+2$). In other words, we might wish to know the value of the 2-step transition probability

(4.2) $$p_{ij}^{(2)} = P\{X_{t+2} = j | X_t = i\}.$$

If we know the basic probabilistic structure of the Markov chain, then we know the values of the (1-step) transition probabilities p_{ij}, for $i = 1, 2$, and $j = 1, 2$. From these 1-step transition probabilities, it is possible to compute the *2-step transition probabilities* $p_{ij}^{(2)}$ in the following manner:

(4.3) $$p_{ij}^{(2)} = p_{i1}p_{1j} + p_{i2}p_{2j}.$$

For example, if $i = 2, j = 1$,

(4.4) $$p_{21}^{(2)} = p_{21}p_{11} + p_{22}p_{21}.$$

The reasoning behind these formulas is as follows. To go from State S_i to State S_j in two time steps, we must first go from State S_i to some state S_l in the first time step (from time t to time $t+1$), and then from State S_l to State S_j in the next time step (from time $t+1$ to time $t+2$). The intermediate state, State S_l, visited at time $t+1$ can either be State S_1 or State S_2, but not both states. The possible transitions which allow the process to go from State S_i to State S_j in two time steps are illustrated in Figure 4.1.

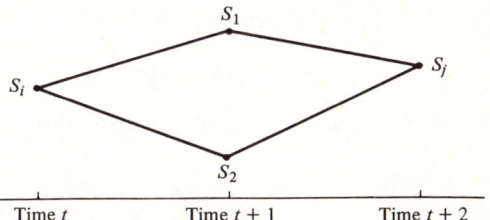

Time t Time $t+1$ Time $t+2$

Figure 4.1 The 2 possible transitions ($S_i \to S_1 \to S_j$ and $S_i \to S_2 \to S_j$) which permit the process to go from State S_i to State S_j in 2 time steps.

Given that at time t the process is in State S_i, let us calculate the conditional probability of the transition $S_i \to S_1 \to S_j$, in which the process moves first from State S_i at time t to State S_1 at time $t+1$, and then from State S_1 at time $t+1$ to State S_j at time $t+2$. We recall (see Chapter 3, Section 4) that

$$P\{X_t = i \text{ and } X_{t+1} = 1 \text{ and } X_{t+2} = j | X_t = i\}$$

(4.5)
$$= \frac{P\{X_t = i \text{ and } X_{t+1} = 1 \text{ and } X_{t+2} = j\}}{P\{X_t = i\}}$$

$$= \frac{P\{X_{t+2} = j | X_{t+1} = 1 \text{ and } X_t = i\}P\{X_{t+1} = 1 | X_t = i\}P\{X_t = i\}}{P\{X_t = i\}}$$

$$= P\{X_{t+2} = j | X_{t+1} = 1 \text{ and } X_t = i\}P\{X_{t+1} = 1 | X_t = i\}.$$

But since the process which we are observing is a Markov chain,
$$P\{X_{t+2}=j|X_{t+1}=1 \text{ and } X_t=i\} = P\{X_{t+2}=j|X_{t+1}=1\}.$$
Further, since the Markov chain has stationary transition probabilities,
$$P\{X_{t+2}=j|X_{t+1}=1\} = p_{1j},$$
$$P\{X_{t+1}=1|X_t=i\} = p_{i1}.$$
Thus, it follows from Equation (4.5) and the above facts that

(4.6) $$P\{X_{t+2}=j \text{ and } X_{t+1}=1|X_t=i\} = p_{i1}p_{1j}.$$

In a similar fashion, we can verify that

(4.7) $$P\{X_{t+2}=j \text{ and } X_{t+1}=2|X_t=i\} = p_{i2}p_{2j}.$$

Since the process under discussion has only two possible states, S_1 and S_2, and since the process is in one and only one of these states at time $t+1$, it follows that

(4.8) $$\{X_{t+2}=j\} = \{X_{t+2}=j \text{ and } X_{t+1}=1\} \cup \{X_{t+2}=j \text{ and } X_{t+1}=2\},$$

and that the events $\{X_{t+2}=j \text{ and } X_{t+1}=1\}$ and $\{X_{t+2}=j \text{ and } X_{t+1}=2\}$ are mutually exclusive. Hence [see Rule 4.3(a) of Chapter 3], from Equations (4.6), (4.7), and (4.8), we conclude that

$$p_{ij}^{(2)} = P\{X_{t+2}=j|X_t=i\} = P\{X_{t+2}=j \text{ and } X_{t+1}=1|X_t=i\}$$
$$+ P\{X_{t+2}=j \text{ and } X_{t+1}=2|X_t=i\} = p_{i1}p_{1j} + p_{i2}p_{2j},$$

which verifies Equation (4.3).

Example 4.1. (Learning Models). There are currently many variations, extensions, and embellishments of the original formulation of statistical learning theory by Estes (1950). Here we describe a simple learning process that is assumed to be a Markov chain with stationary transition probabilities. There are two states:

S_1: the subject is conditioned to make the correct response
(for example, a rat is conditioned to press a bar),
S_2: the subject is not conditioned to make the correct response.

Once a subject has been put in a conditioned state, the subject remains in that state from that time on; that is, at any time t,

$$p_{11} = P\{X_{t+1}=1|X_t=1\} = 1,$$
$$p_{12} = P\{X_{t+1}=2|X_t=1\} = 0.$$

4. CALCULATION OF k-STEP TRANSITION PROBABILITIES

If a subject is in the unconditioned state (S_2) at time t, it may become conditioned (move to state S_1) at time $t+1$ with some probability α, $0 \leq \alpha \leq 1$. Thus,

$$p_{21} = P\{X_{t+1} = 1 | X_t = 2\} = \alpha,$$

and

$$p_{22} = P\{X_{t+1} = 2 | X_t = 2\} = 1 - p_{21} = 1 - \alpha.$$

The transition matrix for this simple Markov chain is shown in Table 4.1.

Table 4.1. Transition probabilities for a 2-state Markov chain learning model.

		State at Time $t+1$	
		S_1	S_2
State at Time t	S_1	1	0
	S_2	α	$1-\alpha$

Suppose that at a given time, time t, the subject is in an unconditioned state (S_2). We might like to know the conditional probability $p_{21}^{(2)}$ that the subject will be in a conditioned state (S_1) two time units from time t (that is, at time $t+2$). From Equation (4.4) and Table 4.1, we find that

$$p_{21}^{(2)} = p_{21}p_{11} + p_{22}p_{21}$$
$$= (\alpha)(1) + (1-\alpha)(\alpha) = 2\alpha - \alpha^2.$$

For example, if $\alpha = \tfrac{1}{2}$,

$$p_{21}^{(2)} = 2\left(\frac{1}{2}\right) - \left(\frac{1}{2}\right)^2 = \frac{3}{4},$$

while if $\alpha = \tfrac{1}{4}$,

$$p_{21}^{(2)} = 2\left(\frac{1}{4}\right) - \left(\frac{1}{4}\right)^2 = \frac{7}{16}.$$

If we want to know the conditional probability that the subject will return to the unconditioned state (S_2) two time units from time t *given* that at time t the subject is in the unconditioned state (S_2), then from Equation (4.3) and Table 4.1,

$$p_{22}^{(2)} = P\{X_{t+2} = 2 | X_t = 2\}$$
$$= p_{21}p_{12} + p_{22}p_{22}$$
$$= (\alpha)(0) + (1-\alpha)(1-\alpha) = (1-\alpha)^2.$$

Alternatively, we can calculate $p_{22}^{(2)}$ by noting that the subject must be in either a conditioned state (S_1) or an unconditioned state (S_2) at time $t+2$.

Thus,
$$\begin{aligned}p_{22}^{(2)} &= P\{X_{t+2}=2\,|\,X_t=2\}\\ &= 1-P\{X_{t+2}=1\,|\,X_t=2\}=1-p_{21}^{(2)}\\ &= 1-(2\alpha-\alpha^2)\\ &= (1-\alpha)^2,\end{aligned}$$
which agrees with the result directly obtained using Equation (4.3).

Suppose that the Markov chain which we are considering has N states, instead of 2 states. In this more general case, we can calculate the 2-step transition probability $p_{ij}^{(2)}$ from the (1-step) transition probabilities by means of the following equation:

(4.9) $$p_{ij}^{(2)} = p_{i1}p_{1j}+p_{i2}p_{2j}+\cdots+p_{iN}p_{Nj}.$$

The verification of Equation (4.9) proceeds in the exact same way as did the verification of Equation (4.3). That is, first we note that for any state $S_l, l=1,2,3,\ldots,N$,

(4.10) $$\begin{aligned}P\{X_{t+2}=j \text{ and } X_{t+1}=l\,|\,X_t=i\} &\\ = P\{X_{t+2}=j\,|\,X_{t+1}=l \text{ and } X_t=i\} \quad &P\{X_{t+1}=l\,|\,X_t=i\}\\ = P\{X_{t+2}=j\,|\,X_{t+1}=l\} \quad &P\{X_{t+1}=l\,|\,X_t=i\}\\ = p_{lj}p_{il},&\end{aligned}$$

where the first equality follows using the definition of conditional probability and the Law of Multiplication (Chapter 3, Rule 4.4), the second equality follows since the process is a Markov chain, and the third equality follows from the definitions of p_{lj} and p_{il} (using the fact that these conditional probabilities are stationary over time). We next observe that

$$\begin{aligned}\{X_{t+2}=j\} &= \{X_{t+2}=j \text{ and } X_{t+1}=1\}\cup\{X_{t+2}=j \text{ and } X_{t+1}=2\}\\ &\quad\cup\cdots\cup\{X_{t+2}=j \text{ and } X_{t+1}=N\},\end{aligned}$$

and that the events on the right-hand side of this equation are mutually exclusive. Consequently,

(4.11) $$\begin{aligned}P\{X_{t+2}=j\,|\,X_t=i\} &= P\{X_{t+2}=j \text{ and } X_{t+1}=1\,|\,X_t=i\}\\ &\quad+P\{X_{t+2}=j \text{ and } X_{t+1}=2\,|\,X_t=i\}+\cdots\\ &\quad+P\{X_{t+2}=j \text{ and } X_{t+1}=N\,|\,X_t=i\}.\end{aligned}$$

We can now verify Equation (4.9) by substituting the results (4.10) into Equation (4.11).

Example 4.2 (Occupational Mobility). The process by which the male

4. CALCULATION OF k-STEP TRANSITION PROBABILITIES

wage-earners in a given family change in occupational status from generation to generation has been of considerable interest both to sociologists and to economists. If, in a given society, sons almost always are observed to follow their father's occupation, or if sons do not rise above (or fall below) the job status of their fathers, then there is good reason for believing that such a society permits individuals little freedom to change the role of life into which they were born. On the other hand, if, in a given society, sons are relatively likely to have a different job or job status as compared to their fathers, and in particular, if the job status of sons tends to be better (higher) than that of their fathers, then the society would seem to offer greater opportunities to talented and ambitious individuals to improve their vocational status.

Berger and Snell (1957) have proposed a Markov chain model for occupational mobility in which the times $0, 1, 2, \ldots$ of observation are generations within families, and the states are levels of occupation. Three occupation levels (states) are distinguished:

S_1: upper level occupations (professional, high administrative, managerial, executive),
S_2: middle level occupations (higher grade supervisory and non-manual, skilled manual and nonmanual),
S_3: lower level occupation (semiskilled manual, unskilled manual).

Glass and Hall (1954) have obtained data on the occupational status of male residents of England and Wales and on the occupational status of their fathers. Using these data, a table (Table 4.2) of transition probabilities for occupational mobility can be obtained (estimated). Here, we are implicitly assuming that the forces acting to cause transitions in occupational levels between generations are the same from generation to generation. This assumption, which is reasonable for a stable economic and social environment such as is found in England and Wales, implies that the Markov chain which models occupational mobility in England and Wales has stationary transition probabilities.

Table 4.2. Transition probabilities for level of occupation.

Occupation Level of Fathers (Time t)		Occupation Level of Sons (Time $t+1$)		
		S_1	S_2	S_3
	S_1	0.45	0.48	0.07
	S_2	0.05	0.70	0.25
	S_3	0.01	0.50	0.49

Suppose we know that the male wage-earner of the (t)th generation of a certain family has a lower level occupation [that is, the process of occupational mobility for this family is at State S_3 in the (t)th generation]. What is the (conditional) probability that this individual's (eldest) grandson [the ($t+2$)nd generation of the family] will be in the highest occupational level (S_1)? We are, in other words, asking for the value of $p_{31}^{(2)}$. From Equation (4.9) and Table 4.2, we find that

$$p_{31}^{(2)} = p_{31}p_{11} + p_{32}p_{21} + p_{33}p_{31}$$
$$= (0.01)(0.45) + (0.50)(0.05) + (0.49)(0.01)$$
$$= 0.0344.$$

Thus, in England and Wales, there is a conditional probability of only 0.0344 that a male wage-earner will have an occupation in the highest occupation level *given* that his grandfather had an occupation in the lowest occupational level. In less formal terms, the probability in England and Wales of a two-generation rise "from rags to riches" is 0.0344.

3-Step Transition Probabilities

We have shown how to compute 2-step transition probabilities $p_{uw}^{(2)}$. A reasonable question to ask next is "How can we calculate the 3-step transition probabilities $p_{ij}^{(3)} = P\{X_{t+3} = j | X_t = i\}$?"

In finding the 2-step transition probabilities $p_{uw}^{(2)}$, we argued that to go from State S_u at time t to State S_w at time $t+2$ the process has to first go from State S_u at time t to some State S_v at one time unit later (time $t+1$), and then from State S_v at time $t+1$ to State S_w one time unit after that (that is, time $t+2$). Thus, one and only one of the transitions $S_u \to S_1 \to S_w, S_u \to S_2 \to S_w, \ldots, S_u \to S_N \to S_w$, must occur if the 2-step transition $S_u \to \to S_w$ is to occur. Thus, $p_{uw}^{(2)}$ equals the sum of probabilities of the form $P\{X_{t+2} = w, X_{t+1} = v | X_t = u\}$, for $v = 1, 2, \ldots, N$. An argument making use of the fact that the process is a Markov chain with stationary transition probabilities shows that $P\{X_{t+2} = w, X_{t+1} = v | X_t = u\} = p_{uv}p_{vw}$, so that

$$p_{uw}^{(2)} = p_{u1}p_{1w} + p_{u2}p_{2w} + \cdots + p_{uN}p_{Nw}.$$

The above argument can be extended to allow us to calculate $p_{ij}^{(3)}$. To go from State S_i at time t to State S_j at time $t+3$, the process must go through two intermediate states, S_h and S_l, at times $t+1$ and $t+2$, respectively. For example, if there are only two possible states S_1 and S_2 for the process, then in order for the process to go from S_i at time t to S_j at time $t+3$, one of the four transitions shown in Figure 4.2 must occur. Thus, using arguments similar to those used to derive Equation (4.3), we can show that

(4.12) $$p_{ij}^{(3)} = p_{i1}p_{11}p_{1j} + p_{i1}p_{12}p_{2j} + p_{i2}p_{21}p_{1j} + p_{i2}p_{22}p_{2j},$$

4. CALCULATION OF k-STEP TRANSITION PROBABILITIES

$$S_i \to S_1 \to S_1 \to S_j$$
$$S_i \to S_1 \to S_2 \to S_j$$
$$S_i \to S_2 \to S_1 \to S_j$$
$$S_i \to S_2 \to S_2 \to S_j$$

Time $\quad t \quad t+1 \quad t+2 \quad t+3$

Figure 4.2. The four possible transitions which permit the process to go from State S_i at time t to State S_j three time units later.

where the first term in the sum on the right-hand side of this equation gives the (conditional) probability that the process undergoes a transition from S_i to S_1, then from S_1 to S_1 and finally from S_1 to S_j, the second term gives the (conditional) probability that the process moves from S_i to S_1, then from S_1 to S_2, and finally from S_2 to S_j, and so on. In general, when the Markov chain has N states, we can show that

(4.13) $\quad p_{ij}^{(3)} = p_{i1}p_{11}p_{1j} + p_{i1}p_{12}p_{2j} + \cdots + p_{i1}p_{1N}p_{Nj} + p_{i2}p_{21}p_{1j}$
$\quad\quad\quad + p_{i2}p_{22}p_{2j} + \cdots + p_{i2}p_{2N}p_{Nj} + \cdots + p_{iN}p_{N1}p_{1j}$
$\quad\quad\quad + p_{iN}p_{N2}p_{2j} + \cdots + p_{iN}p_{NN}p_{Nj}.$

The sum on the right-hand side of Equation (4.13) has N^2 terms, each term consisting of a product of three numbers. If N is even moderately large, calculation of $p_{ij}^{(3)}$ by means of Equation (4.13) can become tedious. In many cases where we are interested in calculating $p_{ij}^{(3)}$, however, we may have already calculated the 2-step transition probabilities $p_{uw}^{(2)}$. Thus, a formula which expresses $p_{ij}^{(3)}$ in terms of the 2-step transition probabilities $p_{uw}^{(2)}$ and the (1-step) transition probabilities p_{hl} would be of considerable help to us in such situations. From Equations (4.9) and (4.13), we see that

$$p_{ij}^{(3)} = p_{i1}(p_{11}p_{1j} + p_{12}p_{2j} + \cdots + p_{1N}p_{Nj})$$
$$+ p_{i2}(p_{21}p_{1j} + p_{22}p_{2j} + \cdots + p_{2N}p_{Nj}) + \cdots$$
$$+ p_{iN}(p_{N1}p_{1j} + p_{N2}p_{2j} + \cdots + p_{NN}p_{Nj})$$
$$= p_{i1}p_{1j}^{(2)} + p_{i2}p_{2j}^{(2)} + \cdots + p_{iN}p_{Nj}^{(2)}.$$

Thus,

(4.14) $\quad p_{ij}^{(3)} = p_{i1}p_{1j}^{(2)} + \cdots + p_{iN}p_{Nj}^{(2)}.$

We can justify Equation (4.14) heuristically by noting that to get from State S_i at time t to State S_j at time $t+3$, the process can move from State S_i at time t to some state, State S_h, one time unit later (that is, at time $t+1$), and then from State S_h at time $t+1$ to State S_j two time steps after that (at time $t+3$). The first term in the sum on the right-hand side of Equation (4.14) gives the conditional probability that the process moves from State S_i to State S_1 in one time step and then from State S_1 to State S_j in two time steps, given that the process is initially in State S_i. Similarly, the last term in the sum on the right-hand side of Equation (4.14) gives the conditional probability that the process moves from State S_i to State S_N in one time step, and then from State S_N to Step S_j in two time steps, given that the process initially is in State S_i. The probability $p_{ij}^{(3)}$ is the sum of these N terms ($p_{i1}p_{1j}^{(2)}$, $p_{i2}p_{2j}^{(2)}$, and so on) since at time $t+1$ the process must be in one and only one of the states $S_1, S_2, ..., S_N$.

Alternatively, we may think of the process going from State S_i at time t to State S_j at time $t+3$ in the following manner: (i) First, the process moves from State S_i to some state, State S_h, in two time steps, and (ii) the process then moves from State S_h to State S_j in one time step. This way of looking at the transition from State S_i at time t to State S_j at time $t+3$ leads to the equation

$$(4.15) \qquad p_{ij}^{(3)} = p_{i1}^{(2)}p_{1j} + p_{i2}^{(2)}p_{2j} + \cdots + p_{iN}^{(2)}p_{Nj}$$

for $p_{ij}^{(3)}$. Verification of Equation (4.15) can be accomplished by grouping appropriate terms in Equation (4.13) and then making use of Equation (4.9). For example, when $N = 2$, we know from Equation (4.13) [or from Equation (4.12)] that

$$p_{ij}^{(3)} = p_{i1}p_{11}p_{1j} + p_{i1}p_{12}p_{2j} + p_{i2}p_{21}p_{1j} + p_{i2}p_{22}p_{2j}$$
$$= (p_{i1}p_{11} + p_{i2}p_{21})p_{1j} + (p_{i1}p_{12} + p_{i2}p_{22})p_{2j}.$$

Thus, making use of Equation (4.9) [or of Equation (4.3)],

$$p_{ij}^{(3)} = p_{i1}^{(2)}p_{1j} + p_{i2}^{(2)}p_{2j}.$$

Example 4.1 (Continued). Recall that in the simple learning model described in this example, there are two states to the learning process:

S_1: the subject is conditioned to make the correct response,
S_2: the subject is not conditioned to make the correct response,

and the learning process is a Markov chain with the following transition matrix:

		State at Time $t+1$	
		S_1	S_2
State at Time t	S_1	1	0
	S_2	α	$1-\alpha$

where α is a number between 0 and 1. If we want to find the conditional probability $p_{21}^{(3)}$ that a subject will be in the conditioned state (S_1) three

4. CALCULATION OF k-STEP TRANSITION PROBABILITIES

time units after a time, time t, in which the subject was observed to be in the unconditioned state (S_2), then from Equation (4.14) with $N=2$, $i=2, j=1$, we see that

$$p_{21}^{(3)} = p_{21}p_{11}^{(2)} + p_{22}p_{21}^{(2)}$$
$$= \alpha p_{11}^{(2)} + (1-\alpha) p_{21}^{(2)}.$$

Earlier, we found that $p_{21}^{(2)} = 2\alpha - \alpha^2$. To obtain $p_{11}^{(2)}$ we can use Equation (4.3), and we find that

$$p_{11}^{(2)} = p_{11}p_{11} + p_{12}p_{21} = (1)(1) + (0)(\alpha) = 1.$$

Thus,

$$p_{21}^{(3)} = (\alpha)(1) + (1-\alpha)(2\alpha - \alpha^2) = 3\alpha - 3\alpha^2 + \alpha^3.$$

If $\alpha = \frac{1}{2}$, then $p_{21}^{(3)} = \frac{7}{8}$.

We could also have computed $p_{21}^{(3)}$ using Equation (4.15). This equation tells us that

$$p_{21}^{(3)} = p_{21}^{(2)}p_{11} + p_{22}^{(2)}p_{21}$$
$$= (p_{21}^{(2)})(1) + (p_{22}^{(2)})(\alpha).$$

Earlier, we found that $p_{21}^{(2)} = 2\alpha - \alpha^2$ and that $p_{22}^{(2)} = (1-\alpha)^2$. Thus,

$$p_{21}^{(3)} = (2\alpha - \alpha^2)(1) + (1-\alpha)^2 \alpha$$
$$= 3\alpha - 3\alpha^2 + \alpha^3,$$

which agrees with the answer for $p_{21}^{(3)}$ obtained using Equation (4.14).

Example 4.2 (Continued). Earlier, we have described a 3-state Markov chain model of occupational mobility in which the states are

S_1: upper occupation level,
S_2: middle occupation level,
S_3: lower occupation level,

and the transition probability matrix is

		State at Time $t+1$		
		S_1	S_2	S_3
State at Time t	S_1	0.45	0.48	0.07
	S_2	0.05	0.70	0.25
	S_3	0.01	0.50	0.49

We previously calculated the conditional probability $p_{31}^{(2)}$ that a son will be in the upper occupation level given that his grandfather was in the lower occupation level. We can now calculate, using Equation (4.14) or (4.15), the conditional probability $p_{31}^{(3)}$ that a son will be in the upper occupation level given that his great-grandfather was in the lower occupation level. For example, using Equation (4.14),

$$p_{31}^{(3)} = p_{31}p_{11}^{(2)} + p_{32}p_{21}^{(2)} + p_{33}p_{31}^{(2)}$$
$$= (0.01)p_{11}^{(2)} + (0.50)p_{21}^{(2)} + (0.49)p_{31}^{(2)}.$$

We already know that $p_{31}^{(2)} = 0.0344$. From Equation (4.9),

$$p_{11}^{(2)} = p_{11}p_{11} + p_{12}p_{21} + p_{13}p_{31}$$
$$= (0.45)(0.45) + (0.48)(0.05) + (0.07)(0.01)$$
$$= 0.2272,$$

and

$$p_{21}^{(2)} = p_{21}p_{11} + p_{22}p_{21} + p_{23}p_{31}$$
$$= (0.05)(0.45) + (0.70)(0.05) + (0.25)(0.01)$$
$$= 0.0600.$$

Thus,

$$p_{31}^{(3)} = (0.01)(0.2272) + (0.50)(0.0600) + (0.49)(0.0344)$$
$$= 0.0491.$$

Note that $p_{31}^{(3)}$ is greater than $p_{31}^{(2)} = 0.0344$. The (conditional) probability that the occupational level of the male wage-earner of a family rises from low occupational level to high occupational level in three generations is greater than the conditional probability that this rise in occupational status takes place in two generations.

General k-Step Transition Probabilities

The kinds of arguments needed to calculate k-step transition probabilities $p_{ij}^{(k)} = P\{X_{t+k} = j \mid X_t = i\}$ are the same as those used to calculate 2-step and 3-step transition probabilities. We may think of the process moving from State S_i at time t to State S_j at time $t+k$ in the following manner: (i) First, the process moves from State i at time t to some state, State S_l, at time $t+1$; (ii) then the process moves from State S_l at time $t+1$ to State S_j at time $t+k$ ($k-1$ time units later).

Note that at time $t+1$, the process must be in one and only one of the N possible states (State S_1, State S_2, ..., State S_N) of the Markov chain. Furthermore, the transition from State S_i to State S_l in one unit of time (time t to time $t+1$), and from State S_l to State S_j in $(k-1)$ units of time (time $t+1$ to time $t+k$) can be shown (by arguments similar to those used

4. CALCULATION OF k-STEP TRANSITION PROBABILITIES

in the case $k = 3$) to be equal to the product $p_{il}p_{lj}^{(k-1)}$ of the 1-step transition probability p_{il} and the $(k-1)$-step transition probability $p_{lj}^{(k-1)}$. Consequently, it follows that

(4.16) $$p_{ij}^{(k)} = p_{i1}p_{1j}^{(k-1)} + p_{i2}p_{2j}^{(k-1)} + \cdots + p_{iN}p_{Nj}^{(k-1)}.$$

We may also think of going from State S_i to State S_j in k time units by: (i) going from State S_i to some state, State S_l, in $(k-1)$ time units, and then (ii) going from State S_l to State S_j in one time unit. Hence, we can show that

(4.17) $$p_{ij}^{(k)} = p_{i1}^{(k-1)}p_{1j} + p_{i2}^{(k-1)}p_{2j} + \cdots + p_{iN}^{(k-1)}p_{Nj}.$$

Using either Equation (4.16) or (4.17), all k-step transition probabilities can be recursively calculated by first calculating all 2-step transition probabilities from the (1-step) transition probabilities p_{uv}, next calculating all 3-step transition probabilities from 1-step and 2-step transition probabilities, next calculating all 4-step transition probabilities from 1-step and 3-step transition probabilities, and so on. Thus, finally, the k-step transition probabilities are calculated from the 1-step and $(k-1)$-step transition probabilities.

Example 4.1 (Continued). In this 2-state Markov chain, the (1-step) transition probabilities are

		State at Time $t+1$	
		S_1	S_2
State at Time t	S_1	1	0
	S_2	α	$1-\alpha$

where $0 \leq \alpha \leq 1$, S_1 is the state in which the subject is conditioned to give a correct response, and S_2 is the state in which the subject is not conditioned to give a correct response. We have already found the 2-step transition probabilities: $p_{11}^{(2)} = 1$, $p_{21}^{(2)} = 2\alpha - \alpha^2$, $p_{22}^{(2)} = (1-\alpha)^2$. It can be shown that the remaining 2-step transition probability $p_{12}^{(2)}$ equals 0.

The 2-step transition probabilities can thus be given in Table 4.3.

Table 4.3. Table of 2-step transition probabilities.

		State at Time $t+2$	
		S_1	S_2
State at Time t	S_1	1	0
	S_2	$2\alpha - \alpha^2$	$(1-\alpha)^2$

The 3-step transition probability $p_{21}^{(3)} = 3\alpha - 3\alpha^2 + \alpha^3$ was also found earlier. The other 3-step transition probabilities can be computed using either Equation (4.14) or Equation (4.15), whichever is more convenient. These 3-step transition probabilities are summarized in Table 4.4.

Table 4.4. Table of 3-step transition probabilities.

		State at Time $t+3$	
		S_1	S_2
State at Time t	S_1	1	0
	S_2	$3\alpha - 3\alpha^2 + \alpha^3$	$(1-\alpha)^3$

Turning next to the 4-step transition probabilities $p_{ij}^{(4)}$, these can be computed from the 1-step and 3-step transition probabilities by use of Equation (4.16), or by use of Equation (4.17). Using Equation (4.16) with $N=2, k=4$,

$$p_{11}^{(4)} = p_{11}p_{11}^{(3)} + p_{12}p_{21}^{(3)} = (1)(1) + (0)(3\alpha - 3\alpha^2 + \alpha^3) = 1,$$
$$p_{12}^{(4)} = p_{11}p_{12}^{(3)} + p_{12}p_{22}^{(3)} = (1)(0) + (0)(1-\alpha)^3 = 0,$$
$$p_{21}^{(4)} = p_{21}p_{11}^{(3)} + p_{22}p_{21}^{(3)} = (\alpha)(1) + (1-\alpha)(3\alpha - 3\alpha^2 + \alpha^3)$$
$$= 4\alpha - 6\alpha^2 + 4\alpha^3 - \alpha^4,$$
$$p_{22}^{(4)} = p_{21}p_{12}^{(3)} + p_{22}p_{22}^{(3)} = (\alpha)(0) + (1-\alpha)(1-\alpha)^3 = (1-\alpha)^4,$$

and these 4-step transition probabilities are summarized in Table 4.5.

Table 4.5. Table of 4-step transition probabilities.

		State at Time $t+4$	
		S_1	S_2
State at Time t	S_1	1	0
	S_2	$4\alpha - 6\alpha^2 + 4\alpha^3 - \alpha^4$	$(1-\alpha)^4$

Because a process must be in some state, State S_l, at time $t+k$, it must always be the case that the row sums of tables of k-step transition probabilities (such as Tables 4.3, 4.4, and 4.5) are all equal to 1. That is, if a Markov chain has N states, $S_1, S_2, \ldots, S_N$, then for any number of time steps k and for any initial state i,

(4.18) $$p_{i1}^{(k)} + p_{i2}^{(k)} + \cdots + p_{iN}^{(k)} = 1.$$

We can verify the truth of Equation (4.18) in Example 4.1. Here, $N=2$, and $k=2, 3$, or 4. Looking at Tables 4.3, 4.4, and 4.5, we see that the

4. CALCULATION OF k-STEP TRANSITION PROBABILITIES

sum of the entries in the first row of each table is

$$p_{11}^{(k)} + p_{12}^{(k)} = 1 + 0 = 1.$$

The sum of the entries in the second row of Table 4.3 and Table 4.4 are

$$p_{21}^{(2)} + p_{22}^{(2)} = (2\alpha - \alpha^2) + (1-\alpha)^2 = 2\alpha - \alpha^2 + 1 - 2\alpha + \alpha^2 = 1,$$
$$p_{21}^{(3)} + p_{22}^{(3)} = (3\alpha - 3\alpha^2 + \alpha^3) + (1-\alpha)^3 = 1,$$

while the sum of the entries in the second row of Table 4.5 is

$$p_{21}^{(4)} + p_{22}^{(4)} = (4\alpha - 6\alpha^2 + 4\alpha^3 - \alpha^4) + (1-\alpha)^4 = 1.$$

Example 4.2 (Continued). In this 3-state Markov chain for occupational mobility, the (1-step) transition probabilities are shown in Table 4.6.

Table 4.6. Table of 1-step transition probabilities for the process of occupational mobility.

		State at Time $t+1$		
		S_1	S_2	S_3
State at Time t	S_1	0.45	0.48	0.07
	S_2	0.05	0.70	0.25
	S_3	0.01	0.50	0.49

We have already found that $p_{11}^{(2)} = 0.2272$, $p_{21}^{(2)} = 0.0600$, and $p_{31}^{(2)} = 0.0344$. Using Equation (4.9), we find that

$$\begin{aligned} p_{12}^{(2)} &= p_{11}p_{12} + p_{12}p_{22} + p_{13}p_{32} \\ &= (0.45)(0.48) + (0.48)(0.70) + (0.07)(0.50) \\ &= 0.5870, \end{aligned}$$

$$\begin{aligned} p_{22}^{(2)} &= p_{21}p_{12} + p_{22}p_{22} + p_{23}p_{32} \\ &= (0.05)(0.48) + (0.70)(0.70) + (0.25)(0.50) \\ &= 0.6390, \end{aligned}$$

and

$$\begin{aligned} p_{32}^{(2)} &= p_{31}p_{12} + p_{32}p_{22} + p_{33}p_{32} \\ &= (0.01)(0.48) + (0.50)(0.70) + (0.49)(0.50) \\ &= 0.5998. \end{aligned}$$

Instead of using Equation (4.9) to compute $p_{31}^{(2)}$, $p_{32}^{(2)}$, and $p_{33}^{(2)}$, we can instead use Equation (4.18), with $k=2$, and obtain

$$p_{13}^{(2)} = 1 - p_{11}^{(2)} - p_{12}^{(2)} = 1 - 0.2272 - 0.5870 = 0.1858,$$
$$p_{23}^{(2)} = 1 - p_{21}^{(2)} - p_{22}^{(2)} = 1 - 0.0600 - 0.6390 = 0.3010,$$

and

$$p_{33}^{(2)} = 1 - p_{31}^{(2)} - p_{32}^{(2)} = 1 - 0.0344 - 0.5998 = 0.3658.$$

Thus, the 2-step transition probabilities of this Markov chain are summarized in Table 4.7.

Table 4.7. Table of 2-step transition probabilities for the process of occupational mobility.

		State at Time $t+2$		
		S_1	S_2	S_3
State at Time t	S_1	0.2272	0.5870	0.1858
	S_2	0.0600	0.6390	0.3010
	S_3	0.0344	0.5998	0.3658

Using Tables 4.6 and 4.7, and either Equation (4.14) or Equation (4.15), we can find the 3-step transition probabilities of this Markov chain and can summarize them in Table 4.8.

Table 4.8. Table of 3-step transition probabilities for the process of occupational mobility.

		State at Time $t+3$		
		S_1	S_2	S_3
State at Time t	S_1	0.133	0.613	0.254
	S_2	0.062	0.627	0.311
	S_3	0.049	0.619	0.332

Using Tables 4.6 and 4.8, and either Equation (4.16) or Equation (4.17), we may compute the 4-step transition probabilities $p_{ij}^{(4)}$, and summarize these probabilities in Table 4.9. Using Tables 4.6 and 4.9, and either Equation (4.16) or Equation (4.17), we can then compute the 5-step transition probabilities. These transition probabilities are summarized in Table 4.10. Although we have not computed the 6-step transition probabilities $p_{ij}^{(6)}$, these probabilities can be obtained from Tables 4.6 and 4.10 by use of either Equation (4.16) or Equation (4.17). Further, from the 6-step transition probabilities, we can obtain the 7-step transition probabilities; from these probabilities we can obtain the 8-step transition probabilities; and so on. Since doing such calculations by hand soon becomes a long and cumbersome task, in practice, k-step transition probabilities are usually obtained through the use of the computer.

4. CALCULATION OF k-STEP TRANSITION PROBABILITIES

Table 4.9. Table of 4-step transition probabilities for the process of occupational mobility.

		State at Time $t+4$		
		S_1	S_2	S_3
State at Time t	S_1	0.093	0.620	0.287
	S_2	0.062	0.624	0.314
	S_3	0.056	0.623	0.321

Table 4.10. Table of 5-step transition probabilities for the process of occupational mobility.

		State at Time $t+5$		
		S_1	S_2	S_3
State at Time t	S_1	0.076	0.622	0.302
	S_2	0.062	0.624	0.314
	S_3	0.060	0.623	0.317

An Alternative Method for Calculating k-Step Transition Probabilities

We have seen that the k-step transition probabilities of a Markov chain can be determined from the $(k-1)$-step and 1-step transition probabilities. Actually, for $k \geq 4$, the k-step transition probabilities can also be determined from the $(k-2)$-step and 2-step transition probabilities. Indeed, if there are two positive integers k_1 and k_2 such that $k_1 + k_2 = k$, then we can obtain the k-step transition probabilities from the following equation:

$$(4.19) \qquad p_{ij}^{(k)} = p_{i1}^{(k_1)} p_{1j}^{(k_2)} + p_{i2}^{(k_1)} p_{2j}^{(k_2)} + \cdots + p_{iN}^{(k_1)} p_{Nj}^{(k_2)}.$$

For example, if $k = 4$ and $k_1 = k_2 = 2$, then

$$(4.20) \qquad p_{ij}^{(4)} = p_{i1}^{(2)} p_{1j}^{(2)} + p_{i2}^{(2)} p_{2j}^{(2)} + \cdots + p_{iN}^{(2)} p_{Nj}^{(2)}.$$

Thus, in Example 4.2, rather than obtain $p_{ij}^{(4)}$ using Tables 4.6 and 4.8 (which is the way Table 4.9 was obtained), we may obtain $p_{ij}^{(4)}$ from the use of Table 4.7 alone. That is, for example, using Equation (4.20),

$$p_{31}^{(4)} = p_{31}^{(2)} p_{11}^{(2)} + p_{32}^{(2)} p_{21}^{(2)} + p_{33}^{(2)} p_{31}^{(2)}$$
$$= (0.0344)(0.2272) + (0.5998)(0.0600) + (0.3658)(0.0344)$$
$$= 0.056.$$

Note that this result for $p_{31}^{(4)}$ agrees with the result for $p_{31}^{(4)}$ given in the third row and first column of Table 4.9.

5. ABSOLUTE OR UNCONDITIONAL PROBABILITIES

In the previous section, we discussed how to calculate k-step transition probabilities $p_{ij}^{(k)}$ for Markov chains. These transition probabilities are *conditional* probabilities, in that $p_{ij}^{(k)}$ gives the probability that the process moves from State S_i to State S_j in k units of time *given* that the process is initially in State S_i at some time t. That is,

$$p_{ij}^{(k)} = P\{X_{t+k} = j \mid X_t = i\},$$

and in particular (when $t = 0$),

(5.1) $$p_{ij}^{(k)} = P\{X_k = j \mid X_0 = i\}.$$

Thus, $p_{ij}^{(k)}$ equals the conditional probability that the process is in State S_j at time k, given that at time 0 the process was in State S_i.

Suppose instead that we are interested in the *unconditional* probability, $P\{X_k = j\}$, that the process is in State S_j at time k. Assume that the process has N possible states, $S_1, S_2, \ldots, S_N$. Recall from Rule 4.6 of Chapter 3 that if B is any event, and $E_1, E_2, \ldots, E_N$ are N mutually exclusive events such that $E_1 \cup E_2 \cup \cdots \cup E_N$ is the whole sample space, then

(5.2) $$P(B) = P(B \mid E_1)P(E_1) + P(B \mid E_2)P(E_2) + \cdots + P(B \mid E_N)P(E_N).$$

We may apply this rule to the events

$$B = \{X_k = j\}, E_1 = \{X_0 = 1\}, E_2 = \{X_0 = 2\}, \ldots, E_N = \{X_0 = N\},$$

obtaining the result

(5.3) $$\begin{aligned} P\{X_k = j\} &= P\{X_k = j \mid X_0 = 1\}P\{X_0 = 1\} \\ &+ P\{X_k = j \mid X_0 = 2\}P\{X_0 = 2\} \\ &+ \cdots + P\{X_k = j \mid X_0 = N\}P\{X_0 = N\}. \end{aligned}$$

Remember that in previous sections, we denoted the initial probabilities $P\{X_0 = i\}$ by a_i; thus,

$$a_i = P\{X_0 = i\}.$$

Substituting this notation into Equation (5.3), and making use of Equation (5.1), we find that

(5.4) $$P\{X_k = j\} = a_1 p_{1j}^{(k)} + a_2 p_{2j}^{(k)} + \cdots + a_N p_{Nj}^{(k)}.$$

Thus, the unconditional probability, $P\{X_k = j\}$, that the process is in State S_j at time k is equal to the sum of N terms, where the ith term $a_i p_{ij}^{(k)}$ of this sum is the product of the initial probability a_i (the probability that the process begins in State S_i at time 0) and the k-step transition probability $p_{ij}^{(k)}$.

5. ABSOLUTE OR UNCONDITIONAL PROBABILITIES

To calculate $P\{X_k = j\}$ for a given Markov chain, when we are given the initial probabilities $a_1, a_2, \ldots, a_N$ and the transition matrix of the process (remember that these quantities determine the probabilistic structure of the process), we proceed as follows: (i) Using the methods of Section 4, we calculate the k-step transition probabilities $p_{ij}^{(k)}$ from the 1-step transition probabilities p_{uv}, (ii) we calculate $P\{X_k = j\}$ from knowledge of $a_1, a_2, \ldots, a_N$ and $p_{1j}^{(k)}, p_{2j}^{(k)}, \ldots, p_{Nj}^{(k)}$ by use of Equation (5.4).

Example 5.1 (Brand Switching). Suppose that there are two brands, Brand 1 and Brand 2, of a given product (say, cigarettes). An individual purchases one of these two brands each day. A variety of factors may affect the individual's choice on any given day. Since the influence of these factors is unpredictable a priori, the individual's brand choice is a random phenomenon, and the sequence of such choices from day to day is a discrete-time stochastic process. One of the factors which influences the individual's choice of brand on a given day is his experience with the brands that he chose on previous days. We assume that individuals have short memory spans concerning the given product, so that the conditional probability that an individual purchases any particular brand on day $t+1$, given the history of his choices of brands on the previous days $0, 1, 2, \ldots, t$, depends only on the individual's choice of brand on the immediately preceding day, day t. Thus, the process of choice of brand is a Markov chain with possible states:

S_1: the individual chooses Brand 1,
S_2: the individual chooses Brand 2.

Let us make the further assumption that the two brands remain constant in quality, and that the other forces (personal taste, social pressures, health, and so on) which influence the individual's choice of brand remain uniform in nature over time. In such a case, we can assume that this choice-of-brand process has stationary transition probabilities.

When an individual first starts purchasing the given product (on day 0), it is not unreasonable to assume that he chooses Brands 1 and 2 each with probability $\frac{1}{2}$. From this assumption it follows that the initial probabilities of the process are

$$a_1 = \tfrac{1}{2}, \qquad a_2 = \tfrac{1}{2}.$$

A survey by a marketing research firm shows that individuals who buy Brand 1 on a given day have equal probability of buying Brand 1 or Brand 2 on the next day, while purchasers of Brand 2 on a given day have probability $\frac{3}{4}$ of switching to Brand 1 on the following day. The transition matrix of this choice-of-brand process is shown in Table 5.1.

Table 5.1. Transition probabilities for choice-of-brand process.

		State at Day $t+1$	
		S_1	S_2
State at Day t	S_1 (Brand 1)	$\frac{1}{2}$	$\frac{1}{2}$
	S_2 (Brand 2)	$\frac{3}{4}$	$\frac{1}{4}$

The market research firm is interested in the unconditional probabilities $P\{X_k = 1\}$, $P\{X_k = 2\}$ that individuals will buy Brand 1 or Brand 2, respectively, k days after they begin to purchase the product of interest.

Using the methods of Section 4, we can determine the 2-step, 3-step, and so on, transition probabilities, and display these transition probabilities in Tables 5.2, 5.3, 5.4, 5.5, and others.

Table 5.2. 2-step transition probabilities for choice-of-brand process.

		State at Day $t+2$	
		S_1	S_2
State at Day t	S_1	10/16	6/16
	S_2	9/16	7/16

Table 5.3. 3-step transition probabilities for choice-of-brand process.

		State at Day $t+3$	
		S_1	S_2
State at Day t	S_1	38/64	26/64
	S_2	39/64	25/64

Table 5.4. 4-step transition probabilities for choice-of-brand process.

		State at Day $t+4$	
		S_1	S_2
State at Day t	S_1	154/256	102/256
	S_2	153/256	103/256

5. ABSOLUTE OR UNCONDITIONAL PROBABILITIES

Table 5.5. 5-step transition probabilities for choice-of-brand process.

| | | State at Day $t+5$ ||
		S_1	S_2
State at Day t	S_1	614/1024	410/1024
	S_2	615/1024	409/1024

Now, from the initial probabilities, Table 5.1, and Equation (5.4), the probability that the brand purchased by an individual on day 1 is Brand 1 is

$$P\{X_1 = 1\} = a_1 p_{11}^{(1)} + a_2 p_{21}^{(1)}$$
$$= a_1 p_{11} + a_2 p_{21}$$
$$= \left(\frac{1}{2}\right)\left(\frac{1}{2}\right) + \left(\frac{1}{2}\right)\left(\frac{3}{4}\right) = \frac{5}{8} = 0.625,$$

while the probability that the brand purchased on day 1 is Brand 2 is

$$P\{X_1 = 2\} = a_1 p_{12} + a_2 p_{22} = \left(\frac{1}{2}\right)\left(\frac{1}{2}\right) + \left(\frac{1}{2}\right)\left(\frac{1}{4}\right) = \frac{3}{8} = 0.375.$$

We may ask for the probability that the individual purchases Brand 1 on day 2. Using the initial probabilities, Table 5.2, and Equation (5.4), we find that

$$P\{X_2 = 1\} = a_1 p_{11}^{(2)} + a_2 p_{21}^{(2)} = \left(\frac{1}{2}\right)\left(\frac{10}{16}\right) + \left(\frac{1}{2}\right)\left(\frac{9}{16}\right) = \frac{19}{32} = 0.594.$$

Similarly,

$$P\{X_2 = 2\} = a_1 p_{12}^{(2)} + a_2 p_{22}^{(2)} = \left(\frac{1}{2}\right)\left(\frac{6}{16}\right) + \left(\frac{1}{2}\right)\left(\frac{7}{16}\right) = \frac{13}{32} = 0.406.$$

Comparable calculations of probabilities can be obtained for days 3, 4, 5, and so on, by making use of Tables 5.3, 5.4, 5.5, the initial probabilities, and Equation (5.4). These results are summarized in Table 5.6.

Table 5.6. Unconditional probabilities (accurate to three decimal places) of brand choice for days 0, 1, 2, 3, 4, and 5.

| Probability | Day k |||||||
	0	1	2	3	4	5	
$P\{X_k = 1\}$	0.500	0.625	0.594	0.602	0.600	0.600	...
$P\{X_k = 2\}$	0.500	0.375	0.406	0.398	0.400	0.400	...

We see from Table 5.6 that the unconditional probability of choosing Brand 1 appears by day 4 to have stabilized at 0.600, and that the unconditional probability of choosing Brand 2 has stabilized at 0.400. That is, from Table 5.6 we are led to anticipate that the probability that an individual chooses Brand 1 k days after he begins purchasing the given product is approximately 0.600 for $k = 4, 5, 6, \ldots$. Hence, if on any given day we observe the choice of brands only of individuals who have been purchasing the given product for 4 or more days, we can expect that approximately 60 percent of these individuals will have purchased Brand 1.

EXERCISES

1. Ship lanes are sufficiently wide so that each ship may vary its direction from time to time to meet various contingencies and still avoid collisions with other ships. Suppose that after each unit of time, a ship chooses its direction by choosing one of 3 states S_1, S_2, S_3, representing specific courses. Further, suppose that the process of choosing such directions is a Markov chain with the following transition probabilities:

		State at Time $t+1$		
		S_1	S_2	S_3
State at Time t	S_1	p_1	p_2	p_3
	S_2	p_3	p_1	p_2
	S_3	p_2	p_3	p_1

   where $p_1 + p_2 + p_3 = 1$. Assume $a_1 = a_2 = a_3 = \frac{1}{3}$. Analyze this model in terms of higher-order (k-step) transition probabilities, and in terms of unconditional probabilities at time k. That is, find formulas for (a) the k-step transition probabilities and (b) the unconditional probabilities at time k for this process. Check your formulas for the case when $p_1 = \frac{1}{2}$, $p_2 = \frac{1}{4}$, and $p_3 = \frac{1}{4}$.

2. In a study of the epidemiology of mental disease, Marshall and Goldhamer (1955) consider various Markov chain models. One of these has as states: S_1 = alive, sane, S_2 = alive, insane (mild), unhospitalized, S_3 = alive, insane (severe), unhospitalized, S_4 = alive, insane, hospitalized, S_5 = dead, outside of mental institution (hospital). The transition probabilities for this 5-state Markov chain are as follows:

		State at Time $t+1$				
		S_1	S_2	S_3	S_4	S_5
State at Time t	S_1	p_{11}	p_{12}	p_{13}	0	p_{15}
	S_2	0	p_{22}	0	p_{24}	p_{25}
	S_3	0	0	p_{33}	p_{34}	p_{35}
	S_4	0	0	0	1	0
	S_5	0	0	0	0	1

(a) Give a rationale for the pattern of transition probabilities shown in the above table. Why, for example, is it necessary that $p_{51} = 0$?
(b) Find the 2-step transition probabilities for this model. In so doing, use a set of numbers for the p_{ij}'s which seems reasonable to you.

A3. The simple 2-state learning model of Example 4.1 may be extended to deal with more complicated experiments. For example, we may have a 3-state learning model in which State S_1 denotes long-term memory, State S_2 denotes short-term memory, and State S_3 denotes guessing. [The details of such models are discussed in Coombs, Dawes, and Tversky (1970), Chapter 9.] The transition matrix of such a 3-state learning model is as follows:

		State at Time $t+1$		
		S_1	S_2	S_3
State at Time t	S_1	1	0	0
	S_2	α	$\bar{\alpha}\bar{\delta}$	$\bar{\alpha}\delta$
	S_3	β	$\bar{\beta}\bar{\delta}$	$\bar{\beta}\delta$

where $\bar{\alpha} = 1-\alpha$, $\bar{\beta} = 1-\beta$, and $\bar{\delta} = 1-\delta$. Initial probabilities a_1, a_2, and a_3 are also given.
(a) Verify that the above table is a valid table of transition probabilities for a Markov chain.
(b) What models do we obtain when (i) $\beta = 0$, (ii) $\delta = 0$, or (iii) $\alpha = \beta = 0$?
(c) If $\alpha = 0.4$, $\beta = 0$, and $\delta = 0.5$, find the 2-step, 3-step, and 4-step transition probabilities. Do you see any pattern developing?
(d) Determine $P\{X_4 = 2|X_3 = 2\}$, $P\{X_4 = 2|X_2 = 2\}$, $P\{X_4 = 2|X_1 = 2\}$, and $P\{X_4 = 2|X_0 = 2\}$.
(e) Determine $P\{X_3 = 2|X_2 = 2\}$, $P\{X_3 = 2|X_1 = 2\}$, $P\{X_3 = 2|X_0 = 2\}$. What is the connection between the answers in parts (d) and (e)?
(f) Determine the unconditional probabilities $P\{X_1 = 2\}$, $P\{X_2 = 2\}$, $P\{X_3 = 2\}$, and $P\{X_4 = 2\}$. Here assume that $a_1 = a_2 = a_3 = \frac{1}{3}$.
(g) How does your answer in (f) change if $a_1 = 1$, $a_2 = 0$, and $a_3 = 0$?
(h) How does your answer in (f) change if $a_1 = 0$, $a_2 = 0$, and $a_3 = 1$?

A4. Suppose we are studying [see Berger and Snell (1957)] intercity population movements, among 3 cities, A, B, and C. Each city sends, within a 1-year time period, certain fractions of its population to itself and to other cities. Here the states represent the cities, and the model for the intercity movement of a randomly chosen individual is assumed to be a Markov chain with transition probabilities:

		State (City) at Time $t+1$		
		A	B	C
State (City) at Time t	A	0.85	0.07	0.08
	B	0.25	0.70	0.05
	C	0.03	0.02	0.95

The initial populations of the 3 cities are 100,000 for city A, 500,000 for city B, and 200,000 for city C.
(a) What are the 2-step transition probabilities of this Markov chain?
(b) Determine the unconditional probabilities for this Markov chain after one unit of time.
(c) Determine the unconditional probabilities after two units of time.

5. Consider the following tables of transition probabilities for a 3-state Markov chain:

State at Time $t+1$

		S_1	S_2	S_3
State at Time t	S_1	1	0	0
	S_2	p	q	0
	S_3	0	p	q

(i)

State at Time $t+1$

		S_1	S_2	S_3
State at Time t	S_1	p	q	0
	S_2	1	0	0
	S_3	0	p	q

(ii)

State at Time $t+1$

		S_1	S_2	S_3
State at Time t	S_1	1	0	0
	S_2	0	1	0
	S_3	0	p	q

(iii)

where $p+q=1$.
(a) In each case [(i), (ii), and (iii)] obtain formulas for all of the 2-step, 3-step, and 4-step transition probabilities in terms of the constants p and q.
(b) In each case, try to derive a general rule (formula) for finding the higher-order transition probabilities.
(c) If the initial probabilities are $a_1 = a_2 = a_3 = \frac{1}{3}$, find the unconditional probabilities $P\{X_3 = 1\}$, $P\{X_3 = 2\}$, and $P\{X_3 = 3\}$ in each of cases (i), (ii), and (iii).
(d) Find these unconditional probabilities if $a_1 = a_2 = \frac{1}{2}$, $a_3 = 0$, and if $a_1 = a_2 = \frac{1}{4}$, $a_3 = \frac{1}{2}$.
(e) Check your answers in (a) through (d) for the case when $p = \frac{1}{4}$ and $q = \frac{3}{4}$.

A6. The following transition probability structures arise in a variety of contexts:

State at Time $t+1$

		S_1	S_2	S_3	S_4
State at Time t	S_1	q	p	0	0
	S_2	q	0	p	0
	S_3	q	0	0	p
	S_4	q	0	0	p

(i)

State at Time $t+1$

		S_1	S_2	S_3	S_4
State at Time t	S_1	p_1	p_2	p_3	p_4
	S_2	0	0	1	0
	S_3	0	0	0	1
	S_4	0	1	0	0

(ii)

	State at Time $t+1$				
	S_1	S_2	S_3	S_4	S_5
S_1	1	0	0	0	0
S_2	p	0	q	0	0
S_3	0	p	0	q	0
S_4	0	0	p	0	q
S_5	0	0	0	0	1

State at Time t

(iii)

where $p+q=1$ and $p_1+p_2+p_3+p_4=1$.

(a) Can you think of a context where such transition probabilities might arise?

(b) In each case, obtain formulas for the 2-step and 3-step transition probabilities in terms of the unspecified constants (that is, p, q, p_1, p_2, p_3, and p_4).

(c) If the initial probabilities are $a_1=a_2=a_3=a_4=\frac{1}{4}$ for (i) and (ii), and $a_1=a_2=a_3=a_4=a_5=\frac{1}{5}$ for (iii), find the unconditional probabilities of the various states at time $t=2$.

(d) Check your results for the case when $p=\frac{1}{4}$, $q=\frac{3}{4}$, $p_1=p_2=\frac{1}{3}$, and $p_3=p_4=\frac{1}{6}$.

7. Markov chain models have been used in a spatial rather than temporal context [see American Geological Institute (1969)]. Suppose we are interested in the study of rock formation, and represent the different rock components as our states: S_1 = sandstone, S_2 = shale, S_3 = siltstone, S_4 = lignite. A vertical strip or section of rock surface, say a cliff or well, is observed upward from its base. Each layer of rock deposit can be thought of as corresponding to the passage of a unit of time. In a particular experiment, a table of transition probabilities is reported (Table E.1).

Table E.1. Transition probabilities for a section in the Oficina formation (miocene) from a well in Venezuela.

	State at Layer $t+1$			
	S_1	S_2	S_3	S_4
S_1	0.79	0.07	0.07	0.07
S_2	0.05	0.79	0.06	0.10
S_3	0.10	0.32	0.43	0.15
S_4	0.18	0.39	0.13	0.30

State at Layer t

(a) If we start our observations (layer 0) with lignite (State S_4), what is the conditional probability that lignite will again be present at layer 2 following immediately after siltstone at layer 1? What is the probability that lignite will again be present at layer 3 following after siltstone at layer 2? In other words, find $P\{X_2=4, X_1=3|X_0=4\}$ and $P\{X_3=4, X_2=3|X_0=4\}$.

(b) Determine the probability that lignite will be present at layer 2 or at layer 3 given that sandstone occurs at layer 1.

(c) It was found that the 40-step transition probabilities of this Markov chain model are:

		State at Layer 40			
		S_1	S_2	S_3	S_4
State at Layer 0	S_1	0.27	0.49	0.12	0.12
	S_2	0.27	0.49	0.12	0.12
	S_3	0.27	0.49	0.12	0.12
	S_4	0.27	0.49	0.12	0.12

Interpret this finding in terms of the given experimental context.

*12

MARKOV CHAINS II: FIRST-PASSAGE TIMES, RECURRENT STATES, AND LONG-RUN PROBABILITIES

Harmony would lose its attractiveness if it did not have a background of discord.

Tehyi Hsieh, Chinese epigrams

Nature speaks in symbols and in signs.
John Greenleaf Whittier to Charles Sumner

In the present chapter, we continue our study of Markov chains. In particular, we attempt to more fully describe the properties of a Markov chain as such a process evolves over time. Thus, our concern in this chapter is with questions such as the following:

(i) Can the process ever reach State S_j if it begins in State S_i?
(ii) How long, on the average, does it take for the process to go from State S_i to State S_j?
(iii) If the process starts in State S_j, will it ever return to that state?
(iv) After the process has been in effect over a long time, does it matter whether the process started in State S_i or in State S_j?

Because Markov chains differ greatly in their properties, no attempt is made here to answer these (or other) questions for every Markov chain. However, answers are obtained that are of use in a variety of practical and scientific contexts.

1. FIRST-PASSAGE TIMES

Suppose that we are observing a discrete-time stochastic process which we have modeled as a Markov chain having N possible states $S_1, S_2, \ldots, S_N$, and having stationary transition probabilities p_{ij}. One state may be of special interest to us, perhaps because this state represents a particularly desirable (or undesirable) happening. For example, in Example

4.1 of Chapter 11 (the learning model), the state of interest may be S_1, the state in which the subject of the learning experiment has been conditioned to make a correct response. In Example 3.3 of Chapter 11 (the genetical breeding experiments) the state S_j of interest may be some desirable (or undesirable) genotype for the offspring. As a last example, in Example 4.2 of Chapter 11 (the occupational mobility process) the state of interest may be S_1, the highest and most desirable occupation level, or it may be S_3, the lowest and least desirable occupation level. Suppose that initially (at time 0), the process is observed to be in some state, State S_i; for the moment, let us assume that the State S_i and the State S_j of interest are not the same state (that is, $i \neq j$). Because State S_j is of special interest to us (or because State S_j is particularly desirable), we may wish to determine how long it will take the process to move from its present state, State S_i, to State S_j. Described another way, given that the process is in State S_i at time 0, we may wish to consider the variable K_{ij}, where K_{ij} is the *first* time after time 0 at which the process is observed to be in State S_j. Such variables K_{ij} are called *first-passage times*, since they tell us the time at which the process first passes into State S_j after having initially been in State S_i.

The observed value of the first-passage time K_{ij} depends upon the particular sequence of states obtained for the process during any particular trial of the random experiment in which the process is observed. Thus, if we observe the process over time during a given trial of the random experiment, and obtain the sequence of states shown in Table 1.1, then $K_{4,9} = 12$, $K_{4,7} = 11$, $K_{4,2} = 1$ and so on. If we were to repeat the random experiment, and if once again S_4 is the state of the process at time 0, but otherwise the sequence of obtained states is as in Table 1.2, then $K_{4,9} = 1$, $K_{4,7} = 3$, $K_{4,2} = 10$, and so on.

Table 1.1. Obtained sequence of states over time for one trial of the random experiment in which a given Markov chain process is observed. Here, the initial state is S_4.

Time	0	1	2	3	4	5	6	7	8	9	10	11	12	13	...
State	S_4	S_2	S_1	S_2	S_5	S_8	S_8	S_6	S_1	S_3	S_4	S_7	S_9	S_5	...

Table 1.2. Obtained sequence of states over time for another trial of the random experiment in which a given Markov chain process is observed, and in which the initial state is S_4.

Time	0	1	2	3	4	5	6	7	8	9	10	11	12	13	...
State	S_4	S_9	S_3	S_7	S_5	S_4	S_3	S_9	S_6	S_8	S_2	S_2	S_1	S_5	...

1. FIRST-PASSAGE TIMES

Because the first-passage times K_{ij} depend on the sequence of states obtained for the random process on any given trial of the random experiment in which this process is observed, the first-passage times K_{ij} are random variables. Recall that the properties of any random variable can be determined from knowledge of its distribution. Thus, we need to obtain the distributions of the first-passage times K_{ij}. Note that the first-passage time K_{ij} is defined for a given process only when the process is known to be initially (at time 0) in State S_i. It therefore seems appropriate to consider the *conditional* distribution of K_{ij} *given* that the process is in State S_i at time 0. Because K_{ij} is a discrete random variable (the possible values of K_{ij} are the times 1, 2, 3, and so on, at which the process can first be in State S_j), the conditional distribution of K_{ij} can be found by determining the values of its (conditional) probability mass function $P\{K_{ij} = k | X_0 = i\}$ for $k = 1, 2, 3$, and so on.

Let k be a given positive integer (that is, $k = 1$, or $k = 2$, or $k = 3$, and so on).

The definition of the first-passage time K_{ij} implies that K_{ij} equals k if the process is observed to be in State S_j for the first time at time k. That is, $K_{ij} = k$ if the process is in State S_j at time k, and the process has *not* been in State S_j at any time (time 1, time 2, ..., time $k-1$) previous to time k. It thus follows from the definition of K_{ij} that the event $\{K_{ij} = k\}$ and the event

$$\{X_0 = i \text{ and } X_1 \neq j \text{ and } X_2 \neq j \text{ and } \cdots \text{ and } X_{k-1} \neq j \text{ and } X_k = j\}$$

are the same event. Hence,

(1.1) $\quad P\{K_{ij} = k | X_0 = i\} = P\{X_0 = i \text{ and } X_1 \neq j \text{ and } X_2 \neq j \text{ and } \cdots$
$\qquad\qquad \text{and } X_{k-1} \neq j \text{ and } X_k = j | X_0 = i\}.$

Adopting a notation similar to that used for k-step transition probabilities, let $f_{ij}^{(k)}$ represent the conditional probability that K_{ij} equals k, given that the process is in State S_i at time 0. It follows from Equation (1.1) that

(1.2) $\quad f_{ij}^{(k)} = P\{K_{ij} = k | X_0 = i\}$
$\qquad = P\{X_0 = i \text{ and } X_1 \neq j \text{ and } \cdots \text{ and } X_{k-1} \neq j \text{ and } X_k = j | X_0 = i\}.$

Although the notations we have used for k-step transition probabilities and first-passage probabilities are similar in form, it is important to note that in general $p_{ij}^{(k)}$ and $f_{ij}^{(k)}$ are not equal. Indeed, it is always the case that

(1.3) $\qquad\qquad\qquad f_{ij}^{(k)} \leq p_{ij}^{(k)}.$

To verify this assertion, note that if the process has entered State S_j for the first time at time k (that is, if the event $\{K_{ij} = k\}$ has occurred), then it must be the case that the process is *in* State S_j at time k (that is, the event $\{X_k = j\}$ has occurred). Thus, the event $\{K_{ij} = k\}$ is included in the event $\{X_k = j\}$, and it follows from the Law of Inclusion for conditional probabilities (see Chapter 3) that $f_{ij}^{(k)} = P\{K_{ij} = k | X_0 = i\}$ is less than or equal to $p_{ij}^{(k)} = P\{X_k = j | X_0 = i\}$. As we see later, for $k > 1$, $f_{ij}^{(k)}$ is in general strictly less than $p_{ij}^{(k)}$.

Calculation of the Probability Distribution of the First-Passage Time K_{ij}

The probability distribution of the first-passage time K_{ij} is determined by the quantities $f_{ij}^{(1)}$, $f_{ij}^{(2)}$, $f_{ij}^{(3)}$, and so on, where for any positive integer k,

$$f_{ij}^{(k)} = P\{K_{ij} = k | X_0 = i\}.$$

As we now show, the quantities $f_{ij}^{(1)}$, $f_{ij}^{(2)}$, $f_{ij}^{(3)}$, and so on, can be computed solely from knowledge of the transition probabilities p_{uv} for the given Markov chain. We illustrate such computations for the first genetical breeding experiment described in Example 3.3 of Chapter 11. Recall that in this experiment, one parent (parent I) in a given mating always has a fixed genotype (here, the fixed genotype is assumed to be the genotype Aa), while the other parent is randomly selected from the offspring of a previous such mating. The state of the mating process at the kth such mating ("time $k-1$") is the genotype of the offspring of this mating which is selected at random for the next [the $(k+1)$st] mating. Thus, the possible states of this process are

S_1: the offspring selected has genotype AA,
S_2: the offspring selected has genotype Aa,
S_3: the offspring selected has genotype aa.

As we observed in Chapter 11, Section 3, the genetical mating process described above is a Markov chain with stationary transition probabilities p_{ij} as summarized in Table 1.3.

Suppose that at the initial mating, both parents have genotype Aa; in other words, suppose that the state of the mating process at time 0 is S_2. If the genotype AA is a particularly desirable genotype, it is of interest (particularly to breeders) to determine the probability distribution of K_{21}, the number of matings after the initial mating that are required to obtain an offspring of genotype AA (State S_1) for the first time. The random variable K_{21} is the first-passage time from State S_2 (genotype Aa) to State S_1 (genotype AA).

To find the conditional probability mass function of K_{21} (that is, to find $f_{21}^{(1)}, f_{21}^{(2)}, f_{21}^{(3)}$, and so on), we proceed as follows. First, we give an argument, true for any 3-state Markov chain (and not just the genetic mating process

Table 1.3. Transition probabilities for genotypes arising from a mating where one parent always has genotype Aa, the other parent is randomly selected from the offspring of a previous such mating, and one offspring is selected at random from their children.

		State at Time $t+1$ (Mating $t+2$)		
		S_1	S_2	S_3
State at Time t (Mating $t+1$)	S_1 (genotype AA)	$\frac{1}{2}$	$\frac{1}{2}$	0
	S_2 (genotype Aa)	$\frac{1}{4}$	$\frac{1}{2}$	$\frac{1}{4}$
	S_3 (genotype aa)	0	$\frac{1}{2}$	$\frac{1}{2}$

described above), that establishes a general formula for $f_{21}^{(1)}, f_{21}^{(2)}, f_{21}^{(3)}$, and so on. Then, using that formula, we find the explicit values of $f_{21}^{(1)}, f_{21}^{(2)}, f_{21}^{(3)}$, and so on, for our example.

The conditional probability that the first-passage time K_{21} from State S_2 to State S_1 is equal to 1, given that the process is initially in State S_2, that is,

$$f_{21}^{(1)} = P\{K_{21} = 1 | X_0 = 2\},$$

is the same as the conditional probability p_{21} that the process is in State S_1 at time 1 given that the process is in State S_2 at time 0. That is,

(1.4) $$f_{21}^{(1)} = P\{X_1 = 1 | X_0 = 2\} = p_{21}.$$

For the genetical mating process, the conditional probability $f_{21}^{(1)}$ that genotype AA (State S_1) is attained for the first time one mating after the initial mating, given that the initial mating is between two parents of genotype Aa (the process is initially in State S_2) can be found from Table 1.3 and Equation (1.4); namely,

$$f_{21}^{(1)} = p_{21} = \tfrac{1}{4}.$$

To find the value of $f_{21}^{(2)}$ we argue: from the definition of $f_{21}^{(2)}, f_{21}^{(2)}$ equals the conditional probability that the process is in State S_1 for the first time at time 2, given that initially (at time 0) the process is in State S_2. For the process to be in State S_1 for the *first* time at time 2, the process cannot have been in State S_1 at time 1. Thus, the conditional probability of the transition from State S_2 at time 0 to State S_1 at time 1, and then to State S_1 at time 2 (that is, the transition $S_2 \to S_1 \to S_1$) cannot be counted when calculating $f_{21}^{(2)}$. Therefore, because the Markov chain we are considering has but three states, S_1, S_2, and S_3 (and one of these states must be in effect at time 1), the only transitions that are compatible with the event that the process passes from State S_2 at time 0 to State S_1 for the first time at time 2 are the transitions $S_2 \to S_2 \to S_1$ and $S_2 \to S_3 \to S_1$. Given that the process is in State S_2 at time 0, we may use the conditional probability arguments of Chapter 11, Section 4 to show that the conditional probability of the transition $S_2 \to S_2 \to S_1$ is equal to

$$P\{S_2 \to S_2 \to S_1 | \text{the process is in State } S_2 \text{ at time } 0\}$$
$$= P\{X_0 = 2 \text{ and } X_1 = 2 \text{ and } X_2 = 1 | X_0 = 2\}$$
$$= p_{22}p_{21}.$$

In a similar fashion, we can show that

$$P\{S_2 \to S_3 \to S_1 | \text{the process is in State } S_2 \text{ at time } 0\}$$
$$= P\{X_0 = 2 \text{ and } X_1 = 3 \text{ and } X_2 = 1 | X_0 = 2\}$$
$$= p_{23}p_{31}.$$

Because the event $\{K_{21} = 2\}$ is the union of the two mutually exclusive events $\{X_0 = 2 \text{ and } X_1 = 2 \text{ and } X_2 = 1\}$ and $\{X_0 = 2 \text{ and } X_1 = 3 \text{ and } X_2 = 1\}$, it follows that

(1.5) $\quad f_{21}^{(2)} = P\{K_{21} = 2 | X_0 = 2\}$
$$= P\{X_0 = 2 \text{ and } X_1 = 2 \text{ and } X_2 = 1 | X_0 = 2\}$$
$$+ P\{X_0 = 2 \text{ and } X_1 = 3 \text{ and } X_2 = 1 | X_0 = 2\}$$
$$= p_{22}p_{21} + p_{23}p_{31}.$$

If we apply Equation (1.5) to our genetical mating process, we find that the conditional probability $f_{21}^{(2)}$ that genotype AA (State S_1) is attained for the first time two matings after the initial mating, given that the initial mating is between two parents of genotype Aa, is equal to

$$f_{21}^{(2)} = p_{22}p_{21} + p_{23}p_{31} = \left(\frac{1}{2}\right)\left(\frac{1}{4}\right) + \left(\frac{1}{4}\right)(0) = \frac{1}{8}.$$

To determine $f_{21}^{(3)}$, we need to find the conditional probability of moving from State S_2 at time 0 to State S_1 for the first time at time 3, given that the process is initially (at time 0) in State S_2. The process moves to State S_1 for the first time at time 3 if any one of the sequences of states over time shown in Figure 1.1 occurs.

$$\begin{array}{cccccccc}
S_2 & \to & S_2 & \to & S_2 & \to & S_1 \\
S_2 & \to & S_2 & \to & S_3 & \to & S_1 \\
S_2 & \to & S_3 & \to & S_2 & \to & S_1 \\
S_2 & \to & S_3 & \to & S_3 & \to & S_1
\end{array}$$

Time 0   Time 1   Time 2   Time 3

Figure 1.1. Any of the above four sequences of states results in the process moving from State S_2 at time 0 to State S_1 for the first time at time 3.

Again making use of the conditional probability arguments of Chapter 11, Section 4, we find that

$$P\{S_2 \to S_2 \to S_2 \to S_1 | \text{the process is in State } S_2 \text{ at time } 0\}$$
$$= P\{X_0 = 2 \text{ and } X_1 = 2 \text{ and } X_2 = 2 \text{ and } X_3 = 1 | X_0 = 2\}$$
$$= p_{22}p_{22}p_{21}.$$

Similarly,

$$P\{S_2 \to S_2 \to S_3 \to S_1 | \text{the process is in State } S_2 \text{ at time } 0\}$$
$$= p_{22}p_{23}p_{31},$$

$$P\{S_2 \to S_3 \to S_2 \to S_1 | \text{the process is in State } S_2 \text{ at time } 0\}$$
$$= p_{23}p_{32}p_{21},$$

and

$$P\{S_2 \to S_3 \to S_3 \to S_1 | \text{the process is in State } S_2 \text{ at time } 0\}$$
$$= p_{23}p_{33}p_{31}.$$

Since the event $\{K_{21} = 3\}$ is the union of the 4 mutually exclusive events $\{S_2 \to S_2 \to S_2 \to S_1\}$, $\{S_2 \to S_2 \to S_3 \to S_1\}$, $\{S_2 \to S_3 \to S_2 \to S_1\}$, and $\{S_2 \to S_3 \to S_3 \to S_1\}$, it follows that

(1.6) $\qquad f_{21}^{(3)} = P\{K_{21} = 3 | X_0 = 2\}$
$$= p_{22}p_{22}p_{21} + p_{22}p_{23}p_{31} + p_{23}p_{32}p_{21} + p_{23}p_{33}p_{31}.$$

For our genetical mating process, use of Equation (1.6) and Table 1.3 allows us to calculate the conditional probability $f_{21}^{(3)}$ that three matings (after the initial mating) are required for us to obtain an offspring with AA genotype for the first time, given that the initial mating involved two parents with Aa genotypes. The value of $f_{21}^{(3)}$ is

$$f_{21}^{(3)} = p_{22}p_{22}p_{21} + p_{22}p_{23}p_{31} + p_{23}p_{32}p_{21} + p_{23}p_{33}p_{31}$$

$$= \left(\frac{1}{2}\right)\left(\frac{1}{2}\right)\left(\frac{1}{4}\right) + \left(\frac{1}{2}\right)\left(\frac{1}{4}\right)(0) + \left(\frac{1}{4}\right)\left(\frac{1}{2}\right)\left(\frac{1}{4}\right) + \left(\frac{1}{4}\right)\left(\frac{1}{2}\right)(0)$$

$$= \frac{3}{32}.$$

Computing $f_{21}^{(k)}$ for $k = 4, 5, \ldots$ by the method illustrated above becomes both tedious and cumbersome as k increases. The conditional probabilities of more and more transitions over longer and longer time periods must be calculated in order to obtain $f_{21}^{(k)}$ as k grows larger. Thus, in the present example of a 3-state Markov chain, the conditional probabilities of the 8 transitions shown in Figure 1.2 must be determined in order to compute $f_{21}^{(4)}$. Any one of these 8 conditional probabilities, in turn, is calculated by taking the product of 4 transition probabilities; for example,

$P\{S_2 \to S_2 \to S_3 \to S_2 \to S_1 |\text{the process is in State } S_2 \text{ at time } 0\}$
$= p_{22}p_{23}p_{32}p_{21}.$

$S_2 \to$	$S_2 \to$	$S_2 \to$	$S_2 \to$	S_1
$S_2 \to$	$S_2 \to$	$S_2 \to$	$S_3 \to$	S_1
$S_2 \to$	$S_2 \to$	$S_3 \to$	$S_2 \to$	S_1
$S_2 \to$	$S_2 \to$	$S_3 \to$	$S_3 \to$	S_1
$S_2 \to$	$S_3 \to$	$S_2 \to$	$S_2 \to$	S_1
$S_2 \to$	$S_3 \to$	$S_2 \to$	$S_3 \to$	S_1
$S_2 \to$	$S_3 \to$	$S_3 \to$	$S_2 \to$	S_1
$S_2 \to$	$S_3 \to$	$S_3 \to$	$S_3 \to$	S_1
Time 0	Time 1	Time 2	Time 3	Time 4

Figure 1.2. Any of the above eight sequences of states results in the process moving from State S_2 at time 0 to State S_1 for the first time at time 4.

To find $f_{21}^{(5)}$ we must find the sum of the conditional probabilities of 16 different transitions, and each of these 16 conditional probabilities is found by taking the product of 5 transition probabilities. In general (still in the context of a 3-state Markov chain), to find $f_{21}^{(k)}$, the conditional probabilities of 2^{k-1} different transitions must be found, and each such conditional probability is the product of k transition probabilities. If the Markov chain we are considering has N possible states, a much larger number of transitions, namely, $(N-1)^{k-1}$ must be considered in order to calculate $f_{21}^{(k)}$. Even if N is only 6 and k is only 5, this means that the conditional probabilities of $(6-1)^{5-1} = 5^4 = 625$ different transitions must be found in order to compute $f_{21}^{(5)}$.

From the above considerations, we are led to look for an alternative way of calculating the first-passage probabilities $f_{21}^{(k)}$ for $k = 1, 2, 3, \ldots$. One way that suggests itself is a recursive method similar to the recursive method used in Chapter 11, Section 4 to compute the k-step transition probabilities $p_{ij}^{(k)}$. To derive such a recursive method, we first show that there exists a certain relationship between first-passage probabilities and transition probabilities.

We earlier pointed out that in general, for $k > 1$, $f_{ij}^{(k)} < p_{ij}^{(k)}$. In particular, for the mating example which we have been considering in the present section, $f_{21}^{(2)} = p_{22}p_{21} + p_{23}p_{31} = \frac{1}{8}$, while

$$p_{21}^{(2)} = p_{21}p_{11} + p_{22}p_{21} + p_{23}p_{31} = \tfrac{2}{8}.$$

The difference between $p_{21}^{(2)}$ and $f_{21}^{(2)}$ is that the transition $S_2 \to S_1 \to S_1$ is counted when computing $p_{21}^{(2)}$, but is not counted when computing $f_{21}^{(2)}$

(since if the transition $S_2 \to S_1 \to S_1$ occurs, then the process reaches State S_1 for the first time at time 1, and not at time 2). Indeed,

$$p_{21}^{(2)} = f_{21}^{(2)} + P\{X_0 = 2 \text{ and } X_1 = 1 \text{ and } X_2 = 1 | X_0 = 2\}$$
$$= f_{21}^{(2)} + p_{21}p_{11},$$

and since we have already seen that $f_{21}^{(1)} = p_{21}$,

(1.7) $$p_{21}^{(2)} = f_{21}^{(2)} + f_{21}^{(1)}p_{11}.$$

We can interpret Equation (1.7) in the following way: To make the transition from State S_2 at time 0 to State S_1 at time 2, the process can either go from State S_2 at time 0 to State S_1 for the *first* time at time 2; or else go from State S_2 at time 0 to State S_1 for the first time at time 1, and then remain in State S_1 at time 2.

Let us try to obtain a relationship similar to Equation (1.7) between the transition probabilities p_{11}, $p_{11}^{(2)}$, and $p_{21}^{(3)}$, and the first-passage probabilities $f_{21}^{(1)}, f_{21}^{(2)}$, and $f_{21}^{(3)}$. To go from State S_2 at time 0 to State S_1 at time 3, one of the following mutually exclusive events must occur:

(i) The process goes from State S_2 at time 0 to State S_1 for the *first* time at time 3. The conditional probability of this event, given that the process is at State S_2 at time 0, is $f_{21}^{(3)}$.

(ii) The process goes from State S_2 at time 0 to State S_1 for the first time at time 2, and then from State S_1 at time 2 to State S_1 once again at time 3. The conditional probability of this event, given that the process is in State S_2 at time 0, is equal to

$$P\{X_2 = 1, X_1 \neq 1 | X_0 = 2\} P\{X_3 = 1 | X_2 = 1 \text{ and } X_1 \neq 1 \text{ and } X_0 = 2\}$$
$$= f_{21}^{(2)} P\{X_3 = 1 | X_2 = 1\}$$
$$= f_{21}^{(2)} p_{11}.$$

To obtain this result we have used the fact that the process is a Markov chain; this fact implies that

$$P\{X_3 = 1 | X_2 = 1 \text{ and } X_1 \neq 1 \text{ and } X_0 = 2\} = P\{X_3 = 1 | X_2 = 1\}.$$

(iii) The process goes from State S_2 at time 0 to State S_1 for the first time at time 1, and then from State S_1 back to State S_1 in two time steps (from time 1 to time 3). The conditional probability of this event, given that the process is in State S_2 at time 0, is equal to

$$P\{X_1 = 1 | X_0 = 2\} P\{X_3 = 1 | X_1 = 1 \text{ and } X_0 = 2\}$$
$$= p_{21}^{(1)} P\{X_3 = 1 | X_1 = 1\}$$
$$= f_{21}^{(1)} p_{11}^{(2)}.$$

Thus, to obtain the conditional probability $p_{21}^{(3)}$ that the process is in State

S_1 at time 3, given that at time 0 the process is in State S_2, we add the conditional probabilities of the events described in (i), (ii), and (iii) above. We obtain the result

(1.8) $$p_{21}^{(3)} = f_{21}^{(3)} + f_{21}^{(2)} p_{11} + f_{21}^{(1)} p_{11}^{(2)}.$$

Using arguments similar to the above we can express the conditional probability $p_{21}^{(k)}$ that the process is in State S_1 at time k, given that the process is in State S_2 at time 0, as the sum of: the conditional probability $f_{21}^{(k)}$ that the process is in State S_1 for the first time at time k; the conditional probability $f_{21}^{(k-1)} p_{11}$ that the process is in State S_1 for the first time at time $k-1$ and then moves from State S_1 back to State S_1 in one time step (from time $k-1$ to time k); the conditional probability $f_{21}^{(k-2)} p_{11}^{(2)}$ that the process is in State S_1 for the first time at time $k-2$ and then moves from State S_1 back to State S_1 in two time steps; and so on. The resulting equation

(1.9) $$p_{21}^{(k)} = f_{21}^{(k)} + f_{21}^{(k-1)} p_{11}^{(1)} + f_{21}^{(k-2)} p_{11}^{(2)} + \cdots + f_{21}^{(1)} p_{11}^{(k-1)}$$

expresses $p_{21}^{(k)}$ in terms of the first-passage probabilities $f_{21}^{(k-1)}, f_{21}^{(k-2)}, \ldots, f_{21}^{(1)}$ and the transition probabilities $p_{11}^{(1)} = p_{11}, p_{11}^{(2)}, p_{11}^{(3)}, \ldots, p_{11}^{(k-1)}$. Alternatively, from Equation (1.9) we obtain the relation

(1.10) $$f_{21}^{(k)} = p_{21}^{(k)} - f_{21}^{(k-1)} p_{11}^{(1)} - f_{21}^{(k-2)} p_{11}^{(2)} - \cdots - f_{21}^{(1)} p_{11}^{(k-1)},$$

which expresses the first-passage probability $f_{21}^{(k)}$ in terms of the k-step transition probability $p_{21}^{(k)}$, the transition probabilities $p_{11}, p_{11}^{(2)}, \ldots, p_{11}^{(k-1)}$, and the first-passage probabilities $f_{21}^{(k-1)}, f_{21}^{(k-2)}, \ldots, f_{21}^{(1)}$. If we have already computed the necessary transition probabilities using the methods of Chapter 11, Section 4, then Equation (1.10) allows us to recursively calculate the first-passage probabilities $f_{21}^{(2)}, f_{21}^{(3)}, f_{21}^{(4)}$, and so on. That is, first we calculate $f_{21}^{(1)} = p_{21}$; then, using the values of $p_{21}^{(2)}$ and p_{11} already obtained, we calculate

(1.11) $$f_{21}^{(2)} = p_{21}^{(2)} - f_{21}^{(1)} p_{11};$$

next, using the values of $p_{21}^{(3)}, p_{11}, p_{11}^{(2)}$, and the values of $f_{21}^{(1)}$ and $f_{21}^{(2)}$ already calculated, we compute $f_{21}^{(3)}$ from

(1.12) $$f_{21}^{(3)} = p_{21}^{(3)} - f_{21}^{(2)} p_{11} - f_{21}^{(1)} p_{11}^{(2)};$$

and so forth.

In our genetical mating example, we have already obtained $f_{21}^{(2)}$ and $f_{21}^{(3)}$ by direct methods. Let us now calculate these same quantities by our recursive method [that is, by the method based on Equation (1.10)]. As a first step we determine all of the 2-step and 3-step transition probabilities for the genetical mating process, using the methods of Chapter 11, Section 4. The 2-step transition probabilities $p_{ij}^{(2)}$ are summarized in Table 1.4, while the 3-step transition probabilities $p_{ij}^{(3)}$ are summarized in Table 1.5.

Table 1.4. The 2-step transition probabilities $p_{ij}^{(2)}$ for the genetical mating process whose transition probabilities are given in Table 1.3.

		State at Time $t+2$ (Mating $t+3$)		
		S_1	S_2	S_3
State at Time t (Mating $t+1$)	S_1	$\frac{3}{8}$	$\frac{1}{2}$	$\frac{1}{8}$
	S_2	$\frac{1}{4}$	$\frac{1}{2}$	$\frac{1}{4}$
	S_3	$\frac{1}{8}$	$\frac{1}{2}$	$\frac{3}{8}$

Table 1.5. The 3-step transition probabilities $p_{ij}^{(3)}$ for the genetical mating process whose transition probabilities are given in Table 1.3.

		State at Time $t+3$ (Mating $t+4$)		
		S_1	S_2	S_3
State at Time t (Mating $t+1$)	S_1	$\frac{5}{16}$	$\frac{1}{2}$	$\frac{3}{16}$
	S_2	$\frac{1}{4}$	$\frac{1}{2}$	$\frac{1}{4}$
	S_3	$\frac{3}{16}$	$\frac{8}{16}$	$\frac{5}{16}$

We are now ready to carry out the computations of $f_{21}^{(2)}$ and $f_{21}^{(3)}$. First, recall from Equation (1.4) that

$$f_{21}^{(1)} = p_{21}^{(1)} = p_{21} = \frac{1}{4}.$$

Next, from Table 1.3 note that $p_{11} = \frac{1}{2}$, and from Table 1.4 note that $p_{21}^{(2)} = \frac{1}{4}$. Thus, using Equation (1.11),

$$f_{21}^{(2)} = p_{21}^{(2)} - f_{21}^{(1)} p_{11} = \frac{1}{4} - \left(\frac{1}{4}\right)\left(\frac{1}{2}\right) = \frac{1}{8},$$

which agrees with the answer for $f_{21}^{(2)}$ obtained by our earlier method. To compute $f_{21}^{(3)}$, we make use of Equation (1.12). Since $p_{11}^{(2)} = \frac{3}{8}$, as can be seen from Table 1.4, and since from Table 1.5, $p_{21}^{(3)} = \frac{1}{4}$, it follows that

$$f_{21}^{(3)} = p_{21}^{(3)} - f_{21}^{(2)} p_{11} - f_{21}^{(1)} p_{11}^{(2)}$$

$$= \frac{1}{4} - \left(\frac{1}{8}\right)\left(\frac{1}{2}\right) - \left(\frac{1}{4}\right)\left(\frac{3}{8}\right)$$

$$= \frac{3}{32}.$$

Again, this result agrees with the answer for $f_{21}^{(3)}$ obtained by our earlier method.

Using Equation (1.10), we can also calculate $f_{21}^{(4)}$, $f_{21}^{(5)}$, $f_{21}^{(6)}$, and so on. For example, since from Table 1.5 we know that $p_{11}^{(3)} = \frac{5}{16}$, and since

$$p_{21}^{(4)} = p_{21}p_{11}^{(3)} + p_{22}p_{21}^{(3)} + p_{23}p_{31}^{(3)}$$
$$= \left(\frac{1}{4}\right)\left(\frac{5}{16}\right) + \left(\frac{1}{2}\right)\left(\frac{8}{32}\right) + \left(\frac{1}{4}\right)\left(\frac{3}{16}\right)$$
$$= \frac{6}{64},$$

it follows from Equation (1.10) and our previous calculations that

$$f_{21}^{(4)} = p_{21}^{(4)} - f_{21}^{(3)}p_{11} - f_{21}^{(2)}p_{11}^{(2)} - f_{21}^{(1)}p_{11}^{(3)}$$
$$= \frac{16}{64} - \left(\frac{3}{32}\right)\left(\frac{1}{2}\right) - \left(\frac{1}{8}\right)\left(\frac{3}{8}\right) - \left(\frac{1}{4}\right)\left(\frac{5}{16}\right)$$
$$= \frac{5}{64}.$$

Equations (1.10), (1.11), and (1.12) provide a method for recursively calculating the first-passage probabilities $f_{21}^{(k)}$, for $k = 1, 2, 3, 4, \ldots$. In deriving these equations, the states S_1 and S_2 were only used for the sake of specificity. Actually, the first-passage probabilities $f_{ij}^{(k)}$, $k = 1, 2, 3, \ldots$, for any two states S_i and S_j could just as well have been the focus of our discussion; and arguments similar to those used above to derive Equations (1.10), (1.11), and (1.12) could have been utilized to obtain the following recursive equations for $f_{ij}^{(k)}$, $k = 1, 2, 3, \ldots$:

(1.13) $\quad f_{ij}^{(1)} = p_{ij},$

$\qquad f_{ij}^{(2)} = p_{ij}^{(2)} - f_{ij}^{(1)} p_{jj},$

$\qquad f_{ij}^{(3)} = p_{ij}^{(3)} - f_{ij}^{(2)} p_{jj} - f_{ij}^{(1)} p_{jj}^{(2)},$

$\qquad \vdots \qquad \vdots \qquad \vdots \qquad \vdots$

$\qquad f_{ij}^{(k)} = p_{ij}^{(k)} - f_{ij}^{(k-1)} p_{jj} - f_{ij}^{(k-2)} p_{jj}^{(2)} - \cdots - f_{ij}^{(2)} p_{jj}^{(k-2)} - f_{ij}^{(1)} p_{jj}^{(k-1)},$

and so on.

Thus, for the genetical mating experiment whose 1-step transition probabilities appear in Table 1.3, we can use Equation (1.13) to recursively calculate the conditional probability mass function

$$f_{23}^{(k)} = P\{K_{23} = k \mid X_0 = 2\}, \qquad k = 1, 2, 3, \ldots,$$

for the first-passage time K_{23}, where K_{23} is the number of matings (after the initial mating) needed to obtain an offspring of genotype aa for the first time. Thus, from Tables 1.3, 1.4, and 1.5, and from Equation (1.13) with $i = 2$, $j = 3$, we find that

$$f_{23}^{(1)} = p_{23} = \frac{1}{4},$$

$$f_{23}^{(2)} = p_{23}^{(2)} - f_{23}^{(1)} p_{33} = \frac{1}{4} - \left(\frac{1}{4}\right)\left(\frac{1}{2}\right) = \frac{1}{8},$$

$$f_{23}^{(3)} = p_{23}^{(3)} - f_{23}^{(2)} p_{33} - f_{23}^{(1)} p_{33}^{(2)} = \frac{1}{4} - \left(\frac{1}{8}\right)\left(\frac{1}{2}\right) - \left(\frac{1}{4}\right)\left(\frac{3}{8}\right) = \frac{3}{32};$$

and since

$$p_{23}^{(4)} = p_{21} p_{13}^{(3)} + p_{22} p_{23}^{(3)} + p_{23} p_{33}^{(3)} = \left(\frac{1}{4}\right)\left(\frac{3}{16}\right) + \left(\frac{1}{2}\right)\left(\frac{1}{4}\right) + \left(\frac{1}{4}\right)\left(\frac{5}{16}\right) = \frac{1}{4},$$

we derive the result

$$f_{23}^{(4)} = p_{23}^{(4)} - f_{23}^{(3)} p_{33} - f_{23}^{(2)} p_{33}^{(2)} - f_{23}^{(1)} p_{33}^{(3)}$$

$$= \frac{1}{4} - \left(\frac{3}{32}\right)\left(\frac{1}{2}\right) - \left(\frac{1}{8}\right)\left(\frac{3}{8}\right) - \left(\frac{1}{4}\right)\left(\frac{5}{16}\right)$$

$$= \frac{5}{64}.$$

We have discussed calculation of the first-passage probabilities $f_{ij}^{(k)}$ only for situations in which the Markov chain under consideration has exactly three possible states. However, note that our derivation of Equation (1.13), or our derivation of Equations (1.10), (1.11), and (1.12), made no use of this fact. Indeed, Equation (1.13) is valid for any Markov chain whatsoever. Assuming that the values of all of the n-step transition probabilities (for $n = 1, 2, 3, \ldots$) are available, it is not difficult to use Equation (1.13) to obtain the first few (say, 4 or 5) first-passage probabilities by hand. However, if the entire conditional probability mass function

(1.14) $\qquad f_{ij}^{(k)} = P\{K_{ij} = k | X_0 = i\}, \qquad k = 1, 2, 3, \ldots$

is desired, it is usually more convenient to make use of computer programs written for this purpose.

Example 1.1. (Political Preference). In a certain state, three different political parties have status for state-wide preferential primaries and delegate selection. Every year, a voter has a chance to register as a member of one of these three parties, and to vote for representatives (delegates) to the state caucus of his party. Since voters change their party registration from year to year in unpredictable fashion, we might model this phenomenon (of political preference) as a discrete-time stochastic process. If, further, we assume that a voter's choice of political party in a given year depends on his past political preferences only in terms of his political choice in the immediately previous year, then we may model the process of political choice as a Markov chain. The states of this process are the three parties, and in addition, since a voter can also choose not to register, a fourth state of being "Not Registered" is required. Thus, the states of the process are the following:

S_1: the voter registers in Party 1,
S_2: the voter registers in Party 2,
S_3: the voter registers in Party 3,
S_4: the voter is not registered.

To a rather gross approximation, the political and social forces which affect political preference can be assumed to act probabilistically in a uniform fashion over time. That is, the process of political choice can be assumed to approximately behave as if it is a Markov chain with stationary transition probabilities.

A political scientist has interviewed a sample of voters in the state, and has obtained from each of these voters a history of their past registrations. Based on this evidence, the political scientist estimates the transition probabilities of the political preference process to be those summarized in Table 1.6.

Table 1.6. Transition probabilities for the political preference process.

		State of Process (Party Preference) in Year $t+1$			
		S_1	S_2	S_3	S_4
State of Process (Party Preference) in Year t	S_1	0.43	0.22	0.10	0.25
	S_2	0.21	0.41	0.11	0.27
	S_3	0.30	0.02	0.33	0.35
	S_4	0.33	0.23	0.10	0.34

If, in a given year, a voter is not registered (State S_4), the length of time (in years) required before the voter registers in Party 1 for the first time may be of interest. In this case, we are interested in the first-passage time K_{41} between State S_4 and State S_1. Using a computer program, the conditional probability mass function

$$f_{41}^{(k)} = P\{K_{41} = k | X_0 = 4\}, \quad k = 1, 2, 3, \ldots$$

has been obtained. This conditional probability mass function is summarized in Table 1.7. Also in Table 1.7, we have summarized the values of $p_{41}^{(k)}$ and $p_{11}^{(k)}$, $k = 1, 2, 3, \ldots$, needed to compute $f_{41}^{(k)}$ by means of Equation

1. FIRST-PASSAGE TIMES

Table 1.7. Values of the conditional probability mass function $f_{41}^{(k)}$ of the first-passage time K_{41} for the political preference process of Example 1.1. The values of $p_{41}^{(k)}$ and $p_{11}^{(k)}$ are also given. All entries are to three significant figures.

k	1	2	3	4	5	6	7	8	9
$f_{41}^{(k)}$	0.330	0.190	0.134	0.097	0.070	0.050	0.036	0.026	0.019
$p_{41}^{(k)}$	0.330	0.332	0.330	0.330	0.330	0.330	0.330	0.330	0.330
$p_{11}^{(k)}$	0.430	0.344	0.331	0.330	0.330	0.330	0.330	0.330	0.330

k	10	11	12	13	14	15	16	17	18	19
$f_{41}^{(k)}$	0.013	0.010	0.007	0.005	0.004	0.003	0.002	0.001	0.001	0.001
$p_{41}^{(k)}$	0.330	0.330	0.330	0.330	0.330	0.330	0.330	0.330	0.330	0.330
$p_{11}^{(k)}$	0.330	0.330	0.330	0.330	0.330	0.330	0.330	0.330	0.330	0.330

(1.13). The values of $p_{41}^{(k)}$ and $p_{11}^{(k)}$ were in turn computed recursively by means of the methods described in Chapter 11, Section 4. All entries in Table 1.7 are given to three significant figures. After $k = 19$, the values of $f_{41}^{(k)}$ are (to three significant figures) equal to 0.

Infinite First-Passage Times

For any finite positive integer k, we showed how to determine the value of the conditional probability mass function, $f_{ij}^{(k)} = P\{K_{ij} = k | X_0 = i\}$, for the first-passage time K_{ij} between two states, State S_i and State S_j, of a Markov chain. However, it is conceivable that if we start the Markov chain process at State S_i, the process never reaches State S_j, no matter how long the process is observed. That is, we can conceive of a sequence of states over time for the process for which the first state of this sequence is State S_i, and for which State S_j never appears in that sequence. In such a situation, the first-passage time K_{ij} between State S_i and State S_j exceeds any finite positive integer k (since for this sequence of states, there is no time k at which the process passes into State S_j for the first time); hence, in this case, we say that K_{ij} is infinite (written $K_{ij} = \infty$). Thus, in general, *the possible values of the first-passage time K_{ij} from State S_i to State S_j are the finite positive integers 1, 2, 3, ..., and the additional value ∞*. We have shown how to compute $P\{K_{ij} = k | X_0 = i\}$ for any finite possible value k ($k = 1, 2, 3, ...$) of the first-passage time K_{ij}, but we have not shown how to compute the conditional probability

$$f_{ij}^{(\infty)} = P\{K_{ij} = \infty | X_0 = i\}$$

that K_{ij} is infinite. Note, however, that if K_{ij} is not equal to one of the finite positive integers 1, 2, 3, ..., then K_{ij} must be infinite. Thus, it follows that

(1.15) $P\{K_{ij} = \infty | X_0 = i\} = 1 - P\{K_{ij} = 1 \text{ or } 2 \text{ or } 3 \text{ or } \cdots | X_0 = i\}$
$= 1 - [P\{K_{ij} = 1 | X_0 = i\} + P\{K_{ij} = 2 | X_0 = i\}$
$+ P\{K_{ij} = 3 | X_0 = i\} + \cdots]$
$= 1 - [f_{ij}^{(1)} + f_{ij}^{(2)} + f_{ij}^{(3)} + \cdots].$

Let

(1.16) $$F_{ij} = f_{ij}^{(1)} + f_{ij}^{(2)} + f_{ij}^{(3)} + \cdots.$$

Then, from Equation (1.15),

(1.17) $$f_{ij}^{(\infty)} = P\{K_{ij} = \infty | X_0 = i\} = 1 - F_{ij}.$$

Since $F_{ij} = P\{K_{ij} = 1 \text{ or } 2 \text{ or } 3 \text{ or } \cdots | X_0 = i\}$ is a probability, it must be the case that

(1.18) $$F_{ij} \leq 1.$$

If $F_{ij} = 1$, then $P\{K_{ij} = \infty | X_0 = i\} = 1 - F_{ij} = 1 - 1 = 0$. Thus, if $F_{ij} = 1$, the event that State S_j cannot be reached by the process, given that the process is initially in State S_i, has a (conditional) probability of 0. In this case, the first-passage time K_{ij} from State S_i to State S_j is finite with probability equal to 1. For example, looking back at Example 1.1, we see (from Table 1.7) that the first-passage time K_{41} is finite with (conditional) probability equal to 1, since $F_{41} = f_{41}^{(1)} + f_{41}^{(2)} + f_{41}^{(3)} + \cdots = 1$.

It is possible, however, to exhibit Markov chains in which there exist two states, S_i and S_j, such that

(1.19) $$F_{ij} < 1.$$

If Inequality (1.19) holds, then the conditional probability that State S_j is never reached, given that the process is initially in State S_i, is a non-zero number. That is, if Inequality (1.19) is true, then

$$P\{K_{ij} = \infty | X_0 = i\} = 1 - F_{ij} > 0,$$

and thus the first-passage time K_{ij} has a positive (conditional) probability of being infinite.

As an example of a Markov chain for which (1.19) holds, consider the Markov chain which has two states S_1 and S_2, and which has transition probabilities

		State at Time $t+1$	
		S_1	S_2
State at Time t	S_1	1	0
	S_2	$\frac{1}{2}$	$\frac{1}{2}$

The learning model of Example 4.1 of Chapter 11 (with $\alpha = \frac{1}{2}$) is an exemplification of such a Markov chain. If this process begins initially in State S_1, then the process cannot leave this state and go to State S_2. To verify this assertion, recall that for any positive integer $k \neq 1$ (that is, $k = 2, 3, 4, \ldots$),

$$p_{12}^{(k)} = p_{11}^{(k-1)} p_{12} + p_{12}^{(k-1)} p_{22}.$$

However, for our particular Markov chain, $p_{12} = 0$ and $p_{22} = \frac{1}{2}$. Thus,

$$p_{12}^{(k)} = \tfrac{1}{2} p_{12}^{(k-1)}.$$

Using this relationship between $p_{12}^{(k)}$ and $p_{12}^{(k-1)}$, we see that since $p_{12} = 0$, then $p_{12}^{(2)} = \frac{1}{2} p_{12} = (\frac{1}{2})(0) = 0$; similarly, $p_{12}^{(2)} = 0$ implies that $p_{12}^{(3)} = \frac{1}{2} p_{12}^{(2)} = 0$; $p_{13}^{(3)} = 0$ implies that $p_{12}^{(4)} = 0$; and so on. Proceeding in this fashion, we can show that $p_{12}^{(k)} = 0$ for all positive integers k. Since $f_{12}^{(k)}$ is a probability, it must be the case that $f_{12}^{(k)} \geq 0$. However, Equation (1.3) tells us that $f_{12}^{(k)} \leq p_{12}^{(k)}$, and we have shown that $p_{12}^{(k)} = 0$, $k = 1, 2, 3, \ldots$. Thus, $f_{12}^{(k)}$ both exceeds and is exceeded by 0, so that $f_{12}^{(k)} = 0$, for all $k = 1, 2, \ldots$. Consequently,

$$\begin{aligned}F_{12} &= f_{12}^{(1)} + f_{12}^{(2)} + f_{12}^{(3)} + \cdots \\ &= 0 + 0 + 0 + \cdots \\ &= 0,\end{aligned}$$

and therefore

$$P\{K_{12} = \infty | X_0 = 1\} = 1 - F_{12} = 1.$$

In words, given that this process is initially in State S_1, then with conditional probability equal to 1, the process never reaches State S_2, and the first-passage time K_{12} between State S_1 and State S_2 is infinite.

For many processes, it is of considerable practical interest to determine the conditional probabilities $f_{ij}^{(\infty)} = P\{K_{ij} = \infty | X_0 = i\}$ for various states S_i and S_j. For example, in the learning process described in Example 4.1 of Chapter 11 (with $\alpha = \frac{1}{2}$), it is of interest to know that (as we have just shown above) once a subject is conditioned (the process is in State S_1), the conditional probability that the subject never becomes unconditioned (that is, the conditional probability that $K_{12} = \infty$) is equal to 1. For this process, it is also of both interest and importance to verify that if a subject starts in an unconditioned state (State S_2), then the first-passage time K_{21} to the conditioned state (S_1) is finite with conditional probability equal to 1. Verification of this latter assertion follows by means of the following argument: First, recall from Equation (4.18) of Chapter 11 that for any $k = 1, 2, 3, \ldots$,

(1.20) $$p_{11}^{(k)} = 1 - p_{12}^{(k)},$$

and that for $k = 2, 3, 4, \ldots,$

(1.21) $$p_{21}^{(k-1)} = 1 - p_{22}^{(k-1)}.$$

We have shown above that for the learning process (with $\alpha = \frac{1}{2}$), $p_{12}^{(k)} = 0$ for $k = 1, 2, 3, \ldots$. Thus, Equation (1.20) implies that $p_{11}^{(k)} = 1$ for all $k = 1, 2, 3, \ldots$. Next, using Equation (1.21) and the fact that

$$p_{21}^{(k)} = p_{21}^{(k-1)} p_{11} + p_{22}^{(k-1)} p_{21}$$
$$= (p_{21}^{(k-1)})(1) + (p_{22}^{(k-1)})(\tfrac{1}{2}),$$

we can show that

$$p_{21}^{(k)} = p_{21}^{(k-1)} + (1 - p_{21}^{(k-1)})(\tfrac{1}{2})$$
$$= \tfrac{1}{2} + \tfrac{1}{2} p_{21}^{(k-1)}.$$

It follows that $p_{21}^{(1)} = \tfrac{1}{2}$, $p_{21}^{(2)} = \tfrac{1}{2} + \tfrac{1}{2} p_{21}^{(1)} = \tfrac{1}{2} + (\tfrac{1}{2})^2 = \tfrac{3}{4}$, $p_{21}^{(3)} = \tfrac{1}{2} + \tfrac{1}{2}[\tfrac{1}{2} + (\tfrac{1}{2})^2] = \tfrac{1}{2} + (\tfrac{1}{2})^2 + (\tfrac{1}{2})^3 = \tfrac{7}{8}$, and in general it can be shown that

$$p_{21}^{(k)} = \tfrac{1}{2} + (\tfrac{1}{2})^2 + (\tfrac{1}{2})^3 + \cdots + (\tfrac{1}{2})^k$$
$$= 1 - (\tfrac{1}{2})^k,$$

for $k = 1, 2, 3, \ldots$. Finally, since $p_{11}^{(m)} = 1$ for $m = 1, 2, 3, \ldots$, we verify that

$$f_{21}^{(k)} = p_{21}^{(k)} - f_{21}^{(k-1)} p_{11} - f_{21}^{(k-2)} p_{11}^{(2)} - \cdots - f_{21}^{(2)} p_{11}^{(k-2)} - f_{21}^{(1)} p_{11}^{(k-1)}$$
$$= 1 - (\tfrac{1}{2})^k - f_{21}^{(k-1)} - f_{21}^{(k-2)} - \cdots - f_{21}^{(2)} - f_{21}^{(1)},$$

or

$$f_{21}^{(1)} + f_{21}^{(2)} + \cdots + f_{21}^{(k)} = 1 - (\tfrac{1}{2})^k.$$

As k grows large, $(\tfrac{1}{2})^k$ gets smaller and smaller, and eventually becomes 0. Therefore,

$$F_{21} = f_{21}^{(1)} + f_{21}^{(2)} + f_{21}^{(3)} + \cdots = 1,$$

and

$$P\{K_{21} = 1 \text{ or } 2 \text{ or } 3 \text{ or } \cdots \mid X_0 = 2\} = P\{K_{21} \text{ is finite} \mid X_0 = 2\}$$
$$= F_{21} = 1.$$

We have thus shown that once a subject starts in an unconditioned state (S_2), the length of time K_{21} required for the subject to enter the conditional State S_1 for the first time is finite with (conditional) probability equal to 1.

For general Markov chains, it is seldom an easy task to determine the conditional probability F_{ij} that the first-passage time K_{ij} between State S_i and State S_j is finite, or even to determine whether $F_{ij} < 1$ or $F_{ij} = 1$. The recursion relations for calculating $f_{ij}^{(k)}$ permit us, in theory, to calculate

$$F_{ij} = f_{ij}^{(1)} + f_{ij}^{(2)} + f_{ij}^{(3)} + \cdots$$

by simply computing $f_{ij}^{(1)}, f_{ij}^{(2)}, f_{ij}^{(3)}$, and so on, recursively, and then adding these quantities. However, unless only a finite number of the $f_{ij}^{(k)}$ are nonzero, computing F_{ij} in this manner requires computation of a infinite number of terms (namely, $f_{ij}^{(1)}$, $f_{ij}^{(2)}$, $f_{ij}^{(3)}$, and so on). For this reason, it becomes necessary in most cases to use mathematical arguments (such as those used above to verify that $F_{12} = 0$ and $F_{21} = 1$ in the learning model example) either to determine the mathematical form of the $f_{ij}^{(k)}$'s, or to directly determine the value of F_{ij}.

2. RECURRENT AND TRANSIENT STATES

In discussing first-passage times K_{ij} from a state S_i to a state S_j, we have up to now assumed that S_i and S_j are not the same state. However, no use of this assumption was made when we verified Equation (1.13). Consequently, if we are interested in the random length K_{jj} of time required for a given state, State S_j, to recur or reappear for the first time, given that the process which we are observing was initially in State S_j, we can repeat the arguments that we used to obtain Equation (1.13), and in this manner derive the following recursive equations for calculating the conditional probability mass function $f_{jj}^{(k)} = P\{K_{jj} = k | X_0 = j\}$ of K_{jj}:

(2.1) $\quad f_{jj}^{(1)} = p_{jj},$

$\qquad f_{jj}^{(2)} = p_{jj}^{(2)} - f_{jj}^{(1)} p_{jj},$

$\qquad f_{jj}^{(3)} = p_{jj}^{(3)} - f_{jj}^{(2)} p_{jj} - f_{jj}^{(1)} p_{jj}^{(2)},$

$\qquad \vdots$

$\qquad f_{jj}^{(k)} = p_{jj}^{(k)} - f_{jj}^{(k-1)} p_{jj} - f_{jj}^{(k-2)} p_{jj}^{(2)} - \cdots - f_{jj}^{(2)} p_{jj}^{(k-2)} - f_{jj}^{(1)} p_{jj}^{(k-1)},$

and so on.

The random variable K_{jj} is a first-passage time between State S_j and itself. Alternatively, we can think of K_{jj} as being a (first) *recurrence time* (or reappearance time) for State S_j. If State S_j is a state of particular interest for a given process, the length K_{jj} of time between appearances of this state will, in many cases, be a random variable of importance. Calculation of the (conditional) probability distribution of this random variable then becomes a basic part of the analysis of the given process.

For example, in the simple genetical mating experiment described in Section 1, suppose that the initial parent (parent II) chosen to mate with the parent (parent I) of fixed genotype Aa has genotype aa (that is, the mating process is in State S_3 at time 0). If genotype aa is an undesirable genotype, we might be interested in the number K_{33} of matings (after the initial mating) that can be performed before an offspring is chosen for mating which again has genotype aa (that is, until State S_3 recurs). From Tables 1.3, 1.4, and 1.5, and from Equation (2.1) with $j = 3$, we can determine the first three values $f_{33}^{(1)}$, $f_{33}^{(2)}$, and $f_{33}^{(3)}$ of the conditional probability mass function $f_{33}^{(k)} = P\{K_{33} = k | X_0 = 3\}$ of K_{33}. These values are

$$f_{33}^{(1)} = p_{33} = \frac{1}{2} = 0.500,$$

$$f_{33}^{(2)} = p_{33}^{(2)} - f_{33}^{(1)} p_{33} = \left(\frac{3}{8}\right) - \left(\frac{1}{2}\right)\left(\frac{1}{2}\right) = \frac{1}{8} = 0.125,$$

and

$$f_{33}^{(3)} = p_{33}^{(3)} - f_{33}^{(2)} p_{33} - f_{33}^{(1)} p_{33}^{(2)} = \left(\frac{5}{16}\right) - \left(\frac{1}{8}\right)\left(\frac{1}{2}\right) - \left(\frac{1}{2}\right)\left(\frac{3}{8}\right) = \frac{1}{16} = 0.0625.$$

The other values of this conditional probability mass function have been determined using a computer. These values are displayed in Table 2.1. From Table 2.1, it appears that

$$F_{33} = f_{33}^{(1)} + f_{33}^{(2)} + f_{33}^{(3)} + \cdots = 1,$$

so that

$$f_{33}^{(\infty)} = P\{K_{33} = \infty | X_0 = 3\} = 1 - F_{33} = 0.$$

However, although Table 2.1 is accurate enough for most practical purposes, the values of $f_{33}^{(k)}$ shown in that table are rounded off to three decimal places, and hence a round-off error could have affected the sum F_{33}. Thus, it is possible that $f_{33}^{(\infty)}$ is actually not 0, but instead is a very small positive number. If it is important to know which is the case, then mathematical analysis is needed to find the exact values of $f_{33}^{(k)}$, $k = 1, 2, 3, \ldots$.

Table 2.1. Values of the conditional probability mass function $f_{33}^{(k)}$ of the (first) recurrence time K_{33} for the genetical mating process.

k	1	2	3	4	5	6	7
$f_{33}^{(k)}$	0.500	0.125	0.063	0.047	0.039	0.033	0.028

k	8	9	10	11	12	13	14	
$f_{33}^{(k)}$	0.024	0.021	0.018	0.015	0.013	0.011	0.009	$\cdots$

In some cases (such as in the above example) the recurrence time K_{jj} (that is, the time required for State S_j to reappear) can never be infinite; that is,

$$f_{jj}^{(\infty)} = P\{K_{jj} = \infty | X_0 = j\} = 0,$$

or, equivalently,

(2.2) $$F_{jj} = f_{jj}^{(1)} + f_{jj}^{(2)} + f_{jj}^{(3)} + \cdots = 1.$$

In this case, the State S_j recurs in a finite length of time with conditional

2. RECURRENT AND TRANSIENT STATES

probability equal to 1, and we say that State S_j is *recurrent*. Thus, in the genetical mating experiment described in Section 1, State S_3 (genotype *aa*) is recurrent, since $F_{33} = 1$ (see Table 2.1). Because a recurrent state S_j has the property that the time K_{jj} until its recurrences is always finite (with conditional probability equal to 1), it follows that once the process has entered State S_j, it must eventually return to that state.

On the other hand, a state, State S_j, is called a *transient* state if, once the process has entered State S_j, there is a positive probability, $P\{K_{jj} = \infty \mid X_0 = j\}$, that the process will never return to State S_j. Put another way, *a state S_j is transient if it is not recurrent*. That is, S_j is a transient state if

(2.3) $$F_{jj} = f_{jj}^{(1)} + f_{jj}^{(2)} + f_{jj}^{(3)} + \cdots < 1,$$

so that

$$f_{jj}^{(\infty)} = P\{K_{jj} = \infty \mid X_0 = j\} = 1 - F_{jj} > 0.$$

Within a given Markov chain, some of the possible states can be recurrent, and some of the possible states can be transient. Classification of the states of a Markov chain in terms of whether they are recurrent or transient is a helpful qualitative way of analyzing such processes. If we know that certain states are recurrent, we know that once we have observed the process to be in one of those states, we will, at a future time, again observe the process to be in that state. On the other hand, once the process has been observed to enter a transient state, we may never again observe that state; transient states are, in other words, transitory. Recurrent states are the "old favorites" or "classics" of the process; once seen, they will be seen again. (Indeed, it can be shown that such states will be seen an arbitrarily large number of times.) Transient states are the "fads" of the process; once seen, they may never be seen again.

[*Remark:* The terms *recurrent* and *transient* are the terms most commonly used to distinguish between states S_j for which $F_{jj}^{(\infty)} = 1$ and states S_i for which $F_{ii}^{(\infty)} < 1$. However, various synonyms have been used in the literature. For example, recurrent states are sometimes called *persistent*, and transient states have been called *unessential*.]

Example 2.1 (Clothing Styles). We have already considered one Markov chain model for consumer preferences in Example 5.1 of Chapter 11. To illustrate a situation in which a Markov chain has both recurrent and transient states, consider the following hypothetical experiment: A longitudinal study of changes in clothing styles for women is being made in the United States. Three alternative styles of dress are identified:

State S_1: miniskirt,
State S_2: midiskirt,
State S_3: maxiskirt.

The choice of dress style for each of several women is recorded daily. (If more than one style is chosen in a day, that style worn "for company" or "for dress-up" is recorded.) The process in which a given woman daily chooses the style of dress which she will wear is assumed to be a Markov chain (that is, the choice of dress style today depends on past choices of dress style only through the choice of dress style worn yesterday). This Markov chain is further assumed to have stationary transition probabilities. Suppose that as a result of the longitudinal study, the transition probabilities shown in Table 2.2 are assigned to the process. Based on this information, let us determine which of the states (styles) S_1, S_2, or S_3 are recurrent, and which are transient.

Table 2.2. Table of transition probabilities for the process of choice of dress style.

		Style on Day $t+1$		
		S_1	S_2	S_3
Style on Day t	S_1 (mini)	$\frac{1}{2}$	$\frac{1}{2}$	0
	S_2 (midi)	$\frac{1}{2}$	$\frac{1}{2}$	0
	S_3 (maxi)	$\frac{1}{4}$	$\frac{1}{4}$	$\frac{1}{2}$

Using the recursion equations for k-step transition probabilities discussed in Chapter 11, Section 4, it can be shown that

(2.4) $\qquad p_{11}^{(k)} = \frac{1}{2}, \qquad p_{12}^{(k)} = \frac{1}{2}, \qquad p_{13}^{(k)} = 0,$

$\qquad p_{21}^{(k)} = \frac{1}{2}, \qquad p_{22}^{(k)} = \frac{1}{2}, \qquad p_{23}^{(k)} = 0,$

and that

(2.5) $\qquad p_{31}^{(k)} = \frac{1}{4} + \frac{1}{2} p_{31}^{(k-1)}, \qquad p_{32}^{(k)} = \frac{1}{4} + \frac{1}{2} p_{32}^{(k-1)}, \qquad p_{33}^{(k)} = \frac{1}{2} p_{33}^{(k-1)},$

for all positive integers $k = 2, 3, 4, \ldots$. Thus, looking first at State S_1 (miniskirt), we find that

$$f_{11}^{(1)} = p_{11} = \frac{1}{2},$$

$$f_{11}^{(2)} = p_{11}^{(2)} - f_{11}^{(1)} p_{11} = \frac{1}{2} - \left(\frac{1}{2}\right)\left(\frac{1}{2}\right) = \frac{1}{4},$$

$$f_{11}^{(3)} = p_{11}^{(3)} - f_{11}^{(2)} p_{11} - f_{11}^{(1)} p_{11}^{(2)} = \frac{1}{2} - \left(\frac{1}{4}\right)\left(\frac{1}{2}\right) - \left(\frac{1}{2}\right)\left(\frac{1}{2}\right) = \frac{1}{8},$$

and in general [making repeated use of Equation (2.1)] it can be shown that

(2.6) $$f_{11}^{(k)} = P\{K_{11} = k | X_0 = 1\} = \left(\frac{1}{2}\right)^k,$$

for all positive integers $k = 1, 2, 3, \ldots$. Now recall that if the discrete random variable X has a geometric distribution with parameter $p = \frac{1}{2}$, then (see Chapter 6, Section 4) the probability mass function $p_X(k) = P_X\{X = k\}$ of X has the form

$$p_X(k) = \frac{1}{2}\left(\frac{1}{2}\right)^{k-1} = \left(\frac{1}{2}\right)^k, \quad k = 1, 2, 3, \ldots.$$

Thus, we see that the recurrence time K_{11} for State S_1 has a (conditional) geometric distribution with parameter $p = \frac{1}{2}$. Because for a geometrically distributed random variable X, it must be the case that

$$p_X(1) + p_X(2) + p_X(3) + \cdots = P\{X = 1 \text{ or } 2 \text{ or } 3 \text{ or } \cdots\}$$
$$= P\{\text{sample space } \mathcal{X}\} = 1,$$

it thus follows that

$$F_{11} = f_{11}^{(1)} + f_{11}^{(2)} + f_{11}^{(3)} + \cdots$$
$$= \left(\frac{1}{2}\right) + \left(\frac{1}{2}\right)^2 + \left(\frac{1}{2}\right)^3 + \cdots$$
$$= 1.$$

Therefore, by definition [see Equation (2.2)], State S_1 is a recurrent state.

Because $p_{11}^{(k)} = p_{22}^{(k)} = \frac{1}{2}$ for all $k = 1, 2, 3, \ldots$, an analysis exactly similar to the above shows that $f_{22}^{(k)} = (\frac{1}{2})^k$, for $k = 1, 2, 3, \ldots$, and thus the recurrence time K_{22} for State S_2 has a geometric distribution with parameter $p = \frac{1}{2}$. Hence,

$$F_{22} = f_{22}^{(1)} + f_{22}^{(2)} + f_{22}^{(3)} + \cdots$$
$$= \left(\frac{1}{2}\right) + \left(\frac{1}{2}\right)^2 + \left(\frac{1}{2}\right)^3 + \cdots$$
$$= 1,$$

and State S_2 (midiskirt) is a recurrent state.

We now turn to State S_3 (maxiskirt). From Equation (2.5), we find that

$$p_{33} = \frac{1}{2},$$
$$p_{33}^{(2)} = \frac{1}{2} p_{33}^{(1)} = \left(\frac{1}{2}\right)\left(\frac{1}{2}\right) = \left(\frac{1}{2}\right)^2,$$
$$p_{33}^{(3)} = \frac{1}{2} p_{33}^{(2)} = \frac{1}{2}\left(\frac{1}{2}\right)^2 = \left(\frac{1}{2}\right)^3,$$

and so on. Thus, $p_{33}^{(k)} = (\tfrac{1}{2})^k$ for $k = 1, 2, 3, \ldots$, and

$$f_{33} = p_{33} = \frac{1}{2},$$
$$f_{33}^{(2)} = p_{33}^{(2)} - f_{33}^{(1)} p_{33} = \frac{1}{4} - \left(\frac{1}{2}\right)\left(\frac{1}{2}\right) = 0,$$
$$f_{33}^{(3)} = p_{33}^{(3)} - f_{33}^{(2)} p_{33} - f_{33}^{(1)} p_{33}^{(2)} = \frac{1}{8} - (0)\left(\frac{1}{2}\right) - \left(\frac{1}{2}\right)\left(\frac{1}{4}\right) = 0,$$

so that in general,

$$f_{33}^{(k)} = 0, \qquad \text{for } k = 2, 3, 4, 5, \ldots.$$

It follows that

$$\begin{aligned} F_{33} &= f_{33}^{(1)} + f_{33}^{(2)} + f_{33}^{(3)} + f_{33}^{(4)} + f_{33}^{(5)} + \cdots \\ &= \frac{1}{2} + 0 + 0 + 0 + 0 + \cdots \\ &= \frac{1}{2} < 1, \end{aligned}$$

and we conclude that State S_3 is a transient state.

In Example 2.1, it is intuitively reasonable that State S_3 should be transient. Looking at Table 2.2, we see that starting in State S_3 at time 0 the process can either return to State S_3 for the first time at time 1 (that is, remain in State S_3), or leave State S_3. If the process leaves State S_3 (and goes to State S_1 or State S_2), it can never return because the probabilities p_{13} and p_{23} of going to State S_3 from State S_1 or State S_2, respectively, are 0. Thus, State S_3 either recurs at time 1 (and this happens with conditional probability $f_{11}^{(1)} = \tfrac{1}{2}$), or State S_3 never recurs. Hence, the (conditional) probability that State S_3 ever recurs is less than 1, and State S_3 is transient.

3. EXPECTED FIRST-PASSAGE TIMES

As we have noted earlier in this section, a first-passage time K_{ij} from a state S_i to a state S_j is a random variable. Although all of the probabilistic properties of K_{ij} can be determined once we know the distribution of K_{ij}, this distribution, as we have seen, can be difficult to obtain. In Chapter 5, we demonstrated that certain descriptive indices of the distribution of a random variable can be used to provide a gross approximate picture of the variational properties of that random variable. In particular, we showed how for a positive random variable X (such as a first-passage time), the approximate magnitudes of the probabilities of certain events could be determined from knowledge of the expected value of X. Considerations such as the above have led probabilists working in the area of Markov

3. EXPECTED FIRST-PASSAGE TIMES

chains to be interested in determining the expected values of the first-passage times K_{ij}.

First-passage times K_{ij} differ from the random variables discussed in Chapters 4 through 10 in that in some cases it is possible (with nonzero conditional probability $f_{ij}^{(\infty)}$) for a first-passage time K_{ij} to be infinite. Although such first-passage times are of great interest to probabilists, methods of dealing with the possibility that K_{ij} can be infinite require the use of fairly complicated mathematical concepts; further, many of the definitions and formulas obtained in Chapters 4 and 5 do not apply to such random variables. Hence, in the following discussion of the expected values of the first-passage times K_{ij}, we will assume that we have already found that

$$f_{ij}^{(\infty)} = P\{K_{ij} = \infty | X_0 = i\} = 0,$$

or, equivalently, that

$$F_{ij} = f_{ij}^{(1)} + f_{ij}^{(2)} + f_{ij}^{(3)} + \cdots = 1,$$

so that

$$f_{ij}^{(k)} = P\{K_{ij} = k | X_0 = i\}, \quad k = 1, 2, 3, \ldots$$

is a probability mass function (see Chapter 4). In this case, we have the following (conditional) probability model for the first-passage time K_{ij}:

k	1	2	3	4	$\cdots$
$P\{K_{ij} = k | X_0 = i\}$	$f_{ij}^{(1)}$	$f_{ij}^{(2)}$	$f_{ij}^{(3)}$	$f_{ij}^{(4)}$	$\cdots$

Using the definition of the expected value of a discrete random variable given in Chapter 5, Equation (2.1), we define the *expected first-passage time* (or *mean first-passage time*) m_{ij} from State S_i to State S_j to be

$$(3.1) \qquad m_{ij} = 1f_{ij}^{(1)} + 2f_{ij}^{(2)} + 3f_{ij}^{(3)} + \cdots + kf_{ij}^{(k)} + \cdots.$$

That is, the expected first-passage time m_{ij} from State S_i to State S_j is the (conditional) expected value $E[K_{ij} | X_0 = i]$ of the first-passage time K_{ij}.

Example 1.1. (Continued). We earlier obtained (see Table 1.7) the conditional probability mass function $f_{41}^{(k)}$, $k = 1, 2, 3, \ldots$, for the number of years K_{41} required until a voter who is not now registered (State S_4) registers for the first time in Party 1 (State S_1). From Table 1.7 and Equation (3.1), the expected number of years m_{41} required before a voter who is not now registered chooses to register for the first time in Party 1 is equal to

$$m_{41} = 1f_{41}^{(1)} + 2f_{41}^{(2)} + 3f_{41}^{(3)} + \cdots + 18f_{41}^{(18)} + 19f_{41}^{(19)} + 0$$
$$= 3.359.$$

Thus, on the average, a voter who is not now registered will take 3.359 years (somewhat over 3 years) before he first chooses to register in Party 1.

The recurrence time K_{jj} for a State S_j is also a first-passage time. Thus, if State S_j is a recurrent state (so that $f_{jj}^{(k)} = P\{K_{jj} = k | X_0 = j\}$ is a probability mass function in the sense defined in Chapter 4), then we can determine the *expected* (or *mean*) *recurrence time* m_{jj} by means of Equation (3.1). That is,

(3.2) $$m_{jj} = 1 f_{jj}^{(1)} + 2 f_{jj}^{(2)} + 3 f_{jj}^{(3)} + \cdots + k f_{jj}^{(k)} + \cdots.$$

The expected recurrence time m_{jj} is thus the (conditional) expected value $E[K_{jj} | X_0 = j]$ of the recurrence time K_{jj}.

Example 2.1 (Continued). In the context of the process of choice of dress style which was described earlier, we found that State S_1 (choice of a miniskirt) is a recurrent state, and we determined [see Equation (2.6)] that the conditional probability mass function $f_{11}^{(k)} = P\{K_{11} = k | X_0 = 1\}$ of the recurrence time K_{11} is given by

$$f_{11}^{(k)} = \left(\frac{1}{2}\right)^k, \qquad k = 1, 2, 3, \ldots.$$

Thus, if we see a miniskirted woman on a given day, we know that the expected number of days m_{11} until for the first time that woman once again chooses to wear a miniskirt is given by the following expression:

$$\begin{aligned} m_{11} &= 1 f_{11}^{(1)} + 2 f_{11}^{(2)} + 3 f_{11}^{(3)} + \cdots \\ &= (1)\left(\frac{1}{2}\right)^1 + (2)\left(\frac{1}{2}\right)^2 + (3)\left(\frac{1}{2}\right)^3 + \cdots. \end{aligned}$$

We recognize this expression as being the same as that for the expected value $E[X]$ of a random variable X having a geometric distribution with parameter $p = \frac{1}{2}$. (This is not surprising since, as we remarked earlier, the conditional distribution of K_{11} is a geometric distribution with parameter $p = \frac{1}{2}$.) Since the expected value $E[X]$ of a random variable X having a geometric distribution with parameter p is equal to $E[X] = 1/p$, and since in this case $p = \frac{1}{2}$, we conclude that

$$m_{11} = \frac{1}{\left(\frac{1}{2}\right)} = 2.$$

Thus, on the average, a miniskirted woman will choose a miniskirt again for the first time 2 days after she initially wears a miniskirt. (Remember, however, that both the Markov chain model and the transition probabilities for this process are hypothetical.)

If we want to find m_{22}, the expected number of days until a midiskirted woman for the first time again wears a midiskirt, identical arguments show that $m_{22} = 2$ days. Because State S_3 (maxiskirt) is a transient state, we cannot compute m_{33} for this state. Note, however, that because S_3 is a transient state, there is a positive conditional probability (namely, $f_{33}^{(\infty)} = 1 - F_{33} = \frac{1}{2}$) that a maxiskirted woman may never again wear a maxiskirt. (This shows how hypothetical our model actually is!)

Alternate Computation for Expected First-Passage Times

The computation of the expected first-passage time m_{ij} (or the expected renewal time m_{jj}) illustrated above requires us to first determine the first-passage probabilities $f_{ij}^{(1)}, f_{ij}^{(2)}, f_{ij}^{(3)}$, and so on (or the renewal probabilities $f_{jj}^{(1)}, f_{jj}^{(2)}, f_{jj}^{(3)}$, and so on) before we can compute the desired expected value. However, one of our motivations for considering the expected first-passage time m_{ij} was to avoid determining the first-passage probabilities $f_{ij}^{(k)}$, $k = 1, 2, \ldots$, since these are often difficult to obtain. Thus, the above method of computation is inappropriate to our needs (at least for most problems). There is, fortunately, an alternative method of computation which determines all of the expected first-passage times m_{ij} simultaneously, and which requires knowledge only of the transition probabilities p_{ij} of the process. This method involves the simultaneous solution of a system of linear equations. Because numerical and algebraic methods are readily available for the solution of linear equations, this method of computation for the expected first-passage times can be quite straightforwardly applied in practice, particularly when a computer is available. For the sake of convenience and specificity, we first show how to apply this method to the calculation of the expected first-passage times for a 3-state Markov chain, and then show how to extend our results to N-state Markov chains.

The 3-State Case. Suppose that we have a Markov chain with three states S_1, S_2, and S_3. Given that we are initially (at time 0) in State S_1, we may be interested in how long, on the average, it takes for the process to first enter State S_j, where j can be 1, 2, or 3. Consider, now, how the process can move from State S_1 to, say, State S_3 for the first time. The process could, of course, go from State S_1 to State S_3 in one time period; this event has (conditional) probability p_{13}. Alternatively, the process could move from State S_1 to State S_1 in one time step, and then move from State S_1 to State S_3 for the first time at some later time (taking, on the average, m_{13} time periods to do so); this event has (conditional) probability p_{11}. Finally, the process could move from State S_1 to State S_2 in one time period, and then move from State S_2 to State S_3 for the first time at some later time (taking, on the average, m_{23} time periods to do so); this event has conditional probability p_{12}. The three possibilities that we have indicated are illustrated in Figure 3.1.

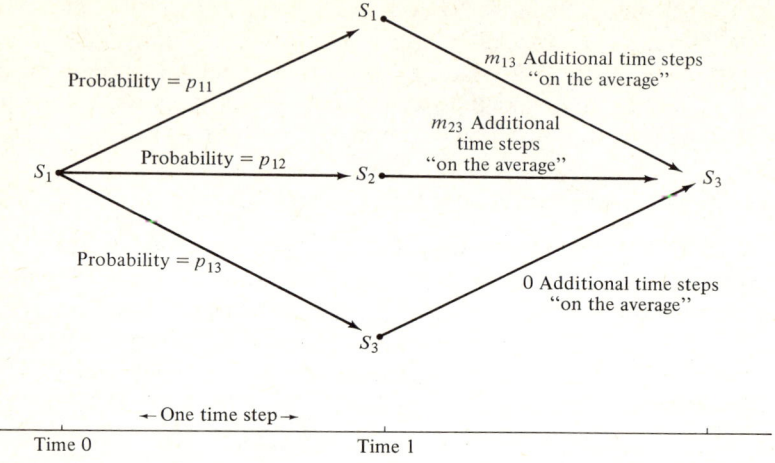

Figure 3.1 Figure illustrating ways that the process can move from State S_1 to State S_3 for the first time, the conditional probabilities of each of these ways, and the number of additional time steps after the first step required "on the average" for each way.

Let W be a random variable defined as follows: If the process goes from State S_1 at time 0 to State S_j at time 1, then W equals the total *additional* length of time required on the average for the process to then go from State S_j to State S_3 for the first time. Thus,

$$W = \begin{cases} 0, & \text{if the process enters State } S_3 \text{ at time 1,} \\ m_{23}, & \text{if the process enters State } S_2 \text{ at time 1,} \\ m_{13}, & \text{if the process remains in State } S_1 \text{ at time 1.} \end{cases}$$

The total length of time required, on the average, for the process to go from State S_1 to State S_3 is then equal to 1 plus the conditional expected value of W, given that the process begins in State S_1. That is,

(3.3) $$m_{13} = 1 + E[W | X_0 = 1].$$

Since

$$P\{W = 0 | X_0 = 1\} = P\{\text{the process is in State } S_3 \text{ at time } 1 | X_0 = 1\}$$
$$= P\{X_1 = 3 | X_0 = 1\} = p_{13},$$
$$P\{W = m_{23} | X_0 = 1\} = P\{X_1 = 2 | X_0 = 1\} = p_{12},$$

and

$$P\{W = m_{13} | X_0 = 1\} = P\{X_1 = 1 | X_0 = 1\} = p_{11},$$

it follows that

$$E[W|X_0 = 1] = (0)P\{W = 0|X_0 = 1\} + (m_{23})P\{W = m_{23}|X_0 = 1\}$$
$$+ (m_{13})P\{W = m_{13}|X_0 = 1\}$$
$$= (0)(p_{13}) + (m_{23})(p_{12}) + (m_{13})(p_{11})$$
$$= p_{11}m_{13} + p_{12}m_{23}.$$

Hence, from Equation (3.3), we conclude that

$$m_{13} = 1 + p_{11}m_{13} + p_{12}m_{23}.$$

We have thus (somewhat heuristically) obtained an equation which relates the expected first-passage times m_{13} and m_{23} to the transition probabilities p_{11} and p_{12}. Proceeding in a similar fashion, we can also argue that

$$m_{23} = 1 + p_{21}m_{13} + p_{22}m_{23},$$

and that

$$m_{33} = 1 + p_{31}m_{13} + p_{32}m_{23}.$$

Summarizing these three equations, we have

(3.4)
$$m_{13} = 1 + p_{11}m_{13} + p_{12}m_{23},$$
$$m_{23} = 1 + p_{21}m_{13} + p_{22}m_{23},$$
$$m_{33} = 1 + p_{31}m_{13} + p_{32}m_{23}.$$

If we are given the values of the transition probabilities p_{11}, p_{12}, p_{21}, p_{22}, p_{31}, and p_{32}, we can solve (3.4) for the unknowns m_{13}, m_{23}, and m_{33}. To see this, note that m_{33} appears only in the last of the three equations shown in (3.4). The first two equations can be rewritten in the form:

(3.5)
$$(1 - p_{11})m_{13} + (-p_{12})m_{23} = 1,$$
$$(-p_{21})m_{13} + (1 - p_{22})m_{23} = 1.$$

Using the general result that if $a_1b_2 - a_2b_1 \neq 0$, the equations

(3.6)
$$a_1x + b_1y = c_1,$$
$$a_2x + b_2y = c_2,$$

have the solutions

(3.7)
$$x = \frac{c_1b_2 - c_2b_1}{a_1b_2 - a_2b_1}, \qquad y = \frac{a_1c_2 - a_2c_1}{a_1b_2 - a_2b_1},$$

for x and y, we conclude from (3.5) that

$$m_{13} = \frac{(1)(1 - p_{22}) - (1)(-p_{12})}{(1 - p_{11})(1 - p_{22}) - (-p_{12})(-p_{21})} = \frac{1 - p_{22} + p_{12}}{1 - p_{11} - p_{22} + p_{11}p_{22} - p_{12}p_{21}},$$

and
$$m_{23} = \frac{1-p_{11}+p_{21}}{1-p_{11}-p_{22}+p_{11}p_{22}-p_{12}p_{21}},$$

provided that $1-p_{11}-p_{22}+p_{11}p_{22}-p_{12}p_{21} \neq 0$. Substituting these results into the last equation of (3.4) permits us to solve for m_{33}.

Arguments similar to those used to derive (3.4) allow us to obtain the following set of three equations for m_{11}, m_{21}, and m_{31} in terms of the transition probabilities $p_{12}, p_{13}, p_{22}, p_{23}, p_{32}$, and p_{33}:

(3.8)
$$m_{11} = 1 + p_{12}m_{21} + p_{13}m_{31},$$
$$m_{21} = 1 + p_{22}m_{21} + p_{23}m_{31},$$
$$m_{31} = 1 + p_{32}m_{21} + p_{33}m_{31}.$$

We may solve the last two of these equations for m_{21} and m_{31} by means of the general solution obtained in (3.6) for any set of two equations in two unknowns of the form (3.5). The answers for m_{21} and m_{31} so obtained can then be substituted into the first equation in (3.8), thus obtaining m_{11}.

Finally, the following set of three equations can be solved for m_{12}, m_{22}, and m_{32}; namely,

(3.9)
$$m_{12} = 1 + p_{11}m_{12} + p_{13}m_{32},$$
$$m_{22} = 1 + p_{21}m_{12} + p_{23}m_{32},$$
$$m_{32} = 1 + p_{31}m_{12} + p_{33}m_{32}.$$

To do this, we may first solve the first and last equations in (3.9) for m_{12} and m_{32}, and then substitute our answers into the middle equation of (3.9) to find m_{22}.

[*Remark:* It should be noted that if $F_{ij} < 1$ for any pair of states S_i and S_j, then not all of Equations (3.4), (3.8), and (3.9) will have finite solutions. Thus, to use these equations for finding the values of the expected first-passage times m_{ij}, it is helpful to know that $F_{ij} = 1$ for all states S_i and S_j (including cases where $i = j$). A condition under which this last result holds (that is, $F_{ij} = 1$ for all i, j) is given in Section 5.]

We have discussed two examples of 3-state Markov chains in this chapter. One of these examples (the clothing style process of Example 2.1) contains a transient state (namely, State S_3: maxiskirts). Thus, $F_{33} < 1$ in that example, and by the above remark, we know that not all of Equations (3.4), (3.8), and (3.9) will have finite solutions. For the other example of a 3-state Markov chain mentioned in this chapter (the genetical mating process whose transition probabilities were given in Table 1.3), it can be shown that $F_{ij} = 1$ for all states S_i and S_j; $i, j = 1, 2, 3$. Thus, this example can serve to illustrate our new method of calculating the expected first-passage times m_{ij}.

3. EXPECTED FIRST-PASSAGE TIMES

Recall that the states of the genetical mating process are

S_1: genotype AA,
S_2: genotype Aa,
S_3: genotype aa.

The transition probabilities for this Markov chain are summarized in Table 1.3. To obtain the expected first-passage times between states S_1 and S_2, states S_2 and S_2, and states S_3 and S_2 (that is, m_{12}, m_{22}, m_{32}), we see from Table 1.3 and Equation (3.4) that we must solve the following set of equations:

$$m_{12} = 1 + (\tfrac{1}{2})m_{12} + (0)m_{22},$$
$$m_{22} = 1 + (\tfrac{1}{4})m_{12} + (\tfrac{1}{4})m_{22},$$
$$m_{32} = 1 + (0)m_{12} + (\tfrac{1}{2})m_{22}.$$

We can solve this set of equations directly [without the need of using the general theory of Equations (3.6) and (3.7)]. From the first equation, we see that $(\tfrac{1}{2})m_{12} = 1$ or $m_{12} = 2$. Substituting this result into the second equation, we obtain the result

$$m_{22} = 1 + (\tfrac{1}{4})(2) + (\tfrac{1}{4})m_{22},$$

or $m_{22} = 2$. Finally, if we substitute the values $m_{12} = 2$, $m_{22} = 2$ into the third equation, we find that

$$m_{32} = 1 + (0)(2) + (\tfrac{1}{2})(3) = \tfrac{5}{2}.$$

Thus, the expected first-passage time (the expected number of matings) required to go from an offspring of genotype AA (State S_1) to an offspring of genotype Aa (State S_2) for the first time is $m_{12} = 2$; the expected number of matings required to go from an offspring of genotype aa (State S_3) to an offspring of genotype Aa (State S_2) for the first time is $m_{22} = 2$; and the expected number of matings required for genotype Aa to recur is $m_{22} = 2$.

To obtain the expected first-passage times m_{11}, m_{21}, m_{31}, we know from Table 1.3 and Equation (3.8) that we must solve the following collection of equations:

(3.10)
$$m_{11} = 1 + (\tfrac{1}{2})m_{21} + (0)m_{31},$$
$$m_{21} = 1 + (\tfrac{1}{2})m_{21} + (\tfrac{1}{4})m_{31},$$
$$m_{31} = 1 + (\tfrac{1}{2})m_{21} + (\tfrac{1}{2})m_{31}.$$

We first solve the last two equations for m_{21} and m_{31}. To do this, we rewrite these equations in the form of Equation (3.6) with $x = m_{21}$, and $y = m_{31}$. Thus, we obtain the equations

$$\tfrac{1}{2}m_{21} + (-\tfrac{1}{4})m_{31} = 1,$$
$$(-\tfrac{1}{2})m_{21} + (\tfrac{1}{2})m_{31} = 1,$$

which [see Equation (3.7)], when solved for m_{21} and m_{31}, yield the solutions

$$m_{21} = \frac{(1)(\frac{1}{2}) - (1)(-\frac{1}{4})}{(\frac{1}{2})(\frac{1}{2}) - (-\frac{1}{2})(-\frac{1}{4})} = 6, \qquad m_{31} = \frac{(\frac{1}{2})(1) - (-\frac{1}{2})(1)}{(\frac{1}{2})(\frac{1}{2}) - (-\frac{1}{2})(-\frac{1}{4})} = 8.$$

Substituting $m_{21} = 6$ and $m_{31} = 8$ into the first equation in (3.10), we find that

$$m_{11} = 1 + (\tfrac{1}{2})m_{21} + (0)m_{31}$$
$$= 1 + (\tfrac{1}{2})(6) + (0)(8) = 4.$$

Thus, $m_{21} = 6$, $m_{31} = 8$, and $m_{11} = 4$.

Finally, to obtain m_{13}, m_{23}, and m_{33}, we know from Table 1.3 and Equation (3.9) that we need to solve the following three equations:

$$m_{13} = 1 + (\tfrac{1}{2})m_{13} + (\tfrac{1}{2})m_{23},$$
$$m_{23} = 1 + (\tfrac{1}{4})m_{13} + (\tfrac{1}{2})m_{23},$$
$$m_{33} = 1 + (0)m_{13} + (\tfrac{1}{2})m_{23}.$$

We first solve the first two equations for m_{13} and m_{23}. To do this, we write these equations in the form of Equation (3.6) with $x = m_{13}$ and $y = m_{23}$, and then solve these equations using (3.7). We obtain

$$m_{13} = 8, \qquad m_{23} = 6.$$

The complete collection of expected first-passage times m_{ij} for the genetical mating process is summarized in Table 3.1.

Table 3.1. Expected first-passage times m_{ij}, from State S_i to State S_j for the genetical mating process whose transition probabilities are given in Table 1.3.

		Final State S_j		
		S_1	S_2	S_3
Initial State S_i	S_1	4	2	8
	S_2	6	2	6
	S_3	8	2	4

The N-State Case. The method that we have just described for finding the expected first-passage times m_{ij} for a 3-state Markov chain can be generalized to more general Markov chain contexts. Suppose that a Markov chain has a finite number N of states $S_1, S_2, \ldots, S_N$. Then, if we are given

3. EXPECTED FIRST-PASSAGE TIMES

the values of the transition probabilities p_{ij} of this Markov chain (and if we know that $F_{ij} = 1$ for all states S_i and S_j, including cases where $i = j$), then, for example, we can find the expected first-passage times $m_{11}, m_{21}, \ldots, m_{N1}$ for this process by solving the following collection of N equations:

(3.11)
$$m_{11} = 1 + p_{12}m_{21} + p_{13}m_{31} + \cdots + p_{1N}m_{N1},$$
$$m_{21} = 1 + p_{22}m_{21} + p_{23}m_{31} + \cdots + p_{2N}m_{N1},$$
$$m_{31} = 1 + p_{32}m_{21} + p_{33}m_{31} + \cdots + p_{3N}m_{N1},$$
$$\vdots$$
$$m_{N1} = 1 + p_{N2}m_{21} + p_{N3}m_{31} + \cdots + p_{NN}m_{N1},$$

for the unknowns $m_{11}, m_{21}, m_{31}, \ldots, m_{N1}$. Note that because the unknown m_{11} appears only in the first equation in (3.11), we may solve this set of equations by first solving the last $N-1$ equations for $m_{21}, m_{31}, \ldots, m_{N1}$, and then substituting our answers into the first equation in order to find m_{11}.

Let us both add and subtract $p_{ii}m_{11}$ from the right-hand side of the (i)th equation in (3.11), for $i = 1, 2, 3, \ldots, N$. Since this operation does not change these equations (we are only adding 0 to each right-hand side), the solutions for $m_{11}, m_{21}, m_{31}, \ldots, m_{N1}$ of the resulting collection of equations

(3.12)
$$m_{11} = (1 - p_{11}m_{11}) + p_{11}m_{11} + p_{12}m_{21} + \cdots + p_{1N}m_{N1},$$
$$m_{21} = (1 - p_{21}m_{11}) + p_{21}m_{11} + p_{22}m_{21} + \cdots + p_{2N}m_{N1},$$
$$m_{31} = (1 - p_{31}m_{11}) + p_{31}m_{11} + p_{32}m_{21} + \cdots + p_{3N}m_{N1},$$
$$\vdots$$
$$m_{N1} = (1 - p_{N1}m_{11}) + p_{N1}m_{11} + p_{N2}m_{21} + \cdots + p_{NN}m_{N1},$$

are the same as the solutions to the set of equations in (3.11). The collection of equations (3.12) exhibits a pattern which permits generalization to a collection of equations solvable for the expected first-passage times $m_{1j}, m_{2j}, m_{3j}, \ldots, m_{Nj}$, $j = 1, 2, 3, \ldots, N$. This general set of equations is the following:

(3.13)
$$m_{1j} = (1 - p_{1j}m_{jj}) + p_{11}m_{1j} + p_{12}m_{2j} + \cdots + p_{1N}m_{Nj},$$
$$m_{2j} = (1 - p_{2j}m_{jj}) + p_{21}m_{1j} + p_{22}m_{2j} + \cdots + p_{2N}m_{Nj},$$
$$m_{3j} = (1 - p_{3j}m_{jj}) + p_{31}m_{1j} + p_{32}m_{2j} + \cdots + p_{3N}m_{Nj},$$
$$\vdots$$
$$m_{Nj} = (1 - p_{Nj}m_{jj}) + p_{N1}m_{1j} + p_{N2}m_{2j} + \cdots + p_{NN}m_{Nj}.$$

For any state S_j, we may solve this set of N equations for the N unknowns $m_{1j}, m_{2j}, \ldots, m_{Nj}$.

As long as N is not more than about 100, a set of N linear equations in N unknowns can be solved quite rapidly for the unknowns by most standard computers. Even the smallest computers can solve up to 30 linear equations in 30 unknowns. Thus, although finding a solution of the collection of equations (3.13) by hand calculation can become tedious when N is even as large as 5, these equations can be conveniently and quickly solved by means of a computer.

4. PROBABILISTIC EQUILIBRIUM OF A MARKOV CHAIN

If the free end of a pendulum is given a strong push, the pendulum will at first swing back and forth in an erratic and irregular fashion. After some time has passed, however, gravitational and frictional forces will tend to regularize the path of the pendulum through space. No matter how we start the pendulum swinging, if we wait a sufficient length of time, the basic dynamic forces (gravity and friction) which work to determine the back-and-forth motion of the pendulum eventually will influence the pendulum to move in the same constant and predictable fashion over time. This basic long-term stability (or reproducibility) of the motion of a pendulum is the reason why pendulums have been used in the construction of timing mechanisms (such as clocks).

The pendulum is an example of a physical process in which there is a "start up" phenomenon that is typically different from what can be expected in the "long run." At the beginning, the performance of such a process depends greatly on the nature (state) of the process at its inception, and on the length of time during which the process has been active. However, after a sufficient length of time has passed, the activity of the process becomes almost completely regulated by balances among the basic dynamic forces that generate the process (and give it its recognizable character), and the process settles down to a consistent and predictable pattern of activity over time. When this phenomenon occurs, we say that the process is in *equilibrium*, or that the process is in a *steady state*. [*Note:* The meaning of the word "state" here differs somewhat from the meaning of this word in the context of Markov chains.] The design and operation of most physical processes (machines, chemical reactors, and so on) depends on the achievement of such a state of equilibrium by the process, since otherwise the process tends to be subject to frequent and unpredictable changes that make it difficult to handle or utilize. Social and biological processes also seem to be most useful when they achieve a state of equilibrium, since when in equilibrium, their activity is predictable, and thus

4. PROBABILISTIC EQUILIBRIUM OF A MARKOV CHAIN

more easily controlled. For these reasons, scientists and technologists are more often concerned with the long-run equilibrium behavior of physical, social, or biological processes than they are with the "start up" behavior of these processes. Indeed, in many cases, a process has already achieved equilibrium when we begin to study it, so that the "start up" behavior of the process may not even be observable.

Corresponding with the notion of a state of equilibrium for processes of a mechanistic type (that is, processes whose dynamics are well described by a mechanistic model), we can conceive of equilibrium behavior for stochastic processes. However, for stochastic processes it is not the observed *performance* over time of such a process which achieves a characteristic and predictable pattern of behavior. Rather, it is the conditional and unconditional *probabilities* of observing the various states of the process which eventually become constant, ceasing to depend in value either on the condition (state) of the process at its inception, or on the length of time that the process has been in operation.

Consider, for example, the simple 2-state Markov chain whose transition probabilities are given in Table 4.1. A 2-state Markov chain with these transition probabilities has previously been used to model the choice-of-brand process described in Section 5 of Chapter 11. Using the

Table 4.1. Table of transition probabilities for a 2-state Markov chain.

		State at Time $t+1$	
		S_1	S_2
State at Time t	S_1	0.500	0.500
	S_2	0.750	0.250

methods of computation described in Section 4 of Chapter 11, we can compute the 2-step, 3-step, 4-step, and so on, transition probabilities of this Markov chain. These transition probabilities are summarized in Table 4.2 (see also Tables 5.2 to 5.5 of Chapter 11). All entries in Table 4.2 are accurate to three decimal places.

If we compare the first two rows of Table 4.2, we can see how the conditional probability of being in State S_1 at time k, given the state, S_1 or S_2, of the process at time 0, depends for its value on the time, time k, and on the initial state. When k is small (the process is just beginning), the probability of reaching State S_1 at time k, given that the process started at State S_1, differs greatly from the probability of reaching State S_1 at time k, given that the process started at State S_2. Thus, for example, $p_{11}^{(1)}$ equals 0.500, but $p_{21}^{(1)} = 0.750$. Again, $p_{11}^{(2)} = 0.625$, but $p_{21}^{(2)} = 0.562$; also, $p_{11}^{(3)} = 0.594$, but $p_{21}^{(3)} = 0.609$. However, for $k = 5, 6, 7$, and for even larger values

Table 4.2. The k-step transition probabilities $p_{ij}^{(k)}$ for the Markov chain whose transition probabilities are given in Table 4.1.

k-Step Transition Probability to:	k							
	1	2	3	4	5	6	7	...
State S_1 $\begin{cases} p_{11}^{(k)} \\ p_{21}^{(k)} \end{cases}$	0.500	0.625	0.594	0.602	0.600	0.600	0.600	...
	0.750	0.562	0.609	0.598	0.600	0.600	0.600	...
State S_2 $\begin{cases} p_{12}^{(k)} \\ p_{22}^{(k)} \end{cases}$	0.500	0.375	0.406	0.398	0.400	0.400	0.400	...
	0.250	0.438	0.391	0.402	0.400	0.400	0.400	...

of k (not explicitly shown in Table 4.2), the conditional probability $p_{11}^{(k)}$ of reaching State S_1 at time k, given that the process started at State S_1, equals the conditional probability $p_{21}^{(k)}$ of reaching State S_1 at time k, given that the process started at State S_2. (Remember, however, that this "equality" is only approximate, since Table 4.2 is accurate only to three decimal places.) That is, after time $k = 5$, the value of the (conditional) probability of reaching State S_1 at time k ceases (at least approximately) to depend on the state of the process at time 0. A similar comparison of rows 3 and 4 of Table 4.2 shows that after time $k = 5$, the value of the (conditional) probability of reaching State S_2 at time k ceases to depend on the state of the process at time 0.

Looking across each row of Table 4.2, we see that for values of k greater than or equal to $k = 5$, the k-step transition probabilities $p_{ij}^{(k)}$ become (at least to three decimal places of accuracy) independent of the time k. Thus,

$$p_{11}^{(5)} = p_{11}^{(6)} = p_{11}^{(7)} = 0.600,$$
$$p_{21}^{(5)} = p_{21}^{(6)} = p_{21}^{(7)} = 0.600,$$
$$p_{12}^{(5)} = p_{12}^{(6)} = p_{12}^{(7)} = 0.400,$$
$$p_{22}^{(5)} = p_{22}^{(6)} = p_{22}^{(7)} = 0.400,$$

and explicit calculation of the transition probabilities $p_{ij}^{(k)}$ for $k = 8, 9, 10$, and so on would show similar independence of the time k.

From this analysis, we can conclude that the conditional probabilities $p_{ij}^{(k)}$ of observing State S_j at time k, given that the process starts in State S_i, eventually (after a long enough length k of time has passed) cease to depend either on the initial state S_i of the process, or on the length k of time that the process has been operating. Hence, after a sufficiently long time of operation, the probabilities of this Markov chain process exhibit equilibrium behavior, and we say that the Markov chain is in *probabilistic* (or *stochastic*) *equilibrium*.

Not all Markov chains can achieve probabilistic equilibrium. However, if a Markov chain *does* achieve probabilistic equilibrium, then regardless

4. PROBABILISTIC EQUILIBRIUM OF A MARKOV CHAIN

of the state S_i in which the process begins, the conditional probability of observing a given state S_j at time k has (approximately) the same value π_j for all sufficiently distant times k. That is, for all initial states S_i and all sufficiently large values of k,

(4.1) $$p_{ij}^{(k)} \cong \pi_j,$$

where we recall that the symbol "$\cong$" means "approximately equal to." This common value π_j of the k-step transition probabilities $p_{ij}^{(k)}$ is called the *long-run* (or *steady-state*, or *equilibrium*) *probability* of State S_j. If we know that a Markov chain with states $S_1, S_2, ..., S_N$ can achieve probabilistic equilibrium, then we know that there is a probability distribution assigning probability π_j to State S_j, for $j = 1, 2, ..., N$, such that for all sufficiently large values of k, a table of the k-step transition probabilities of the Markov chain has the form shown in Table 4.3.

Table 4.3. The k-step transition probabilities of an N-state Markov chain, when k is large enough so that the Markov chain is in probabilistic equilibrium.

		State at Time k			
		S_1	S_2	...	S_N
State at Time 0	S_1	π_1	π_2	...	π_N
	S_2	π_1	π_2	...	π_N
	⋮	⋮	⋮		⋮
	S_N	π_1	π_2	...	π_N

The probability distribution that assigns probability π_j to State S_j is called the *long-run* (or *steady-state*, or *equilibrium*) *distribution* of the Markov chain. Knowledge of this long-run distribution eliminates the need to calculate tables of k-step transition probabilities for large values of k, since all such tables have the form of Table 4.3.

Provided a long-run distribution exists for a given Markov chain (that is, provided the Markov chain can achieve probabilistic equilibrium), knowledge of this long-run distribution also enables us to calculate the values of the absolute probabilities $P\{X_k = j\}$ for all sufficiently large values of k. Recall from Chapter 11, Section 5, that

(4.2) $$P\{X_k = j\} = a_1 p_{1j}^{(k)} + a_2 p_{2j}^{(k)} + \cdots + a_N p_{Nj}^{(k)},$$

where $a_1, a_2, ..., a_N$ are the initial probabilities (the probabilities at time 0) of the states $S_1, S_2, ..., S_N$, respectively. However, from Equation (4.1) we know that for k sufficiently large,

(4.3) $$p_{1j}^{(k)} \cong p_{2j}^{(k)} \cong \cdots \cong p_{Nj}^{(k)} \cong \pi_j.$$

Substituting Equation (4.3) into Equation (4.2), and remembering that $a_1 + a_2 + \cdots + a_N = 1$, we obtain

(4.4)
$$\begin{aligned} P\{X_k = j\} &\cong a_1\pi_j + a_2\pi_j + \cdots + a_N\pi_j \\ &= \pi_j(a_1 + a_2 + \cdots + a_N) \\ &= \pi_j. \end{aligned}$$

Thus, for all sufficiently large values of k, the unconditional probability that the process is in State S_j at time k is (approximately) equal to the long-run probability π_j of State S_j. Note that this result is true *regardless of the initial probabilities* $a_1, a_2, \ldots, a_N$ *of the states* $S_1, S_2, \ldots, S_N$ *of the Markov chain*. Equation (4.4) is thus another illustration of the characteristic property of probabilistic equilibria mentioned earlier; namely, that the probabilities of the states $S_1, S_2, \ldots, S_N$ in the long run cease to depend on the condition of the Markov chain at its start.

From the above discussion, we can see that if a Markov chain can achieve probabilistic equilibrium, much can be learned about the properties of this process from knowledge of its long-run distribution. Thus, in studying a Markov chain, it is useful to: (i) determine whether or not the process can achieve probabilistic equilibrium, and (ii) find the long-run distribution of the process in cases where the Markov chain can achieve probabilistic equilibrium.

In the remaining pages of this chapter, we describe a large class of Markov chains that can achieve probabilistic equilibrium, and show how to find the long-run distributions for such processes. In the course of our presentation, we also introduce certain new concepts concerning Markov chains which not only are of importance to our study of probabilistic equilibrium, but are also of independent interest. Our study will not identify *all* Markov chains that achieve probabilistic equilibrium; however, the new concepts and techniques that we present can be (and have been) used to accomplish that goal.

5. CLASSES OF STATES

In studying probabilistic equilibria and long-run probability distributions for Markov chains, we will find it useful to classify the states of a Markov chain according to various criteria. We have already discussed one such classification for the states of a Markov chain in Section 2; namely, the classification of states according to whether such states are recurrent or transient. Before considering other, equally useful, classifications of states, let us make clear what we mean by a *class of states*, and briefly indicate one important property that certain classes of states possess: the property of being *closed*.

5. CLASSES OF STATES

In general, a *class* $\mathscr{C}$ of states of a Markov chain is any well-defined collection of such states. A class $\mathscr{C}$ of states can be defined by listing those states belonging to the class, or by indicating some property (or criterion) only met by all states in the class. Thus, in the Markov chain described in Example 2.1 (the process of choice of skirt style), one class of states is the class $\mathscr{C}$ containing the states S_1 (miniskirts) and S_2 (midiskirts). This class could be described as the class of states corresponding to styles in skirts in which the hem of the skirt falls above the ankle; or, as we verified in Section 2, this class can be described as the class of recurrent states.

Suppose that we are interested in a particular class $\mathscr{C}$ of states for a given Markov chain. We say that such a class is *closed* (or *absorbing*) if it is the case that once the process is in one of the states belonging to class $\mathscr{C}$ at time t, it can never leave the states in class $\mathscr{C}$ at any later time: $t+1$, $t+2$, $t+3$, and so on. For example, if a Markov chain has four states S_1, S_2, S_3, and S_4, and if the class $\mathscr{C}$ consists of states S_1 and S_2, then $\mathscr{C}$ is closed if once the process enters State S_1, it can then never go either to State S_3 or to State S_4; and if once the process enters State S_2, it then cannot go to States S_3 or S_4.

It can be shown that a class $\mathscr{C}$ of states is closed if for each state S_i belonging to class $\mathscr{C}$, and for every state S_j not belonging to class $\mathscr{C}$, the 1-step transition probability p_{ij} is equal to 0. Thus, if the Markov chain has four states S_1, S_2, S_3, and S_4, and if the class $\mathscr{C}$ consists of states S_1 and S_2, then $\mathscr{C}$ is closed if

$$p_{13} = p_{14} = p_{23} = p_{24} = 0,$$

and otherwise $\mathscr{C}$ is not closed. If the 4-state Markov chain has transition probabilities of the form shown in Table 5.1(a), then the class $\mathscr{C}$ con-

Table 5.1(a). Table of transition probabilities for a Markov chain in which class $\mathscr{C}$ is closed. Here, an "x" in any position in the table represents a transition probability which may or may not be equal to 0. Class $\mathscr{C}$ consists of States S_1 and S_2.

		State at Time $t+1$			
		S_1	S_2	S_3	S_4
State at Time t	S_1	x	x	0	0
	S_2	x	x	0	0
	S_3	x	x	x	x
	S_4	x	x	x	x

sisting of states S_1 and S_2 is closed. On the other hand, if the Markov chain has transition probabilities of the form shown in Table 5.1(b), then class $\mathscr{C}$ is not closed, since $p_{13} \neq 0$.

Table 5.1(b). Table of transition probabilities for a Markov chain in which class $\mathscr{C}$ is not closed. Class $\mathscr{C}$ consists of States S_1 and S_2.

		State at Time $t+1$			
		S_1	S_2	S_3	S_4
State at Time t	S_1	x	x	$\frac{1}{4}$	0
	S_2	x	x	0	0
	S_3	x	x	x	x
	S_4	x	x	x	x

Classes of Communicating States

Two states, S_1 and S_j, of a Markov chain process are said to *communicate* if it is possible for the process to go from State S_i to State S_j in a finite number of time periods, and if it is possible for the process to go from State S_j to State S_i in a finite number of time periods. Formally, a state S_j *can be reached* from another state S_i if, given that the process starts at State S_i, there is a nonzero conditional probability that the process will be in State S_j after a finite number of time periods. Two states, S_i and S_j, communicate if each can be reached from the other. To demonstrate that State S_j can be reached from State S_i, all we have to show is that there is a finite integer n such that $p_{ij}^{(n)}$ is not 0. Thus, to verify that two states, S_i and S_j, communicate with each other, we need to show that there is a finite integer n such that $p_{ij}^{(n)} \neq 0$, and another finite integer m such that $p_{ji}^{(m)} \neq 0$. To show that two states, S_i and S_j, do *not* communicate, we can try to show either that $p_{ij}^{(k)} = 0$ for $k = 1, 2, 3$, and so on, or that $p_{ji}^{(k)} = 0$ for $k = 1, 2, 3$, and so on. However, in some cases it is easier to show that one of the two states, say, S_i, belongs to a closed class to which the other state, S_j, does not belong, since then, by the definition of a closed class, we know that S_j cannot be reached from S_i.

For example, consider the 4-state Markov chain which has the transition probabilities p_{ij} shown in Table 5.2. Let us first decide whether states S_1 and S_2 communicate. From Table 5.2, we see that $p_{12} = \frac{1}{2} \neq 0$, and thus State S_2 can be reached from State S_1 in $n = 1$ time steps. Also from Table 5.2, we note that $p_{21} = 0$. Hence, the process cannot go from State S_2 to State S_1 in one time step. However, we may use the methods of computation described in Chapter 11, Section 4, to compute the 2-step transi-

5. CLASSES OF STATES

Table 5.2. Transition probabilities for a 4-state Markov chain.

		\multicolumn{4}{c}{State at Time $t+1$}

		S_1	S_2	S_3	S_4
State at Time t	S_1	$\frac{1}{2}$	$\frac{1}{2}$	0	0
	S_2	0	$\frac{1}{2}$	$\frac{1}{2}$	0
	S_3	$\frac{1}{2}$	0	$\frac{1}{2}$	0
	S_4	$\frac{1}{4}$	$\frac{1}{4}$	$\frac{1}{4}$	$\frac{1}{4}$

tion probabilities $p_{ij}^{(2)}$; these 2-step transition probabilities are summarized in Table 5.3. From Table 5.3, we see that $p_{21}^{(2)} = \frac{1}{4} \neq 0$, and thus State S_1 can be reached from State S_2 in $m = 2$ time steps. From this analysis, we conclude that State S_2 can be reached from State S_1, and that State S_1 can be reached from State S_2. Hence, States S_1 and S_2 communicate.

Table 5.3. Two-step transition probabilities for the 4-state Markov chain whose transition probabilities appear in Table 5.2.

		S_1	S_2	S_3	S_4
State at Time t	S_1	$\frac{1}{4}$	$\frac{1}{2}$	$\frac{1}{4}$	0
	S_2	$\frac{1}{4}$	$\frac{1}{4}$	$\frac{1}{2}$	0
	S_3	$\frac{1}{2}$	$\frac{1}{4}$	$\frac{1}{4}$	0
	S_4	$\frac{5}{16}$	$\frac{5}{16}$	$\frac{5}{16}$	$\frac{1}{16}$

A similar analysis, using Tables 5.2 and 5.3, shows that States S_2 and S_3 communicate (since State S_3 can be reached from State S_2 in $n = 1$ time steps, and since State S_2 can be reached from State S_3 in $m = 2$ time steps). We could also use Tables 5.2 and 5.3 to prove that States S_1 and S_3 communicate. However, intuitively, this result should follow from the fact that State S_1 communicates with State S_2, and the fact that State S_2 communicates with State S_3. For, if States S_1 and S_2 communicate, then State S_2 can be reached from State S_1. In turn, the fact that States S_2 and S_3 communicate means that State S_3 can be reached from State S_2. Thus, starting from State S_1, the process can first reach State S_2; and from State S_2, the process can then reach State S_3. Hence, we seem to have shown that State S_3 can be reached from State S_1. A similar argument, using the fact that State S_1 can be reached from State S_2, and that State S_2 can be reached from State S_3, seems to indicate that State S_1 can be reached from State S_3. Hence, States S_1 and S_3 communicate.

In general (for any Markov chain) it is true that *if State S_i communi-*

cates with State S_j, and if State S_j communicates with State S_h, then States S_i and S_h must communicate with each other. To verify this assertion, recall from Chapter 11, Section 4, that for any two integers n_1 and n_2 such that $n = n_1 + n_2$,

(5.1) $\quad p_{ih}^{(n)} = p_{i1}^{(n_1)}p_{1h}^{(n_2)} + p_{i2}^{(n_1)}p_{2h}^{(n_2)} + \cdots + p_{ij}^{(n_1)}p_{jh}^{(n_2)} + \cdots + p_{iN}^{(n_1)}p_{Nh}^{(n_2)}$.

Thus, if State S_j can be reached from State S_i in n_1 time steps (that is, $p_{ij}^{(n_1)} \neq 0$), and State S_h can be reached from State S_j in n_2 time steps (that is, $p_{jh}^{(n_2)} \neq 0$), then $p_{ij}^{(n_1)}p_{jh}^{(n_2)} \neq 0$, and thus $p_{ih}^{(n)} \neq 0$, so that State S_h can be reached from State S_i in $n = n_1 + n_2$ time steps. [*Note:* State S_h may be reached from State S_i in fewer than n time steps. This possibility does not, however, need to trouble us. To show that State S_h can be reached from State S_i we need only show that $p_{ih}^{(n)} \neq 0$ for *some* n, and not necessarily the smallest such n.] Using the equation

(5.2) $\quad p_{hi}^{(m)} = p_{h1}^{(m_1)}p_{1i}^{(m_2)} + p_{h2}^{(m_1)}p_{2i}^{(m_2)} + \cdots + p_{hj}^{(m_1)}p_{ji}^{(m_2)} + \cdots + p_{hN}^{(m_1)}p_{Ni}^{(m_2)}$,

we can also show that if State S_j can be reached from State S_h in m_1 time steps, and if State S_i can be reached from State S_j in m_2 time steps, then $p_{hj}^{(m_1)}p_{ji}^{(m_2)} \neq 0$, so that $p_{hi}^{(m)} \neq 0$, and thus State S_i can be reached from State S_h in m time steps. Hence, if States S_i and S_j communicate (so that S_j can be reached from S_i in n_1 time steps and S_i can be reached from S_j in m_2 time steps), and if States S_j and S_h communicate (so that S_h can be reached from S_j in n_2 time steps; S_j can be reached from S_h in m_1 time steps), then States S_i and S_h each can be reached from the other, and thus must communicate.

Turning back to our 4-state Markov chain example, we see that we have yet to determine whether or not any of states S_1, S_2, or S_3 communicate with State S_4. However, note from Table 5.2 that

$$p_{14} = p_{24} = p_{34} = 0.$$

It thus follows that the class $\mathscr{C}$ consisting of states S_1, S_2 and S_3 is a closed class to which State S_4 does not belong. Hence, State S_4 cannot be reached from any of the states S_1, S_2, or S_3, and therefore none of the pairs of states S_1 and S_4, S_2 and S_4, or S_3 and S_4 communicate with one another. It is worth noting, however, that although State S_4 cannot be reached from states S_1, S_2, or S_3, each of the states S_1, S_2, and S_3 can be reached in one time step from State S_4 (that is, $p_{41} \neq 0$, $p_{42} \neq 0$, $p_{43} \neq 0$). This fact demonstrates that even though two states, S_i and S_j, do not communicate with each other, it may be possible for one of these states to be reached from the other state.

A class $\mathscr{C}$ of states which has the property that every pair of states in the class communicates with one another, and no state in the class communicates with any state outside of the class, is called a *communicating class*. A communicating class which is also closed is called a *closed, communicating class*. In a Markov chain that has a finite number N of states, communicating classes (and closed, communicating classes) have the following two important properties:

(i) The states in a communicating class are either all recurrent states, or all transient states. The states in a closed, communicating class are all recurrent.

(ii) For every pair of states, S_i and S_j, in a closed, communicating class, the conditional probability F_{ij} that the first-passage time K_{ij} between State S_i and State S_j is finite is equal to 1. That is, for every pair of states S_i and S_j in a closed, communicating class, $F_{ij} = 1$, and

$$f_{ij}^{(\infty)} = P\{K_{ij} = \infty | X_0 = i\} = 1 - F_{ij} = 0.$$

In most cases, it is quite straightforward to find the communicating classes of a Markov chain and to verify whether or not each such communicating class is closed. Thus, for example, in the 4-state Markov chain whose transition probabilities are given in Table 5.2, we quite easily verified that the class $\mathscr{C}$ which consists of states S_1, S_2, and S_3 is a closed, communicating class. Hence, from the above results, we know that these three states (S_1, S_2, and S_3) are recurrent, and that $F_{12} = F_{13} = F_{21} = F_{23} = F_{31} = F_{32} = 1$ (that is, the first-passage times between any pair of the states S_1, S_2, and S_3 are finite with probability equal to 1). If we attempted to obtain these same results using the methods outlined in Section 1 of this chapter, we would have to first compute all of the k-step first-passage probabilities $f_{ij}^{(k)}$ for $k = 1, 2, 3, \ldots$. Then we would have to form the sums

$$F_{ij} = f_{ij}^{(1)} + f_{ij}^{(2)} + f_{ij}^{(3)} + \ldots; \quad i = 1, 2, 3, \quad j = 1, 2, 3,$$

and check for each one of these nine sums (that is, F_{11}, F_{12}, F_{13}, F_{21}, F_{22}, F_{23}, F_{31}, F_{32}, F_{33}) whether or not the sum is equal to, or less than 1. However, using the properties of a closed, communicating class mentioned above, we immediately obtain the result that all nine of these sums are equal to 1. Thus, this example clearly illustrates the great utility of the concept of a closed, communicating class in the analysis of the first-passage times of a Markov chain.

There is one special case of a closed, communicating class which should be mentioned. Suppose that *all* of the N states of an N-state Markov chain communicate with one another. In this case, all of the states of the Markov chain form one communicating class. Since there are *no* states of the Markov chain outside of this class, it follows that once the process enters this class (as it must, since the process must enter one of the states at time 0), it can never leave this class. Thus, in this case, all of the states of the Markov chain form one big, closed, communicating class. It therefore follows, from the properties of a closed, communicating class quoted above, that every state of this Markov chain is recurrent, and that every pair of states, S_i and S_j, have a first-passage time K_{ij} which is finite with (conditional) probability $F_{ij} = 1$.

A Markov chain which has the property that every pair of its states communicate is called an *irreducible Markov chain*. Note that the 2-state Markov chain whose transition probabilities are summarized in Table 4.1 is an irreducible Markov chain (States S_1 and S_2 communicate since $p_{12} = 0.500 \neq 0$, and since $p_{21} = 0.750 \neq 0$). It was this Markov chain which we used earlier to exhibit the possibility that a Markov chain can attain probabilistic equilibrium.

6. APERIODIC, IRREDUCIBLE MARKOV CHAINS

In the previous section, we noted that irreducible Markov chains have some nice properties: namely, all states of such a process are recurrent, and every state S_j can be reached from every other state S_i in a finite number of time steps with (conditional) probability F_{ij} equal to 1. From such considerations, it follows that we can use the "alternative method of computation" described in Section 3 to calculate the expected first-passage times m_{ij} between all states, S_i and S_j, of the process.

We also noted in the last section that the 2-state Markov chain whose transition probabilities are summarized in Table 4.1 is an irreducible Markov chain, and that this Markov chain achieves a probabilistic equilibrium. Although it is tempting to generalize from this single example and state that all irreducible Markov chains achieve a probabilistic equilibrium (as defined in Section 4), this assertion is, unfortunately, false. Consider, for example, the 2-state Markov chain whose transition probabilities are summarized in Table 6.1. Since $p_{12} = p_{21} = 1 \neq 0$, it follows that States S_1 and S_2 of this Markov chain communicate. Since the Markov chain only has the two states S_1 and S_2, it is by definition irreducible. However, inspection of Table 6.1 reveals that if the process starts in State S_1, then

Table 6.1. Transition probabilities for a 2-state, irreducible Markov chain that does not achieve probabilistic equilibrium.

		State at Time $t+1$	
		S_1	S_2
State at Time t	S_1	0	1
	S_2	1	0

6. APERIODIC, IRREDUCIBLE MARKOV CHAINS

it must go to State S_2 at time 1, return to State S_1 at time 2, go to State S_2 at time 3, return to State S_1 at time 4, and in this manner alternate between State S_1 (at times 0, 2, 4, 6, 8, 10, and so on) and State S_2 (at times 1, 3, 5, 7, 9, and so on) at all future time points. Similarly, if the process begins in State S_2, then the process must visit State S_2 at all even times (times 0, 2, 4, 6, 8, and so on), and must visit State S_1 at all odd times (times 1, 3, 5, 7, 9, and so on). We conclude from this analysis that a table of k-step transition probabilities for this Markov chain has the form

		State at Time $t+k$	
		S_1	S_2
State at Time t	S_1	1	0
	S_2	0	1

when k is even ($k = 2, 4, 6, 8$, and so on), and that the table of k-step transition probabilities has the form

		State at Time $t+k$	
		S_1	S_2
State at Time t	S_1	0	1
	S_2	1	0

when k is odd. We can summarize these results in a table (Table 6.2) of k-step transition probabilities similar to Table 4.2.

From Table 6.2, we see that no matter how large k becomes, the k-step transition probability $p_{ij}^{(k)}$ always depends on the time k and the initial state S_i. Thus, the irreducible Markov chain whose 1-step transition probabilities appear in Table 6.1 does not achieve probabilistic equilibrium. Consequently, we are forced to conclude that not all irreducible Markov chains achieve a probabilistic equilibrium.

However, in examining Table 6.2, we see that the k-step transition

Table 6.2. The k-step transition probabilities $p_{ij}^{(k)}$ for the Markov chain whose transition probabilities are given in Table 4.1.

k-Step Transition Probability to:		k								
		1	2	3	4	5	6	7	8	9	...
State S_1	$p_{11}^{(k)}$	0	1	0	1	0	1	0	1	0	...
	$p_{21}^{(k)}$	1	0	1	0	1	0	1	0	1	...
State S_2	$p_{12}^{(k)}$	1	0	1	0	1	0	1	0	1	...
	$p_{22}^{(k)}$	0	1	0	1	0	1	0	1	0	...

probabilities (and the Markov chain itself) do exhibit a "back-and-forth" regularity, somewhat akin to the motion of a pendulum. The k-step transition probabilities $p_{ij}^{(k)}$ go from 0 (or 1) to 1 (or 0), and then back to 0 (or 1), periodically. That is, as a function of the time k, $p_{ij}^{(k)}$ has the same value for every other value of k; for example, $p_{11}^{(k)}$ has the same value, 1, for $k = 2, 4, 6, 8, 10$, and so on. Looking at this phenomenon in another way, we see that if the process begins in State S_1 at time 0, it has nonzero (conditional) probability of returning to State S_1 at time k only when k is divisible by 2. Thus, the process can only *periodically* return to State S_1. Similarly, if the process starts at State S_2, the probability that it returns to State S_2 at time k is nonzero only if k is divisible by 2, and we see that the process can only periodically return to State S_2. These considerations motivate definition of the period (and periodicity) of a state, S_j, of a Markov chain.

Periodic States and Periodic Chains

A state S_j of a Markov chain is *periodic* if there exists an integer h which is greater than 1 (that is, $h > 1$) such that $p_{jj}^{(k)} = 0$ whenever k is not an integer multiple of h (that is, whenever k/h is not an integer). If a state is periodic, then given that the process has entered State S_j at some time t, the process has nonzero conditional probability $p_{jj}^{(k)}$ of returning to State S_j only if k is an integer multiple of some fixed integer $h > 1$. (Note, however, that we are not asserting that $p_{jj}^{(k)}$ is *always* nonzero when k is an integer multiple of h.) The *period* T_j of a periodic state S_j is the *smallest* integer h ($h > 1$) such that $p_{jj}^{(k)} = 0$ whenever k is not an integer multiple of h. The period of any periodic state is unique; that is, a periodic state can have one and only one period.

For the 2-state Markov chain whose transition probabilities are summarized in Table 6.1, we have already shown that $p_{11}^{(k)} = 0$ whenever k is not an integer multiple of the positive integer $h = 2$ (that is, whenever k is odd). Since $h = 2$ is also the smallest integer that is greater than 1, the period T_1 of State S_1 must be equal to 2. Similarly, we can show that State S_2 has period 2.

As another example of a periodic state, consider the 4-state Markov chain whose transition probabilities p_{ij} are summarized in Table 6.3. Using the methods of computation described in Chapter 11, Section 4, it can be shown for this Markov chain that

$$p_{11}^{(k)} = \begin{cases} \frac{1}{2}, & \text{if } k = 2, 4, 6, 8, 10, \ldots, \\ 0, & \text{if } k = 1, 3, 5, 7, 9, \ldots. \end{cases}$$

Since $p_{11}^{(k)} = 0$ whenever k is not an integer multiple of $h = 2$, it follows that State S_1 is periodic, and that the period T_1 of State S_1 is equal to 2.

6. APERIODIC, IRREDUCIBLE MARKOV CHAINS

Table 6.3. Table of transition probabilities for a 4-state Markov chain.

		State at Time $t+1$			
		S_1	S_2	S_3	S_4
State at Time t	S_1	0	$\frac{1}{2}$	0	$\frac{1}{2}$
	S_2	$\frac{1}{2}$	0	$\frac{1}{2}$	0
	S_3	0	$\frac{1}{2}$	0	$\frac{1}{2}$
	S_4	$\frac{1}{2}$	0	$\frac{1}{2}$	0

It can be shown that if any single state in a communicating class $\mathscr{C}$ is periodic and has period T, then all states in that class are periodic and have period T. Thus, it is possible to talk about the periodicity (and the period) of a communicating class. A communicating class is periodic of period T if any one state in the class is periodic and has period T. From this fact it follows that if a Markov chain is irreducible, and if one state of that Markov chain is periodic with period T, then *all* states of that irreducible Markov chain are periodic with period T. If any state of an irreducible Markov chain is periodic with period T, then we say that the Markov chain itself is periodic, and that the period of the Markov chain is equal to T. For example, for the 4-state Markov chain whose transition probabilities are summarized in Table 6.3, we already know that State S_1 is periodic with period $T = 2$. Since for this Markov chain, $p_{12} = p_{21} = \frac{1}{2} \neq 0$, $p_{23} = p_{32} = \frac{1}{2} \neq 0$, and $p_{34} = p_{43} = \frac{1}{2} \neq 0$, it follows that States S_1 and S_2 communicate, that States S_2 and S_3 communicate, and that States S_3 and S_4 communicate. Thus (see the discussion in Section 5), every pair of States, S_i and S_j, communicate, and the Markov chain is irreducible. Since State S_1 has period 2, it follows that all of the states (S_1, S_2, S_3, and S_4) of the Markov chain have period $T = 2$, and thus the Markov chain itself is, by definition, periodic with period $T = 2$.

Irreducible Markov chains which are periodic do not achieve probabilistic equilibrium. We have already exhibited an example of this fact when we discussed the 2-state, periodic, irreducible Markov chain whose transition probabilities are given in Table 6.1. However, even though periodic, irreducible, Markov chains do not achieve a probabilistic equilibrium in the sense defined in Section 4, such Markov chains do exhibit certain long-run probabilistic regularities. Because discussion of such regularities can become rather complicated, we leave the description and investigation of such properties to more advanced textbooks [see, for example, Feller (1968)].

Aperiodic States

If a state S_j of a Markov chain is not periodic, we say that such a state is an *aperiodic state*. A state S_j is not periodic if for every integer h greater than 1, there is another integer k such that k is not an integer multiple of h (that is, k/h is not an integer), and such that $p_{jj}^{(k)} \neq 0$. That is, State S_j is not periodic if for every integer h that could be the period for State S_j, it is possible for the process to return to State S_j after a length k of time has passed which is not an integer multiple of h. Although this definition of an aperiodic state may seem awkward and difficult to verify, in most cases it is straightforward to determine whether or not a state S_j is aperiodic. One quick test for the aperiodicity of a state S_j is to look at the 1-step transition probability p_{jj}. *If p_{jj} is not zero* (that is, $p_{jj} > 0$), *then State S_j is aperiodic*. This assertion follows from the definition of an aperiodic state; if p_{jj} is positive, then for every integer h greater than 1, there is another integer k (namely, $k = 1$) which is not an integer multiple of h (since 1 cannot be divided without remainder by any integer h larger than 1), and for which $p_{jj}^{(k)}$ is not 0. Thus, for example, a look at Table 4.1 reveals that both State S_1 and State S_2 are aperiodic since $p_{11} = \frac{1}{2} > 0$, and $p_{22} = \frac{1}{4} > 0$.

Another test for the aperiodicity of a state S_j is the following: If we can find two positive integers k_1 and k_2 such that $p_{jj}^{(k_1)} > 0$ and $p_{jj}^{(k_2)} > 0$, and such that the only positive integer which divides both k_1 and k_2 without remainder is the integer 1, then State S_j is aperiodic. Thus, in particular, if State S_j has the property that $p_{jj}^{(2)} > 0$ and $p_{jj}^{(3)} > 0$, then State S_j is aperiodic, since $k_1 = 2$ and $k_2 = 3$ cannot both be divided without remainder by any positive integer other than the integer 1. For example, if we are interested in the 3-state Markov chain with transition probabilities as shown in Table 6.4, then even though $p_{11} = p_{22} = p_{33} = 0$, we still are able to show that each of the states S_1, S_2, and S_3 is aperiodic. To do so, we compute the 2-step and 3-step transition probabilities of this Markov chain (these appear in Tables 6.5 and 6.6, respectively), and then note

Table 6.4. Table of transition probabilities for a Markov chain for which all three states, S_1, S_2, S_3, are aperiodic.

		State at Time $t+1$		
		S_1	S_2	S_3
State at Time t	S_1	0	$\frac{2}{3}$	$\frac{1}{3}$
	S_2	$\frac{1}{4}$	0	$\frac{3}{4}$
	S_3	$\frac{1}{2}$	$\frac{1}{2}$	0

6. APERIODIC, IRREDUCIBLE MARKOV CHAINS

Table 6.5. Table of 2-step transition probabilities for the Markov chain whose transition probabilities appear in Table 6.4.

		State at Time $t+2$		
		S_1	S_2	S_3
State at Time t	S_1	$\frac{1}{3}$	$\frac{1}{6}$	$\frac{1}{2}$
	S_2	$\frac{3}{8}$	$\frac{13}{24}$	$\frac{1}{12}$
	S_3	$\frac{1}{8}$	$\frac{1}{3}$	$\frac{13}{24}$

Table 6.6. Table of 3-step transition probabilities for the 3-state Markov chain whose transition probabilities are given in Table 6.4.

		State at Time $t+3$		
		S_1	S_2	S_3
State at Time t	S_1	$\frac{21}{72}$	$\frac{34}{72}$	$\frac{17}{72}$
	S_2	$\frac{17}{96}$	$\frac{28}{96}$	$\frac{51}{96}$
	S_3	$\frac{17}{48}$	$\frac{17}{48}$	$\frac{14}{48}$

that $p_{11}^{(2)} = 1/3 > 0$, $p_{11}^{(3)} = 21/72 > 0$; that $p_{22}^{(2)} = 13/24 > 0$, $p_{22}^{(3)} = 28/96 > 0$; and that $p_{33}^{(2)} = 13/24 > 0$, $p_{33}^{(3)} = 14/48 > 0$. Applying our test for aperiodicity, we thus conclude that states S_1, S_2, and S_3 are all aperiodic.

If one state, say State S_i, in a class $\mathscr{C}$ of communicating states is aperiodic, then all of the states in the class $\mathscr{C}$ must be aperiodic. For, if it were the case that some state, State S_j, is in class $\mathscr{C}$, and that State S_j is periodic of period T, then (as was stated earlier) all of the states in $\mathscr{C}$ would have to be periodic with period T. Thus, since State S_i belongs to class $\mathscr{C}$, State S_i would have to be periodic. Since State S_i is known to be aperiodic (not periodic), we have a contradiction caused by assuming that some state in $\mathscr{C}$ is periodic. Thus, no state in $\mathscr{C}$ can be periodic; or, in other words, every state in $\mathscr{C}$ is aperiodic, thus proving our original assertion. If a communicating class $\mathscr{C}$ contains one aperiodic state, then we say that $\mathscr{C}$ is an *aperiodic communicating class*.

The above fact offers us another way of determining whether or not a given state is aperiodic: *If State S_i belongs to an aperiodic communicating class $\mathscr{C}$, then State S_i is aperiodic.* Thus, in the Markov chain whose transition probabilities appear in Table 6.3, we could have shown that states S_2 and S_3 are aperiodic by first noting that State S_2 and State S_3 each communicate with State S_1 (since p_{12}, p_{21}, p_{13}, and p_{31} are all nonzero), and then verifying that State S_1 is aperiodic (by showing that $p_{11}^{(2)} \neq 0$ and $p_{11}^{(3)} \neq 0$, or by some other method). Note that since State S_1

communicates with State S_2 and with State S_3, this Markov chain is irreducible (since all of its states belong to one communicating class). *An irreducible Markov chain which has one aperiodic state is called an aperiodic irreducible Markov chain.* All of the states of an aperiodic irreducible Markov chain are aperiodic states.

Looking back at the Markov chain whose transition probabilities appear in Table 4.1, we see that this Markov chain is irreducible and aperiodic. Recall again from Section 4 that this Markov chain achieves a probabilistic equilibrium. Although, as we have demonstrated earlier, not all irreducible Markov chains achieve probabilistic equilibrium, perhaps it is the case that all *aperiodic* irreducible Markov chains achieve an equilibrium. This is indeed the case, and in fact the following assertions are true for any aperiodic irreducible Markov chain:

(i) The Markov chain achieves probabilistic equilibrium.
(ii) The long-run probability π_j of State S_j of the Markov chain is equal to $1/m_{jj}$, where we recall that m_{jj} is the expected (first) recurrence time of State S_j, $j = 1, 2, ..., N$.

Recall that an N-state Markov chain achieves probabilistic equilibrium if we can find probabilities $\pi_1, \pi_2, ..., \pi_N$ for the states $S_1, S_2, ..., S_N$, respectively, such that for every pair of states, S_i and S_j, and for all sufficiently large values of k.

(6.1) $$p_{ij}^{(k)} \cong \pi_j.$$

If we can show that a given Markov chain is aperiodic and irreducible, then Assertion (i) tells us that the long-run probabilities $\pi_1, \pi_2, ..., \pi_N$ can be found (but does not tell us their values). Armed with the information that the long-run probabilities $\pi_1, \pi_2, ..., \pi_N$ do exist, we could, of course, attempt to determine the values of $\pi_1, \pi_2, ..., \pi_N$ by recursively calculating the k-step transition probabilities $p_{ij}^{(k)}$ for $k = 1, 2, 3$, and so on, stopping when the rows of a table of k-step transition probabilities, for some value of k, are all the same. If our computations of the k-step transition probabilities are perfectly accurate (to any number of decimal places), then we can be sure that when we stop calculation, the first entry in any row of the table is π_1, the second entry in that row is π_2, and so forth (see Table 4.3).

However, since our calculation of k-step transition probabilities cannot usually be done with perfect accuracy (we have to "round-off" in the course of our calculations), we cannot be certain that the values of the long-run probabilities $\pi_1, \pi_2, ..., \pi_N$ obtained by this method are sufficiently accurate. Further, even if we are able to calculate with sufficient accuracy, we may have to calculate tables of k-step transition probabilities for

6. APERIODIC, IRREDUCIBLE MARKOV CHAINS

many values of k before we can stop calculation. Assertion (ii) provides us with an alternative method of computing the long-run probabilities $\pi_1, \pi_2, \ldots, \pi_N$. Using this approach, we first calculate the expected (first) recurrence times $m_{11}, m_{22}, \ldots, m_{NN}$ by one of the methods described in Section 3. Once we have obtained the values of $m_{11}, m_{22}, \ldots, m_{NN}$, we can then obtain the long-run probabilities $\pi_1, \pi_2, \ldots, \pi_N$ by means of the equation

(6.2) $$\pi_j = \frac{1}{m_{jj}}, \quad j = 1, 2, \ldots, N.$$

Thus, for example, consider the 2-state Markov chain whose transition probabilities appear in Table 4.1. As we have noted several times, this aperiodic, irreducible Markov chain achieves a probabilistic equilibrium; from Table 4.2, the long-run probabilities of the states S_1 and S_2 appear to be $\pi_1 = 0.600$ and $\pi_2 = 0.400$, respectively. To verify that 0.600 and 0.400 are indeed the long-run probabilities of states S_1 and S_2, we first use the alternative method for computing the expected first-passage times m_{ij}, described in Section 3, to compute m_{11} and m_{22}. From Equation (3.13), with $j = 1$,

$$m_{11} = (1 - p_{11}m_{11}) + p_{11}m_{11} + p_{12}m_{21} = 1 + (\tfrac{1}{2})m_{21},$$
$$m_{21} = (1 - p_{21}m_{11}) + p_{21}m_{11} + p_{22}m_{21} = 1 + (\tfrac{1}{4})m_{21}.$$

Thus, from the second of these two equations, $m_{21} = \tfrac{4}{3}$, and therefore $m_{11} = 1 + (\tfrac{1}{2})(\tfrac{4}{3}) = \tfrac{5}{3}$. Again, from Equation (3.13), with $j = 2$,

$$m_{12} = (1 - p_{12}m_{22}) + p_{11}m_{12} + p_{12}m_{22} = 1 + (\tfrac{1}{2})m_{12},$$
$$m_{22} = (1 - p_{22}m_{22}) + p_{21}m_{12} + p_{22}m_{22} = 1 + (\tfrac{3}{4})m_{12},$$

so that $m_{12} = 2$ and $m_{22} = 1 + (\tfrac{3}{4})(2) = \tfrac{5}{2}$. Hence, from Equation (6.2),

$$\pi_1 = \frac{1}{m_{11}} = \frac{1}{(\tfrac{5}{3})} = \frac{3}{5} = 0.600,$$
$$\pi_2 = \frac{1}{m_{22}} = \frac{1}{(\tfrac{5}{2})} = \frac{2}{5} = 0.400.$$

As another example of an aperiodic, irreducible Markov chain, consider the genetical mating process whose transition probabilities appear in Table 1.3. Recall that the states of this process are S_1: genotype AA, S_2: genotype Aa, and S_3: genotype aa. From Table 1.4, we see that $p_{12}^{(2)} = \tfrac{1}{2} > 0$, $p_{21}^{(2)} = \tfrac{1}{4} > 0$, and that $p_{23}^{(2)} = \tfrac{1}{2} > 0$, $p_{32}^{(2)} = \tfrac{1}{2} > 0$. Thus, from the discussion in Section 5, we conclude that the three states S_1, S_2, and S_3 form a communicating class. Since the genetical mating process has *only* three states, it follows that this Markov chain is irreducible. Finally, since $p_{11} = \tfrac{1}{2} > 0$ (see Table 1.3), State S_1 is aperiodic, and it follows that the genetical mating process is an aperiodic, irreducible Markov chain. From Assertion (i), we conclude that this Markov chain achieves a probabilistic equilibrium.

The expected first-passage times m_{ij} of this genetical mating process have been obtained in Section 3, and are summarized in Table 3.2. From Table 3.2, we see that $m_{11} = 4$, $m_{22} = 2$, and $m_{33} = 4$. Hence, from Equation (6.2), the long-run probabilities of this process are

$$\pi_1 = \text{the long-run probability of genotype } AA = \frac{1}{m_{11}} = \frac{1}{4},$$

$$\pi_2 = \text{the long-run probability of genotype } Aa = \frac{1}{m_{22}} = \frac{1}{2},$$

$$\pi_3 = \text{the long-run probability of genotype } aa = \frac{1}{m_{33}} = \frac{1}{4}.$$

Thus, suppose we run a genetical mating experiment of the kind described in Example 3.3 of Chapter 11 (the first genetical mating experiment described there), and suppose that the parent (parent I) of fixed genotype has genotype Aa. Then, from our analysis above, we know that after many matings have been performed genotype AA has probability $\pi_1 = \frac{1}{4}$ of being observed, genotype Aa has probability $\pi_2 = \frac{1}{2}$ of being observed, and genotype aa has probability $\pi_3 = \frac{1}{4}$ of being observed as the result of any given mating, regardless of the genotype of the other parent (parent II) in the initial mating.

Steady-State Equations

Useful as it is, Equation (6.2) still is a somewhat indirect way of calculating the long-run probabilities $\pi_1, \pi_2, \ldots, \pi_N$ of an aperiodic irreducible Markov chain. Since we know that the nature of the probabilistic equilibrium of a Markov chain is determined solely by the probabilistic dynamics (that is, the 1-step transition probabilities) that generate the Markov chain, we should be able to obtain the long-run probabilities $\pi_1, \pi_2, \ldots, \pi_N$ of a Markov chain directly from knowledge of the 1-step transition probabilities p_{ij} (rather than indirectly, through calculation of the expected first-passage times). As we now demonstrate, this is indeed possible.

Suppose that we are observing an N-state Markov chain which we know achieves probabilistic equilibrium. However, instead of having observed this process from its beginning, we start observing the process at a time t at which the process is already in probabilistic equilibrium. Because the process is in probabilistic equilibrium, the unconditional probability, $P\{X_t = j\}$, that the process is in State S_j at time t is (at least approximately) equal to π_j. That is,

(6.3) $$P\{X_t = j\} = \pi_j, \quad j = 1, 2, ..., N.$$

However, since the Markov chain is in probabilistic equilibrium, we also know that at the next time, time $t+1$, at which we observe the process, the unconditional probability, $P\{X_{t+1} = j\}$, that the process is in State S_j is the same as it is at time t; that is,

(6.4) $$P\{X_{t+1} = j\} = \pi_j, \quad j = 1, 2, ..., N.$$

If we think of the Markov chain process as beginning (for us) at time t, then we can think of the unconditional probabilities in (6.3) as if they were initial probabilities for the process, starting at time t. That is, we can think of time t as if it were time 0 on a new clock for the process: a clock which starts when we start observing the process. On this new clock, time $t+1$ (measured on the old clock) becomes time 1, time $t+2$ becomes time 2, and so forth. The initial probabilities for the process now become $a_1 = \pi_1, a_2 = \pi_2, ..., a_N = \pi_N$. From Chapter 11, Section 5, we know that the unconditional probability of State S_j at time 1 on the new clock (time $t+1$ on the old clock) can be found from the initial probabilities $a_1, a_2, ..., a_N$, and the 1-step transition probabilities p_{ij} through the equation

(6.5) $$P\{X_1 = j\} = a_1 p_{1j} + a_2 p_{2j} + \cdots + a_N p_{Nj}.$$

But time 1 on the new clock is time $t+1$ on the old clock, and from Equation (6.4) we know that the unconditional probability that the process is in a given state, say State S_j, at time $t+1$ (on the old clock) is π_j. Thus, substituting π_j for $P\{X_1 = j\}$, and $\pi_1, \pi_2, ..., \pi_N$ for $a_1, a_2, ..., a_N$, respectively, in Equation (6.5), we conclude that

(6.6) $$\pi_j = \pi_1 p_{1j} + \pi_2 p_{2j} + \cdots + \pi_N p_{Nj}.$$

Equation (6.6) is true for each state S_j; consequently,

(6.7)
$$\pi_1 = \pi_1 p_{11} + \pi_2 p_{21} + \cdots + \pi_N p_{N1},$$
$$\pi_2 = \pi_1 p_{12} + \pi_2 p_{22} + \cdots + \pi_N p_{N2},$$
$$\vdots \qquad \vdots \qquad \vdots \qquad \vdots$$
$$\pi_j = \pi_1 p_{1j} + \pi_2 p_{2j} + \cdots + \pi_N p_{Nj},$$
$$\vdots \qquad \vdots \qquad \vdots \qquad \vdots$$
$$\pi_N = \pi_1 p_{1N} + \pi_2 p_{2N} + \cdots + \pi_N p_{NN}.$$

Hence, if we know the transition probabilities of the given Markov chain, Equation (6.7) provides us with a set of linear equations which may be solved for the values of the unknowns $\pi_1, \pi_2, ..., \pi_N$. Since in the physical sciences a condition of equilibrium is frequently called a "steady state," and since the equations in (6.7) were obtained by using the fact that the unconditional probabilities of the states of a Markov chain which is in probabilistic equilibrium are steady (constant) in value over time, equations (6.7) are called the *steady-state equations* for the Markov chain.

As an example of the use of these steady-state equations, let us obtain the long-run probabilities π_1 and π_2 of the 2-state choice-of-brand Markov chain whose transition probabilities are given in Table 4.1. From Equation (6.7),

(6.8) $$\pi_1 = \pi_1 p_{11} + \pi_2 p_{21} = \pi_1 \left(\frac{1}{2}\right) + \pi_2 \left(\frac{3}{4}\right),$$

$$\pi_2 = \pi_1 p_{12} + \pi_2 p_{22} = \pi_1 \left(\frac{1}{2}\right) + \pi_2 \left(\frac{1}{4}\right).$$

Thus, from the first of these equations, $(1-\frac{1}{2})\pi_1 = (\frac{3}{4})\pi_2$, or

(6.9) $$\pi_1 = \left(\frac{3}{2}\right)\pi_2.$$

However, from the second of these equations we also obtain the result that $\pi_1 = (\frac{3}{2})\pi_2$. In other words, one of the two equations in (6.8) is redundant. A similar fact is true for the more general set of equations in (6.7): *We can obtain no more information about the long-run probabilities $\pi_1, \pi_2, \ldots, \pi_N$ from the collection of all N equations in (6.7) than we can by using any subcollection of $N-1$ of these equations.* The choice of which subcollection of $N-1$ equations in (6.7) to use can be made at our convenience.

Turning back to Equation (6.9), it appears that there are many values π_1 and π_2 that satisfy this equation. For example, $\pi_1 = -1$ and $\pi_2 = -\frac{2}{3}$ satisfy Equation (6.9); so also do $\pi_1 = \frac{1}{2}$ and $\pi_2 = \frac{1}{3}$. However, two requirements that must be satisfied by the long-run probabilities π_1 and π_2 have not yet been used. First, since they are probabilities, both π_1 and π_2 must be nonnegative (that is, $\pi_1 \geq 0$, $\pi_2 \geq 0$). Second, since at any time, time t, one of the states, S_1 or S_2, of the process must be observed, it follows that the sum of π_1 and π_2 must equal 1; that is,

(6.10) $$\pi_1 + \pi_2 = 1.$$

Substituting $(\frac{3}{2})\pi_2$ for π_1 in (6.10), we find that

$$\left(\frac{3}{2}\right)\pi_2 + \pi_2 = 1,$$

or $\pi_2 = \frac{2}{5}$. Thus, from (6.10), $\pi_1 = 1 - \pi_2 = \frac{3}{5}$. We conclude that

$$\pi_1 = \frac{3}{5} = 0.600, \qquad \pi_2 = \frac{2}{5} = 0.400.$$

This is the same result obtained earlier by use of Equation (6.2).

In general, if we add to any $N-1$ of the N steady-state equations in (6.7) the additional requirement that

$$\pi_1 + \pi_2 + \cdots + \pi_N = 1, \tag{6.11}$$

then the resulting N equations in the N unknowns $\pi_1, \pi_2, \ldots, \pi_N$ may be solved for the unknowns $\pi_1, \pi_2, \ldots, \pi_N$. The unique solutions of these equations will automatically be nonnegative numbers adding to 1.

For example, for the 3-state Markov chain (the genetical mating process) whose transition probabilities appear in Table 1.3, we can find the long-run probabilities π_1, π_2, π_3 by taking any two of the steady-state equations:

$$\pi_1 = \left(\frac{1}{2}\right)\pi_1 + \left(\frac{1}{4}\right)\pi_2 + (0)\pi_3,$$

$$\pi_2 = \left(\frac{1}{2}\right)\pi_1 + \left(\frac{1}{2}\right)\pi_2 + \left(\frac{1}{2}\right)\pi_3,$$

$$\pi_3 = (0)\pi_1 + \left(\frac{1}{4}\right)\pi_2 + \left(\frac{1}{2}\right)\pi_3,$$

and solving these equations for the unknowns π_1, π_2, π_3, under the additional requirement that

$$\pi_1 + \pi_2 + \pi_3 = 1.$$

From the first of the steady-state equations, we see that $\pi_2 = 2\pi_1$. From the second of the steady-state equations, we see that $\pi_3 = \pi_1$; thus, $\pi_3 = \pi_1 = \frac{1}{2}\pi_2$. Substituting $2\pi_1$ for π_2 and π_1 for π_3 in the equation $\pi_1 + \pi_2 + \pi_3 = 1$, we find that

$$\pi_1 + 2\pi_1 + \pi_1 = 1,$$

or $\pi_1 = \frac{1}{4}$. Hence, $\pi_3 = \pi_1 = \frac{1}{4}$. The resulting solutions

$$\pi_1 = \frac{1}{4}, \qquad \pi_2 = \frac{1}{2}, \qquad \pi_3 = \frac{1}{4},$$

are the same as those obtained earlier for this process by use of Equation (6.2).

7. ABSORBING MARKOV CHAINS

If we observe an irreducible Markov chain over time, the process typically jumps from one state to another and only rarely stays in one state for very long. In contrast, there exist Markov chains that have one or more states, called *absorbing* states, such that if the process enters one of these states, then the process can never leave that state (it is "absorbed" into that state). For example, consider the simple learning process first discussed in Example 4.1 of Chapter 11. This 2-state Markov chain has transition probabilities of the form shown in Table 7.1.

In studying Table 7.1, recall that State S_1 is the state in which a subject under observation is conditioned to make a correct response (that is, the subject is in a conditioned state), while State S_2 is the state in which the subject is still unconditioned. Once the process enters State S_1 (the subject is conditioned), it remains in State S_1 forever. We verified this assertion in Section 1 (for the case where α, in Table 7.1, equals $\frac{1}{2}$) by showing that $p_{12}^{(k)} = 0$ for all $k = 1, 2, 3, \ldots$. That is, given that the process is in State S_1 at some time, time t, the conditional probability of leaving State S_1 and going to State S_2 at any future time, time $t+k, k = 1, 2, 3, \ldots$, is equal to 0. Thus, for the learning process, State S_1 is an absorbing state. On the other hand, State S_2 is not an absorbing state when $\alpha \neq 0$, since in this case $p_{21} \neq 0$, and the process can leave State S_2 and go to State S_1 one time unit after a time in which the process has been in State S_2.

Table 7.1. Table of transition probabilities for a simple learning process. The constant α is a number between 0 and 1 (that is, $0 \leq \alpha \leq 1$).

		State at Time $t+1$	
		S_1	S_2
State at Time t	S_1	1	0
	S_2	α	$1-\alpha$

In general, a state, State S_i, of a Markov chain is *absorbing* if the 1-step transition probability p_{ii} between State S_i and itself is equal to 1. That is, State S_i is absorbing if $p_{ii} = 1$. If p_{ii} is not equal to 1, then since, from Equation (2.6) of Chapter 11, we know that

(7.1) $$p_{i1} + p_{i2} + \cdots + p_{ii} + \cdots + p_{iN} = 1,$$

it follows that at least one of the 1-step transition probabilities $p_{ij}, j \neq i$, must be nonzero. Hence, if $p_{ii} \neq 1$, it is possible for the process to leave State S_i, and hence, State S_i cannot be absorbing. We conclude that *State S_i is an absorbing state if, and only if, $p_{ii} = 1$.*

It follows directly from the definition of an absorbing state that every absorbing state is also a recurrent state. If the process enters an absorbing state at time t, it can never leave that state afterwards. Thus, the process must return to (remain in) that state at the very next time, time $t+1$. Consequently, the (first) recurrence time for that absorbing state is finite (indeed, is equal to 1) with probability equal to 1, and hence, by definition, it follows that the absorbing state is a recurrent state.

If a Markov chain has one or more absorbing states, it is called an *absorbing Markov chain*. If an absorbing Markov chain has just one absorbing state, and if this absorbing state can be reached (see Section 5) from every other state of the Markov chain, then the Markov chain has exactly one recurrent state (namely, the one absorbing state) and all of the other states of the process are transient. Such a Markov chain achieves a probabilistic equilibrium in which the long-run probability of the single absorbing state is equal to 1, and the long-run probabilities of all of the other states are equal to 0.

For example, the learning process whose transition probabilities are tabled in Table 7.1 is (when $\alpha \neq 0$) an absorbing Markov chain with a single absorbing state, State S_1. Thus, State S_1 is a recurrent state and State S_2 is a transient state (we proved this fact by a direct method in Section 2). Further, this process achieves a probabilistic equilibrium in which State S_1 has long-run probability π_1 equal to 1 (that is, $\pi_1 = 1$), and in which the long-run probability π_2 of State S_2 is equal to 0 (that is, $\pi_2 = 0$). Hence, after a long enough period of time, k, has passed, the unconditional probability $P\{X_k = 1\}$ that the process is in State S_1 is approximately equal to 1, and the unconditional probability that the process is in State S_2 is approximately equal to 0. (Indeed, when $k = 13$, the assertion that $P\{X_k = 1\} = 1.000$ and $P\{X_k = 0\} = 0.000$ is accurate to three decimal places.)

It is possible for an absorbing Markov chain to have just one absorbing state, and yet have many recurrent states. For example, suppose that a Markov chain of interest to us has three states S_1, S_2, and S_3, and suppose that the transition probabilities of this process are those shown in Table 7.2. Since State S_1 is absorbing ($p_{11} = 1$), this Markov chain is an absorbing Markov chain. Although neither State S_2 nor State S_3 are absorbing ($p_{22} = p_{33} = \frac{1}{2} \neq 1$, these two states together form a closed, communicating class (closed since $p_{21} = p_{31} = 0$; communicating since $p_{23} \equiv 0$

Table 7.2. Table of transition probabilities for an absorbing Markov chain which has one absorbing state and three recurrent states.

		State at Time $t+1$		
		S_1	S_2	S_3
State at Time t	S_1	1	0	0
	S_2	0	$\frac{1}{2}$	$\frac{1}{2}$
	S_3	0	$\frac{1}{2}$	$\frac{1}{2}$

$p_{32} \neq 0$. Hence, from Assertion (i) of Section 5, states S_2 and S_3 are recurrent. We have thus exhibited an example of an absorbing Markov chain in which there is only one absorbing state, but in which there are three recurrent states. Note that this example also illustrates the fact that although all absorbing states are recurrent, not all recurrent states are absorbing.

Suppose that we are interested in an absorbing Markov chain that has more than one absorbing state. The conformity process described in Example 3.2 of Chapter 11 is an example of such an absorbing process. Recall that this process has four states:

S_1: The subject in the conformity experiment is motivated to answer correctly, independent of the answers of the group, at all trials of the experiment.
S_2: The subject is motivated to answer correctly this time, but is still indecisive about whether or not to conform in the future.
S_3: The subject is motivated to conform to the incorrect answer of the group, but is still indecisive about future conformity.
S_4: The subject is motivated to conform to the group's answer on all trials.

The transition probabilities of such a process are of the form shown in Table 7.3. From Table 7.3, we see that State S_1 and State S_4 are absorbing states (since $p_{11} = 1$, $p_{44} = 1$). Once the process enters one of these states, it can never leave that state.

Table 7.3. Transition probabilities for a conformity model.

		State at Time $t+1$			
		S_1	S_2	S_3	S_4
State at Time t	S_1	1	0	0	0
	S_2	p_{21}	p_{22}	p_{23}	0
	S_3	0	p_{32}	p_{33}	p_{34}
	S_4	0	0	0	1

Suppose, however, that a subject is initially undecided between answering correctly or conforming (that is, the conforming process is either in State S_2 or in State S_3). If p_{21} and p_{34} are not equal to 0, then it can be shown that States S_2 and S_3 are transient states, and that the process eventually must leave both of these states and be absorbed either in State S_1 or in State S_4.

Suppose that the subject in the conformity experiment starts by answer-

7. ABSORBING MARKOV CHAINS

ing incorrectly, but is undecided whether or not to remain in conformity with the incorrect answers of the group at future trials of the experiment. That is, suppose that the conformity process initially is in State S_3. We know that eventually the process will leave State S_3 to be absorbed in one of the two states, S_1 or S_4. However, we do not know which state, S_1 or S_4, will be the state that "absorbs" the process. Thus, it is of interest to find the conditional probability A_{3j} that the process will be absorbed in State S_j, for $j = 1$ and $j = 4$, given that the process initially is in State S_3. Similarly, given that the process starts in State S_2, it is of interest to find the conditional probability A_{2j} that the process is absorbed in State S_j, when $j = 1$ or $j = 4$. The quantities A_{31}, A_{34}, A_{21}, and A_{24} are called *absorption probabilities* for the Markov chain. In general, if State S_j is an absorbing state of an (absorbing) Markov chain, and if State S_j can be reached from a nonabsorbing state, State S_i, then the absorption probability A_{ij} equals the conditional probability that the process will eventually be absorbed in State S_j, given that the process starts at State S_i. Symbolically,

(7.2) $A_{ij} = P\{\text{the process is eventually absorbed in State } S_j | X_0 = i\}.$

Although the symbol A_{ij} is mnemonically useful for remembering that the quantity (7.2) is an absorption probability, we have discussed this quantity before under another name. Note that once the process reaches the absorbing state, State S_j, for the *first* time, it is absorbed in State S_j. Consequently, the absorption probability A_{ij} and the probability F_{ij} of a finite first-passage time from State S_i to State S_j (see Section 1) are the same quantity. That is, if State S_j is an absorbing state, then

(7.3) $$A_{ij} = F_{ij}.$$

Thus, when feasible, we may calculate the absorption probability A_{ij} using the methods of computation described in Section 1.

There is, however, a more direct method for calculating the absorption probabilities A_{ij}, which is usually easier to apply. Suppose that an absorbing Markov chain has N states, $S_1, S_2, \ldots, S_N$, and that State S_j is an absorbing state that can be reached from at least one of the nonabsorbing states of the Markov chain. Since State S_j can never be reached from any (other) absorbing state, let us adopt the convention that $A_{hj} = 0$ whenever State S_h is an absorbing state, $h \neq j$. Further, since if the process starts in the absorbing state, State S_j, it is already absorbed in State S_j, it follows that $A_{jj} = 1$. Under these notational conventions, it can be shown that for every nonabsorbing state S_i,

(7.4) $$A_{ij} = p_{i1}A_{1j} + p_{i2}A_{2j} + \cdots + p_{iN}A_{Nj}.$$

If there are M nonabsorbing states, $S_{i_1}, S_{i_2}, \ldots, S_{i_M}$ ($M < N$), in the Markov chain, we can write down M simultaneous linear equations [one equation of the form (7.4) for each nonabsorbing state], and solve these equations for the M unknowns $A_{i_1 j}, A_{i_2 j}, \ldots, A_{i_M j}$.

For example, in the conformity process described earlier in this section, State S_2 and State S_3 are nonabsorbing states (that is, $M = 2$, $i_1 = 2$, $i_2 = 3$). If we want the absorption probabilities A_{21} and A_{31} (the probabilities of being absorbed in State S_1), then we can solve the pair of linear equations:

$$A_{21} = p_{21}A_{11} + p_{22}A_{21} + p_{23}A_{31} + p_{24}A_{41},$$

$$A_{31} = p_{31}A_{11} + p_{32}A_{21} + p_{33}A_{31} + p_{34}A_{41},$$

for the unknowns A_{21} and A_{31}. Recalling that under our notational conventions $A_{11} = 1$ and $A_{41} = 0$, this pair of linear equations becomes

(7.5) $$A_{21} = p_{21} + p_{22}A_{21} + p_{23}A_{31},$$

$$A_{31} = p_{31} + p_{32}A_{21} + p_{33}A_{31}.$$

In one of the conformity experiments (see Example 3.2 of Chapter 11) conducted by Cohen (1958), the unspecified transition probabilities in Table 7.3 were found to be

(7.6) $\quad\quad p_{21} = 0.06, \quad\quad p_{22} = 0.76, \quad\quad p_{23} = 0.18,$

$\quad\quad\quad\quad\quad p_{31} = 0.00, \quad\quad p_{32} = 0.27, \quad\quad p_{33} = 0.69,$

and $p_{34} = 0.04$. Substituting these values into (7.5), we see that we must solve the pair of linear equations

(7.7) $$A_{21} = 0.06 + (0.76)A_{21} + (0.18)A_{31},$$

$$A_{31} = 0.00 + (0.27)A_{21} + (0.69)A_{31},$$

for the unknowns A_{21} and A_{31}. From the second of these two equations we find that $(1 - 0.69)A_{31} = (0.27)A_{21}$, or

$$A_{31} = \frac{27}{31} A_{21}.$$

Substituting $(27/31)A_{21}$ for A_{31} in the first equation in (7.7), we obtain

$$A_{21} = 0.06 + (0.76)A_{21} + (0.18)\left(\frac{27}{31} A_{21}\right),$$

or

$$\left[1 - 0.76 - \frac{27(0.18)}{31}\right] A_{21} = 0.06.$$

Thus, $A_{21} = 0.721$, and $A_{31} = (27/31)A_{21} = 0.629$. Hence, in a conformity experiment with transition probabilities given by (7.6), a subject who starts the experiment undecided between conformity and nonconformity has a conditional probability of 0.721 of eventually becoming permanently nonconforming (State S_1), given that his initial response is to give the correct answer (State S_2). Similarly, he has a (conditional) probability of 0.629 of eventually becoming permanently nonconforming, given that his initial response is to conform and give an incorrect answer (State S_3).

NOTES AND REFERENCES

The theory of Markov chains is quite extensive. In this chapter and in Chapter 11, we have contented ourselves with providing coverage of some of the main concepts of this theory. Our treatment of these concepts, however, has by no means been exhaustive. For example, although we have mostly considered N-state Markov chains, the concepts (and many of the methods and results) introduced in our discussion are meaningful and valid for Markov chains that have an infinite number of possible states. Further, in our analysis of the long-run behavior of N-state Markov chains, we confined our investigations to Markov chains that are either irreducible or absorbing. However, the Markov chain described in Example 2.1 of this chapter, namely, the choice-of-dress process whose transition probabilities are summarized in Table 8.1, is an example of a Markov chain that is neither irreducible nor absorbing. The long-run behavior of such a process can be studied by using some of the concepts which we have introduced in this chapter (for example, first-passage times, closed, communicating classes, and long-run probabilities). However, somewhat different methods of analysis are required.

Table 8.1. Transition probabilities for a Markov chain.

		State at Time $t+1$		
		S_1	S_2	S_3
State at Time t	S_1	$\frac{1}{2}$	$\frac{1}{2}$	0
	S_2	$\frac{1}{2}$	$\frac{1}{2}$	0
	S_3	$\frac{1}{4}$	$\frac{1}{4}$	$\frac{1}{2}$

The development of the theory of Markov chains continues to be of interest to probabilists. In the last few years, many novel approaches have been suggested, and many new results have been obtained. For example, considerable progress has occurred in that area dealing with "limit theorems" for the k-step transition probabilities of Markov chains. Another area of importance has been the study of the long-run behavior of Markov chains whose 1-step transition probabilities are not stationary over time. A third area of continuing interest is the use of Markov chains to approximate the behavior of stochastic processes that are not Markov chains.

Further details concerning the theory of Markov chains may be found in the following textbooks, each of which requires a more comprehensive background in probability theory and mathematics than has been assumed in the present volume.

1. Bartholomew, D. J. (1967). *Stochastic Models for Social Processes*. John Wiley & Sons, Inc., New York.
2. Chung, Kai-Lai (1960). *Markov Chains with Stationary Transition Probabilities*. Springer-Verlag, Berlin.
3. Feller, W. (1968). *An Introduction to Probability Theory and Its Applications* (3rd edition). John Wiley & Sons, Inc., New York.
4. Kemeny, John G., and Snell, J. Laurie (1960). *Finite Markov Chains*. D. Van Nostrand Company, Inc., Princeton, New Jersey.
5. Parzen, E. (1962). *Stochastic Processes*. Holden-Day, Inc., San Francisco.

Treatments of the theory of Markov chains at a less mathematical level are the following:

1. Bartos, Otomar J. (1967). *Simple Models of Group Behavior*. Columbia University Press, New York.
2. Kemeny, John G., Snell, J. Laurie, and Thompson, Gerald L. (1957). *Introduction to Finite Mathematics*. Prentice-Hall, Inc., Englewood Cliffs, New Jersey.
3. Suppes, Patrick, and Atkinson, Richard C. (1960). *Markov Learning Models for Multiperson Interactions*. Stanford University Press, Stanford, California.

EXERCISES

1. In Washington D.C., taxicab fares are based on zones arranged in a pattern of concentric circles. A taxicab may start the day in one zone. The zone of destination of the first passenger then determines the zone in which the taxicab driver cruises for his next fare. The zone of destination of his next passenger then determines a new cruising zone in which the driver looks for his third fare, and so on. The process that we have just described might be modeled as a Markov chain. Suppose that there are 4 fare zones: $S_1 =$ the center zone, S_2, S_3, and $S_4 =$ the outer zones. Suppose that the transition probabilities among the zones are stationary over time, and that these

transition probabilities are as follows:

| | | \multicolumn{4}{c}{Zone of Destination of $(t+1)$st Fare} |
		S_1	S_2	S_3	S_4
Zone of Destination of (t)th Fare	S_1	0.80	0.14	0.05	0.01
	S_2	0.60	0.20	0.18	0.02
	S_3	0.50	0.40	0.05	0.05
	S_4	0.30	0.30	0.30	0.10

(a) If a taxicab driver lives in the center of town (State S_1) and starts looking for his first fare in State S_4 (the outermost zone), what is the (conditional) probability that he returns to his home zone (State S_1) for the first time after he has driven 4 fares to their destinations?

(b) If the driver starts looking for fares in the outermost zone (State S_4), what is the conditional probability that he must carry 5 or more passengers before he reaches his home zone for the first time?

(c) If the driver starts in the central zone (State S_1), what is the conditional probability that he will return to this zone for the first time by driving his second passenger to that passenger's zone of destination? That is, what is the value of $f_{11}^{(2)}$? What is the value of $f_{11}^{(3)}$?

(d) Given that the driver picks up his first passenger in the central zone, what is the expected number of passengers that the driver must take to their destinations before he returns to the central zone again for the first time? (Central zone = State S_1.)

(e) Does the above Markov chain process achieve a probabilistic equilibrium? Support your assertion.

(f) The taxicab driver generally stops for the day after he has driven 30 passengers. What is the (approximate) unconditional probability that he will stop for the day in his home zone (State S_1), and thus not have to drive far to reach his house? Assume that the initial probabilities for the zones in which the driver finds his first passenger of the day are $a_1 = 0.75$, $a_2 = 0.10$, $a_3 = 0.10$, and $a_4 = 0.05$.

A2. In a study on the pattern of diseased and healthy trees, Pielou (1965) uses Markov chain models. As we walk along a randomly chosen path (transect) that leads through the trees in a given forested area, we successively encounter trees that are either healthy or diseased. Healthy trees may themselves be part of a sub-area of noninfested, healthy trees, called a *gap*, or they may be part of a sub-area of trees containing both diseased and healthy trees, called a *patch*. It is assumed that every sub-area of trees can be uniquely labeled either as a patch or as a gap, and that every tree that we encounter can be assigned to one (and only one) of these sub-areas of trees. We model the succession of trees that we encounter as a Markov chain with stationary transition probabilities. The states of the Markov chain are: S_1 = diseased tree, S_2 = healthy patch tree, S_3 = healthy gap tree. From theoretical considerations, the transition probabilities of this Markov chain have the form:

		State of $(t+1)$st Tree Encountered		
		S_1	S_2	S_3
State of (t)th Tree Encountered	S_1	αv	$\alpha(1-v)$	$1-\alpha$
	S_2	αw	$\alpha(1-w)$	$1-\alpha$
	S_3	γw	$\gamma(1-v)$	$1-\gamma$
		$1-v+w$	$1-v+w$	

(a) Show that for the entries in the table to be transition probabilities, we must have $0 \leq \alpha \leq 1$, $0 \leq v < 1$, $0 \leq \gamma \leq 1$, and $0 \leq w \leq 1$. Do we also need the condition $0 \leq 1-v+w$?

(b) Suppose for a certain species of tree, $\alpha = 0.4$, $\gamma = 0.5$, $v = 0.8$, and $w = 0.2$. Find the first-passage probabilities $f_{11}^{(2)}, f_{12}^{(2)}, f_{13}^{(2)}, f_{21}^{(2)}, f_{22}^{(2)}, f_{23}^{(2)}, f_{31}^{(2)}, f_{32}^{(2)}$, and $f_{33}^{(2)}$.

(c) When $\alpha = 0.4$, $\gamma = 0.5$, $v = 0.8$, and $w = 0.2$, find the expected first-passage times $m_{11}, m_{12}, m_{13}, m_{21}, m_{22}, m_{23}, m_{31}, m_{32}$, and m_{33}.

(d) When $\alpha = 0.4$, $\gamma = 0.5$, $v = 0.8$, and $w = 0.2$, show that this Markov chain is irreducible. Does the Markov chain achieve probabilistic equilibrium? Is the Markov chain an absorbing chain? Support your assertions.

(e) If $\gamma = 0.0$, $\alpha = 0.4$, $v = 0.8$, and $w = 0.2$, find $f_{33}^{(1)}, f_{33}^{(2)}, f_{33}^{(3)}$, and $f_{33}^{(4)}$. Also find $f_{13}^{(1)}, f_{13}^{(2)}, f_{13}^{(3)}$, and $f_{13}^{(4)}$.

3. In certain Markov chains not only do the row sums of a table of transition probabilities all equal 1, but each of the column sums of such a table equals 1. For example, consider the 3-state Markov chain with the transition probabilities

		State at Time $t+1$		
		S_1	S_2	S_3
State at Time t	S_1	$\frac{6}{8}$	$\frac{2}{8}$	0
	S_2	$\frac{1}{8}$	$\frac{4}{8}$	$\frac{3}{8}$
	S_3	$\frac{1}{8}$	$\frac{2}{8}$	$\frac{5}{8}$

Note that each column has a sum of 1.

(a) Show that a Markov chain with the above transition matrix is irreducible and aperiodic. Thus, show that such a Markov chain achieves a probabilistic equilibrium and find the long-run probabilities π_1, π_2, and π_3.

(b) Show that any irreducible 3-state Markov chain having a table of transition probabilities in which all columns sum to 1 must achieve a probabilistic equilibrium. Further, show that the long-run probabilities

of any such Markov chain are the same as the long-run probabilities found in part (a).
(c) Use the result in part (b) above to analyze the long-run behavior of the Markov chain described in Exercise 1 of Chapter 11.

In each of the following Markov chains analyze the chain according to the concepts developed in Chapter 12. In particular, identify recurrent and transient states, identify absorbing states, and find the period of each state. Identify communicating classes and state whether each such class is closed or not. Verify whether the chain is aperiodic and irreducible. If it is, find the expected first-passage times and long-run probabilities. If it is not, see if it is an absorbing chain, and calculate the absorption probabilities A_{ij} where these probabilities are appropriate.

4. The genetical mating experiment whose transition probabilities appear in Table 3.8 of Chapter 11.

A5. The occupational mobility process described in Example 4.2 of Chapter 11.

6. The political preference process described in Example 1.1 of Chapter 12.

A7. The learning process described in Exercise 3 of Chapter 11 when $\alpha = 0.4$, $\beta = 0.0$, and $\delta = 0.5$.

8. The intercity population mobility process described in Exercise 4 of Chapter 11.

9. Coleman (1964, p. 168) considers a Markov chain model in which there are two attributes A and B, to which we may respond with a "+" or "−". These responses may mean "present-absent" in one context, "agree-disagree" in another, and so on. We then have 4 states of the system representing the responses on attributes A and B, respectively:

S_1: response is (+,+), S_3: response is (−,+),
S_2: response is (+,−), S_4: response is (−,−).

Thus, State S_2 is the state in which attribute A is present (+) and attribute B is absent (−). In one study of the social system of adolescents in 10 high schools, 3260 (female) students were asked in a questionnaire about the "leading crowd." The attribute A was whether the individual was a member of the leading crowd, and B was whether she agreed or disagreed with the sentence: "If a girl wants to be part of the leading crowd around here, she sometimes has to go against her principles." The resulting (estimated) transition probabilities are given as follows:

		State at Time 2 (May 1958)			
		S_1	S_2	S_3	S_4
State at Time 1 (October 1957)	S_1	0.676	0.130	0.149	0.045
	S_2	0.376	0.369	0.101	0.154
	S_3	0.103	0.032	0.610	0.255
	S_4	0.075	0.076	0.307	0.542

10. In a second study considered by Coleman (see Exercise 9 above), the effects on sales of a favorable attitude toward a brand of a grocery item were studied. Here attribute A referred to the respondent's attitude toward the brand, and attribute B referred to whether the particular item was the respondent's "usual brand." A total of 1633 respondents were interviewed, yielding the following (estimated) transition probabilities:

		State at Time $t+1$			
		S_1	S_2	S_3	S_4
State at Time t	S_1	0.848	0.033	0.055	0.064
	S_2	0.200	0.167	0.033	0.600
	S_3	0.638	0.056	0.153	0.153
	S_4	0.047	0.104	0.025	0.824

BIBLIOGRAPHY

Abramowitz, M., and Stegun, S. A. (Editors) (1965). *Handbook of Mathematical Functions*. New York: Dover Publications, Inc.

Adams, J. D. (1962). Failure time distribution estimation. *Semiconductor Reliability*, vol. 2, pp. 41–52.

Ahrens, L. H. (1954). The lognormal distribution of the elements. *Geochimica et Cosmochimica Acta*, vol. 5, pp. 49–73.

Ahrens, L. H. (1954). The lognormal distribution of the elements. *Geochimica et Cosmochimica Acta*, vol. 6, pp. 121–131.

Ahrens, L. H. (1957). The lognormal distribution of the elements. *Geochimica et Cosmochimica Acta*, vol. 11, pp. 205–212.

Aitchison, J., and Brown, J. A. C. (1963). *The Lognormal Distribution*. Cambridge: Cambridge University Press.

American Geological Institute (1969). *Models of Geologic Processes—An Introduction to Mathematical Geology*. Washington, D.C.: American Geological Institute.

Asch, S. E. (1952). *Social Psychology*. Englewood Cliffs, New Jersey: Prentice-Hall, Inc.

Ayer, A. J. (1965). Chance. *Scientific American*, vol. 213, October, p. 44.

Bartholomew, D. J. (1967). *Stochastic Models for Social Processes*. New York: John Wiley & Sons, Inc.

Bartos, Otomar J. (1967). *Simple Models of Group Behavior*. New York: Columbia University Press.

Berger, J., and Snell, J. L. (1957). On the concept of equal exchange. *Behavioral Science*, vol. 2, pp. 111–118.

Bernoulli, J. (1966). Translations from James Bernoulli, by B. Sung, with a preface by A. P. Dempster, Technical Report No. 2, Harvard University.

Birnbaum, Z. W., and Saunders, S. C. (1958). A statistical model for the life-length of materials. *Journal of the American Statistical Association*, vol. 53, pp. 151–160.

Bliss, C. I. (1934). The method of probits. *Science*, vol. 79, pp. 38–39, 409–410.

Borel, Émile (1962). *Probabilities and Life*. (Translated from the French by M. Baudin). New York: Dover Publications, Inc.

Burgess, E. W., and Cottrell, L. S., Jr. (1955). The prediction of adjustment in marriage, in Lazarsfeld, P. F., and Rosenberg, M. (1955). *The Language of Social Research*. Glencoe, Illinois: The Free Press, pp. 267–276.

Campbell, Angus, and Kahn, Robert (1952). *The People Elect a President*. Ann Arbor, Michigan: University of Michigan, Survey Research Center.

Chatfield, C. (1970). Discrete distributions in market research, in Patil, G. P. (Editor). *Random Counts in Physical Science, Geo Science, and Business*, University Park, Pennsylvania: Pennsylvania University Press.

Chayes, F. (1954). The lognormal distribution of the elements: A discussion. *Geochimica et Cosmochimica Acta*, vol. 6, pp. 119–120.

Chung, Kai-Lai (1960). *Markov Chains with Stationary Transition Probabilities*. Berlin: Springer-Verlag.

Clarke, R. D. (1946). An application of the Poisson distribution. *Journal of the Institute of Actuaries*, vol. 72, p. 481.

Cohen, B. (1958). A probability model for conformity. *Sociometry*, vol. 21, pp. 69–81.

Coleman, James S. (1964). *Introduction to Mathematical Sociology*. London: Free Press of Glencoe, Collier-Macmillan Ltd., pp. 367–375.

Coombs, Clyde H., Dawes, Robyn M., and Tversky, Amos (1970). *Mathematical Psychology, An Elementary Introduction*. Englewood Cliffs, New Jersey: Prentice-Hall, Inc.

David, F. N. (1962). *Games, Gods, and Gambling*. New York: Hafner Publishing Company.

Davis, D. J. (1952). An analysis of some failure data. *Journal of the American Statistical Association*, vol. 47, pp. 113–150.

Davis, H. T. (1933, 1935). *Tables of the Higher Mathematical Functions*, 2 vols. Bloomington, Indiana: Principia Press.

Detlefsen, J. A. (1918). Fluctuations of sampling in a Mendelian population. *Genetics*, vol. 3, pp. 597–607.

Dhrymes, Phoebus J. (1962). On devising unbiased estimators for the parameters of the Cobb-Douglas production function. *Econometrica*, vol. 30, pp. 297–304.

Dwass, Meyer (1967). *First Steps in Probability*. New York: McGraw-Hill Book Company, Inc.

Epstein, B. (1947). The mathematical description of certain breakage mechanisms leading to the logarithmico-normal distribution. *Journal of the Franklin Institute*, vol. 244, pp. 471–477.

Epstein, B. (1948). Statistical aspects of fracture problems. *Journal of Applied Physics*, vol. 19, pp. 140–147.

Estes, W. K. (1950). Toward a statistical theory of learning. *Psychological Review*, vol. 57, pp. 94–107.

Fechner, G. T. (1897). *Kollektivmasslehre*. Leipzig: W. Engelmann.

Feinlieb, M. (1960). A method of analyzing log-normally distributed survival data with incomplete follow-up. *Journal of the American Statistical Association*, vol. 55, pp. 534–545.
Feller, W. (1968). *An Introduction to Probability Theory and Its Applications* (3rd ed.). New York: John Wiley & Sons, Inc.
Fieller, E. C. (1932). The distribution of the index in a normal bivariate population. *Biometrika*, vol. 24, pp. 428–440.
Fienberg, S. E. (1971). Randomization and social affairs: The 1970 draft lottery. *Science*, vol. 171, pp. 255–261.
Finkelstein, M. O. (1966). The application of statistical decision theory to the jury discrimination cases. *Harvard Law Review*, vol. 80, pp. 338–376.
Fisher, R. A. (1950). *Statistical Methods for Research Workers* (11th ed.). Edinburgh: Oliver & Boyd, Ltd.
Fisher, R. A., and Mather, K. (1936). A linkage test with mice. *Annals of Eugenics*, vol. 7, pp. 265–280.
Freedle, R. O., and Lewis, M. (1971). Application of Markov processes to the concept of state. *Research Bulletin*, 71–34. Princeton, New Jersey: Educational Testing Service.
Froggatt, P. (1970). Application of discrete distribution theory to the study of noncommunicable events in medical epidemiology, in Patil, G. P. (Editor). *Random Counts in Biomedical and Social Sciences*, vol. 2. University Park, Pennsylvania: Pennsylvania State University Press.
Gaddum, J. H. (1945). Lognormal distributions. *Nature*, vol. 156, pp. 463–466.
Galton, Francis (1879). The geometric mean in vital and social statistics. *Proceedings of the Royal Society, London*, vol. 29, pp. 365–367.
Galton, Francis (1889), *Natural Inheritance*, London: Macmillan & Co., Ltd.
Galton, Francis (1892). *Finger Prints*, London: Macmillan & Co., Ltd.
Glass, D. V., and Hall, J. R. (1954). A study of inter-generation changes in status, in Glass, D. V. (Editor). *Social Mobility in Britain*. Glencoe, Illinois: The Free Press, pp. 177–241.
Godwin, H. J. (1955). On generalizations of Tchebychef's inequality. *Journal of the American Statistical Association*, vol. 50, pp. 923–945.
Goldberg, S. (1960). *Probability, An Introduction*. Englewood Cliffs, New Jersey: Prentice-Hall, Inc.
Goldthwaite, L. R. (1961). Failure rate study for the lognormal lifetime model. *Proceedings of the Seventh National Symposium on Reliability and Quality Control in Electronics*, pp. 208–213.
Greenwood, M., Jr. (1904). A first study of the weight, variability, and correlation of the human viscera, with special reference to the healthy and diseased heart. *Biometrika*, vol. 3, pp. 63–83.
Greenwood, M., and Yule, G. U. (1920). An inquiry into the nature of frequency distributions of multiple happenings with particular reference to the occurrence of multiple attacks of disease or repeated accidents. *Journal of the Royal Statistical Society*. Series A, vol. 83, pp. 255–279.
Gregory, S. (1963). *Statistical Methods and the Geographer*. London: Longmans, Green & Co., Ltd.

Griffiths, John C. (1960). Frequency distributions in accessory mineral analysis. *The Journal of Geology*, vol. 68, pp. 353–365.

Grundy, P. M. (1951). The expected frequencies in a sample of an animal population in which the abundances of species are log-normally distributed. I. *Biometrika*, vol. 38, pp. 427–434.

Guenther, William C. (1968). *Concepts of Probability*. New York: McGraw-Hill Book Company, Inc.

Gutenberg, B., and Richter, C. F. (1944). Frequency of earthquakes in California. *Bulletin of the Seismological Society of America*, vol. 34, pp. 185–188.

Hagstroem, K.-G. (1960). Remarks on Pareto distributions. *Skandinavisk Aktuarietidskrift*, vol. 43, pp. 59–71.

Haight, F. A. (1967). *Handbook of the Poisson Distribution*. New York: John Wiley & Sons, Inc.

Haight, F. A. (1970). Group size distributions, with applications to vehicle occupancy, in Patil, G. P. (Editor). *Random Counts in Physical Science, Geo Science, and Business*, vol. 3. University Park, Pennsylvania: Pennsylvania State University Press, pp. 95–105.

Hatch, T., and Choute, S. P. (1929). Statistical description of the size properties of non-uniform particles. *Journal of the Franklin Institute*, vol. 207, pp. 369–380.

Herdan, G. (1958). The relation between the dictionary distribution and the occurrence distribution of word length and its importance for the study of quantitative linguistics. *Biometrika*, vol. 45, pp. 222–228.

Herdan, G. (1960). *Small Particle Statistics* (2nd ed.). London: Butterworth & Co., Ltd.

Herdan, G. (1964). *Quantitative Linguistics*. London: Butterworth & Co., Ltd.

Herdan, G. (1966). *The Advanced Theory of Language as Choice and Chance*. New York: Springer-Verlag.

Hodges, J. L., Jr., and Lehmann, E. L. (1970). *Elements of Finite Probability* (2nd ed.). San Francisco: Holden-Day, Inc.

Hogg, Jane M. (1965). The effect of some climatological variations of the incidence and spread of fires in buildings in England and Wales from 1951 to 1961. *Journal of the Royal Statistical Society, Series C, Applied Statistics*, vol. 14, pp. 140–161.

Hollingshead, A. de B. (1949). *Elmtown's Youth: The Impact of Social Classes on Adolescents*. New York: John Wiley & Sons, Inc.

Indow, Tarow (1971). Models for responses of customers with a varying rate. *Journal of Marketing Research*, vol. 8, pp. 78–84.

Jaffe, Joseph, and Feldstein, Stanley (1970). *Rhythms of Dialogue*. New York: Academic Press, Inc.

James, John (1953). The distribution of free-forming small group size. *American Sociological Review*, vol. 18, p. 569.

Johnson, Norman L., and Kotz, Samuel (1969). *Distributions in Statistics: Discrete Distributions*. Boston: Houghton Mifflin Company.

Johnson, Norman L., and Kotz, Samuel (1970). *Distributions in Statistics: Continuous Univariate Distributions-1*. Boston: Houghton Mifflin Company.

Johnson, Norman L., and Kotz, Samuel (1970). *Distributions in Statistics: Continuous Univariate Distributions-2.* Boston: Houghton Mifflin Company.

Kac, M. (1964). Probability. *Scientific American,* vol. 211, September, p. 92.

Kemeny, John G., and Snell, J. Laurie (1960). *Finite Markov Chains,* Princeton, New Jersey: D. Van Nostrand Company, Inc.

Kemeny, John G., Snell, J. Laurie, and Thompson, Gerald L. (1957). *Introduction to Finite Mathematics.* Englewood Cliffs, New Jersey: Prentice-Hall, Inc.

Kendall, M. G. (1963). Isaac Todhunter's history of the mathematical theory of probability. *Biometrika,* vol. 50, pp. 204–205.

Kendall, M. G., and Buckland, W. R. (1960). *A Dictionary of Statistical Terms.* London: Oliver & Boyd, Ltd.

Kitagawa, T. (1951). *Tables of Poisson Distribution.* Tokyo: Baifukan.

Klein, M. (1962). *Mathematics: A Cultural Approach.* Reading, Massachusetts: Addison-Wesley Publishing Company, Inc.

Koch, G. S., Jr., and Link, R. F. (1971). *Statistical Analysis of Geological Data,* vol. II. New York: John Wiley & Sons, Inc.

Kolmogorov, A. N. (1933). *Foundations of the Theory of Probability* (Translation of the original). New York: Chelsea Publishing Company, 1950, 1956.

Krumbein, W. C. (1936). Application of logarithmic moments to size frequency distributions of sediments. *Journal of Sedimentary Petrology,* vol. 6, pp. 35–47.

Krumbein, W. C. (1954). Applications of statistical methods to sedimentary rocks. *Journal of the American Statistical Association,* vol. 49, pp. 51–66.

Kyburg, H. E., and Smokler, H. E. (1964). *Studies in Subjective Probability.* New York: John Wiley & Sons, Inc.

Laplace, P. S. de (1951). *A Philosophical Essay on Probabilities* (Translation with an introduction by E. T. Bell). New York: Dover Publications, Inc.

Latter, O. H. (1901). The egg of *Cuculus Canorus. Biometrika,* vol. 1, pp. 164–176.

Lazarsfeld, Paul F., and Thielens, Wagner, Jr. (1958). *The Academic Mind.* Glencoe, Illinois: The Free Press.

Lieberman, G. J., and Owen, D. B. (1961). *Tables of the Hypergeometric Distribution.* Stanford, California: Stanford University Press.

Life Insurance Fact Book (1967). New York: Institute of Life Insurance.

Macdonell, W. R. (1902). On criminal anthropometry and the identification of criminals. *Biometrika,* vol. 1, pp. 177–227.

Marshall, Andrew W., and Goldhamer, Herbert (1955). An application of Markov processes to the study of the epidemiology of mental disease. *Journal of the American Statistical Association,* vol. 50, pp. 99–129.

Masuyama, M., and Kuroiwa, Y. (1952). Table for the likelihood solutions of gamma distribution and its medical applications. *Reports of Statistical Application Research, (JUSE),* vol. 1, pp. 18–23.

McAlister, D. (1879). The law of the geometric mean. *Proceedings of the Royal Society of London,* vol. 29, pp. 367–375.

Menzerath, P., and Meyer-Eppler, W. (1954). *Die Architektonik des Deutschen Wortschatzes,* Bonn: Dummler.

Miller, R. L., and Kahn, J. S. (1962). *Statistical Analysis in the Geological Sciences*. New York: John Wiley & Sons, Inc.

Molina, E. C. (1940). *Poisson's Exponential Binomial Limit*. New York: D. Van Nostrand Company, Inc.

Mosteller, F. (1965). *Fifty Challenging Problems in Probability*. Reading, Massachusetts: Addison-Wesley Publishing Company, Inc.

Mosteller, F., Rourke, R. E. K., and Thomas, G. B. (1961). *Probability with Statistical Applications*. Reading, Massachusetts: Addison-Wesley Publishing Company, Inc.

Mueller, C. G. (1950). Theoretical relationships among some measures of conditioning. *Proceedings of the National Academy of Sciences*, vol. 56, pp. 123–134.

National Bureau of Standards (1950). *Tables of the Binomial Probability Distribution, Applied Mathematics Series 6*. Washington, D.C.: U.S. Government Printing Office.

National Bureau of Standards (1951). *Tables of n! and $\Gamma(n+1/2)$ for the First Thousand Values of n, Applied Mathematics Series 16*. Washington, D.C.: U.S. Government Printing Office.

National Bureau of Standards (1959). *Tables of the Bivariate Normal Distribution Function and Related Functions, Applied Mathematics Series 50*. Washington, D.C.: U.S. Government Printing Office.

Newman, James R. (1956). *The World of Mathematics*. New York: Simon and Schuster, Inc.

Neyman, J. (1950). *A First Course in Probability and Statistics*. New York: Holt, Rinehart and Winston, Inc.

Ore, Oyestein (1953). *Cardano, the Gambling Scholar*. Princeton, New Jersey: Princeton University Press.

Ore, Oyestein (1960). Pascal and the invention of probability theory. *American Mathematical Monthly*, vol. 67, pp. 409–419.

Parzen, E. (1962). *Stochastic Processes*. San Francisco: Holden-Day, Inc.

Patil, G. P. (Editor) (1965). *Classical and Contagious Distributions*. Calcutta: Statistical Publishing Society.

Pearl, R. (1905). Biometrical studies on man. I. Variation and correlation in brain-weight. *Biometrika*, vol. 4, pp. 13–104.

Pearson, E. S., and Hartley, H. O. (Editors) (1958). *Biometrika Tables for Statistics*, vol. I. Cambridge: Cambridge University Press.

Pearson, E. S., and Kendall, M. G. (1970). *Studies in the History of Statistics and Probability*. New York: Hafner Publishing Company.

Pearson, K. (1903). Craniological notes. *Biometrika*, vol. 2, pp. 338–347.

Pearson, K. (1924). On a certain double hypergeometrical series and its representation by continuous frequency surfaces. *Biometrika*, vol. 16, pp. 172–188.

Pearson, K. (Editor) (1934). *Tables of the Incomplete Beta-Function*. London: The "Biometrika" Office, University College.

Pearson, K., and Lee, A. (1903). On the laws of inheritance in man. I. Inheritance of physical characters. *Biometrika*, vol. 2, pp. 357–462.

Pielou, E. C. (1965). The concept of segregation pattern in ecology: Some discrete distributions applicable to the run lengths of plants in narrow transects, in Patil, G. P. (Editor), *Classical and Contagious Discrete Distributions*. Calcutta, India: Statistical Publishing Society, pp. 410–418.

Pretorius, S. J. (1930). Skew bivariate frequency surfaces, examined in the light of numerical illustrations. *Biometrika*, vol. 22, pp. 109–223.

Proschan, F. (1963). Theoretical explanation of observed decrease failure rate. *Technometrics*, vol. 5, pp. 375–384.

Rasch, G. (1960). *Probabilistic Models for Some Intelligence and Attainment Tests*. Studies in Mathematical Psychology I. Copenhagen, Denmark: Nielson and Lydiche.

Richardson, Lewis F. (1944). Distribution of wars in time. *Journal of the Royal Statistical Society*, vol. 107, pp. 242–250.

Romig, H. G. (1953). *50–100 Binomial Tables*. New York: John Wiley & Sons, Inc.

Rutherford, Ernest, Chadwick, James, and Ellis, C. D. (1930). *Radiations from Radioactive Substances*. New York: The Macmillan Company.

Rutherford, Ernest, and Geiger, Hans (1910). The probability variations in the distribution of a particle. *Philosophical Magazine*, vol. 20, pp. 698–707.

Sartwell, P. E. (1950). The distribution of incubation periods of infectious diseases. *American Journal of Hygiene*, vol. 51, pp. 310–318.

Savage, I. R. (1961). Probability inequalities of the Tchebycheff type. *Journal of Research of National Bureau of Standards*, vol. 65B, pp. 211–222.

Savage, L. J. (1954). *The Foundations of Statistics*. New York: John Wiley & Sons, Inc.

Siddiqui, M. M., and Weiss, George H. (1963). Families of distributions for hourly median power and instantaneous power of received radio signals. *Journal of Research of the National Bureau of Standards*, vol. 67D, pp. 753–762.

Simon, H. A., and Bonini, C. P. (1958). *The Size Distribution of Business Firms*. Pittsburgh, Pennsylvania: Reprint 20 of the Graduate School of Business Administration, Carnegie Institute of Technology.

Simpson, G. G., and Roe, A. (1939). *Quantitative Zoology*. New York: McGraw-Hill Book Company, Inc.

Simpson, G. G., Roe, A., and Lewontin, R. C. (1960). *Quantitative Zoology*. New York: Harcourt, Brace and World, Inc.

Slack, H. A., and Krumbein, W. C. (1955). Measurement and statistical evaluation of low-level radioactivity in rocks. *Transactions, American Geophysical Union*, vol. 36, pp. 460–464.

Smith, E., and Suchman, E. A. (1955). Do people know why they buy? Lazarsfeld, P. F., and Rosenberg, M. (Editors). *The Language of Social Research*. Glencoe, Illinois: The Free Press, pp. 404–410.

Spencer, H. (1877). *The Principles of Sociology*. New York: Appleton and Co.

Suppes, Patrick, and Atkinson, Richard C. (1960). *Markov Learning Models for Multiperson Interactions*. Stanford, California: Stanford University Press.

Svedberg, Theodor (1912). *Existenz der Moleküle*. Leipzig: Akademische Verlagsgesellschaft m.b.H.

Terman, L. M. (1919). *The Intelligence of School Children*. Boston: Houghton Mifflin Company.

Thorndike, F. (1926). Applications of Poisson's probability summation. *Bell System Technical Journal*, vol. 5, pp. 604–624.

Todd, G. F. (Editor) (1966). *Statistics of Smoking in the United Kingdom* (4th ed.). London: Tobacco Research Council.

Todhunter, Isaac (1865). *A History of the Mathematical Theory of Probability from the Time of Pascal to That of Laplace* (Reprint). New York: Chelsea Publishing Company, 1949.

Trumpler, Robert J., and Weaver, Harold F. (1953). *Statistical Astronomy*. Berkeley: University of California Press.

U.S. Bureau of the Census, Statistical Abstract of the United States: 1959 (80th ed.). Washington, D.C.

U.S. Bureau of the Census, Statistical Abstract of the United States: 1965 (86th ed.). Washington, D.C.

Urban, F. M. (1909). Die psychophysischen Massmethoden als Grundlage empirischer Messungen. *Archiv für die gesamte Psychologie*, vol. 15, pp. 261–415.

Von Mises, Richard (1957). *Probability, Statistics, and Truth* (second revised English edition prepared by Hilda Geiringer). New York: The Macmillan Company.

Weaver, W. (1950). Probability. *Scientific American*, vol. 183, October, p. 44.

Weaver, W. (1952). Statistics. *Scientific American*, vol. 186, January, p. 60.

Weibull, W. (1951). A statistical distribution function of wide applicability. *Journal of Applied Mechanics*, vol. 18, pp. 293–297.

Whitaker, L. (1914). On the Poisson law of small numbers. *Biometrika*, vol. 10, pp. 36–71.

Wilk, M. B., Gnanadesikan, R., and Huyett, Marilyn J. (1962). Estimation of parameters of the gamma distribution using order statistics. *Biometrika*, vol. 49, pp. 525–546.

Williams, C. B. (1940). A note on the statistical analysis of sentence length as a criterion of literary style. *Biometrika*, vol. 31, pp. 356–361.

Williams, C. B. (1956). Studies in the history of probability and statistics. IV. A note on an early statistical study of literary style. *Biometrika*, vol. 43, pp. 248–256.

Williamson, E., and Bretherton, M. K. (1963). *Tables of the Negative Binomial Probability Distribution*. New York: John Wiley & Sons, Inc.

Wright, S. (1968). *Genetic and Biometric Foundations*, vol. I. Chicago: University of Chicago Press.

Zipf, G. K. (1949). *Human Behavior and the Principle of Least Effort*. Reading, Massachusetts: Addison-Wesley Publishing Company, Inc.

TABLES

Table T.1 Individual Terms, $p_Z(k) = \binom{n}{k} p^k (1-p)^{n-k}$, of the probability Mass Function of a Binomially Distributed Random Variable Z

Table T.2 Values for the Probability Mass Function of the Hypergeometric Distribution for $N = 2(1)9$ and Selected Values of n and p

Table T.3 Individual Terms, $p_X(k)$, of the Probability Mass Function of the Poisson Distribution with Parameter λ

Table T.4 Individual Terms $p_Z(k) = p(1-p)^{k-1}$, of the Probability Mass Function of a Random Variable Z Having a Geometric Distribution

Table T.5 Values of $p_Y(k)$ for the Negative Binomial Distribution with $r = 2, 3, 4, 5$, and $p = 0.2, 0.4, 0.5, 0.6, 0.8$

Table T.6 Table of the Cumulative Distribution Function of a Standard Normal Random Variable

Table T.7 Tail Probabilities, $P\{\mathcal{N}(Z;0,1) > k\}$, for the Standard Normal Distribution

Table T.8 Values of the Exponential Function e^{-x}, for Negative Arguments

Table T.9 Values of the Incomplete Gamma Function, $I_r(\tau)$, for Use in the Computation of the Cumulative Gamma Distribution Function

Table T.10 Values of the Incomplete Beta Function to Aid in the Graph of the Cumulative Beta Distribution for $r = 0.5, 1, 2, 3, 4, 5$, and $s = 2, 3, 4$, and 5

Table T.11 Values of the Probability Content of an Upper Quadrant, $P\{X > h, Y > k\}$ for the Standard Bivariate Normal Distribution with Correlation ρ.

Table T.1 Individual Terms, $p_Z(k) = \binom{n}{k} p^k (1-p)^{n-k}$, of the Probability Mass Function of a Binomially Distributed Random Variable Z

											p								
n	k	.01	.02	.03	.04	.05	.10	.15	.20	.25	.30	.35	.40	.45	.50
1	0	.99000	.98000	.97000	.96000	.95000	.90000	.85000	.80000	.75000	.70000	.65000	.60000	.55000	.50000
	1	.01000	.02000	.03000	.04000	.05000	.10000	.15000	.20000	.25000	.30000	.35000	.40000	.45000	.50000
2	0	.98010	.96040	.94090	.92160	.90250	.81000	.72250	.64000	.56250	.49000	.42250	.36000	.30250	.25000
	1	.01980	.03920	.05820	.07680	.09500	.18000	.25500	.32000	.37500	.42000	.45500	.48000	.49500	.50000
	2	.00010	.00040	.00090	.00160	.00250	.01000	.02250	.04000	.06250	.09000	.12250	.16000	.20250	.25000
3	0	.97030	.94119	.91267	.88474	.85738	.72900	.61413	.51200	.42188	.34300	.27463	.21600	.16638	.12500
	1	.02940	.05762	.08468	.11059	.13538	.24300	.32513	.38400	.42188	.44100	.44363	.43200	.40838	.37500
	2	.00030	.00118	.00262	.00461	.00712	.02700	.05737	.09600	.14063	.18900	.23888	.28800	.33413	.37500
	3	.00000	.00001	.00003	.00006	.00013	.00100	.00337	.00800	.01563	.02700	.04287	.06400	.09113	.12500
4	0	.96060	.92237	.88529	.84935	.81451	.65610	.52201	.40960	.31641	.24010	.17851	.12960	.09151	.06250
	1	.03881	.07530	.10952	.14156	.17148	.29160	.36848	.40960	.42188	.41160	.38447	.34560	.29948	.25000
	2	.00059	.00230	.00508	.00885	.01354	.04860	.09754	.15360	.21094	.26460	.31054	.34560	.36754	.37500
	3	.00000	.00003	.00010	.00025	.00047	.00360	.01148	.02560	.04688	.07560	.11148	.15360	.20048	.25000
	4	.00000	.00000	.00000	.00000	.00001	.00010	.00051	.00160	.00391	.00810	.01501	.02560	.04101	.06250
5	0	.95099	.90392	.85873	.81537	.77378	.59049	.44371	.32768	.23730	.16807	.11603	.07776	.05033	.03125
	1	.04803	.09224	.13279	.16997	.20363	.32805	.39150	.40960	.39551	.36015	.31239	.25920	.20589	.15625
	2	.00097	.00376	.00821	.01416	.02143	.07290	.13818	.20480	.26367	.30870	.33642	.34560	.33691	.31250
	3	.00001	.00008	.00025	.00059	.00113	.00810	.02438	.05120	.08789	.13230	.18115	.23040	.27565	.31250
	4	.00000	.00000	.00000	.00001	.00003	.00045	.00215	.00640	.01465	.02835	.04877	.07680	.11277	.15625
	5	.00000	.00000	.00000	.00000	.00000	.00001	.00008	.00032	.00098	.00243	.00525	.01024	.01845	.03125
6	0	.94148	.88584	.83297	.78276	.73509	.53144	.37715	.26214	.17798	.11765	.07542	.04666	.02768	.01563
	1	.05705	.10847	.15457	.19569	.23213	.35429	.39933	.39322	.35596	.30253	.24366	.18662	.13589	.09375
	2	.00144	.00553	.01195	.02038	.03054	.09842	.17613	.24576	.29663	.32801	.32801	.31104	.27648	.23438
	3	.00002	.00015	.00049	.00113	.00214	.01458	.04145	.08192	.13184	.18522	.23549	.27648	.30322	.31250
	4	.00000	.00000	.00001	.00004	.00008	.00121	.00549	.01536	.03296	.05953	.09510	.13824	.18607	.23438
	5	.00000	.00000	.00000	.00000	.00000	.00005	.00031	.00154	.00439	.01021	.02048	.03686	.06089	.09375
	6	.00000	.00000	.00000	.00000	.00000	.00000	.00001	.00006	.00024	.00073	.00184	.00410	.00830	.01563

Table T.1 (*continued*)

n	k	.01	.02	.03	.04	.05	.10	.15	.20	.25	.30	.35	.40	.45	.50
7	0	.93207	.86813	.80798	.75145	.69834	.47830	.32058	.20972	.13348	.08235	.04902	.02799	.01522	.00781
	1	.06590	.12402	.17492	.21917	.25728	.37201	.39601	.36700	.31146	.24706	.18478	.13064	.08719	.05469
	2	.00200	.00759	.01623	.02740	.04062	.12400	.20965	.27525	.31146	.31765	.29848	.26127	.21402	.16406
	3	.00003	.00026	.00084	.00190	.00356	.02296	.06166	.11469	.17303	.22689	.26787	.29030	.29185	.27344
	4	.00000	.00001	.00003	.00008	.00019	.00255	.01088	.02867	.05768	.09724	.14424	.19354	.23878	.27344
	5	.00000	.00000	.00000	.00000	.00001	.00017	.00115	.00430	.01154	.02500	.04660	.07741	.11722	.16406
	6	.00000	.00000	.00000	.00000	.00000	.00001	.00007	.00036	.00128	.00357	.00836	.01720	.03197	.05469
	7	.00000	.00000	.00000	.00000	.00000	.00000	.00000	.00001	.00006	.00022	.00064	.00164	.00374	.00781
8	0	.92274	.85076	.78374	.72139	.66342	.43047	.27249	.16777	.10011	.05765	.03136	.01680	.00837	.00391
	1	.07457	.13890	.19392	.24046	.27933	.38264	.38469	.33554	.26697	.19765	.13726	.08958	.05481	.03125
	2	.00264	.00992	.02099	.03507	.05146	.14880	.23760	.29360	.31146	.29648	.25869	.20902	.15695	.10938
	3	.00005	.00040	.00130	.00292	.00542	.03307	.08386	.14680	.20764	.25412	.27859	.27869	.25683	.21875
	4	.00000	.00001	.00005	.00015	.00036	.00459	.01850	.04588	.08652	.13614	.18751	.23224	.26266	.27344
	5	.00000	.00000	.00000	.00001	.00002	.00041	.00261	.00918	.02307	.04668	.08077	.12386	.17192	.21875
	6	.00000	.00000	.00000	.00000	.00000	.00002	.00023	.00115	.00385	.01000	.02175	.04129	.07033	.10938
	7	.00000	.00000	.00000	.00000	.00000	.00000	.00001	.00008	.00037	.00122	.00335	.00786	.01644	.03125
	8	.00000	.00000	.00000	.00000	.00000	.00000	.00000	.00000	.00002	.00007	.00023	.00066	.00168	.00391
9	0	.91352	.83375	.76023	.69253	.63025	.38742	.23162	.13422	.07508	.04035	.02071	.01008	.00461	.00195
	1	.08305	.15314	.21161	.25970	.29854	.38742	.36786	.30199	.22525	.15565	.10037	.06047	.03391	.01758
	2	.00336	.01250	.02618	.04328	.06285	.17219	.25967	.30199	.30034	.26683	.21619	.16124	.11099	.07031
	3	.00008	.00060	.00189	.00421	.00772	.04464	.10692	.17616	.23360	.26683	.27162	.25082	.21188	.16406
	4	.00000	.00002	.00009	.00026	.00061	.00744	.02830	.06606	.11680	.17153	.21939	.25082	.26004	.24609
	5	.00000	.00000	.00000	.00001	.00003	.00083	.00499	.01652	.03893	.07351	.11813	.16722	.21276	.24609
	6	.00000	.00000	.00000	.00000	.00000	.00006	.00059	.00275	.00865	.02100	.04241	.07432	.11605	.16406
	7	.00000	.00000	.00000	.00000	.00000	.00000	.00004	.00029	.00124	.00386	.00979	.02123	.04069	.07031
	8	.00000	.00000	.00000	.00000	.00000	.00000	.00000	.00002	.00010	.00041	.00132	.00354	.00832	.01758
	9	.00000	.00000	.00000	.00000	.00000	.00000	.00000	.00000	.00000	.00002	.00008	.00026	.00076	.00195
10	0	.90438	.81707	.73742	.66483	.59874	.34868	.19687	.10737	.05631	.02825	.01346	.00605	.00253	.00098
	1	.09135	.16675	.22807	.27701	.31512	.38742	.34743	.26844	.18771	.12106	.07249	.04031	.02072	.00977
	2	.00415	.01531	.03174	.05194	.07463	.19371	.27590	.30199	.28157	.23347	.17565	.12093	.07630	.04395
	3	.00011	.00083	.00262	.00577	.01048	.05740	.12983	.20133	.25028	.26683	.25222	.21499	.16648	.11719
	4	.00000	.00003	.00014	.00042	.00096	.01116	.04010	.08808	.14600	.20012	.23767	.25082	.23837	.20508
	5	.00000	.00000	.00001	.00002	.00006	.00149	.00849	.02642	.05840	.10292	.15357	.20066	.23403	.24609
	6	.00000	.00000	.00000	.00000	.00000	.00014	.00125	.00551	.01622	.03676	.06891	.11148	.15957	.20508
	7	.00000	.00000	.00000	.00000	.00000	.00001	.00013	.00079	.00309	.00900	.02120	.04247	.07460	.11719
	8	.00000	.00000	.00000	.00000	.00000	.00000	.00001	.00007	.00039	.00145	.00428	.01062	.02289	.04395
	9	.00000	.00000	.00000	.00000	.00000	.00000	.00000	.00000	.00003	.00014	.00051	.00157	.00416	.00977
	10	.00000	.00000	.00000	.00000	.00000	.00000	.00000	.00000	.00000	.00001	.00003	.00010	.00034	.00098

Table T.2 Values for the probability mass function of the hypergeometric distribution for $N = 2(1)9$ and selected values of n and p.

N	n	Np	k	$p_X(k)$	N	n	Np	k	$p_X(k)$
2	1	1	0	0.5000				2	0.6000
			1	0.5000				3	0.1000
3	1	1	0	0.6667		4	1	0	0.2000
			1	0.3333				1	0.8000
						4	2	1	0.4000
	2	1	0	0.3333				2	0.6000
			1	0.6667		4	3	2	0.6000
	2	2	1	0.6667				3	0.4000
			2	0.3333		4	4	3	0.8000
								4	0.2000
4	1	1	0	0.7500					
			1	0.2500	6	1	1	0	0.8333
								1	0.1667
	2	1	0	0.5000					
			1	0.5000		2	1	0	0.6667
	2	2	0	0.1667				1	0.3333
			1	0.6667		2	2	0	0.4000
			2	0.1667				1	0.5333
								2	0.0667
	3	1	0	0.2500					
			1	0.7500		3	1	0	0.5000
	3	2	1	0.5000				1	0.5000
			2	0.5000		3	2	0	0.2000
	3	3	2	0.7500				1	0.6000
			3	0.2500				2	0.2000
						3	3	0	0.0500
5	1	1	0	0.8000				1	0.4500
			1	0.2000				2	0.4500
								3	0.0500
	2	1	0	0.6000					
			1	0.4000		4	1	0	0.3333
	2	2	0	0.3000				1	0.6667
			1	0.6000		4	2	0	0.0667
			2	0.1000				1	0.5333
								2	0.4000
	3	1	0	0.4000		4	3	1	0.2000
			1	0.6000				2	0.6000
	3	2	0	0.1000				3	0.2000
			1	0.6000		4	4	2	0.4000
			2	0.3000				3	0.5333
	3	3	1	0.3000				4	0.0667

Table T.2 (*continued*)

N	n	Np	k	$p_X(k)$	N	n	Np	k	$p_X(k)$
6	5	1	0	0.1667	7	5	1	0	0.2857
			1	0.8333				1	0.7143
	5	2	1	0.3333		5	2	0	0.0476
			2	0.6667				1	0.4762
	5	3	2	0.5000				2	0.4762
			3	0.5000		5	3	1	0.1429
	5	4	3	0.6667				2	0.5714
			4	0.3333				3	0.2857
	5	5	4	0.8333		5	4	2	0.2857
			5	0.1667				3	0.5714
								4	0.1429
7	1	1	0	0.8571		5	5	3	0.4762
			1	0.1429				4	0.4762
								5	0.0476
	2	1	0	0.7143					
			1	0.2857		6	1	0	0.1429
	2	2	0	0.4762				1	0.8571
			1	0.4762		6	2	1	0.2857
			2	0.0476				2	0.7143
						6	3	2	0.4286
	3	1	0	0.5714				3	0.5714
			1	0.4286		6	4	3	0.5714
	3	2	0	0.2857				4	0.4286
			1	0.5714		6	5	4	0.7143
			2	0.1429				5	0.2857
	3	3	0	0.1143		6	6	5	0.8571
			1	0.5143				6	0.1429
			2	0.3429					
			3	0.0286	8	1	1	0	0.8750
								1	0.1250
	4	1	0	0.4286					
			1	0.5714		2	1	0	0.7500
	4	2	0	0.1429				1	0.2500
	4	2	1	0.5714		2	2	0	0.5357
			2	0.2857				1	0.4286
	4	3	0	0.0286				2	0.0357
			1	0.3429					
			2	0.5143		3	1	0	0.6250
			3	0.1143				1	0.3750
	4	4	1	0.1143		3	2	0	0.3571
			2	0.5143				1	0.5357
			3	0.3429				2	0.1071
			4	0.0286		3	3	0	0.1786

Table T.2 (*continued*)

N	n	Np	k	$p_X(k)$	N	n	Np	k	$p_X(k)$
8	3	3	1	0.5357				2	0.5357
			2	0.2679				3	0.3571
			3	0.0179		6	4	2	0.2143
								3	0.5714
	4	1	0	0.5000				4	0.2143
			1	0.5000		6	5	3	0.3571
	4	2	0	0.2143				4	0.5357
			1	0.5714				5	0.1071
			2	0.2143		6	6	4	0.5357
	4	3	0	0.0714				5	0.4286
			1	0.4286				6	0.0357
			2	0.4286					
			3	0.0714	8	7	1	0	0.1250
	4	4	0	0.0143				1	0.8750
			1	0.2286		7	2	1	0.2500
			2	0.5143				2	0.7500
			3	0.2286		7	3	2	0.3750
			4	0.0143				3	0.6250
						7	4	3	0.5000
	5	1	0	0.3750				4	0.5000
			1	0.6250		7	5	4	0.6250
	5	2	0	0.1071				5	0.3750
			1	0.5357		7	6	5	0.7500
			2	0.3571				6	0.2500
	5	3	0	0.0179		7	7	6	0.8750
			1	0.2679				7	0.1250
			2	0.5357					
			3	0.1786	9	1	1	0	0.8889
	5	4	1	0.0714				1	0.1111
			2	0.4286					
			3	0.4286		2	1	0	0.7778
			4	0.0714				1	0.2222
	5	5	2	0.1786		2	2	0	0.5833
			3	0.5357				1	0.3889
			4	0.2679				2	0.0278
			5	0.0179					
						3	1	0	0.6667
	6	1	0	0.2500				1	0.3333
			1	0.7500		3	2	0	0.4167
	6	2	0	0.0357				1	0.5000
			1	0.4286				2	0.0833
			2	0.5357		3	3	0	0.2381
	6	3	1	0.1071				1	0.5357

Table T.2 (*continued*)

N	n	Np	k	$p_X(k)$	N	n	Np	k	$p_X(k)$
			2	0.2143				1	0.5000
			3	0.0119				2	0.4167
					9	6	3	0	0.0119
9	4	1	0	0.5556				1	0.2143
			1	0.4444				2	0.5357
	4	2	0	0.2778				3	0.2381
			1	0.5556		6	4	1	0.0476
			2	0.1667				2	0.3571
	4	3	0	0.1190				3	0.4762
			1	0.4762				4	0.1190
						6	5	2	0.1190
	4	3	2	0.3571				3	0.4762
			3	0.0476				4	0.3571
	4	4	0	0.0397				5	0.0476
			1	0.3175		6	6	3	0.2381
			2	0.4762				4	0.5357
			3	0.1587				5	0.2143
			4	0.0079				6	0.0119
	5	1	0	0.4444		7	1	0	0.2222
			1	0.5556				1	0.7778
	5	2	0	0.1667		7	2	0	0.0278
			1	0.5556					
			2	0.2778		7	2	1	0.3889
	5	3	0	0.0476				2	0.5833
			1	0.3571		7	3	1	0.0833
			2	0.4762				2	0.5000
			3	0.1190				3	0.4167
	5	4	0	0.0079		7	4	2	0.1667
			1	0.1587				3	0.5556
			2	0.4762				4	0.2778
			3	0.3175		7	5	3	0.2778
			4	0.0397				4	0.5556
	5	5	1	0.0397				5	0.1667
			2	0.3175		7	6	4	0.4167
			3	0.4762				5	0.5000
			4	0.1587				6	0.0833
			5	0.0079		7	7	5	0.5833
								6	0.3889
	6	1	0	0.3333				7	0.0278
			1	0.6667		8	1	0	0.1111
	6	2	0	0.0833				1	0.8889

Table T.2 (*continued*)

N	n	Np	k	$p_X(k)$	N	n	Np	k	$p_X(k)$
9	8	2	1	0.2222	9			5	0.4444
			2	0.7778		8	6	5	0.6667
	8	3	2	0.3333				6	0.3333
			3	0.6667		8	7	6	0.7778
	8	4	3	0.4444				7	0.2222
			4	0.5556		8	8	7	0.8889
	8	5	4	0.5556				8	0.1111

Table T.3 Individual Terms, $p_X(x)$, of the Probability Mass Function of the Poisson Distribution with Parameter λ

x	0.1	0.2	0.3	0.4	λ 0.5	0.6	0.7	0.8	0.9	1.0
0	.9048	.8187	.7408	.6703	.6065	.5488	.4966	.4493	.4066	.3679
1	.0905	.1637	.2222	.2681	.3033	.3293	.3476	.3595	.3659	.3679
2	.0045	.0164	.0333	.0536	.0758	.0988	.1217	.1438	.1647	.1839
3	.0002	.0011	.0033	.0072	.0126	.0198	.0284	.0383	.0494	.0613
4	.0000	.0001	.0003	.0007	.0016	.0030	.0050	.0077	.0111	.0153
5	.0000	.0000	.0000	.0001	.0002	.0004	.0007	.0012	.0020	.0031
6	.0000	.0000	.0000	.0000	.0000	.0000	.0001	.0002	.0003	.0005
7	.0000	.0000	.0000	.0000	.0000	.0000	.0000	.0000	.0000	.0001

x	1.1	1.2	1.3	1.4	λ 1.5	1.6	1.7	1.8	1.9	2.0
0	.3329	.3012	.2725	.2466	.2231	.2019	.1827	.1653	.1496	.1353
1	.3662	.3614	.3543	.3452	.3347	.3230	.3106	.2975	.2842	.2707
2	.2014	.2169	.2303	.2417	.2510	.2584	.2640	.2678	.2700	.2707
3	.0738	.0867	.0998	.1128	.1255	.1378	.1496	.1607	.1710	.1804
4	.0203	.0260	.0324	.0395	.0471	.0551	.0636	.0723	.0812	.0902
5	.0045	.0062	.0084	.0111	.0141	.0176	.0216	.0260	.0309	.0361
6	.0008	.0012	.0018	.0026	.0035	.0047	.0061	.0078	.0098	.0120
7	.0001	.0002	.0003	.0005	.0008	.0011	.0015	.0020	.0027	.0034
8	.0000	.0000	.0001	.0001	.0001	.0002	.0003	.0005	.0006	.0009
9	.0000	.0000	.0000	.0000	.0000	.0000	.0001	.0001	.0001	.0002

x	2.1	2.2	2.3	2.4	λ 2.5	2.6	2.7	2.8	2.9	3.0
0	.1225	.1108	.1003	.0907	.0821	.0743	.0672	.0608	.0550	.0498
1	.2572	.2438	.2306	.2177	.2052	.1931	.1815	.1703	.1596	.1494
2	.2700	.2681	.2652	.2613	.2565	.2510	.2450	.2384	.2314	.2240
3	.1890	.1966	.2033	.2090	.2138	.2176	.2205	.2225	.2237	.2240
4	.0992	.1082	.1169	.1254	.1336	.1414	.1488	.1557	.1622	.1680
5	.0417	.0476	.0538	.0602	.0668	.0735	.0804	.0872	.0940	.1008
6	.0146	.0174	.0206	.0241	.0278	.0319	.0362	.0407	.0455	.0504
7	.0044	.0055	.0068	.0083	.0099	.0118	.0139	.0163	.0188	.0216
8	.0011	.0015	.0019	.0025	.0031	.0038	.0047	.0057	.0068	.0081
9	.0003	.0004	.0005	.0007	.0009	.0011	.0014	.0018	.0022	.0027
10	.0001	.0001	.0001	.0002	.0002	.0003	.0004	.0005	.0006	.0008
11	.0000	.0000	.0000	.0000	.0000	.0001	.0001	.0001	.0002	.0002
12	.0000	.0000	.0000	.0000	.0000	.0000	.0000	.0000	.0000	.0001

x	3.1	3.2	3.3	3.4	λ 3.5	3.6	3.7	3.8	3.9	4.0
0	.0450	.0408	.0369	.0334	.0302	.0273	.0247	.0224	.0202	.0183
1	.1397	.1304	.1217	.1135	.1057	.0984	.0915	.0850	.0789	.0733
2	.2165	.2087	.2008	.1929	.1850	.1771	.1692	.1615	.1539	.1465
3	.2237	.2226	.2209	.2186	.2158	.2125	.2087	.2046	.2001	.1954
4	.1734	.1781	.1823	.1858	.1888	.1912	.1931	.1944	.1951	.1954
5	.1075	.1140	.1203	.1264	.1322	.1377	.1429	.1477	.1522	.1563
6	.0555	.0608	.0662	.0716	.0771	.0826	.0881	.0936	.0989	.1042
7	.0246	.0278	.0312	.0348	.0385	.0425	.0466	.0508	.0551	.0595
8	.0095	.0111	.0129	.0148	.0169	.0191	.0215	.0241	.0269	.0298
9	.0033	.0040	.0047	.0056	.0066	.0076	.0089	.0102	.0116	.0132

Table T.3 (*continued*)

x	3.1	3.2	3.3	3.4	λ 3.5	3.6	3.7	3.8	3.9	4.0
10	.0010	.0013	.0016	.0019	.0023	.0028	.0033	.0039	.0045	.0053
11	.0003	.0004	.0005	.0006	.0007	.0009	.0011	.0013	.0016	.0019
12	.0001	.0001	.0001	.0002	.0002	.0003	.0003	.0004	.0005	.0006
13	.0000	.0000	.0000	.0000	.0001	.0001	.0001	.0001	.0002	.0002
14	.0000	.0000	.0000	.0000	.0000	.0000	.0000	.0000	.0000	.0001

x	4.1	4.2	4.3	4.4	λ 4.5	4.6	4.7	4.8	4.9	5.0
0	.0166	.0150	.0136	.0123	.0111	.0101	.0091	.0082	.0074	.0067
1	.0679	.0630	.0583	.0540	.0500	.0462	.0427	.0395	.0365	.0337
2	.1393	.1323	.1254	.1188	.1125	.1063	.1005	.0948	.0894	.0842
3	.1904	.1852	.1798	.1743	.1687	.1631	.1574	.1517	.1460	.1404
4	.1951	.1944	.1933	.1917	.1898	.1875	.1849	.1820	.1789	.1755
5	.1600	.1633	.1662	.1687	.1708	.1725	.1738	.1747	.1753	.1755
6	.1093	.1143	.1191	.1237	.1281	.1323	.1362	.1398	.1432	.1462
7	.0640	.0686	.0732	.0778	.0824	.0869	.0914	.0959	.1002	.1044
8	.0328	.0360	.0393	.0428	.0463	.0500	.0537	.0575	.0614	.0653
9	.0150	.0168	.0188	.0209	.0232	.0255	.0280	.0307	.0334	.0363
10	.0061	.0071	.0081	.0092	.0104	.0118	.0132	.0147	.0164	.0181
11	.0023	.0027	.0032	.0037	.0043	.0049	.0056	.0064	.0073	.0082
12	.0008	.0009	.0011	.0014	.0016	.0019	.0022	.0026	.0030	.0034
13	.0002	.0003	.0004	.0005	.0006	.0007	.0008	.0009	.0011	.0013
14	.0001	.0001	.0001	.0001	.0002	.0002	.0003	.0003	.0004	.0005
15	.0000	.0000	.0000	.0000	.0001	.0001	.0001	.0001	.0001	.0002

x	5.1	5.2	5.3	5.4	λ 5.5	5.6	5.7	5.8	5.9	6.0
0	.0061	.0055	.0050	.0045	.0041	.0037	.0033	.0030	.0027	.0025
1	.0311	.0287	.0265	.0244	.0225	.0207	.0191	.0176	.0162	.0149
2	.0793	.0746	.0701	.0659	.0618	.0580	.0544	.0509	.0477	.0446
3	.1348	.1293	.1239	.1185	.1133	.1082	.1033	.0985	.0938	.0892
4	.1719	.1681	.1641	.1600	.1558	.1515	.1472	.1428	.1383	.1339
5	.1753	.1748	.1740	.1728	.1714	.1697	.1678	.1656	.1632	.1606
6	.1490	.1515	.1537	.1555	.1571	.1584	.1594	.1601	.1605	.1606
7	.1086	.1125	.1163	.1200	.1234	.1267	.1298	.1326	.1353	.1377
8	.0692	.0731	.0771	.0810	.0849	.0887	.0925	.0962	.0998	.1033
9	.0392	.0423	.0454	.0486	.0519	.0552	.0586	.0620	.0654	.0688
10	.0200	.0220	.0241	.0262	.0285	.0309	.0334	.0359	.0386	.0413
11	.0093	.0104	.0116	.0129	.0143	.0157	.0173	.0190	.0207	.0225
12	.0039	.0045	.0051	.0058	.0065	.0073	.0082	.0092	.0102	.0113
13	.0015	.0018	.0021	.0024	.0028	.0032	.0036	.0041	.0046	.0052
14	.0006	.0007	.0008	.0009	.0011	.0013	.0015	.0017	.0019	.0022
15	.0002	.0002	.0003	.0003	.0004	.0005	.0006	.0007	.0008	.0009
16	.0001	.0001	.0001	.0001	.0001	.0002	.0002	.0002	.0003	.0003
17	.0000	.0000	.0000	.0000	.0000	.0000	.0001	.0001	.0001	.0001

Table T.3 (*continued*)

x	6.1	6.2	6.3	6.4	6.5	6.6	6.7	6.8	6.9	7.0
0	.0022	.0020	.0018	.0017	.0015	.0014	.0012	.0011	.0010	.0009
1	.0137	.0126	.0116	.0106	.0098	.0090	.0082	.0076	.0070	.0064
2	.0417	.0390	.0364	.0340	.0318	.0296	.0276	.0258	.0240	.0223
3	.0848	.0806	.0765	.0726	.0688	.0652	.0617	.0584	.0552	.0521
4	.1294	.1249	.1205	.1162	.1118	.1076	.1034	.0992	.0952	.0912
5	.1579	.1549	.1519	.1487	.1454	.1420	.1385	.1349	.1314	.1277
6	.1605	.1601	.1595	.1586	.1575	.1562	.1546	.1529	.1511	.1490
7	.1399	.1418	.1435	.1450	.1462	.1472	.1480	.1486	.1489	.1490
8	.1066	.1099	.1130	.1160	.1188	.1215	.1240	.1263	.1284	.1304
9	.0723	.0757	.0791	.0825	.0858	.0891	.0923	.0954	.0985	.1014
10	.0441	.0469	.0498	.0528	.0558	.0588	.0618	.0649	.0679	.0710
11	.0245	.0265	.0285	.0307	.0330	.0353	.0377	.0401	.0426	.0452
12	.0124	.0137	.0150	.0164	.0179	.0194	.0210	.0227	.0245	.0264
13	.0058	.0065	.0073	.0081	.0089	.0098	.0108	.0119	.0130	.0142
14	.0025	.0029	.0033	.0037	.0041	.0046	.0052	.0058	.0064	.0071
15	.0010	.0012	.0014	.0016	.0018	.0020	.0023	.0026	.0029	.0033
16	.0004	.0005	.0005	.0006	.0007	.0008	.0010	.0011	.0013	.0014
17	.0001	.0002	.0002	.0002	.0003	.0003	.0004	.0004	.0005	.0006
18	.0000	.0001	.0001	.0001	.0001	.0001	.0001	.0002	.0002	.0002
19	.0000	.0000	.0000	.0000	.0000	.0000	.0000	.0001	.0001	.0001

x	7.1	7.2	7.3	7.4	7.5	7.6	7.7	7.8	7.9	8.0
0	.0008	.0007	.0007	.0006	.0006	.0005	.0005	.0004	.0004	.0003
1	.0059	.0054	.0049	.0045	.0041	.0038	.0035	.0032	.0029	.0027
2	.0208	.0194	.0180	.0167	.0156	.0145	.0134	.0125	.0116	.0107
3	.0492	.0464	.0438	.0413	.0389	.0366	.0345	.0324	.0305	.0286
4	.0874	.0836	.0799	.0764	.0729	.0696	.0663	.0632	.0602	.0573
5	.1241	.1204	.1167	.1130	.1094	.1057	.1021	.0986	.0951	.0916
6	.1468	.1445	.1420	.1394	.1367	.1339	.1311	.1282	.1252	.1221
7	.1489	.1486	.1481	.1474	.1465	.1454	.1442	.1428	.1413	.1396
8	.1321	.1337	.1351	.1363	.1373	.1382	.1388	.1392	.1395	.1396
9	.1042	.1070	.1096	.1121	.1144	.1167	.1187	.1207	.1224	.1241
10	.0740	.0770	.0800	.0829	.0858	.0887	.0914	.0941	.0967	.0993
11	.0478	.0504	.0531	.0558	.0585	.0613	.0640	.0667	.0695	.0722
12	.0283	.0303	.0323	.0344	.0366	.0388	.0411	.0434	.0457	.0481
13	.0154	.0168	.0181	.0196	.0211	.0227	.0243	.0260	.0278	.0296
14	.0078	.0086	.0095	.0104	.0113	.0123	.0134	.0145	.0157	.0169
15	.0037	.0041	.0046	.0051	.0057	.0062	.0069	.0075	.0083	.0090
16	.0016	.0019	.0021	.0024	.0026	.0030	.0033	.0037	.0041	.0045
17	.0007	.0008	.0009	.0010	.0012	.0013	.0015	.0017	.0019	.0021
18	.0003	.0003	.0004	.0004	.0005	.0006	.0006	.0007	.0008	.0009
19	.0001	.0001	.0001	.0002	.0002	.0002	.0003	.0003	.0003	.0004
20	.0000	.0000	.0001	.0001	.0001	.0001	.0001	.0001	.0001	.0002
21	.0000	.0000	.0000	.0000	.0000	.0000	.0000	.0000	.0001	.0001

Table T.3 (*continued*)

x	8.1	8.2	8.3	8.4	8.5	8.6	8.7	8.8	8.9	9.0
0	.0003	.0003	.0002	.0002	.0002	.0002	.0002	.0002	.0001	.0001
1	.0025	.0023	.0021	.0019	.0017	.0016	.0014	.0013	.0012	.0011
2	.0100	.0092	.0086	.0079	.0074	.0068	.0063	.0058	.0054	.0050
3	.0269	.0252	.0237	.0222	.0208	.0195	.0183	.0171	.0160	.0150
4	.0544	.0517	.0491	.0466	.0443	.0420	.0398	.0377	.0357	.0337
5	.0882	.0849	.0816	.0784	.0752	.0722	.0692	.0663	.0635	.0607
6	.1191	.1160	.1128	.1097	.1066	.1034	.1003	.0972	.0941	.0911
7	.1378	.1358	.1338	.1317	.1294	.1271	.1247	.1222	.1197	.1171
8	.1395	.1392	.1388	.1382	.1375	.1366	.1356	.1344	.1332	.1318
9	.1256	.1269	.1280	.1290	.1299	.1306	.1311	.1315	.1317	.1318
10	.1017	.1040	.1063	.1084	.1104	.1123	.1140	.1157	.1172	.1186
11	.0749	.0776	.0802	.0828	.0853	.0878	.0902	.0925	.0948	.0970
12	.0505	.0530	.0555	.0579	.0604	.0629	.0654	.0679	.0703	.0728
13	.0315	.0334	.0354	.0374	.0395	.0416	.0438	.0459	.0481	.0504
14	.0182	.0196	.0210	.0225	.0240	.0256	.0272	.0289	.0306	.0324
15	.0098	.0107	.0116	.0126	.0136	.0147	.0158	.0169	.0182	.0194
16	.0050	.0055	.0060	.0066	.0072	.0079	.0086	.0093	.0101	.0109
17	.0024	.0026	.0029	.0033	.0036	.0040	.0044	.0048	.0053	.0058
18	.0011	.0012	.0014	.0015	.0017	.0019	.0021	.0024	.0026	.0029
19	.0005	.0005	.0006	.0007	.0008	.0009	.0010	.0011	.0012	.0014
20	.0002	.0002	.0002	.0003	.0003	.0004	.0004	.0005	.0005	.0006
21	.0001	.0001	.0001	.0001	.0001	.0002	.0002	.0002	.0002	.0003
22	.0000	.0000	.0000	.0000	.0001	.0001	.0001	.0001	.0001	.0001

x	9.1	9.2	9.3	9.4	9.5	9.6	9.7	9.8	9.9	10
0	.0001	.0001	.0001	.0001	.0001	.0001	.0001	.0001	.0001	.0000
1	.0010	.0009	.0009	.0008	.0007	.0007	.0006	.0005	.0005	.0005
2	.0046	.0043	.0040	.0037	.0034	.0031	.0029	.0027	.0025	.0023
3	.0140	.0131	.0123	.0115	.0107	.0100	.0093	.0087	.0081	.0076
4	.0319	.0302	.0285	.0269	.0254	.0240	.0226	.0213	.0201	.0189
5	.0581	.0555	.0530	.0506	.0483	.0460	.0439	.0418	.0398	.0378
6	.0881	.0851	.0822	.0793	.0764	.0736	.0709	.0682	.0656	.0631
7	.1145	.1118	.1091	.1064	.1037	.1010	.0982	.0955	.0928	.0901
8	.1302	.1286	.1269	.1251	.1232	.1212	.1191	.1170	.1148	.1126
9	.1317	.1315	.1311	.1306	.1300	.1293	.1284	.1274	.1263	.1251
10	.1198	.1210	.1219	.1228	.1235	.1241	.1245	.1249	.1250	.1251
11	.0991	.1012	.1031	.1049	.1067	.1083	.1098	.1112	.1125	.1137
12	.0752	.0776	.0799	.0822	.0844	.0866	.0888	.0908	.0928	.0948
13	.0526	.0549	.0572	.0594	.0617	.0640	.0662	.0685	.0707	.0729
14	.0342	.0361	.0380	.0399	.0419	.0439	.0459	.0479	.0500	.0521
15	.0208	.0221	.0235	.0250	.0265	.0281	.0297	.0313	.0330	.0347
16	.0118	.0127	.0137	.0147	.0157	.0168	.0180	.0192	.0204	.0217
17	.0063	.0069	.0075	.0081	.0088	.0095	.0103	.0111	.0119	.0128
18	.0032	.0035	.0039	.0042	.0046	.0051	.0055	.0060	.0065	.0071
19	.0015	.0017	.0019	.0021	.0023	.0026	.0028	.0031	.0034	.0037

Table T.3 (continued)

x	9.1	9.2	9.3	9.4	9.5 λ	9.6	9.7	9.8	9.9	10
20	.0007	.0008	.0009	.0010	.0011	.0012	.0014	.0015	.0017	.0019
21	.0003	.0003	.0004	0004	.0005	.0006	.0006	.0007	.0008	.0009
22	.0001	.0001	.0002	.0002	.0002	.0002	.0003	.0003	.0004	.0004
23	.0000	.0001	.0001	.0001	.0001	.0001	.0001	.0001	.0002	.0002
24	.0000	.0000	.0000	.0000	.0000	.0000	.0000	.0001	.0001	.0001

x	11	12	13	14	15 λ	16	17	18	19	20
0	.0000	.0000	.0000	.0000	.0000	.0000	.0000	.0000	.0000	.0000
1	.0002	.0001	.0000	.0000	.0000	.0000	.0000	.0000	.0000	.0000
2	.0010	.0004	.0002	.0001	.0000	.0000	.0000	.0000	.0000	.0000
3	.0037	.0018	.0008	.0004	.0002	.0001	.0000	.0000	.0000	.0000
4	.0102	.0053	.0027	.0013	.0006	.0003	.0001	.0001	.0000	.0000
5	.0224	.0127	.0070	.0037	.0019	.0010	.0005	.0002	.0001	.0001
6	.0411	.0255	.0152	.0087	.0048	.0026	.0014	.0007	.0004	.0002
7	.0646	.0437	.0281	.0174	.0104	.0060	.0034	.0018	.0010	.0005
8	.0888	.0655	.0457	.0304	.0194	.0120	.0072	.0042	.0024	.0013
9	.1085	.0874	.0661	.0473	.0324	.0213	.0135	.0083	.0050	.0029
10	.1194	.1048	.0859	.0663	.0486	.0341	.0230	.0150	.0095	.0058
11	.1194	.1144	.1015	.0844	.0663	.0496	.0355	.0245	.0164	.0106
12	.1094	.1144	.1099	.0984	.0829	.0661	.0504	.0368	.0259	.0176
13	.0926	.1056	.1099	.1060	.0956	.0814	.0658	.0509	.0378	.0271
14	.0728	.0905	.1021	.1060	.1024	.0930	.0800	.0655	.0514	.0387
15	.0534	.0724	.0885	.0989	.1024	.0992	.0906	.0786	.0650	.0516
16	.0367	.0543	.0719	.0866	.0960	.0992	.0963	.0884	.0772	.0646
17	.0237	.0383	.0550	.0713	.0847	.0934	.0963	.0936	.0863	.0760
18	.0145	.0256	.0397	.0554	.0706	.0830	.0909	.0936	.0911	.0844
19	.0084	.0161	.0272	.0409	.0557	.0699	.0814	.0887	.0911	.0888
20	.0046	.0097	.0177	.0286	.0418	.0559	.0692	.0798	.0866	.0888
21	.0024	.0055	.0109	.0191	.0299	.0426	.0560	.0684	.0783	.0846
22	.0012	.0030	.0065	.0121	.0204	.0310	.0433	.0560	.0676	.0769
23	.0006	.0016	.0037	.0074	.0133	.0216	.0320	.0438	.0559	.0669
24	.0003	.0008	.0020	.0043	.0083	.0144	.0226	.0328	.0442	.0557
25	.0001	.0004	.0010	.0024	.0050	.0092	.0154	.0237	.0336	.0446
26	.0000	.0002	.0005	.0013	.0029	.0057	.0101	.0164	.0246	.0343
27	.0000	.0001	.0002	.0007	.0016	.0034	.0063	.0109	.0173	.0254
28	.0000	.0000	.0001	.0003	.0009	.0019	.0038	.0070	.0117	.0181
29	.0000	.0000	.0001	.0002	.0004	.0011	.0023	.0044	.0077	.0125
30	.0000	.0000	.0000	.0001	.0002	.0006	.0013	.0026	.0049	.0083
31	.0000	.0000	.0000	.0000	.0001	.0003	.0007	.0015	.0030	.0054
32	.0000	.0000	.0000	.0000	.0001	.0001	.0004	.0009	.0018	.0034
33	.0000	.0000	.0000	.0000	.0000	.0001	.0002	.0005	.0010	.0020
34	.0000	.0000	.0000	.0000	.0000	.0000	.0001	.0002	.0006	.0012
35	.0000	.0000	.0000	.0000	.0000	.0000	.0000	.0001	.0003	.0007
36	.0000	.0000	.0000	.0000	.0000	.0000	.0000	.0001	.0002	.0004
37	.0000	.0000	.0000	.0000	.0000	.0000	.0000	.0000	.0001	.0002
38	.0000	.0000	.0000	.0000	.0000	.0000	.0000	.0000	.0000	.0001
39	.0000	.0000	.0000	.0000	.0000	.0000	.0000	.0000	.0000	.0001

SOURCE: From *Handbook of Tables for Probability and Statistics*, 2nd ed., William H. Beyer, ed. (Cleveland: The Chemical Rubber Co.) pp. 206–211. Reprinted by permission of The Chemical Rubber Co.

Table T.4 Individual Terms, $p_Z(k) = p(1-p)^{k-1}$, of the Probability Mass Function of a Random Variable z Having a Geometric Distribution

p \ k	1	2	3	4	5	6	7	8	9	10	11	12	13	14	15
0.10	.100	.090	.081	.073	.066	.059	.053	.048	.043	.039	.035	.031	.028	.025	.023
0.12	.120	.106	.093	.082	.072	.063	.056	.049	.043	.038	.033	.029	.026	.023	.020
0.14	.140	.120	.104	.089	.077	.066	.057	.049	.042	.036	.031	.027	.023	.020	.017
0.16	.160	.134	.113	.095	.080	.067	.056	.047	.040	.033	.028	.024	.020	.017	.014
0.18	.180	.148	.121	.099	.081	.067	.055	.045	.037	.030	.025	.020	.017	.014	.011
0.20	.200	.160	.128	.102	.082	.066	.052	.042	.034	.027	.021	.017	.014	.011	.009
0.22	.220	.172	.134	.104	.082	.064	.050	.039	.030	.024	.018	.014	.011	.009	.007
0.24	.240	.182	.139	.105	.080	.061	.046	.035	.027	.020	.015	.012	.009	.007	.005
0.26	.260	.192	.142	.105	.078	.058	.043	.032	.023	.017	.013	.009	.007	.005	.004
0.28	.280	.202	.145	.105	.075	.054	.039	.028	.020	.015	.010	.008	.005	.004	.003
0.30	.300	.210	.147	.103	.072	.050	.035	.025	.017	.012	.008	.006	.004	.003	.002
0.32	.320	.218	.148	.101	.068	.047	.032	.022	.015	.010	.007	.005	.003	.002	.001
0.34	.340	.224	.148	.098	.065	.043	.028	.019	.012	.008	.005	.004	.002	.002	.001
0.36	.360	.230	.147	.094	.060	.039	.025	.016	.010	.006	.004	.003	.002	.001	.001
0.38	.380	.236	.146	.091	.056	.035	.022	.013	.008	.005	.003	.002	.001	.001	.001
0.40	.400	.240	.144	.086	.052	.031	.019	.011	.007	.004	.002	.001	.001	.001	
0.42	.420	.244	.141	.082	.048	.028	.016	.009	.005	.003	.002	.001	.001		
0.44	.440	.246	.138	.077	.043	.024	.014	.008	.004	.002	.001	.001			
0.46	.460	.248	.134	.072	.039	.021	.011	.006	.003	.002	.001	.001			
0.48	.480	.250	.130	.067	.035	.018	.009	.005	.003	.001	.001				
0.50	.500	.250	.125	.063	.031	.016	.008	.004	.002	.001	.001				
0.52	.520	.250	.120	.058	.028	.013	.006	.003	.002	.001					
0.54	.540	.248	.114	.053	.024	.011	.005	.002	.001						
0.56	.560	.246	.108	.048	.021	.009	.004	.002	.001						
0.58	.580	.244	.102	.043	.018	.008	.003	.001	.001						
0.60	.600	.240	.096	.038	.015	.006	.002	.001							
0.62	.620	.236	.090	.034	.013	.005	.002	.001							
0.64	.640	.230	.083	.030	.011	.004	.001	.001							
0.66	.660	.224	.076	.026	.009	.003	.001								
0.68	.680	.218	.070	.022	.007	.002	.001								
0.70	.700	.210	.063	.019	.006	.002	.001								
0.72	.720	.202	.056	.016	.004	.001									
0.74	.740	.192	.050	.013	.003	.001									
0.76	.760	.182	.044	.011	.003	.001									
0.78	.780	.172	.038	.008	.002										
0.80	.800	.160	.032	.006	.001										
0.82	.820	.148	.027	.005	.001										
0.84	.840	.134	.022	.003	.001										
0.86	.860	.120	.017	.002											
0.88	.880	.106	.013	.002											
0.90	.900	.090	.009	.001											

Table T.4 (*continued*)

p \ k	16	17	18	19	20	21	22	23	24	25	26	27	28	29	30
0.10	.021	.019	.017	.015	.014	.012	.011	.010	.009	.008	.007	.006	.006	.005	.005
0.12	.018	.016	.014	.012	.011	.009	.008	.007	.006	.006	.005	.004	.004	.003	.003
0.14	.015	.013	.011	.009	.008	.007	.006	.005	.004	.004	.003	.003	.002	.002	.002
0.16	.012	.010	.008	.007	.006	.005	.004	.003	.003	.002	.002	.002	.001	.001	.001
0.18	.009	.008	.006	.005	.004	.003	.003	.002	.002	.002	.001	.001	.001		
0.20	.007	.006	.005	.004	.003	.002	.002	.001	.001	.001					
0.22	.005	.004	.003	.002	.002	.002	.001	.001							
0.24	.004	.003	.002	.002	.001	.001									
0.26	.003	.002	.002	.001	.001										
0.28	.002	.001	.001	.001											
0.30	.001	.001	.001												
0.32	.001	.001													
0.34	.001														

Table T.5. Values of $p_Y(k)$ for the negative binomial distribution with $r = 2, 3, 4, 5$ and $p = 0.2, 0.4, 0.5, 0.6, 0.8$.

p \ k	r = 2				
	0.2	0.4	0.5	0.6	0.8
0	0.0400	0.1600	0.2500	0.3600	0.6400
1	0.0640	0.1920	0.2500	0.2880	0.2560
2	0.0768	0.1728	0.1875	0.1728	0.0768
3	0.0819	0.1382	0.1250	0.0922	0.0205
4	0.0819	0.1037	0.0781	0.0461	0.0051
5	0.0786	0.0746	0.0469	0.0221	0.0012
6	0.0734	0.0523	0.0273	0.0103	0.0003
7	0.0671	0.0358	0.0156	0.0047	0.0001
8	0.0604	0.0242	0.0088	0.0021	
9	0.0537	0.0161	0.0049	0.0009	
10	0.0472	0.0106	0.0027	0.0004	
11	0.0412	0.0070	0.0015	0.0002	
12	0.0357	0.0045	0.0008	0.0001	
13	0.0308	0.0029	0.0004		
14	0.0264	0.0019	0.0002		
15	0.0225	0.0012	0.0001		
16	0.0191	0.0008	0.0001		
17	0.0162	0.0005			
18	0.0137	0.0003			
19	0.0115	0.0002			
20	0.0097	0.0001			
21	0.0081	0.0001			
22	0.0068				
23	0.0057				
24	0.0047				
25	0.0039				
26	0.0033				
27	0.0027				
28	0.0022				
29	0.0019				
30	0.0015				

Table T.5 (*continued*)

k \ p	r = 3				
	0.2	0.4	0.5	0.6	0.8
0	0.0080	0.0640	0.1250	0.2160	0.5120
1	0.0192	0.1152	0.1875	0.2592	0.3072
2	0.0307	0.1382	0.1875	0.2074	0.1229
3	0.0410	0.1382	0.1562	0.1382	0.0410
4	0.0492	0.1244	0.1172	0.0829	0.0123
5	0.0551	0.1045	0.0820	0.0464	0.0034
6	0.0587	0.0836	0.0547	0.0248	0.0009
7	0.0604	0.0645	0.0352	0.0127	0.0002
8	0.0604	0.0484	0.0220	0.0064	0.0001
9	0.0591	0.0355	0.0134	0.0031	
10	0.0567	0.0255	0.0081	0.0015	
11	0.0536	0.0181	0.0048	0.0007	
12	0.0500	0.0127	0.0028	0.0003	
13	0.0462	0.0088	0.0016	0.0002	
14	0.0422	0.0060	0.0009	0.0001	
15	0.0383	0.0041	0.0005		
16	0.0345	0.0028	0.0003		
17	0.0308	0.0019	0.0002		
18	0.0274	0.0012	0.0001		
19	0.0242	0.0008	0.0001		
20	0.0213	0.0005			
21	0.0187	0.0004			
22	0.0163	0.0002			
23	0.0142	0.0002			
24	0.0123	0.0001			
25	0.0106	0.0001			
26	0.0091				
27	0.0079				
28	0.0067				
29	0.0058				
30	0.0049				

Table T.5 (*continued*)

p k	\multicolumn{5}{c}{r = 4}				
	0.2	0.4	0.5	0.6	0.8
0	0.0016	0.0256	0.0625	0.1296	0.4096
1	0.0051	0.0614	0.1250	0.2074	0.3277
2	0.0102	0.0922	0.1562	0.2074	0.1638
3	0.0164	0.1106	0.1562	0.1659	0.0655
4	0.0229	0.1161	0.1367	0.1161	0.0229
5	0.0294	0.1115	0.1094	0.0743	0.0073
6	0.0352	0.1003	0.0820	0.0446	0.0022
7	0.0403	0.0860	0.0586	0.0255	0.0006
8	0.0443	0.0709	0.0403	0.0140	0.0002
9	0.0472	0.0568	0.0269	0.0075	
10	0.0491	0.0443	0.0175	0.0039	
11	0.0500	0.0338	0.0111	0.0020	
12	0.0500	0.0254	0.0069	0.0010	
13	0.0493	0.0187	0.0043	0.0005	
14	0.0479	0.0136	0.0026	0.0002	
15	0.0459	0.0098	0.0016	0.0001	
16	0.0436	0.0070	0.0009	0.0001	
17	0.0411	0.0049	0.0005		
18	0.0383	0.0035	0.0003		
19	0.0355	0.0024	0.0002		
20	0.0327	0.0017	0.0001		
21	0.0299	0.0011			
22	0.0272	0.0008			
23	0.0246	0.0005			
24	0.0221	0.0004			
25	0.0198	0.0002			
26	0.0177	0.0002			
27	0.0157	0.0001			
28	0.0139	0.0001			
29	0.0123				
30	0.0108				

Table T.5 (*continued*)

k \ p	r = 5				
	0.2	0.4	0.5	0.6	0.8
0	0.0003	0.0102	0.0312	0.0778	0.3277
1	0.0013	0.0307	0.0781	0.1555	0.3277
2	0.0031	0.0553	0.1172	0.1866	0.1966
3	0.0057	0.0774	0.1367	0.1742	0.0918
4	0.0092	0.0929	0.1367	0.1393	0.0367
5	0.0132	0.1003	0.1230	0.1003	0.0132
6	0.0176	0.1003	0.1025	0.0669	0.0044
7	0.0221	0.0946	0.0806	0.0420	0.0014
8	0.0266	0.0851	0.0604	0.0252	0.0004
9	0.0307	0.0738	0.0436	0.0146	0.0001
10	0.0344	0.0620	0.0305	0.0082	
11	0.0375	0.0507	0.0208	0.0045	
12	0.0400	0.0406	0.0139	0.0024	
13	0.0419	0.0318	0.0091	0.0012	
14	0.0431	0.0246	0.0058	0.0006	
15	0.0436	0.0187	0.0037	0.0003	
16	0.0436	0.0140	0.0023	0.0002	
17	0.0431	0.0104	0.0014	0.0001	
18	0.0422	0.0076	0.0009		
19	0.0408	0.0055	0.0005		
20	0.0392	0.0040	0.0003		
21	0.0373	0.0028			
22	0.0353	0.0020			
23	0.0332	0.0014			
24	0.0309	0.0010			
25	0.0287	0.0007			
26	0.0265	0.0005			
27	0.0243	0.0003			
28	0.0223	0.0002			
29	0.0203	0.0002			
30	0.0184	0.0001			

Table T.6 Table of the Cumulative Distribution Function of a Standard Random Variable

k	.0	.01	.02	.03	.04	.05	.06	.07	.08	.09
0.0	.50000	.50399	.50798	.51197	.51595	.51994	.52392	.52790	.53188	.53586
0.1	.53983	.54380	.54776	.55172	.55567	.55962	.56356	.56749	.57142	.57535
0.2	.57926	.58317	.58706	.59095	.59483	.59871	.60257	.60642	.61026	.61409
0.3	.61791	.62172	.62552	.62930	.63307	.63683	.64058	.64431	.64803	.65173
0.4	.65542	.65910	.66276	.66640	.67003	.67364	.67724	.68082	.68439	.68793
0.5	.69146	.69497	.69847	.70194	.70540	.70884	.71226	.71566	.71904	.72240
0.6	.72575	.72907	.73237	.73565	.73891	.74215	.74537	.74857	.75175	.75490
0.7	.75804	.76115	.76424	.76730	.77035	.77337	.77637	.77935	.78230	.78524
0.8	.78814	.79103	.79389	.79673	.79955	.80234	.80511	.80785	.81057	.81327
0.9	.81594	.81859	.82121	.82381	.82639	.82894	.83147	.83398	.83646	.83891
1.0	.84134	.84375	.84614	.84849	.85083	.85314	.85543	.85769	.85993	.86214
1.1	.86433	.86650	.86864	.87076	.87286	.87493	.87698	.87900	.88100	.88298
1.2	.88493	.88686	.88877	.89065	.89251	.89435	.89617	.89796	.89973	.90147
1.3	.90320	.90490	.90658	.90824	.90988	.91149	.91309	.91466	.91621	.91774
1.4	.91924	.92073	.92220	.92364	.92507	.92647	.92785	.92922	.93056	.93189
1.5	.93319	.93448	.93574	.93699	.93822	.93943	.94062	.94179	.94295	.94408
1.6	.94520	.94630	.94738	.94845	.94950	.95053	.95154	.95254	.95352	.95449
1.7	.95543	.95637	.95728	.95818	.95907	.95994	.96080	.96164	.96246	.96327
1.8	.96407	.96485	.96562	.96638	.96712	.96784	.96856	.96926	.96995	.97062
1.9	.97128	.97193	.97257	.97320	.97381	.97441	.97500	.97558	.97615	.97670
2.0	.97725	.97778	.97831	.97882	.97932	.97982	.98030	.98077	.98124	.98169
2.1	.98214	.98257	.98300	.98341	.98382	.98422	.98461	.98500	.98537	.98574
2.2	.98610	.98645	.98679	.98713	.98745	.98778	.98809	.98840	.98870	.98899
2.3	.98928	.98956	.98983	.99010	.99036	.99061	.99086	.99111	.99134	.99158
2.4	.99180	.99202	.99224	.99245	.99266	.99286	.99305	.99324	.99343	.99361
2.5	.99379	.99396	.99413	.99430	.99446	.99461	.99477	.99492	.99506	.99520
2.6	.99534	.99547	.99560	.99573	.99585	.99598	.99609	.99621	.99632	.99643
2.7	.99653	.99664	.99674	.99683	.99693	.99702	.99711	.99720	.99728	.99736
2.8	.99744	.99752	.99760	.99767	.99774	.99781	.99788	.99795	.99801	.99807
2.9	.99813	.99819	.99825	.99831	.99836	.99841	.99846	.99851	.99856	.99861
3.0	.99865	.99869	.99874	.99878	.99882	.99886	.99889	.99893	.99896	.99900
3.1	.99903	.99906	.99910	.99913	.99916	.99918	.99921	.99924	.99926	.99929
3.2	.99931	.99934	.99936	.99938	.99940	.99942	.99944	.99946	.99948	.99950
3.3	.99952	.99953	.99955	.99957	.99958	.99960	.99961	.99962	.99964	.99965
3.4	.99966	.99968	.99969	.99970	.99971	.99972	.99973	.99974	.99975	.99976
3.5	.99977	.99978	.99978	.99979	.99980	.99981	.99981	.99982	.99983	.99983
3.6	.99984	.99985	.99985	.99986	.99986	.99987	.99987	.99988	.99988	.99989
3.7	.99989	.99990	.99990	.99990	.99991	.99991	.99992	.99992	.99992	.99992
3.8	.99993	.99993	.99993	.99994	.99994	.99994	.99994	.99995	.99995	.99995
3.9	.99995	.99995	.99996	.99996	.99996	.99996	.99996	.99996	.99997	.99997
4.0	.99997	.99997	.99997	.99997	.99997	.99997	.99998	.99998	.99998	.99998

Table T.7 Tail Probabilities, $P\{\mathcal{N}(Z;0,1) > k\}$, for the Standard Normal Distribution

k	.0	.01	.02	.03	.04	.05	.06	.07	.08	.09
0.0	.50000	.49601	.49202	.48803	.48405	.48006	.47608	.47210	.46812	.46414
0.1	.46017	.45620	.45224	.44828	.44433	.44038	.43644	.43251	.42858	.42465
0.2	.42074	.41683	.41294	.40905	.40517	.40129	.39743	.39358	.38974	.38591
0.3	.38209	.37828	.37448	.37070	.36693	.36317	.35942	.35569	.35197	.34827
0.4	.34458	.34090	.33724	.33360	.32997	.32636	.32276	.31918	.31561	.31207
0.5	.30854	.30503	.30153	.29806	.29460	.29116	.28774	.28434	.28096	.27760
0.6	.27425	.27093	.26763	.26435	.26109	.25785	.25463	.25143	.24825	.24510
0.7	.24196	.23885	.23576	.23270	.22965	.22663	.22363	.22065	.21770	.21476
0.8	.21186	.20897	.20611	.20327	.20045	.19766	.19489	.19215	.18943	.18673
0.9	.18406	.18141	.17879	.17619	.17361	.17106	.16853	.16602	.16354	.16109
1.0	.15866	.15625	.15386	.15151	.14917	.14686	.14457	.14231	.14007	.13786
1.1	.13567	.13350	.13136	.12924	.12714	.12507	.12302	.12100	.11900	.11702
1.2	.11507	.11314	.11123	.10935	.10749	.10565	.10383	.10204	.10027	.09853
1.3	.09680	.09510	.09342	.09176	.09012	.08851	.08691	.08534	.08379	.08226
1.4	.08076	.07927	.07780	.07636	.07493	.07353	.07215	.07078	.06944	.06811
1.5	.06681	.06552	.06426	.06301	.06178	.06057	.05938	.05821	.05705	.05592
1.6	.05480	.05370	.05262	.05155	.05050	.04947	.04846	.04746	.04648	.04551
1.7	.04457	.04363	.04272	.04182	.04093	.04006	.03920	.03836	.03754	.03673
1.8	.03593	.03515	.03438	.03362	.03288	.03216	.03144	.03074	.03005	.02938
1.9	.02872	.02807	.02743	.02680	.02619	.02559	.02500	.02442	.02385	.02330
2.0	.02275	.02222	.02169	.02118	.02068	.02018	.01970	.01923	.01876	.01831
2.1	.01786	.01743	.01700	.01659	.01618	.01578	.01539	.01500	.01463	.01426
2.2	.01390	.01355	.01321	.01287	.01255	.01222	.01191	.01160	.01130	.01101
2.3	.01072	.01044	.01017	.00990	.00964	.00939	.00914	.00889	.00866	.00842
2.4	.00820	.00798	.00776	.00755	.00734	.00714	.00695	.00676	.00657	.00639
2.5	.00621	.00604	.00587	.00570	.00554	.00539	.00523	.00508	.00494	.00480
2.6	.00466	.00453	.00440	.00427	.00415	.00402	.00391	.00379	.00368	.00357
2.7	.00347	.00336	.00326	.00317	.00307	.00298	.00289	.00280	.00272	.00264
2.8	.00256	.00248	.00240	.00233	.00226	.00219	.00212	.00205	.00199	.00193
2.9	.00187	.00181	.00175	.00169	.00164	.00159	.00154	.00149	.00144	.00139
3.0	.00135	.00131	.00126	.00122	.00118	.00114	.00111	.00107	.00104	.00100
3.1	.00097	.00094	.00090	.00087	.00084	.00082	.00079	.00076	.00074	.00071
3.2	.00069	.00066	.00064	.00062	.00060	.00058	.00056	.00054	.00052	.00050
3.3	.00048	.00047	.00045	.00043	.00042	.00040	.00039	.00038	.00036	.00035
3.4	.00034	.00032	.00031	.00030	.00029	.00028	.00027	.00026	.00025	.00024
3.5	.00023	.00022	.00022	.00021	.00020	.00019	.00019	.00018	.00017	.00017
3.6	.00016	.00015	.00015	.00014	.00014	.00013	.00013	.00012	.00012	.00011
3.7	.00011	.00010	.00010	.00010	.00009	.00009	.00008	.00008	.00008	.00008
3.8	.00007	.00007	.00007	.00006	.00006	.00006	.00006	.00005	.00005	.00005
3.9	.00005	.00005	.00004	.00004	.00004	.00004	.00004	.00004	.00003	.00003
4.0	.00003	.00003	.00003	.00003	.00003	.00003	.00002	.00002	.00002	.00002

Table T.8. Values of the exponential function, e^{-x}, for negative arguments.

	0.00	0.01	0.02	0.03	0.04	0.05	0.06	0.07	0.08	0.09
0.0	1.00000	0.99005	0.98020	0.97045	0.96079	0.95123	0.94176	0.93239	0.92312	0.91393
0.1	0.90484	0.89583	0.88692	0.87810	0.86936	0.86071	0.85214	0.84366	0.83527	0.82696
0.2	0.81873	0.81058	0.80252	0.79453	0.78663	0.77880	0.77105	0.76338	0.75578	0.74826
0.3	0.74082	0.73345	0.72615	0.71892	0.71177	0.70469	0.69768	0.69073	0.68386	0.67706
0.4	0.67032	0.66365	0.65705	0.65051	0.64404	0.63763	0.63128	0.62500	0.61878	0.61263
0.5	0.60653	0.60050	0.59452	0.58860	0.58275	0.57695	0.57121	0.56553	0.55990	0.55433
0.6	0.54881	0.54335	0.53794	0.53259	0.52729	0.52204	0.51685	0.51171	0.50662	0.50158
0.7	0.49659	0.49164	0.48675	0.48191	0.47711	0.47237	0.46767	0.46301	0.45841	0.45384
0.8	0.44933	0.44486	0.44043	0.43605	0.43171	0.42741	0.42316	0.41895	0.41478	0.41066
0.9	0.40657	0.40252	0.39852	0.39455	0.39063	0.38674	0.38289	0.37908	0.37531	0.37158

	0.0	0.1	0.2	0.3	0.4	0.5	0.6	0.7	0.8	0.9
1.0	0.36788	0.33287	0.30119	0.27253	0.24660	0.22313	0.20190	0.18268	0.16530	0.14957
2.0	0.13534	0.12246	0.11080	0.10026	0.09072	0.08208	0.07427	0.06721	0.06081	0.05502
3.0	0.04979	0.04505	0.04076	0.03688	0.03337	0.03020	0.02732	0.02472	0.02237	0.02024
4.0	0.01832	0.01657	0.01500	0.01357	0.01228	0.01111	0.01005	0.00910	0.00823	0.00745
5.0	0.00674	0.00610	0.00552	0.00499	0.00452	0.00409	0.00370	0.00335	0.00303	0.00274
6.0	0.00248	0.00224	0.00203	0.00184	0.00166	0.00150	0.00136	0.00123	0.00111	0.00101

Table T.9 Values of the incomplete gamma function $I_r(\tau)$ for use in the computation of the cumulative gamma distribution function.

τ \ r	1	2	3	4	5
0.2	0.18127	0.01752	0.00115	0.00006	0.00000
0.4	0.32968	0.06155	0.00793	0.00078	0.00006
0.6	0.45119	0.12190	0.02312	0.00336	0.00039
0.8	0.55067	0.19121	0.04742	0.00908	0.00141
1.0	0.63212	0.26424	0.08030	0.01899	0.00366
1.2	0.69881	0.33737	0.12051	0.03377	0.00775
1.4	0.75340	0.40817	0.16650	0.05372	0.01425
1.6	0.79810	0.47507	0.21664	0.07881	0.02368
1.8	0.83470	0.53716	0.26938	0.10871	0.03641
2.0	0.86466	0.59399	0.32332	0.14288	0.05265
2.2	0.88920	0.64543	0.37729	0.18065	0.07250
2.4	0.90928	0.69156	0.43029	0.22128	0.09587
2.6	0.92573	0.73262	0.48157	0.26400	0.12258
2.8	0.93919	0.76892	0.53055	0.30806	0.15232
3.0	0.95021	0.80085	0.57681	0.35277	0.18474
3.2	0.95924	0.82880	0.62010	0.39748	0.21939
3.4	0.96663	0.85316	0.66026	0.44164	0.25582
3.6	0.97268	0.87431	0.69725	0.48478	0.29356
3.8	0.97763	0.89262	0.73110	0.52652	0.33216
4.0	0.98168	0.90842	0.76190	0.56653	0.37116
4.2	0.98500	0.92202	0.78976	0.60460	0.41017
4.4	0.98772	0.93370	0.81486	0.64055	0.44882
4.6	0.98995	0.94371	0.83736	0.67429	0.48677
4.8	0.99177	0.95227	0.85746	0.70577	0.52374
5.0	0.99326	0.95957	0.87535	0.73497	0.55951
5.2	0.99448	0.96580	0.89121	0.76193	0.59387
5.4	0.99548	0.97109	0.90524	0.78671	0.62669
5.6	0.99630	0.97559	0.91761	0.80938	0.65785
5.8	0.99697	0.97941	0.92849	0.83004	0.68728
6.0	0.99752	0.98265	0.93803	0.84880	0.71494
6.2	0.99797	0.98539	0.94638	0.86577	0.74082
6.4	0.99834	0.98770	0.95368	0.88108	0.76493
6.6	0.99864	0.98966	0.96003	0.89485	0.78730
6.8	0.99889	0.99131	0.96556	0.90719	0.80797
7.0	0.99909	0.99270	0.97036	0.91823	0.82701
7.2	0.99925	0.99388	0.97453	0.92808	0.84448
7.4	0.99939	0.99487	0.97813	0.93685	0.86047
7.6	0.99950	0.99570	0.98124	0.94463	0.87506
7.8	0.99959	0.99639	0.98393	0.95152	0.88833
8.0	0.99966	0.99698	0.98625	0.95762	0.90037

Table T.9 (*continued*)

τ \ r	1	2	3	4	5
8.5	0.99980	0.99807	0.99072	0.96989	0.92564
9.0	0.99988	0.99877	0.99377	0.97877	0.94504
9.5	0.99993	0.99921	0.99584	0.98514	0.95974
10.0	0.99995	0.99950	0.99723	0.98966	0.97075
10.5	0.99997	0.99968	0.99817	0.99285	0.97891
11.0	0.99998	0.99980	0.99879	0.99508	0.98490
11.5	0.99999	0.99987	0.99920	0.99664	0.98925
12.0	0.99999	0.99992	0.99948	0.99771	0.99240
12.5	1.00000	0.99995	0.99966	0.99845	0.99465
13.0	1.00000	0.99997	0.99978	0.99895	0.99626
13.5	1.00000	0.99998	0.99986	0.99929	0.99740
14.0	1.00000	0.99999	0.99991	0.99953	0.99819
14.5	1.00000	0.99999	0.99994	0.99968	0.99875
15.0	1.00000	1.00000	0.99996	0.99979	0.99914

τ \ r	6	7	8	9	10
1.0	0.00059	0.00008	0.00001		
1.2	0.00150	0.00025	0.00004	0.00000	
1.4	0.00320	0.00062	0.00011	0.00002	
1.6	0.00604	0.00134	0.00026	0.00005	0.00001
1.8	0.01038	0.00257	0.00056	0.00011	0.00002
2.0	0.01656	0.00453	0.00110	0.00024	0.00005
2.2	0.02491	0.00746	0.00198	0.00047	0.00010
2.4	0.03567	0.01159	0.00334	0.00086	0.00020
2.6	0.04904	0.01717	0.00533	0.00149	0.00038
2.8	0.06511	0.02441	0.00813	0.00243	0.00066
3.0	0.08392	0.03351	0.01190	0.00380	0.00110
3.2	0.10541	0.04462	0.01683	0.00571	0.00176
3.4	0.12946	0.05785	0.02307	0.00829	0.00271
3.6	0.15588	0.07327	0.03079	0.01167	0.00402
3.8	0.18444	0.09089	0.04011	0.01598	0.00580
4.0	0.21487	0.11067	0.05113	0.02136	0.00813
4.2	0.24686	0.13254	0.06394	0.02793	0.01113
4.4	0.28009	0.15635	0.07858	0.03580	0.01489
4.6	0.31424	0.18197	0.09505	0.04507	0.01953
4.8	0.34899	0.20920	0.11333	0.05582	0.02514
5.0	0.38404	0.23782	0.13337	0.06809	0.03183
5.2	0.41909	0.26761	0.15508	0.08193	0.03967

Table T.9 (*continued*)

τ \ r	6	7	8	9	10
5.4	0.45387	0.29833	0.17834	0.09735	0.04875
5.6	0.48814	0.32974	0.20302	0.11432	0.05913
5.8	0.52169	0.36161	0.22897	0.13281	0.07084
6.0	0.55432	0.39370	0.25602	0.15276	0.08392
6.2	0.58589	0.42579	0.28398	0.17409	0.09838
6.4	0.61626	0.45767	0.31268	0.19669	0.11420
6.6	0.64533	0.48916	0.34192	0.22044	0.13136
6.8	0.67302	0.52008	0.37151	0.24523	0.14982
7.0	0.69929	0.55029	0.40129	0.27091	0.16950
7.2	0.72410	0.57964	0.43106	0.29733	0.19035
7.4	0.74744	0.60804	0.46067	0.32435	0.21226
7.6	0.76932	0.63538	0.48996	0.35181	0.23515
7.8	0.78975	0.66159	0.51879	0.37956	0.25889
8.0	0.80876	0.68663	0.54704	0.40745	0.28338
8.5	0.85040	0.74382	0.61440	0.47689	0.34703
9.0	0.88431	0.79322	0.67610	0.54435	0.41259
9.5	0.91147	0.83505	0.73134	0.60818	0.47817
10.0	0.93291	0.86986	0.77978	0.66718	0.54207
10.5	0.94962	0.89837	0.82149	0.72059	0.60287
11.0	0.96248	0.92139	0.85681	0.76801	0.65949
11.5	0.97227	0.93973	0.88627	0.80941	0.71121
12.0	0.97966	0.95418	0.91050	0.84497	0.75761
12.5	0.98518	0.96543	0.93017	0.87508	0.79857
13.0	0.98927	0.97411	0.94597	0.90024	0.83419
13.5	0.99227	0.98075	0.95852	0.92100	0.86474
14.0	0.99447	0.98577	0.96838	0.93794	0.89060
14.5	0.99606	0.98955	0.97606	0.95162	0.91224
15.0	0.99721	0.99237	0.98200	0.96255	0.93015
15.5	0.99803	0.99446	0.98654	0.97121	0.94481
16.0	0.99862	0.99599	0.99000	0.97801	0.95670
16.5	0.99903	0.99712	0.99261	0.98331	0.96626
17.0	0.99933	0.99794	0.99457	0.98741	0.97388

τ \ r	11	12	13	14	15
4.0	0.00284	0.00091	0.00027	0.00008	0.00002
4.5	0.00667	0.00240	0.00081	0.00025	0.00007
5.0	0.01370	0.00545	0.00202	0.00070	0.00023
5.5	0.02525	0.01099	0.00445	0.00169	0.00060

Table T.9 (*continued*)

τ \ r	11	12	13	14	15
6.0	0.04262	0.02009	0.00883	0.00363	0.00140
6.5	0.06684	0.03388	0.01603	0.00710	0.00296
7.0	0.09852	0.05335	0.02700	0.01281	0.00572
7.5	0.13776	0.07924	0.04267	0.02156	0.01026
8.0	0.18411	0.11192	0.06380	0.03418	0.01726
8.5	0.23664	0.15134	0.09092	0.05141	0.02743
9.0	0.29401	0.19699	0.12423	0.07385	0.04147
9.2	0.31797	0.21682	0.13926	0.08438	0.05999
9.4	0.34236	0.23743	0.15524	0.09581	0.05590
9.6	0.36705	0.25876	0.17212	0.10815	0.06428
9.8	0.39195	0.28072	0.18988	0.12139	0.07346
10.0	0.41696	0.30322	0.20844	0.13554	0.08346
10.2	0.44197	0.32618	0.22777	0.15055	0.09429
10.4	0.46687	0.34951	0.24779	0.16641	0.10596
10.6	0.49159	0.37310	0.26843	0.18309	0.11847
10.8	0.51603	0.39687	0.28963	0.20054	0.11318
11.0	0.54011	0.42073	0.31130	0.21871	0.14596
11.2	0.56376	0.44459	0.33337	0.23756	0.16090
11.4	0.58690	0.46837	0.35576	0.25702	0.17661
11.6	0.60949	0.49198	0.37839	0.27703	0.19305
11.8	0.63146	0.51535	0.40117	0.29754	0.21019
12.0	0.65277	0.53840	0.42403	0.31846	0.22798
12.2	0.67338	0.56108	0.44690	0.33974	0.24637
12.4	0.69327	0.58331	0.46968	0.36130	0.26531
12.6	0.71239	0.60504	0.49232	0.38307	0.28474
12.8	0.73075	0.62623	0.51475	0.40498	0.30462
13.0	0.74832	0.64684	0.53690	0.42696	0.32487
13.2	0.76510	0.66681	0.55870	0.44893	0.34543
13.4	0.78108	0.68614	0.58012	0.47084	0.36625
13.6	0.79627	0.70478	0.60110	0.49262	0.38725
13.8	0.81068	0.72273	0.62158	0.51421	0.40838
14.0	0.82432	0.73996	0.64154	0.53555	0.42956
14.2	0.83720	0.75647	0.66094	0.55659	0.45075
14.4	0.84934	0.77225	0.67975	0.57728	0.47188
14.6	0.86076	0.78731	0.69793	0.59756	0.49289
14.8	0.87149	0.80164	0.71549	0.61741	0.51373
15.0	0.88154	0.81525	0.73239	0.63678	0.53435
15.5	0.90388	0.84622	0.77173	0.68292	0.58459
16.0	0.92260	0.87301	0.80688	0.72549	0.63247
16.5	0.93813	0.89593	0.83790	0.76426	0.67746
17.0	0.95088	0.91533	0.86498	0.79913	0.71917

Table T.9 (*continued*)

τ \ r	11	12	13	14	15
17.5	0.96126	0.93160	0.88835	0.83013	0.75736
18.0	0.96963	0.94511	0.90833	0.85740	0.79192
18.5	0.97635	0.95624	0.92525	0.88114	0.82286
19.0	0.98168	0.96533	0.93944	0.90160	0.85025
19.5	0.98589	0.97269	0.95125	0.91908	0.87427
20.0	0.98919	0.97861	0.96099	0.93387	0.89514
20.5	0.99176	0.98335	0.96897	0.94630	0.91310
21.0	0.99375	0.98710	0.97545	0.95664	0.92843
21.5	0.99528	0.99005	0.98069	0.96520	0.94141
22.0	0.99645	0.99237	0.98488	0.97222	0.95231
22.5	0.99735	0.99418	0.98823	0.97794	0.96140
23.0	0.99802	0.99557	0.99088	0.98257	0.96893
23.5	0.99853	0.99665	0.99297	0.98630	0.97512
24.0	0.99892	0.99748	0.99460	0.98928	0.98018
24.5	0.99920	0.99811	0.99587	0.99166	0.98428

Table T.10. Values of the incomplete beta function to aid in the graph of the cumulative beta distribution for $r = 0.5, 1, 2, 3, 4, 5$ and $s = 2, 3, 4,$ and 5.

	\multicolumn{6}{c}{$s = 2$}					
y \ r	0.5	1	2	3	4	5
0.05	32982	09750	00725	00048	00006	00000
0.10	45853	19000	02800	00370	00046	00006
0.15	55190	27750	06075	01198	00223	00040
0.20	62610	36000	10400	02720	00672	00160
0.25	68750	43750	15626	05078	01562	00464
0.30	73943	51000	21600	08370	03078	01094
0.35	78388	57750	28175	12648	05402	02232
0.40	82219	64000	35200	17920	08704	04096
0.45	85530	69750	42525	24148	13122	06920
0.50	88388	75000	50000	31250	18750	10938
0.55	90848	79750	57475	39098	25622	16357
0.60	92952	84000	64800	47520	33696	23328
0.65	94732	87750	71825	56298	42842	31908
0.70	96216	91000	78400	65170	52822	42018
0.75	97428	93750	84375	73828	63281	53394
0.80	98387	96000	89600	81920	73728	65536
0.85	99110	97750	93925	89048	83521	77648
0.90	99612	99000	97200	94770	91854	88574
0.95	99905	99750	99275	98598	97741	96723

	\multicolumn{6}{c}{$s = 3$}					
y \ r	0.5	1	2	3	4	5
0.05	40550	14262	01402	00116	00009	00001
0.10	55458	27100	05230	00856	00127	00018
0.15	65683	38588	10952	02661	00589	00122
0.20	73343	48800	18080	05792	01696	00467
0.25	79297	57812	26172	10352	03760	01288
0.30	84007	65700	34830	16308	07047	02880
0.35	87761	72538	43702	23517	11742	05561
0.40	90757	78400	52480	31744	17920	09626
0.45	93139	83362	60902	40687	25526	15293
0.50	95017	87500	68750	50000	34375	22656
0.55	96480	90888	75852	59313	44152	31644
0.60	97599	93600	82080	68256	54432	41990
0.65	98435	95712	87352	76483	64709	53228
0.70	99040	97300	91630	83692	74431	64707
0.75	99458	98438	94922	89648	83057	75641
0.80	99729	99200	97280	94208	90112	85197
0.85	99888	99662	98802	97339	95266	92623
0.90	99968	99900	99630	99144	98415	97431
0.95	99996	99988	99952	99884	99777	99624

Table T.10 (*continued*)

$s = 4$

y \ r	0.5	1	2	3	4	5
0.05	46539	18549	02259	00223	00019	00004
0.10	62663	34390	08146	01585	00273	00043
0.15	73116	47799	16479	04734	01210	00285
0.20	80498	59040	26272	09888	03334	01041
0.25	85889	68359	36719	16943	07056	02730
0.30	89878	75990	47178	25569	12604	05797
0.35	92839	82149	57158	35291	19985	10609
0.40	95026	87040	66304	45568	28979	17367
0.45	96627	90849	74378	55848	39171	26038
0.50	97780	93750	81250	65625	50000	36328
0.55	98592	95899	86878	74474	60829	47696
0.60	99148	97440	91296	82080	71021	59409
0.65	99515	98499	94598	88258	80015	70640
0.70	99746	99190	96922	92953	87396	80590
0.75	99880	99609	98438	96240	92944	88618
0.80	99952	99840	99328	98304	96666	94372
0.85	99985	99949	99777	99411	98790	97865
0.90	99997	99990	99954	99873	99727	99498
0.95	1.00000	99999	99997	99991	99981	99963

$s = 5$

y \ r	0.5	1	2	3	4	5
0.05	51521	22622	03277	00376	00037	00003
0.10	68336	40951	11426	02569	00502	00089
0.15	78644	55629	22352	07377	02135	00563
0.20	85507	67232	34464	14803	05628	01958
0.25	90215	76270	46606	24359	11382	04893
0.30	93474	83193	57982	35293	19410	09881
0.35	95726	88397	68092	46772	29360	17172
0.40	97268	92224	76672	58010	40591	26657
0.45	98305	94967	83643	68356	52304	37858
0.50	98988	96875	89062	77344	63672	50000
0.55	99424	98155	93080	84707	73962	62142
0.60	99691	98976	95904	90374	82633	73343
0.65	99846	99475	97768	94439	89391	82828
0.70	99931	99757	98906	97120	94203	90119
0.75	99973	99902	99536	98712	97270	95107
0.80	99991	99968	99840	99533	98959	98042
0.85	99998	99992	99960	99878	99715	99437
0.90	1.00000	99999	99994	99982	99957	99911
0.95	1.00000	1.00000	1.00000	99999	99996	99997

Table T.11. Values of the Probability Content of an Upper Quadrant, $P\{X > h, Y > k\}$, for the Standard Bivariate Normal Distribution with Correlation ρ.

$$\rho = 0.500$$

h \ k	0.0	0.1	0.2	0.3	0.4	0.5	0.6	0.7	0.8
0.0	333333	312961	291886	270344	248589	226878	205468	184605	164512
0.1	312961	294422	275161	255392	235345	215260	195377	175927	157126
0.2	291886	275161	257709	239718	221397	202965	184644	166650	149190
0.3	270344	255392	239718	223488	206888	190114	173370	156858	140769
0.4	248589	235345	221397	206888	191979	176847	161676	146649	131946
0.5	226878	215260	202965	190114	176847	163320	149694	136139	122816
0.6	205468	195377	184644	173370	161676	149694	137570	125451	113486
0.7	184605	175927	166650	156858	146649	136139	125451	114718	104071
0.8	164512	157126	149190	140769	131946	122816	113486	104071	094686
0.9	145388	139168	132448	125281	117733	109882	101819	093640	085448
1.0	127398	122215	116586	110550	104159	097477	090578	083546	076465
1.1	110671	106398	101733	096704	091350	085722	079882	073896	067839
1.2	095297	091814	087990	083844	079407	074718	069825	064784	059656
1.3	081329	078522	075421	072042	068404	064539	060485	056284	051988
1.4	068785	066546	064061	061337	058388	055237	051913	048451	044891
1.5	057646	055882	053912	051740	049377	046836	044142	041320	038401
1.6	047867	046493	044950	043238	041365	039340	037180	034905	032539
1.7	039379	038321	037126	035793	034325	032729	031018	029204	027307
1.8	032095	031290	030375	029348	028211	026968	025627	024198	022695
1.9	025912	025307	024615	023834	022963	022006	020968	019854	018677
2.0	020724	020274	019756	019169	018510	017782	016987	016130	015218
2.1	016417	016087	015704	015268	014776	014228	013627	012974	012276
2.2	012882	012642	012363	012043	011679	011272	010823	010332	009804
2.3	010011	009840	009638	009406	009141	008842	008510	008145	007750
2.4	007706	007585	007441	007275	007083	006867	006624	006357	006065
2.5	005875	005790	005689	005571	005435	005280	005105	004911	004698
2.6	004436	004377	004307	004225	004129	004019	003894	003755	003602
2.7	003317	003277	003229	003172	003105	003029	002941	002842	002733
2.8	002457	002430	002397	002358	002312	002259	002198	002130	002053
2.9	001802	001784	001762	001736	001705	001669	001627	001579	001526
3.0	001309	001297	001283	001265	001244	001220	001192	001159	001123
3.1	000942	000934	000925	000913	000899	000883	000864	000842	000817
3.2	000671	000666	000660	000652	000644	000633	000620	000606	000589
3.3	000473	000470	000466	000462	000456	000449	000441	000431	000420
3.4	000331	000329	000326	000323	000320	000315	000310	000304	000297
3.5	000229	000228	000226	000224	000222	000219	000216	000212	000207
3.6	000157	000156	000155	000154	000153	000151	000149	000146	000143
3.7	000107	000106	000106	000105	000104	000103	000102	000100	000098
3.8	000072	000071	000071	000071	000070	000069	000069	000068	000066
3.9	000048	000048	000047	000047	000047	000046	000046	000045	000045
4.0	000031	000031	000031	000031	000031	000031	000030	000030	000030

Table T.11 (*continued*)

$$\rho = 0.500$$

h \ k	0.9	1.0	1.1	1.2	1.3	1.4	1.5	1.6	1.7
0.0	145388	127398	110671	095297	081329	068785	057646	047867	039379
0.1	139168	122215	106398	091814	078522	066546	055882	046493	038321
0.2	132448	116586	101733	087990	075421	064061	053912	044950	037126
0.3	125281	110550	096704	083844	072042	061337	051740	043238	035793
0.4	117733	104159	091350	079407	068404	058388	049377	041365	034325
0.5	109882	097477	085722	074718	064539	055237	046836	039340	032729
0.6	101819	090578	079882	069825	060485	051913	044142	037180	031018
0.7	093640	083546	073896	064784	056284	048451	041320	034905	029204
0.8	085448	076465	067839	059656	051988	044891	038401	032539	027307
0.9	077344	069426	061786	054504	047649	041275	035421	030110	025349
1.0	069426	062514	055812	049394	043323	037651	032418	027648	023353
1.1	061786	055812	049991	044388	039062	034063	029428	025184	021345
1.2	054504	049394	044388	039545	034920	030556	026490	022749	019349
1.3	047649	043323	039062	034920	030942	027170	023639	020373	017391
1.4	041275	037651	034063	030556	027170	023944	020907	018085	015494
1.5	035421	032418	029428	026490	023639	020907	018323	015909	013681
1.6	030110	027648	025184	022749	020373	018085	015909	013864	011969
1.7	025349	023353	021345	019349	017391	015494	013681	011969	010372
1.8	021134	019534	017915	016297	014701	013146	011652	010233	008902
1.9	017447	016178	014888	013591	012304	011044	009826	008663	007566
2.0	014260	013266	012249	011221	010196	009186	008204	007261	006367
2.1	011539	010769	009977	009171	008363	007563	006780	006024	005304
2.2	009242	008654	008043	007420	006790	006163	005546	004947	004373
2.3	007328	006883	006419	005941	005457	004971	004490	004021	003568
2.4	005751	005418	005069	004708	004339	003968	003598	003234	002882
2.5	004468	004222	003962	003692	003415	003134	002852	002574	002303
2.6	003435	003255	003065	002865	002659	002449	002237	002027	001820
2.7	002613	002484	002346	002200	002049	001894	001736	001579	001424
2.8	001968	001876	001777	001672	001562	001449	001333	001217	001102
2.9	001467	001402	001332	001257	001178	001097	001013	000928	000843
3.0	001082	001037	000988	000935	000879	000821	000761	000700	000639
3.1	000789	000759	000725	000688	000649	000608	000566	000522	000478
3.2	000570	000549	000526	000501	000474	000446	000416	000385	000354
3.3	000408	000394	000378	000361	000343	000323	000303	000281	000260
3.4	000288	000279	000269	000257	000245	000232	000218	000203	000188
3.5	000202	000196	000189	000181	000173	000164	000155	000145	000135
3.6	000140	000136	000132	000127	000121	000115	000109	000102	000096
3.7	000096	000093	000091	000087	000084	000080	000076	000072	000067
3.8	000065	000064	000062	000060	000057	000055	000052	000049	000046
3.9	000044	000043	000042	000040	000039	000037	000036	000034	000032
4.0	000029	000028	000028	000027	000026	000025	000024	000023	000022

Table T.11 (*continued*)

$$\rho = 0.500$$

h\k	1.8	1.9	2.0	2.1	2.2	2.3	2.4	2.5	2.6
0.0	032095	025912	020724	016417	012882	010011	007706	005875	004436
0.1	031290	025307	020274	016087	012642	009840	007585	005790	004377
0.2	030375	024615	019756	015704	012363	009638	007441	005689	004307
0.3	029348	023834	019169	015268	012043	009406	007275	005571	004225
0.4	028211	022963	018510	014776	011679	009141	007083	005435	004129
0.5	026968	022006	017782	014228	011272	008842	006867	005280	004019
0.6	025627	020968	016987	013627	010823	008510	006624	005105	003894
0.7	024198	019854	016130	012974	010332	008145	006357	004911	003755
0.8	022695	018677	015218	012276	009804	007750	006065	004698	003602
0.9	021134	017447	014260	011539	009242	007328	005751	004468	003435
1.0	019534	016178	013266	010769	008654	006883	005418	004222	003255
1.1	017915	014888	012249	009977	008043	006419	005069	003962	003065
1.2	016297	013591	011221	009171	007420	005941	004708	003692	002865
1.3	014701	012304	010196	008363	006790	005457	004339	003415	002659
1.4	013146	011044	009186	007563	006163	004971	003968	003134	002449
1.5	011652	009826	008204	006780	005546	004490	003598	002852	002237
1.6	010233	008663	007261	006024	004947	004021	003234	002574	002027
1.7	008902	007566	006367	005304	004373	003568	002882	002303	001820
1.8	007671	006546	005531	004626	003830	003138	002544	002041	001620
1.9	006546	005608	004758	003996	003322	002733	002225	001793	001429
2.0	005531	004758	004053	003418	002853	002358	001928	001560	001249
2.1	004626	003996	003418	002895	002427	002014	001654	001344	001081
2.2	003830	003322	002853	002427	002043	001703	001405	001146	000926
2.3	003138	002733	002358	002014	001703	001426	001181	000968	000785
2.4	002544	002225	001928	001654	001405	001181	000983	000809	000659
2.5	002041	001793	001560	001344	001146	000968	000809	000669	000548
2.6	001620	001429	001249	001081	000926	000785	000659	000548	000451
2.7	001273	001127	000989	000859	000740	000630	000532	000444	000367
2.8	000989	000879	000775	000676	000584	000500	000424	000355	000295
2.9	000760	000678	000600	000526	000457	000393	000334	000282	000235
3.0	000578	000518	000460	000405	000353	000305	000261	000221	000185
3.1	000434	000391	000349	000308	000270	000234	000201	000171	000144
3.2	000323	000292	000262	000232	000204	000178	000154	000131	000111
3.3	000238	000216	000194	000173	000153	000134	000116	000099	000084
3.4	000173	000157	000142	000127	000113	000099	000086	000075	000064
3.5	000124	000114	000103	000093	000083	000073	000064	000055	000047
3.6	000088	000081	000074	000067	000060	000053	000047	000040	000035
3.7	000062	000057	000052	000048	000043	000038	000034	000029	000025
3.8	000043	000040	000037	000033	000030	000027	000024	000021	000018
3.9	000030	000028	000025	000023	000021	000019	000017	000015	000013
4.0	000020	000019	000017	000016	000015	000013	000012	000010	000009

Table T.11 (*continued*)

$$\rho = 0.500$$

h\k	2.7	2.8	2.9	3.0	3.1	3.2	3.3	3.4	3.5
0.0	003317	002457	001802	001309	000942	000671	000473	000331	000229
0.1	003277	002430	001784	001297	000934	000666	000470	000329	000228
0.2	003229	002397	001762	001283	000925	000660	000466	000326	000226
0.3	003172	002358	001736	001265	000913	000652	000462	000323	000224
0.4	003105	002312	001705	001244	000899	000644	000456	000320	000222
0.5	003029	002259	001669	001220	000883	000633	000449	000315	000219
0.6	002941	002198	001627	001192	000864	000620	000441	000310	000216
0.7	002842	002130	001579	001159	000842	000606	000431	000304	000212
0.8	002733	002053	001526	001123	000817	000589	000420	000297	000207
0.9	002613	001968	001467	001082	000789	000570	000408	000288	000202
1.0	002484	001876	001402	001037	000759	000549	000394	000279	000196
1.1	002346	001777	001332	000988	000725	000526	000378	000269	000189
1.2	002200	001672	001257	000935	000688	000501	000361	000257	000181
1.3	002049	001562	001178	000879	000649	000474	000343	000245	000173
1.4	001894	001449	001097	000821	000608	000446	000323	000232	000164
1.5	001736	001333	001013	000761	000566	000416	000303	000218	000155
1.6	001579	001217	000928	000700	000522	000385	000281	000203	000145
1.7	001424	001102	000843	000639	000478	000354	000260	000188	000135
1.8	001273	000989	000760	000578	000434	000323	000238	000173	000124
1.9	001127	000879	000678	000518	000391	000292	000216	000157	000114
2.0	000989	000775	000600	000460	000349	000262	000194	000142	000103
2.1	000859	000676	000526	000405	000308	000232	000173	000127	000093
2.2	000740	000584	000457	000353	000270	000204	000153	000113	000083
2.3	000630	000500	000393	000305	000234	000178	000134	000099	000073
2.4	000532	000424	000334	000261	000201	000154	000116	000086	000064
2.5	000444	000355	000282	000221	000171	000131	000099	000075	000055
2.6	000367	000295	000235	000185	000144	000111	000084	000064	000047
2.7	000300	000242	000194	000153	000120	000093	000071	000054	000040
2.8	000242	000197	000158	000126	000099	000077	000059	000045	000034
2.9	000194	000158	000128	000102	000081	000063	000049	000037	000028
3.0	000153	000126	000102	000082	000065	000051	000040	000030	000023
3.1	000120	000099	000081	000065	000052	000041	000032	000025	000019
3.2	000093	000077	000063	000051	000041	000032	000025	000020	000015
3.3	000071	000059	000049	000040	000032	000025	000020	000016	000012
3.4	000054	000045	000037	000030	000025	000020	000016	000012	000009
3.5	000040	000034	000028	000023	000019	000015	000012	000009	000007
3.6	000030	000025	000021	000017	000014	000011	000009	000007	000006
3.7	000022	000018	000015	000013	000011	000009	000007	000005	000004
3.8	000016	000013	000011	000009	000008	000006	000005	000004	000003
3.9	000011	000010	000008	000007	000006	000005	000004	000003	000002
4.0	000008	000007	000006	000005	000004	000003	000003	000002	000002

Table T.11 (*continued*)

$$\rho = 0.500$$

h \ k	3.6	3.7	3.8	3.9	4.0
0.0	000157	000107	000072	000048	000031
0.1	000156	000106	000071	000048	000031
0.2	000155	000106	000071	000047	000031
0.3	000154	000105	000071	000047	000031
0.4	000153	000104	000070	000047	000031
0.5	000151	000103	000069	000046	000031
0.6	000149	000102	000069	000046	000030
0.7	000146	000100	000068	000045	000030
0.8	000143	000098	000066	000045	000030
0.9	000140	000096	000065	000044	000029
1.0	000136	000093	000064	000043	000028
1.1	000132	000091	000062	000042	000028
1.2	000127	000087	000060	000040	000027
1.3	000121	000084	000057	000039	000026
1.4	000115	000080	000055	000037	000025
1.5	000109	000076	000052	000036	000024
1.6	000102	000072	000049	000034	000023
1.7	000096	000067	000046	000032	000022
1.8	000088	000062	000043	000030	000020
1.9	000081	000057	000040	000028	000019
2.0	000074	000052	000037	000025	000017
2.1	000067	000048	000033	000023	000016
2.2	000060	000043	000030	000021	000015
2.3	000053	000038	000027	000019	000013
2.4	000047	000034	000024	000017	000012
2.5	000040	000029	000021	000015	000010
2.6	000035	000025	000018	000013	000009
2.7	000030	000022	000016	000011	000008
2.8	000025	000018	000013	000010	000007
2.9	000021	000015	000011	000008	000006
3.0	000017	000013	000009	000007	000005
3.1	000014	000011	000008	000006	000004
3.2	000011	000009	000006	000005	000003
3.3	000009	000007	000005	000004	000003
3.4	000007	000005	000004	000003	000002
3.5	000006	000004	000003	000002	000002
3.6	000004	000003	000003	000002	000001
3.7	000003	000003	000002	000001	000001
3.8	000003	000002	000002	000001	000001
3.9	000002	000001	000001	000001	000001
4.0	000001	000001	000001	000001	000000

Table T.11 (*continued*)

$$\rho = -0.500$$

h\k	0.0	0.1	0.2	0.3	0.4	0.5	0.6	0.7	0.8
0.0	1666667	1472109	1288543	1117443	0959897	0816598	0687848	0573588	0473431
0.1	1472109	1295818	1130216	0976550	0835700	0708178	0594141	0493418	0405553
0.2	1288543	1130216	0982164	0845419	0720665	0608253	0508211	0420282	0343956
0.3	1117443	0976550	0845419	0724876	0615434	0517301	0430398	0354399	0288762
0.4	0959897	0835700	0720665	0615434	0520367	0435548	0360817	0295794	0239929
0.5	0816598	0708178	0608253	0517301	0435548	0362982	0299375	0244321	0197268
0.6	0687848	0594141	0508211	0430398	0360817	0299375	0245803	0199680	0160471
0.7	0573588	0493418	0420282	0354399	0295794	0244321	0199680	0161454	0129134
0.8	0473431	0405553	0343956	0288762	0239929	0197268	0160471	0129134	0102785
0.9	0386718	0329853	0278526	0232782	0192530	0157559	0127561	0102154	0080912
1.0	0312570	0265442	0223135	0185637	0152821	0124469	0100285	0079918	0062984
1.1	0249952	0211319	0176829	0146428	0119973	0097244	0077966	0061823	0048478
1.2	0197727	0166407	0138601	0114231	0093142	0075127	0059935	0047286	0036890
1.3	0154710	0129603	0107439	0088123	0071504	0057388	0045553	0035756	0027751
1.4	0119721	0099821	0082354	0067219	0054273	0043340	0034227	0026728	0020636
1.5	0091615	0076023	0062416	0050694	0040726	0032357	0025422	0019748	0015167
1.6	0069322	0057246	0046769	0037796	0030210	0023879	0018663	0014421	0011017
1.7	0051861	0042617	0034644	0027856	0022150	0017417	0013541	0010408	0007909
1.8	0038356	0031363	0025367	0020292	0016052	0012556	0009710	0007424	0005610
1.9	0028042	0022814	0018359	0014610	0011497	0008945	0006881	0005232	0003932
2.0	0020265	0016403	0013132	0010396	0008138	0006298	0004818	0003644	0002723
2.1	0014474	0011656	0009283	0007310	0005692	0004381	0003334	0002507	0001864
2.2	0010217	0008185	0006484	0005079	0003934	0003011	0002279	0001704	0001260
2.3	0007128	0005680	0004476	0003487	0002686	0002045	0001539	0001144	0000841
2.4	0004913	0003895	0003053	0002365	0001812	0001372	0001027	0000759	0000555
2.5	0003347	0002639	0002058	0001586	0001208	0000909	0000677	0000498	0000362
2.6	0002253	0001767	0001370	0001050	0000795	0000596	0000441	0000323	0000233
2.7	0001498	0001168	0000901	0000687	0000517	0000385	0000283	0000206	0000148
2.8	0000984	0000763	0000585	0000444	0000332	0000246	0000180	0000130	0000093
2.9	0000639	0000493	0000376	0000283	0000211	0000155	0000113	0000081	0000058
3.0	0000409	0000314	0000238	0000179	0000132	0000097	0000070	0000050	0000035
3.1	0000259	0000198	0000149	0000111	0000082	0000060	0000043	0000030	0000021
3.2	0000162	0000123	0000092	0000068	0000050	0000036	0000026	0000018	0000013
3.3	0000100	0000076	0000057	0000041	0000030	0000022	0000015	0000011	0000008
3.4	0000061	0000046	0000034	0000025	0000018	0000013	0000009	0000006	0000004
3.5	0000037	0000028	0000020	0000015	0000011	0000008	0000005	0000004	0000003
3.6	0000022	0000016	0000012	0000009	0000006	0000004	0000003	0000002	0000001
3.7	0000013	0000010	0000007	0000005	0000004	0000002	0000002	0000001	0000001
3.8	0000007	0000005	0000004	0000003	0000002	0000001	0000001	0000001	0000000
3.9	0000004	0000003	0000002	0000002	0000001	0000001	0000001	0000000	0000000
4.0	0000002	0000002	0000001	0000001	0000001	0000000	0000000	0000000	0000000

Table T.11 (*continued*)

$$\rho = -0.500$$

h \ k	0.9	1.0	1.1	1.2	1.3	1.4	1.5	1.6	1.7
0.0	0386718	0312570	0249952	0197727	0154710	0119721	0091615	0069322	0051861
0.1	0329853	0265442	0211319	0166407	0129603	0099821	0076023	0057246	0042617
0.2	0278526	0223135	0176829	0138601	0107439	0082354	0062416	0046769	0034644
0.3	0232782	0185637	0146428	0114231	0088123	0067219	0050694	0037796	0027856
0.4	0192530	0152821	0119973	0093142	0071504	0054273	0040726	0030210	0022150
0.5	0157559	0124469	0097244	0075127	0057388	0043340	0032357	0023879	0017417
0.6	0127561	0100285	0077966	0059935	0045553	0034227	0025422	0018663	0013541
0.7	0102154	0079918	0061823	0047286	0035756	0026728	0019748	0014421	0010408
0.8	0080912	0062984	0048478	0036890	0027751	0020636	0015167	0011017	0007909
0.9	0063376	0049085	0037587	0028455	0021294	0015751	0011515	0008319	0005940
1.0	0049085	0037823	0028814	0021699	0016152	0011884	0008641	0006209	0004409
1.1	0037587	0028814	0021836	0016357	0012111	0008863	0006409	0004580	0003234
1.2	0028455	0021699	0016357	0012188	0008975	0006532	0004698	0003339	0002345
1.3	0021294	0016152	0012111	0008975	0006574	0004758	0003403	0002405	0001680
1.4	0015751	0011884	0008863	0006532	0004758	0003425	0002436	0001712	0001189
1.5	0011515	0008641	0006409	0004698	0003403	0002436	0001723	0001204	0000831
1.6	0008319	0006209	0004580	0003339	0002405	0001712	0001204	0000837	0000574
1.7	0005940	0004409	0003234	0002345	0001680	0001189	0000831	0000574	0000392
1.8	0004190	0003093	0002257	0001627	0001159	0000815	0000567	0000389	0000264
1.9	0002921	0002144	0001556	0001115	0000790	0000553	0000383	0000261	0000176
2.0	0002012	0001469	0001059	0000755	0000532	0000370	0000254	0000173	0000116
2.1	0001369	0000994	0000713	0000505	0000354	0000245	0000167	0000113	0000075
2.2	0000920	0000664	0000474	0000334	0000232	0000160	0000108	0000073	0000048
2.3	0000611	0000439	0000311	0000218	0000151	0000103	0000070	0000046	0000031
2.4	0000401	0000286	0000201	0000140	0000097	0000066	0000044	0000029	0000019
2.5	0000260	0000184	0000129	0000089	0000061	0000042	0000028	0000018	0000012
2.6	0000166	0000117	0000082	0000056	0000039	0000026	0000017	0000011	0000007
2.7	0000105	0000074	0000051	0000035	0000024	0000016	0000010	0000007	0000004
2.8	0000066	0000046	0000031	0000021	0000014	0000010	0000006	0000004	0000003
2.9	0000040	0000028	0000019	0000013	0000009	0000006	0000004	0000002	0000002
3.0	0000025	0000017	0000012	0000008	0000005	0000003	0000002	0000001	0000001
3.1	0000015	0000010	0000007	0000005	0000003	0000002	0000001	0000001	0000001
3.2	0000009	0000006	0000004	0000003	0000002	0000001	0000001	0000000	0000000
3.3	0000005	0000003	0000002	0000002	0000001	0000001	0000000	0000000	0000000
3.4	0000003	0000002	0000001	0000001	0000001	0000000	0000000	0000000	0000000
3.5	0000002	0000001	0000001	0000000	0000000	0000000	0000000	0000000	0000000
3.6	0000001	0000001	0000000	0000000	0000000	0000000	0000000	0000000	0000000
3.7	0000001	0000000	0000000	0000000	0000000	0000000	0000000	0000000	0000000
3.8	0000000	0000000	0000000	0000000	0000000	0000000	0000000	0000000	0000000
3.9	0000000	0000000	0000000	0000000	0000000	0000000	0000000	0000000	0000000
4.0	0000000	0000000	0000000	0000000	0000000	0000000	0000000	0000000	0000000

Table T.11 (*continued*)

$$\rho = -0.500$$

h\k	1.8	1.9	2.0	2.1	2.2	2.3	2.4	2.5	2.6
0.0	0038356	0028042	0020265	0014474	0010217	0007128	0004913	0003347	0002253
0.1	0031363	0022814	0016403	0011656	0008185	0005680	0003895	0002639	0001767
0.2	0025367	0018359	0013132	0009283	0006484	0004476	0003053	0002058	0001370
0.3	0020292	0014610	0010396	0007310	0005079	0003487	0002365	0001586	0001050
0.4	0016052	0011497	0008138	0005692	0003934	0002686	0001812	0001208	0000795
0.5	0012556	0008945	0006298	0004381	0003011	0002045	0001372	0000909	0000596
0.6	0009710	0006881	0004818	0003334	0002279	0001539	0001027	0000677	0000441
0.7	0007424	0005232	0003644	0002507	0001704	0001144	0000759	0000498	0000323
0.8	0005610	0003932	0002723	0001864	0001260	0000841	0000555	0000362	0000233
0.9	0004190	0002921	0002012	0001369	0000920	0000611	0000401	0000260	0000166
1.0	0003093	0002144	0001469	0000994	0000664	0000439	0000286	0000184	0000117
1.1	0002257	0001556	0001059	0000713	0000474	0000311	0000201	0000129	0000082
1.2	0001627	0001115	0000755	0000505	0000334	0000218	0000140	0000089	0000056
1.3	0001159	0000790	0000532	0000354	0000232	0000151	0000097	0000061	0000039
1.4	0000815	0000553	0000370	0000245	0000160	0000103	0000066	0000042	0000026
1.5	0000567	0000383	0000254	0000167	0000108	0000070	0000044	0000028	0000017
1.6	0000389	0000261	0000173	0000113	0000073	0000046	0000029	0000018	0000011
1.7	0000264	0000176	0000116	0000075	0000048	0000031	0000019	0000012	0000007
1.8	0000177	0000117	0000077	0000050	0000032	0000020	0000012	0000008	0000005
1.9	0000117	0000077	0000050	0000032	0000020	0000013	0000008	0000005	0000003
2.0	0000077	0000050	0000032	0000021	0000013	0000008	0000005	0000003	0000002
2.1	0000050	0000032	0000021	0000013	0000008	0000005	0000003	0000002	0000001
2.2	0000032	0000020	0000013	0000008	0000005	0000003	0000002	0000001	0000001
2.3	0000020	0000013	0000008	0000005	0000003	0000002	0000001	0000001	0000000
2.4	0000012	0000008	0000005	0000003	0000002	0000001	0000001	0000000	0000000
2.5	0000008	0000005	0000003	0000002	0000001	0000001	0000000	0000000	0000000
2.6	0000005	0000003	0000002	0000001	0000001	0000000	0000000	0000000	0000000
2.7	0000003	0000002	0000001	0000001	0000000	0000000	0000000	0000000	0000000
2.8	0000002	0000001	0000001	0000000	0000000	0000000	0000000	0000000	0000000
2.9	0000001	0000001	0000000	0000000	0000000	0000000	0000000	0000000	0000000
3.0	0000001	0000000	0000000	0000000	0000000	0000000	0000000	0000000	0000000
3.1	0000000	0000000	0000000	0000000	0000000	0000000	0000000	0000000	0000000
3.2	0000000	0000000	0000000	0000000	0000000	0000000	0000000	0000000	0000000
3.3	0000000	0000000	0000000	0000000	0000000	0000000	0000000	0000000	0000000
3.4	0000000	0000000	0000000	0000000	0000000	0000000	0000000	0000000	0000000
3.5	0000000	0000000	0000000	0000000	0000000	0000000	0000000	0000000	0000000
3.6	0000000	0000000	0000000	0000000	0000000	0000000	0000000	0000000	0000000
3.7	0000000	0000000	0000000	0000000	0000000	0000000	0000000	0000000	0000000
3.8	0000000	0000000	0000000	0000000	0000000	0000000	0000000	0000000	0000000
3.9	0000000	0000000	0000000	0000000	0000000	0000000	0000000	0000000	0000000
4.0	0000000	0000000	0000000	0000000	0000000	0000000	0000000	0000000	0000000

Table T.11 (*continued*)

$$\rho = -0.500$$

h\k	2.7	2.8	2.9	3.0	3.1	3.2	3.3	3.4	3.5
0.0	0001498	0000984	0000639	0000409	0000259	0000162	0000100	0000061	0000037
0.1	0001168	0000763	0000493	0000314	0000198	0000123	0000076	0000046	0000028
0.2	0000901	0000585	0000376	0000238	0000149	0000092	0000057	0000034	0000020
0.3	0000687	0000444	0000283	0000179	0000111	0000068	0000041	0000025	0000015
0.4	0000517	0000332	0000211	0000132	0000082	0000050	0000030	0000018	0000011
0.5	0000385	0000246	0000155	0000097	0000060	0000036	0000022	0000013	0000008
0.6	0000283	0000180	0000113	0000070	0000043	0000026	0000015	0000009	0000005
0.7	0000206	0000130	0000081	0000050	0000030	0000018	0000011	0000006	0000004
0.8	0000148	0000093	0000058	0000035	0000021	0000013	0000008	0000004	0000003
0.9	0000105	0000066	0000040	0000025	0000015	0000009	0000005	0000003	0000002
1.0	0000074	0000046	0000028	0000017	0000010	0000006	0000003	0000002	0000001
1.1	0000051	0000031	0000019	0000012	0000007	0000004	0000002	0000001	0000001
1.2	0000035	0000021	0000013	0000008	0000005	0000003	0000002	0000001	0000000
1.3	0000024	0000014	0000009	0000005	0000003	0000002	0000001	0000001	0000000
1.4	0000016	0000010	0000006	0000003	0000002	0000001	0000001	0000000	0000000
1.5	0000010	0000006	0000004	0000002	0000001	0000001	0000000	0000000	0000000
1.6	0000007	0000004	0000002	0000001	0000001	0000000	0000000	0000000	0000000
1.7	0000004	0000003	0000002	0000001	0000001	0000000	0000000	0000000	0000000
1.8	0000003	0000002	0000001	0000001	0000000	0000000	0000000	0000000	0000000
1.9	0000002	0000001	0000001	0000000	0000000	0000000	0000000	0000000	0000000
2.0	0000001	0000001	0000000	0000000	0000000	0000000	0000000	0000000	0000000
2.1	0000001	0000000	0000000	0000000	0000000	0000000	0000000	0000000	0000000
2.2	0000000	0000000	0000000	0000000	0000000	0000000	0000000	0000000	0000000

$$\rho = -0.500$$

h\k	3.6	3.7	3.8	3.9	4.0
0.0	0000022	0000013	0000007	0000004	0000002
0.1	0000016	0000010	0000005	0000003	0000002
0.2	0000012	0000007	0000004	0000002	0000001
0.3	0000009	0000005	0000003	0000002	0000001
0.4	0000006	0000004	0000002	0000001	0000001
0.5	0000004	0000002	0000001	0000001	0000000
0.6	0000003	0000002	0000001	0000001	0000000
0.7	0000002	0000001	0000001	0000000	0000000
0.8	0000001	0000001	0000000	0000000	0000000
0.9	0000001	0000001	0000000	0000000	0000000
1.0	0000001	0000000	0000000	0000000	0000000
1.1	0000000	0000000	0000000	0000000	0000000
1.2	0000000	0000000	0000000	0000000	0000000
1.3	0000000	0000000	0000000	0000000	0000000
1.4	0000000	0000000	0000000	0000000	0000000
1.5	0000000	0000000	0000000	0000000	0000000
1.6	0000000	0000000	0000000	0000000	0000000
1.7	0000000	0000000	0000000	0000000	0000000
1.8	0000000	0000000	0000000	0000000	0000000
1.9	0000000	0000000	0000000	0000000	0000000
2.0	0000000	0000000	0000000	0000000	0000000
2.1	0000000	0000000	0000000	0000000	0000000
2.2	0000000	0000000	0000000	0000000	0000000

Table T.11 (*continued*)

$$\rho = 0.85$$

h\k	0.0	0.1	0.2	0.3	0.4	0.5	0.6	0.7	0.8
0.0	411699	390507	367028	341657	314893	287297	259454	231930	205241
0.1	390507	372323	351791	329205	304980	279618	253671	227700	202237
0.2	367028	351791	334234	314547	293048	270158	246373	222227	198251
0.3	341657	329205	314547	297774	279096	258844	237439	215365	193127
0.4	314893	304980	293048	279096	263239	245704	226826	207018	186743
0.5	287297	279618	270158	258844	245704	230869	214580	197166	179027
0.6	259454	253671	246373	237439	226826	214580	200848	185873	169978
0.7	231930	227700	222227	215365	207018	197166	185873	173295	159674
0.8	205241	202237	198251	193127	186743	179027	169978	159674	148275
0.9	179823	177753	174935	171220	166474	160597	153539	145316	136013
1.0	156018	154636	152704	150090	146664	142314	136961	130575	123180
1.1	134073	133179	131894	130110	127710	124585	120640	115817	110097
1.2	114136	113575	112748	111568	109938	107758	104936	101397	097097
1.3	096270	095929	095413	094657	093583	092109	090151	087631	084490
1.4	080464	080264	079952	079483	078798	077832	076514	074772	072546
1.5	066651	066537	066355	066073	065649	065036	064176	063010	061478
1.6	054719	054656	054553	054388	054135	053758	053215	052458	051437
1.7	044525	044492	044435	044343	044196	043972	043639	043163	042503
1.8	035911	035894	035864	035813	035731	035602	035405	035115	034702
1.9	028707	028699	028684	028657	028612	028540	028428	028257	028006
2.0	022746	022742	022734	022720	022697	022659	022596	022499	022352
2.1	017863	017861	017857	017850	017839	017818	017785	017731	017648
2.2	013903	013902	013900	013897	013891	013881	013864	013835	013789
2.3	010724	010723	010722	010721	010718	010713	010705	010690	010666
2.4	008197	008197	008196	008196	008195	008193	008188	008181	008169
2.5	006210	006210	006209	006209	006209	006208	006206	006202	006196
2.6	004661	004661	004661	004661	004661	004660	004659	004658	004655
2.7	003467	003467	003467	003467	003467	003467	003466	003465	003464
2.8	002555	002555	002555	002555	002555	002555	002555	002555	002554
2.9	001866	001866	001866	001866	001866	001866	001866	001866	001865
3.0	001350	001350	001350	001350	001350	001350	001350	001350	001350
3.1	000968	000968	000968	000968	000968	000968	000968	000968	000968
3.2	000687	000687	000687	000687	000687	000687	000687	000687	000687
3.3	000483	000483	000483	000483	000483	000483	000483	000483	000483
3.4	000337	000337	000337	000337	000337	000337	000337	000337	000337
3.5	000233	000233	000233	000233	000233	000233	000233	000233	000233
3.6	000159	000159	000159	000159	000159	000159	000159	000159	000159
3.7	000108	000108	000108	000108	000108	000108	000108	000108	000108
3.8	000072	000072	000072	000072	000072	000072	000072	000072	000072
3.9	000048	000048	000048	000048	000048	000048	000048	000048	000048
4.0	000032	000032	000032	000032	000032	000032	000032	000032	000032

Table T.11 (*continued*)

$$\rho = 0.85$$

h \ k	0.9	1.0	1.1	1.2	1.3	1.4	1.5	1.6	1.7
0.0	179823	156018	134073	114136	096270	080464	066651	054719	044525
0.1	177753	154636	133179	113575	095929	080264	066537	054656	044492
0.2	174935	152704	131894	112748	095413	079952	066355	054553	044435
0.3	171220	150090	130110	111568	094657	079483	066073	054388	044343
0.4	166474	146664	127710	109938	093583	078798	065649	054135	044196
0.5	160597	142314	124585	107758	092109	077832	065036	053758	043972
0.6	153539	136961	120640	104936	090151	076514	064176	053215	043639
0.7	145316	130575	115817	101397	087631	074772	063010	052458	043163
0.8	136013	123180	110097	097097	084490	072546	061478	051437	042503
0.9	125791	114870	103518	092030	080696	069787	059532	050104	041619
1.0	114870	105798	096173	086238	076254	066476	057135	048420	040472
1.1	103518	096173	088208	079814	071209	062623	054275	046359	039031
1.2	092030	086238	079814	072894	065650	058276	050969	043917	037278
1.3	080696	076254	071209	065650	059701	053516	047262	041110	035213
1.4	069787	066476	062623	058276	053516	048456	043230	037982	032855
1.5	059532	057135	054275	050969	047262	043230	038971	034602	030244
1.6	050104	048420	046359	043917	041110	037982	034602	031055	027441
1.7	041619	040472	039031	037278	035213	032855	030244	027441	024518
1.8	034133	033376	032398	031178	029703	027976	026016	023860	021559
1.9	027652	027167	026525	025701	024679	023451	022022	020411	018648
2.0	022138	021838	021428	020890	020203	019356	018345	017174	015863
2.1	017523	017343	017090	016749	016302	015735	015041	014216	013269
2.2	013719	013614	013463	013254	012972	012605	012142	011579	010914
2.3	010627	010568	010481	010357	010184	009954	009656	009283	008831
2.4	008148	008116	008068	007996	007894	007754	007568	007328	007030
2.5	006185	006169	006142	006102	006044	005962	005849	005700	005510
2.6	004650	004641	004627	004606	004574	004527	004461	004371	004253
2.7	003462	003457	003450	003439	003422	003396	003359	003307	003236
2.8	002553	002551	002547	002542	002533	002519	002499	002469	002428
2.9	001865	001864	001862	001860	001855	001848	001837	001821	001798
3.0	001349	001349	001348	001347	001345	001341	001336	001327	001315
3.1	000967	000967	000967	000966	000965	000964	000961	000957	000950
3.2	000687	000687	000687	000687	000686	000685	000684	000682	000679
3.3	000483	000483	000483	000483	000483	000483	000482	000481	000479
3.4	000337	000337	000337	000337	000337	000337	000336	000336	000335
3.5	000233	000233	000233	000233	000233	000232	000232	000232	000232
3.6	000159	000159	000159	000159	000159	000159	000159	000159	000159
3.7	000108	000108	000108	000108	000108	000108	000108	000108	000108
3.8	000072	000072	000072	000072	000072	000072	000072	000072	000072
3.9	000048	000048	000048	000048	000048	000048	000048	000048	000048
4.0	000032	000032	000032	000032	000032	000032	000032	000032	000032

Table T.11 (*continued*)

$$\rho = 0.85$$

h\k	1.8	1.9	2.0	2.1	2.2	2.3	2.4	2.5	2.6
0.0	035911	028707	022746	017863	013903	010724	008197	006210	004661
0.1	035894	028699	022742	017861	013902	010723	008197	006210	004661
0.2	035864	028684	022734	017857	013900	010722	008196	006209	004661
0.3	035813	028657	022720	017850	013897	010721	008196	006209	004661
0.4	035731	028612	022697	017839	013891	010718	008195	006209	004661
0.5	035602	028540	022659	017818	013881	010713	008193	006208	004660
0.6	035405	028428	022596	017785	013864	010705	008188	006206	004659
0.7	035115	028257	022499	017731	013835	010690	008181	006202	004658
0.8	034702	028006	022352	017648	013789	010666	008169	006196	004655
0.9	034133	027652	022138	017523	013719	010627	008148	006185	004650
1.0	033376	027167	021838	017343	013614	010568	008116	006169	004641
1.1	032398	026525	021428	017090	013463	010481	008068	006142	004627
1.2	031178	025701	020890	016749	013254	010357	007996	006102	004606
1.3	029703	024679	020203	016302	012972	010184	007894	006044	004574
1.4	027976	023451	019356	015735	012605	009954	007754	005962	004527
1.5	026016	022022	018345	015041	012142	009656	007568	005849	004461
1.6	023860	020411	017174	014216	011579	009283	007328	005700	004371
1.7	021559	018648	015863	013269	010914	008831	007030	005510	004253
1.8	019177	016780	014438	012213	010155	008300	006671	005274	004103
1.9	016780	014858	012937	011073	009314	007698	006253	004992	003919
2.0	014438	012937	011403	009880	008412	007035	005780	004665	003700
2.1	012213	011073	009880	008668	007473	006329	005264	004299	003448
2.2	010155	009314	008412	007473	006526	005599	004716	003900	003166
2.3	008300	007698	007035	006329	005599	004867	004154	003480	002862
2.4	006671	006253	005780	005264	004716	004154	003594	003052	002544
2.5	005274	004992	004665	004299	003900	003480	003052	002628	002221
2.6	004103	003919	003700	003448	003166	002862	002544	002221	001903
2.7	003144	003027	002885	002716	002523	002309	002080	001841	001600
2.8	002373	002302	002212	002103	001975	001829	001668	001496	001319
2.9	001766	001724	001669	001601	001518	001422	001313	001193	001066
3.0	001297	001273	001241	001199	001148	001086	001014	000933	000845
3.1	000940	000927	000909	000884	000853	000815	000769	000716	000657
3.2	000674	000666	000656	000642	000624	000601	000573	000539	000501
3.3	000477	000473	000468	000460	000450	000436	000419	000399	000374
3.4	000334	000332	000329	000325	000319	000312	000302	000290	000275
3.5	000231	000230	000229	000227	000224	000220	000214	000207	000198
3.6	000158	000158	000157	000156	000155	000153	000150	000146	000141
3.7	000108	000107	000107	000107	000106	000105	000103	000101	000098
3.8	000072	000072	000072	000072	000071	000071	000070	000069	000067
3.9	000048	000048	000048	000048	000048	000047	000047	000046	000046
4.0	000032	000032	000032	000032	000031	000031	000031	000031	000031

Table T.11 (*continued*)

$$\rho = 0.85$$

h \ k	2.7	2.8	2.9	3.0	3.1	3.2	3.3	3.4	3.5
0.0	003467	002555	001866	001350	000968	000687	000483	000337	000233
0.1	003467	002555	001866	001350	000968	000687	000483	000337	000233
0.2	003467	002555	001866	001350	000968	000687	000483	000337	000233
0.3	003467	002555	001866	001350	000968	000687	000483	000337	000233
0.4	003467	002555	001866	001350	000968	000687	000483	000337	000233
0.5	003467	002555	001866	001350	000968	000687	000483	000337	000233
0.6	003466	002555	001866	001350	000968	000687	000483	000337	000233
0.7	003465	002555	001866	001350	000968	000687	000483	000337	000233
0.8	003464	002554	001865	001350	000968	000687	000483	000337	000233
0.9	003462	002553	001865	001349	000967	000687	000483	000337	000233
1.0	003457	002551	001864	001349	000967	000687	000483	000337	000233
1.1	003450	002547	001862	001348	000967	000687	000483	000337	000233
1.2	003439	002542	001860	001347	000966	000687	000483	000337	000233
1.3	003422	002533	001855	001345	000965	000686	000483	000337	000233
1.4	003396	002519	001848	001341	000964	000685	000483	000337	000232
1.5	003359	002499	001837	001336	000961	000684	000482	000336	000232
1.6	003307	002469	001821	001327	000957	000682	000481	000336	000232
1.7	003236	002428	001798	001315	000950	000679	000479	000335	000232
1.8	003144	002373	001766	001297	000940	000674	000477	000334	000231
1.9	003027	002302	001724	001273	000927	000666	000473	000332	000230
2.0	002885	002212	001669	001241	000909	000656	000468	000329	000229
2.1	002716	002103	001601	001199	000884	000642	000460	000325	000227
2.2	002523	001975	001518	001148	000853	000624	000450	000319	000224
2.3	002309	001829	001422	001086	000815	000601	000436	000312	000220
2.4	002080	001668	001313	001014	000769	000573	000419	000302	000214
2.5	001841	001496	001193	000933	000716	000539	000399	000290	000207
2.6	001600	001319	001066	000845	000657	000501	000374	000275	000198
2.7	001365	001142	000936	000752	000593	000458	000347	000257	000188
2.8	001142	000969	000806	000658	000526	000412	000316	000238	000175
2.9	000936	000806	000681	000564	000458	000364	000283	000216	000161
3.0	000752	000658	000564	000474	000391	000315	000249	000193	000146
3.1	000593	000526	000458	000391	000327	000268	000215	000169	000130
3.2	000458	000412	000364	000315	000268	000223	000182	000145	000113
3.3	000347	000316	000283	000249	000215	000182	000151	000122	000097
3.4	000257	000238	000216	000193	000169	000145	000122	000101	000081
3.5	000188	000175	000161	000146	000130	000113	000097	000081	000067
3.6	000134	000127	000118	000108	000098	000087	000075	000064	000054
3.7	000095	000090	000085	000079	000072	000065	000057	000050	000042
3.8	000066	000063	000060	000056	000052	000048	000043	000037	000032
3.9	000045	000043	000042	000039	000037	000034	000031	000028	000024
4.0	000030	000029	000028	000027	000026	000024	000022	000020	000018

Table T.11 (continued)

$$\rho = 0.85$$

h\k	3.6	3.7	3.8	3.9	4.0
0.0	000159	000108	000072	000048	000032
0.1	000159	000108	000072	000048	000032
0.2	000159	000108	000072	000048	000032
0.3	000159	000108	000072	000048	000032
0.4	000159	000108	000072	000048	000032
0.5	000159	000108	000072	000048	000032
0.6	000159	000108	000072	000048	000032
0.7	000159	000108	000072	000048	000032
0.8	000159	000108	000072	000048	000032
0.9	000159	000108	000072	000048	000032
1.0	000159	000108	000072	000048	000032
1.1	000159	000108	000072	000048	000032
1.2	000159	000108	000072	000048	000032
1.3	000159	000108	000072	000048	000032
1.4	000159	000108	000072	000048	000032
1.5	000159	000108	000072	000048	000032
1.6	000159	000108	000072	000048	000032
1.7	000159	000108	000072	000048	000032
1.8	000158	000108	000072	000048	000032
1.9	000158	000107	000072	000048	000032
2.0	000157	000107	000072	000048	000032
2.1	000156	000107	000072	000048	000032
2.2	000155	000106	000071	000048	000031
2.3	000153	000105	000071	000047	000031
2.4	000150	000103	000070	000047	000031
2.5	000146	000101	000069	000046	000031
2.6	000141	000098	000067	000046	000031
2.7	000134	000095	000066	000045	000030
2.8	000127	000090	000063	000043	000029
2.9	000118	000085	000060	000042	000028
3.0	000108	000079	000056	000039	000027
3.1	000098	000072	000052	000037	000026
3.2	000087	000065	000048	000034	000024
3.3	000075	000057	000043	000031	000022
3.4	000064	000050	000037	000028	000020
3.5	000054	000042	000032	000024	000018
3.6	000044	000035	000027	000021	000016
3.7	000035	000028	000023	000017	000013
3.8	000027	000023	000018	000014	000011
3.9	000021	000017	000014	000012	000009
4.0	000016	000013	000011	000009	000007

Table T.11 (*continued*)

$$\rho = -0.85$$

h\k	0.0	0.1	0.2	0.3	0.4	0.5	0.6	0.7	0.8
0.0	0883009	0696651	0537127	0404314	0296851	0212402	0147991	0100334	0066146
0.1	0696651	0539919	0408547	0301542	0216905	0151935	0103559	0068639	0044211
0.2	0537127	0408547	0303123	0219192	0154352	0105767	0070478	0045639	0028705
0.3	0404314	0301542	0219192	0155166	0106889	0071604	0046616	0029477	0018094
0.4	0296851	0216905	0154352	0106889	0071984	0047113	0029950	0018483	0011068
0.5	0212402	0151935	0105767	0071604	0047113	0030109	0018680	0011246	0006566
0.6	0147991	0103559	0070478	0046616	0029950	0018680	0011306	0006637	0003777
0.7	0100334	0068639	0045639	0029477	0018483	0011246	0006637	0003797	0002106
0.8	0066146	0044211	0028705	0018094	0011068	0006566	0003777	0002106	0001137
0.9	0042377	0027658	0017526	0010778	0006428	0003717	0002083	0001131	0000595
1.0	0026368	0016797	0010383	0006226	0003620	0002039	0001113	0000589	0000301
1.1	0015927	0009897	0005966	0003487	0001975	0001084	0000576	0000297	0000148
1.2	0009335	0005656	0003323	0001893	0001044	0000558	0000289	0000145	0000070
1.3	0005306	0003134	0001794	0000995	0000535	0000278	0000140	0000068	0000032
1.4	0002924	0001682	0000938	0000507	0000265	0000134	0000066	0000031	0000014
1.5	0001561	0000875	0000475	0000250	0000127	0000063	0000030	0000014	0000006
1.6	0000808	0000441	0000233	0000119	0000059	0000028	0000013	0000006	0000003
1.7	0000405	0000215	0000111	0000055	0000027	0000012	0000006	0000002	0000001
1.8	0000196	0000102	0000051	0000025	0000012	0000005	0000002	0000001	0000000
1.9	0000092	0000046	0000023	0000011	0000005	0000002	0000001	0000000	0000000
2.0	0000042	0000021	0000010	0000005	0000002	0000001	0000000	0000000	0000000
2.1	0000018	0000009	0000004	0000002	0000001	0000000	0000000	0000000	0000000
2.2	0000008	0000004	0000002	0000001	0000000	0000000	0000000	0000000	0000000
2.3	0000003	0000002	0000001	0000000	0000000	0000000	0000000	0000000	0000000
2.4	0000001	0000001	0000000	0000000	0000000	0000000	0000000	0000000	0000000
2.5	0000001	0000000	0000000	0000000	0000000	0000000	0000000	0000000	0000000
2.6	0000000	0000000	0000000	0000000	0000000	0000000	0000000	0000000	0000000

h\k	0.9	1.0	1.1	1.2	1.3	1.4	1.5	1.6	1.7
0.0	0042377	0026368	0015927	0009335	0005306	0002924	0001561	0000808	0000405
0.1	0027658	0016797	0009897	0005656	0003134	0001682	0000875	0000441	0000215
0.2	0017526	0010383	0005966	0003323	0001794	0000938	0000475	0000233	0000111
0.3	0010778	0006226	0003487	0001893	0000995	0000507	0000250	0000119	0000055
0.4	0006428	0003620	0001975	0001044	0000535	0000265	0000127	0000059	0000027
0.5	0003717	0002039	0001084	0000558	0000278	0000134	0000063	0000028	0000012
0.6	0002083	0001113	0000576	0000289	0000140	0000066	0000030	0000013	0000006
0.7	0001131	0000589	0000297	0000145	0000068	0000031	0000014	0000006	0000002
0.8	0000595	0000301	0000148	0000070	0000032	0000014	0000006	0000003	0000001
0.9	0000303	0000149	0000071	0000033	0000015	0000006	0000003	0000001	0000000
1.0	0000149	0000072	0000033	0000015	0000006	0000003	0000001	0000000	0000000
1.1	0000071	0000033	0000015	0000007	0000003	0000001	0000000	0000000	0000000
1.2	0000033	0000015	0000007	0000003	0000001	0000001	0000000	0000000	0000000
1.3	0000015	0000006	0000003	0000001	0000001	0000000	0000000	0000000	0000000
1.4	0000006	0000003	0000001	0000001	0000000	0000000	0000000	0000000	0000000
1.5	0000003	0000001	0000001	0000000	0000000	0000000	0000000	0000000	0000000
1.6	0000001	0000000	0000000	0000000	0000000	0000000	0000000	0000000	0000000
1.7	0000000	0000000	0000000	0000000	0000000	0000000	0000000	0000000	0000000
1.8	0000000	0000000	0000000	0000000	0000000	0000000	0000000	0000000	0000000
1.9	0000000	0000000	0000000	0000000	0000000	0000000	0000000	0000000	0000000
2.0	0000000	0000000	0000000	0000000	0000000	0000000	0000000	0000000	0000000

Table T.11 (*continued*)

$$\rho = -0.85$$

h\k	1.8	1.9	2.0	2.1	2.2	2.3	2.4	2.5	2.6
0.0	0000196	0000092	0000042	0000018	0000008	0000003	0000001	0000001	0000000
0.1	0000102	0000046	0000021	0000009	0000004	0000002	0000001	0000000	0000000
0.2	0000051	0000023	0000010	0000004	0000002	0000001	0000000	0000000	0000000
0.3	0000025	0000011	0000005	0000002	0000001	0000000	0000000	0000000	0000000
0.4	0000012	0000005	0000002	0000001	0000000	0000000	0000000	0000000	0000000
0.5	0000005	0000002	0000001	0000000	0000000	0000000	0000000	0000000	0000000
0.6	0000002	0000001	0000000	0000000	0000000	0000000	0000000	0000000	0000000
0.7	0000001	0000000	0000000	0000000	0000000	0000000	0000000	0000000	0000000
0.8	0000000	0000000	0000000	0000000	0000000	0000000	0000000	0000000	0000000
0.9	0000000	0000000	0000000	0000000	0000000	0000000	0000000	0000000	0000000
1.0	0000000	0000000	0000000	0000000	0000000	0000000	0000000	0000000	0000000
1.1	0000000	0000000	0000000	0000000	0000000	0000000	0000000	0000000	0000000

ANSWERS

CHAPTER 1

2. (a) Yes, it satisfies definition given in Section 3.
   (b) Assume a coin has two distinguishable faces—called "head" and "tail." Simple results are "visible face of the coin is a head" and "visible face of coin is a tail" (or simply "head" and "tail"). A simple result is one of a collection of different possible descriptions of the outcome of an experiment, where the descriptions are detailed enough to provide all information of interest about the outcome of the experiment, and where one of these descriptions applies to every possible happening of the experiment. This definition is meaningful for any experiment; a random experiment has simple results which are unpredictable in advance.
   (c) Identify the simple result of having a boy in the sex-choice random experiment with one face (say "heads") of the coin in the coin tossing experiment. Identify "girl" with the other face of the coin ("tails"). If the long-run relative frequency of "boy" in repeated births equals the long-run relative frequency of "heads" in repeated coin tosses, both random experiments can be modeled by the same probability model.
4. (a) 0,1,2,3,4, ... (in hundred dollar units). That is, the simple results are nonnegative integers;
   (b) r.f.(E) = 13/49, the simple results 085,086,087, ..., 099,100 are contained in E.
6. (a) 309/2246; (b) S.
8. (a) Let J represent a Jack, Q a Queen, K a King. Let C represent clubs, D represent diamonds, H represent hearts, and S represent spades. A simple result is given by writing the value (1,2,3, ..., 9,10,J,Q,K) first and the suit

717

second. The possible simple results (there are 52 in all) are:

1C,2C,3C, ..., 9C,10C,JC,QC,KC,   1D,2D,3D, ..., 9D,10D,JD,QD,KD,
1H,2H,3H, ..., 9H,10H,JH,QH,KH,   1S,2S,3S, ..., 9S,10S,JS,QS,KS.

(b) 1H,2H,3H,4H,5H,6H,7H,8H,9H,10H,JH,QH,KH;
(c) $p = \frac{1}{4}$. Since shuffling is well done, every card has equal chance to appear. There are 52 cards and 13 hearts, so that intuition suggests r.f.$(F_1) = \frac{13}{52} = \frac{1}{4}$ (see Chapter 3, Section 1);
(d) JC,QC,KC,JD,QD,KD,JH,QH,KH,JS,QS,KS;
(e) $p = \frac{12}{52} = \frac{3}{13}$ by same reasoning given in part (c).

11. (a)

k	0	1	2	3	4	5	6	7	8	9
Frequency	12	9	9	15	13	13	7	8	7	7
Relative Frequency	0.12	0.09	0.09	0.15	0.13	0.13	0.07	0.08	0.07	0.07

(b) The experiment in Question 9 since both experiments have the same 10 simple results: 0,1,2,3,4,5,6,7,8,9. If we assume that the 10 simple results 0,1,2,...,8,9 in the experiment of Question 9 are equally likely, the two experiments also are identical in that the relative frequencies of any simple result all eventually stabilize at $\frac{1}{10}$.

14. (a) The experiment is to choose a patient, have him independently diagnosed by the computer and by the team of physicians. If the computer and the physicians agree on a diagnosis an A (for agreement) is recorded; otherwise, a D (for disagreement) is recorded. We can denote the simple results as "A" and "D." There are 2 simple results.
(b) "Neither A nor B is recorded," "A," "B," "one of A or B is recorded." There are 4 composite results.
(c) We need to show that the relative frequencies of "neither A nor B is recorded," "A," "B," and "one of A or B is recorded" stabilize. Let these composite results be symbolized by E_1, E_2, E_3, and E_4, respectively. Since r.f.(E_1) is always 0 and r.f.$\{A,B\}$ is always 1 (either A or B must be the simple result at each trial), the relative frequencies of these two composite results stabilize at 0 and 1, respectively. To show that r.f.(E_2) and r.f.(E_3) stabilize we use the given data and graph r.f.(E_2) and r.f.(E_3) against the number of trials (as in Figures 3.1–3.3). We find that r.f.(E_1) stabilizes at around $68/80 = 0.85$, and r.f.(E_2) stabilizes around $12/80 = 0.15$. Thus, the random experiment can be modeled using probability theory.

CHAPTER 2

2. (a) $\omega_1, \omega_2, \ldots, \omega_{12}, \omega_{13}, \omega_{24}, \omega_{25}, \omega_{26}, \omega_{37}, \omega_{38}, \omega_{39}, \omega_{50}, \omega_{51}, \omega_{52}$;
(b) $\omega_{11}, \omega_{12},$ and ω_{13};
(c) $\omega_{24}, \omega_{25}, \omega_{26}, \omega_{37}, \omega_{38}, \omega_{39}, \omega_{50}, \omega_{51}, \omega_{52}$;

ANSWERS   719

(d) $\omega_1, \omega_2, \ldots, \omega_9, \omega_{10}$;
(e) Problem 8 of the exercises to Chapter 1.

4. (a) $E_1 \cup E_2 \cup E_3$;
(b) $(E_1 \cap E_2^c \cap E_3^c) \cup (E_1^c \cap E_2 \cap E_3^c) \cup (E_1^c \cap E_2^c \cap E_3)$.
(c) $(E_1 \cap E_2 \cap E_3^c) \cup (E_1 \cap E_2^c \cap E_3) \cup (E_1^c \cap E_2 \cap E_3)$;
(d) $(E_1 \cap E_2 \cap E_3)^c$;
(e) $E_1 \cap E_2 \cap E_3$; (f) $E_1^c \cap E_2^c \cap E_3^c$.

5. (a) Let ω be any outcome in E_1. Since ω is in E_1 and $E_1 \subset E_2$, ω is in E_2. Since ω is in E_2 and $E_2 \subset E_3$, ω is in E_3. Thus, any outcome in E_1 is also in E_3; and hence by definition $E_1 \subset E_3$.

7. The events in (b).

10. (a) 107,981; (b) 96,672; (c) 20,928; (d) 87,355; (e) 98,709; (f) 112,110; (g) 721.

12. (a) $P(C)$; (b) $P(D)$; (c) $P(D^c)$; (d) $P(C^c \cap D)$; (e) $P(C^c \cap D^c)$.

13. (b) Since $(A \cap B)$ and $(A \cap B^c)$ are mutually exclusive, and since $(A \cap B) \cup (A \cap B^c) = A$, it follows from Axiom 3 that $(P(A) = P(A \cap B) + P(A \cap B^c)$, or $P(A) - P(A \cap B) = P(A \cap B^c)$.

14. (a) False, $P(A \cap B) = 0.5$; (b) true; (c) false, $P(A \cap B^c) = 0.1$; (d) true; (e) false, $P(B^c) = 0.6$; (f) true.

18. $P(A) = 0.3, P(B) = 0.5, P(A \cup B) = 0.7$.

19. (a) 0.17; (b) 0.32; (c) 0.20; (d) 0.31; (e) 0.51.

22. (b) Yes.
(c) Use Rule 4.5 to find $P(M)$ from the probabilities of the events E_i. Use Rule 4.2 to find $P(F)$ from $P(M)$.
(d) The probabilities $P(A_1), P(A_2), P(A_3), P(A_4)$ must sum to 1 since the events A_1, A_2, A_3, A_4 are pairwise mutually exclusive and $A_1 \cup A_2 \cup A_3 \cup A_4$ is the entire sample space.
(f) To find $P(M \cup A_1 \cup B_1)$ make use of Rule 4.4.
(g) A_1 and M are not mutually exclusive. Hence $P(M \cup A_1) \neq P(M) + P(A_1)$, unless $P(M \cap A_1) = 0$ under your probability model.
(h) The events M and $M^c \cap A$, are mutually exclusive. Thus, $P(M \cup (M^c \cap A)) = P(M) + P(M^c \cap A)$ for any probability model.
(i) $M \cap A_1, M \cap A_2, M \cap A_3, M \cap A_4$ are pairwise mutually exclusive and $M = (M \cap A_1) \cup (M \cap A_2) \cup (M \cap A_3) \cup (M \cap A_4)$. Now use Rule 4.5 to show that $P(M) = P(M \cap A_1) + P(M \cap A_2) + P(M \cap A_3) + P(M \cap A_4)$.

23. (a and b) We may assign outcomes to events according to the following table:

	Red Eyes		Pale Eyes	
	Long Antennae	Short Antennae	Long Antennae	Short Antennae
Marked Wings	ω_1	ω_2	ω_3	ω_4
Unmarked Wings	ω_5	ω_6	ω_7	ω_8

Using a Venn diagram show that $\{\omega_1\} = E_1 \cap E_2 \cap E_3$, $\{\omega_2\} = E_1 \cap E_2 \cap E_3^c$, $\{\omega_3\} = E_1 \cap E_2^c \cap E_3$, $\{\omega_4\} = E_1 \cap E_2^c \cap E_3^c$, $\{\omega_5\} = E_1^c \cap E_2 \cap E_3$, $\{\omega_6\} = E_1^c \cap E_2 \cap E_3^c$, $\{\omega_7\} = E_1^c \cap E_2^c \cap E_3$, $\{\omega_8\} = E_1^c \cap E_2^c \cap E_3^c$

Then

$$P(\{\omega_8\}) = P(E_1^c \cap E_2^c \cap E_3^c),$$
$$P(\{\omega_7\}) = P(E_1^c \cap E_2^c) - P(\{\omega_8\}) = 1 - P(E_1 \cup E_2) - P(\{\omega_8\}),$$
$$P(\{\omega_6\}) = P(E_1^c \cap E_3^c) - P(\{\omega_8\}) = 1 - P(E_1 \cup E_3) - P(\{\omega_8\}),$$
$$P(\{\omega_4\}) = 1 - P(E_2 \cup E_3) - P(\{\omega_8\}),$$
$$P(\{\omega_5\}) = P(E_1^c) - P(\{\omega_6\}) - P(\{\omega_7\}) - P(\{\omega_8\})$$
$$= 1 - P(E_1) - P(\{\omega_6\}) - P(\{\omega_7\}) - P(\{\omega_8\}),$$
$$P(\{\omega_3\}) = 1 - P(E_2) - P(\{\omega_4\}) - P(\{\omega_7\}) - P(\{\omega_8\}),$$
$$P(\{\omega_2\}) = 1 - P(E_3) - P(\{\omega_4\}) - P(\{\omega_6\}) - P(\{\omega_8\}),$$
$$P(\{\omega_1\}) = 1 - P(\{\omega_2\}) - P(\{\omega_3\}) - \cdots - P(\{\omega_8\}).$$

(c) Use Rule 4.5 to verify that the probability of an event E is the sum of the probabilities of the events $\{\omega_j\}$ corresponding to outcomes ω_j in E.
(d) $P(\{\omega_1\}) = P(\{\omega_2\}) = \cdots = P(\{\omega_7\}) = P(\{\omega_8\}) = \frac{1}{8}$.

24. There are many ways of showing that the researcher made a mistake. For example, $P(\{\omega_6\}) = 1 - P(E_1 \cup E_2) - P(\{\omega_8\}) = 1 - \frac{2}{3} - \frac{1}{2} = -\frac{1}{6} < 0$. But all probabilities must be nonnegative by Axiom 1.

CHAPTER 3

1. (a) 9/19; (b) 6/19; (c) 3/19; (d) 12/19; (e) 9/19.
3. (a) 1/145; (b) 27/145; (c) 117/145.
5. $P_J = 1/2, P_H = 5/9, P_J - P_H = -(1/18)$.
7. 1/33; 6/33 = 2/11; 3; and 7/33.
9. (a) 1000; (b) 8; (c) 512; (d) 488; (e) 0.008; (f) 0.512; (g) 0.488.
11. (a) 0.902; (b) 0.852; the conditional probability is greater than the unconditional probability, but they are close to one another $(0.902 - 0.852 = 0.050)$; the small difference between the two probabilities is due to the fact that the new sample space for the conditional probability has unconditional probability very close to 1.
13. (a) $P(A|A \cap B) = 1$ since $A \cap (A \cap B) = A \cap B$; $P(A|A^c) = 0$ since if A^c occurs, then A cannot occur; (b) If $A \subset B$, then $A \cap B = A$, and thus $P(B|A) = P(A \cap B)/P(A) = P(A)/P(A) = 1$; (c) $P(B|A) = 0 = P(A|B)$ since if A occurs, then B cannot, and vice-versa.
17. (a) 3/10; (b) 3/40; (c) 1/20; (d) 3/40; (e) 1/5; (f) 1/5; (g) 17/20.
19. (a) 0.51; (b) 0.49.
21. (a) 0.56; (b) 0.995; (c) 0.551, 0.361; (d) 0.002.
23. (a) 0.24; (b) No. $P\{\text{both vote}\} \neq P\{\text{husband votes}\}P\{\text{wife votes}\}$; (c) 0.686.
27. (a) 0.03; (b) 0.68; (c) 0.12; (d) 0.17.
30. $P(A \text{ and } B|B) = P(A)$.
33. (a) There are 21 different ways in which the circuit can conduct electricity; (b) 0.621; (c) 0.819; (d) 0.758; (e) Multiply answers to (c) and (d) since total circuit conducts if both subcircuits conduct, and subcircuits conduct independently. $P\{\text{entire circuit conducts}\} = (0.819)(0.758) = 0.621$; (f) $(1 - 0.9)/(1 - 0.621) = 0.264$.
35. We would expect, for example, that the relative frequency of "Arches for A

children *and* Loops for *B* children" equals (at least approximately) the product of the relative frequency of "Arches for *A* Children" and the relative frequency of "Loops for *B* children." Thus, we would expect "Arches for *A* children *and* Loops for *B* children," to have a frequency of 101(22/101) × (34/101) = 7.4 for nonfraternal children and (105)(10/105)(61/105) = 5/8 for fraternal children. The frequency distributions which we would expect are summarized in the following tables:

Nonfraternal Children

		A Children			
		Arches	Loops	Whorls	Totals
B Children	Arches	5.4	12.4	7.2	25
	Loops	7.4	16.8	9.8	34
	Whorls	9.2	20.8	12.0	42
	Totals	22	50	29	101

Fraternal Children

		A Children			
		Arches	Loops	Whorls	Totals
B Children	Arches	1.8	12.3	4.9	19
	Loops	5.8	39.5	15.7	61
	Whorls	2.4	16.2	6.4	25
	Totals	10	68	27	105

36. 1/2.
38. 12! = 479,001,600.
40. (a) $(26)^3 = 17{,}576$; (b) $(26)^3 + (26)^2 + 26 = 18{,}278$; (c) $(26)^4 + (26)^3 + (26)^2 + 26 = 475{,}254$.
43. (a) $\binom{6}{3} = 20$; (b) $\binom{6}{3} = 20$; (c) It does not matter whether the blocks are arranged in 1 row or 2 rows.
45. (a) $9!/3!3!3! = 1680$; (b) $\binom{6}{3} = 20$.
46. $25!/7!5!6!7! = 7{,}067{,}582{,}121{,}600$.
49. (a) $(10)_{(7)} = 604{,}800$; (b) $(10)^7 = 10{,}000{,}000$.
52. $(12)_{(5)}/(12)^5 = 0.382$.
54. (a) $\binom{8}{3}\binom{5}{2} = 560$; (b) $\binom{8}{0}\binom{5}{5} + \binom{8}{1}\binom{5}{4} + \binom{8}{2}\binom{5}{3} = 321$.
55. (a) $\binom{165}{6}\binom{135}{6} / \binom{300}{12} = 0.2163$.

(b) $\dfrac{1}{\binom{300}{12}}\left[\binom{165}{7}\binom{135}{5} + \binom{165}{8}\binom{135}{4} + \binom{165}{9}\binom{135}{3} + \binom{165}{10}\binom{135}{2} + \binom{165}{11}\binom{135}{1} + \binom{165}{12}\binom{135}{0}\right] = 0.5272$.

CHAPTER 4

2. (a and b) Write (i,j,k,l) to mean that the first player shows i fingers and guesses j, and the second player shows k fingers and guesses l. Let X be the amount of money won by the first player. The 16 possible outcomes and the corresponding values of X are listed below:

Outcome	X	Outcome	X	Outcome	X	Outcome	X
(1,1,1,1)	0	(1,2,1,1)	−2	(2,1,1,1)	3	(2,2,1,1)	0
(1,1,1,2)	2	(1,2,1,2)	0	(2,1,1,2)	0	(2,2,1,2)	−3
(1,1,2,1)	−3	(1,2,2,1)	0	(2,1,2,1)	0	(2,2,2,1)	4
(1,1,2,2)	0	(1,2,2,2)	3	(2,1,2,2)	−4	(2,2,2,2)	0

(c)

x	−4	−3	−2	0	2	3	4
Probability	$\frac{1}{16}$	$\frac{2}{16}$	$\frac{1}{16}$	$\frac{8}{16}$	$\frac{1}{16}$	$\frac{2}{16}$	$\frac{1}{16}$

;

  (d) We are now restricted to the outcomes (1,1,1,1), (1,1,2,2), (2,2,1,1), (2,2,2,2) for each of which $X = 0$. Hence, $P\{X = 0\} = 1$.

4. (c).
5. (a) and (d).
8. (a) Discrete; (b) 8.0, 9.0, 10.0, 11.0, and 12.5;

  (c)

x	8.0	9.0	10.0	11.0	12.5
Probability	0.1	0.3	0.3	0.2	0.1

;

  (d) (i) 0.4, (ii) 0.3, (iii) 0.5, (iv) 0.8.

10. (a) Continuous; (b) $0 \leq Y \leq 1$; (c) 1/2; (d) 3/4; (e and f) $F_Y(\tau) = P_Y\{Y \leq \tau\}$ is less than $F_X(\tau) = P_X\{X \leq \tau\}$ for all values of τ strictly between 0 and 1 (and in particular for $\tau = 0.2, 0.4, 0.6, 0.8$). $F_Y(\tau) = F_X(\tau)$ for $\tau \leq 0$ or $\tau \geq 1$ (and in particular for $\tau = -1.0, 0.0, 1.0, 2.0$). In this sense, the answer depends upon τ. On the other hand, we can say that $F_Y(\tau) \geq F_X(\tau)$ for any value of τ.

11. (a) Mixed discrete and continuous; the graph of the cumulative distribution function follows the horizontal axis for $\tau < 0$, rises in a straight line for $0 \leq \tau < \frac{1}{2}$, jumps at $\tau = \frac{1}{2}$, rises in another straight line for $\frac{1}{2} < \tau < 1$, and then is flat along the line $F_W(\tau) = 1$ for $\tau > 1$;
  (b) $0 \leq W \leq 1$; (d) $P_W\{W = \frac{1}{2}\} = \frac{1}{4}$, $P_W\{W > \frac{1}{2}\} = \frac{1}{4}$, $P\{W < \frac{1}{4}\} = \frac{1}{4}$, $P_W\{\frac{1}{3} \leq W < \frac{2}{3}\} = \frac{1}{2}$.

13. $F_X(\tau)$ is equal to 0 when $\tau < 20'7\frac{3}{8}''$, changes values only at the τ-values indicated below, and is equal to 1 for $\tau \geq 20'8\frac{1}{2}''$. The graph of $F_X(\tau)$ is a step-function.

$\tau - 20'7''$	$\frac{3}{8}$	$\frac{4}{8}$	$\frac{5}{8}$	$\frac{6}{8}$	$\frac{7}{8}$	$\frac{8}{8}$	$\frac{9}{8}$	$\frac{10}{8}$	$\frac{11}{8}$	$\frac{12}{8}$
$F_X(\tau)$	0.04	0.08	0.12	0.24	0.36	0.56	0.72	0.84	0.92	1.00

.

ANSWERS   723

16. (a)

τ	0	1	2	3	4
$p_X(\tau)$	6/36	10/36	10/36	6/36	4/36

;

(b) $20/36 = 5/9$; (c) graphs are left to reader;

(d)

τ	0	1	2	3	4
$p_X(\tau)$	4/24	5/24	6/24	5/24	4/24

.

17. (a)

τ	0	1	4	5	6	9
$p_Y(\tau)$	0.1	0.2	0.2	0.1	0.2	0.2

;

(b)

τ	0	1	4	5	6	9
$p_Z(\tau)$	0.1	0.4	0.0	0.1	0.4	0.0

;

(c)

τ	0	1	4	5	6	9
$p_W(\tau)$	0.1	0.4	0.0	0.1	0.4	0.0

.

(d) No. Since the distributions of Z and W are the same, so will be the distribution of W and any random variable obtained by any further stage of the process.

18. (a)–(c). See Exercise 4 of Chapter 5;

(d)

τ	0	1	2	3	4	5
$p_{X_4}(\tau)$	44/120	45/120	20/120	10/120	0/120	1/120

.

19. (d) 7/31, 14/31, 28/31.
22. (b) 0.6; (c) 0.3; (d) 0.3; (e) 0.6; (f) 0.4.
24. (b) 1/8, 6/8, 1/8; (c) 0.75. The correct answer is $6/8 = 0.75$, so the approximation is exact. If the number of intevals is even, the approximation is exactly correct.
26. (a) The graph of the cumulative distribution function $F_X(\tau)$ lies along the horizontal axis when $\tau < 22.47$, then rises in steps for $22.47 \leq \tau < 36.50$. For $\tau \geq 36.50$, $F_X(\tau) = 1$. The jumps of the graph (each of height 1/30) occur at the observed X-values (22.47, 24.01, 24.87, and so on).
(b)

x	22.0	24.0	26.0	28.0	30.0	32.0	34.0	36.0
r.f.$\{x - 1.00 \leq X \leq x + 1.00\}$	1/30	2/30	10/30	6/30	5/30	2/30	3/30	1/30

(c) Strategy (iii) is slightly better than strategy (ii), and both strategies are considerably better than strategy (i). However, the conclusions do not necessarily hold for other groupings or other data. Examples to show this are fairly easily constructed.

CHAPTER 5

2. (a)

τ	1	2	3	4
$p_X(\tau)$	0.1	0.4	0.2	0.3

;

 (b) 2.7; (c) 2.5; (d) 2.0; (e) 2, the probability of a correct guess is 0.4; (f) 1.01.
4. $E(X_2) = E(X_3) = E(X_4) = 1$, $\text{Var}(X_2) = \text{Var}(X_3) = \text{Var}(X_4) = 1$.
6. The probability mass function of X is

τ	0	1	2	3	4	5
$p_X(\tau)$	0	0.09	0.36	0.42	0.12	0.01

,

 and thus $E(X) = 2.6$.
9. (a) 1.60; (b) 0.92; (c) $\text{Med}(X) = 1.0$, $Q_{0.25}(X) = 1.0$, $Q_{0.75}(X) = 2.0$; (d) $\hat{\mu}_X + \hat{\sigma}_X = 2.559$, $\hat{\mu}_X + 2\hat{\sigma}_X = 3.518$.
12. (b) Yes, the mode is 0. The expected value and median are both to the right of (that is, greater than) the mode.
 (d) Solve $1 - (2-\tau)^3/8 = 0.5$ for τ. The solution $\tau = 0.413$ is the median. Similarly, the solution $\tau = 1.072$ of the equation $1 - (2-\tau)^3/8 = 0.9$ is $Q_{0.90}(X)$.
14. $a = \frac{1}{2}, b = \frac{1}{2}$.
15. (a) 7; (b) 10; (c) 16; (d) cannot find without knowing either the variance or the entire distribution of X.
16. $E(X^2) = 7/6$, $E(X^3) = 11/16$, $E(X^4) = 19/6$.
18. (a) $c = (715/650)$. Thus, the largest possible probability is $1/c = 0.91$;
 (b) $c = 780/650$. Largest possible probability is $1/c = 0.83$;
 (c) On the basis of our answer to (b) our answer must be "Yes" since $0.30 < 0.83$. On the basis of published facts about the S.A.T., the answer is "No." The inequality is too gross to help.
20. (a)

d	1.1	1.5	1.8	2.0	2.2
Smallest value $1 - (1/d^2)$	0.83	0.40	0.31	0.25	0.21

,

 (b)

b	0.5	0.7	1.0	1.2	1.5
Largest value $1/(1+b^2)$	0.80	0.67	0.50	0.41	0.31

.

CHAPTER 6

2. (a) 0.8; (b) 0.000004; (c) 0.0008; (d) Z has a binomial distribution with parameters $n = 10$ and $p = 0.8$ (see Table 1.2);
 (e) $\mu_Z = 8.0$ and $\sigma_Z^2 = 1.6$; (f) 0.9936. Yes, it is worth studying since probability of passing increases from 0.0328 to 0.9936.

ANSWERS 725

4. (a) 0.0128; (b) Use tables of Romig (1953). The desired probability is 0.1410; (c) 0.0184.
6. Assume that alleles received by one offspring in a litter are statistically independent of the alleles received by other offspring in the litter.

Number, x of mice with dominant character	0	1	2	3	4	5
Theoretical relative frequency	0.0312	0.1562	0.3125	0.3125	0.1562	0.0312
Theoretical frequency	10.296	51.546	103.125	103.125	51.546	10.296

8. (a) binomial distribution with parameters $n = 8$ and $p = 0.5$ (see Table 1.4);
   (b)

τ	0	1	2	3	4	5	6	7	8
$p_Z(\tau)$	0.0001	0.0012	0.0100	0.0467	0.1361	0.2541	0.2965	0.1977	0.0576

   (c) If she is guessing, $P\{6 \text{ correct}\} = 0.1094$, while if she is skilled, $P\{6 \text{ correct}\} = 0.2965$. Thus, we tentatively would conclude that she is skilled. (You can also argue comparing probabilities of $\{6$ or more correct$\}$.) If we repeated this experiment several times and each time she got 6 (or more) correct, we would definitely conclude that she is skilled.

9. (a) (i) and (ii); (b) 0.6172; (c) 300; (d) 210; (e) 0.0720.

11.

Probability, P, of selecting a single non-white venireman	Probability, P*, of selecting a single venire of 30 with 5 or fewer non-whites	Probability of selecting 30 such venires = (P*)^{30}
0.20	0.42751	0.00000
0.10	0.92681	0.10226
0.05	0.99672	0.90614

13. Desired probabilities are (i) $671/1296 = 0.518$, (ii) $1 - (35/36)^{24} = 0.491$. The probability of the first event (i) is larger, but only by 0.027.
16. 5.
17. (a)

τ	0	1	2	3
$p_X(\tau)$	0.8836	0.1128	0.0036	0.000

   (b) 0; 0; 0.12; (c) 0.0036;

(d) No, the new probability mass function is

τ	0	1	2	3
$p_X(\tau)$	0.856	0.138	0.006	0.000

and thus $P\{X > 1\} = 0.006$.

19. (a) $N = 9.0, p = 0.2$; (b) $N = 7.0, p = 0.4$;
    (c) $N = 5[\sigma_Z^2 - \mu_Z(5 - \mu_Z)]/[5\sigma_Z^2 - \mu_Z(5 - \mu_Z)]$, provided that the result is an integer;
    (d) No, Yes ($N = [\sigma_Z^2 - 25p(1-p)]/[\sigma_Z^2 - 5p(1-p)]$);
    (e) Yes, from $n = 5$ and μ_Z you get $p = (\mu_Z/5)$. From Np and p you get N.
21. (b and c)

x	0	1	2	3	4	5 or more
Theoretical frequency $\lambda = 0.70$	214.53	150.16	52.57	12.27	2.16	0.30
Theoretical frequency $\lambda = 0.69$	216.69	149.52	51.70	11.88	2.03	0.32

23. Since $\hat{\mu}_X = 0.53$ use $\lambda = 0.53$. The theoretical frequencies are:

x	0	1	2	3	4 or more
Theoretical frequency	58.86	31.20	8.27	1.46	0.21

25. (a) 1.5, 1.5 "on the average";

    (b)

τ	0	1	2	3	4	5
$p_X(\tau)$	0.2231	0.3347	0.2510	0.1255	0.0471	0.0141
$F_X(\tau)$	0.2231	0.5578	0.8088	0.9343	0.9814	0.9955

τ	6	7	8	9	...
$p_X(\tau)$	0.0035	0.0008	0.0001	0.0000	...
$F_X(\tau)$	0.9990	0.9998	0.9999	0.9999	...

(c) $\text{Med}(X) = 1$, $\text{Mode}(X) = 1, 1$;
(d) $e^{-1.5} = 0.2231$, 133.86 or 134.

27. (a) Binomial distribution with parameters $n = 100, p = 0.01$;
    (b) $P\{X = 0\} = (0.99)^{100} = 0.366$, $P\{X \geq 1\} = 0.634$;
    (c) Poisson distribution with parameter $\lambda = 100(0.01) = 1$, $P\{X = 0\} \cong e^{-1} = 0.368$, $P\{X \geq 1\} = 0.632$;
    (d) Yes.

ANSWERS 727

28. (a) Hypergeometric distribution with parameters $N = 500$, $n = 50$, $p = 0.02$; (b) Finite population factor is 0.902, so that binomial distribution is a good approximation (since 0.902 is close to 1); (c) Since n is large and p is small, Poisson distribution with parameter $\lambda = np = 1$ is good approximation; (d) 0.0803; Poisson distribution with $\lambda = np$ can be used to approximate hypergeometric distribution with parameters N, n, and p when n is large, p is small, and $(N - n)/(N - 1)$ is close to 1.

29. (a) $e^{-5} = 0.0067$; (b) Y has binomial distribution with parameters $n = 300$ and $p = 0.0067$. Thus, Y approximately has a Poisson distribution with $\lambda = 300(0.0067) = 2.01$. Use Table 3.1 with $\lambda = 2$ to get $P\{Y = 0\} \cong 0.1353$. Hence, the desired answer is $P\{Y \geqslant 1\} \cong 0.8647$.

30. $R = 116$ raisins. Since $R/25 = \lambda$ is moderate in size, Poisson approximations should be fairly good.

33. (a) $p = 0.4$, $\text{Med}(X) = 2$, $\text{Mode}(X) = 1$;
    (b) $p = 0.6$, $\text{Med}(X) = 1 = \text{Mode}(X)$;
    (c) $p = 0.8$, $\text{Med}(X) = 1 = \text{Mode}(X)$;
    All distributions are positively skewed.

35. Since $\hat{\mu}_X = (2282/1469)$, let $p = 1/\hat{\mu}_X = 0.6437$. Theoretical frequencies are

x	1	2	3	4	5	6 or more
Theoretical frequency	945.6	337.0	120.0	42.7	15.3	8.4

37. (a) $(0.4)^3 = 0.064$; (b) $1/(0.6)(0.3) = 5.56$.
39. (a) 0.207; (b) 0.269; (c) 0.276; (d) 0.710; (e) 0.6.
41. $\hat{\mu}_X = 2.794872$, $\hat{\sigma}_X^2 = 4.501512$. Thus $r = 4.5770$ and $p = 0.621$. The theoretical frequencies for this r and p are

x	0	1	2	3	4	5	6
Theoretical frequency	22.04	38.22	40.38	33.56	24.10	15.67	9.49

x	7	8	9	10 or more
Theoretical frequency	5.42	2.98	1.58	1.58

45. (a) Mode = 0, Med = 1;
    (b) Mode = 0, Med = 0;
    (c) Mode = 1, Med = 2;
    (d) Mode = 0, Med = 0.
47. 0.5582
48. In general $P\{Y = k\} = \left[\binom{800}{k}\binom{200}{2}\bigg/\binom{1000}{k+2}\right]\left[\dfrac{198}{998-k}\right]$.
    (a) 0.0079; (b) 0.0190; (c) 0.0305; (d) 0.0408; (e) 0.0491; (f) 0.00476. Using negative binomial approximation, we get (a) 0.0080; (b) 0.0192; (c) 0.0307; (d) 0.0410; (e) 0.0492; (f) 0.00491. The negative binomial provides a good approximation for the probabilities of all events listed. This is not

surprising since in cases (a)–(f) the total population of fish ($N = 1000$) is very much larger than the number of fish sampled.

50. (a) $19/37 = 0.5135$; (b) $10/37 = 0.2703$; (c) $13/37 = 0.3514$; X has a generalized discrete uniform distribution with parameters $c = -1$, $h = 1$, $K = 37$.

CHAPTER 7

2. (a) 0.3759; (b) 0.4589; (c) 0.6166; (d) 0.2182; (e) 0.4755; (f) 0.5805.
3. 65.
4. (a) $(0.9772)^3 = 0.9331$; (b) $1-(0.8413)^3 = 0.4045$; (c) 0.0915; (d) 0.0000.
6. (a) 0.0227; (b) $Q_{0.90}(X) = 55.82$, $Q_{0.95}(X) = 59.45$, $Q_{0.99}(X) = 66.27$;
   (c) $Q_{0.50}(X) = 43$, $\text{Mode}(X) = 43$;
   (d) $\frac{1}{2}[Q_{0.75}(X) - Q_{0.25}(X)] = \frac{1}{2}[49.75 - 36.25] = 6.75$.
8. No, we can solve the following 2 equations for μ and σ: $(87-\mu)/\sigma = 1.282$, $(96-\mu)/\sigma = 1.645$. The results are $\mu = 55.22$ and $\sigma = 24.79$ (or $\sigma^2 = 614.54$).
9. (b) $\mu_X = -21.0$, $\sigma_X^2 = 259.0$. Hence the theoretical relative frequencies and theoretical frequencies are:

Interval of velocities	Theoretical relative frequency	Theoretical frequency
Less than -80	0.0001	0.008
-80 to -70	0.0011	0.088
-70 to -60	0.0065	0.520
-60 to -50	0.0281	2.248
-50 to -40	0.0832	6.656
-40 to -30	0.1691	13.528
-30 to -20	0.2366	18.928
-20 to -10	0.2281	18.248
-10 to 20	0.1512	12.096
0 to 10	0.0690	5.520
10 to 20	0.0216	1.728
20 to 30	0.0046	0.368
30 to 40	0.0007	0.056
more than 40	0.0001	0.008
	1.0000	80.000

(d) $P\{X > 0\} = P\{[X-(-21)]/\sqrt{259.0} > 1.3049\} = 0.9040$.

11. (a) $\text{Med}(X) = 6.9$ and thus $P_Y\{Y > 6.9\} = 0.5235$;
    (b) $\text{Med}(Y) = 7.0$ and thus $P_X\{X > 7.0\} = 0.4751$;
    (c) $Q_{0.90}(X) = 8.95 = Q_{0.87}(Y)$;
    (d) $Q_{0.90}(Y) = 9.18 = Q_{0.92}(X)$.
14. (a) If $Y = 16{,}000$, $P_X\{X > 5000\} = P\{\mathcal{N}(Z;0,1) > [5000-(0.34)(16{,}000)]/\sqrt{(0.01)(16{,}000)^2}\} = 0.6083 < 0.99$. Thus, 16,000 pounds is too much weight.
    (b) To find the weight limit, find Y so that $P_X\{X > 5000\} = P\{\mathcal{N}(Z;0,1) >$

$[5000-(0.34)Y]/\sqrt{(0.01)Y^2}\} = 0.99$. That is, solve the equation $[5000-(0.3)Y]/(0.1)Y = 2.33$ for $Y > 0$. The answer $Y = 8726.00$ is the desired weight limit.

16. (a) $P\{\hat{\mu}_X \geq 5.25\} = P\{\mathcal{N}(Z;0,1) \geq 2.5\} = 0.0062$; (b) Underapproximation, since distribution of X is positively skewed, and thus large values of X (and of $\hat{\mu}_X$) are more probable than would be the case under a normal distribution; (c) Upper bound to probability $P\{\mu_X \geq 4.90 + b\sqrt{(0.98/50)}\}$ is $1/(1+b^2)$. Here, $5.25 = 4.90 + b(0.14)$, so that $b = 2.5$ and $1/(1+b^2) = 0.138$. Hence our approximation here is 0.138, a value much larger than the approximation in part (a). Since the distribution of X is only slightly skewed, the value 0.0062 is probably closer to the true probability $P\{\hat{\mu}_X \geq 5.25\}$ than is 0.138. The value 0.138 is, however, certain to exceed $P(\hat{\mu}_X \geq 5.25)$ no matter what distribution X has, so choosing 0.138 never overstates the probability.

CHAPTER 8

Remark. The notation exp $[x]$ means e^x.

1. (a) $\hat{\mu}_X = 618.50/1050 = 0.589$, $\hat{\theta} = 1/\hat{\mu}_X = 1.6977$.
(b) Compute $F_X(x) = 1 - \exp[-1.6977x]$.

x	$F_X(x)$	Midpoint	Theoretical probability	Theoretical frequency
0.50	0.5721	0.25	0.5721	600.7
1.00	0.8169	0.75	0.2448	257.0
1.50	0.9216	1.25	0.1047	109.9
2.00	0.9665	1.75	0.0449	47.1
2.50	0.9856	2.25	0.0191	20.1
3.00	0.9939	2.75	0.0083	8.7
3.50	0.9974	3.25	0.0035	3.7
		> 3.50	0.0026	2.7

(c) $P\{M > 6\} = P\{X > 2.76\} = \exp[-2.75(1.6977)] = 0.0094$.
(d) $\hat{\mu}_Y = 0.5058$, $\hat{\theta} = 1.9769$. Compute $F_Y(y) = 1 - \exp[-1.9769y]$.

y	$F_Y(y)$	Midpoint	Theoretical probability	Theoretical frequency
0.50	0.6278	0.25	0.6278	295.7
1.00	0.8615	0.75	0.2337	110.1
1.50	0.9485	1.25	0.0870	41.0
2.00	0.9808	1.75	0.0323	15.2
2.50	0.9929	2.25	0.0121	5.7
3.00	0.9973	2.75	0.0044	2.1
3.50	0.9990	> 3.25	0.0027	1.3

2. (a) $P\{X > 1\} = \exp[-2] = 0.1353$,
   (b) $F_X(x) = 0.5 = 1 - \exp[-2x]$, so that $x = 0.3466 = Q_{0.5}(x)$,
   $F_X(x) = 0.9 = 1 - \exp[-2x]$, so that $x = 1.1513 = Q_{0.9}(x)$.
   (c) $P\{Y < 1\} = F(1) = 1 - \exp[-\theta] = 0.75$, so that $\theta = 1.3863$.
5. (a) $P\{2 \leq X \leq 4\} = F_X(4) - F_X(2) = \exp[-0.2(2)] - \exp[-0.2(4)]$
   $= 0.2210$.
   (b) $P\{X > 1.5\} = \exp[-(1.5)(0.2)] = 0.7408$.
   (c) $P\{X < 3\} = 1 - \exp[-3(0.2)] = 0.4512$
7. (a) $P\{X < 1\} = 1 - \exp[-1/48] = 0.0206$.
   (b) Expectation $= 1/\theta = 48$ hours.
8. (a) $P\{X > 5, Y > 5\} = P\{X > 5\}P\{Y > 5\} = \exp[-0.1(5)(2)] = 0.3679$.
   (b) $P\{X > \tau, Y > \tau\} = \exp[-2\tau\theta]$.
   (c) $P\{Z \leq \tau\} = 1 - \exp[-2\tau\theta]$, so Z has an exponential distribution with parameter 2θ.
   (d) Expectation $= 1/2\theta = 5$.
11. (c) $\hat{\theta} = 1400.91/151{,}618.36 = 0.00924$, $r = (1400.91)^2/151{,}618.36 = 12.94$.
    Use $r = 13$ as an approximation.

Endpoint y	$F_Y(y) = I_r(\theta y)$	Midpoint	Theoretical probability	Theoretical frequency	Observed frequency
550	0.00241	460	0.0024	0.2	1
730	0.02141	640	0.0190	1.9	2
910	0.08592	820	0.0645	6.5	8
1090	0.21536	1000	0.1294	13.1	11
1270	0.39373	1180	0.1784	18.0	18
1450	0.57991	1360	0.1862	18.8	14
1630	0.73720	1540	0.1573	15.9	20
1810	0.85005	1720	0.1129	11.4	11
1990	0.92145	1900	0.0714	7.2	10
2170	0.96180	2080	0.0404	4.1	3
2350	0.98248	2260	0.0207	2.1	2
2530	0.99246	2440	0.0100	1.0	1
> 2530	1.00000		0.0074	0.7	0
			1.0000	100.9	101

13. (a) $\theta = 4/8 = 0.5$, $r = 16/8 = 2$.
    (b) $P\{1 \leq X \leq 4\} = F_X(4) - F_X(1) = I_2(2) - I_2(0.5) = 0.59399 - 0.09173$
    $= 0.502$.
    (c) $0.5 = P\{X < X_{0.50}\} = I_2(X_{0.50}/2)$, so $X_{0.50}/2 = 1.68$ and $Q_{0.50}(X) = 3.36$.
15. (a) $P\{X \geq 14\} = \exp[-(14-0.2)/10^2] = \exp[-1.44] = 0.237$.
    (b) $P\{X < 0\} = 0$.
    (c) Expectation $= 2 + 10\Gamma(1.5) = 10.862$.
    (d) Solve $0.5 = 1 - \exp[-(X-2)/10^2]$ for $Q_{0.50}(X) = 10.33$.
    (e) 11, which is answer for both mean and median.
17. (a) $P\{X > 8\} = \exp\{-[(8-1.5)/6.41]^{2.298}\} = 0.356$.
    (b) $P\{6 < X < 10\} = \exp[-((6-1.5)/6.41)^{2.298}]$
    $- \exp[-((10-1.5)/6.41)^{2.298}] = 0.494$.

ANSWERS   731

(c) $P\{X < 4\} = 1 - \exp[-((4-1.5)/6.41)^{2.298}] = 1 - 0.892 = 0.108$.
Solve $0.5 = \exp[-((X-1.5)/6.41)^{2.298}]$ for X;
   $X_{0.50} = 6.9650$.

19. (b) Solve $0.25 = 1 - \exp[-((X-1.0)/7360)^{0.98}]$ obtain $Q_{0.25}(X) = 2065$.
Solve $0.75 = 1 - \exp[-((X-1.0)/7360)^{0.98}]$ to obtain $Q_{0.75}(X) = 10{,}272$.
(c) $P\{X < 7\} = 1 - \exp[-((7-1.0)/7360)^{0.98}] = 1 - 0.99906 = 0.00094$.

21. (a) $P\{10 < X < 40 \text{ or } 60 < X < 90\} = 30/100 + 30/100 = 0.6$.
(b) $P\{0 < X < 15 \text{ or } 35 < X < 40 \text{ or } 60 < X < 65 \text{ or } 85 < X < 100\} = 0.4$,
so the probability of travelling less than 10 miles is smaller.

24. (a) Set $x = F_W(w) = 1 - \exp[-((w-1.6)/3.0)^4]$, and solve for w.
(b) Solve $0.77 = 1 - \exp[-((w-1.6)/3.0)^4]$ to obtain $w = 4.90$.

26. The solution is given for both (i) $r = 0.5$, $s = 3.5$ and (ii) $r = 0.5766$, $s = 3.7$. In case (i) we use single interpolation or we may find the values directly from more extensive tables. In case (ii) we require double interpolation.

	$r = 0.5$, $s = 3.5$		$r = 0.5766$, $s = 3.7$		
Midpoint	Theoretical probability	Theoretical frequency	Theoretical probability	Theoretical frequency	Observed frequency
0.02	0.3941	394.1	0.3793	379.3	209
0.06	0.1451	145.1	0.1445	144.5	264
0.10	0.0997	99.7	0.1008	100.8	170
0.14	0.0750	75.0	0.0766	76.6	105
0.18	0.0587	58.7	0.0604	60.4	58
0.22	0.0468	46.8	0.0485	48.5	53
0.26	0.0377	37.7	0.0394	39.4	33
0.30	0.0306	30.6	0.0320	32.0	22
0.34	0.0247	24.7	0.0260	26.0	12
0.38	0.0200	20.0	0.0225	22.5	11
0.42	0.0161	16.1	0.0168	16.8	11
0.46	0.0129	12.9	0.0133	13.3	10
0.50	0.0102	10.2	0.0107	10.7	4
0.54	0.0080	8.0	0.0082	8.2	5
0.58	0.0061	6.1	0.0063	6.3	4
0.62	0.0046	4.6	0.0048	4.8	8
0.66	0.0034	3.4	0.0035	3.5	1
0.70	0.0024	2.4	0.0025	2.5	5
0.74	0.0016	1.6	0.0016	1.6	4
0.78	0.0011	1.1	0.0011	1.1	5
0.82	0.0006	0.6	0.0006	0.6	2
0.86	0.0003	0.3	0.0003	0.3	1
0.90	0.0001	0.1	0.0001	0.1	1
0.94	0.0000	0.0	0.0000	0.0	0
0.98	0.0000	0.0	0.0000	0.0	2
	0.9998	999.8	0.9998	999.8	1000

$P\{\text{Pb} > 2\text{Cu}\} = P\{X < \tfrac{1}{3}\} = I_{0.333}(0.5766, 3.7) = 0.88853$.

28. (a) $r = 0.6[(0.6)(0.4) - 0.048]/0.048 = 2.4$
$s = 0.4[(0.6)(0.4) - 0.048]/0.048 = 1.6$.

(b) $P\{0.2 \leq X \leq 1.0\} = I_{1.0}(3,2) - I_{0.2}(3,2) = 1 - 0.02720 = 0.97280$.
From the Chebyshev Inequality, the probability is at least 0.750. For the normal distribution, the probability is 0.9544.
(c) Solve $0.50 = I_x(3,2) - I_0(3,2)$ for x. Since $I_0(3,2) = 0$, and $I_{0.60}(3,2) = 0.47520$, $I_{0.65}(3,2) = 0.56298$, we find $Q_{0.50}(x) = 0.614$.

31. Transform the data to logarithms, and then fit a normal distribution with mean 3.4014, and standard deviation 0.00805.

Height at endpoint x	Logarithmic transformation $y = \log x$	Theoretical cumulative Probability	Height at midpoint	Theoretical probability	Theoretical frequency	Observed frequency
30.80	3.428	0.9996				
30.70	3.424	0.9979	30.75	0.0017	3.3	1
30.60	3.421	0.9934	30.65	0.0045	8.8	1
30.50	3.418	0.9826	30.55	0.0108	21.1	10
30.40	3.414	0.9463	30.45	0.0363	70.8	32
30.30	3.411	0.8925	30.35	0.0538	105.0	111
30.20	3.408	0.8078	30.25	0.0847	165.2	214
30.10	3.405	0.6915	30.15	0.1163	226.9	386
30.00	3.401	0.5000	30.05	0.1915	373.6	365
29.90	3.398	0.3557	29.95	0.1443	281.5	288
29.80	3.395	0.2266	29.85	0.1291	251.9	199
29.70	3.391	0.1075	29.75	0.1191	232.4	129
29.60	3.388	0.0537	29.65	0.0538	105.0	86
29.50	3.384	0.0174	29.55	0.0363	70.8	62
29.40	3.381	0.0066	29.45	0.0108	21.1	26
29.30	3.378	0.0021	29.35	0.0045	8.8	17
29.20	3.374	0.0004	29.25	0.0017	3.3	10
29.10	3.371	0.0001	29.15	0.0003	0.6	9
29.00	3.367	0.0000	29.05	0.0001	0.2	2
28.90	3.364					2
28.80						1
					1950.3	1951

CHAPTER 9

1. (a) The probability mass functions of (X,Y), (X,Z), (Y,Z) all have the form:

(x,y)	$(1,0)$	$(0,1)$	$(0,0)$	$(1,1)$
$p(x,y)$	$\frac{1}{4}$	$\frac{1}{4}$	$\frac{1}{4}$	$\frac{1}{4}$

;

(b) The probability mass functions of X, of Y, and of Z all have the form:

x	0	1
$P(x)$	$\frac{1}{2}$	$\frac{1}{2}$

;

(c) X and Y are independent; Y and Z are independent; X and Z are independent;
(d) $\rho_{X,Y} = 0$. No, since independence implies zero correlation.

(e) No. $P\{(X,Y,Z)=(1,1,1)\}=\frac{1}{4}$, whereas $P\{X=1\}P\{Y=1\}P\{Z=1\}=\frac{1}{8}$.

(f)

x	0	1
$p_{X|Y=1}(x)$	$\frac{1}{2}$	$\frac{1}{2}$

;

(g) Yes; because X and Y are independent, $p_{X|Y=1}(x) = p_X(x)$;

(h)

(x,y)	$(0,0)$	$(1,1)$
$p_{(X,Y)|Z=1}(x,y)$	$\frac{1}{2}$	$\frac{1}{2}$

.

3. (a)

x	-2	-1	0	1	2
$p_X(x)$	0.092	0.129	0.161	0.203	0.415

,

y	0	1	2	3	4	5
$p_Y(y)$	0.075	0.044	0.094	0.143	0.244	0.400

;

(b)

y	0	1	2	3	4	5
$p_{Y|X=-2}(y)$	0.304	0.098	0.196	0.120	0.120	0.163
$p_{Y|X=-1}(y)$	0.225	0.171	0.186	0.217	0.116	0.085
$p_{Y|X=0}(y)$	0.056	0.056	0.137	0.205	0.261	0.286
$p_{Y|X=1}(y)$	0.034	0.010	0.074	0.143	0.360	0.379
$p_{Y|X=2}(y)$	0.005	0.005	0.036	0.101	0.248	0.605

;

(c)

x	-2	-1	0	1	2
$E(Y|X=x)$	2.145	2.083	3.419	3.922	4.397

;

(d) $\mu_X=0.720$, $\mu_Y=3.637$, $\sigma_X^2=1.842$, $\sigma_Y^2=2.383$, $\sigma_{XY}=1.162$, $\rho_{X,Y}=0.56$;
(e) Yes.

5. (a, b)

y \ x	0	1	2	3	4	5	$p_Y(y)$
0	0	0	0	0	0	0	0
1	0	0	0.02304	0.03456	0	0	0.0576
2	0	0.01536	0.13824	0.20736	0.05184	0	0.4128
3	0	0.03072	0.06912	0.10368	0.10368	0	0.3072
4	0	0.03072	0	0	0.10368	0	0.1344
5	0.01024	0	0	0	0	0.07776	0.0880
$p_X(x)$	0.01024	0.0768	0.2304	0.3456	0.2592	0.07776	1.0000

,

$\mu_Y=2.7824$, $\sigma_Y^2=1.0823$

734   ANSWERS

(c)

x	1	2	3	4
$P_{X|Y=3}(x)$	0.1000	0.2250	0.3375	0.3375

;

(d)

y	1	2	3
$P_{Y|X=3}(y)$	0.1	0.6	0.3

;

(e) $P\{X > 3 | Y \leq 2\} = 0.05184/0.47040 = 0.11$.
(f) $P\{Y \geq 3 | X = 3\} = 0.10368/0.34560 = 0.30$.

7. (a) $\rho_{X,Y} = 0.50$ (b) $\sigma_{ZW} = 0.50$ (c) $\rho_{Z,W} = 0.5$;
(d) $\rho_{X,Y} = \rho_{Z,W}$;   (e) $\sigma_{ZY} = (\frac{1}{2}) \sigma_{XY} = 1.50$;
$\sigma_{XW} = (\frac{1}{3}) \sigma_{XY} = 1.00$; (f) $\rho_{Z,Y} = 0.50 = \rho_{X,W}$.

9. (a) Each pair of numbers shown has probability 1/40;
(b)

x	1.75	2.00	2.25	2.50	2.75	3.00	3.25	3.50	3.75	4.00	4.25	4.50
$p(x)$	$\frac{3}{40}$	$\frac{7}{40}$	$\frac{1}{40}$	$\frac{1}{40}$	0	$\frac{1}{40}$	0	$\frac{5}{40}$	$\frac{5}{40}$	$\frac{11}{40}$	$\frac{3}{40}$	$\frac{3}{40}$

y	45	47	48	50	51	52	53	54	56	61	63	65
$p_Y(y)$	$\frac{1}{40}$	$\frac{1}{40}$	$\frac{1}{40}$	$\frac{1}{40}$	$\frac{3}{40}$	$\frac{1}{40}$	$\frac{1}{40}$	$\frac{1}{40}$	$\frac{1}{40}$	$\frac{1}{40}$	$\frac{2}{40}$	$\frac{2}{40}$

y	68	69	70	71	72	73	74	76	77	79	83
$p_Y(y)$	$\frac{2}{40}$	$\frac{1}{40}$	$\frac{4}{40}$	$\frac{3}{40}$	$\frac{3}{40}$	$\frac{2}{40}$	$\frac{2}{40}$	$\frac{1}{40}$	$\frac{3}{40}$	$\frac{2}{40}$	$\frac{1}{40}$

(c) $\mu_X = 3.3375$, $\sigma_X^2 = 0.8611$, $\mu_X = 65.775$, $\sigma_Y^2 = 107.224$,
$\sigma_{XY} = 8.382$, $\rho_{X,Y} = 0.87$;

(d)

y	63	68	70	71	76	77	79
$p_{Y|X=}(y)$	$\frac{2}{11}$	$\frac{1}{11}$	$\frac{2}{11}$	$\frac{2}{11}$	$\frac{1}{11}$	$\frac{2}{11}$	$\frac{1}{11}$

;

(e) If use mode, predict 63, 70, 71, or 77; if use mean, predict 71.4, if use median, predict 71. If $X = 2$, then

y	45	50	51	52	53	61
$p_{Y|X=2}(y)$	$\frac{1}{7}$	$\frac{1}{7}$	$\frac{2}{7}$	$\frac{1}{7}$	$\frac{1}{7}$	$\frac{1}{7}$

,

so that the prediction is different.

10. (a) $X^* = 0.9083X - 2.5250$, $Y^* = 0.3869Y - 2.8231$,
(b) $\rho_{X^*,Y^*} = \rho_{X,Y} = 0.8448$. Probability contour approximates contour in Figure 5.4 with $\rho = 0.8$;

(c) $P\{X > 5.5, Y > 16.5\} = P\{X^* > 2.47, Y^* > 3.56\}$.
Using $\rho = 0.85$, we obtain 0.000172. The interpolation is carried out in two steps using the National Bureau of Standards tables:

k \ h	2.4	2.5
3.5	0.000214	0.000207
3.56	0.000176	0.000170
3.6	0.000150	0.000146

k \ h	2.4	2.47	2.5
3.56	0.000176	0.000172	0.000170

;

(d) $P\{X \leq 3.5, Y \leq 5.5\} = P\{X^* \geq -0.654, Y^* \geq 0.695\}$
$= P\{\mathcal{N}(Z;0,1) \geq 0.695\} - P\{V \geq 0.654, W > 0.695\} = 0.2435 - 0.0005$
$= 0.2430$.
Note: Interpolation from National Bureau of Standards table for $\rho = -0.85$:

k \ h	0.6	0.7
0.6	0.0011306	0.0006637
0.695	0.0006870	0.0003939
0.7	0.0006637	0.0003797

k \ h	0.6	0.654	0.7
0.695	0.000687	0.0005287	0.0003939

;

(e) r.f. $\{X > 5.5, Y > 16.5\} = 55/20453 = 0.002689$, compared with 0.000172
r.f. $\{X \leq 3.5, Y \leq 5.5\} = 5287/20453 = 0.258495$, compared with 0.2430.

13. Combined length is $U = X + Y$, which has a normal distribution with mean 6.50 and variance $2(0.05)^2(1 + \rho)$.
   (a) $P\{\mathcal{N}(U; 6.5, (0.07)^2) < 6.6\} = 0.9207$,
   (b) $P\{\mathcal{N}(U; 6.5, (0.0866)^2) < 6.6\} = 0.8749$,
   (c) $P\{\mathcal{N}(U; 6.5, (0.05)^2) < 6.6\} = 0.9772$,
   (d) $P\{\mathcal{N}(U; 6.5, (0.1)^2) < 6.6\} = 0.8413$,
   (e) $P\{\mathcal{N}(U; 6.5, 0) < 6.6\} = 1.0000$.

CHAPTER 10

2. (a) $p_Y(\tau) = e^{-\lambda}\lambda^{(1/5)\tau}/(\tfrac{1}{5}\tau)!$ for $\tau = 0, 5, 10, 15, \ldots$, and $p_Y(\tau) = 0$ otherwise.
   (b) $\mu_Y = 5\lambda$. (c) $\sigma_Y^2 = 25\lambda$. (d) No, compare the forms of the probability mass functions. (e) $P_Y\{Y \leq 7\} = P_X\{5X \leq 7\} = P_X\{X = 0 \text{ or } X = 1\} = 0.3679 + 0.3679 = 0.7358$.
4. $U = X + Y$ has a normal distribution with expected value $\mu = 70$ and variance $\sigma^2 = 9 + 16 + 2(0.7)(3)(4) = 41.8$.
6. (a) Binomial distribution with parameters $n = 100$ and p.
   (b) Binomial distribution with parameters $n = 300$ and p.

(c) $W = X + Y$ has a binomial distribution with parameters $n = 400$ and p.

(d) $P_W\{W \geq 50\} = P_W\left\{\dfrac{W - 400p}{\sqrt{400p(1-p)}} \geq \dfrac{50 - 400p}{\sqrt{400p(1-p)}}\right\}$

$\cong P\left\{\mathcal{N}(Z;0,1) \geq \dfrac{50 - 400p}{\sqrt{400p(1-p)}}\right\}$

from Central Limit Theorem. But

$P\left\{\mathcal{N}(Z;0,1) \geq \dfrac{50 - 400p}{\sqrt{400p(1-p)}}\right\} = P\{\mathcal{N}(Z;0,1) \geq 1.67\} = 0.048$

when $p = 0.1$.

8. $W = \frac{1}{5}(X_1 + X_2 + X_3 + X_4 + X_5)$ has a gamma distribution with parameters $r = 5$ and $5\theta = 2.5$. Also $\mu_W = 2.0$, $\sigma_W^2 = 0.8$. $P_W\{W > 2.5\} = 1 - P_W\{W \leq 2.5\} = 1 - I_5((2.5)(2.5)) = 0.25315$.

10. (a) $P_X\{X \geq 5.0\} \cong P\left\{\mathcal{N}(Z;0,1) \geq \dfrac{(5.0 - 1.5)}{\sqrt{50(0.03)(0.97)}}\right\} = 0.0019$.

(b) $P_X\{X \geq 5.0\} \cong P\left\{\mathcal{N}(Z;0,1) \geq \dfrac{(5.0 - 0.5 - 1.5)}{\sqrt{50(0.03)(0.97)}}\right\} = 0.0064$.

(c) Correct value of $P_X\{X \geq 5.0\}$ is 0.01681. Continuity correction gives closer approximation, but is not fully satisfactory.

12. Without continuity correction,

$P_X\{X \geq 3\} \cong P\left\{\mathcal{N}(Z;0,1) \geq \dfrac{(3.0 - 1.0)}{\sqrt{100(0.01)(0.99)}}\right\} = 0.0222$.

With continuity correction,

$P_X\{X \geq 3\} = P\left\{\mathcal{N}(Z;0,1) \geq \dfrac{3.0 - 0.5 - 1.0}{\sqrt{100(0.01)(0.99)}}\right\} = 0.0658$.

Can also use Poisson approximation with $\lambda = 1$ and obtain $P_X\{X \geq 3\} \cong 1 - 0.3679 - 0.3679 - 0.1839 = 0.0803$. All of these approximations either assume that cans are sampled with replacement, or that the number N of cans in the factory output is very much larger than $n = 100$. Alternatively, we can assume that the defectiveness or nondefectiveness of the cans are mutually statistically independent events. The exact value of $P_X\{X \geq 3\}$ is 0.07937, so that the Poisson approximation seems best in this case. The normal approximations, however, are often easier to compute.

13. Let $q = 1 - p$. By Central Limit Theorem, $P\{-0.005 \leq \hat{p} - p \leq 0.005\} \cong P\{-0.005/\sqrt{pq/n} \leq \mathcal{N}(Z;0,1) \leq 0.005/\sqrt{pq/n}\} = 2P\{\mathcal{N}(Z;0,1) \leq 0.005/\sqrt{pq/n}\} - 1$. The probability $P\{\mathcal{N}(Z;0,1) \leq 0.005/\sqrt{pq/n}\}$ is smallest when $0.005/\sqrt{pq/n}$ is smallest, or equivalently when pq/n is largest. If we graph $pq/n = p(1-p)/n$ as a function of P, pq/n is largest when $p = \frac{1}{2}$, and that this largest value is $(1/4)n$. Thus, $\hat{p}$ is within 0.005 of p with probability at least 0.90 for all values of p when $2P\{\mathcal{N}(Z;0,1) \leq 0.005/\sqrt{1/4n}\} - 1 \geq$

0.90. Now $2P\{\mathcal{N}(Z;0,1) \leq 1.645\} - 1 = 0.90$ and thus if $1.645 \leq 0.005/\sqrt{1/4n}$, we achieve the desired goal. Hence, we want $n \geq [1.645/2(0.005)]^2 = 27{,}060$

15. (a) $E(\hat{\mu}_X - \hat{\mu}_Y) = \mu - \nu = \Delta$, $\text{Var}(\hat{\mu}_X - \hat{\mu}_Y) = \sigma^2(1/n + 1/m)$.
    (b) From Inequality (6.8) of Chapter 5, we have

$$P\left\{-d\sigma\sqrt{\frac{1}{m}+\frac{1}{n}} \leq \hat{\mu}_X - \hat{\mu}_Y - \Delta \leq d\sigma\sqrt{\frac{1}{m}+\frac{1}{n}}\right\} \geq 1 - \frac{1}{d^2}.$$

As m and n increase the term $\sqrt{1/m + 1/n}$ gets smaller and thus $d\sigma\sqrt{1/m+1/n}$ gets closer to zero. Another way to see this is to let $k = d\sqrt{1/m + 1/n}$. Then $P\{\Delta - k\sigma \leq \mu_X - \mu_X \leq \Delta + k\sigma\} \geq 1 - (m+n)/mnk^2$. Suppose $k = 2$, then for $m = n = 10$ we get 0.95, for $m = n = 100$ we get 0.995, and for $m = n = 10{,}000$ we get 0.99995. Thus the probability is getting closer to 1.

(c) $\mathcal{N}(\hat{\mu}_X - \hat{\mu}_Y; \Delta, \sigma^2(\frac{1}{m}+\frac{1}{n}))$.

(d)
$$P\{-1 \leq \hat{\mu}_X - \hat{\mu}_Y - \Delta \leq 1\} = P\left\{-\frac{1}{0.2828} \leq \frac{\hat{\mu}_X - \hat{\mu}_Y - \Delta}{0.28} \leq \frac{1}{0.2828}\right\}$$
$$\doteq P\{\mathcal{N}(Z;0,1) \leq 3.536\} - P\{(Z;0,1) \leq -3.536\}$$
$$= 0.9996.$$

(e) If we assume normality, then we require

$$\frac{1}{\sigma\sqrt{\frac{1}{m}+\frac{1}{n}}} = \frac{1}{2\sqrt{\frac{2}{n}}} \geq 1.96,$$

so that $n \geq (1.96)^2 4(2) = 30.7$. Hence, we need $n \geq 31$. If we do not assume a normal distribution, we may use the Bienaymé-Chebyshev inequality. Now we let $k\sigma = 1$, so that $k = \frac{1}{2}$, and solve $1 - (2/n)/k^2 \geq 0.95$. Thus $n \geq 8/0.05 = 160$. [*Note:* We expect to require more observations if we do not know the form of the distribution.]

CHAPTER 11

3. (a) The entries in the table (transition matrix) are all nonnegative numbers, and the sum of the entries in each row equals 1.
   (b)

$\beta = 0$

	S_1	S_2	S_3
S_1	1	0	0
S_2	α	$\bar{\alpha}\bar{\delta}$	$\bar{\alpha}\bar{\delta}$
S_3	0	$\bar{\delta}$	δ

$\delta = 0$

	S_1	S_2	S_3
S_1	1	0	0
S_2	α	0	$\bar{\alpha}$
S_3	β	$\bar{\beta}$	0

$\alpha = \beta = 0$

	S_1	S_2	S_3
S_1	1	0	0
S_2	0	δ	$\bar{\delta}$
S_3	0	$\bar{\delta}$	δ

(c)

State at Time $t+2$

		S_1	S_2	S_3
State at Time t	S_1	1	0	0
	S_2	0.52	0.24	0.24
	S_3	0.20	0.40	0.40

State at Time $t+3$

		S_1	S_2	S_3
State at Time t	S_1	1	0	0
	S_2	0.616	0.192	0.192
	S_3	0.360	0.320	0.320

State at Time $t+4$

		S_1	S_2	S_3
State at Time t	S_1	1	0	0
	S_2	0.6928	0.1536	0.1536
	S_3	0.4880	0.2560	0.2560

Patterns that emerge are that $p_{11}^{(k)} = 1$, $p_{12}^{(k)} = p_{13}^{(k)} = 0$, $p_{22}^{(k)} = p_{23}^{(k)}$, $p_{32}^{(k)} = p_{33}^{(k)}$ for $k = 1, 2, 3, \ldots$. It can also be shown that $p_{21}^{(k)} = 1 - (0.6)(0.8)^{k-1}$ and $p_{31}^{(k)} = 1 - (0.8)^{k-1}$.

(d and e)
$P\{X_4 = 2 | X_3 = 2\} = P\{X_3 = 2 | X_2 = 2\} = p_{22} = 0.3$,
$P\{X_4 = 2 | X_2 = 2\} = P\{X_3 = 2 | X_1 = 2\} = p_{22}^{(2)} = 0.24$,
$P\{X_4 = 2 | X_1 = 2\} = P\{X_3 = 2 | X_0 = 2\} = p_{22}^{(3)} = 0.192$,
$P\{X_4 = 2 | X_0 = 2\} = p_{22}^{(4)} = 0.1536$.
(f) $P\{X_1 = 2\} = 0.27$, $P\{X_2 = 2\} = 0.213$, $P\{X_3 = 2\} = 0.171$, $P\{X_4 = 2\} = 0.137$.
(g) $P\{X_1 = 2\} = P\{X_2 = 2\} = P\{X_3 = 2\} = P\{X_4 = 2\} = 0$.
(h) $P\{X_1 = 2\} = 0.5$, $P\{X_2 = 2\} = 0.4$, $P\{X_3 = 2\} = 0.32$, $P\{X_4 = 2\} = 0.256$.

4.

(a)

State at Time $t+2$

		S_1	S_2	S_3
State at Time t	S_1	0.7424	0.1101	0.1475
	S_2	0.3890	0.5085	0.1025
	S_3	0.0590	0.0351	0.9059

(b and c) From the assumptions, $a_1 = 100{,}000/800{,}000 = \frac{1}{8}$, $a_2 = \frac{5}{8}$, and $a_3 = \frac{2}{8}$. Thus, the unconditional probabilities after one and two units of time are respectively:

x	1	2	3
$P\{X_1 = x\}$	0.270	0.451	0.279

x	1	2	3
$P\{X_2 = x\}$	0.351	0.340	0.309

6. (a) (i) A learning situation (or a task) in which each stage of solution depends upon accomplishing the previous stage of solution, and in which failure at any stage forces learning (or the task) to begin at the start (for example, building a house with playing cards).

(ii) A physical system in which there is a resting state (S_1) which has a chance of being excited to higher energy levels (S_2, S_3, S_4). However, once any higher energy level is reached the system cycles among these levels forever.

(iii) A game between two players in which they initially divide 4 chips between themselves. Then they play the same game (in which one player has probability p of winning and the other player has probability $q = 1 - p$ of winning) over and over, wagering one chip for each game, until one of the players loses all of his chips and cannot wager any more.

(b)

State at Time $t+2$

	S_1	S_2	S_3	S_4
S_1	q	qp	p^2	0
S_2	q	qp	0	p^2
S_3	q	qp	0	p^2
S_4	q	qp	0	p^2

State at Time t

(i)

State at Time $t+2$

	S_1	S_2	S_3	S_4
S_1	p_1^2	$p_1p_2 + p_4$	$p_1p_3 + p_2$	$p_1p_4 + p_3$
S_2	0	0	0	1
S_3	0	1	0	0
S_4	0	0	1	0

(ii)

State at Time $t+2$

	S_1	S_2	S_3	S_4	S_5
S_1	1	0	0	0	0
S_2	p	pq	0	q^2	0
S_3	p^2	0	$2pq$	0	q^2
S_4	0	p^2	0	pq	q
S_5	0	0	0	0	1

State at Time t

(iii)

State at Time $t+3$

	S_1	S_2	S_3	S_4
S_1	q	qp	qp^2	p^3
S_2	q	qp	qp^2	p^3
S_3	q	qp	qp^2	p^3
S_4	q	qp	qp^2	p^3

State at Time t

(i)

State at Time $t+3$

	S_1	S_2	S_3	S_4
S_1	p_1^3	$p_1^2 p_2 + p_1 p_4 + p_3$	$p_1^2 p_3 + p_1 p_2 + p_4$	$p_1^2 p_4 + p_1 p_3 + p_2$
S_2	0	1	0	0
S_3	0	0	1	0
S_4	0	0	0	1

(ii)

740 ANSWERS

State at Time $t+3$

		S_1	S_2	S_3	S_4	S_5
	S_1	1	0	0	0	0
State	S_2	$p+p^2q$	0	$2pq^2$	0	q^3
at Time	S_3	p^2	$2p^2q$	0	$2pq^2$	q^2
t	S_4	p^3	0	$2p^2q$	0	pq^2+q
	S_5	0	0	0	0	1

(iii)

(c)

x	1	2	3	4
$P\{X_2 = x\}$	q	qp	$\frac{1}{4}p^2$	$\frac{3}{4}p^2$

(i)

x	1	2	3	4
$P\{X_2 = x\}$	$\frac{1}{4}p_1^2$	$\frac{1}{4}(1+p_1p_2+p_4)$	$\frac{1}{4}(1+p_1p_3+p_2)$	$\frac{1}{4}(1+p_1p_4+p_3)$

(ii)

x	1	2	3	4	5
$P\{X_2 = x\}$	$\frac{1}{5}(1+p+p^2)$	$\frac{1}{5}p$	$\frac{2}{5}pq$	$\frac{1}{5}q$	$\frac{1}{5}(1+q+q^2)$

(iii)

CHAPTER 12

2. (a) $P = \begin{pmatrix} 0.32 & 0.08 & 0.60 \\ 0.08 & 0.32 & 0.60 \\ 0.25 & 0.25 & 0.50 \end{pmatrix}$, $P^{(2)} = \begin{pmatrix} 0.2588 & 0.2012 & 0.5400 \\ 0.2012 & 0.2588 & 0.5400 \\ 0.2250 & 0.2250 & 0.5500 \end{pmatrix}$,

$f_{11}^{(2)} = p_{11}^{(2)} - f_{11}p_{11} = 0.2588 - (0.32)(0.32) = 0.1564,$

$f_{12}^{(2)} = p_{12}^{(2)} - f_{12}p_{22} = 0.2012 - (0.8)(0.32) = 0.1756,$

$f_{13}^{(2)} = p_{13}^{(2)} - f_{13}p_{33} = 0.5400 - (0.60)(0.50) = 0.2400,$

$f_{21}^{(2)} = p_{21}^{(2)} - f_{21}p_{11} = 0.2012 - (0.08)(0.32) = 0.1756,$

$f_{31}^{(2)} = 0.1450$, $f_{32}^{(2)} = 0.1450$, $f_{33}^{(2)} = 0.300$, $f_{22}^{(2)} = 0.1564$, $f_{23}^{(2)} = 0.2400$.

(b) Solve the equations

$m_{11} = 1 + p_{12}m_{21} + p_{13}m_{31} = 1 + 0.08m_{21} + 0.60m_{31},$

$m_{21} = 1 + p_{22}m_{21} + p_{23}m_{31} = 1 + 0.32m_{21} + 0.60m_{31},$

$m_{31} = 1 + p_{32}m_{21} + p_{33}m_{31} = 1 + 0.25m_{21} + 0.50m_{31}.$

to obtain $m_{21} = 5.79$, $m_{31} = 4.89$, $m_{11} = 4.40$.

Solve the equations
$$m_{12} = 1 + 0.32m_{12} + 0.60m_{32},$$
$$m_{22} = 1 + 0.08m_{12} + 0.60m_{32},$$
$$m_{32} = 1 + 0.25m_{12} + 0.50m_{32},$$

to obtain $m_{12} = 5.79$, $m_{32} = 4.89$, $m_{22} = 4.40$.
Solve the equations
$$m_{13} = 1 + 0.32m_{13} + 0.08m_{23},$$
$$m_{23} = 1 + 0.08m_{13} + 0.32m_{23},$$
$$m_{33} = 1 + 0.25m_{13} + 0.25m_{23},$$

to obtain $m_{13} = m_{23} = 1.67$, $m_{33} = 1.83$.
(c) The chain is irreducible since all the transition probabilities are positive (not zero). The Markov chain achieves equilibrium. The chain is not absorbing since there are no absorbing states.
(d)

$$P = \begin{pmatrix} 0.32 & 0.08 & 0.60 \\ 0.08 & 0.32 & 0.60 \\ 0.00 & 0.00 & 1.00 \end{pmatrix}, \quad P^{(2)} = \begin{pmatrix} 0.1088 & 0.0512 & 0.8400 \\ 0.0512 & 0.1088 & 0.8400 \\ 0.0000 & 0.0000 & 1.0000 \end{pmatrix},$$

$$P^{(3)} = \begin{pmatrix} 0.0389 & 0.0251 & 0.9360 \\ 0.0251 & 0.0389 & 0.9360 \\ 0.0000 & 0.0000 & 1.0000 \end{pmatrix}, \quad P^{(4)} = \begin{pmatrix} 0.0145 & 0.0111 & 0.9744 \\ 0.0111 & 0.0145 & 0.9744 \\ 0.0000 & 0.0000 & 1.0000 \end{pmatrix}.$$

$f_{33}^{(1)} = p_{33} = 1$, $\quad f_{33}^{(2)} = p_{33}^{(2)} - f_{33}p_{33} = 0$,

$f_{33}^{(3)} = p_{33}^{(3)} - f_{33}^{(2)}p_{33} - f_{33}p_{33}^{(2)} = 0$, $\quad f_{33}^{(4)} = 0$,

$f_{13}^{(1)} = p_{13} = 0.60$, $\quad f_{13}^{(2)} = p_{13}^{(2)} - f_{13}p_{33} = 0.84 - (0.60)(0.60) = 0.24$,

$f_{13}^{(3)} = p_{13}^{(3)} - f_{13}^{(2)}p_{33} - f_{13}^{(1)}p_{33}^{(2)} = 0.9360 - 0.24 - 0.60 = 0.0960$,

$f_{13}^{(4)} = p_{13}^{(4)} - f_{13}^{(3)}p_{33} - f_{13}^{(2)}p_{33}^{(2)} - f_{13}p_{33}^{(3)} = 0.9744 - 0.0960 - 0.24 - 0.60$

$= 0.0384$.

5.
$$P = \begin{pmatrix} 0.45 & 0.48 & 0.07 \\ 0.05 & 0.70 & 0.25 \\ 0.01 & 0.50 & 0.49 \end{pmatrix}.$$

All the states communicate, so that there is a closed communicating class. All the states are aperiodic, there are no absorbing states, and we have an aperiodic irreducible chain.

$m_{21} = 27.14$, $\quad m_{31} = 28.57$, $\quad m_{11} = 16.03$,

$m_{12} = 2.07$, $\quad m_{32} = 2.00$, $\quad m_{22} = 1.60$,

$m_{13} = 5.53$, $\quad m_{23} = 4.26$, $\quad m_{33} = 3.18$.

$\pi_1 = 1/m_{11} = 0.06$, $\quad \pi_2 = 1/m_{22} = 0.63$, $\quad \pi_3 = 1/m_{33} = 0.31$.

7.
$$P = \begin{pmatrix} 1.0 & 0.0 & 0.0 \\ 0.4 & 0.3 & 0.3 \\ 0.0 & 0.5 & 0.5 \end{pmatrix}.$$

State 1 is absorbing, states 2 and 3 communicate, states 2 and 3 are transient.

$$P^{(k)} = \begin{pmatrix} 1 & 0 & 0 \\ 1-0.6(0.8)^{k-1} & 0.3(0.8)^{k-1} & 0.3(0.8)^{k-1} \\ 1-(0.8)^{k-1} & 0.5(0.8)^{k-1} & 0.5(0.8)^{k-1} \end{pmatrix}$$

and we get $\pi_1 = 1$, $\pi_2 = 0$, $\pi_3 = 0$.

NAME INDEX

Abramowitz, M., 108, 663
Adams, J.D., 413, 663
Ahrens, L. H., 413, 663
Aitchison, J., 418, 663
Asch, S. E., 563, 663
Atkinson, R. C., 658, 669
Ayer, A. J., 54, 663

Bacon, F., 65
Bartholomew, D. J., 658, 663
Bartos, O. J., 658, 663
Baudin, M., 664
Bayes, T., 95, 96, 114
Bell, E. T., 16, 667
Berger, J., 577, 593, 663
Bernoulli, J., 3, 4, 247, 274, 276, 280, 663
Bertrand, J., 4
Bienayme, I. J., 227, 340
Birnbaum, Z. W., 422, 663
Bliss, C. I., 413, 664
Bonini, C. P., 298, 299, 669
Borel, E., 664
Bretherton, M. K., 292, 293, 670
Brown, J. A. C., 418, 663
Buckland, W. R., 4, 667
Bunyan, J., 18, 19
Burgess, E. W., 511, 664
Bush, R. R., 360

Campbell, A., 116, 664
Cardano, G., 16
Carroll, L., 25, 519
Carter, B., 114
Cauchy, A., 414, 415
Chadwick, J., 272, 669
Chatfield, C., 290, 303, 664
Chayes, F., 413, 664
Chebyshev, P. L., 227, 340, 356
Choute, S. P., 413, 666
Chung, K. L., 658, 664

Clarke, R. D., 270, 664
Cohen, B., 564, 565, 664
Coleman, J. S., 662, 664
Coombs, C. H., 593, 664
Cottrell, L. S., Jr., 511, 664

Darwin, C., 358
David, F. N., 15, 664
Davis, D. J., 160, 664
Davis, H. T., 664
Dawes, R. M., 593, 664
de Moivre, A., 3, 326
de Morgan, A., 38
Dempster, A. P., 4, 663
Detlefsen, J. A., 307, 664
Dewey, T. A., 116
Dhrymes, P. J., 413, 664
Dwass, M., 54, 664

Eddington, A. S., 550
Einstein, A., 4, 8, 519
Ellis, C. D., 272, 669
Epstein, B., 413, 664
Estes, W. K., 574, 664

Fechner, G. T., 413, 664
Feinlieb, B. M., 413, 665
Feldstein, S., 419, 666
Feller, W., 55, 643, 658, 665
Fermat, P. de, 3
Fieller, E. C., 418, 665
Fienberg, S. E., 295, 665
Finkelstein, M. O., 309, 665
Fisher, R. A., 254, 255, 306, 665
Freedle, R. O., 559, 562, 563, 665
Froggatt, P., 320, 665

Gaddum, J. H., 413, 665
Galanter, E., 360
Galton, F., 120, 324, 326, 330, 413, 665

Gauss, C. F. W., 4, 8, 326, 345
Geiger, H., 271, 272, 669
Geiringer, H., 55, 670
Geissler, A., 306
Gide, A., 175
Glass, D. V., 577, 665
Gnanadesikan, R., 372, 373, 670
Godwin, H. J., 233, 665
Goethe, J. W. von, 242
Goldberg, S., 54, 665
Goldhamer, H., 592, 667
Goldthwaite, L. R., 413, 665
Graunt, J., 52
Greenwood, M., Jr., 287, 288, 330, 665
Gregory, S., 111, 173, 665
Griffiths, J. C., 290, 666
Grundy, P. M., 413, 666
Guenther, W. C., 54, 666
Gutenberg, B., 418, 419, 666

Hagstroem, K. G., 243, 666
Haight, F. A., 274, 278, 317, 666
Hall, J. R., 577, 665
Halley, E., 52
Hartley, H. O., 256, 668
Hasofer, A. M., 15, 295
Hatch, T., 413, 666
Heisenberg, W., 7
Herdan, G., 18, 19, 413, 515, 666
Hodges, J. L., Jr., 54, 666
Hogg, J. M., 666
Hollingshead, A. B., 81, 666
Huyett, M. J., 372, 373, 670
Hsieh, T., 597

Indow, T., 424, 666

Jaffe, J., 419, 666
James, J., 302, 666

743

NAME INDEX

Johnson, N.L., 304, 410, 418, 666, 667
Johnson, S., 241

Kac, M., 54, 667
Kahn, J. S., 110, 119, 668
Kahn, R., 116, 664
Kemeny, J. G., 658, 667
Kendall, M. G., 4, 15, 16, 667, 668
Keynes, J. M., 16
Kitagawa, T., 274, 667
Klein, M., 15, 667
Koch, G. S., Jr., 427, 667
Kolmogorov, A. N., 55, 667
Kotz, S., 304, 410, 418, 666, 667
Krumbein, W. C., 313, 370, 413, 667, 669
Kuroiwa, Y., 421
Kyburg, H. E., 55, 667

Laplace, P. S. de, 3, 4, 8, 16, 326, 667
Latter, O. H., 330, 667
Lazarsfeld, P. F., 112, 664, 667, 669
Lee, A., 327, 328, 495, 496, 668
Lehmann, E. L., 54, 666
Lewis, M., 559, 562, 563, 665
Lewontin, R. C., 306, 669
Lieberman, G. J., 264, 667
Link, R. F., 427, 667
Luce, R. D., 360

Macdonell, W. R., 330, 667
Markov, A. A., 224, 556
Mann, T., 432
Marshall, A. W., 592, 667
Masuyama, M., 421, 667
Mather, K., 254, 255, 665
Maxwell, J. C., 4
Mc Alister, D., 413, 667
Mc Gill, W. J., 360
Mendel, G. J., 566
Menzerath, P., 133, 667
Meré, C. de, 3, 16, 310
Meyer-Eppler, W., 113, 667
Miller, R. L., 110, 119, 668
Molina, E. C., 274, 668
Mosteller, F., 54, 115, 668

Mueller, C. G., 359, 365, 668

Nagel, E., 16
Newman, J. R., 16, 668
Newton, I., 7
Neyman, J., 308, 668

Ore, O., 16, 668
Owen, D. B., 264, 667

Pareto, V., 298
Parzen, E., 658, 668
Pascal, B., 3, 16, 104, 283, 310
Patil, G. P., 278, 303, 304, 320, 664, 665, 666, 668, 669
Pearl, R., 331, 668
Pearson, E. S., 15, 256, 495, 496, 668
Pearson, K., 263, 264, 327, 328, 329, 330, 409, 495, 496, 668
Peirce, C. S., 16
Pielou, E. C., 659, 668
Pliny the Elder, 358
Poincaré, H., 4, 16
Poisson, S. D., 265, 299
Pope, A., 247
Pretorius, S. J., 429, 431, 669
Proschan, F., 361, 669

Quetelet, L. A. J., 326, 327

Rasch, G., 419, 669
Richardson, L. F., 312, 669
Richter, C. F., 418, 419, 666
Riemann, G. F. B., 297, 298
Roe, A., 267, 269, 306, 669
Romig, H. G., 256, 669
Rosenberg, M., 664, 669
Rourke, R. E. K., 54, 668
Rutherford, E., 271, 272, 669

Saunders, S. C., 422, 663
Sartwell, P. E., 413, 669
Savage, I. R., 233, 669
Savage, L. J., 55, 669
Shaw, G. B., 25
Siddiqui, M. M., 428, 669
Simon, H. A., 229, 298, 299, 669

Simpson, G. G., 267, 269, 306, 669
Slack, H. A., 370, 669
Smith, E., 114, 669
Smokler, H. E., 55, 667
Snell, J. L., 577, 593, 658, 663, 667
Spencer, H., 5, 669
Stegun, I. A., 108, 663
Suchman, E. A., 114, 669
Sumner, C., 597
Sung, B., 4, 663
Suppes, P., 658, 669
Svedberg, T., 181, 669

Terence (Publius Terentius), 247
Terman, L. M., 341, 342, 670
Thackeray, W. M., 149, 150
Thielens, W., Jr., 112, 667
Thomas, G. B., 54, 668
Thompson, G. L., 658, 667
Thorndike, F., 271, 313, 670
Todd, G. F., 45, 670
Todhunter, I., 16, 667, 670
Truman, H. S., 116
Trumpler, R. J., 353, 670
Tversky, A., 593, 664

Urban, F. M., 670

Venn, J., 29
Viëtor-Meyer, 133, 515, 516
Virgil, 432
Von Mises, R., 55, 670

Weaver, H. F., 353, 670
Weaver, W., 54, 670
Weibull, W., 378, 387, 388, 389
Weiss, G. H., 428, 669
Weldon, W. F. R., 330
Whitaker, L., 314, 670
Whitehead, A. N., 550
Whittier, J. G., 597
Wiener, N., 4
Wilk, M. B., 372, 373, 670
Williams, C. B., 149, 413, 670
Williamson, E., 292, 293, 670
Wright, S., 307, 670

Yule, G. U., 287, 288, 665

Zipf, G. K., 298, 670

SUBJECT INDEX

Absolute k-step probabilities, 588
Absolute probabilities, 588
Absorbing Markov chains, 651
Absorbing states, 635, 651
Absorption probabilities, 655
Accident data, 288, 320
Actuarial tables, 5, 52
Addition, law of, 45, 84, 140
Age data, 430
Airplane failure data, 362
Aperiodic Markov chain, 640
Aperiodic state, 644
Ars conjectandi, 4
Astronomical data, 138, 142, 353
Automobile occupancy data, 278, 291
Average, arithmetic, 134
Axioms of probability theory, 41

Bayes' rule, 95
Bernoulli distribution, 148, 249
Bernoulli random variable, 248
Bernoulli trial, 247
Beta distribution, 359, 398
Beta distribution, fitting of, 404
Beta distribution, graphs of, 400
Beta function, incomplete, 406
Beta function, incomplete, tables of, 698
Bienaymé-Chebyshev inequality, 227
Bimodal distribution, 203
Binomial coefficient, 103
Binomial distribution, 247, 249, 251
Binomial distribution, approximation of, 272
Binomial distribution, normal approximation to, 535
Binomial distribution, tables of, 252, 673
Binomial expansion, 103
Binomial random variables, sum of, 526

Bivariate distribution, conditional, 506
Bivariate distribution, continuous, 475
Bivariate normal distribution, 433, 486, 494, 527
Bivariate normal distribution, contours of, 490
Bivariate normal distribution, standard, 493
Bivariate normal distribution, tables of, 701
Bivariate probability mass function, 436
Bombing hits, Poisson distribution fit to, 270
Brand switching data, 589
Breeding experiment, 566
Brownian motion, 181

Card hands, distribution of, 264
Cauchy and normal distributions, comparison of, 416
Cauchy distribution, 359, 414
Cauchy distribution, applications of, 418
Cauchy random variables, sum of, 527
Central limit theorem, 345, 349, 356, 522, 533
Central moments, 217, 221
Chains, periodic, 642
Chance, games of, 3, 4, 10, 241, 264
Chance phenomena, 3
Chance variable, 126
Characterization of the normal distribution, 488
Chebyshev inequality, 227, 233, 356, 531
Classes of states, 634, 636
Closed states, 634
Cloud-seeding experiment, 21
Cobb-Douglas distribution, 413
College Board Examination data, 352, 517
Combinations, 106
Combinatorial analysis, 69, 71, 100

745

SUBJECT INDEX

Communicating class of states, 639
Communicating states, 636, 645
Complement of event sets, 31
Complementation, law of, 43, 84, 134, 140, 228
Component random experiments, 74
Composite random experiments, 65, 72, 75, 99, 433
Composite results, 25
Conditional bivariate distribution, 506
Conditional distribution, 446, 506
Conditional expected value, 451
Conditional probability, 65, 79, 83
Conditional probability density functions, 482
Conditional statistical independence, 509
Conditional variance, 488
Conformity data, 563, 654
Contingency table, 434, 503
Continuity correction for central limit theorem, 539
Continuous random variable, 131, 139, 186, 202
Continuous-time stochastic process, 552
Correlation coefficient, 464
Correlation coefficient, partial, 506
Correlation and independence, 469, 489
Countable collection, 39
Counting rules, 100
Counts using Poisson distribution, 267
Covariance, 485, 527
Criminology data, 129
Cumulative distribution function, 135, 140, 144

Death notice data, 314
Deciles, 203
Density function, probability, 146, 150
Denumerable, 39
Dependence and mutually exclusive, 98
Dependence, statistical, 472
Dependent variable, 451
Discrete bivariate distribution, 433
Discrete random variable, 130, 202
Discrete-time stochastic process, 552
Discrete uniform distribution, 293
Disjoint event sets, 34
Dispersion, measure of, 175, 210
Distributions, shape of, 205, 222
Dow-Jones average, 85
Dynamic probability model, 551

Earthquake data, 418
Empirical cumulative distribution function, 144
Empirical probability mass function, 233
Epidemiology data, 592
Equality of event sets, 28
Equilibrium of Markov chain, 630
Equilibrium, probabilistic, 630, 646
Estimation of expected value, 233
Estimation of initial probabilities, 563
Estimation of transition probabilities, 560
Estimation of variance, 235
Event, null, 27
Event sets, 26, 29, 51
Events, disjoint, 34
Events, inclusion of, 29
Events, intersection of, 30, 33
Events, mutually exclusive, 34
Events, nested, 89
Events, simple, 66
Events, union of, 30, 33
Expected first-passage times, 620
Expected value, 177, 186, 193, 209, 457, 484, 531
Expected value, conditional, 451
Expected value as a measure of location, 191
Expected value of a sum, 527
Experiment, sequential, 79
Exponential distribution, fitting of, 361
Exponential distribution, graph of, 361
Exponential function, tables of, 693
Exponential random variables, sum of, 527

Failure data, 362, 372
Finite population factor, 262
Finger print data, 120
Fire incidence data, 425
First moment, 217
First-passage times, 597, 600
First-passage times, distribution of, 600
First-passage times, expected, 620
First-passage times, infinite, 611
Fishery data, 281
Fitting a probability model, 51
Flood data, 111
Forestry data, 659
Fourth central moment as measure of kurtosis, 222
Fractile of a distribution, 203
Frequency distribution, 237

SUBJECT INDEX 747

Frequency histogram, 163
Frequency interpretation of probability, 80
Frequency ratio, 5
Functions of a random variable, 127

Games of chance, 3, 4, 10, 241, 264
Gamma density function, standard, 370
Gamma distribution, 359, 369
Gamma distribution, relation to exponential distribution, 370
Gamma function, 285, 286
Genetic experiment, 254, 566
Geological data, 119, 120, 314, 595
Geometric distribution, 274, 619
Geometric distribution, fitting of, 277
Geometric distribution, tables of, 685
Goodness-of-fit tests, 255
Grouped data, 167

Higher-order transition probabilities, 571, 582
Histograms, 160, 162
Hypergeometric distribution, 256, 264
Hypergeometric distribution, approximation of, 262
Hypergeometric distribution, tables of, 261, 675
Hypergeometric distribution, uses of, 262

Identically distributed trials, 100
Inclusion of event sets, 29
Inclusion, law of, 42, 84
Income distribution, 206
Incomplete beta function, 406
Incomplete beta function, tables of, 698
Incomplete gamma function, tables of, 694
Independence, 97
Independence and correlation, 469, 489
Independence, mutual statistical, 99
Independence, physical, 99
Independence, statistical, 469, 483, 508
Independence, stochastic, 65, 97
Independent random variables, 451
Independent trials, 100
Independent variable, 451
Industrial absence data, 319
Inequalities, probability, 223, 227
Inequality, Bienaymé-Chebyshev, 227, 232
Inequality, Markov, 227, 232
Initial probabilities, estimation of, 563
Initial probabilities for a Markov chain, 557
Intelligence quotient data, 341

Interpolation, linear, 196
Intersection of event sets, 31
Irreducible Markov chain, 640, 646

Joint distribution, 433
Joint probability mass function, 433
Jury choices, 106, 308

k-step transition probabilities, 571, 582, 587
Kurtosis, 177, 227

Latency distribution, 360, 365
Law of addition, 45, 84, 140
Law of addition, conditional probabilities, 84
Law of complementation, 43, 84, 134, 140, 228
Law of complementation, conditional probabilities, 84
Law of errors, 326
Law of inclusion, 42, 84
Law of inclusion, conditional probabilities, 84
Law of large numbers, 530
Law of multiplication, 86, 89
Law of multiplication for nested events, 89
Learning models, 375, 574, 593, 658
Length of life and Weibull distribution, 385
Life insurance, 183
Life tables, 52, 63, 79
Lifetime data, 158, 422
Linear combination of normal variables, 488
Linear interpolation, 196, 200
Linguistics, statistical, 18, 133, 137, 142, 149, 241, 242, 515
Location, change of, 519
Location, measure of, 175, 205
Lognormal distribution, 359, 410
Lognormal distribution, applications of, 413
Lognormal distribution, relation to normal 411
Long-run probability, 597, 633, 646
Lost article data, 313
Lottery, 8, 73, 294

Marginal distributions, 439, 505
Marginal probabilities, 80, 436
Marginal probability density functions, 479
Markov chain, 550, 556
Markov chains, absorbing, 651
Markov chains, aperiodic, 640
Markov chains, applications of, 559

748 SUBJECT INDEX

Markov chains, equilibrium of, 630
Markov chains, irreducible, 640, 646
Markov chains, transition probabilities of, 557
Markov inequality, 224
Mass function, probability, 146, 149, 179, 180
Matches, 170, 240
Mean, 177, 193, 208
Mean of a distribution, 193
Mean, first-passage time, 621
Mean, joint, 453
Mean, sample, 182
Mean vector, 453
Measure of covariation, 455
Measure of dispersion, 175, 486, 521
Measure of location, 175, 205, 485, 521
Measurement error, 8, 326
Median, 177, 193, 198, 203, 207
Mendelian laws, 306
Meteorological data, 429
Migration data, 593
Mode, 177, 203, 208, 453
Model, fitting of a probability, 51
Model, probability, 4
Model, stochastic, 4, 432
Moment of distribution, 193, 217, 221
Moment-generating function, 541
Moment-generating function of a sum, 542
Moments, 217
Monte Carlo simulation, 397
Morra, two-finger, 165
Mortality table, 184
Multinomial distribution, 509
Multiplication, law of, 85, 89
Multivariate distribution, 433
Mutual independence, 99
Mutually exclusive event sets, 34

Negative binomial distribution, 280, 283
Negative binomial distribution, derivation of, 222
Negative binomial distribution, fitting of, 287
Negative binomial distribution, generalized, 284
Negative binomial distribution, tables of, 687
Newtonian mechanics, 6
Nested events, 89
Normal and Cauchy distributions, comparison of, 416

Normal distribution, 324
Normal distribution, fit of, 341
Normal distribution, standard, 333
Normal distribution, tables of, 691, 692
Normal ogive, 336
Normal random variables, sum of, 526
n-tuples, 74
Null event set, 27

Objective probabilities, 55
Occupational mobility, 576
Ogive, normal, 336
Old Faithful data, 514
Ordered sample, 69
Ordered sample, 69
Ordered sample with replacement, 71
Orderings, 100, 104
Outcomes, 25, 75

Pairwise statistical independence, 508
Parameters of bivariate distribution, 487
Pareto distribution, 243, 298
Partial correlation coefficients, 506
Partitions, 93
Pascal distribution, 283
Pascal's triangle, 104
Peakedness, 177
Percentile, 203
Periodic chains, 642
Periodic states, 642
Permutations, 100
Persistent state, 617
Poisson distribution, 54, 265
Poisson distribution, applications of, 269
Poisson distribution, fitting of, 266
Poisson distribution, tables of, 266, 680
Poisson random variables, sum of, 526
Political preference data, 609
Population density data, 513
Probabilistic equilibrium, 632, 646
Probability, absolute, 588
Probability, absorption, 655
Probability, conditional, 80
Probability density function, 146
Probability, frequency interpretation of, 80
Probability, long-run, 597, 633
Probability, marginal, 80
Probability mass function, 146, 149, 179, 180
Probability model, 4, 51
Probability model, finite, 65, 66
Probability model, fitting, 9, 51

SUBJECT INDEX 749

Probability model, uniform, 65, 68
Probability, objective, 55
Probability, steady-state, 633
Probability, subjective, 55
Probability, unconditional, 80, 588
Problem of 'matches', 170, 240
Problem of 'rencontres', 170, 240

Quality control, 10
Quantile, 177, 198, 203
Quantile as measure of location, 201
Quartile, 203

Radioactive emissions, Poisson distribution fit to, 271
Radioactivity and gamma distribution, 370
Random experiment, 9, 25, 125
Random number generators, 396
Random sample, 8, 69, 73
Random sampling without replacement, 256
Random variable, 126, 140
Random variable, continuous, 131, 139, 186, 202
Random variable, discrete, 130, 202
Random variables, standardized, 519
Random variables, sum of, 523
Random walk, 517, 548
Reading time data, 417
Rectangular distribution, 391
Rectangular probability approximations, 153, 159, 188
Recurrence time, 615
Recurrent states, 597, 615, 639, 653
Regression function, 450, 488, 508
Relative frequency, 11, 21, 25
Relative frequency histogram, 162
Relative frequency, stability of, 4, 14
Relative mode, 204
Rencontre, 170, 240
Repeated trial, 10
Rescaling techniques, 460
Riemann zeta function, 297
Roulette, 67, 241, 275

Sample average, 234, 531
Sample cumulative distribution function, 144
Sample mean, 182
Sample space, 26
Sample survey, 9, 65
Sample survival function, 36
Sample universe, 26

Sample variance, 235
Sampling, random, 69
Sampling with replacement, 70, 98, 540
Sampling without replacement, 70, 105, 256
Sampling, survey, 65
Scale, change of, 519
Scaling, 460
Semi-interdecile range, 216
Semi-interquartile range, 215
Sequential experiment, 79
Set, event, 26
Set theory, 27
Simple events, 66
Simple random sample, 69, 70, 72
Simulation, 396
Size of particles data, 385
Skewed distribution, 221
Social classes, 81
Social groups data, 302
Spare parts distribution, 369
Standard bivariate normal distribution, 493
Standard deviation, 210
Standard gamma density function, 370
Standard normal distribution, 333
State, persistent, 617
States, absorbing, 635
States, aperiodic, 644
States, closed, 634
States, communicating, 636, 645
States, periodic, 642
States of process, 553
States, recurrent, 597, 615
States, transient, 615
Stationary transition probabilities, 557
Statistical dependence, 472
Statistical independence, 65, 97, 469, 483, 508
Statistical linguistics, 18, 133, 137, 142, 149, 241, 242, 515
Statistical regularity, 14
Stature data, 327, 495
Steady state, 630
Steady-state equations, 648
Steel production data, 298
Step-function, 137
Strength of steel data, 389
Stochastic, 4, 65
Stochastic equilibrium, 632
Stochastic matrix, 558
Stochastic model, 432
Stochastic process, 552, 658
Stochastic variable, 126

SUBJECT INDEX

Subjective probability, 55
Surgical consultation data, 322
Survey, telephone, 114
Survival function, 365, 389

Tail probabilities, 336
Tea-tasting experiment, 308
Telephone connections, Poisson distribution fit to, 270
Telephone survey data, 114
Third central moment, 221
Traffic data, 317
Transient states, 615
Transition matrix, 558
Transition probabilities, calculation of, 587
Transition probabilities, higher-order, 571, 607
Transition probabilities for Markov chain, 553
Transition probabilities, three-step, 578, 607
Transition probabilities, two-step, 572, 607
Trials, independent, 99
Trials, repeated, 10
Trivariate distribution, 433, 501
Truncated Poisson distribution, 299
Two-finger Morra, 165

Unconditional probabilities, 80, 588
Uniform distribution, 359, 390
Uniform distribution, applications of, 396
Uniform distribution, discrete, 293
Uniform probability model, 68, 148
Unimodal distribution, 203
Union of event sets, 30
Univariate marginal distribution, 505
Upper quantile, 203

Variance, 177, 210
Variance of a sum, 527
Venn diagram, 29
Vocalization data, 559

Wars, number of, 312
Weibull distribution, 359, 378
Weibull distribution, fitting of, 385
Wholesale price data, 404
Wine-tasting experiment, 435, 502

Zeta distribution, 296
Zipf's law, 298